Introduction to Protozoology

INTRODUCTION TO PROTOZOOLOGY

REGINALD D. MANWELL

PROFESSOR (EMERITUS) OF ZOOLOGY
AND RESEARCH ASSOCIATE,
SYRACUSE UNIVERSITY

SECOND REVISED EDITION

DOVER PUBLICATIONS, INC. NEW YORK

Published in Canada by General Publishing Company, Ltd., 30 Lesmill Road, Don Mills, Toronto, Ontario.
Published in the United Kingdom by Constable and Company, Ltd., 10 Orange Street, London WC 2.

This Dover edition, first published in 1968, is a revised version of the work originally published in 1961. It is published by special arrangement with St. Martin's Press, Inc., 175 Fifth Avenue, New York City 10010.

The publisher is indebted to the authors and journals who have granted permission to use illustrations from their works. Acknowledgment is also gratefully made to the following book publishers who have permitted the use of copyrighted illustrations: Academic Press for Fig. 7.1; Balliere, Tindall & Cox for Figs. 20.2, 25.5, 25.7; Burgess for Fig. 25.4; Columbia University Press for Fig. 9.1; Harvard University Press for Figs. 3.10 and 15.9; Iowa State University Press for Figs. 25.1 and 25.4B; Lea & Febiger for Fig. 16.7; Masson & Cie for Figs. 14.2, 14.7, 14.9, 22.5, 23.12; McGraw-Hill for Fig. 12.2; Princeton University Press for Fig. 15.11; Charles C Thomas for Fig. 9.2; John Wiley for Figs. 9.5, 9.6, 9.7, 22.9. The illustrations in color on the inside covers are reproduced from *Practical Malariology* by Russell *et al.,* by permission of the Oxford University Press.

Standard Book Number: 486-61897-8
Library of Congress Catalog Card Number: 68-26130

Manufactured in the United States of America
Dover Publications, Inc.
180 Varick Street
New York, N. Y. 10014

To Dr. Harold H. Plough and Dr. Robert Hegner who first instilled in me a love for the microorganisms that constitute the subject of this book, and to all my students, with whom learning about the Protozoa has been a cooperative enterprise.

Preface to the Dover Edition

In a world already flooded with books, there should always be good reason for publishing a new one or even reprinting an old one. The favorable reception of the first edition of this book both here and abroad convinced the author of its value to many readers, and he therefore welcomed the opportunity afforded by the publishers of the present edition to make it conform to present knowledge, even though the changes which could be made were of limited scope. Several typographical errors in the first edition have been corrected, and certain sections have been rewritten. References cited in this revised edition are marked by an asterisk (*); these are listed after the references at the ends of the chapters under the heading ADDITIONAL REFERENCES.

Protozoology is an important branch of biology, and with the introduction of new techniques and the advent of a new generation of protozoologists qualified to use them, knowledge of the Protozoa is rapidly growing. Although closely related to microorganisms more clearly in the animal and plant kingdoms than some of the Protozoa appear to be, they are nevertheless in many ways unique in nature, and the study of them should be a part of all curricula in biology. It is my hope that the book will prove useful to students and biologists, and cause the tiny organisms we know as Protozoa to seem as interesting to the reader as they have always seemed to me.

REGINALD D. MANWELL

December, 1967
Syracuse University
Syracuse, New York

Preface to the First Edition

For three hundred years, since Leeuwenhoek chanced to see *Vorticella* in rain water, we have been amassing information about the Protozoa, and in recent years at an increasing rate. Now more than ever there is a need for an introductory text geared to the beginning student, and I have tried to meet that need with a sound teaching instrument.

This book, therefore, stresses principles. But often neglected aspects of protozoology—the antiquity and evolution of the Protozoa, the history of the subject itself—are emphasized. I give special attention to parasitism, for many protozoan parasites cause serious disease in man and animals. And because what we do not know is as significant as our achievement, I take pains to discuss the gaps in our knowledge in the hope that the student will be challenged, perhaps even to attempting some research of his own.

I wish to acknowledge with gratitude the invaluable criticism of Dr. John Corliss of the University of Illinois and Dr. Elizabeth S. Hobbs of Smith College. Dr. Adolph Spandorf of St. Lawrence University and Dr. Margaret Weiss of Wayne State University read the manuscript with kindly eyes and made helpful suggestions; so also has Dr. William Balamuth of the University of California (Berkeley). I owe much to my wife, Dr. Elizabeth M. Manwell, who has gone through the entire manuscript with care.

REGINALD D. MANWELL

Syracuse University
Syracuse, New York

Contents

Introduction to Protozoology

1

Protozoa: Nature, Variety, Habitat

"The minute living Animals exhibited in . . . . this Work, will excite a considerate Mind to admire in how small a Compass Life can be contained, what various Organs it can actuate, and by what different Means it can subsist . . . ." Henry Baker, *Employment for the Microscope*, 1753.

To most of us nothing seems as peaceful as a shaded woodland pool, or a little pond shimmering in the sunshine. But could we only see it, this quiet conceals life being lived to the full by myriad minute beings. Many of these are Protozoa. Though they may complete their brief life spans in a few hours or less, their experiences are those of all living things. Youth, maturity, and even old age are passed in quick succession by the fortunate few, but more often there is no old age for these organisms either because they fall victims to forms speedier or hungrier than themselves or have the vigor of youth again conferred on them by the internal reorganization accompanying reproduction of one kind or another. All of them are engaged in a continuous struggle for existence in which nothing is asked nor given. This is the drama which goes on before our unseeing eyes, but luckily the microscope makes it visible.

The Protozoa rival anything in the animal kingdom for uniqueness and novelty. Could they be magnified to the size of vertebrates we are most familiar with, some might even be terrifying. Their habits would be equally frightening; some Protozoa manage to capture and swallow animals larger than themselves; others are cannibals. Many are unconscionably greedy, gorging themselves twenty-four hours a day. Many, however, are content to bask near the surface of a stagnant pool or ride the ocean currents, using the sun's radiant energy to manufacture their food.

Where are the Protozoa found? Among the most adaptable of animals, they have established themselves in almost every conceivable environmental niche. They are most abundant in moist situations, however—swamps, ponds, lakes, and streams. Many species find the warm surface currents of the ocean ideal, while others are happy in the ooze on the ocean floor. A dozen or so species live in the Great Salt Lake, in which the salt concentration far exceeds that of the ocean. Some are characteristic of the mosses and lichens found on trees. Certain euglenids have been found in great numbers in the petroleum-covered pools about oil wells—seemingly a most improbable environment. Still others live in the upper few inches of the soil. Perhaps the most popular habitat of all is the bodies of animals and even of plants, where food lies within easy reach. A few species of Protozoa have found even the bodies of other Protozoa convenient havens. Only the air has proved unsuitable as a medium in which growth and reproduction are possible. It is therefore easy to see why most species of Protozoa are found the world over.

Protozoa are much easier to define than to describe. Most biologists content themselves with saying that these little beings are simply "unicellular animals." But even here there is ample room for difference of opinion.

Protozoa vary enormously in size, shape, color, and in virtually every morphological characteristic. They also differ in habit, behavior, and life cycles. Biologists who know them best have often questioned whether the group should not be regarded as a subkingdom rather than a phylum. We might also reasonably ask, "Why not a kingdom?", for in the protozoan world things are not always what they seem. The boundaries between the plant and animal kingdoms tend to disappear among the Protozoa.

Although most Protozoa are microscopic, many are not. There are such relatively gigantic species among the ciliates as *Spirostomum ambiguum*, which may be as long as 3,000μ (3 mm.); a fossil *Nummulites* (foraminiferidan) reached the almost unbelievable diameter (for a protozoan) of 18 cm., or about seven inches, and a single living species is known which may somewhat exceed this size. At the other end of the scale are such tiny amoebae as *Endolimax nana*, a parasite of man often only a few micra in diameter, and flagellates like the Rhyncomonads, which are even smaller.

Unfortunately for the neophyte protozoologist, there is little about a protozoan that makes it immediately recognizable. Only experience allows us to tell a protozoan from a variety of other animals and even plants.

How Protozoa Move

In general, protozoa are minute and active organisms that move by the use of organelles* known as flagella, pseudopodia, or cilia. Flagella (singular *flagellum*, Latin for "whip") are filamentous structures, usually arising at the anterior end of the animal, which pull it along by actively lashing about. There may be only one or a few flagella, as in most free-living forms, or

*Since the term "organ" is usually defined as a structure serving a specific purpose (or, better, performing a given function), and made up of tissues, it is clearly not properly applicable to the corresponding parts of a protozoan, which is generally regarded as a single cell. The term "organelle" is therefore used instead.

perhaps hundreds, as in certain species parasitic in termites and wood roaches.

Pseudopodia (from two Greek words meaning "false feet," singular *pseudopodium*) are temporary extensions of protoplasm which in their simplest form are more or less fingerlike. The organism may be said to move by flowing into itself. Such pseudopods are known as lobopodia. A second type, the "filopodium," is like the first, except that it is filamentous and may be branched. A third type differs from the second in that its branches may fuse, or anastomose, on contact. It is called a "rhizopodium" (from two Greek words meaning "root" and "feet"), since it has a rootlike appearance under the microscope. There is still a fourth type, known as an "axopodium," because it has an axial filament. Axopodia are usually extended radially and resemble flagella in some ways. For this reason, it has been thought that pseudopods may have evolved from flagella, although recent research with the electron microscope has shown that the finer structure of axopodia and flagella is quite different. Nevertheless, amoeboid organisms may include flagellated stages in their life cycles and for this and other reasons they are generally thought to be less primitive than the flagellates from which, perhaps, they evolved.

Cilia (*cilium*, Latin for "eyelid") are individually similar to flagella, but are relatively smaller and more numerous. A typical ciliate is covered by cilia much as a cat is covered with hair. An individual cilium functions like a flagellum, but since there are generally many cilia, some means of coordination is necessary. This alone requires a complex internal organization, which is invisible in life. And of course the mechanics of ciliary motion are different, since the flagellum is usually at the anterior end and in the great majority of cases pulls the organism, whereas the cilia operate more like oars.

The Major Groups

For convenience, free-living Protozoa may be divided into three groups according to the type of locomotor organelles. In a few cases this method of classification presents a problem, since species are known which have both flagella and pseudopodia, sometimes simultaneously and sometimes at different stages of the life cycle. Fortunately no protozoan has yet been discovered which has all three types.

Forms with flagella are called Mastigophora, or sometimes Flagellata, the first term from the Greek and the second from the Latin. If the organisms make their way about with pseudopodia, they are said to be Sarcodina. Those using cilia for progression are designated Ciliophora.

Not every animal possessing flagella, pseudopodia, or cilia, however, is a protozoan. Cells with one or more of these three types of locomotor organelles have been incorporated into the bodies of many kinds of living things. Among such multicellular forms are some very minute organisms as small or smaller than many Protozoa which may occur in the same environment. To distinguish between them is often not at first easy.

All free-living Protozoa and many of the parasitic ones belong to one or another of the three great groups. There is also a fourth one, however, the Sporozoa,* all the members of which are parasites. The group receives its name from its method of reproduction, which involves the formation of many progeny from a single parent cell. This process of multiple division is often called "sporulation," or sometimes "sporogony." Other kinds of Protozoa usually multiply by binary fission, though there is variety even here.

Organization

If the first essential for the recognition of a protozoan is determining its method of movement, the second is to observe carefully its type of organization. It can be argued that the simpler multicellular animals (e.g., *Hydra*) show a rather less intricate organization than some of the more highly evolved Protozoa (as, for example, *Troglodytella*, a genus of ciliates inhabiting the colon of anthropoid apes, or *Diplodinium*, a cattle ciliate). But the organization of the protozoan is on a different level: parts of the cell (if a protozoan should be regarded as such), rather than the cells themselves, have become specialized for the performance of specific functions.

The protozoan is therefore often highly complex even in its grosser aspects. There may be one or many nuclei, or even (in the ciliates) two kinds of nuclei, with one or many of each. The individual nucleus may be almost any shape. During cell division protozoan nuclei (except the macronuclei of ciliates) undergo mitosis, passing through the usual stages and showing a definite number of chromosomes. Details of the process vary much, however, in different species and groups. In many species there is a cell mouth ("cytostome"), usually with a short gullet, at the base of which temporary stomachs ("food vacuoles") form. About the mouth there are often structures that move food particles toward and into it. Some ciliates also have a cell anus, or "cytopyge." In free-living fresh-water species, excess water and soluble wastes are secreted into a vacuole which voids its wastes periodically, and is therefore called a "contractile vacuole."

Within the cell there are also other structures, though they vary greatly with the group and the species. Associated with organelles of locomotion, especially in ciliates, there may be a system of fibrils and granules, believed to have as one of its functions coordination, and which therefore has been called a neuromotor apparatus. Not infrequently there are also contractile fibrils, or "myonemes," reminiscent of muscle fibers of the Metazoa and possibly functioning in similar fashion.

Many species of Protozoa are beautifully colored. In some the pigment

*A detailed classification of the Protozoa follows consideration of each of the main groups. See Chapters 14, 15, 16, 17.

is soluble and diffused throughout the cytoplasm; in others, particularly the plantlike flagellates (Phytomastigina), it is aggregated into masses known as "chromatophores" (or, sometimes, "chloroplasts"). These same flagellates also often possess a light-sensitive organelle known as an eyespot, or "stigma." It appears as a conspicuous orange-red body in the anterior portion of the cell. Some of these flagellates wear a cellulose jacket masking the green of the chlorophyll within, as the common brown *Trachelomonas* and the exquisitely marked dinoflagellates do.

Skeletal elements of other kinds, internal and external, occur in many Protozoa. These vary all the way from coats of mail, more or less like the armor of the knights of old, to one or more skeletal elements stiffening some cytoplasmic organelle. A jacket of ectoplasmic plates is worn by the ubiquitous ciliate *Coleps*, and the vaselike garment called a lorica protects the flowerlike, stalked ciliate *Cothurnia* from the perils of life. Other ciliates, such as *Chilodonella*, have developed supporting structures around the mouth, perhaps to minimize the possibly dangerous consequences of excessive greediness. These are known as trichites, and collectively constitute the "pharyngeal basket."

Shelled Forms

The less actively motile Protozoa often possess shells (also termed "tests" or "loricae"). The majority of such fresh-water forms belong to the rhizopod order Testacida. Their shells may be secreted by the organism, in which case they are likely to be of pseudochitin (perhaps colored by mineral salts), as in the case of *Arcella vulgaris*, or of foreign bodies picked up from the environment, such as diatom shells or grains of sand, held together by some organic cement. Some forms utilize only the shells of certain kinds of diatoms to make their own. Still others, like the rather common bottom-dwelling *Euglypha*, secrete a shell of minute siliceous plates.

But the great majority of shell-bearing species are marine, and most of them belong to one or the other of the two large rhizopod orders, Foraminiferida and Radiolarida. Foraminiferida usually build their shells of calcium carbonate, whereas the Radiolarida utilize either silica or strontium sulfate. The shells of the former are often coiled in a single plane, rather like certain molluscs (which they were indeed once thought to be), and are pierced with many minute openings through which pseudopodia can be extruded. Most radiolaridan shells are more or less spherical in form, consisting of two concentric capsules, latticed and radiating spines. They are certainly among the most beautiful objects in all of nature.

Sessile Types

Though free-living Protozoa are generally capable of independent motion throughout their existence (except in the cyst stage), at least one group is motile only during the embryonic period. These are the Suctorida, a class of Ciliophora. As adults they remain at one spot, attached to the substrate by a stalk, while lying in wait for chance protozoan passersby to become entangled and paralyzed by their radiating tentacles. The vitals of the hapless prey are then quite literally sucked out, its cytoplasm slowly passing down the tentacles and into the body of its captor.

Numerous other examples of flagellate and ciliate species have adopted a semi-sedentary existence. Among the most familiar to students of fresh-water Protozoa are the stentors and various vorticellids. The first protozoan to be seen and accurately described by man was a *Vorticella*. When Leeuwenhoek wrote down what he had seen for the Royal Society, he referred to it as a "poor little creature," and remarked that when "striving to disentangle [its] tayle" (as he called the stalk), its "whole body lept back towards the globul of the tayle, which then rolled together serpent-like." Less often observed are the tiny little flagellates of the genus *Amphimonas*, and others of the same family, which may be seen either free-swimming or attached by stalks, consisting of threads of cytoplasm, to vegetation and debris.

Communal and Colonial Living

Perhaps it is only a step from a sedentary but solitary existence to a social or communal one, somewhat as in the evolution of human society, where hunting and the life of the nomad gradually gave way to that of the village dweller. At any rate, some species of *Vorticella* are commonly found in communities of many individuals, and closely related forms may develop colonies so large that they are quite easily detected by the naked eye.

The same evolutionary tendency has been at work among the flagellates. A remarkable example is the species known as *Rhipidodendron splendidum* (belonging to the same family as *Amphimonas*). This species, though not often seen, was first studied by Friedrich Stein, a pioneer in protozoology. Almost eighty years ago he published a description and figure of this "splendid" form in his great monograph *Infusionsthiere;* so good was it that the eminent English naturalist, W. Saville Kent, referred to it in his own *Manual of the Infusoria* as the "gem of the entire series."

Rhipidodendron splendidum is a colonial species that builds itself a structure resembling a minute shrub. At the end of each tiny twig is a little tube with a flagellate inside, free to come and go, but living a solitary and celibate existence (for sexual stages are rare in the life cycles of the Zoomastigina).

The advantages to the organism of living communally are not easily understood, since there is no specialization of labor in individuals of these species. But some forms, such as the Volvocidae, have made a beginning in this direction. These plantlike flagellates form large colonies consisting of as many as 15,000 cells. It is said that cytoplasmic connections exist between adjacent cells, at least in some species. Possibly some functions affecting the colony as a whole, such as movement of the flagella, are coordinated. Thus, though most of these cells are alike, individual cells may not be wholly independent. In each colony, however, a few cells differ from the rest, and function only in reproduction. These are either male or female, though development is in some cases parthenogenetic.

Some biologists interested in evolution feel that perhaps Protozoa such as these represent an early stage in the origin of multicellular animals, and colonies of *Volvox* have often been compared to the blastula stage of the higher Metazoa. Of course it is impossible to suppose that *Volvox* or any other living organisms could represent the actual ancestral types from which multicellular animals arose. But conceivably these spherical colonies of cells, a few of them specialized for reproduction, may correspond in a general way to such hypothetical ancestral types.

Are Protozoa Cells?

At this point, a very fundamental question may be asked. What is the real nature of a protozoan? Is it actually a cell, in the sense usually attached to the word? Is the conventional definition of Protozoa as "unicellular animals" a sound one?

The great English protozoologist Clifford Dobell, who preferred not to consider Protozoa by themselves but lumped them with such other "unicellular" forms as Bacteria and Algae in a group called the "Protista," answered this query with an unqualified "no."

The term "cell" was first proposed by the English physician Robert Hooke for the structural units he saw in cork viewed under the microscope. Later it was applied by Schleiden and Schwann and others to the structural units of plant and animal tissues. But these units are part of an organism, rather than organisms in themselves, as are the Protozoa and other Protista, and this difference seemed very important to Dobell. The cells of the multicellular organism are mutually interdependent, whereas the protozoan is independent.

Furthermore, the protozoan has evolved along lines completely different from those followed by the Metazoa, often developing highly specialized organelles and a complex life cycle whose stages represent equally great specialization of another sort. Although the physiological mechanisms of protozoan and metazoan cells are essentially alike, the former are usually much more self-sufficient. This means that they are likely to be more rather than less complex than the latter. The protozoan "cell" also retains the ability to reproduce through most of its life cycle (except, possibly, in the cyst stage), while the fully differentiated metazoan cell has usually lost it.* It can also be argued that the protozoan exhibits more variety in important structural characteristics than the metazoan cell. The wide variation in

*Obviously there are exceptions to this rule, such as epithelial cells of various kinds, leucocytes and their precursors, etc.

nuclear condition has already been mentioned. The morphological changes that often characterize the different stages of the protozoan life cycle are even more varied.

Clearly the conventional concept of the cell as the fundamental structural unit of living things must be somewhat modified if it is applied to the protozoan organism. Possibly this concept has less meaning today than it once did. Modern discoveries of molecular and protoplasmic structure, and of the fundamental nature of viruses and the gene, have tended to emphasize the chemical and physiological unity of life, thus making a unity based on cellular organization seem less important—and less real.

Therefore, for convenience and to avoid the confusion of a new definition of the Protozoa based on the view that they are "noncellular," the usual definition of Protozoa as "unicellular animals" will be here adopted. But this solves no problems, as we shall see. Nature evolved Protozoa before man evolved definitions. There are not only difficulties in regarding Protozoa as unicellular, but there are frequently convincing objections to calling many of them animals.

Parasitism Among the Protozoa

It was stated earlier that Protozoa are extremely adaptable and are found in almost every environmental niche, including the bodies of animals and plants. But of those which live in other living things, not all harm their hosts. It is not in their interest to do so (of what use is a dead host to its parasites?) and parasitologists have long recognized that the relationship between host and parasite is usually one of mutual toleration.

We must of course suppose that parasitic species arose from free-living ones. Yet the parasitic mode of life must be very ancient—perhaps almost as ancient as life itself. Sandon (1932) remarked: "The animal kingdom presumably had its origin when some unicellular organisms, previously accustomed to nourishing themselves after the manner of plants, began to eat the bodies (either living or dead) of their neighbors." Forms which preferred to eat their victims alive were probably not only the first predators, but the ancestors of truly parasitic species.

That parasitism must indeed be old is also indicated by the large number of species which have adopted it as a convenient method of making a living. The number of known protozoan species, parasitic and free-living, has been variously estimated. Weisz (1954) puts the number at "50,000 plus"; others make the number of described species as small as 15,000 (e.g., Rogers, *et al.*, 1942). What is almost certain, though, is that there are more parasitic species than free-living ones, although no one can say with assurance how many of either there really are, for new ones are being described all along.

The evidence for the assertion that more parasitic species exist than free-living ones is that most parasites are restricted to a single (or perhaps to several closely related) species of host. Man, for example, has some 30 species of protozoan parasites, and there is no reason to think that nature has here been exceptionally generous. If we assume that each of the 5,000 or so species of mammals are equally well provided with protozoan parasites, the total number of such species, in mammals alone, would number 150,000.

Among the protozoan parasites of man are those causing malaria, amoebic dysentery, African sleeping sickness, and the leishmanias (espundia, kala-azar, and Oriental sore). Collectively, these diseases have a nearly world-wide distribution and afflict at one time probably a quarter of mankind. In addition to the pathogenic types there are many harmless species and some which are only occasionally injurious to man.

But Protozoa need not be human parasites to be of great importance. Many species cause serious disease in domestic animals, and others may inflict significant injury on game. Some parasitic species play an important role in maintaining biological balances and in limiting the population of various species in nature. A few species of Protozoa have become so indispensable to their hosts that without them the host could not live. Such are those flagellates occurring in the gut of termites and wood roaches, insects which have lost the ability to digest the wood on which they feed. When deprived of their parasite fauna experimentally they continue to feed actively, but literally starve to death in the midst of plenty.

Importance to Man

Protozoa also affect human life in a number of other ways. Many free-living species are one end of a food chain whose other end is man. These Protozoa (particularly the plantlike flagellates, or Phytomastigina) are important elements in the diet of animals somewhat larger than themselves, such as the small crustacea often known as Entomostraca. The latter are in turn eaten by small fish and other animals until we reach species eaten by man. Microorganisms such as these Protozoa are in part the ultimate sources of some vitamins, such as B_{12} and D. Occasionally they and the blue-green algae are also sources of substances poisonous to man, with the result that molluscs and other food species which consume such organisms may in their turn be rendered dangerous for human consumption.

Marine and fresh-water Protozoa may well become significant to man in a most harmful way in the future. Some of them have the ability to concentrate as much as a thousandfold radioactive substances present in water and derived from nuclear fission and fusion. These substances are always produced by the explosion of atomic bombs and as by-products in the manufacture of nuclear power. Isotopes such as these, after incorporation into the protoplasm of the protozoan, would then be passed along the food chain until they eventually reach man. Any considerable increase in the concentration of radioactive substances in human tissue would be a very grave matter, for it would increase somatic and genetic mutations, and this would entail a sharp rise in leukemia, bone cancer, and hereditable defects of many kinds.

There are also certain other ways in which Protozoa touch human life, less important than those just discussed, but still not to be discounted. One is the occurrence occasionally of unpleasant tastes and odors in domestic water supplies. For this one or more of about a dozen species of plantlike flagellates are usually responsible. The water is usually said to have a "fishy" taste, but aromatic oils in the bodies of these Protozoa, much more often than fish, cause the mischief.

Another way in which Protozoa affect human welfare is the mortality among fish resulting from excessive multiplication of certain species of dinoflagellates (a group of plantlike forms, chiefly marine, characterized especially by their beautifully sculptured cellulose jackets and, in another group, their delicate pastel tints). The species chiefly responsible is believed to be in most instances *Gymnodinium brevis*. Occasionally these Protozoa become so numerous that many square miles of ocean surface may be colored red (hence the name "red tide"), and many thousand dead fish pile up on the beaches. It has even been suggested that certain large accumulations of fish fossils originated from similar catastrophes in bygone ages.

Protozoa are also very useful research tools. Because of the rapidity with which they reproduce and the ease with which some free-living species can be cultivated, they lend themselves especially well to studies on genetics. It is no exaggeration to say that more is known about inheritance in certain Protozoa than in man. Protozoa have also been much used in the study of problems of sex, senescence, and populations.

Recently Protozoa, particularly some of the flagellates and the ciliate *Tetrahymena*, have been extensively employed in fundamental nutritional research, with the result that a good deal has been learned of what might be called physiological evolution. It appears, for example, that the basic nutritional requirements of such ciliate species as *Tetrahymena* and man are nearly the same. Vitamins needed by different species differ, but it appears likely that most species need B_{12}. Indeed some flagellate species are so sensitive to the lack of certain vitamins that cultures of these forms are used experimentally and commercially for vitamin assay. Among such are species of *Euglena*. They have also been used in the study of antibiotics, the effects of radiation, photosynthesis, and other aspects of cell physiology (*Wolken, 1961). *Tetrahymena pyriformis* has recently been useful as an assay organism for nicotinic acid (Baker, *et al.*, 1960). Evidently there has been much less change in the physiology of living things than in their morphology and life cycles. The basic machinery of energy production and metabolism is much the same whether living things are great or small, unicellular or multicellular. So also is that of growth, reproduction, and inheritance.

REFERENCES

Baker, H., Frank, O., Pasher, I., Sobotka, H. 1960. Nicotinic acid assay in blood and urine. *J. Protozool.*, 7(Suppl.): 12–13.

Calkins, G. N. 1933. *Biology of the Protozoa.* Philadelphia: Lea & Febiger.

Conn, H. W. 1905. *Protozoa of the Fresh Waters of Connecticut.* Hartford Press.

Dobell, C. 1911. Principles of Protistology. *Arch. f. Protistenk.*, 23: 269–310.

——— 1932. *Antony van Leeuwenhoek and his Little Animals.* New York: Harcourt, Brace.

Dujardin, F. 1841. *Histoire naturelle des zoophytes. Infusoires.* Paris.

Edmondson, C. H. 1906. The Protozoa of Iowa. *Proc. Davenport Acad. Sci.*, 11: 1–124.

Hamilton, L. D., Hutner, S. H., and Provasoli, L. 1952. The use of Protozoa in analysis. *Analyst*, 77: 618–28.

Jahn, T. L. and Jahn, F. F. 1949. *How to Know the Protozoa.* Dubuque: Wm. C. Brown.

Kent, W. S. 1880–81. *A Manual of the Infusoria.* London: David Bogue.

Kudo, R. R. 1954. *Protozoology*, 4th ed. Springfield: Charles C Thomas.

Leidy, J. 1879. *Fresh-water Rhizopods of North America.* U. S. Geol. Surv., Vol. 12.

Minchin, E. A. 1912. *Introduction to the Study of the Protozoa.* London: Edward Arnold.

Rogers, J. S., Hubbell, T. H., and Byers, C. F. 1942. *Man and the Biological World.* New York: McGraw-Hill.

Sandon, H. 1927. *Composition and Distribution of the Protozoan Fauna of the Soil.* London: Oliver and Boyd.

——— 1932. *The Food of Protozoa.* Cairo: Egyptian University.

Stein, F. R. v. 1859–83. *Der Organismus der Infusionsthiere.* Leipzig.

Stokes, A. C. 1886–88. Fresh-water Infusoria. *J. Trenton Nat. Hist. Soc.*, 1: 71–344.

Ward, H. B. and Whipple, G. C. 1959. *Fresh-water Biology*, 2d ed. New York: Wiley.

Weisz, P. B. 1954. *Biology.* New York: McGraw-Hill.

Wenrich, D. H. 1952. Protozoa as material for biological research. *Bios*, 23: 126–45.

West, L. S. 1952. Protozoa of the Upper Michigan Peninsula. *Papers of the Michigan Acad. Sci., Arts, Letters*, 88: 269–84.

Woodruff, L. L. 1938. Philosophers in little things. *Univ. Okla. Bull., New Series*, 739: 21–33.

ADDITIONAL REFERENCE

Wolken, J. J. 1961. *Euglena. An Experimental Organism for Biochemical and Biophysical Studies.* New Brunswick: Institute of Microbiology, Rutgers University.

2

The Microscope and Protozoology

"By this (the microscope) the Earth it selfe, which lyes so neer us, under our feet, shews quite a new thing to us, and in every *little particle* of its matter, we now behold almost as great a variety of Creatures, as we were able to reckon up in the whole UNIVERSE it selfe." R. Hooke, *Micrographia*, 1665.

Man has never devised a key which unlocks as many doors as has the microscope. Combinations of magnifying lenses resulting in the compound microscope have revealed to man the structure of his own body, the varied fauna and flora which make of him almost a botanical and zoological garden, and the infinitely numerous and varied population of a microenvironment whose existence was for so many centuries completely hidden. Algae, bacteria, yeasts and molds, rotifers, Protozoa, and much else people that world.

No one knows just when, or by whom, the magnifying glass was invented. It is said that the Assyrians used lenses of rock crystal, for they have been found in the ruins of the palace of Nimrud at Nineveh. The Romans knew that glass bowls filled with water had a magnifying effect, for in 63 A.D. the historian and philosopher Seneca mentions their use. But many centuries were yet to pass before such knowledge was put to real use. More than a thousand years later, in the thirteenth century, Roger Bacon wrote about simple lenses and the optical theory behind them. He even suggested their use as spectacles, though persons with defective vision had to wait another century for such relief. Credit for this device, which has added so much to our comfort, goes to Amata of Florence and de Spina of Pisa.

The science of optics, as distinguished from the use of lenses, was meanwhile developing even more slowly, though it began with Euclid. Two centuries after Roger Bacon, the great and versatile Leonardo da Vinci continued the study of the properties of lenses. Descartes also became interested in their effects on light, and in 1637 he tried mounting them on stands, thus adding a good deal to the convenience of the user.

But the era of microscopy really begins with Leeuwenhoek (1632–1723), a somewhat younger contemporary of Descartes. Leeuwenhoek, a Dutchman

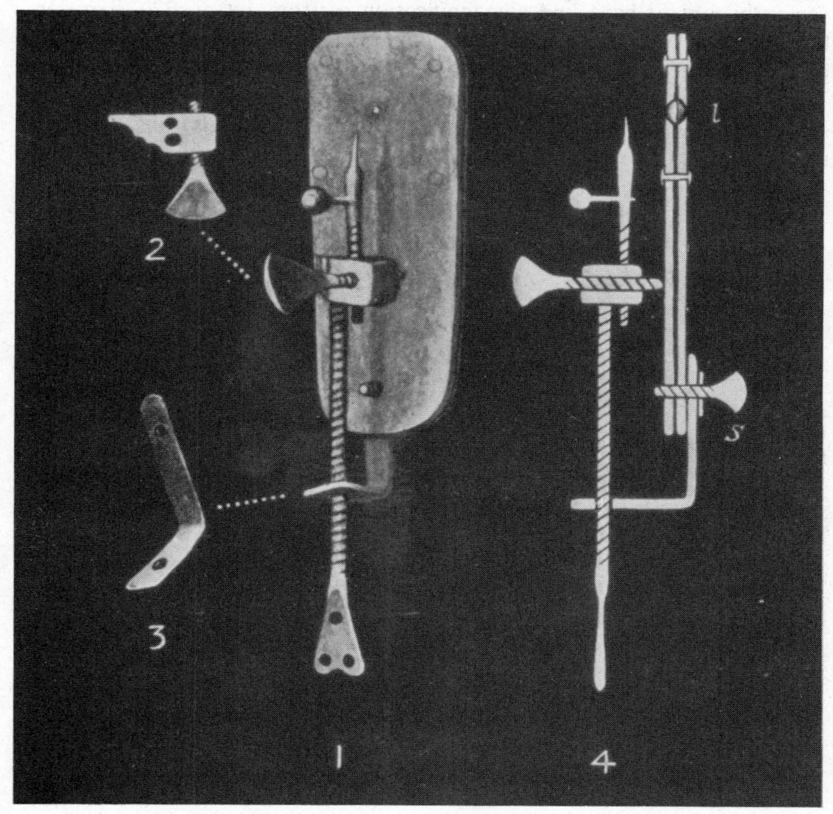

FIG. 2.1. Leeuwenhoek's simple microscope (after Dobell). The object rather than the optical system was moved to secure good focus. "*l*" indicates the lens, a finely ground double-convex one capable of a magnification of several hundred diameters. The object was mounted on the pin (shown just back of the lens in *4*), and manipulated with the screws at its base. Focal adjustment was secured by turning the thumbscrew indicated in *2*. The instrument was held close to the eye (as with a simple hand magnifier), and probably utilized a system of dark-field illumination.

with little formal education, was originally a draper or cloth merchant, but after a few years succeeded in getting himself placed on the payroll of his native town of Delft. Apparently his duties required little time, for he spent most of the remaining days of a long life in grinding lenses and using them to study almost every conceivable object (Fig. 2.1). The number of his discoveries is large. He was the first to see bacteria, Protozoa, and liver flukes. Among the Protozoa he observed and figured were Foraminiferida, the flagellates *Anthophysa* and *Volvox*, and the ciliates *Vorticella*, *Coleps*, and *Cothurnia* (see Fig. 2.2). He also saw and accurately described the cellular structure of wood. With his simple lenses, the most powerful of which is variously said to have had a magnification of from 160 to 270 diameters, he even observed human sperm and demonstrated the capillary connection between arteries and veins. His curiosity seems to have had no bounds, for —not satisfied with seeing germ cells and fertilization—he tried crossing rabbits in order to discover how coat color is inherited. He thus influenced many biological sciences, but for us his chief contribution is his observations

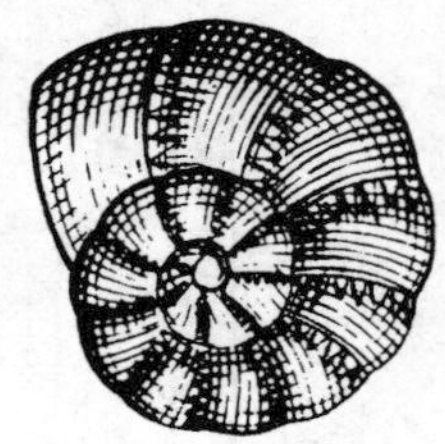

FIG. 2.2. *Left*. The ciliate *Coleps* in a drawing by Leeuwenhoek. *Right*. The shell of a foraminiferidan, perhaps *Polystomella*, in a drawing by Leeuwenhoek. (From Dobell's *Antony van Leeuwenhoek and His Little Animals*.)

of Protozoa. Detailed accounts of his discoveries are still extant in some of his numerous letters to the Royal Society of London, and it is clear that he was the first to see a number of species, free-living and parasitic (Fig. 2.3).

Leeuwenhoek's letter to the Royal Society bearing the first news of his discovery is worth quoting, both for its historical importance and its quaint language. The portion given below is taken from a translation by Dobell.

In the year 1675, about half-way through September (being busy with studying air, when I had much compressed it by means of water), I discovered living creatures in rain, which had stood but a few days in a new tub, that was painted blue within. This observation provoked me to investigate this water more narrowly; and especially because these little animals were, to my eye, more than ten thousand times smaller than the animalcule which Swammerdam has portrayed, and called by the name of water flea, or water louse, which you can see alive and moving in water with the naked eye.

Of the first sort that I discovered in the said water, I saw, after divers observations, that the bodies consisted of 5, 6, 7, or 8 very clear globules, but without being able to discern any membrane or skin that held these globules together, or in which they were inclosed. When these animalcules bestirred 'emselves, they sometimes stuck out two little horns, which were continually moved, after the fashion of a horse's ears. The part between these little horns was flat, their body else being roundish, save only that it ran somewhat to a point at the hind end; at which pointed end it had a tail, near four times as long as the whole body, and looking as thick, when viewed through my microscope, as a spider's web. At the end of this tail there was a pellet, of the bigness of one of the globules of the body; and this tail I could not perceive to be used by them for their movements in very clear water. These little animals were the most wretched creatures that I have ever seen; for when, with the pellet, they did but hit on any particles or little filaments (of which there are many in water, especially if it hath but stood some days), they stuck intangled in them; and then pulled their body out into an oval, and did struggle, by strongly stretching themselves, to get their tail loose; whereby their whole body then sprang back towards the pellet of the tail, and their tails then coiled up serpent-wise, after the fashion of a copper or iron wire that, having been wound close about a round stick, and then taken off, kept all its windings. This motion, of stretching out and pulling together the tail, continued; and I have seen several hundred animalcules, caught fast by one another in a few filaments, lying within the compass of a coarse grain of sand.

FIG. 2.3. *Above*, a seal used by Leeuwenhoek in his earlier years bearing his initials. *Below*, the seal used in Leeuwenhoek's later years appears on many of his letters to the Royal Society. The portrait is of Leeuwenhoek himself, and is said by Dobell to have been made in 1686 by Verkolje. (From Dobell's *Antony van Leeuwenhoek and His Little Animals*.)

To anyone who has studied the organisms of stagnant water under the microscope, Leeuwenhoek's description instantly brings to mind that fascinating infusorian which we today call *Vorticella*. His reference to the "two little horns" (the peristomial cilia), moving "after the fashion of a horse's ears" shows that he had imagination and a gift for pungent expression, as well as a scientist's instinct for accurate description and careful observation. Five years later Leeuwenhoek added another discovery to his already long list. In a sample of his own feces he saw the remarkable flagellate that today bears the name *Giardia lamblia.*

To the neophyte just beginning a study of protozoan intestinal parasites, the diagnosis of different species usually seems to call for wonderful skill. That Leeuwenhoek could detect these forms, seeing them well enough with his single lenses and crude illumination to give a description recognizable today, seems almost incredible. Yet observations of *Vorticella* and *Giardia* are not his only discoveries in the field of protozoology and his collected works in microscopy fill four large volumes.

Further studies of the Protozoa awaited the development of the compound microscope, which came slowly and represented the work of many men. It is said that another Dutchman, Janssen the elder, chanced to notice that a combination of lenses gave better results than a single lens. In any event, Janssen and his son about 1590 constructed a crude compound microscope. Galileo however seems to have been the first to study scientifically the optical laws involved and the first to use such an instrument effectively. He built one for his own use in 1609, but it is not on record that he made any significant observations of Protozoa with it.

The first compound microscopes were fitted with negative eyepieces, but within a few years better ones of the positive type were invented by Fontana; and O. F. Müller, using an instrument of the Janssen type and the new eyepieces, studied and described (1773–86) the remarkable group of flagellates which we know today as the Dinoflagellata, and which constitute so important a fraction of the protistan population of the sea.

But there were others who, using microscopes of a similar or even more primitive type, made very significant observations even before those of Müller. Included among them were Harris, who in 1696 first saw and described *Euglena* (although more than a century was to pass before it received its present name at the hands of the pioneer German protozoologist, Ehrenberg). Harris tells of seeing in "some Puddle-water which stood in my Yard . . . Animals of several Shapes and Magnitudes; But the most remarkable were those which . . . gave the Water that Green Color, and were Oval Creatures, whose middle part was Grass Green, but each end Clear and Transparent. They would contract and dilate themselves, tumble over and over many times together, and then shoot away like fishes." What he used for a microscope is not clear, but it may have been a simple lens, such as Leeuwenhoek's, since he also remarks that his "Glass did not magnify very much."

The first really thorough study of the Protozoa was that of Joblot, published in 1718. A Parisian, he not only described numerous new types of organisms and the contractile vacuole of Protozoa, but a number of new microscopes. He should likewise be remembered because he was the first to attempt naming the tiny organisms his microscopes revealed, and even more for his experiments on spontaneous generation. This theory he regarded as completely disproved, although his work antedated by more than fifty years the more frequently mentioned studies of Spallanzani.

Dr. John Hill (1716–75) was another pioneer in both protozoology and microscopy. His first important work, the *History of Animals*, was published in London in 1752. In it he undertook the classification of Protozoa, remarking that he had "arranged them into a regular method, and [had given] them denominations," one of his "denominations" being the well-known *Paramecium.* Probably no student of the Protozoa had as many interests as "the versatile Sir John Hill," as Woodruff calls him. A physician by vocation, and almost everything else by avocation, he achieved some distinction in nearly every field of his interest—and made a number of vigorous enemies among the bigwigs of the day. He was a prolific author with some eighty publications to his credit, many of them large volumes. The third volume in his *Natural History* series dealt with animals, particularly with those "visible only by the assistance of microscopes," for many of which he coined names, remarking that no one should object since they bore none before. Fossils, insects, plants, and minerals also interested Dr. Hill, but his contributions (though especially noteworthy in botany) failed to impress his contemporaries. One of them expressed his opinion rather pungently in the epigram (unearthed by Woodruff):

> Hill puffs himself, forbear to chide!
> An insect vile and mean
> Must first, he knows, be magnified
> Before it can be seen!

A contemporary of Hill's, the famous Linnaeus, was also sufficiently interested in the Protozoa to try to classify them, but he probably never studied them himself. He created three genera to contain all the forms then known: *Volvox*, *Furia*, and *Chaos;* only the first is still in general use.

Spallanzani was a contemporary of both Hill and Linnaeus. Originally trained for the law, he later entered the clergy, and then became a professor of Greek and philosophy; nevertheless his chief interests were in natural science. His contributions to embryology and regeneration were considerable, but he is remembered by protozoologists for his painstaking experiments (published in 1777) to refute the theory of spontaneous generation, which had recently been given renewed support by Needham. Joblot, a half-century earlier, had originated the method of boiling infusions to kill any living organisms in his experiments on spontaneous generation, but Spallanzani seems to have been unaware of this, and in any case he improved on Joblot's methods by sealing his flasks before applying heat. When this was done Spallanzani showed clearly that no microscopic life appeared thereafter.

Meanwhile attempts to improve the microscope continued. The chief difficulty was chromatic aberration. Objects viewed with lenses of ordinary glass, especially if no iris was used, would show fringes of color. No less a physicist than Newton called the problem insoluble, but eventually Samuel Klingenstierna, a professor of physics at the famous old Swedish university of Upsala, discovered how it might be done, and the English craftsman Dollond made some achromatic lenses under his direction. But, though used previously in the making of telescopes, not until the 1820's was the invention adapted to the construction of microscopes. The Italian Amici and the Frenchman Chevalier are usually given the credit, and instruments of their manufacture were soon quite widely used in Europe. Very soon afterwards an American, Charles Spencer, was making even better objectives than those produced abroad. His son, Herbert Spencer, and Robert B. Tolles, who began as Charles Spencer's apprentice, also made important improvements in microscope manufacture. Spencer the elder and later Tolles were able to construct objectives with greater numerical aperture (and hence of higher resolving power) than the best made abroad, and they also pioneered in the use of fluorspar as material for lensmaking. Probably it was in the middle fifty years of the nineteenth century that progress in microscope manufacture was most rapid.

At about the time Amici and Chevalier were introducing achromatic objectives, other important innovations were being made in England by J. J. Lister (father of Joseph Lister, who introduced antiseptic surgery). Microscopes were a hobby with J. J. Lister, although he was in the wine business. Lister devised a method of correcting spherical aberration of objectives by using two achromatic lenses, the positive aberration of one being equal to the negative aberration of the other. He pioneered in dark-field microscopy, and he was the inventor of the so-called "Lister limb," mounting the draw tube with its adjustments on the same portion of the stand as the stage and mirror, the whole being hinged to the foot as it is today.

Thus really good microscopes were available by the 1830's, and there were good biologists to use them. Ehrenberg, a native of Leipzig, was one of the greatest of the early protozoologists. Although he was trained in medicine, the biological sciences were much more to his liking, and his treatise *Die Infusionsthierchen als vollkommene Organismen* is one of the landmarks in protozoology. Published in 1838, it was the product of fourteen years of exacting study, and it contained beautifully executed figures of some hundreds of forms. Many of the names he gave these forms are still in use. Ehrenberg's main thesis, that the Protozoa are organized along the same lines as the Metazoa, has of course been disproved, but his conception of these minute animals as "complete organisms" (*vollkommene Organismen*) is certainly true. He was also mistaken in lumping together a great variety of microscopic forms which we know today have little or no relationship; the marvel is not that he made any mistakes but that he made so few, especially since the microscopes he used lacked immersion objectives, condensers, dependable illumination, and many other refinements that every college biology student today takes as a matter of course.

Another pioneer in protozoology was Félix Dujardin, a contemporary of Ehrenberg. Dujardin's studies of certain of the Foraminiferida (then regarded as minute Mollusca, similar to the chambered *Nautilus*), some of the Testacida, the amoebae, and certain ciliates (among them *Paramecium bursaria*) convinced him that Ehrenberg erred in thinking they possessed organs and organ systems like those of the multicellular animals, and he published in 1841 a treatise of his own under the impressive and awesome title *Histoire Naturelle des Zoophytes. Infusoires, comprenant La Physiologie et la Classification de ces Animaux, et la manière de les étudier à l'aide du microscope.* Like Müller and Ehrenberg, Dujardin is also remembered for the names he gave a number of new species, and his work too is notable for its many beautiful figures. Dujardin was the first to distinguish the three types of free-living forms, "les flagellés, les rhizopods, et les ciliés," and he also suggested that the protoplasm of organisms such as the amoebae was of an especially simple type, for which he proposed the term "sarcode." Bütschli, many years later, coined the term "Sarcodina" in Dujardin's honor. Dujardin's view soon prevailed, and names such as Ehrenberg's "Polygastrica" (based on the idea that gastric vacuoles were so many tiny stomachs) speedily went out of use.

The infant science of protozoology now grew rapidly, and further improvements in the microscope accelerated its progress. The study of parasitology, particularly parasitism in animals, began to receive attention. Leeuwenhoek had seen the oöcysts of the rabbit coccidian, *Eimeria stiedae*, in 1674 but his observations were not published, and Hake's account of them in 1839 was the first to be printed. Leeuwenhoek was also the first to see a protozoan parasite of man, observing *Giardia* in his own stool in 1681. Gregarines, that large and important group of Sporozoa, were described by Kölliker in a memoir published in 1848. Trypanosomes, which cost man so much in terms of human and animal disease, were first seen by Gruby in the blood of a frog in 1843, when he discovered *Trypanosoma rotatorium*. The first of the half-dozen or so species of amoebae known to parasitize man, *Entamoeba gingivalis* of the mouth, was found by Gros in 1849.

The broader aspects of protozoology were also being noted, although the emphasis continued for a long time to be on morphology. That Protozoa were cells, like the cells of Metazoa, was urged by Barry in 1843, and supported in much more thoroughgoing fashion by von Siebold two years later. Like many early biologists, Siebold had studied medicine, but found natural science more to his taste. He not only recognized that the Infusoria and Rhizopoda were single-celled animals, to which he limited Protozoa (a term previously coined by Goldfuss in 1817 to include a larger group, among them the coelenterates), but he recognized their essentially simple structure and, for that reason, put the rotifers into another group.

While Siebold was carrying on his studies at the German universities of Erlangen, Breslau, and Munich, Friedrich Stein (1818–85), at the Czech university of Prague, was busily engaged in researches on the free-living Protozoa. His four beautifully illustrated volumes under the title *Der Organismus der Infusionsthiere* are classics which every protozoologist would like to have on his bookshelf. Stein was later knighted, and the list of his honors—meticulously given in seven lines of fine print on the cover of Volume II of the *Infusionsthiere*—comprises memberships in many academies of natural science, including one in Philadelphia. Stein's first notable work on the Protozoa, appearing in 1848, dealt with the life history of a gregarine, and the last volume of the *Infusionsthiere* appeared in 1883.

During this time the microscope continued to be improved. The water immersion objective was introduced in the 60's and some of these objectives were of high quality. After a few years of wide use, however, they were superseded for most purposes by the oil immersion objective, invented by the German Abbe, to whom all microscopists owe an incalculable debt. Among Abbe's many other contributions to microscope design are the condenser, without which no immersion objective would be of much use; apochromatic objectives, using in their manufacture fluorite and new and finer optical glasses; and the grinding of lenses from previous calculations of their optics, thus eliminating much needless trial and error. It has been said that microscopy owes more to Abbe and Lister than to anyone else, and of the two, more to Abbe.

But others, too, have been innovators in microscope design. Christian Huygens, a Dutch physicist, who lived at the time of Leeuwenhoek, invented the so-called Huygenian eyepiece, consisting of two lenses, the field lens being below the diaphragm. He is said to have introduced dark-field observation, and, incidentally, made epochal improvements in pendulum clocks, and observed the planet Saturn with a telescope of his own manufacture. Binocular microscopes were introduced by Wenham, an Englishman, but an American named Riddell, resident in New Orleans, devised the first really successful design in 1851. Yet not until 1913 were the refinements in the prism system necessary to make such microscopes really practical developed, and these were the work of Dr. F. Jentsche of the great German firm of E. Leitz, Inc. Mechanically, the microscope has changed relatively little since Lister's time, except for the coarse and fine adjustments. The diagonal rack, adding much to the smooth operation of the coarse adjustment, was invented and introduced by the English firm, J. Swift and Son, in 1880. The fine adjustment has also benefited from many improvements, but unlike most other microscope features, it is still not standardized.

A major development in microscopy is, of course, the electron microscope. By the use of a stream of electrons and suitable magnetic fields to act as lenses, magnifications reaching the almost incredible figure of several hundred thousand can be obtained. But since the image formed by the electrons is invisible to the eye it must either be recorded on photographic film or made visible by the use of a luminescent screen. Another limitation is that the objects to be viewed must be placed in a vacuum. Most biological work with the electron microscope has been done on the viruses, bacteria, and certain tissues or skeletal structures of a few of the higher animals. But with refinements in technique it has become possible to study Protozoa also, and some very interesting facts about their finer structure have already come to light.

Even more recently the method of using the ordinary light microscope known as "phase microscopy" has been introduced. The principle underlying phase (or phase contrast) microscopy is that light waves are altered in phase, or retarded, when they pass through a medium in which the optical path is different. Since objects being viewed are not likely to be homogeneous a microscope equipped with suitable condenser, diffraction plate, and objective lens system, can detect such differences, causing them to appear as areas of varying brightness or color. Thus the method often reveals in the living cell structures which even the best staining processes might not otherwise uncover, and it is equally well applied to the study of stained preparations themselves.

Meanwhile studies of the "infinitely small" continued. An early problem—where to draw the line between plants and animals—is by no means solved even now, for the botanists and zoologists still differ strongly (though quite amicably) about the correct position of the plantlike flagellates (*Phytomastigina*). A related problem concerned the real nature of Protozoa which owed their green color to the presence of symbiotic algae, for it was only when the two organisms—host and symbiont—could be separated that one could be sure that each had an identity of its own. Symbiosis of this sort was not established until Cohn did so in 1851.

Our knowledge of the Foraminiferida again begins with Leeuwenhoek, who saw the forms we know as *Polystomella* in the stomach of a shrimp, but these organisms were generally ignored until d'Orbigny in 1826 placed them in the order Foraminiferida along with the Cephalopods, to which the shells of some of them have a strong resemblance. Dujardin corrected this mistake nine years later, and put them with the Rhizopoda where they remain today.

The other great group of marine Protozoa, the Radiolarida, were discovered at almost the same time (1834) by Meyen, who described two species, but not until 1858 were they placed with the Rhizopoda by J. Müller. But the greatest discoveries in the field of these remarkable marine organisms were those of the great German zoologist Haeckel. A student of Müller's, he began his microscopic studies as a boy, having been given a microscope by his father, who hoped it would start his son toward a medical career. But it seems to have given the boy a consuming interest in microbiology instead, and while still in his twenties, young Haeckel published a monumental work on the Radiolarida, *Die Radiolarien*, in which he described more than 4,000 species, most of them new.

While Haeckel was doing his elaborate studies of the Radiolarida in

Germany, Émile Francois Maupas engaged in equally important researches on the ciliates. Maupas' chief interest was in their life cycles and the significance of the changes taking place in conjugation. Balbiani and Stein had regarded the process as sexual but they misunderstood the cytological changes involved. The former had thought the chromosomes in the dividing micronucleus to be spermatozoa, and believed the macronucleus an ovary. Bütschli, who may well be regarded as one of the fathers of modern protozoology (despite the fact that he was trained in mineralogy), showed that the micronucleus was essentially reproductive, and that the macronucleus was ultimately derived from it in the process of conjugation. In studies published in 1888–89, Maupas proved to his own satisfaction that conjugation is a rejuvenating process and described in detail the nuclear changes involved. There is no doubt of the accuracy of Maupas' observations or of the thoroughness of his experiments, but more recent studies with refined techniques by Calkins, Woodruff, and others have tended to modify Maupas' conclusions considerably. It is now clear that species differ greatly and that this extends to the effects of conjugation. Yet Maupas' interpretations of ciliate behavior as in part a manifestation of youth, maturity, and old age are now believed to be largely correct.

Few things, however, are as simple as they often seem, and conjugation among the ciliates has turned out to be a problem with many ramifications. These have become more and more evident with the revelation of mating types, pioneered by the brilliant work of Sonneborn, and the varieties, or physiological species, recently dubbed "syngens." All this has made of ciliate genetics a complex subject indeed.

The last quarter of the nineteenth century saw important work in protozoology prosecuted in both Europe and America. Kent, in England, published his famous *Manual of the Infusoria* in 1880–81, still a standard reference work for protozoologists studying the taxonomy of the free-living species. Leidy pioneered in the investigation of both free-living and parasitic Protozoa on this side of the Atlantic, in addition to important work in other fields of natural science. He was a physician, but his interest in natural history was great. Leidy was the first to recognize that parasitic amoebae were sufficiently different from free-living forms to justify a distinct genus, *Endamoeba*, which he created in 1879 originally for one of the amoebae of the cockroach. (The first such amoeba to be seen—by Gros in 1849—seems to have been *Entamoeba gingivalis*). But Leidy's greatest contribution to protozoology was his monograph *Fresh-water Rhizopods of North America*, magnificently illustrated with his own meticulously accurate drawings, partly in color.

Another American physician who pioneered in somewhat similar fashion was Alfred C. Stokes. His interests were chiefly in the ciliates, of which he described numerous new species in a monographic article published in 1888. Somewhat surprisingly, though Leidy and Stokes were trained in medicine, the possible relationship of Protozoa to disease appears not to have impressed them. This may have been because protozoan diseases are most prevalent in warmer climates, and, therefore, have been relatively unimportant in the greater part of the United States. Furthermore, we have never had a colonial empire with extensive areas in the tropics and resulting health problems due to parasite-caused diseases. Nevertheless, malaria and amoebiasis were common enough over much of the United States in the nineteenth century.

The amoeba of amoebic dysentery was first seen by Loesch (1875) in a dysentery patient in St. Petersburg, and the plasmodia of malaria were discovered in 1880 by the French army surgeon Laveran. The transmission of malaria by mosquitoes was demonstrated first by the English army doctor, Ronald Ross (later knighted) in bird malaria, and confirmed by the Italian Grassi, in the human disease a year later.

Other discoveries in malariology (some of them erroneous) were made at about the same time by the young German scientist Schaudinn (1871–1906), although he is probably best known for his discovery of the syphilis spirochete. He also carried out significant researches on the life cycles of Foraminiferida, and on such Sporozoa as *Monocystis* of the earthworm, and the very important group known as *Coccidia.* Death came to him at the early age of thirty-five—it is said from an amoebic abscess which resulted from a successful attempt to infect himself. Goldschmidt has recently said of Schaudinn, "Present-day protozoology is not imaginable without him."

It may also be said that knowledge of the role of Protozoa as causes of disease increased more rapidly in the years just prior to and just after the turn of the century than at any other time, before or since. That human sleeping sickness is caused by trypanosomes was shown by Forde in 1901, and its transmission by tsetse flies was proved two years later by Bruce and Nabarro. These studies received their initial inspiration from the work of Sir Patrick Manson who, in 1877, found that the microfilaria from cases of human filariasis would develop in culicine mosquitoes, and also from the proof by the American scientists Smith and Kilbourne in 1893 that Texas cattle fever (a protozoan disease then threatening the cattle industry of the Southwest) is spread by infected ticks.

Since 1900 our knowledge of both free-living and parasitic Protozoa has increased at an ever-accelerating rate, until protozoology has become recognized as a branch of biology quite equal in rank to bacteriology, helminthology, entomology, and the many other life science specialities. It is not possible, without seeming to do injustice to many of its devotees, to attempt to recognize the many who have contributed to its development. Yet no summary, incomplete as it may be, should omit mention of S. von Prowazek and Reichenow in Germany, Dogiel in Russia, Brumpt and Penard in France, Wenyon in England, and Hegner and Kofoid in the United States. Science does not recognize national boundaries, and scientists come from all nations. Protozoology is still young, but through the efforts of many men in many countries it is rapidly coming of age.

REFERENCES

Baker, H. 1753. *Employment for the Microscope*. London.

Belling, J. 1930. *The Use of the Microscope*. New York: McGraw-Hill.

Clay, R. S. 1938. A review of the mechanical improvements of microscopes in the last forty years. *J. Roy. Mic. Soc.*, 58: 1–29.

Cole, F. J. 1926. *History of Protozoology*. University of London Press.

Dobell, C. 1932. *Antony van Leeuwenhoek and his Little Animals*. New York: Harcourt, Brace.

Gage, S. H. 1943. Brief History of lenses and microscopes. In *The Microscope*. Ithaca, New York: Comstock.

Goldschmidt, R. B. 1956. *Portraits from Memory. Recollections of a Zoologist*. Seattle: University of Washington Press.

Kent, S. 1880. *A Manual of the Infusoria*. London: David Bogue.

Nordenskiold, E. 1929. *The History of Biology*. New York: Knopf.

Three American Microscope Builders. 1945. American Optical Co.

Wenrich, D. H. 1956. Some American Pioneers in Protozoology. *J. Protozool.*, 3: 1-7.

Woodruff, L. L. 1919. Hooke's *Micrographia*. *Am. Nat.*, 53: 247–64.

——— 1937. Louis Joblot and the Protozoa. *Sci. Monthly*, 44: 41–7.

——— 1939. Some pioneers in microscopy, with special reference to protozoology. *Trans. N. Y. Acad. Sci.*, 1: 1–4.

——— 1940. Microscopy before the nineteenth century. *Biol. Symp.*, 1: 5–36.

3

Antiquity of the Protozoa

"About six years ago, being in England, out of curiosity, and seeing the great chalk cliffs and chalky lands at Gravesend and Rochester, it ofttimes set me a-thinking; and at the same time I also tried to penetrate the parts of the chalk. At last I observed that chalk consisteth of very small transparent particles. . . ."* Leeuwenhoek, Letter 6, September 7, 1674.

No group can boast of greater antiquity than the Protozoa. Far older than the pyramids of the great Egyptian Pharaohs are the rocks filled with protozoan fossils of which they are made. The fossils are of Foraminiferida, inhabitants of the seas for more than half a billion years, and silent reminders of man's relatively recent arrival on the earth.

To the biologist fossils are extremely significant, for they are the best evidence of evolution; the rocks containing them provide a prehistoric textbook, with many pages missing of course, but open for all to read who can. From them we may learn not only about the story of life on earth, but even of the climates and geography of the past. Fortunately, though many kinds of Protozoa have left no fossil remains, others are quite abundantly represented in ancient rocks.

The minute and intricately marked skeletons of the dinoflagellates are often found as fossils in the flint concretions associated with chalk and limestones. Certain species, notably *Peridinium conicum* (Fig. 3.1), a dinoflagellate known from Cretaceous flints, are still abundant in the Atlantic and Pacific oceans. *Peridinium* seems not to have changed for at least 125 million years. The eminent French protozoologist Deflandre remarks that we owe much to these tiny organisms. Prehistoric man started his fires and made his weapons using the flint to which their fossil remains contributed. Modern man uses chalk and limestone to manufacture the cement to build his great

*Some of these particles must have been fossil Foraminiferida. (Quotes from Leeuwenhoek are in Dobell's translation.)

skyscrapers, his fortifications, and his highways. Even the hydrocarbons, of which petroleum and its products consist, may have come in part from the carbohydrates synthesized by these microorganisms in the sunshine falling on the ancient seas.

Protozoan fossils, like those of other living things, may tell us something of the climatic cycles of antiquity. The Foraminiferida especially are sensitive to environmental temperatures and both the occurrence and abundance of different species are indices of the degree of warmth of the seas in which

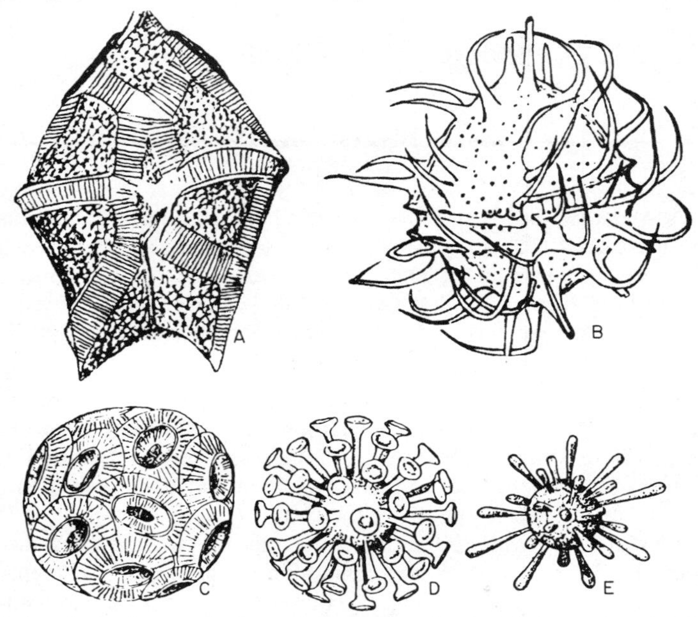

FIG. 3.1. Fossil flagellates. *A*. Theca of the dinoflagellate *Peridinium conicum* beautifully preserved in Cretaceous flint of Denmark. *B*. Another fossil dinoflagellate, *Hystrichodinium*, found in Cretaceous flint from the neighborhood of Paris. *C*. A coccolith, *Coccolithus pelagicus*, very common in the Cretaceous seas. *D and E*. Two other fossil coccoliths. (From Deflandre, *Nature*.)

they live. The temperatures of ocean waters change gradually with the passing of time, and there is every reason to think that such oscillations are correlated with long-term fluctuations in climate, although much remains to be learned of these relationships and little is yet known of their ultimate causes. It is fairly certain that the upper layers of the oceans were considerably colder during the Ice Ages than they are now. New techniques of dating, based on the amounts of the isotopes of carbon (C_{14}) and oxygen (O_{18}) in fossils, have made it possible to set the age of relatively recent fossils quite accurately. Fossil Foraminiferida so dated from sea bottom deposits in the Arctic, Pacific, and Atlantic Oceans and in the Caribbean Sea, seem to show that sea water temperatures began to drop about 80,000 years ago, and continued to do so for 65,000 years. Presumably this period corresponded to the last Ice Age. In the 15,000 years since there has been a marked temperature rise in ocean water. One of the more sensitive indicators of these changes is the foraminiferidan genus *Globigerinoides*. Two species, *G. sacculifera* and *G. rubra*, have proved especially useful.

Likewise, study of the distribution of fossil Foraminiferida has apparently shed light on the changing geography of the world at different periods in the past. Comparison of fossil types found in the rocks of the Gulf States and the coastal regions of Ecuador shows similarities which can be plausibly explained only by postulating a connection between the two oceans in late Eocene times, perhaps 50 million years ago. Fossil Foraminiferida have even been used to elucidate the geologic history of islands such as Guam and the Marianas. The remains of these Protozoa show that there was a period of maximum volcanic activity in this island neighborhood in Eocene times, and it is this which is thought to have established their present geography.

Protozoa Found as Fossils

By far the largest proportion of protozoan fossils consist of the remains of Foraminiferida and Radiolarida (Fig. 3.2), but those of flagellates (particularly the plantlike forms, or Phytomastigina) are abundant in rocks of some parts of the world. Even the fossilized tests of a few species of ciliates have been discovered.

We are tempted to think of these very ancient Protozoa as both simple and primitive, as our contemporary, the humble amoeba, is still generally considered. In truth they were neither. The remains of fossil Foraminiferida show that they closely resemble modern types, both in shell structure and life cycle (Figs. 3.7,8,9). We know something about the latter because of the occurrence of two types of fossil shells differing in size. The smaller is said to be "megalospheric" and the larger "microspheric," because the initial chamber, or "proloculum," is larger in the smaller shell, and vice versa. These two shell types are evidence of an alternation of generations which occurred many millions of years ago, just as a similar alternation does today. Truly simple protozoan forms must have existed before these, complex in life cycle and structure, could have developed, but we know nothing of what they were like or when they lived.

Though Foraminiferida are undoubtedly better represented as fossils than any other group of Protozoa, the Radiolarida are also abundant. The former are typically dwellers on the sea bottom, perhaps because of the weight of their many-chambered shells, but the latter are mostly inhabitants of the open ocean, carried freely by the waves and wind-induced currents. Although a few Foraminiferida are known in fresh-water habitats, or in brackish waters, all the Radiolarida are marine. Most of them have beautiful and intricately designed skeletons, more or less spherical, of silica or (in Acantharida) strontium sulfate (Fig. 3.2). Although their shells are heavy, many Radiolarida can regulate their buoyancy through a system of hydrostatic vacuoles and thus can float or sink to a considerable depth. The principle is exactly that of the submarine. Radiolaridan fossils are among the oldest known, for they have

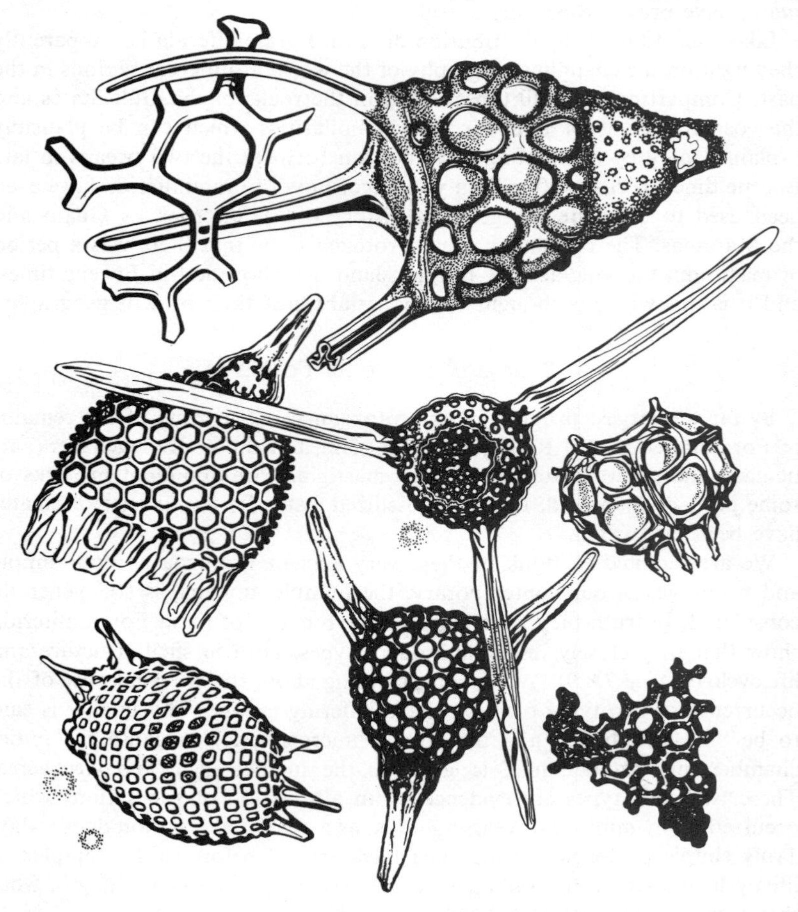

FIG. 3.2. Siliceous skeletons of half a dozen species of recent (now living) Radiolarida. Note especially the perfect symmetry and the intricate patterns of the shells.

been found in pre-Cambrian rocks. But even in that almost unimaginably remote time, when these deposits were being laid down at the bottom of some primordial sea, we may be sure that protozoan life was already hoary with age for these ancient remains indicate an already high degree of skeletal specialization.

Dinoflagellates, like Radiolarida, are typically residents of the open ocean and their fossils have been found in rocks formed from marine sediments laid down in regions as far apart as France and Australia (Fig. 3.1). Although less ancient than the Foraminiferida and Radiolarida, dinoflagellate remains have been identified in rocks of upper Jurassic age and thus these Protozoa existed much as they do today at least 150 million years ago. Even more remarkable than their antiquity is the nearly perfect preservation of their skeletons in the ancient limestones entombing these tiny organisms. Evidently the tough and elaborately carved cellulose envelopes which enclose many of these organisms and the silica-impregnated shells which cover a few of the rarer forms make for easy fossilization, and they also make identification relatively simple, since they resemble nothing else in the protozoan world. Still it seems remarkable that the dinoflagellates were among the first fossil Protozoa to be thoroughly studied, and that that pioneer protozoologist, Ehrenberg, who had so many other interests, was the one to do it. For the micropalaeontologist of today uses refined techniques entirely unknown a century and a quarter ago when the German scholar did his work, and his modern counterpart has far better optical equipment.

Several other groups of flagellates (although they are often regarded as algae by the botanists and the micropalaeontologists) are known in the fossil state. One such group is the Coccolithina or Coccolithidae (there is little agreement about its taxonomic rank), so called because of the calcareous platelets or coccoliths making up the skeleton (Fig. 3.1). Coccolithic remains are often associated in great numbers with fossil Foraminiferida such as *Globigerina*, a genus which itself goes back well over a hundred million years. But the organisms which left us their coccoliths, or armor plates, first appeared in the upper Cambrian, which makes them at least 450 million years old. The relative position of this and other geologic periods is shown in Table 3.1. The coccolith-bearing flagellates which have survived to the present day are not a very numerous group, and like the dinoflagellates are almost wholly restricted to salt water, although a few live in brackish water and a handful of aberrant forms have acclimated themselves to life in fresh waters.

A second group well represented as fossils is that of the silicoflagellates. Like the wearers of the coccoliths, the silicoflagellates are regarded by some as constituting an order and by others as a family. Their preservation as fossils is due to the siliceous skeleton they produce, often of bizarre shape and elaborately decorated with bristling spines. Fossil types are known from Cretaceous rocks, and their owners were therefore contemporaries of the flying reptiles and toothed birds which filled the skies of that strange world of 150 million years ago. Living silicoflagellates are almost exclusively marine and pelagic in habit.

The Ebriedians, which numerous fossil remains indicate have been contemporaries of the silicoflagellates during much of their long history, constitute a third group (Figs. 3.3 and 3.4). They possess an internal siliceous skeleton built about a central spicule having three or four symmetrically branched spines, and often elaborated to a high degree of complexity. Most of these flagellates are now extinct, and only their skeletal characteristics are known, but one may suppose that they had two flagella and a nucleus of the dinokaryon type, like the few surviving species, and conceivably they lacked chlorophyll and were holozoic (animal-like) in their nutrition. And they were doubtless all marine. Living Ebriedians reproduce by a form of binary fission resembling budding, although, as Deflandre remarks, the skeletal complexities of some fossil types suggest that they may have had more complicated life cycles. Deflandre, whose studies are responsible for most of what we know of the group, divides it into four families having a

TABLE 3.1

GEOLOGICAL TIME SCALE
AND THE SUCCESSION OF LIVING THINGS

Era	*Period*	*Years Since Beginning (millions of years)*	*Genesis of Animal Types**
Cenozoic	Recent		
	Pleistocene	1	Man
	Pliocene	15	
	Miocene	32	
	Oligocene	47	
	Eocene	68	
Mesozoic	Cretaceous	140	Flying reptiles; toothed birds; snakes
	Jurassic	167	First birds
	Triassic	196	First mammals (marsupials)
Palaeozoic	Permian	220	
	Carboniferous	275	First true reptiles
	Devonian	318	First amphibia; first snails (land); insects
	Silurian	350	
	Ordovician	430	Ostracoderms (armored fishes); corals; echinoderms; molluscs; trilobites; sponges; Foraminiferida
	Cambrian	510	
Archeozoic	Laurentian	?	Radiolarida

*It must be borne in mind that the time of appearance given for the different animal groups is based only on the occurrence of the oldest known fossils. The actual first appearance of each of these groups must therefore have been even earlier.

total of 18 genera, the oldest of which dates back some 75 million years. Fossil Ebriedians have been collected in many parts of the world, among them Russia, Hungary, Jutland, the United States, Japan, and New Zealand.

Nor are the flagellate groups listed above all. There are, for example, the Archaeomonads (Fig. 3.5), and the Hystrichosphaeridians (Fig. 3.6), curious looking forms with spiny shells sometimes seemingly as complex in structure as the names they bear. They have been found as fossils in Cretaceous rocks in Europe and Australia, and doubtless occur elsewhere.

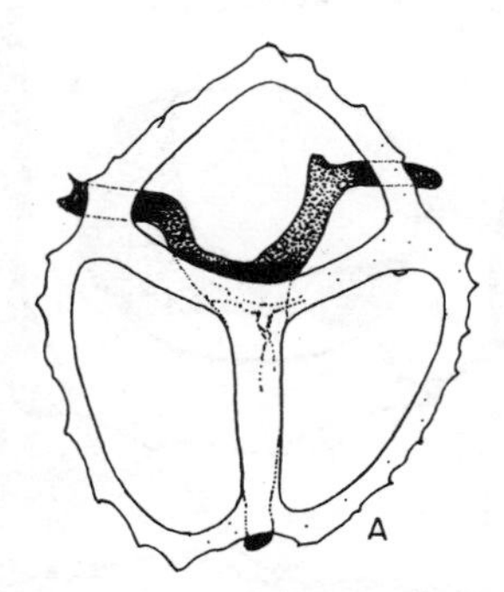

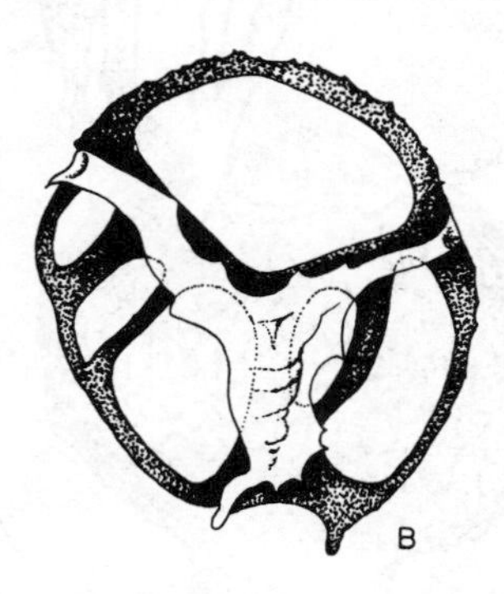

FIG. 3.3. Two views of ebriedian *Ebriopsis narbonensis*, from Oligocene deposits of southern France. *A*, ventral view; *B*, dorsal view. (After Deflandre and Gageonnet.)

Not long ago, too, plant remains were described from the extremely ancient rocks of the Canadian Shield of Southern Ontario, dating back possibly two billion years—almost far enough, one might think, to glimpse the dawn of life. Yet the organisms found were not really primitive, for they included two types of fossil algae, two of fungi, and one provisionally identified as perhaps a species of the marine flagellate genus *Discoaster*. These last are planktonic organisms with calcareous shells, perhaps related to the Coccolithidae.

And there are, finally, a few test-bearing ciliates that have left us a record in ancient rocks of their existence. One of them is a species of *Calpionella*, thought to be related to the familiar and ubiquitous *Halteria*, the fossils of which have been found in Alpine limestones 150 million years old.

Of the other major groups of Protozoa, none is known to occur as fossils. This is not surprising as far as the naked forms are concerned, but it might reasonably be expected that those with tests in these groups (and there are many such species) would have left fossil evidence of their history. Certainly the degree of structural differentiation, and especially the highly complex life cycles of many such forms, indicate an origin far back in antiquity. Were it not for their minuteness, fossils of such Protozoa might indeed have been discovered long since, for some are so abundant in many kinds of aquatic environments as to make it almost certain that fossil evidence of their existence in the remote past exists. One must suppose that discovery awaits the

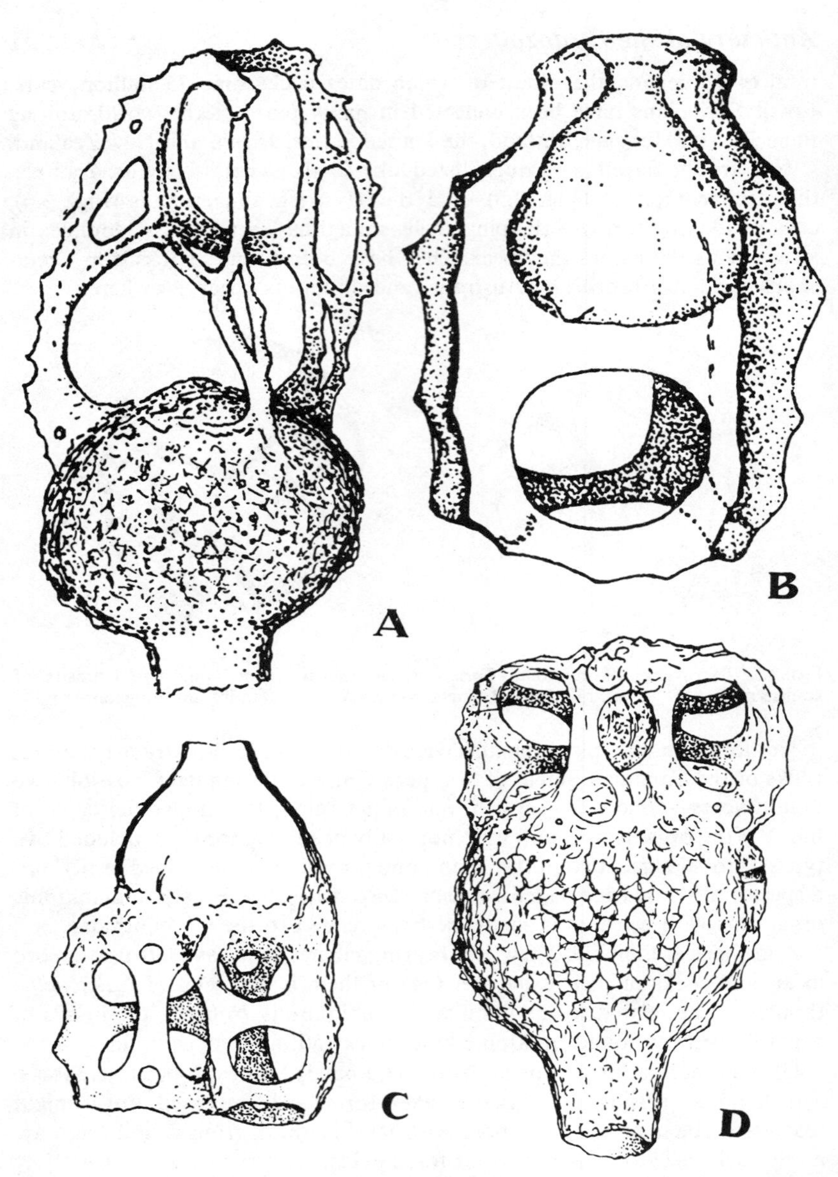

FIG. 3.4. *A*. Lorica of the fossil ebriedian flagellate, *Pedamphora elgeri*. *B*. Fossilized cyst of the ebriedian *Ammodochium rectangulare*. *C*, *D*. Two variant fossil loricae of *Ammodochium ampulla*. (After Deflandre.)

efforts of a researcher with the requisite patience and mastery of some of the recent refined techniques. These include methods of cutting rock into thin and nearly transparent sections, easily examined under the microscope, and, when necessary, treating the fossil-containing specimens with acid to dissolve out extraneous material, sifting and even staining the residue, with final mounting on a slide for microscopic study. Fortunately for the fossil

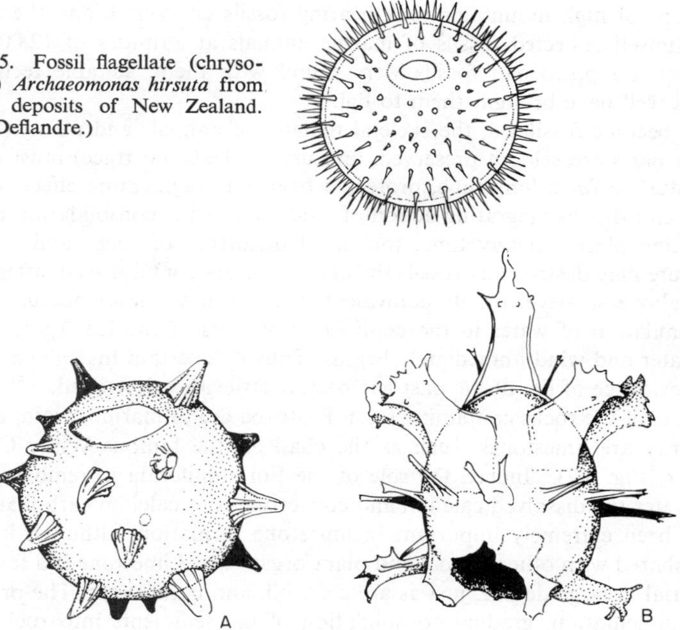

FIG. 3.5. Fossil flagellate (chrysomonad) *Archaeomonas hirsuta* from marine deposits of New Zealand. (After Deflandre.)

FIG. 3.6. Fossil flagellates of the hystrichosphaerids, perhaps related to the dinoflagellates. These fossils have a wide occurrence; those illustrated are from Australia. *A*, *Hystrichosphaeridium striatoconus*, upper Cretaceous; *B*, *Hystrichokolpoma rigaudi*, Eocene. (After Deflandre and Cookson.)

hunter, a few forms, such as certain Foraminiferida, are very easily seen with the naked eye, although to the layman it often seems a marvel that fossils of Protozoa were ever discovered at all.

HOW FOSSILS ARE FORMED

As generation followed generation, the bodies of shell- and armor-bearing Protozoa settled on the sea bottom until, with other sediments, layers many feet in thickness were formed. Such sediments are called "ooze" and cover most of the ocean floor. Sedimentation is not confined to the seas, but layers so formed in bodies of fresh water are thinner, for the life of a lake or pond is relatively short. In the bottom silt the protoplasm of these tiny organisms speedily disintegrated, but the skeletons remained intact for long periods of time. Eventually the strata of accumulated sediment became rock (usually some variety of limestone) and, with changes in the contour of the earth's crust or in sea level, some of this material was raised from the ocean floor to become dry land, often at a considerable elevation. Even at or near the tops of high mountains rock bearing fossils are found, and the author has himself collected fossils of marine animals at altitudes of 12,000 feet. Though no protozoan fossils were found with them, suitable techniques might well have brought them to light.

To become fossilized, the skeletal remains of animals and plants (for the softer parts are seldom preserved, and usually leave no trace) must remain undisturbed for a long time, protected from the fragmenting effects of currents and the disintegrating action of oxidation, while consolidation to rock is taking place. At any time, too, local distortion of rocks and resulting pressure may destroy any fossils the affected layers contain. And, after elevation above sea-level or, its equivalent, a drop in sea level because of the accumulation of water in the continental glaciers of the Ice Ages, erosion by water and wind immediately begins. Thus it is certain that much invaluable evidence of life of the past has been irretrievably destroyed.

Most of the rocks containing fossil Protozoa are of marine origin, and the majority are limestones, such as the chalk of the famous White Cliffs of Dover (Fig. 3.7). Indeed the role of the Foraminiferida in removing from sea water the dissolved calcium and converting it to calcium carbonate must have been extremely important in limestone formation, although it was a role shared with other animal and plant organisms. Limestone is a fossil raw material in large degree, just as are coal, oil, and natural gas. The processes of sedimentation, gradual consolidation of the sediments into rock, uplift through local deformation of the earth's crust, and erosion by wind, water, and successive freezing and thawing, are continuous and of course still going on. So also may be the formation of hydrocarbons, for it has been shown that "recent" sediments (those having an age of about 12,000 years) from the floor of the Gulf of Mexico contain appreciable amounts of such substances. Their source is believed to be "the hydrocarbon remains of many forms of marine life," and of such life it is certain that the Protozoa are an important part (Smith, 1952).

A like role, but probably a relatively less important one, was played by the Radiolarida in fixing dissolved silica, and thus making it available for the formation of some rocks. Radiolaridan ooze, however, accumulates only at depths exceeding 12,000 feet, and uplifts of the earth's crust are seldom sufficient to bring the rock which may ultimately form from such sediments to the surface or above it. Thus rocks rich in radiolaridan fossils are less often discovered. Most silicate rocks are not the product of living organisms

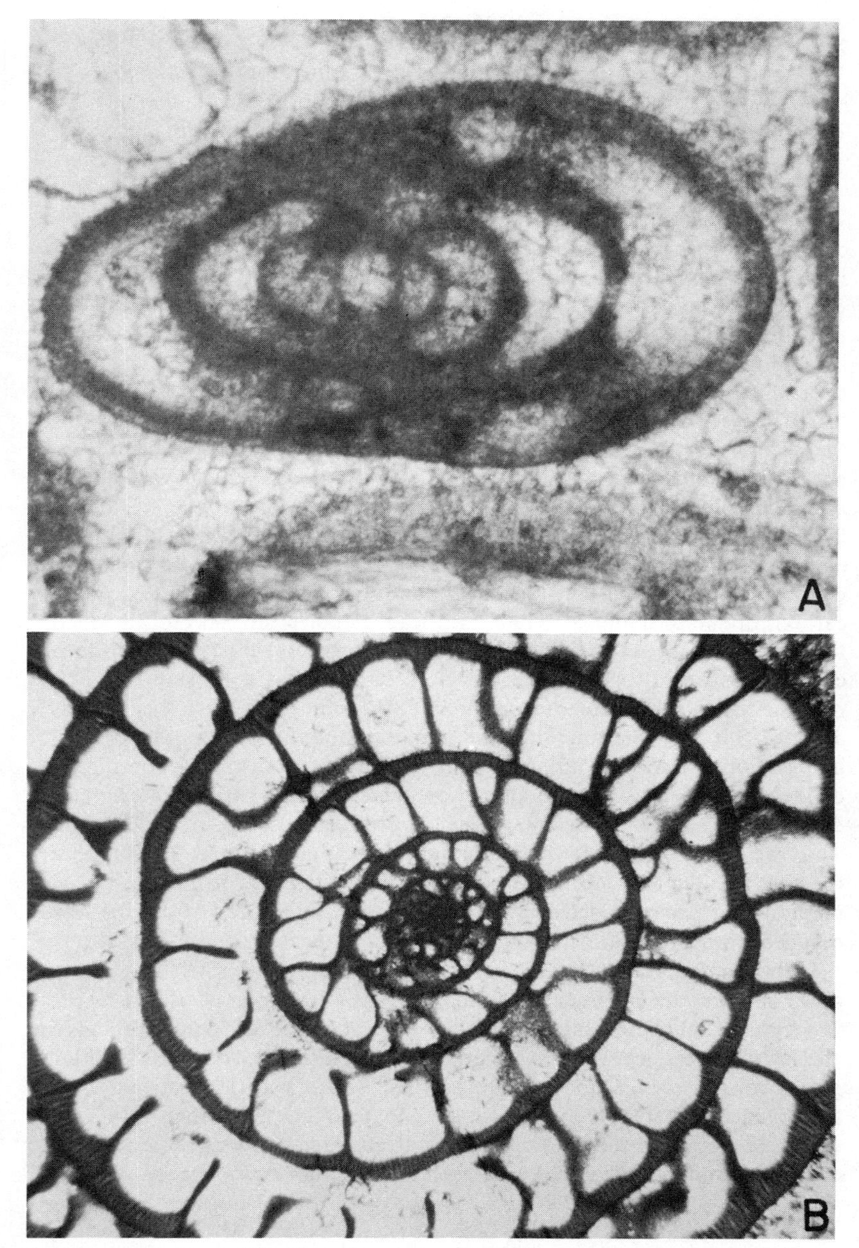

FIG. 3.7. Fossil Foraminiferida in limestone. *A*. Perhaps a species of *Discorbis* from Matlock, England. Cross section, ×176. *B*. Fossil form from China (place and age unknown). It may be a species of *Rotalia*. Although the magnification is only ×40, the shell was nevertheless too large to be included in its entirety in a single field. Sagittal section.

as are the limestone, although flint, to the formation of which dinoflagellates have sometimes contributed, is a partial exception.

Determining the Age of Fossils

Fossils constitute the best evidence of evolution, but the value of such evidence is largely dependent on knowing how old they are, and their age is measured in a number of ways. The best method, when it can be used, depends on the fact that certain elements, such as uranium and thorium, or for very recent remains, carbon 14, decay at a constant rate. The products of the decay of uranium and thorium eventually are several isotopes of lead; of carbon 14, the stable and usual form of the element, carbon 12. If the amount of isotopic lead in a piece of rock can be determined with exactness, and checked against the uranium or thorium remaining, the resulting ratio will be a measure of age. This method gives fairly accurate determinations even for very ancient rocks, since the half-lives of uranium and thorium are about 4.5 and 14 billion years respectively. Thus the amounts of these elements originally present will have been reduced by half after these lengths of time have elapsed. Similarly, the half-life of carbon 14 is 5,568 years. This form of carbon, unlike uranium, thorium, and lead, is not a normal constituent of the earth's crust, but is formed in the upper atmosphere by the action of cosmic rays on nitrogen. Since this process occurs at a constant rate, the ratio of the two types of carbon dioxide (the one containing C_{12} and the other C_{14}) is also constant; the ratio applies to the carbon compounds of all living things, which are ultimately derived in part from the carbon dioxide of the atmosphere. After the animal or plant dies, the carbon 14 gradually disintegrates, and the amount present is a measure of age of the remains. Thus we can get accurate determinations of age up to about 25,000 years; for ages greater than 40,000 the method is of little use. It is only by this method that accurate determinations of the ages of cores of marine ooze dredged up from various depths can be made.

Indirect methods of age determination may also be used. If the age of a given stratum can be measured, quite obviously layers below it will (except in very unusual circumstances) be older and those above younger. If a given species of animal or plant is known to have existed only within a certain geological time range, the age of any fossil found with it must fall within that range. Sometimes it is possible to date a given stratum with some exactness simply because it contains an assemblage of certain fossil species, even when individual species among them could not be so used.

Practical Utility

Unquestionably, the greatest stimulus to the study of protozoan fossils has been the information they supply the oil driller about the identity and

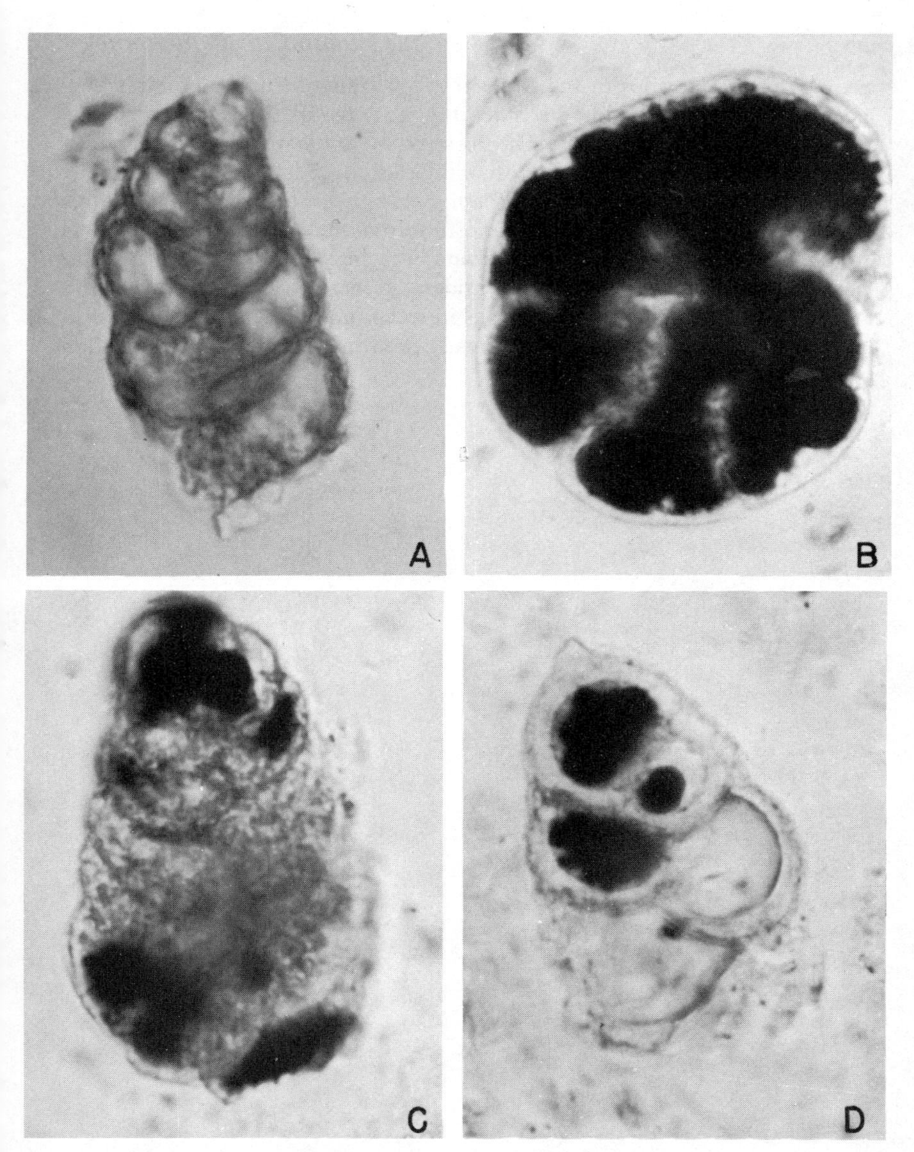

FIG. 3.8. Fossil Foraminiferida from Texas oil well cores, ×676. *A*, *C*, *D*. Complete or nearly complete shells, probably species of *Textularia* from the Eocene, and thus fifty or more million years old. *B*. Perhaps a species of *Ceratobulimina*, also probably of Eocene age. The black inclusions so conspicuous in this shell, and less so in the shells shown in *C* and *D*, represent the carbonaceous remains of organic material.

geological relationships of the strata through which he drills. For this purpose, fossil Foraminiferida are much the most useful. Recognition of this fact is largely due to the late Dr. Joseph A. Cushman, who established a laboratory for the study of Foraminiferida at Sharon, Massachusetts. Figure 3.8 shows several types of fossil Foraminiferida from oil-well cores collected in Texas. The use of these tiny fossilized organisms as markers for the identification of different strata was introduced into Japan after World War II and resulted in a considerable increase in oil production.

The value of foraminiferidan fossils in recent marine sediments as indicators of past climatic changes has already been mentioned, and information

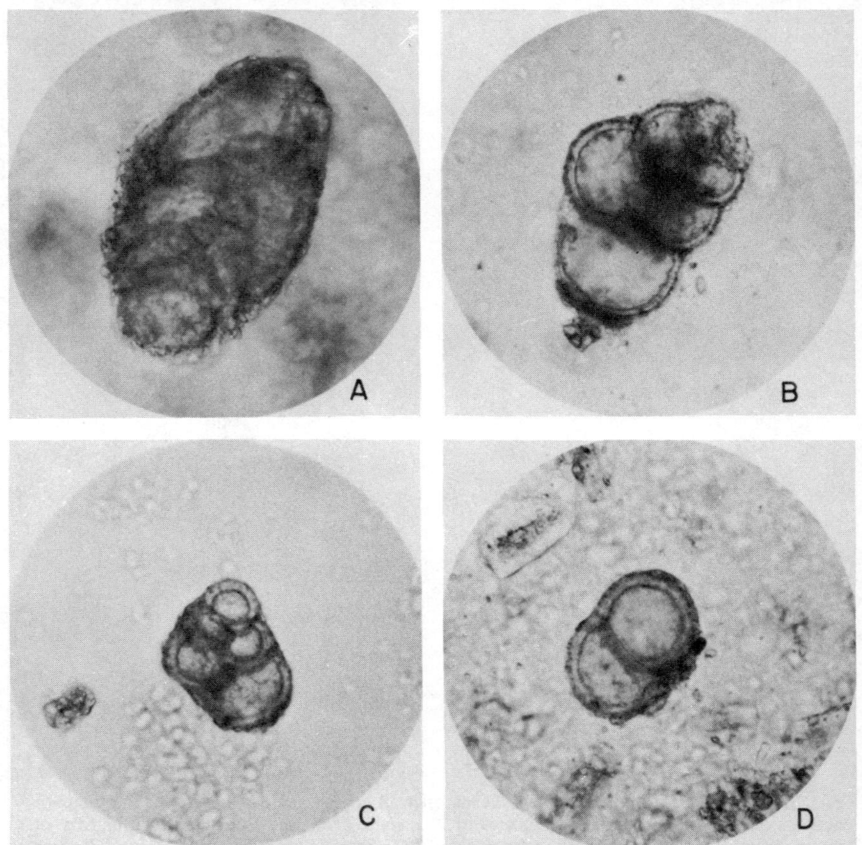

FIG. 3.9. Fossil Foraminiferida in blackboard chalk, ×500. *A.* Larger form with fairly numerous chambers. *B.* Multichambered shell, perhaps of *Heterohelix americana.* The family to which this belongs, the Heterohelicidae, is first known from the Cretaceous, some 100 million years ago. Certain genera persist today in the Indo-Pacific oceans. *C.* Smaller shell, probably also *Heterohelix*. *D.* Two-chambered shell (fragment?), probably of the same genus. (Charles Reiner.)

so obtained has even been used as a basis for the prediction of corresponding future climatic fluctuations. Competent geologists see in such data ground for thinking that we may now be in the middle of a short interglacial period, the last recession of the great continental ice sheet only some 10,000 years behind us and perhaps another advance to be looked for some 10,000 years hence.

One other practical use of fossil Foraminiferida, more relevant to the present, may be mentioned. They and their fragments make up the greater part of chalk, and blackboards would be of much less use without them. Some typical forms from a piece of ordinary chalk are shown in Figure 3.9.

Protozoan Fossils and Evolution

What do protozoan fossils tell us about protozoan evolution? Rather surprisingly perhaps, especially when we recall their abundance in some rocks, they tell us less than has been learned from the palaeontology of some vertebrate groups. This is partly because the fossil record of Protozoa is spotty, largely due to the absence of skeletons and shells in so many of them, and partly because fewer scholars have been interested in this field of study.

But we do know that arenaceous Foraminiferida (those which make their shells of foreign bodies) are much older than those which have learned to use calcium carbonate. Indeed, almost a quarter of a billion years apparently elapsed before this new material for test building was widely adopted, for these lime-secreting Protozoa did not become abundant until the Jurassic period, which began about 167 million years ago. There were, to be sure, some families (e.g., the Lagenidae) producing calcareous tests which appeared earlier, but it was a long time before they became numerous.

The evolution of certain families in the order Foraminiferida can be traced, however, as Cushman did for the Polymorphinidae, which go back at least 160 million years (Fig. 3.10). Similar pedigrees can undoubtedly be worked out for other families of the Foraminiferida when their fossil remains are as intensively studied, and we can safely predict that some of these families will prove to be among the oldest known in the animal kingdom. Fossils belonging to the genus *Textularia* have been found in Cambrian rocks, perhaps 500 million years old.

We also know that the largest Foraminiferida, like the largest species of insects and vertebrates, did not persist for long and are now extinct. Only one living species of Foraminiferida reaches the size of that giant in the protozoan world, a fossil species of *Nummulites*, fully three inches in diameter. Yet for a variety of reasons, some of them not hard to guess, nature seems to have found great size a disadvantage. With large size goes of course greater pressure on the food supply, and it usually means slower reproduction, longer growth periods, and hence smaller populations. The smaller the population

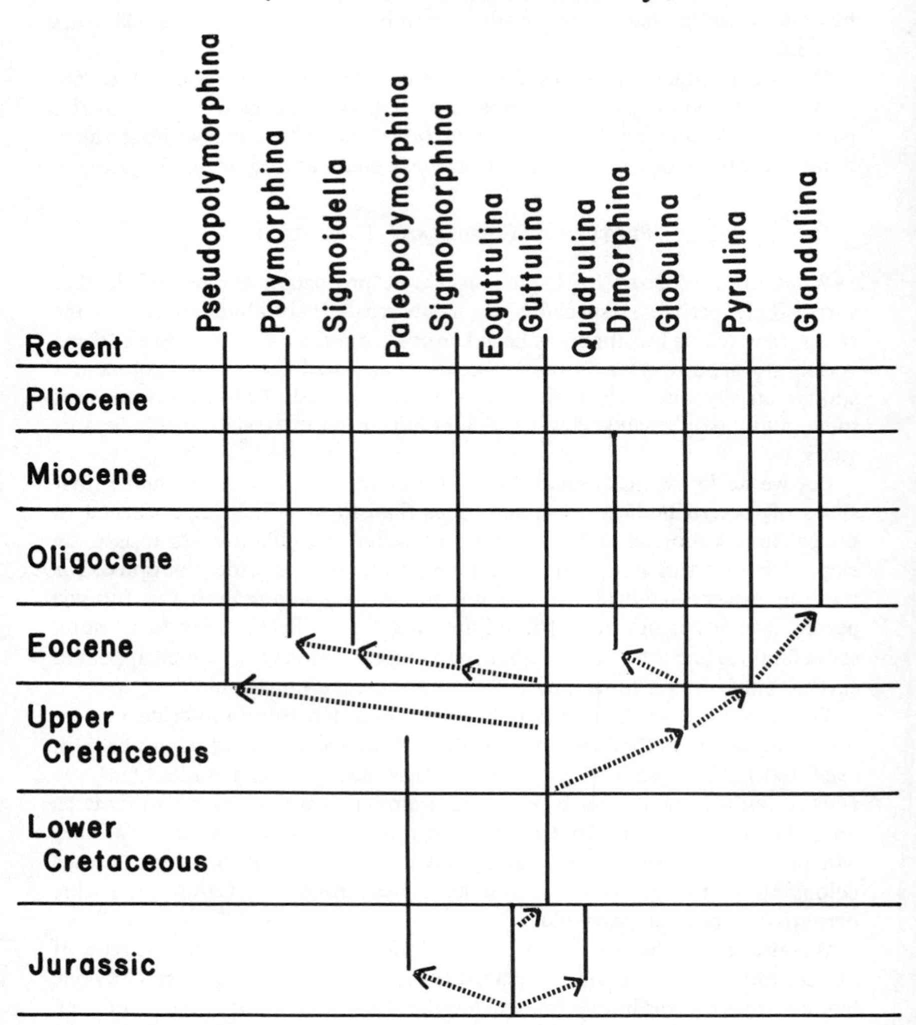

FIG. 3.10. Relationships and geologic distribution of the Polymorphinidae (a foraminiferidan family.) (After Cushman, *Foraminifera.*)

the fewer are likely to be the mutating individuals, and mutation is nature's chief weapon in coping with the exigencies of changing environment.

Less can be said about the evolutionary history of other groups of Protozoa, although fossil remains of a few are quite abundant. The calcium-secreting flagellates (Coccolithidae) often occur in great numbers with fossil Foraminiferida in layers of chalk in France and elsewhere. Deflandre has estimated that a cubic millimeter of such material, an amount one could quite easily balance on a thumbnail, may contain as many as 10 million coccolithic skeletons. Indeed, it seems likely that these organisms were much more numerous in the Cretaceous seas, some hundred millions of years ago, than they are in the oceans of today, although no one can yet explain why. Evidently the Coccolithidae missed some evolutionary opportunities, possibly because of overspecialization, as the very intricate architecture of the shells of some suggest.

The silicoflagellates are forms, once more numerous than now, that appear to have undergone rather little evolutionary change. The living genus *Dictyocha* is fundamentally like its remote ancestors of 150 million years ago. It is not that nature made no attempts to evolve new types, for, as Deflandre remarks, a number of them appeared at different times, including some very large ones. But these for some reason soon became extinct, while *Dictyocha*, which was apparently the parent stock, persisted.

The Ebriedians, like the silicoflagellates, reached the zenith of their development many million of years ago, and only a few survive in modern seas. Little can yet be said about their evolution, except that it seems to differ strikingly from that of the silicoflagellates. Deflandre states that the so-called triode type, which is represented by the genus *Ebria*, appeared rather suddenly in Miocene times, and that no forms from which it might have evolved are known from older rocks. Relationships of these flagellates to both the dinoflagellates and the Radiolarida (to which their skeletons bear some resemblance) have been suggested, but convincing fossil evidence of either possibility is lacking.

Facts such as these raise intriguing questions. Why should extinction be the lot of some species and even of larger groups, and not of others? Why do certain stocks retain the capacity for evolutionary change while others lose it? Sometimes, especially for terrestrial organisms, profound changes in the environment occur to which evolutionary adaption is apparently no longer possible. But for marine types, such as the Protozoa just discussed, there is every reason to believe that the environment has remained stable for hundreds of millions of years, except perhaps for the occasional appearance of new species of parasites or of predators, or of new competitors for the same food supply. To questions such as these present knowledge has no answers.

Fossils of course tell us nothing about either the antiquity or the evolution of the parasitic Protozoa, although other kinds of evidence will be discussed when specific forms or groups are treated. Yet in at least one instance ectoparasitic Foraminiferida have been found as fossils on sea-lilies (crinoids) preserved in Silurian rocks. To what extent the relation had any of the elements of dependence involved in true parasitism we cannot tell, but in any case this is an association of a protozoan and a metazoan dating back at least 350 million years. Protozoa still live in this fashion on Metazoa of various kinds. In some, as with the Vorticellids, which often almost cover the bodies of small Crustacea, it is hard to know whether they are not simply stealing a ride. With others, such as the ciliate *Kerona* often seen scampering about the body of *Hydra*, the relation is certainly parasitic. It is logical to suppose that associations such as this might have started as simple matters of convenience and gradually become first habit and then necessity.

Since relatively few protozoan groups are represented by fossils, evidence of actual evolutionary relationships within the phylum and to other phyla must be sought elsewhere, and it is at best speculative.

*Dillon (1962), on the basis of a comparison of cell structure (nuclear, mitotic spindle, mitochondrial, and flagellar) of a number of animal and plant types, concluded that blue-green algae are probably the most primitive forms now living; others have reached similar conclusions. From them arose the sulfur bacteria and—a little higher on the scale—true bacteria and yeasts. Flagellated Protozoa come next, followed by amoebae and protociliates (opalinids). Well up in rank are the true ciliates and even higher the flagellates symbiotic in termites and wood roaches. Each of these groups appears as a branch of the genealogical tree, rather than an actual stage in evolution. No close relationship is postulated between green algae and green flagellates, such as "euglenoids" and dinoflagellates. Most students of evolution would probably agree that the main relationships suggested by Dillon are the correct ones, but there is ample room for difference of opinion on how close or distant they may be.

Relationships within the phylum are discussed by *Corliss (1959). He proposes a phylogenetic tree with two major branches, the one culminating in the green flagellates and the other splitting at intervals to give rise to the Sarcodina, Sporozoa, animal-like flagellates (Zoomastigina), and ciliates (Ciliophora). Opalinids are shown as evolving from the same stem as "complex symbionts" (flagellates of termites and wood roaches). Ciliates, according to the tree, developed from a similar stock which diverged somewhat earlier. A point of interest is the suggestion that Metazoa and the higher plants (via "certain algae") may have evolved from the same primitive stock ancestral to the green flagellates or Phytomastigina.

Thus it appears that as a phylum the Protozoa are extremely old and may indeed be the oldest of the great animal groups. They go back certainly more than half a billion years, and they are neither simple nor recently evolved. Yet we still know little of protozoan evolution. We cannot say with assurance either that the phylum arose from a single ancestral stock, or that it came from several. Nevertheless, the relationship among and within certain groups is undoubted, and in some cases it is close, as we shall see in later chapters. Although the parasitic forms must be supposed to have arisen from free-living ancestors, apparently this divergence occurred in the far distant past. So much evolutionary change in the morphology of both free-living and parasitic types has occurred that true biological relations are often extremely difficult to determine.

REFERENCES

Cushman, J. A. 1948. *The Foraminifera.* Cambridge: Harvard University Press.

Cloud, P. E. and Cole, W. S. 1953. Eocene Foraminifera from Guam, and their implications. *Science,* 117: 323–4.

Deflandre, G. 1937. Les microfossiles de la craie et des silex. *La Nature* No. 3010, pp. 314–20.

——— 1951. Recherches sur les Ebriédiens. Paléobiologie. Evolution. Systématique. *Bull. Biol. de la France et de la Belgique.* 85: 1–84. Fasc. 1.

Glaessner, M. F. 1945. *Principles of Micropalaeontology.* Melbourne: Melbourne University Press.

Huxley, T. H. "On a Piece of Chalk." In Huxley's *Collected Essays,* and in *The Autobiography of Science,* Moulton and Schifferes (eds.). New York: Doubleday, Doran.

Jones, D. J. 1956. *Introduction to Microfossils.* New York: Harper.

Moore, R. C. 1954. *Treatise on Invertebrate Palaeontology.* Part D. Protista. Geol. Soc. America and University of Kansas Press.

Shepherd, G. 1937. *Geology of Southwest Ecuador.* London: Thomas Murby.

Shrock, R. R and Twenhofel, W. H. 1953. *Principles of Invertebrate Palaeontology.* New York: McGraw-Hill.

Smith, P. V. 1952. The occurrence of hydrocarbons in the recent sediments from the Gulf of Mexico. *Science,* 116: 437–9.

Suess, H. E. 1956. Absolute chronology of the last glaciation. *Science,* 123: 355–7.

Tyler, S. A. and Barghoorn, E. S. 1954. Occurrence of structurally preserved plants in pre-Cambrian rocks of the Canadian Shield. *Science,* 119: 606.

ADDITIONAL REFERENCES

Corliss, J. O. 1959. Comments on the systematics and phylogeny of the Protozoa. *Systematic Zool.,* 8: 169–90.

Dillon, L. S. 1962. Comparative cytology and the evolution of life. *Evolution,* 16: 102–17.

4

Ecology

> "Nothing is simpler than to prepare infusions and to watch infusoria develop in them; but nothing is harder than to produce like results in two infusions prepared under apparently the same conditions. . . ."
> Felix Dujardin, *Histoire Naturelle des Zoophytes. Infusoires.* 1841.

The Protozoa—like man—are found everywhere, and for much the same reasons. Both the Protozoa and man are highly adaptable, although of the two man comes out a poor second. For the climate in which he must live, despite all his air conditioning, is essentially that between his clothing and his skin. He is, therefore, really restricted to tropical or subtropical conditions.

Another advantage that people and Protozoa share is ease of travel. Partly because of their small size, Protozoa are readily transported from place to place. The tiny rivulets that follow a summer shower, the mighty river winding for hundreds or thousands of miles on its way to the oceans, and the globe-girdling currents of the seven seas, all carry their myriads of Protozoa. Air currents waft the minute cysts of many species from one area to another, often for long distances. Hitchhiking as cysts in the guts of animals, from the smallest to the largest, is in the protozoan world an entirely respectable way of getting about, whether the species is free living or parasitic.

Adaptability

Protozoa could scarcely have populated the world as they have, with all their voyaging, except for their great adaptability. The same species are often found widely separated in space and even, as we have seen, in time. A few are known to have spanned geologic eras of a hundred or more million years. Indeed there are virtually no environmental niches, except the air, where if life is possible at all Protozoa have not learned to live. Individual species have their preferred niches, even to the characteristic fauna of the utricles of the pitcher plant (Hegner, 1926). But as a biological group Protozoa have excelled most, if not all others, in their ability to adapt themselves to widely varying habitats. Protozoa exist in the Arctic snow, in hot springs, in the soil, in the sands of the ocean beaches, in fresh water and salt lakes, in oceans and inland seas, at high altitudes and below sea level, in wells and sewage, on the sun-flecked surface of the ocean and in the ooze at its bottom, within the bodies and on the surfaces of many animals and some plants—in sum, almost everywhere.

DETERMINING FACTORS

Despite their wide distribution, the range of environmental conditions in which many species of Protozoa can live and multiply successfully is relatively narrow, and there is always an optimum, less sharply defined for free-living species. For parasitic species, however, even small deviations from the normally encountered environment may be fatal. The reason, no doubt, is the relative constancy of environmental factors within a host as compared with the continual change occurring in the outside world. When the need for adaptation ceases, organisms usually lose their capacity to adapt.

Of all limiting factors, the most important is the presence of moisture. Existence is possible for many species, for a time at least, during periods of extreme dessication, but reproduction is not. Obviously the presence of food and some source of oxygen (not necessarily free molecular oxygen) are also necessary. So also is a hydrogen-ion concentration, or pH, falling within a certain range. Usually, Protozoa are less sensitive to an increased than a decreased pH; a point close to neutrality, or a little on the alkaline side, favors most free-living and many parasitic species. In nature, variations in pH are normally rather unimportant in governing the size of protozoan populations. For most species, growth and multiplication are favored by certain temperatures; light, too, is needed by some, but not too much or too little. Probably all Protozoa require mineral substances of a variety of kinds, and it is certain that many must have a diet supplemented by vitamins.

Temperature

To changes in temperature, Protozoa react in many different ways. For most free-living species, the optimum probably falls within the range of 10° C. and 25°C., or 50°–80°F. But species have adapted themselves to life in the snow in the far north, or at high altitudes, or even in hot springs. Joseph Leidy, an American physician of the last century who pioneered in the study of Protozoa, noted that many of the species found around his home in Philadelphia also occurred in the High Rockies, where summers are short and winter temperatures very low. On the whole, a sharp drop in temperature is less injurious than a moderate rise. Many species can endure prolonged freezing, although probably most of them do so in the encysted state. A similar adaptation is seen in many parasitic species. Parasitic amoebae (members of the genus *Entamoeba*), so closely alike as to be almost indistinguishable morphologically, occur in both cold-blooded and mammalian hosts (cf. *Entamoeba ranarum* of the frog and *E. histolytica* of man and other primates). The biological relationship must be close, but that they are distinct species is shown by failure of either species to infect the other host.

The most obvious effect of a rise in temperature might be expected to be a rise in metabolic rate. That this is indeed the case is suggested by the greater frequency of pulsation of the contractile vacuole often observed under such conditions. The rate of contraction in *Paramecium*, for example, increases from almost zero at 1° or 2°C. to ten or twelve times a minute at 25°C. (Hance, 1917).

An increase in temperature of only a few degrees beyond the optimum for the species is usually fatal, although if the rise is sufficiently gradual it may be withstood. As long ago as 1887 Dallinger found that certain species of flagellates could be acclimated to temperatures as high as 70°C., but his experiments required several years. Species of *Amoeba* and *Oxytricha* have been found in Japanese hot springs where the temperature exceeded 50°C. (Uyemura, 1936). When such acclimatization develops in nature the temperature change may also have been gradual. Perhaps certain individuals by chance have mutated in the direction of increased tolerance to high temperatures. But most parasitic or free-living Protozoa, except encysted individuals, find temperatures exceeding 55°C. almost immediately lethal.

Resistance to the effects of rising temperatures may result from factors other than mutations. The age of the culture used for experiment is, with some species at least, an important consideration. Doudaroff (1936) found that cultures of a clone of *Paramecium multimicronucleatum* showed a gradually greater resistance to increased temperature, followed by a period of constancy, and finally again a diminishing resistance. The effects of high temperatures also seem to vary somewhat with the amount of food present, as will be seen. It is clear, too, that temperature changes must indirectly affect Protozoa by favoring or hindering the multiplication of organisms used as food.

In any case, and for whatever reason, temperatures seem to have a definite relationship to the occurrence and abundance of certain species of Protozoa in nature. The common ciliate *Coleps hirtus* is a good example of a form preferring warmer waters whereas the hardier *Trochilia palustris* thrives under cooler conditions. Similar tastes on the part of certain Foraminiferida have made it possible to use the relative abundance of some species in marine sediments as indicators of temperature oscillations in ocean waters of past ages, and hence to date the glacial epochs.

In contrast to the deleterious effects of high temperatures, in some cases lethal even at 42°C. (as for *Paramecium multimicronucleatum*), cold is usually relatively harmless, unless actual freezing results. It is believed that the organism dies under freezing conditions because of the formation of microscopic ice crystals in its protoplasm, with subsequent disruption of cell organization. But as in most other natural phenomena the explanation may not be this simple. For one thing, the cyst stages are much more resistant to freezing than are the vegetative forms. As a practical consequence, it is important to remember that waters polluted with sewage containing cysts of *Entamoeba histolytica* (the cause of amoebic dysentery) and other pathogenic microorganisms are not rendered safe simply because they have been frozen. Numerous species of Protozoa may remain alive in the frozen state for long periods provided the freezing process is rapid enough.

The great resistance of Protozoa to freezing may be easily demonstrated in the laboratory by exposing a tube containing a few drops from a culture to a mixture of dry ice and alcohol. Quick immersion of the tube results in almost instantaneous freezing of its contents, and subsequent thawing can be accomplished by placing the tube in tepid water. The author has kept a number of species of malaria parasites and trypanosomes in a dry-ice refrigerated cabinet for many months, and found them quite capable of producing typical infections; at $-170°$ C (as in liquid nitrogen) viability would probably endure indefinitely.

The fact that deep-frozen organisms remain alive for many months and even years raises the fundamental question of what life is. In this condition the organisms do not eat, do not grow, do not move, and of course do not reproduce. In sum, metabolism is at a virtual standstill; none of the normal activities is being carried on. The eminent protozoologist Calkins may have been right in saying that life is perhaps after all simply a matter of organization of protoplasmic components, visible, microscopic, and submicroscopic. As long as these components are intact and their mutual relationships maintained, life continues—even though physiological activities may approach zero.

Food

The nutritional state of the organism influences its resistance to adverse environmental factors, among them temperatures outside the optimum range suggested above, and it has numerous other effects. Starvation is often a stimulus to encystment, although there are species in which an overly large meal has the same effect. On the other hand, an ample food supply may stimulate excystment.

Obviously, too little food is certain to limit protozoan populations as inevitably as it does human population. Lowell Noland, studying the effects of a variety of environmental factors on more than 65 species of fresh-water ciliates some years ago, concluded that "the nature and amount of available food" had more to do with their abundance and distribution than any other single factor. However, temperature may in turn determine the abundance of food, as we have seen. A moderate degree of warmth not only stimulates the multiplication of such food organisms as the bacteria and algae, but it often favors the rapid reproduction of those Protozoa which are themselves eaten by others.

Laboratory Cultures

Although it is easy to maintain laboratory cultures of Protozoa, it is—as Dujardin observed more than a century ago—difficult to grow them with uniform and reproducible results and even harder to interpret these results so that they can help us in understanding what takes place in nature. Nevertheless, carefully controlled experiments have taught us a great deal about factors influencing protozoan multiplication. A culture medium definite in composition and suitable for growth of a given species is always necessary. Preferably the medium should enable the experimental organism to grow and reproduce in the absence of any other form of life (such cultures are said to be "axenic," meaning "without a stranger"). As indicated elsewhere (Chap. 27), axenic cultivation is now possible for a number of species of Protozoa, although so far only a small beginning has been made.

From such experiments it appears that conditions governing the growth of protozoan populations are often highly complex. Reproduction may be more rapid when many individuals are present than when there are few (the "allelocatalytic" effect); apparently Protozoa love at least a moderate amount of company. Yet growth is slowed after a certain population density is reached, perhaps because of the accumulation of deleterious waste products, or the pressure on the food supply. Rather oddly, though, the same waste products that inhibit the species producing them may stimulate the growth of other species. Certain essential nutrients, for example, the B vitamins, may be elaborated piecemeal, one species synthesizing one fraction of a vitamin molecule and another species another fraction, each of the two then "borrowing" from the other that molecular fraction which it itself cannot make. Thus species may be interdependent in a very complex fashion. Yet the extent to which relationships of this kind exist in nature remains highly uncertain, and we can hardly say that laboratory studies of protozoan populations have as yet shed much light on the ecology of Protozoa in their natural habitats.

Inorganic Salts

Unquestionably important in determining the character and abundance of protozoan population are the minerals in natural waters. The significance of different substances, however, is highly variable. Relative hardness of the water is undoubtedly of consequence, although numerous species of Protozoa seem able to live about equally well—other things being the same—in very soft and very hard waters. Contained mineral salts do not seem to affect the size of a protozoan population, but they may have other consequences. The shell of *Arcella vulgaris* is either colorless or some shade of brown, depending on the amount of iron present.

Dissolved mineral salts exert their effects chiefly in three ways: they may nourish the organisms or the microorganisms constituting its food, they are important in determining the pH, and—perhaps most important of all—they affect the osmotic relationships of the organism with its environment. For example, in fresh-water protozoans water tends to pass through the semipermeable membrane of the cell into its protoplasm, thus diluting its contents, but in marine types and parasitic forms such water accumulation takes place at a much lower rate or not at all.

Most free-living Protozoa, as we have seen, are tolerant of a fairly wide range of pH, particularly on the alkaline side. This tolerance also holds, though to a lesser extent, for salinity. Species normally occurring in fresh water are often found in brackish water too. A few may even tolerate the change from fresh to sea water, at least in the laboratory. Survival is facilitated if the change is gradual; doubtless this is also true in nature. It is not known how many species are common to both fresh and sea water, but Kudo (1954) lists 24 and suggests that there are probably many others. We might expect that at least some parasitic species of Protozoa are tolerant enough of change to occur in the free-living condition, but this double life seems to be rarely if ever true. Species of *Tetrahymena* (a genus of usually free-living ciliates) however are apparently parasitic and free living at different times.

That so many species can tolerate considerable changes in pH and salinity is surprising when we consider the small size of Protozoa, and how profound a problem in physiological water balance so great an environmental change must entail. Yet there are species capable of solving much more difficult problems. The marine rhizopod *Amoeba* (*Flabellula*) *mira* comes rightly enough by its species name (from the Latin *mirari*, "to be wondered at"). This remarkable form, though ordinarily a sea dweller, can apparently live in anything from distilled water to sea water evaporated to one-tenth its volume. Furthermore, according to Mast and Hopkins (1941) these amoebae adapt themselves to the change within as short a period as 90 minutes. There must be few other animals, protozoan or multicellular, which can match this amoeba in its tolerance to so rapidly changing an environment.

Additional evidence that such adaptability is indeed exceptional is Pack's statement (1919) that the Great Salt Lake, with its salinity of about 33 per cent (335 grams salts per liter, as compared to about 42 grams per liter for sea water), contains "probably less than fifty forms of life." Two of these are ciliates, one a species of *Prorodon* and the other a *Uroleptus;* like *Amoeba mira*, they proved able to withstand greatly lowered salinity, but only when the change was gradual. In a medium of reduced density (it was finally brought down to a density of 1.015) these organisms showed greatly increased activity and reached a considerably larger size. Since both these ciliates normally make their living by hunting, their ability to adapt to the nearly saturated salt solution of the Great Salt Lake may be in part the result of acquiring symbiotic algae. They would then be much less dependent for food on the capture of microorganisms, the amount and variety of which is limited in so salty a habitat. Along with physiological change doubtless developed modifications in behavior, for light has a marked attraction for these ciliates (a reaction obviously facilitating photosynthesis and known as positive phototropism, or phototaxis).*

These ciliates are the lonely survivors of a once much more varied lacustrine fauna; for the Great Salt Lake is itself the tiny remnant of the ancient Pleistocene Lake Bonneville, which during the much wetter climate of glacial times was a truly majestic inland sea more than a thousand feet deep and nearly 20,000 square miles in area. It is worth remembering that fresh-water Protozoa themselves probably evolved from sea-dwelling ancestors, although of this we may never be quite certain. In any event, most fossil Protozoa, as well as the oldest known types among them, are of marine origin. There are other reasons, too, for thinking that the oceans were the ancestral home of the Protozoa. Here food must have been more abundant. Here still reside the greatest variety of species. The seas are thought to have been the cradle of life. Thus, species such as the two found in Great Salt Lake must have achieved an evolutionary feat seldom performed, for they have managed to adapt themselves first to the change from a marine to a fresh-water existence, and then back to a habitat more than seven times saltier than the sea in which their ancestors probably once lived.

Relative Abundance

Marine Protozoa are themselves numerous as far as species are concerned, and often almost incredibly abundant. A study of the protozoan fauna of the summer coastal waters of Woods Hole, Cape Cod, showed 264 species (Lackey, 1936), and the same author remarked that a population of 300,000 Protozoa per liter of sea water would "still be a relatively sparse one." In the light of a statement made by Gran (quoted by Deflandre, 1937) that he had found "five to six millions of *Pontosphaera Huxleyi* (a species of the flagellate group Coccolithophoridae) per liter of sea water" in the fjord of Oslo, perhaps Lackey's claim should be regarded as a model of understatement. We must also remember that the relative abundance of protozoan marine fauna, at least in the case of certain groups (e.g., the dinoflagellates) varies greatly with the season. Both species distribution and number of individuals are greatly influenced by temperature and sunlight; warm currents such as the Gulf Stream support an especially large protozoan population. Under particularly favorable conditions, the density of such populations may be very great. The author has a rock fragment with an area of about

*The tendency of an organism to react in a definite way to a stimulus of any sort is often called a "taxis" or "tropism," positive if the movement is toward the source of the stimulus, negative otherwise.

two square inches into which are crowded fossil foraminiferidan shells lying upon one another much like wheat grains, which they closely resemble in both shape and size.

Fresh-water species may also become extremely numerous. The chlorophyll-bearing types sometimes multiply to such an extent that in late summer the surface of a small pond becomes green (or even red, since certain species possess the red pigment haematochrome). Such an accumulation, known as a "bloom" (discussed in Chap. 14), is due to the light and warmth characteristic of the season.

The population density of parasitic species may be equally great. The colon of an ordinary grass frog is often crowded with trichomonads, and may also support large numbers of other Protozoa, such as the ubiquitous Opalinas. The blood of a malaria-infected bird is frequently found to have more parasites than red cells; this is possible because a single red cell often has within its cytoplasm several plasmodia. But parasites, especially when pathogenic, by their very presence usually stimulate the host to produce antibodies and then their environment may quite suddenly become very adverse. A precipitous drop in parasite count promptly results, although in protozoan infections a few of the parasites usually survive and, if conditions change, may cause a relapse.

No one can yet make an accurate estimate of the number of protozoan species, whether marine, fresh water, or parasitic. New ones are continually being described, and new ones may be evolving just as they have been in all the millions of years since the first protozoan appeared, perhaps some time in the Archeozoic Era.

Cyclical Changes

One of the most striking phenomena about populations of free-living Protozoa is the way they fluctuate in both seasonal and irregularly recurring cycles. The relation of variations in temperature and daylight to the abundance of chlorophyll-containing forms and the Protozoa preying upon them is clear. But the explanation of such population changes is far from simple. Especially in lakes and in protected and largely landlocked bays or lagoons, seasonal changes in water temperature are not confined to the surface layers and they never affect the entire body of water, unless it is very shallow. Instead there is a spring and fall inversion of the upper and lower water layers. With the advent of shorter and cooler fall days, the upper layer cools until it reaches the point of greatest density at 4°C. and then sinks to the bottom, displacing the warmer (and hence lighter) water below. Eventually all the water reaches this temperature, and further cooling results in expansion so that the colder water remains at the surface and finally freezes. Months later, as the days lengthen and the warm spring sunshine melts the ice, the thawed water again reaches the critiical temperature of 4°C., at which it is heaviest and sinks, the deeper layers having been warmed by their contact with the earth. Such changes, of course, can only happen in the temperate zones. The water in tropical lakes and ponds never becomes cool enough for this type of circulation, and in polar regions it is never warm enough. But elsewhere there are seasonal periods of relative stagnation of the surface and bottom strata.

Such cyclical changes have profound consequences for aquatic life of all kinds. The surface waters* become warm enough, and so remain above for a sufficient length of time to permit the algae and chlorophyll-bearing Protozoa to become very numerous. Their abundance in turn creates conditions favorable for that vast assortment of organisms, microscopic and near microscopic, which depend on the plantlike types for food, and thus we get the essentially self-sufficient assemblage of living things collectively called "plankton" (from the Greek *planktos* meaning "drifting" or "floating"). Most of these forms are concentrated near the surface, since below ten meters even in clear water the intensity of illumination is usually too low for active photosynthesis; if the water is turbid, light penetration is even less.

Ultimately, most of the larger animals in fresh and salt water are dependent on the plankton for food, and even the famous voyagers of the *Kon-Tiki* tried concentrating it, making a kind of soup. Whether the recipe could be recommended for general use is questionable, although these modern Argonauts claimed their plankton broth to be quite palatable. If the human population of the world continues to increase at its present rate, the time may yet come when we shall have to turn to this crop of the sea for subsistence, but we are not sure that it will equal in attractiveness the present products of our popular cuisines.

The seasonal inversions of which we have just spoken do not occur to any extent in the seas, because there tides and winds maintain a nearly continuous circulation. Equatorial heating and polar cooling, aided by factors such as coast topography and the earth's rotation, are also responsible for the great currents such as the Gulf Stream.

In fresh waters the seasonal sinking of the upper layers doubtless carries with them most of the Protozoa. When the latter reach the bottom the majority of them probably encyst because of the darkness, cold, and relative lack of food, and in this condition they spend the winter. With the coming of spring and the vernal inversion of bottom and surface layers, the warmer water with its greater oxygen content may stimulate excystment. Though the water movement may be too gentle actually to dislodge and carry to the surface the protozoan cysts, the excysted forms reacting negatively to gravity and aided by the gentle upward current again reach the surface. The perennial thermal mixing, assisted by wind and wave motion, also influences the abun-

*These layers are mainly those above a rather sharply delineated stratum known as the "thermocline," at which point there is an especially sharp temperature gradient.

dance of plankton life by helping to prevent local accumulation or depletion of dissolved and suspended substances.

Just as striking as these seasonal cycles, and much more dramatic, are the marine phenomena known as "red tides." These occur in many parts of the world. In the United States they are especially frequent along the California coast and in the Gulf of Mexico. The Florida coasts are notorious because of them. When such a "tide" occurs vast areas of red or brown sea water

FIG. 4.1. The havoc caused by a "red tide" in the Gulf of Mexico. (Collier and Hutner, *Sci. Am.*)

suddenly cover hundreds of square miles. Something in that water is lethal to fish, whose corpses float by the million on the surface (Fig. 4.1), and may be thrown up on neighboring beaches in grisly rows. For a long time the cause of such catastrophes was unknown. Unlike the seasonal cycles just described, they recur at irregular intervals and seem to have little connection with other events. Though the ultimate cause is still uncertain, the immediate cause appears to be the uncontrolled multiplication of a tiny dinoflagellate known as *Gymnodinium brevis*. Other species in this same group may also be involved; *Gonyaulax catenella* has been mentioned, and so has the luminous flagellate *Noctiluca* (Connell and Cross, 1950; Perlmutter, 1952). Just what is really responsible for the mass killing of so many fish is still uncertain. It has been suggested that the flagellates elaborate a fish-killing toxin. Another possibility is that up-welling currents bring to the surface lethal concentrations of hydrogen sulfide, a gas product of the multiplication of certain bacteria often numerous in the oxygen-poor bottom ooze. Nitrates and phosphates, which may promote the rapid multiplication of dinoflagellates, may also be brought to the surface. Another and more probable suggestion is that unusually large amounts of vitamin B_{12}, believed to be essential to the nutrition of all Protozoa, condition the appearance of these flagellates in such vast numbers. This vitamin, which animals apparently cannot elaborate, comes ultimately from the metabolic activities of bacteria and perhaps also from the blue-green algae. It is supposed that in this case unusually heavy rainfall and resultant floods wash out accumulations of the vitamin from bogs or other sites of active bacterial growth, and carry it to the sea. Red tides have sometimes been observed to coincide with a heavy influx of flood water into coastal areas of the sea.

Protozoa and Drinking Water

Under some conditions the fresh-water protozoan fauna may also become of considerable direct importance. During the summer certain species of plantlike flagellates may become numerous enough in lakes and reservoirs to give the water an objectionable taste, often described as "fishy." Certain ciliates, such as *Bursaria*, as well as a number of species of algae and diatoms, may share the guilt. The tastes and odors they produce apparently stem from the liberation of certain essential oils contained within the bodies of the organisms. Not all such odors are disagreeable: one of the common chlorophyll-bearing flagellates, *Cryptomonas*, whose color varies from green to some shade of brown, is said to cause an odor resembling "candied violets." Control of these organisms is relatively simple: it consists simply in adding minute amounts of copper sulfate to the water, often done by dragging a sack of the chemical behind a rowboat. Devices for automatically feeding the substance into the water are also sometimes used.

Much more important than these Protozoa are the numerous species that play a role in the natural purification of sewage-polluted waters; their function is ever more essential because the growth of human population results in increasing pollution. The commonest method of sewage disposal is its discharge into streams and lakes. Since these bodies of water more often than not are sources of supply for nearby municipalities downstream, natural or artificial purification is an obvious necessity.

Natural purification depends primarily on the activities of a variety of microorganisms plus such agencies as oxidation, reduction, and sunlight. The Protozoa make a dual contribution: some, such as the bacteria feeders, tend to reduce the bacterial content; others promote oxidation by the liberation of oxygen as a by-product of photosynthesis. Among the most active consumers of bacteria are probably the vorticellids. Purdy (according to Whipple, 1927) found the population of these ciliates in a sample of grossly polluted water to be 141,000 per ml.

Since it has been observed that the nature of the protozoan population differs according to the amount of organic material present in the environment, Protozoa have been classified into four general types. Forms predominating where pollution is high, decomposition active, and little oxygen available, are "polysaprobic." Where there is more oxygen and less decomposition of nitrogenous substances the dominant Protozoa are termed "mesosaprobic." If oxidation is the chief type of chemical change and there is little or no nitrogenous decomposition, the protozoan population is chiefly of "oligosaprobic" forms. At the end of this spectrum, in water rich in oxygen and otherwise relatively pure, are the "katharobic" Protozoa.

A good example of a polysaprobic protozoan population is found in the Imhoff tanks of sewage disposal plants. These tanks are really digestion chambers into which screened raw sewage is fed in order to undergo bacterial decomposition for a period of time. Since the bacteria mainly responsible for the dissolution of the organic matter are either facultative or obligate anaerobes,* the Protozoa are of corresponding physiological types. More than 70 species were identified in the contents of these tanks by Lackey (1925), the large majority being colorless or animal-like flagellates. One of the commonest was *Bodo*, a minute bacterial feeder with two flagella, one trailing. Many of these species were the so-called coprozoic forms, living and multiplying readily in diluted human or animal excrement. It is not surprising that a number of them have at various times been mistaken for intestinal parasites.

What role these Protozoa play, if indeed they play any, in the series of complex chemical degradations occurring in Imhoff tanks is so far totally unknown. The final result of the combined activities of bacteria and Protozoa is a sludge that can be used as fertilizer, and an almost odorless effluent that can be fed into natural bodies of water with little risk either to health or the esthetic sensibilities of people in the neighborhood.

It is likely that most of the Protozoa encountered in sewage come originally from soil. Some occur commonly enough, though in small numbers, in lakes and streams which serve as public water supplies. It is probable that cysts of these types can withstand ordinary chlorination; for example, cysts of *Entamoeba histolytica* are not always killed even by five times the concen-

*"Anaerobic" refers to an environment in which there is no free oxygen; an obligate anaerobe is one which must live in such conditions, whereas a facultative anaerobe is less rigid in its requirements.

tration of chlorine normally reached in the treatment of waters destined for drinking. We also know that the cysts of many species of Protozoa often pass unharmed through the gut of man and animals. Such cysts are doubtless frequently present on the leaves of salad plants, on tomatoes and strawberries, in short on all fruits and vegetables grown on the ground and eaten raw.

Soil Protozoa

Fortunately, as long as these soil-derived protozoan cysts are of normally free-living species, there is no health hazard. But where human manure is used as fertilizer, as in parts of Europe and Asia, the dangers to health are great. This is a practice which the pressure of a dense human population on the food supply makes almost necessary. Methods are known for the previous treatment of such material which make it safe, but they involve expense and imply a level of sanitation not yet reached in these areas. It is unlikely that either disease-producing bacteria or the cysts of pathogenic Protozoa long survive the conditions within an Imhoff tank, but modern sewage disposal systems are extremely expensive and seldom used except in rather large cities. Thus it happens that vegetables grown in many parts of the world are very likely to be contaminated with dangerous pathogens.

That some Protozoa are always to be found in the soil was first noted by Ehrenberg, although it is very probable that Leeuwenhoek was the first actually to observe such forms. Perhaps the most thorough study of the protozoan fauna of the soil is Sandon's (1927). He examined many types of soil samples from all over the world, identifying some 250 species, including flagellates, amoeboid forms, and ciliates. Many were coprozoic, an expected finding when the promiscuous nature of animal defecation is considered. A surprising discovery was that, of all samples examined, Greenland's was richest in Protozoa. Garden soils, doubtless because of the frequent application of manure as fertilizer, gave the highest average protozoan count. Yet, as Sandon remarked, "even the most barren soil" had its population of these microorganisms.

The relative importance of various environmental factors in determining the protozoan population of a given soil remains unknown, and Sandon's study shed little light on it. But clearly rainfall, temperature, pH, and amount of organic matter are all of significance. So also is the soil type. Rhizopods are, for example, especially numerous in acid peaty soils. The amount of free oxygen undoubtedly plays a part, since many of the species are anaerobes or facultative anaerobes. Another factor is depth: the protozoan population is largest at from 10 to 12 centimeters below the surface.

Since Protozoa are present in all soil, and apparently most numerous where it is intensively tilled, it is reasonable to suppose that they affect soil fertility. But, here again, we can say nothing about just what role they play.

The size of the protozoan population must have some effect on both the number and type of bacteria since different species of Protozoa are known to have preferences in their bacterial diet. Some species of bacteria are highly important in soil fertility because of their ability to fix atmospheric nitrogen, others because they break down nitrogenous substances; so also are other bacterial species of significance in cycles involving other elements, such as carbon and sulfur. These elements, and of course numerous others, are essential constituents of the protoplasm of all living things. Since their supply is limited, it is necessary that they be available for reuse over and over again. Although not all decay is the result of bacterial activity, the greater part is; thus their role in the economy of nature is of the most outstanding importance. To the extent that Protozoa limit bacterial increase, they also become vitally important and are always potentially a factor in fertility.

Perhaps one of the most interesting groups of soil Protozoa are the ciliates inhabiting the tidal sands of the ocean beaches. The eminent French protozoologist Fauré-Fremiet found numerous sand species there, some of which seem to occur in no other ecological niche. Apparently the size of the sand grains (and hence of the spaces between) is the chief environmental factor governing the occurrence of these infusoria in a given littoral. Since the oceans of the world are all interconnected, it is not surprising that these Protozoa have a cosmopolitan distribution. Fauré-Fremiet remarks, ". . . sand-living species are generally characterized by their very small or medium size, their flattened shape, and their jerky motion, with alternately very fast swimming and sudden stopping when in contact with solid surfaces." In a world of such sharply restricted spaces and as full of obstacles as sand grains the ability for "sudden stopping" might be very important indeed. Certainly these ciliates have become highly adapted to their environment, both in morphology and behavior.

Environmental Relationships of Parasitic Protozoa

So far only the ecology of the free-living Protozoa has been considered. But that of the parasitic species, particularly of those living in man and domestic animals, is perhaps more important.

Basically, the factors governing the occurrence, abundance, and distribution of parasitic Protozoa are the same as those affecting free-living forms. But the interrelationship of these factors seems more complex, for the parasite not only has its own immediate environment, the "internal environment" of the host, but the very conditions within the host to some degree reflect the latter's surroundings.

The most obvious needs of the parasite are food and oxygen, though equally significant are other physical chemical factors: the proper pH, temperature, carbon dioxide tension, and perhaps certain vitamins, to name a few only. Species of parasites differ greatly in their requirements, and the ability (or inability) of potential host species to meet them doubtless determines that more or less rigid host-parasite relationship which we call host specificity.

The greatest difficulty confronting parasites is getting from one host to another. In this nature has displayed great ingenuity. What are the ecological factors which stimulate *Entamoeba histolytica* to produce the cyst, which is its infective stage? What is the stimulus which causes some malaria parasites in the host's blood to become gametocytes, the only stage capable of survival and development in its next host, the mosquito? We are still far from solutions to these problems, and others like them.

Even with food to be had for the taking, parasites are nevertheless usually selective, and they generally confine themselves to a given habitat within the host and to a given food. The least choosy, on the whole, are species living in the intestinal tract. The intestinal amoebae, for example, live for the most part on the bacteria and other microorganisms which throng the colon, although *Entamoeba histolytica* (the cause, as we have seen, of amoebic dysentery in man) has developed a strong predilection for erythrocytes and the ability to digest its way into the submucosa where a good supply is almost guaranteed. In marked contrast are the malaria parasites which can live and multiply, in their vertebrate hosts, only in the blood cells (with certain partial exceptions to be discussed in Chap. 24). Most protozoan parasites are at least facultative anaerobes when they dwell in vertebrates, since little free oxygen is available in the body. We are still quite ignorant of the requirements of protozoan parasites of invertebrates. Many of their hosts are small in size and hence difficult to study, but the environment they furnish to their protozoan guests may well be quite different from that supplied by vertebrates to theirs.

Even though the environment of a parasite within its hosts may be relatively constant, it is not entirely so; the host itself is subject to a changing environment, and such changes are reflected by internal physiological fluctuations. It is clear that these may thus affect the parasite. Certain species of malaria plasmodia and their avian hosts exemplify such a relationship. The reproductive cycle of the plasmodia is sharply marked, reaching a peak each day about a given time. This cycle is the parasite's response to a diurnal cycle in the physiological activities of its host. Thus if the latter are exposed to lengthened periods of light and darkness (or what we might call "artificial days") of, say, 36 hours instead of 24, after a short initial lag the parasites respond by slowing their own reproductive cycle until it is also 36 hours in length instead of the normal 24. Some species of malaria plasmodia react in a similar way to the cyclic physiological activities of their mammalian hosts.

Much more valuable to the parasite than its ability to make these adjustments (it is hard to see what advantage accrues to the parasite in this case) is its capacity to survive in its continuous struggle with its host. Somewhat effective even at the start, the resisting mechanism of the host becomes increasingly efficient as the infection progresses. In protozoan infections, the mechanism is primarily cellular. After an initial lag of variable length, the so-called fixed phagocytic cells, chiefly the macrophages of the spleen and liver, increase in number and exhibit vastly greater activity in destroying the parasites. Humoral mechanisms are also involved, including the elaboration of substances capable of killing the parasites directly (as in many trypanosome infections) or occasionally of inhibiting parasite reproduction. Humoral components may extinguish the infection entirely, or the parasites may become partially or completely immune to their effects. Such resistance is especially likely with the trypanosomes, which react in a similar way to trypanocidal drugs when the initial dosage is too small. For such drug resistance the term "drug-fastness" is often used. In general the immunity which vertebrate hosts develop toward protozoan parasites is of a special type known as "premunition." The parasites do not completely disappear from the body but persist in small numbers, so what really results is an unstable equilibrium between host and parasite. When and if the parasites disappear entirely the host is likely to lose its immunity and be as susceptible to reinfection as if it had never experienced infection in the first place. This susceptibility contrasts strongly with the lifelong immunity frequently following recovery from bacterial and virus-induced infections. In such infections the immunity is primarily humoral (although, especially in bacterial diseases, the role of leucocytes is important), yet the cellular immune mechanism of protozoan infections is highly effective in keeping the parasites under control, and hence in keeping the host clinically well.

Parasites are also confronted with the need to adjust to a different environment when they leave the host or exchange a host of one species for another (e.g., the alternation of hosts when the malaria plasmodium transfers its residence from man to the mosquito). Sometimes the problem is simply one of survival until chance permits entry into a fresh host, and then their prospects are likely to be dim unless the parasites have been forehanded enough to encyst. Many species have adapted to an alternation of hosts, so that the specter of perishing in a hard outer world has been permanently banished. Nevertheless, the need to adjust to a different environment remains with a change of host, for the host species are usually widely separated biologically (for example, the tsetse fly and the trypanosomes causing sleeping sickness in Africa). The protozoan responds by a change in morphology and often in reproductive phase. When the mosquito vector (which must be of a susceptible species) of malaria takes a meal of infected blood, only the gametocytes of the parasites survive. In the gut of the mosquito they are somehow stimulated to undergo maturation into gametes, and the male and female gametes then unite to form the fertilized egg, or oökinete. The stimulus causing maturation is said to be the increased carbon dioxide tension of the drawn blood, and it is as effective on a slide under the microscope as in the mosquito.

Thus parasitic Protozoa have not entirely lost their ability to respond to changes in the environment, even though their free-living ancestors were undoubtedly more versatile. It is as proper to speak of the ecology of the parasitic Protozoa as of their free-living, though rather distant, relatives. Unfortunately as yet we know little of the ecology of either. We can fairly easily analyze the physical components of the environments to which Protozoa are exposed in that microcosmos we call a laboratory culture, and we can even determine the effects of most of the components on these organisms as well as the nature of their reactions. But it is a far more difficult task to study the relative importance of corresponding physical factors in nature. And it is a task which protozoologists have scarcely begun.

REFERENCES

Cutler, D. W., and Crump, L. M. 1935. *Problems in Soil Microbiology*. London: Longmans, Green.

Dallinger, W. H. 1887. The president's address. *J. Roy. Mic. Soc.*, 7: 185–99.

Deflandre, G. 1937. Les microfossiles de la craie et des silex. *La Nature*, No. 3010, pp. 314–20.

Doudoroff, M. 1936. Studies in thermal death in *Paramecium. J. Exp. Zool.*, 72: 369–85.

Fauré-Fremiet, E. 1950. Ecologie des Ciliés psammophiles littoraux. *Bull. Biol. de la France et de la Belgique*, 84: 35–75.

——— 1951. Marine sand-dwelling ciliates of Cape Cod. *Biol. Bull.*, 100: 59–70.

Hance, R. T. 1917. Studies on a race of *Paramoecium* possessing extra contractile vacuoles. I. An account of the morphology, physiology, genetics, and cytology of the new race. *J. Exp. Zool.*, 23: 287–333.

Hegner, R. W. 1926. The Protozoa of the pitcher plant, *Sarracenia purpurea. Biol. Bull.*, 50: 271–6.

——— 1926. Homologies and analogies between free-living and parasitic Protozoa. *Amer. Nat.*, 60: 516–25.

Hutchison, G. E. 1941. Ecological aspects of succession in natural populations. *Biol. Symp.*, 4: 8–20.

Hutner, S. H. and McLaughlin, J. J. A. 1958. Poisonous tides. *Sci. Am.*, 199: 42.

Kofoid, C. A. The distribution of the pelagic ciliates in the eastern tropical Pacific. *Fifth Pacific Science Congress*, pp. 2159–61 A5. 31.

Kudo, R. R. 1954. *Protozoology*, 4th ed. Springfield: Charles C Thomas.

Lackey, J. B. 1925. Studies on the biology of sewage disposal. The fauna of Imhoff tanks. *Bull.* 417, *N. J. Exp. Stations*. September.

——— 1936. Occurrences and distribution of the marine protozoan species in the Woods Hole area. *Biol. Bull.*, 70: 264–78.

Loefer, J. B. 1939. Acclimation of fresh-water ciliates to media of higher osmotic pressure. *Physiol. Zool.*, 12: 161–72.

Mast, S. O. and Hopkins, D. L. 1941. Regulation of water content of *Amoeba mira* and adaptation to changes in the osmotic concentration of the surrounding medium. *J. Cell. and Comp. Physiol.*, 17: 31–48.

Needham, J. G. and Lloyd, J. T. 1916. *The Life of Inland Waters*. Ithaca: Comstock.

Noland, L. E. 1925. Factors influencing the distribution of fresh-water ciliates *Ecology*, 6: 437–52.

Pack, D. A. 1919. Two ciliata of Great Salt Lake. *Biol. Bull.*, 36: 273–82.

Perlmutter, A. 1952. Mystery on Long Island. *N. Y. State Conservationist*, Feb. March, p. 11.

Sandon, H. 1927. *Composition and Distribution of the Protozoan Fauna of the Soil*. London: Oliver and Boyd.

Uyemura, M. 1936. Biological studies of thermal waters in Japan. *IV Ecol. St.*, 2: 171.

Whipple, G. C. 1927. *The Microscopy of Drinking Water* (rev. Fair and Whipple). New York: Wiley.

Zobell, C. E. 1946. *Marine Microbiology*. Waltham, Mass: Chronica Botanica Co.

5

Morphology

> ". . . we find, almost without Exception, in those *Specks of Life* whose Minuteness renders them almost imperceptible to the Eye of Man, a greater Number of Members to be put in Motion, more Wheels and Pullies to be kept going, a greater Variety of Machinery, an Apparatus more complex and curious, a Plan seemingly of deeper Contrivance; in short, more Elegance and *Workmanship* (if the term may be excused) in the Composition, more Beauty and Ornament in the Finishing, than are seen in the enormous Bulk of the Elephant, the Crocodile, and the Whale. . . ."
> Henry Baker, *Employment for the Microscope*, 1753.

The study of Protozoa, like most attempts to understand nature, began as a descriptive science. The early microscopists were intrigued by the complexities of structure and behavior of the "little animals." Inevitably the growth of a real understanding of either was limited by the inability of contemporary microscopes to reveal the finer physical features of these most minute members of the animal kingdom, and perhaps also by the natural tendency of the human mind (including those of some biologists) to suppose that what cannot be seen must therefore not exist. As the late and eminent nematologist N. A. Cobb was fond of saying, "No amount of not finding a thing proves that it does not exist." But now it is universally conceded that only the grosser aspects of protozoan morphology are visible with the ordinary light microscope. Even to make the most of such observations requires the use of highly refined staining processes and often of complicated special techniques. Yet patient observation of living Protozoa with any good microscope will reveal much more than most students see. It is easy to forget that all the older observations, many of them astonishingly accurate, were made with instruments which by modern standards would be regarded as primitive, and on living organisms, without resort to special illumination, narcotics to dull protozoan ambition, viscous media to impede their progress, or other such devices. Such were the studies of Stein, the first of which appeared in 1859.

The recent development of the phase contrast and electron microscopes has added greatly to our ability to explore the world of the "infinitely small." As applied to Protozoa, the use of both is essentially a product of the last two decades. Phase microscopy has the advantage of making visible in the living organism much that would otherwise require staining to reveal. The great advantage of the electron microscope is of course that extremely high magnifications are possible. But it cannot be employed for viewing living organisms, because objects to be studied must first be placed in an evacuated chamber through which the electron beam is passed.

Size and Form

Although the great majority of Protozoa are invisible to the naked eye, some are relatively large. Certain species of the foraminiferidan genus *Cyclo-clypaeus* may reach or even exceed a diameter of two inches (50 mm. plus). Much smaller, but still easily perceived, are the larger ciliates such as species of *Spirostomum*, which are often a millimeter or more in length. Some giant amoebae are almost as large. So also are the colonies of some flagellates, e.g., *Volvox*. Largest of all living Protozoa, if we concede that they are members of the phylum, as most protozoologists now do, are certain slime molds which in the plasmodial stage may reach a diameter of a few inches or more.

The smallest Protozoa are somewhat harder to pick out for special mention, for in the life cycle of many species there are stages of grossly different size. For example, the spores of many of the Microsporida may be little larger than bacteria. Some amoebae, even at their largest, are still only a few micra in diameter. Many species of animal flagellates, notably those of the Bodonidae, are also extremely small.

Variety of form equals that in size. Indeed, it is likely that in no other phylum of the animal kingdom is there greater diversity. Some Protozoa are asymmetrical; others exhibit bilateral, radial, or homaxonic symmetry. Species lacking a definite outer envelope ("pellicle") or exoskeleton (often called a "test," "lorica," or "shell"—terms that are nearly synonymous), such as amoebae, may change in shape from one instant to the next. Those with a limiting pellicle must of course retain a more or less constant form, but there is almost infinite variety among groups and species. Probably an oval or spherical shape is nearest that of the ancestral protozoan stock, whatever the latter may have been like in other respects.

Bilaterally symmetrical forms originated perhaps as the result of abortive division, the fission process in some ancestral individual having been arrested after duplication of the nuclear and cytoplasmic structures.

The radially symmetrical Protozoa are predominantly sessile types, like the Vorticellas and their numerous relatives collectively known as the Peritrichida. Since division in this group is superficially longitudinal (often with the production of unequal daughters), rather than transverse as in most other ciliates, the oral disc or broad end of the "bell" may correspond to the dorsal aspect of other ciliates. If such is the case, it has involved a rather drastic relocation of the cytostome from its original position at the anterior end.

Homaxonic symmetry is best seen in the Radiolarida, an ancient and essentially pelagic group of marine Sarcodina. The body in most species is spherical and usually supported by a skeleton of exquisite delicacy of design (Fig. 3.2), consisting of a lattice-like central capsule of pseudochitin (also called "tectin"), within which are contained the nucleus and other of the more vital parts of the organism. The minute pores, which are variously arranged in the capsular membrane, permit communication between the intracapsular protoplasm and the outer capsular envelope of cytoplasm, which is concerned with pseudopod formation, food assimilation, and the control of buoyancy by change in vacuolar number and size. Lying outside the central capsule, and concentric with it, is usually a delicate structure of siliceous lattices, interspersed with radiating spines and spicules. In some Radiolarida these structures are of strontium sulfate, although in a few there is no skeleton at all. There are also Radiolarida in which the symmetry is bilateral, and others in which it is called "monaxonic," the basally spherical body being modified in various ways about a central axis.

Tests

Testate Protozoa exhibit an originality of plan and choice of building materials almost rivalling that of our own species, although radially symmetrical "houses" are distinctly more popular among them. Such are the cuplike dwellings of the loricate (sessile) vorticellids and those of the testate rhizopods (e.g., the various species of *Arcella*). All but a few species of the Foraminiferida build many-chambered shells of calcium carbonate, often resembling those of cephalopod molluscs, such as the beautiful chambered *Nautilus*. Indeed the Foraminiferida were once thought to be minute species of cephalopods. Perforating the foraminiferidan shell at many points are tiny apertures through which pseudopodia may be extruded.

Materials for test construction, other than the minerals mentioned above, include foreign bodies such as sand grains and diatom shells. These are picked up from the environment, often quite selectively. Highly polymerized proteins secreted by the animal itself cement the particles together, or they may constitute the entire shell, additionally reinforced in certain species by silica and iron salts. Some species have been thought to make their tests from chitin (a complex carbohydrate), but there has been some doubt recently cast on this belief. Many plantlike flagellates construct jackets of cellulose, such as the beautifully sculptured creations of the dinoflagellates (Fig. 5.1).

Such coverings are often called "thecae" (singular "theca," from the Greek word *theke*, "case").

Shells of the testate rhizopods (a fresh-water group) are all of organic material, presumably polymerized proteins. To this may be fastened a covering of sand grains (as in the common genus *Difflugia*) or particles of other sorts. Many of these species are finely depicted in Leidy's famous monograph *Fresh-water Rhizopods of North America*, published almost a century

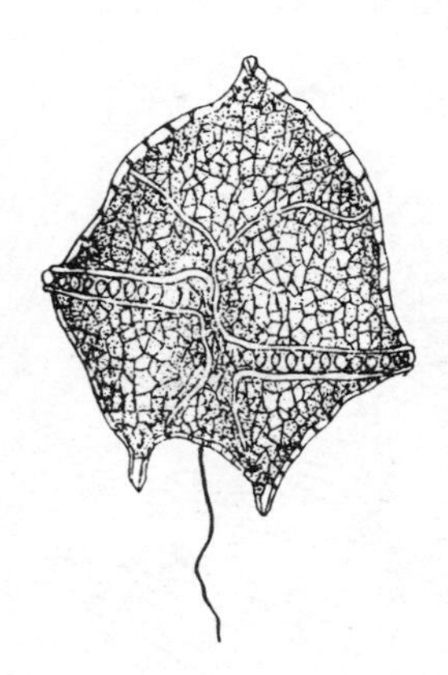

FIG. 5.1. *Protoperidinium limbatum*, a moss-dwelling fresh-water dinoflagellate. Note the elaborately marked theca; the girdle is somewhat spiral. (After Stokes.)

ago. In some species the tests are of one piece, while others, such as *Euglypha*, some dinoflagellates, and some coccoliths (Coccolithidae), construct them of many minute platelets, arranged like shingles on a house.

Colony Formation

Some test-dwelling Protozoa are gregarious and build the protozoan equivalent of apartment houses. Such are the zooflagellates of certain genera of Amphimonadidae. Perhaps the most remarkable of these species is *Rhipidodendron splendidum* (Fig. 5.2), sometimes referred to as the "organ-pipe animalcule" because its complexly branched colonies resemble somewhat an organ-builder's nightmare. Each little monad in the colony lives in a tubule at the end of a branch, with only its two flagella projecting into the surrounding fresh water. The substance of which the colony is made is gelatinous or (according to Kent) horny material.

Other colonial flagellates live in groups, sometimes of only a few individuals and sometimes of many thousand (*e.g.*, *Volvox*). But here the individuals of the colony are bound together by a common matrix, often of a gelatinous substance, and retain little or no independence. They are thus hardly comparable to species like *Rhipidodendron splendidum*. Their colonies may be flat, the members forming a platelike aggregation (*e.g.*, *Gonium*), or radially arranged as are the chrysomonad flagellates of the genus *Synura*, or spherical like *Volvox*.

Among the most interesting of colonial Protozoa are the slime molds. These, as the name suggests, are in many respects similar to fungi and have often been regarded as such. They are still included in some standard botany textbooks, such as Fuller and Tippo's (1955), who call them Myxomycophyta and remark, "Doubt still exists as to their affinities with other groups of living organisms. . . . Most biologists, however, now consider them to be plants."

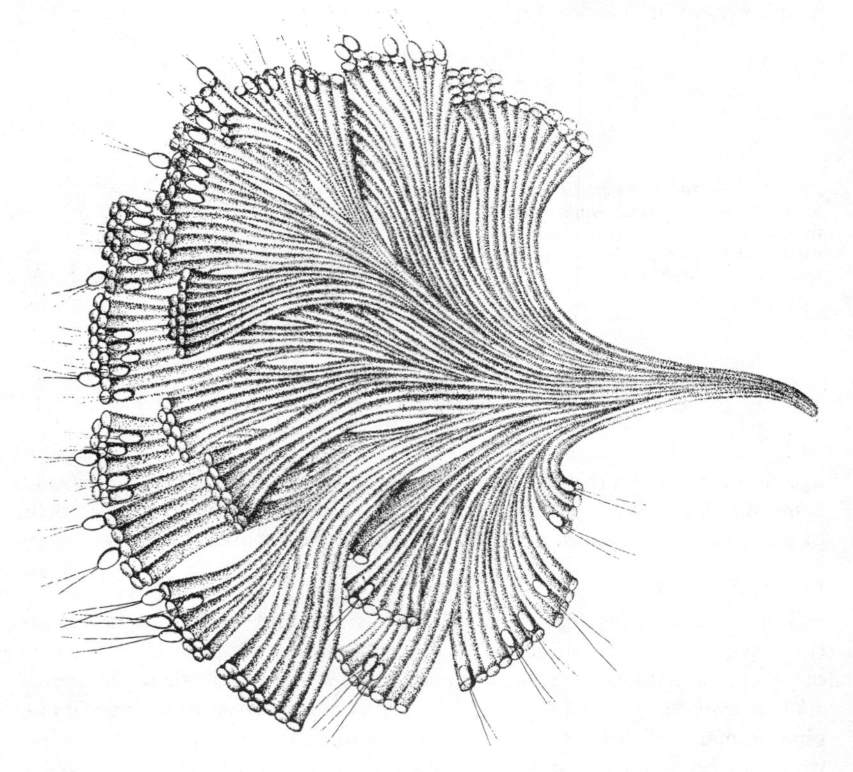

FIG. 5.2. A colony of *Rhipidodendron splendidum*, first described by Stein. Each tiny branch is occupied by a small oval flagellate with two flagella of equal length. The individual organisms are about 13μ long, the full-grown colony about 350μ in height. These forms are restricted to fresh water. (After Kent.)

Most protozoologists are inclined to think of them as Protozoa and assign them to the order Mycetozoida. Included here are four smaller groups, of which the Myxomycetes are the largest; probably the groups have little or no relationship to one another (Bonner, 1959). All of them, however, are alike in having a highly complex life history, in one stage of which aggregation occurs with the formation of large multicellular masses (Fig. 5.3) termed "plasmodia" (singular, plasmodium). Such masses are usually said to reach

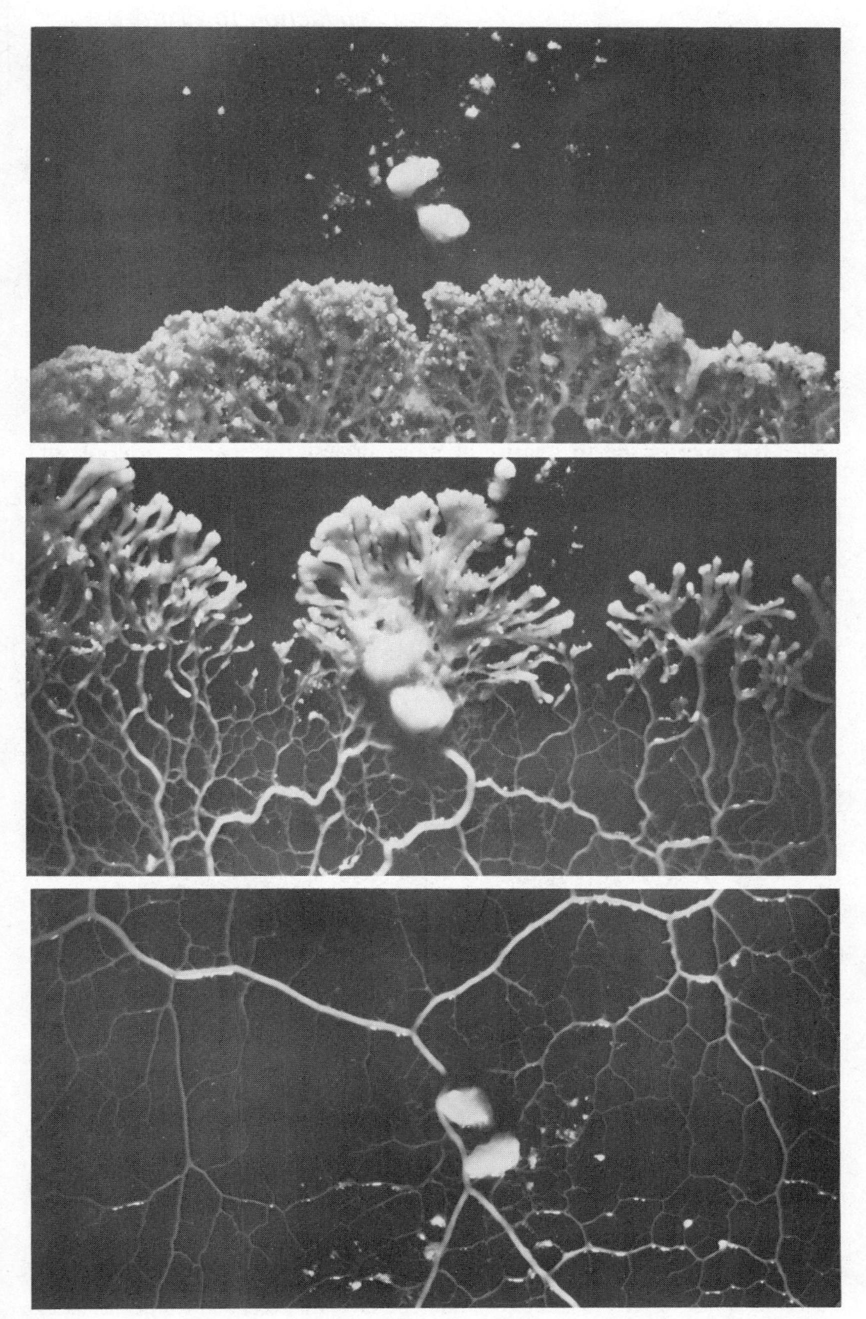

FIG. 5.3. Photographs of a slime mold plasmodium (*Physarum polycephalum*) approaching, engulfing, and digesting food particles. (After Lonert, General Biological Supply House.)

a diameter of a few inches, but Crowder (1926) has described colonies of almost a square yard. They occur on damp objects, such as old logs lying in swamps, where they often appear as brightly colored masses "like the pile of a bright Persian rug" (Crowder). From these eventually develop myriads of dustlike spores, from which, in many species under favorable conditions of warmth and moisture combined with dense shade or darkness, emerge biflagellated swarm cells. These in turn become amoebulae behaving as gametes and, after fusion, giving rise to the plasmodium though the details of the process differ greatly according to group and species.

Ectoplasm

Living Protozoa, if naked so that the body proper is not concealed, appear under the microscope to consist of more or less transparent protoplasm. The protoplasm in many cases shows a definite differentiation into two areas: a clear "hyaline" outer envelope of ectoplasm, and an internal, less transparent region of endoplasm. The differentiation is particularly noticeable in the amoeboid forms and in some Sporozoa. The ectoplasm of the ciliates may contain structures such as trichocysts, easily seen in *Paramecium*, and it also bears the cilia and derived organelles. The trichocysts of *Paramecium* have been regarded by some as anchoring devices, enabling the animal to

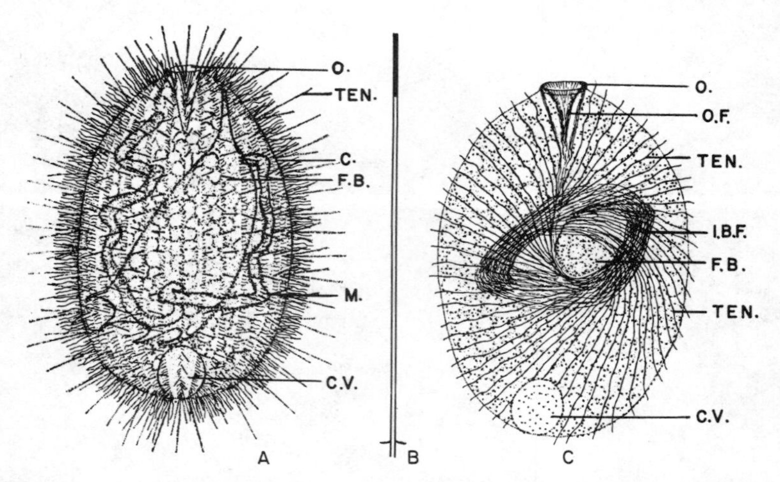

Fig. 5.4. *A*. The ciliate *Actinobolina vorax*, showing structure and the outlines of an ingested rotifer, *Anurea*. *B*. A single tentacle bearing a toxicyst at its end. *C*. Side view, showing inner extensions of tentacles and their association with the skein of internal fibrils. At its maximum extension a tentacle protrudes as much as twice the diameter of the animal. The toxicyst at its tip paralyzes any prey it touches. (*Actinobolus* ranges between 100μ and 200μ.) Key: C., cilia; C.V., contractile vacuole; F.B., food body; I.B.F., inner bundle of fibrils; M., macronucleus; O., oral opening; O.F., oral fibrils; Ten., tentacles. (After Wenrich, *Biol. Bull.*)

remain in one spot while feeding, although the more widely held view is that they are defensive. Other types of trichocysts occur, such as the toxicysts of the ciliate *Actinobolina* (Fig. 5.4). This vicious little animal uses its toxicyst-tipped tentacles to paralyze its victims. Even rotifers succumb quickly to the poison within these tiny weapons.

Trichocysts similar to those of *Paramecium* occur also in some dinoflagellates and even in certain cryptomonad flagellates such as the common

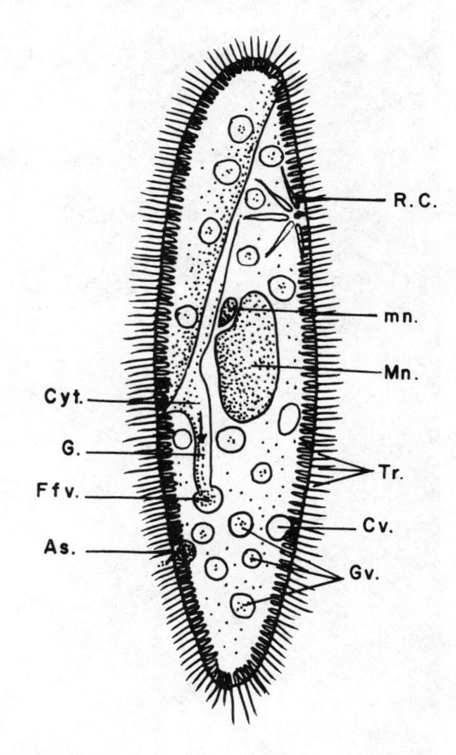

FIG. 5.5. *Paramecium caudatum.* As. anal spot; C.v., contractile vacuole; Cyt., cytostome; F.f.v., forming food vacuole; G., gullet; G.v., gastric vacuole; Mn., macronucleus; mn., micronucleus; R.C., radial canals; Tr., trichocysts (shown in a row just under cilia). (After Wenrich, *Trans. Am. Micr. Soc.*)

Chilomonas. There are also the so-called mucoid trichocysts of ciliates and some flagellates, paratrichocysts, discobolocysts, and others. Except for the toxicysts, the functions of trichocysts remain unknown.

Often also regarded as trichocysts are the nematocysts of some dinoflagellates, especially species of *Polykrikos*. Their occurrence is thought by some to have phylogenetic significance, especially as they strongly resemble not only the nematocysts of coelenterates but the cnidocysts of the Cnidosporidia.

Trichocysts differ considerably in structure and may be quite complicated. They were first seen in *Paramecium* by Ellis some two centuries ago, according to Wichterman (1953), and are easily observed under the ordinary light microscope despite their relatively small size (4μ by 2μ). Among the ciliates trichocysts occur rather commonly, but by no means universally. In *Paramecium* they lie in great numbers just under the surface (Fig. 5.5). Their appearance when discharged, viewed under the electron microscope, is shown in Figure 5.6. An individual trichocyst consists of a finely striated filament or shaft bearing a spearlike tip. Discharge requires only two or three milli-

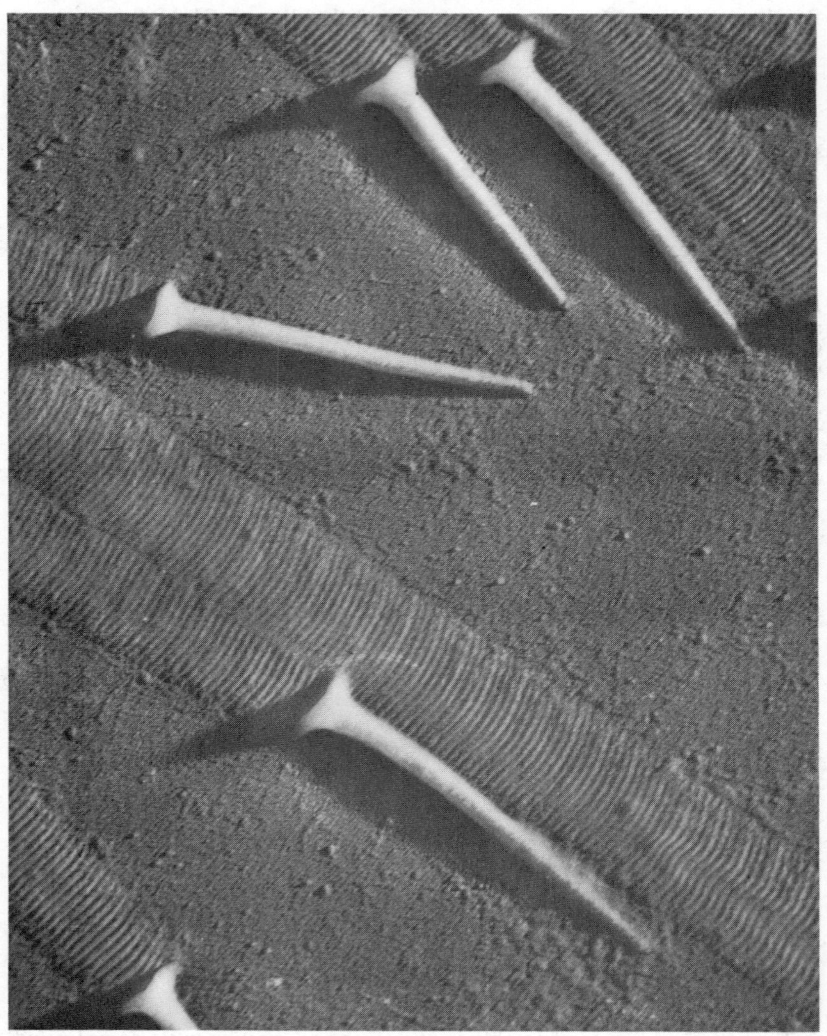

FIG. 5.6. Trichocysts of *Paramecium*, as shown by electron microscopy in shadow-cast chromium preparations, ×18,144. Note the finely striated trichocyst shaft. (Jakus and Hall, *Biol. Bull.*)

seconds and may be triggered by a variety of stimuli: chemical, mechanical, and even electrical. Its exact mechanism is unknown, but it may be that the trichocyst contains some substance capable of taking up water suddenly and swelling with almost explosive violence, much as if a tiny submarine were shooting a microscopic torpedo.

Of other ectoplasmic elements, perhaps none is stranger than the tentacles of the Suctorida, those peculiar predatory ciliates which capture their prey

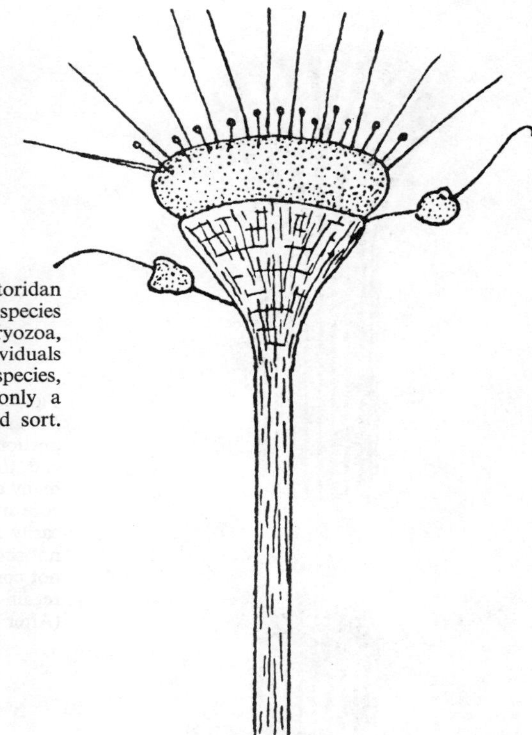

FIG. 5.7. *Ephelota plana*, a suctoridan with two types of tentacles. A species often found attached to Bryozoa, attached to it here are two individuals of another suctoridan species, *Ophryocephalus*, which has only a single tentacle of the knobbed sort. (After Wailes.)

by lying in wait for it, like the spider for the fly. When a protozoan of acceptable sort wanders by and chances to touch an outstretched tentacle it is promptly paralyzed by some secreted toxin, while the protoplasm of the luckless victim is then sucked through the tentacle into the body of the captor. The process is a fascinating one to watch under the microscope. The mechanism by which it is accomplished remains uncertain. A suggestion made many years ago by Sand (1901) and still sometimes urged is that a negative pressure within the body of the suctoridan may be produced by an accelerated water elimination through the contractile vacuole. A sixfold increase in vacuolar activity during feeding has actually been observed in the suctoridan *Tokophrya infusionum* by Rudzinska and Chambers (1951). Not all tentacles of the Suctorida, however, are alike. A second type, less universally distributed, seems to have a prehensile function. The first, or sucking, type is tipped with a tiny knob, while the second type of tentacle resembles an axopodium, with an inner filament extending deep into the cytoplasm until it terminates close to the nucleus (Fig. 5.7).

Also characteristic of most Suctorida, and apparently of ectoplasmic

FIG. 5.8. Stalk or pedicel of *Tokophrya lemnarum*. The funnel-shaped upper portion bears the body and the lower end the basal disk. Within are as many as 24 delicate rods and an inner core of homogeneous substance not easily demonstrated by staining and not shown in the figure. The stalk is not contractile, but is elastic and can regain its position after bending. (After Noble.)

origin, is the terminal disc and stalk by which the organisms attach themselves to the substrate (Fig. 5.8). The stalk, both of the Suctorida and the vorticellids (Peritrichida), is secreted by the disc ("scopula"), and this is in turn of ciliary origin.

Locomotor Organelles

Movement is probably what first attracts the attention of the microscopist interested in "little animals." The organelles by which Protozoa move (flagella, pseudopodia, cilia) are not always easily visible. The grosser characteristics of these structures, whose versatility enables them to be used not only for travel, but for food-getting and perhaps also for tactile and sensory purposes, have already been described.

Ciliary arrangement varies greatly with the group and the species. Typically, the cilia form rows originating in the region of the mouth and proceeding as meridians or spirals toward the posterior end. Underlying each cilium is a basal granule or "kinetosome"; the relation of the latter to cilium formation will be discussed later (p. 87). The cilia around the mouth are often modified in a variety of ways to facilitate food capture. Some of the larger species of Holotricha may have more than 10,000 cilia which, in this order, are typically arranged uniformly about the organism.

Among the more highly evolved ciliates, the cilia are often fewer in number and their distribution restricted. This kind of change often occurs in the course of evolution, and is best exemplified by the Hypotrichida, where ciliation is for practical purposes limited to the ventral surface. Here small groups of adjacent cilia have fused to form massive locomotor organelles known as "cirri" (singular, cirrus), which are used for creeping, walking, and even for the protozoan equivalent of jumping. That cilia have indeed originated by the fusion of neighboring cilia is proved by the numerous basal granules underlying them, for a single cilium has but one basal granule. It is also possible to fragment cirri into their component cilia by special treatment.

Many ciliates also possess bristles. These appear to have essentially the same structure as cilia, but lack the fibrillar ("neuromotor") connections of cilia used for locomotion. Since such bristles are not capable of independent movement, their independence of the neuromotor system is no more than one would expect. Their function is unknown, though it has been suggested that they are tactile or sensory; this is, however, a difficult theory to prove experimentally. The most conspicuous of such bristles are the long caudal ones possessed by *Uronema* and some of its holotrich relatives.

Although an individual cilium looks different from a flagellum and functions in a somewhat different manner, recent studies with the electron microscope have shown that the structure of both is remarkably complex and very similar. Each consists of a cytoplasmic sheath with what appears to be a filament within, but the latter is made up of eleven longitudinal fibrils, nine of which are peripheral and two central (Fig. 5.9). By some accounts, the former are each double. Flagella may have lateral branches, however; a flagellum of this type is known as a "flimmer" or ciliary flagellum (Fig. 5.10).

By comparison, pseudopods have a much simpler structure, although there are the four general types (lobopodia, filopodia, rhizopodia, and axopodia or actinopodia) mentioned elsewhere (pp. 141, 320). Pseudopodia are

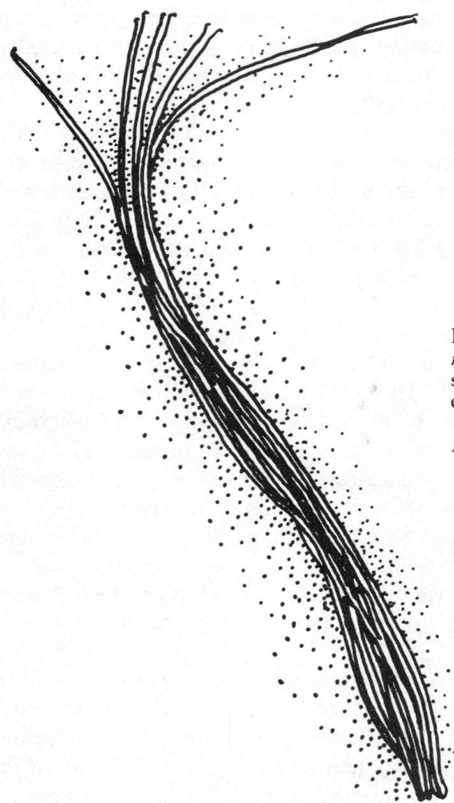

FIG. 5.9. A frayed-out cilium of *Paramecium* seen under the electron microscope, consisting of 11 minute fibrils, each about 400 A in diameter, $\times$11,000. (After Jakus and Hall, *Biol. Bull.*)

often highly specific in form and mode of formation, and their characteristics aid in identifying species. They differ in the relative amounts of ectoplasm and endoplasm, as well as in the manner of extrusion and withdrawal; they also differ in use. Filopodia, for example, contain almost no endoplasm, whereas lobopodia are made up of both. Rhizopodia have a rootlike appearance (as the name suggests), branching profusely. But unlike roots, the branches fuse or anastomose on contact. The lobopodium is the most commonly observed type among the Sarcodina, since it is characteristic of amoebae and many Testacida, but most Foraminiferida, perhaps the largest order of the class, produce rhizopodia.

Axopodia are unique in structure and appearance. They consist of a cytoplasmic sheath enveloping an axial filament, the former being almost fluid in consistency. Granules in the sheath may often clearly be seen moving up on one side of the central filament and back into the body of the animal on the other. Structurally, flagella and axopodia are superficially alike, but the latter, though fibrillar, lack the typical "9+2" pattern of minute fibrils

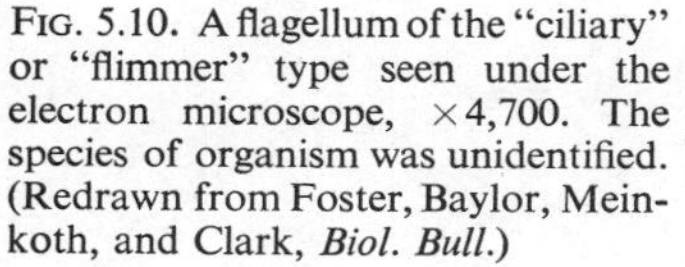

FIG. 5.10. A flagellum of the "ciliary" or "flimmer" type seen under the electron microscope, ×4,700. The species of organism was unidentified. (Redrawn from Foster, Baylor, Meinkoth, and Clark, *Biol. Bull.*)

revealed in the former by electron microscopy. Axopodia, however, are sometimes capable of a slow vibratile motion, as in some Radiolarida. The internal filament of the axopodium extends to a central body, sometimes contained within the nucleus. For all its appearance of rigidity, the axopodium is capable of being bent or quickly withdrawn.

Cytostome and Associated Organelles

Since food-getting is the first requirement of any living thing, it is not surprising that locomotor organs of any type are commonly used for this purpose, and they are very often associated with the cytostome. This organelle lies at or close to the anterior end in many flagellates and the more primitive ciliates; in the more advanced ciliates it is often found anywhere in the anterior two-thirds of the body and, in a few species such as certain parasitic ciliates of molluscs, at the posterior end. In still others of this interesting group (known as thigmotrichs) it is completely absent. The Opalinids, parasitic in frogs and toads and perhaps more closely related to the flagellates than to the ciliates, also lack a mouth (see Chap. 22). Thus the position of the mouth, when one is present, is a very important factor in identifying species.

In the immediate neighborhood of the cytostome (the actual opening into the body) is the area known as the "peristome" (literally, "about the mouth"), also called the "buccal cavity." In the peristome lie, in many species, compound ciliary organelles of many kinds used in feeding (Fig. 5.11); their function is primarily to create a current of water that drives food particles into the mouth. The Italian monk Spallanzani remarked in a monograph published in Italy in 1777, "The oscillating filaments make a whirlpool, which draws into the opening or mouth of the animalcule objects which may be swimming about in the infusion, and the little animal then chooses the more dainty bits for its foods, or at least those which suit it best." He

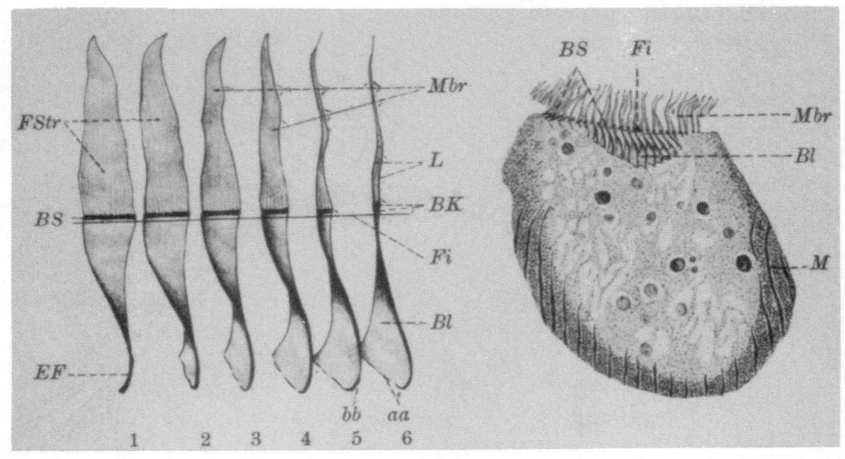

FIG. 5.11. Structure of the membranelles of *Stentor coeruleus* (as reconstructed by Dierks, 1926). aa, lower edge of the basal plate; bb, posterior edge of the sheetlike basal plate end; FStr, longitudinal striation of the membranelles; BS, basal junction; EF, terminal filaments; Bl, basal plate; Fi, area of conducting fibrils in peristome; BK, basal bodies; L, two fused ciliary plates; Mbr, membranelle.

also commented, "One is forced to believe that, penetrating the body of the animal (from the mouth) is some invisible canal."

More highly evolved ciliates may have associated with the cytostome and peristome such cilia-derived structures as the ciliary plates ("membranelles") representing the fusion of two, or even three, adjacent and short ciliary rows. Membranelles also occur in the Holotricha, in which are included the more primitive ciliates. Ciliary rows may form membranes, often appearing of relatively great size (Fig. 5.12) but actually more gossamer-like than the finest spider's web. Their constant motion is like that of a sail in the wind, even though their function is so different. In some species several such undulating membranes are present, or there may be a single undulating membrane and several membranelles, as in the much studied genus *Tetrahymena* (although here the generic name is somewhat misleading, since its literal meaning is "four-membraned").

The margin of the peristome, in most ciliates, bears a fringe of membranelles or membranes known as the *adoral zone.* This is also important in identification, since it may wind right or left toward the mouth.

Among the flagellates the mouth, when present, and adjacent area exhibit relatively little specialization, and there is no hint of the complex and elaborate food-getting mechanisms often found in the ciliates. In many flagellates there is no mouth at all, and when one is present (as in euglenids) it appears that it may not function for the ingestion of food. In Sarcodina, as in Sporozoa, the mouth is, of course, completely absent.

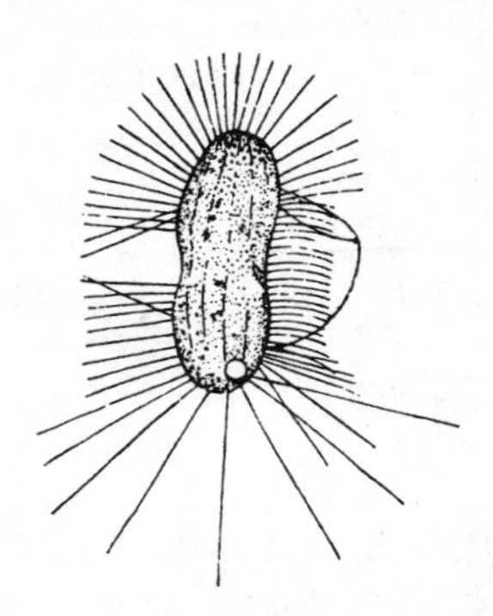

FIG. 5.12. *Cyclidium litomesum.* A ciliate with a very large undulating membrane shown on the right; its width equals that of the body. Note the subterminal contractile vacuole and the very long caudal bristles. (After Stokes.)

ENDOPLASM

Within the endoplasm of living Protozoa a great variety of inclusions may be discerned. The plantlike flagellates in many species have a conspicuous orange-red eyespot ("stigma") and, of course, chlorophyll bodies (variously known as "chloroplasts," "chromatophores," or "chromoplasts"), differing according to species in shape, size, and number. Their characteristics are important in species identification. Symbiotic bacteria and algae, as well as certain green flagellates, occur in the cytoplasm of many Protozoa, and are especially common in the Radiolarida. *Paramecium bursaria*, almost always encountered by any beginning student, has its cytoplasm so loaded with a species of the alga *Chlorella* that it is always a bright green. Bodies of paramylum in the chlorophyll-containing flagellates, and of glycogen (animal starch) in numerous species, among them the parasitic amoebae, are commonly seen and are often of characteristic shape and size. They constitute reserve food material, and disappear with the passage of time in starved and encysted individuals.

In some Protozoa (e.g., the ciliates *Chilodonella* and *Dileptus*, Figs. 5.13, 5.14, 11.8), skeletal elements such as the trichites supporting the mouth are easily observed. The ciliate *Coleps* (Fig. 5.15) has an armorlike jacket of minute plates embedded in its cytoplasm. Most remarkable of all, perhaps, are the exquisitely intricate and beautiful skeletons of the Radiolarida. Among the ciliates are the curiously elaborate internal skeletons of certain cattle ciliates and their close relatives (e.g., *Epiplastron* of antelopes).

Food Vacuoles

Food vacuoles are likely to be conspicuous in Protozoa that take their food in particulate form. They arise wherever food is ingested. In ciliates this is, of course, at the lower end of the gullet, a structure which is often highly

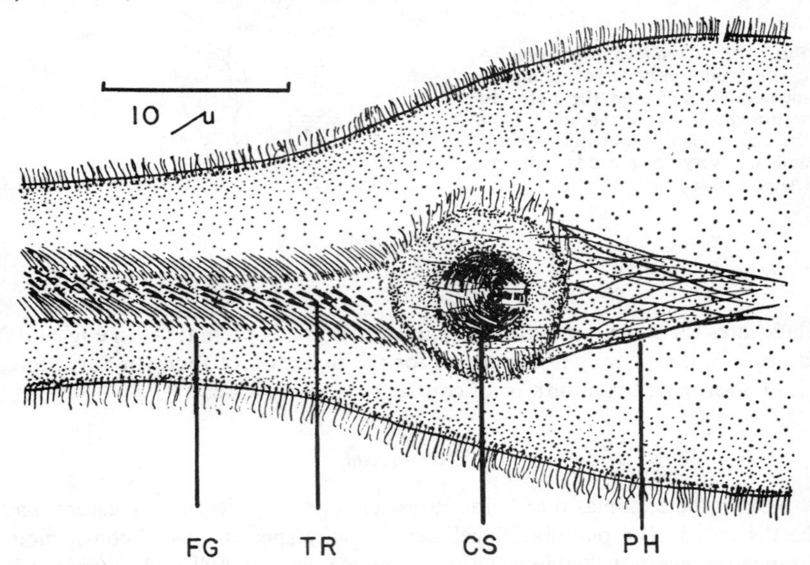

FIG. 5.13. Cytostome of *Dileptus monilatus* PH, pharyngeal basket; CS, cytostome; TR, trichocysts; FG, feeding groove. (After Jones and Beers, *J. Elisha Mitchell Sci. Soc.*)

complicated. Figure 5.16 shows the mouth and esophagus of *Paramecium*, as depicted by Mast (1947). From this point the food vacuoles move about in the cytoplasm, generally following a rather definite path, presumably determined by cyclosis. Close observation of their contents will show them to be filled with prey of various sorts: bacteria, yeasts, fungi, algae, diatoms, other Protozoa, and occasionally even rotifers and the smaller crustacea.

In many species of ciliates the course followed by the food vacuoles terminates at a definite point in the posterior portion of the animal, where there appears to be an opening known as the cell anus ("cytopyge" or "cytoproct"), through which any undigested or indigestible residue from the last meal is voided. Cosmovici (1933) claimed that in some ciliates there is a protozoan analogue of the alimentary tract in higher animals, consisting of a minute more or less convoluted canal. Though his experiments were ingeniously done his conclusions have not been generally accepted.

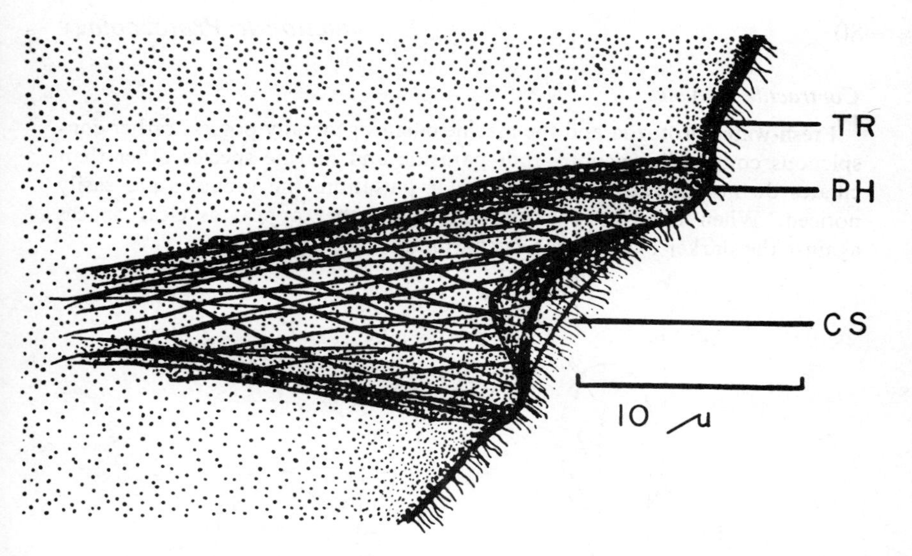

FIG. 5.14. Pharyngeal basket of *Dileptus monilatus*. TR, trichocysts; PH, pharyngeal basket; CS, cytostome. (After Jones and Beers, *J. Elisha Mitchell Sci. Soc.*)

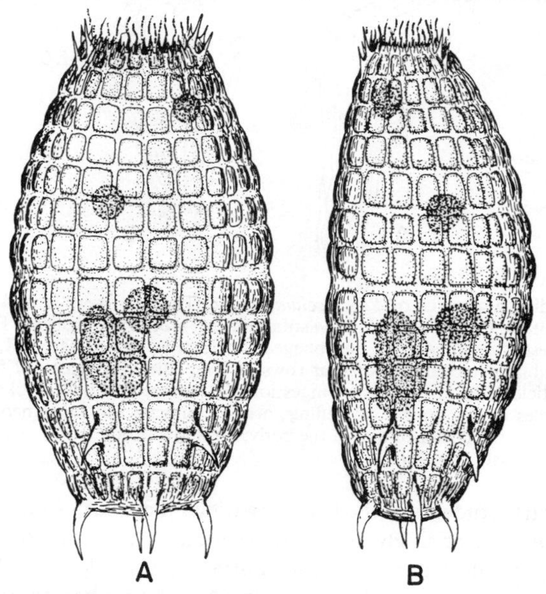

FIG. 5.15. *Coleps octospinus*, a ciliate with armor consisting of plates, each of which is further divided into small rectangular areas. It is these areas, rather than the individual plates, which one usually observes in the animals. Cilia project between the plates and the mouth occupies most of the anterior end. At the posterior end are spines, the number and arrangement of which are of taxonomic value. *A* shows a dorsal view and *B* a side view of the animal. (After Noland, *Trans. Am. Micr. Soc.*)

Contractile Vacuoles

Fresh-water Protozoa and the marine ciliates generally possess rather conspicuous contractile vacuoles, but parasitic and marine species other than ciliates do not. Because of their regular pulsation, the vacuoles are easily noticed. When completely or partially filled, they appear as clear spaces against the darker cytoplasmic background.

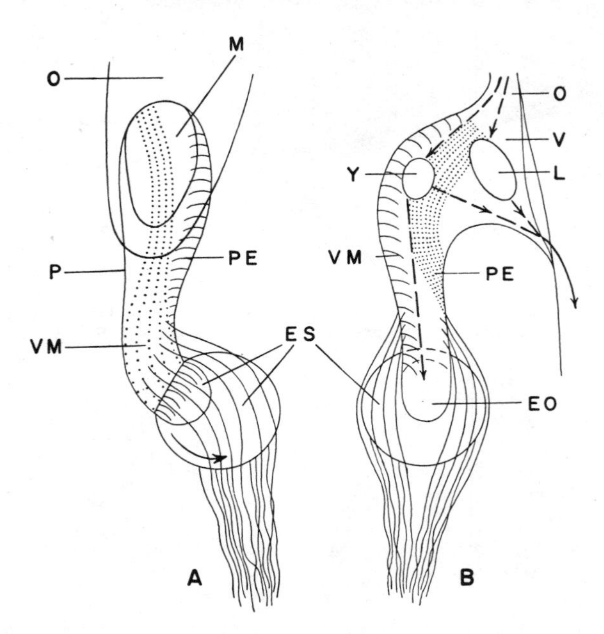

FIG. 5.16. Feeding apparatus of *Paramecium aurelia*. The two views *A* and *B* are at right angles to each other. O, oral groove; V, vestibulum; M, mouth; P, pharynx; EO, esophageal opening; ES, esophageal sac; EF, esophageal fibers; PE, peniculus; VM, "Vierermembran" (a "membrane" consisting of four rows of long cilia); Y, yeast cell being ingested; L, a large particle, also in process of ingestion; S, surface of body; broken lines, paths taken by particles during course of feeding; arrows, direction of movement. Fibrils, said to extend nearly to the posterior end of the body, are shown projecting from the wall of the pharynx. (After Mast, *Biol. Bull.*)

Even the early students of the Protozoa noted these curious little organelles, and made some surprisingly acute observations. Among these pioneers was Spallanzani, who described both the contractile vacuoles and the collecting canals that in many species lead to them: ". . . Another organ I have discovered in the course of recent observations, and which I suspect is concerned with respiration; it is composed of two stars which have at their centers globules. . . . These two stars are always in motion, whether the animals are moving or at rest, but the motion is regular and alternate. Every three or four seconds, the two little central globules fill up like bladders and become three or four times larger; finally they contract, and their emptying, like their filling, is carried out very slowly; one may see the same timing in the rays of these stars, but with this difference: when the globules fill, the rays empty, and when the latter fill, the former empty. . . ." It seems almost certain that Spallanzani's "animalcules" were either species of *Paramecium* or of some other genus with similar contractile vacuoles. One suspects from Spallanzani's own accounts of his experiments that the new world revealed by the microscope fascinated him far more than did the ancient Greek world of the classics and metaphysics about which he lectured to the students of the universities of Modena and Pavia.

The contractile vacuoles of different species of Protozoa vary greatly in detail even though they are alike in function. Among the ciliates there may be one, two, or more vacuoles, although in a given species the number and position are usually constant. The vacuole is often located at the posterior end or near the pharynx. When a number of contractile vacuoles are present as in *Loxophyllum* they are likely to be marginally placed.

In ciliates, in which contractile vacuoles have received more study than those of other Protozoa, collecting canals may or may not be present, and when present vary in number even in different individuals of the same species. King (1935) found that in *Paramecium multimicronucleatum* even the canals were of rather complex structure (Fig. 5.17), being divided into

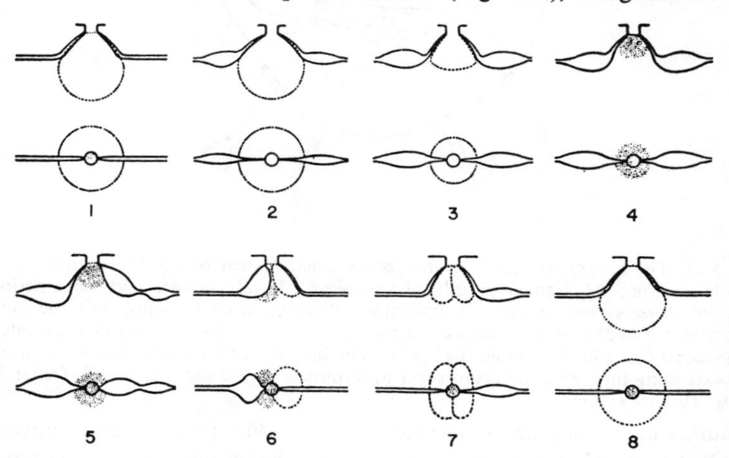

FIG. 5.17. Diagrammatic sketch of the successive stages in the operation of the contractile vacuole of *Paramecium multimicronucleatum*. The first and third rows show the vacuole, with a pair of feeder canals, in a "horizontal" view; the second and fourth are in a plane at right angles to this. *1*, full diastole, with membrane closing pore; *2*, beginning of systole, with pore open; *3*, middle of systole, with ampullae filled; *4*, end of systole, with membrane closing pore; *5*, contents of ampullae passing into injection canal; *6*, formation of vesicles from injection canals; *7*, fusion of vesicles to form central vesicle; *8*, central vesicle in full diastole. (After King, *J. Morphol.*, 58: 555.)

three clearly differentiated parts. The vacuole emptied to the exterior through a tubule ending in a pore on the surface, the two constituting a permanent organelle, but the vacuole itself was formed anew after each contraction ("systole") by the fusion of several vesicles. An electron microscope study of the suctoridan *Tokophrya infusionum* by Rudzinska (1957) revealed that the contractile vacuole has a similar structure, with a permanent voiding canal. Apparently the latter is provided with a closure mechanism consisting

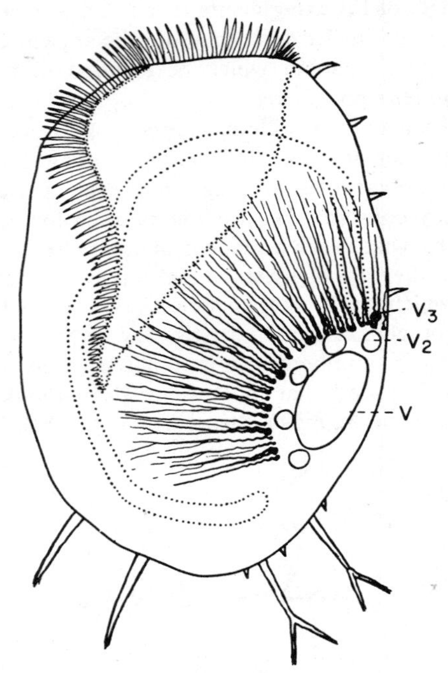

FIG. 5.18. The contractile vacuole and feeder canal system of *Euplotes patella*. V, contractile vacuole; V2, smaller peripheral vacuoles; V3, still smaller and more peripheral vacuoles. After systole and the disappearance of the contractile vacuole (V), the vacuoles lying about it (V2) coalesce and take its place, and the smaller vacuoles (V3) outside take the place of those in the middle ring (V2). The finer collecting canals (shown as irregular rays extending into the cytoplasm) seem to terminate in the vacuoles of V3. (After King, *Trans. Am. Micr. Soc.*)

of numerous minute fibrils surrounding its middle portion, an arrangement strikingly reminiscent of the sphincter at the mouth of the mammalian bladder.

In *Euplotes*, another ciliate studied by King (1933), the vacuolar arrangement is essentially like that of *Paramecium*, except that the feeding canals are much more numerous and minute, and two supplementary sets of vacuoles surround the main one (Fig. 5.18). The first of these sets is fed by the collecting tubules, and the second, or intermediate, set is formed by coalescence of vacuoles of the first. The latter then contribute to the filling of the main vacuole.

The contractile vacuoles of flagellates appear to be simpler. They are often arranged about the gullet (frequently referred to as the "reservoir"), into which they empty. In amoebae and other Sarcodina their position is naturally variable, and they are ordinarily absent in the Sporozoa.

Because the chief function of the contractile vacuole is now thought to be hydrostatic (see p. 124), or osmoregulatory (i.e., governing the water balance), some believe it corresponds to the "vacuome" (vacuolar system) of many plant and animal cells, which is also concerned with the control of protoplasmic water content. This may perhaps be true in the green flagellates, among which a special problem may exist because of the production of water in photosynthesis. Here such vacuoles are present even in marine types. Ordinarily, however, the protozoan vacuome is regarded as a system of granules, stainable with neutral red, the presence of which has been demonstrated in numerous species. In some they seem to be associated with the food vacuoles, and it has therefore been suggested that they may be secretory in function. But opinion differs as to their distribution among species and even within the organism, and as to their identity and function. Unlike the contractile vacuole, they are visible only after coloration with suitable dyes.

Nuclei

The nuclei of Protozoa are seldom visible in the normal living organism, although there are numerous exceptions to this rule. Since chloroplasts occur in the cytoplasm, the nucleus of the green flagellates may often be seen as a clear space, usually in the central part of the organism. (Do not confuse it with the paramylum bodies, which are also colorless and frequently of comparable size.) The nuclei of the larger amoebae are also often visible, as for example in *Amoeba proteus*, in which it has a disclike form.

Protozoan nuclei vary a great deal in size, shape, and number. They may be round, oval, ribbonlike, of horseshoe shape, or beaded. Many species have two, or even many, nuclei and even in forms ordinarily uninucleate there may be multinucleate stages in the life cycle, as in the Foraminiferida and Myxosporida. Such a stage is called a "plasmodium" or "syncytium." *Arcella vulgaris* is a common testate rhizopod with two oppositely placed nuclei. Arranged irregularly about the periphery may also be seen a ring of "chromidia," bodies which stain heavily with haemotoxylin and which, for that reason, were long thought to represent dispersed chromatin material. Since they do not take a Feulgen stain they are now regarded as perhaps secretory granules of some sort.

In gross structure, the nuclei of Protozoa are often said to be either "vesicular" or "compact," although the distinction is arbitrary and perhaps without much real significance. In nuclei of the former type there is a lesser amount of chromatin and usually a more or less conspicuous "karyosome" (also termed "endosome"). Vesicular nuclei are especially characteristic of the Sarcodina and Mastigophora. The position and relative size of the karyosome may be of considerable diagnostic importance in the case of the intestinal amoebae, but staining is usually needed to make such structures clearly visible.

The ciliates (except the Opalinas and their relatives) possess nuclei of two sorts, and in this respect they are unique. The macronucleus, the larger of the two, appears to be chiefly concerned with control of the vegetative activities of the organism, but it also determines phenotypic characteristics and governs regeneration after injury. It looks almost structureless under an ordinary microscope, but staining shows it to be composed of a great number of minute particles that reproduce themselves at the time of division. Electron microscope studies, so far made on only a few species, show the nuclear membrane to be apparently double and perforated by myriad tiny pores. The material within the membrane seems to consist of numerous Feulgen-negative nucleoli, interspersed with much smaller Feulgen-positive bodies which Dippell (1957) thinks may be actual chromosomes. These appear against a background of "wispy material," often as short chains, although they may be in reality coiled, and it is said that they show premonitory changes indicative of approaching division even before anything of the sort can be seen in the micronucleus. We realize how minute these bodies are when we know that they were observed in sections of only 0.02 micron thickness.

The micronucleus in most species seems to have genetic functions only, and amicronucleate individuals that appear able to live and grow and divide normally, although they cannot survive conjugation, have often been seen. Like the macronucleus, the micronucleus apparently is enveloped by a very delicate double membrane. Species differ in the number of nuclei of each kind, but usually there are not more than two. There are, however, exceptions such as *Dileptus*, in which there are fifty or more of each.

In form, macronuclei vary greatly, but micronuclei are typically either round or oval. The latter are almost never visible in the living ciliate, but the former may sometimes be seen, particularly in moribund individuals confined too long under the cover slip.

The two types of nuclei vary greatly in behavior during division, conjugation, and encystment. During division only the micronucleus exhibits mitosis; the macronucleus divides amitotically (one of the few examples of amitosis normally occurring among the Protozoa). The details of nuclear changes during conjugation and encystment differ with the species (see Chaps. 10, 11). In general, the macronucleus disintegrates at the time of conjugation while the micronucleus is dividing several times, two of the divisions being meiotic, the first of the latter being reductional and the second equational. Thus the process is like metazoan gametogenesis. All but two of the products of these divisions in each conjugant degenerate, and a mutual exchange of one of the nuclei of each pair ensues, followed by fusion of the stationary and wandering nuclei. The normal nuclear complex for the species is restored by successive nuclear divisions, some of the products becoming micronuclei and others macronuclei. Thus, since each kind of nucleus has a common origin, it is easy to understood how the macronucleus may determine phenotypic characters and control regeneration.

Cysts

It is not easy to generalize about what happens to the nuclei during encystment because of differences among species and the existence of a variety of cyst types. But there often appears to be rather extensive nuclear and cytoplasmic reorganization, and in some cases actual multiplication occurs within the cyst (Chap. 10). The cysts themselves vary in appearance as much as their makers. Some are spherical and smooth externally, some spherical and wrinkled, or spherical and knobby. They may be oval or quite irregular in form. Usually there is an inner and outer wall, and there may also be a middle one ("endocyst," "ectocyst," and "mesocyst" respectively). In some species the cyst is provided with an emergence pore, often of a diameter so much smaller than even the smallest dimension of the occupant that egress must be a major task. The first step in the process of escape from such cysts is forcing open the lid or operculum closing the aperture. In the cysts of numerous species avenues of escape are apparently lacking, and egress requires either the solution of the cyst wall by enzyme action or its rupture. In view of the microscopic dimensions of the cyst, a considerable pressure in terms of pounds per square inch must be required, achieved in some forms by the imbibition of water, final exit being aided by the efforts of the organism itself.

Neuromotor Apparatus

Invisible without special staining techniques, yet perhaps the central part of that mechanism of microscopic "pullies and wheels" envisaged two centuries ago by Baker, is the elaborate fibrillar system of the ciliates commonly known as the "neuromotor apparatus." As described originally by Sharp (1914) in the cattle ciliate *Epidinium* (then known as *Diplodinium*), it consists of a central portion he called the "motorium" from which radiates a complex arrangement of minute fibrils terminating at the bases of the various motor organelles. Since Sharp, similar systems have been described in species of other ciliate genera, among them *Euplotes*, *Paramecium*, and *Telotrochidium* (all free living), and *Nyctotherus* and *Balantidium* (both parasitic). *Metopus* may also be added to the latter two, although it is a kind of parasite-by-courtesy in the digestive tract of the sea urchin, since its environment there is much like sea water, and it does its host no harm.

Although the details of the neuromotor system in different species vary greatly, in many, the motorium is closely associated with the pharynx or gullet, but in others, it is variously placed. From the motorium fibrils extend through the cytoplasm to the organelles about the mouth, apparently constituting a mechanism for coordinating membranelles, undulating membranes, and other cilia-derived structures. Fibrils serve the cilia and cirri of the various body regions, and, presumably, the contractile filaments or myonemes when these are present. Figure 22.2 shows the system as originally described by Sharp. He remarked, "Every mobile territory is supplied by strands from the central mass (*motorium*) and especially are the bases of the membranelles, both dorsal and adoral, well supplied by these fibers," a plausible but insufficient reason for assigning the system a motor-coordinating function.

More recently, Kofoid and Rosenberg (1940) worked out the anatomy of the neuromotor apparatus of the vorticellid *Telotrochidium henneguyi* (formerly *Opisthonecta henneguyi*). It is of particular interest because fibrils from the motorium to the sphincter myonemes by which the organism folds its oral end are clearly visible (Fig. 5.19).

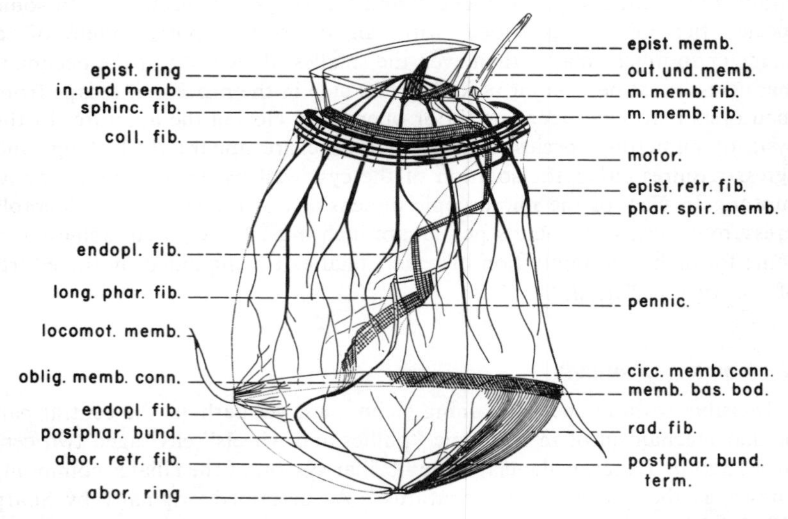

FIG. 5.19. Neuromotor system of the ciliate *Telotrochidium henneguyi*, showing its complex structure. In the figure the radial fibrils and the basal bodies of the membranelles are omitted from one side of the organism for clarity. Key: *abor. retr. fib.*, aboral retractor fibril; *abor. ring*, aboral ring; *circ. memb. conn.*, circular membranelle connective; *coll. fib.*, collarette fibril; *endopl. fib.*, endoplasmic fibrils; *epist. memb.*, epistomal membrane; *epist. retr. fib.*, epistome-retractor fibril; *epist. ring*, epistomal ring; *in. und. memb.*, inner undulating membrane; *locomot. memb.*, locomotor membranelles; *long. phar. fib.*, longitudinal pharyngeal fibril; *memb. bas. bod.*, membranelle basal bodies; *m. memb. fibs.*, motomembrano fibrils; *motor.*, motorium; *obliq. memb. conn.*, oblique membranelle connective; *out. und. memb.*, outer undulating membrane; *pennic.*, penniculus; *phar. spir. memb.*, pharyngeal spiral membrane; *postphar. bund.*, postpharyngeal bundle; *postphar. bund. term.*, postpharyngeal bundle terminal; *rad. fib.*, radial fibril; *sphinct. fib.*, sphincter fibril. (After Kofoid and Rosenberg.)

Despite the fact that fibrillar systems such as these undoubtedly exist in many species of ciliates, we may still question whether their function is really neuromotor. It has been argued that these minute strands of apparently specialized cytoplasm may have a supporting function in addition to, if not instead of, that of coordination, or that they may be contractile or sensory in nature. This is a difficult problem to solve experimentally. Taylor (1920) attempted to answer the question by the techniques of microdissection, severing the neuromotor fibrils of *Euplotes* to observe the effects on their connected cirri. He concluded that the disorganization of movement that resulted was decisive evidence of a coordinating function for the fibrils. But recent work by *Jahn (1961), *Pitelka (1961), and others suggests that what has been called the neuromotor system may not have this function, though these authors do not try to explain what this fibrillar system does do. Jahn believes that electrical phenomena alone may account for ciliary activation and reversal.

The relation of the neuromotor apparatus to the complex pattern of surface markings and ectoplasmic fibrils revealed by Klein's method of silver impregnation has also proved difficult to determine. It appears that the two systems are probably distinct since the techniques by which they can be demonstrated are quite different. For delineating the neuromotor fibrils Mallory's triple stain and Heidenhain's haematoxylin have usually proved best. As the result of painstaking studies of the two systems in *Paramecium*, Lund (1933) concluded that "Klein's method fails completely to demonstrate the most intricate part of the neuromotor complex. . . ."

Kinetosomes

Although the fibrils, granules, and other ectoplasmic structures revealed by silver impregnation have often been referred to as the "silver-line system," it seems fairly certain that they do not constitute a real system at all. But the technique has been used by Chatton and Lwoff (1935) in an ingenious way to study and interpret the behavior of the basal granules of the ciliates. These they have found to be self-duplicating elements, comparable in this respect to genes, and capable of giving rise not only to cilia, but to fibrils, trichocysts, trichites, and even to the mouth (in some forms) and its associated structures. Flagella likewise have their basal granules (often termed "blepharoplasts"), and it is an interesting and significant fact that they are found not only in the flagellate Protozoa but in other types of flagellated cells, such as spermatozoa. Similarly, the cilia of the ciliated epithelia of Metazoa have their basal granules.

Since these granules seem to be essential to the movement of both cilia and flagella, and may indeed themselves move about in the cytoplasm on occasion, they have been termed "kinetosomes" (from the Greek *kinematos* and *soma*, meaning "movement" and "body"). A row of these tiny elements with their cilia has been named "kinety."

Not the least remarkable fact about the kinetosomes is their persistence (easily demonstrated by silver impregnation) in some flagellates and ciliates even during stages in the life cycle when locomotor organelles are absent, as in the adults of species of Suctorida. More remarkable still is the ability of the kinetosomes to give rise to different organelles at different times. The problem is very much like that of explaining the fate of different cells in a metazoan embryo.

Associated with the kinetosomes are certain other structures, such as the granules known as "kinetoplasts" (perhaps really the enlarged basal ends of the cilia), and a minute longitudinal fibril called "kinetodesma" lying to the right of and underneath each row. Thus the whole complex appears to form a system for which Lwoff and Chatton, who were pioneers in its study, coined the term "infraciliature."

Differentiation

The Apostomatida, on which much of the research on infraciliature was originally done, are a small and curious group of ciliates. Nearly all of them are parasites of Crustacea, with whose life cycles their own have become intricately intertwined. Yet the system functions in much the same way in other ciliates. As more is learned, it is becoming apparent that its behavior is a key to solving the puzzle of cellular differentiation, a problem both fascinating and vexing. Experiments in regeneration of ciliates, performed by Paul Weisz and others, have shown that differentiation, even in organisms as minute and (as many have naively supposed) as simple as the Protozoa, is a very complex process indeed. It seems to depend on three factors: macronucleus (and hence genes), kinetosomes, and the proximity of the gullet, which in some fashion influences macronuclear ability to control regeneration. Although the gulf separating Protozoa and metazoa seems great, the problem of differentiation in the developing organism may be fundamentally similar in both. Weisz remarks, "Thus the embryologist reflects, his embryonic cells and his Protozoa share the same basic secret" (Weisz, 1953).

Division in the common and highly evolved hypotrich *Euplotes* is a good illustration. Here, as in most ciliates, fission is heralded by the appearance of a new mouth just back of the old one, destined for the posterior (the "opisthe") of the two individuals-to-be. This new organelle becomes evident long before there are other noticeable changes in the nucleus and the cytoplasm, apparently as the result of the influence of a single kinetosome originally lying close to the mouth of the parent.

Lwoff puts the matter succinctly when he says, "It is possible to describe the morphology of ciliates in terms of kinetosomes and their derivatives. The morphogenesis of a ciliate is essentially the multiplication, distribution, and organization of populations of kinetosomes and of the organelles which are the result of their activity. Kinetosomal order is controlled, at least partially, by the kinetodesmas" (Lwoff, 1950).

But the complexities of internal organization do not end here. There must, of course, be some control over the activities of the kinetosomes. Although we know the macronucleus is involved, there is still the question *How*. A suggestion made by Weisz and others, although of little help presently, is that kinetosomal behavior is somehow mediated by the immediate molecular environment.

Mitochondria

Among other very minute structures of the protozoan organism are the "mitochondria" (singular mitochondrion). The name is derived from two Greek words, *mitos* and *chondros*, meaning "thread" and "grain" respectively, but it is a misleading term, since these tiny organelles show much diversity of form. They are cytoplasmic structures, spherical, rodlike, cylindrical, or dumbbell-shaped—even sometimes in the same cell. To

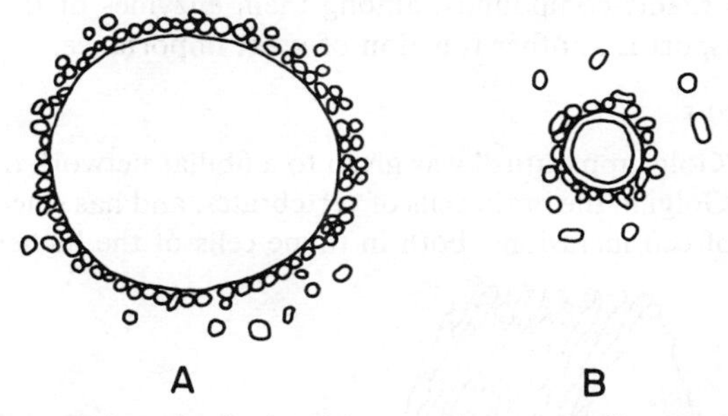

FIG. 5.20. *A*. Accumulation of mitochondria around a contractile vacuole of *Pelomyxa carolinensis*. *B*. Mitochondria about a food vacuole of *Pelomyxa carolinensis*. (After Torch, *J. Protozool.*, 2: 170.)

demonstrate them, vital dyes or permanent preparations made after osmium fixation are most commonly employed. In size, mitochondria are often extremely minute, but in some species, such as the giant amoeba *Pelomyxa carolinensis*, they may reach a length of seven or eight micra (Torch, 1955). Here they resemble relatively large bacteria.

The internal structure of the mitochondria of two species of *Paramecium* has been recently studied in very thin sections by Powers, Ehret, and Roth (1954). In this genus, the mitochondria are tiny rods about a micron long, and contain a number of extremely minute tubular spaces about 0.014μ in diameter. Cytochemically they were found to resemble kinetosomes. Still more recently, the electron microscope has been turned on the mitochondria of the common and relatively gigantic ciliate *Spirostomum*. According to Randall (1957), these organelles are very numerous in the ectoplasm, and appear as short rods of two types. One apparently consists of sheets about 150 Å in thickness (10^8 Å = 1 cm.), between which are sandwiched equally thin layers of some other material. Minute as the mitochondria are, there is nevertheless enough room within to accommodate a total area of 1.25 $cm.^2$ of these almost infinitely tenuous membranes. The second type seems to be somewhat like the mitochondria of *Paramecium*, with minute tubular cavities, except that these may lie in or between irregularly spaced membranes known as "cristae." It thus appears that mitochondria, infinitesimal though they seem, are big enough to be complexly organized.

Mitochondria are very common in both plant and animal cells, and their presence in the Protozoa has been known for many years. Unfortunately, however, they have received a variety of names (e.g., "chrondriosomes") at the hands of different investigators, and it is even probable that there has been confusion in their identification.

The role mitochondria play in metazoan cells is extremely important, and there is no reason to think it different in Protozoa. They are involved in oxidation (in the broad sense), including the synthesis of ATP, and in the synthesis of organic compounds, among them enzymes of the Krebs cycle. Electron transport is another function of great importance.

Golgi Apparatus

The term "Golgi apparatus" was given to a fibillar network observed many years ago by Golgi in the brain cells of vertebrates, and has since been applied to a variety of cell inclusions, both in tissue cells of the higher animals and

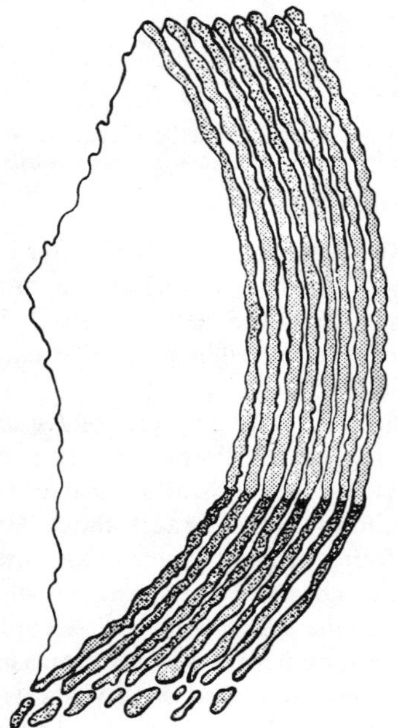

FIG. 5.21. Dictyosomes of a zooflagellate. These bodies ("saccules") are considered part of the Golgi apparatus, along with the vesicles, a few of which are shown breaking away at one of the corners. Both dictyosomes and vesicles are said to have the same structure as the parabasal apparatus in flagellates such as those of termites and woodroaches. (Grassé and Carasso, *Nature*.)

in Protozoa. As with mitochondria, treatment with osmic fixatives has often been employed to make them visible. But it is likely that such bodies have often been of many kinds, and quite unrelated, and Baker (1957) goes as far as to say, "It follows [from histochemical studies] that *there is no such thing as a Golgi substance.*" But Grassé and Carasso (1957), basing their opinion on an electron microscope study of very thin sections of both certain metazoan cells and the zooflagellates, conclude that "the Golgi apparatus, like the mitochondria, is an autonomous cell element and also has its own structure." As this implies, these elements are also self-replicating. Grassé and Carasso describe this apparatus as consisting of "very flat *saccules* . . . so flattened that one could interpret them as being double membranes" (Fig. 5.21). The Golgi apparatus is usually thought to be secretory in function. Although commonly seen in animals it is absent in most plants, including the algae and perhaps also the plantlike flagellates.

Myonemes

But whatever the truth may be as to the universality of a Golgi apparatus in Protozoa, there is no debate as to the existence of the contractile fibrils termed "myonemes," or "muscle threads" (Fig. 5.22). These minute cytoplasmic

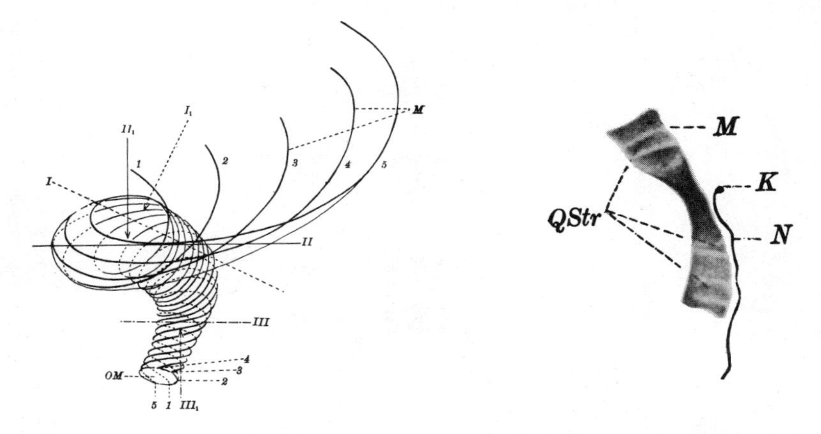

FIG. 5.22. *Left*. Diagram of myoneme system of mouth and esophagus of *Stentor coeruleus*. Note that there are five myonemes. M, myoneme; OM, lumen of esophagus. *Right*. Myoneme with terminal knob, as seen in highly magnified preparation from *Stentor coeruleus*. M, myoneme; N, neuroid; K, end organelle or "knob"; QStr, transverse striation of myoneme. (After Dierks.)

filaments occur in many Protozoa, and those of *Spirostomum* have recently been observed, again by Randall, under the electron microscope. Here they are extremely small, being close to the minimum diameter visible under an ordinary optical microscope, but electron microscopy shows them to be composed of a number of superimposed fibrillar sheets, each of which is

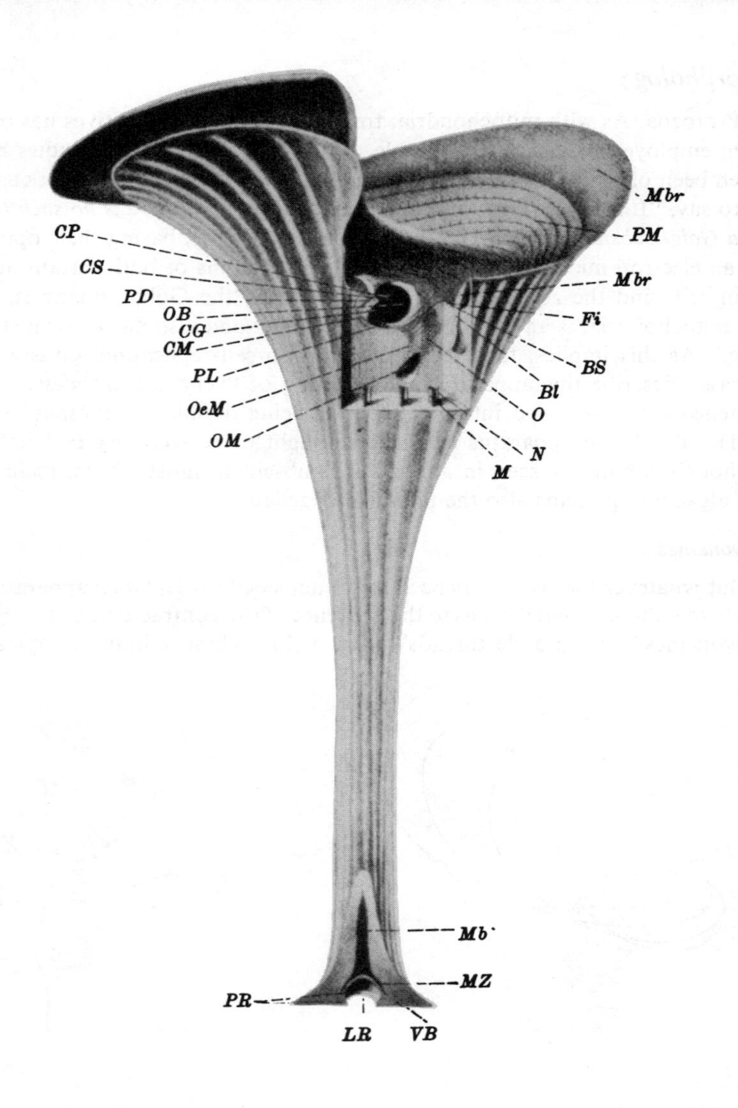

FIG. 5.23. Structure of *Stentor coeruleus*. A schematic reconstruction of an organism in extended condition, as worked out using conventional staining techniques and the ordinary optical microscope. BS, basal edge; Bl, basal plate; CG, cytopharynx; CS, cytostomal margin; Fi, conductile fibrils of the peristomal field; LR, air cavity under the myoneme umbrella; M, myoneme of body wall; Mb, myoneme bundle; Mbr, membranelle; MZ, myoneme umbrella; N, "neuroid" fibril; O, esophagus; OB, beginning of esophagus; OeM, myoneme encircling esophagus; OM, lumen of esophagus; PD, cytopharyngeal roof; PL, protoplasmic boundary separating cytopharynx and esophagus; PM, peristomal myoneme field; PR, protoplasmic ring; VB, pedal disk of *Stentor* (or "foot"). (After Dierks.)

about 0.0000025 cm. in thickness. Fauré-Fremiet *et al.* found the myonemes of certain Urceolaria (parasitic peritrichous ciliates) to have a similar structure, and also remark that they resemble the myonemes of *Stentor* (Fig. 5.23). In all these forms, however, the myonemes are of ectoplasmic origin, except that in peritrichs both ectoplasm- and endoplasm-derived contractile fibrils may occur; the latter are said to lack a clear relation to the ciliary rows and to resemble smooth muscle fibrils, even to being aggregated in bundle fashion.

Whether the myonemes of all Protozoa are of essentially similar structure no one can yet say, and it is also true that contractility does not necessarily imply the presence of specialized cytoplasmic strands of this sort. In general it may be said that myonemes are especially characteristic of ciliates, notably forms like *Stentor* and the vorticellids, and of the gregarines (a rather large group of Sporozoa restricted mostly to insect hosts). In the latter group there are said to be both longitudinal and circular myonemes, the action of which is coordinated so as to produce the peculiar gliding movements of these bizarre little animals. The occurrence of myonemes in flagellates is doubtful. Few of these organisms are capable of the sudden changes in form often seen among ciliates such as *Spirostomum*, *Trachelocerca*, and *Lacrymaria*, or the vorticellids.

Other Organelles

But if such contractile fibrils are absent among the Mastigophora, other organelles associated with movement are not. Such are the undulating membranes of the trypanosomes and the trichomonads, although these membranes are essentially simply pellicular folds enclosing the flagellum. Nevertheless, such membranes may have a complicated structure. Anderson (1955) states that in *Trichomonas muris* it consists of a series of very thin lamellae of about the same order of thickness as the sheets found in the myonemes of *Spirostomum.*

Characteristic of the trypanosomes and leishmanias is a conspicuous "kinetoplast" (formerly frequently called the "parabasal body") lying near the base of the flagellum. It takes ordinary stains much as does the nucleus, and was once thought to correspond to the micronucleus of the ciliates. What its actual function is remains obscure. Evidently it is not essential since some species of trypanosomes (e.g., *T. equinum*) have lost it. Minute fibrils, or rhizoplasts, connect the kinetosome to the blepharoplast.

The kinetoplast, blepharoplast, connecting fibrils, and other structures associated with the flagella of zooflagellates are often referred to as the "mastigont system," a system little less complex than the corresponding locomotor apparatus of the ciliates. Among these other structures is the so-called true parabasal body of such flagellates as those of termites. These bodies vary greatly in form and are of uncertain function; unlike blepharoplasts and kinetoplasts they do not usually divide independently when the organism divides.

A variety of other structures, almost all of uncertain function, have been described in different flagellates. Among them are the "costa" and "axostyle" of the trichomonads, and the "pelta" and the "cresta" which may occur in this group and in the descovinid flagellates of termites; according to Kirby (1947) the two flagellate groups are related, the latter representing "the flowering of the trichomonad type of polymastigote." (The Polymastigida are interesting forms with from three to eight flagella; the majority are parasites. See Chap. 18). The costa is a rodlike affair, and the axostyle is an elongated tubule extending the length of the animal, usually with a projection at the posterior end (Fig. 5.24). Perhaps both organelles serve for support,

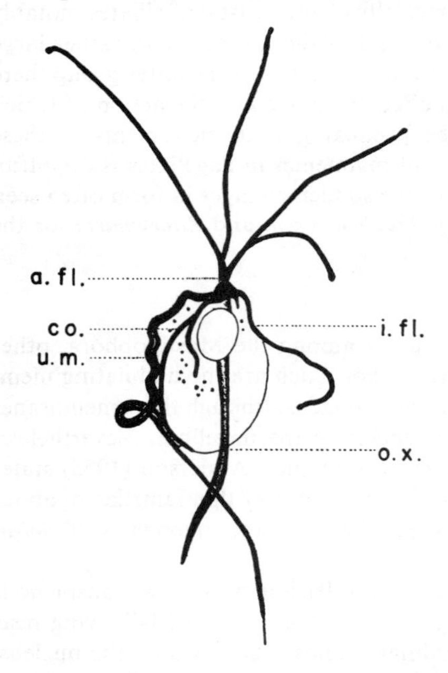

FIG. 5.24. Structure of the human trichomonad *Trichomonas hominis*. a.fl., anterior flagella; o.x., axostyle; co., costa; i.fl., independent flagellum; u.m., undulating membrane. (After Kirby, *J. Parasitol.*)

but the axostyle has also been regarded as a "powerful motor organ" (Kofoid and Swezy, 1915), used when the animal wends its way about on a solid surface such, possibly, as the mucosa of its host's colon. The pelta is a crescent-shaped membranous organelle situated anteriorly, usually on the right side, close to the base of the flagella. It stains readily with silver. Its form differs considerably in different species. The cresta is also a membrane and similarly situated, but often visible during life and able to move independently.

Indispensable, of course, to the functioning of any motor system is the ability of the organism to become aware of changes in its environment of a sort to make movement desirable. But this does not always imply the presence of specialized receptors, for in many Protozoa sensitivity to a variety of physical agents is highly developed, and yet sensory organelles seem entirely absent. How, for example, does an amoeba become aware of its prey? How does a malaria merozoite detect and enter an erythrocyte, rather than a leucocyte? Or a reticulocyte in preference to a mature red cell? We may say, in the case of a normally intracellular parasite like the malaria plasmodium, that it is a matter of greater hospitality on the part of the "preferred" host cells, or that they are more easily invaded. Yet this can hardly be the whole story. Organelles of perception are known to occur in certain Protozoa. The nonmotile bristles to which such a function has been ascribed in the ciliates have already been mentioned. Vacuoles containing a minute particle free to roll about, thus perhaps informing the animal of changes in position, have been described for a few Protozoa. The eyespot (stigma) of the euglenids and certain other plantlike flagellates, as well as of some colorless forms, has also been remarked. It is usually bright orange because of the presence of haematochrome. Mast (1928), who made an intensive study of the stigma in a number of species, found the pigment to lie behind a minute transparent lens. Thus the structure is much like that of a vertebrate eye, although it is certain that no image is formed. Some dinoflagellates also possess an eyespot (here called an "ocellus") of even more complex structure. That of *Erythropsis* (Fig. 5.25) Kofoid and Swezy (1921)

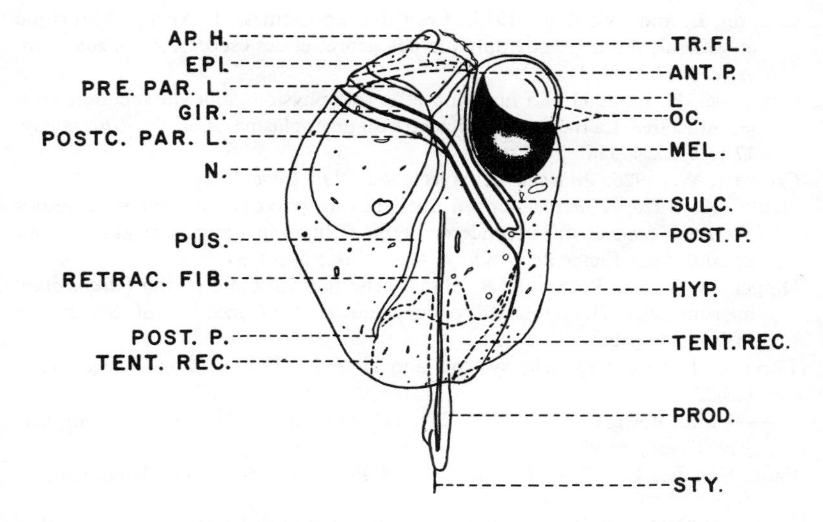

FIG. 5.25. Unarmored dinoflagellate *Erythropsis cornuta*. Note the complex structure, and particularly the very large eyespot or *ocellus* (oc.), with its lens (l.) and pigment or *melanosome* (mel.). Key: *ant. p.*, antapical pore; *ap. h.*, apical horn; *epi.*, epicone; *gir.*, girdle; *hyp.*, hypocone; *l.*, lens; *mel.*, melanosome; *n.*, nucleus; *oc.*, ocellus; *post. p.*, posterior pore; *post. par. l.*, postcingular paradinial lines; *pre. par. l.*, precingular paradinial lines; *prod.*, prod or tentacle; *pus.*, pusule; *retrac. fib.*, retractor fibrillae; *sulc.*, sulcus; *sty.*, stylet; *tent. rec.*, recess of prod or tentacle; *tr. fl.*, transverse flagellum. (After Kofoid and Swezy.)

call the "most efficient optical organ among the Dinoflagellida and the Protozoa as a whole." It is relatively of great size, its diameter equaling one-third that of the entire body, and consists of a large lens embedded in an equally large mass of dark pigment.

We have not come to the end of the story of protozoan morphology; nor even the end of what we know, for a much longer chapter would be required to include our present knowledge. But enough has been said to show how remarkably complex are these little animals which Leeuwenhoek, Spallanzani, Ehrenberg, and their many successors have found so fascinating. They do, indeed, exhibit that "elegance and workmanship" which so impressed Henry Baker two centuries ago. They are among nature's masterpieces. And the many intricate details of that workmanship will continue to be revealed as techniques of research become progressively more refined.

REFERENCES

Anderson, E. 1955. The electron microscopy of *Trichomonas muris*. *J. Protozool.*, 2: 114–24.

Baker, J. R. 1957. The Golgi controversy. Symposia Soc. Exp. Biol., No. 10., pp. 1–10.

Chatton, E. and Lwoff, A. 1935. Les Ciliés apostomes. 1. Aperçu historique et général; étude monographique des genres et des espèces. *Arch. zool. exp. et gén.*, 77: 1–453.

Cosmovici, N. L. 1933. La nutrition et le rôle physiologique du vacuome chez les infusoires. La théorie canaliculaire du protoplasma. *Ann. Sci. Univ. Jassy*, 17 (3/4): 294–336.

Crowder, W. 1926. Marvels of Mycetozoa. *Nat. Geog. Mag.*, 49: 421–43.

Dierks, K. 1926. Untersuchungen über die Morphologie und Physiologie des *Stentor coeruleus*, mit besonderer Berücksichtigung seiner kontraktilien und konduktilien Elemente. *Arch. f. Protistenk.*, 54: 1–91.

Dippell, R. V. and Porter, K. R. 1957. The fine structure of the *Paramecium* micronucleus. (Unpublished paper read at 1957 meeting of Society of Protozoologists.)

Ehret, C. F. 1960. Organelle systems and biological organization. *Science*, 132: 115-23.

——— and Powers, E. L. 1959. The cell surface of *Paramecium*. *Internat. Rev. Cytol.*, 8: 97-133

Fauré-Fremiet, E. 1953. Morphology of Protozoa. *Ann. Rev. Microbiol.*, 7: 1–18.

——— 1954. Les problèmes de la différenciation chez les Protistes. *Bull. Soc. Zool. France*, 79: 311–29.

——— 1956. Microscopie électronique de quelques Ciliés. *Bull. Soc. Zool. France*, 81; 9–11.

——— 1959. Le cortex de la vacuole contractile et son ultrastructure chez les Ciliés. *J. Protozool.*, 6: 29–36.

——— and Rouiller, C. 1955. Microscopie électronique des structures ecto-

plasmiques chez les Ciliés du genre *Stentor*. *C. Acad. Sci.*, 241: 678–80.
———, ———, and Gauchery, M. 1956. L'appareil squellettique et myoide des Urceolaires: étude au microscope électronique. *Bull. Soc. Zool. France*, 81: 77–84.
———, ———, ——— 1956. Les structures myoides chez les Ciliés. Etude au microscope électronique. *Arch. d'Anat. Microscop. et de Morph. Exp.*, 45: 139–61.

Fuller, H. J. and Tippo, O. 1955. *College Botany*. New York: Holt.

Gelei, J. v. 1934. Die feinere Bau des Cytopharynx von *Paramecium* und seine systematische Bedeutung. *Arch. f. Protistenk.*, 82: 331–62.

Grassé, P.-P., and Carasso, N. 1957. Ultrastructure of the Golgi apparatus in Protozoa and Metazoa (somatic and germinal cells). *Nature*, 179: 31–33.

Grell, C. G. 1952. Der Stand unserer Kenntnisse über den Bau der Protistenkerne. Freiburg: Verhand. Deutsch. Zool. Gesellsch.
——— 1956. Morphologie und Entwicklungsgeschichte der Protozoen. *Fortsch. Zool.*, 10: 1–32.

Kent, W. S. 1880–81. *A Manual of the Infusoria*. London: David Bogue.

King, R. L. 1933. The contractile vacuole of *Euplotes*. *Trans. Am. Mic. Soc.*, 52: 103–6.
——— 1935. The contractile vacuole of *Paramecium multimicronucleatum*. *Trans. Am. Mic. Soc.*, 58: 555–72.

Kitching, J. A. 1956. Contractile vacuoles of Protozoa. *Protoplasmatologia*, D3a. 3: 1–45.
——— 1956. Food vacuoles. *Protoplasmatologia*, D3b. 3: 1–54.

Kirby, H. 1947. Flagellate and host relationships of trichomonad flagellates. *J. Parasitol.*, 33: 214–28.

Kofoid, C. A. and Rosenberg, L. E. 1940. The neuromotor system of *Opisthonecta henneguyi* (Fauré-Fremiet). *Proc. Am. Phil. Soc.*, 82: 421–36.
——— and Swezy, O. 1915. Mitosis and multiple fission in trichomonad flagellates. *Proc. Am. Acad. Arts and Sci.*, 51: 289–378.
——— and ——— 1921. The free-living unarmored *Dinoflagellata*. *Memoirs*, vol. 5. University of California Press.

Leidy, J. 1879. *Fresh-water Rhizopods of North America*. U.S. Geol. Surv., Vol. 12.

Lund, E. E. 1933. A correlation of the silverline and neuromotor systems of *Paramecium*. *Univ. Calif. Pub. Zool.*, 39: 35–76.
——— 1935. The neuromotor system of *Oxytricha*. *J. Morphol.*, 58: 257–78.
——— 1941. The feeding mechanisms of various ciliated Protozoa. *J. Morphol.*, 69: 563–73.

Lwoff, A. 1950. *Problems of Morphogenesis in Ciliates*. New York: Wiley.

Mast, S. O. 1928. Structure and function of the eyespot in unicellular and colonial organisms. *Arch. f. Protistenk.*, 60: 197–220.
——— 1947. The food vacuole in *Paramecium*. *Biol. Bull.*, 92: 31–72.

Metz, C. B. and Westfall, J. A. 1954. The fibrillar systems of ciliates as revealed by the electron microscope. II. *Tetrahymena*. 107: 106–22.

Nanney, D. L. and Rudzinska, M. A. 1960. Protozoa. In *The Cell*, Brachet and Mirsky (eds.). New York: Academic Press.

Powers, E. L., Ehret, C. F., and Roth, L. E. 1954. Morphology of the mitochondrion and its relationship to other structures in *Paramecium*. *J. Protozool.*, 1(Supp.): 5.

Randall, J. T. 1957. The fine structure of the protozoan *Spirostomum ambiguum*. *Symposia Soc. Exp. Biol.*, No. 10. pp. 185–96.

Rudzinska, M. A. 1957. Mechanisms involved in the function of the contractile vacuole in *Tokophrya infusionum* as revealed by electron microscopy. *J. Protozool.*, 4(Supp.): 9.

——— and Chambers, R. 1951. The activity of the contractile vacuole in a suctorian (*Tokophrya infusionum*). *Biol. Bull.*, 100: 49–58.

Sand, R. 1901. Etude monographique sur le groupe des Infusoires Tentaculifères. *Ann. Soc. Belge Microsc.* 24, 25, 26: 1–441.

Sharp, R. G. 1914. *Diplodinium ecaudatum*, with an account of its neuromotor apparatus. *Univ. Calif. Pub. Zool.*, 13: 43–122.

Spallanzani, L. 1786. *Observations et Expériences Faites sur les Animalcules des Infusions*, (1st ed. 1777), Geneva.

Taylor, C. V. 1920. Demonstration of the function of the neuromotor apparatus in *Euplotes* by the method of microdissection. *Univ. Calif. Pub. Zool.*, 19: 404–70.

Torch, R. 1955. Cytological studies on *Pelomyxa carolinensis.*, with special reference to the mitochondria. *J. Protozool.*, 2: 167–77.

Weisz, P. B. 1953. The embryologist and the protozoon. *Sci. Am.*, 188: 68–75.

——— 1954. Morphogenesis in Protozoa. *Quart. Rev. Biol.*, 29: 207–29.

Wenrich, D. H. 1947. Morphology and cytology of the Protozoa. *Ann. Rev. Microbiol.*, pp. 1–18.

Wichterman, R. 1953. *The Biology of Paramecium*. Philadelphia: Blakiston.

Wolharth-Botterman, K. E. 1953. Experimentelle und elektronenoptische Untersuchungen zur Funktion der Trichocysten von *Paramecium caudatum*. *Zool. Inst. Univ. Munster. Arch. f. Protistenk.*, 98: 171–226.

Wolken, J. J. 1956. A molecular morphology of *Euglena gracilis* var. *bacillaris*. *J. Protozool.*, 3: 211–21.

Worley, L. G., Fischbein, E., and Shapiro, J. E. 1953. The structure of ciliated epithelial cells as revealed by the electron microscope. *J. Morphol.*, 92: 545–78.

ADDITIONAL REFERENCES

Jahn, T. L. 1961. The mechanism of ciliary movement. 1. Ciliary reversal and activation by electric current; the Ludloff phenomenon in terms of core and volume conductors. *J. Protozool.*, 8: 369–80.

Pitelka, D. R. 1961. Fine structure of the silverline and fibrillar systems of three tetrahymenid ciliates. *J. Protozool.*, 8: 75–89.

6

Nutrition

"... when they stick their tails out again, they displace the water round about them, and being thus come into *different water*, they can get fresh food out of it.... This being so, we are faced once more with the mysteries, and unconceivable order, which such tiny creatures (which quite escape one's naked eye) are endowed with." Leeuwenhoek, Letter 150, February 5, 1703.

Physiology, like protozoology, is a relatively new branch of biology and to many people it means study of the functioning of only the human organism. Yet the subject is much broader than this: physiology deals with the life processes of animals and plants generally. The word itself is derived from two Greek roots meaning "a knowledge of nature"; it therefore covers all that lives.

It is hard to realize how little was known even about the physiology of man fifty years ago. Yet today standard texts on human physiology may approach in size an unabridged dictionary. By contrast, it may seem that little could be learned about the internal functioning of a microscopic organism such as a protozoan. Yet the basic living processes of the latter are little different from those of our own bodies. If anything, the former are somewhat more complex, because of the division of labor among groups of cells in a multicellular animal. The protozoan cell, it must never be forgotten, is a complete organism, and must do all that is essential for the maintenance of its life and of the species. Any given type of human body cell has much done for it by other types of cells, and—by the same token—has lost the ability itself to perform these other functions.

Much the same thing has happened to many parasitic species of Protozoa: they have learned to depend on their hosts for the performance of numerous vital functions their free-living ancestors had to carry on for themselves, and in the process they have lost much of what might be called physiological versatility.

Whatever the mode of life, all organisms occupy themselves with the same basic physiological tasks: the getting of food, its digestion and assimilation, the utilization of certain of its products for growth and repair or for energy production, and the excretion of metabolic products. These activities inevitably involve respiration. Much variety exists in the ways in which such tasks are accomplished, even among the Protozoa. Like the descendants of Adam, all are condemned to work for a living.

FOOD-GETTING

Although many Protozoa make a "great stir in the water" in their active search for food, as Leeuwenhoek watched them doing more than two centuries ago, numerous other ways are also utilized. Some species synthesize their own food, others have it made for them by algae living in their cytoplasm, and still others wait passively until food comes within reach, capturing it almost by stealth. Still others live as parasites, usually doing little or no harm to the host, but occasionally causing serious disease, or sometimes becoming so useful that the host cannot live without them.

No matter which method of food-getting is used (and they are often combined), food is absorbed of necessity either in solution or in solid form. For the first of these the term "osmotrophy" has been proposed, and for the second "phagotrophy." Some organisms can take in food by only one of these two methods, and are therefore called "compulsory osmotrophs" or "compulsory phagotrophs." Others, less restricted in their abilities, may take their food in both ways, though not necessarily at the same time; these would be the "facultative osmotrophs" or "facultative phagotrophs," depending on which was the preferred method. Many parasitic organisms are compulsory osmotrophs, as of course any protozoan must be if nature failed to endow it with a mouth or with the ability to ingest food particles with pseudopods. Perhaps there are few, if any, organisms wholly unable to absorb in dissolved form at least some substances which might be classed as food. Many of the chlorophyll-bearing flagellates take in needed substances in this way, even though such materials are often of a chemically simple sort; rather surprisingly, others, like some dinoflagellates, though they supply most of their nutritional requirements by photosynthesis, supplement their diet by the pseudopodial ingestion of other organisms.

Free-living Protozoa that absorb foodstuffs in solution and take in no solid food are often called "saprophytic" or "saprozoic." They possess of course no food vacuoles. Species such as the green flagellates, which because of their ability to carry on photosynthesis are able to meet most of their food requirements, are said to be "autotrophic" or "holophytic." Hunters among the Protozoa are called "holozoic," whether their food consists of captured animals or plants, or both. Protozoa combining any of these methods are often termed "heterotrophic." Parasitic nutrition is sometimes regarded as a fifth method of securing food, but it actually does not differ in any significant way from the holozoic and saprophytic methods already mentioned, for parasitic species take food into the body either by absorbing it through the surface or ingesting it in the form of solid food particles obtained from the host. In some cases the process may be complicated by previous digestion of the host's tissues by enzymes secreted by the parasite, as with *Entamoeba histolytica*, which in this manner erodes its way into the intestinal lining while it doubtless absorbs and utilizes for its own nutrition certain of the digested products.

Autotrophy

Probably the most primitive of all these methods is autotrophy, but the process of photosynthesis involved can hardly be called primitive. It occurs in the presence of light in all the plantlike flagellates, and of course in all green plants from the simplest alga to the giant redwood. In all these forms chlorophyll is essential for photosynthesis, but there are two groups of bacteria (Chlorobacteriaceae and Thiorhodaceae) that can also carry on photosynthesis, utilizing other pigments in place of chlorophyll, and H_2S instead of H_2O. This must be regarded as further evidence of the great antiquity of this process in nature, a process which is the primary source of food and fuel for all animals, for without plants to serve as food they would all starve in a short time. It is likely that, for the earth as whole, the algae and their close relatives the green flagellates fix more carbon than all the terrestrial plants together, for much more of the earth's surface is covered by water than by dry land.

Photosynthesis involves not only energy from sunlight, but carbon dioxide and water as raw materials, chlorophyll, and doubtless numerous enzymes. Indeed respiration, as far as carbohydrate metabolism is concerned, is the reverse of photosynthesis, and requires some twenty reactions, for each of which enzymes are needed. Dextrose is the main product and oxygen a by-product of photosynthesis; from the dextrose starch or (in some of the green flagellates) paramylum may be formed. The latter substance is a complex polysaccharide related to starch as its name suggests, and is especially characteristic of the euglenid flagellates, although something very like it occurs in the walls of yeast cells. Because of the basic importance of photosynthesis in nature, much research has gone into an understanding of its chemistry, but much remains unknown. The over-all equation may be written as follows:

$$6\,CO_2 + 6\,H_2O \underset{\text{oxidation}}{\overset{\text{photosynthesis}}{\rightleftharpoons}} 6\,O_2 + C_6H_{12}O_6$$

Chlorophyll, the active agent in photosynthesis, exists in two slightly different forms known as "chlorophyll *a*" and "chlorophyll *b*"; the formula of the first has been found to be $C_{55}H_{72}O_5N_4Mg$, and gives some indication of the complexity of this remarkable substance. Other variants of chlorophyll have also been reported, but on the whole the chlorophylls of many different kinds of plants are similar. Which form of chlorophyll occurs in the chlorophyll-bearing Protozoa may be open to question; in many cases there is doubtless more than one, as in the euglenids and phytomonads. The chrysomonads however are said to contain only chlorophyll *a*. Associated with the chlorophylls are often other pigments, such as the carotenes and the xanthophylls, which may or may not play a role in photosynthesis. Indeed the precise role of chlorophyll itself is not entirely certain, although one of its chief functions is the trapping of the radiant energy required for carbohydrate synthesis. It is an interesting and rather remarkable fact that in this respect photosynthesis is only about 20 per cent efficient. Somewhat surprisingly, too, not all the reactions involved in this complicated process require light; some can proceed only in the dark.

Of course no organism can subsist on a carbohydrate diet alone, and photosynthesis must therefore be supplemented by the synthesis of fats and proteins. For such synthesis, and doubtless for other uses, mineral salts and vitamins are required, and these must be secured from the medium in which the protozoan lives. Nitrates and sometimes ammonium salts may serve as nitrogen sources, and B_{12} seems to be usually, if not universally, essential, although occasionally related cobalt-containing compounds will take its place. The vitamin thiamin is also a requirement for most species so far studied.

Of all the plantlike flagellates so far investigated, the dietary requirements of the minute brown-pigmented forms called "chrysomonads" seem the simplest. These Protozoa are abundant in both salt and fresh water, and in the soil. Though they normally carry on photosynthesis they are also phagotrophic, eating any small food particles—bacteria, algae, and even luckless neighbors of their own kind. Curiously enough, they are apparently unable to synthesize enough carbohydrate to supply their own needs, for it was necessary to add sugar or glycerol to the medium in which certain strains were grown in the laboratory. In addition, vitamin B_{12}, thiamin, biotin, and certain acids of the citric acid cycle were required. Given such a medium, the organisms grew as well in darkness as in light (Hutner, Provasoli, and Filfus, 1953). Thus, they are only partially phototrophic or autotrophic.

Oddly some species of *Euglena*, such as *E. gracilis*, and doubtless of other genera, are able to use acetate as a carbon source although other species of the genus cannot.* They flourish in an acetate-containing medium in the dark, without need of photosynthesis, although they lose chlorophyll in the process; normally the chlorophyll is regained when they are again exposed

*Other organic substances may also be used on occasion. One strain of *Euglena gracilis* can make do with ethyl alcohol alone (Wilson, 1959). Apparently life for a laboratory-bred protozoan need not always be dull.

to light. Many colorless species of flagellates, similar to the green forms except for the absence of chlorophyll, may have arisen from mutants able to use acetate and devoid of pigment. Since chloroplasts are self-duplicating in division, any failure of duplication might result in colorless offspring. Experimentally, chloroplasts may also be destroyed by treatment with streptomycin; such loss is irreversible. The interesting suggestion has been made by Hutner and Provasoli (1951) that colorless species may originate naturally in the latter fashion, since streptomycin-producing actinomycetes are common in soils. *Astasia*, a colorless flagellate but otherwise very similar to *Euglena* and common in soil and fresh water, might have arisen in one of these two ways.

Such facts as these emphasize the folly of attempting to separate some organisms from the rest on the ground that they must be either plants or animals, and therefore fall into the botanist's or the zoologist's realm. Nature created life, but she did not create classifications. These are all manmade. Could we but witness the genesis of species, and even of the larger groups, as the eons pass, it is likely that they would look like scenes of a play, gradually unfolding from previously depicted events. In the course of evolution, nature may have done with organisms such as these what researchers have recently done with the mold *Neurospora*, and even with green flagellates such as *Chlamydomonas*, knocking out a gene here and one there, so that they become progressively less able to carry on the complex syntheses that normal living requires.

Some Protozoa, though lacking chlorophyll of their own manufacture and therefore unable to carry on autotrophy on their own, nevertheless have managed to secure the advantage of an unearned income for themselves by "domesticating" certain algae. One of the most commonly encountered species of this kind is *Paramecium bursaria*, the green *Paramecium*, whose cytoplasm contains the alga *Chlorella vulgaris;* this animal can survive starvation for a long time if only light is available for photosynthesis. If conditions become too severe, it ungratefully digests its algae, apparently imperiling its future should the environment later improve (Parker, 1926). But fortunately enough of the plant cells usually survive to restore the algal population to normal. Symbiotic relationships of a like sort are also common among the Foraminiferida living in the shallow waters of tropical coral reefs, and in the Radiolarida.

Heterotrophy

Although clear enough in theory, the distinction between heterotrophy and autotrophy is in practice difficult to make. The reasons for this are already evident. Many species using photosynthesis as a means of food manufacture also take in some part of their diet in dissolved form by osmotrophy or by phagotrophy, even to the extent of capturing other organisms. Some, like *Euglena gracilis*, may lose their chlorophyll and adopt a typically animal-like pattern of nutrition if condemned to life in darkness, provided only that suitable organic substances are present. Others like *Peranema*, obviously a close relative, have lost their chlorophyll permanently, and may subsist on living organisms such as yeast. One of the most epicurean of such species is *Peranema tricophorum*, which Storm and Hutner (1953) found to flourish in a medium of cream supplemented by a variety of amino acids, mineral salts, and vitamins (in addition to B_{12} and thiamin, riboflavin was required). Eventually it was found that cholesterol and lecithin could replace the cream, and then that linoleate could substitute for most of the lecithin. This last finding was of special interest, because some highly evolved animals, such as insects, birds, and mammals, also require linoleate. Although so far the complete nutritional requirements for only a few of the many species of animal-like flagellates are known, it is already certain that they are varied.

Most Protozoa, however, must work for a living. They can depend on no inquiring protozoologist to play the role of fairy godmother, and laboriously work out the diet they most like. Most of them subsist by hunting. Some live on bacteria, fungal spores, and algae, others capture such hapless Protozoa and rotifers as they can, and many species will ingest anything edible that comes their way. How the distinction is made between objects suitable for food and those that are not we do not know; experiments have shown that inedible particles within a certain size range are sometimes ingested, but such mistakes are not common. Some Protozoa are extremely fastidious in their tastes, insisting on certain kinds of food and passing up all others; *Actinobolina radians* ordinarily eats only the ciliate *Halteria grandinella*. Nothing but paramecia will satisfy *Didinium nasutum*. These two fussy ciliates are relatively rare, for obviously they can exist only where their required food species occur. It would be interesting to know how such a selective diet evolved, for it would seem a serious handicap.

Protozoa are very greedy creatures as a rule. Most of them take in food whenever they can get it, and they feed as actively at night as during the day. Sleep is a luxury indulged in only by the higher animals. As among humans, some protozoan groups are not above cannibalism. The writer has often watched *Pleurotricha lanceolata* ingest its own kind. *Blepharisma* is also exceedingly cannibalistic, and may exhibit double chains of macronuclei. Since these ciliates may also grow until they become almost giants in their tiny world, one might expect them to strike terror among their kind if indeed this were possible. And perhaps one could say that most Protozoa are as careless in their manners as they are ruthless in their habits. For they swallow whole anything that will go into the mouth, regardless of size. As a rule, no supporting structures about the "cytostome" (cell mouth) or gullet limit ingestion. Some species, however, such as those of the genus *Chilodonella*, have a kind of skeletal framework, or "pharyngeal basket," around the mouth limiting the size of what they swallow.

Methods of Phagotrophy

The manner in which Protozoa take in solid food naturally varies with the species. The Sarcodina engulf it with their pseudopodia as do some Mastigophora. When a cell mouth is present, food particles are wafted toward it by water currents produced by cilia or flagella, as Leeuwenhoek first correctly observed so many years ago in the case of *Vorticella.*

Amoebae ingest their food in various ways. This was first recognized by Rhumbler (1910), who distinguished four methods: "import," "circumfluence," "circumvallation," and "invagination." Differences between these methods (defined in the glossary) seem less important today.

Another method, known as pinocytosis, has recently received considerable attention. First described by *Lewis (1930) for mammalian cells, and then by *Mast and *Doyle (1934) in amoebae, it is often called "cell drinking," but it is really more than that. It has been intensively studied by *Holter (1959), among others. Apparently certain substances, particularly some of rather large molecular weight such as certain proteins, must be present to stimulate the amoeba to ingest them. This it does, much as in invagination, by the formation of tiny cytoplasmic tubules. These are often too small for detection by the light microscope. Ingestion itself is said to occur by "sips and gulps" rather than continuously, and the immediate stimulus may involve differences in charge between cell surface and exciting substances in the solute.

Other Sarcodina may capture their food in somewhat different ways, relying more on patience and craft. The Heliozoida ("sun animals," e.g., *Actinophrys sol*) are always ready to hold and pull in any other luckless protozoan that comes within reach. Even such relatively large and active forms as rotifers are on occasion ingested by amoebae. Apparently their pseudopodial embrace is one of considerable power, for amoebae have been seen actually to pinch a *Paramecium* in two. Diatoms are a favorite food for many *Sarcodina.* Leidy was so much impressed with the appetite for diatoms of one species that he named it *Ouramoeba vorax* ("greedy amoeba"). Sometimes a diatom much larger than its would-be attacker is tackled (Fig. 6.1). After ingestion, the still-living prey may often be observed struggling to escape from the freshly formed gastric vacuole. But after a few seconds of desperate swimming the cilia or flagella become motionless, and digestion soon begins. The killing agent is as yet undetermined; it may be an acid secreted into the vacuole to furnish the proper medium for a digestive enzyme, or it may be asphyxiation.

The Foraminiferida produce delicate reticulopodia that branch and anastomose extensively forming a net that Leidy likened to a spider's web, and indeed it serves a similar purpose. For the protoplasm on the surface of each tiny thread is viscous and in constant movement, flowing away from the organism on one side and toward it on the other, so that any tiny particle, such as an alga or tiny flagellate, sticks to the filament on contact and is then dragged to the central mass of the organism to be engulfed.

The flagellates may engulf solid food by pseudopodia or ingest it by means of a cell mouth, although what has been called a "cytostome" in some species (e.g. of *Euglena*) does not always function as such. The flagellum often functions not only to propel the animal but to bring small food particles toward the mouth or toward an area from which pseudopods can be extended.

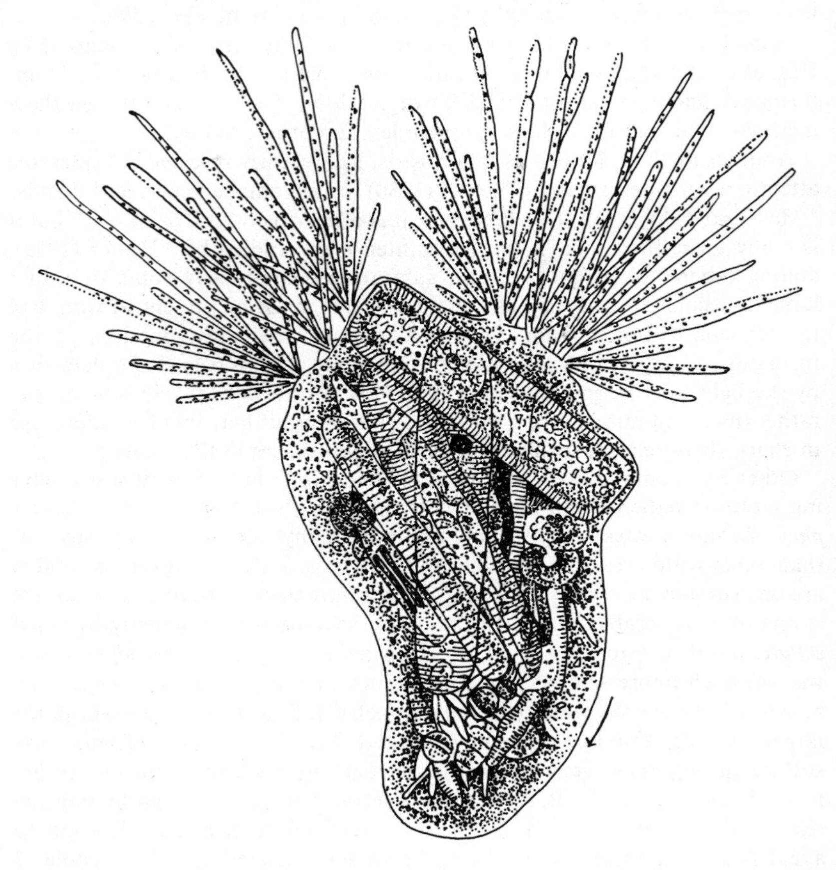

FIG. 6.1. *Ouramoeba vorax* Leidy, about 300μ long, though often smaller. The cytoplasm is loaded with diatoms and algae. The caudal filaments are flexible tubes, "perfectly passive, neither retractile or extensile." Arrow shows direction of movement. (After Leidy.)

Ciliates have brought the science of food capture to its highest point. Many of them ingest other ciliates, as does *Enchelys* (Fig. 6.2). In some the motion of the organism through the water is the chief factor in bringing food to the mouth. This is especially true for organisms having the mouth at or close to the anterior end and lacking strong membranelles or other structures in its neighborhood. In others special cilia cause a strong current of water to flow toward the oral region (Fig. 6.3). In still others these cilia have fused to become membranelles or undulating membranes creating water currents that may actually assist in carrying the prey to the mouth or even down the gullet toward the forming food vacuole. In the more specialized ciliates the rows of cilia originate in the oral region, showing the close relationship of the ciliary system to the mouth. *Holophrya simplex* is a good example of a

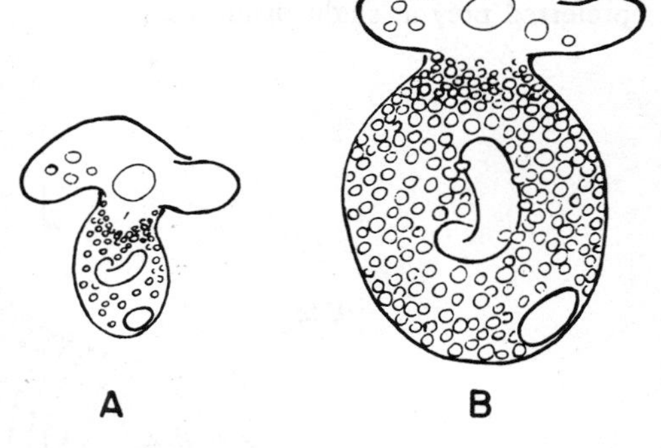

FIG. 6.2. Capture of *Colpidium* by *Enchelys mutans*, a holotrichous ciliate which feeds chiefly on other ciliates. It occurs commonly in cultures in two forms: *A*, a small, typically banana-shaped individual, and *B*, a much larger and somewhat pear-shaped organism, which may be called the *trophont*. (After Fauré-Fremiet.)

more primitive type of ciliate with the mouth at the anterior end without any special cilia surrounding it. A groove with strong cilia sweeps food particles toward the mouth of *Paramecium*. *Lembadion conchoides* eats very well aided by its relatively enormous undulating membrane sweeping nearly the length of the body. The hypotrichous ciliates, standing near the top of the ciliate evolutionary scale, are provided with large adoral zones bearing a row of huge membranelles, supplying the protozoan equivalent of plank steaks. Vegetables, in the form of bacteria and algae, help vary their diet. The number of bacteria ingested by the larger ciliates may be enormous. It has been estimated that a single *Paramecium caudatum* may devour from two to five million *Escherichia coli* (the common colon bacillus) in 24 hours, and it is easy to see how such Protozoa aid in the natural purification of polluted waters.

Didinium nasutum is not only a fastidious feeder but a rather original one. It has a proboscis with which it jabs any luckless paramecium in its neighborhood. With this proboscis, well armed with trichocysts and also an efficient seizing organ, it may capture as many as eight paramecia daily, each larger than itself. Occasionally it attacks other species of ciliates, but it is said that only a diet of paramecia is completely adequate for continued reproduction.

Actinobolina radians is, if possible, even greedier and more particular in its tastes. Tentacles richly provided with trichocysts trap *Halteria grandinella*, sometimes at the rate of one every two minutes. Other species swim by unnoticed, but whenever one of the ubiquitous Halterias wanders past a tentacle is thrust out, and the hapless protozoan, paralyzed by the action of the trichocysts, is slowly drawn toward the body and swallowed whole.

How these ciliates, and other Protozoa which appear to have some power of food selection, "choose" their prey is unknown. In many cases what appears to be selection is not. If *Paramecium*, for example, takes only relatively small organisms as food it may be because its mouth is too small to accommodate bigger forms. Species such as *Actinobolina radians* and *Didinium nasutum* probably are stimulated to capture their prey by some sort of chemical attraction. A few species of ciliates can apparently even detect their preferred prey at slight distances.

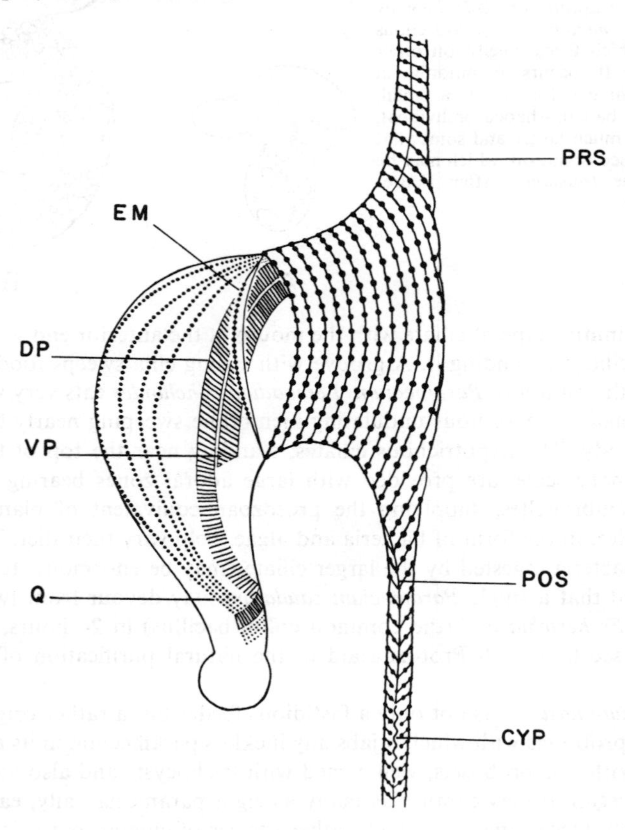

FIG. 6.3. The buccal apparatus of *Paramecium* shown diagrammatically from the right ventral side. PRS, pre-oral suture; POS, post-oral suture; CYP, cytoproct; EM, endoral membrane; VP, ventral peniculus; DP, dorsal peniculus; Q, quadrulus. The peniculus and quadrulus are compound ciliary organelles characteristic of many of the hymenostomes. (After Yusa, *J. Protozool.*, 4: 128.)

Food capture by the use of tentacles occurs also among the Suctorida (Fig. 6.4). These uncommon forms possess cilia only during their embryonic stages, and develop tentacles and lead a sedentary life thereafter. Then they lie in wait, like a duck hunter in a shooting box, ready to ambush any small protozoan coming their way. Yet some of them seem to use a certain amount of selection, for one may often watch wandering ciliates and flagellates approach the innocent-looking suctoridan and even swim through its outstretched tentacles. Any suitable food species that comes into contact with

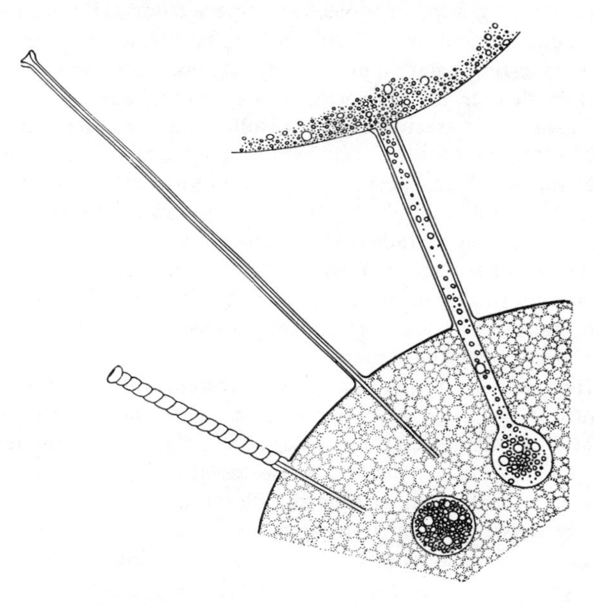

FIG. 6.4. Tentacles of the suctoridan *Tokophrya lemnarum* in the act of feeding (top), fully extended (middle), and retracted (bottom). The tentacles appear to be tubular. (After Noble.)

the forest of tentacles is promptly paralyzed by some toxin (termed "hypnotoxin"), and is then sucked into the body of the captor and digested at leisure. Yet it would seem that these remarkable organisms are rather derelict in making full use of their opportunities, for they often take in only the cytoplasm of their prey, rejecting the nucleus, from a nutritional point of view perhaps the most valuable part.

Nutrition of Parasites

The evolutionary gap separating forms such as *Actinobolina* and *Didinium*, so rigid in their food requirements, from obligate parasites may not be very great. Obligate parasites are usually restricted to life within a given species of host, or a small group of closely related host species, and this is but another way of saying that their nutritional needs can be adequately satisfied by substances characteristic of the tissues of such hosts. Parasites, however, differ greatly in the strictness of their host specificity and in their food habits. On the whole, intracellular parasites such as the malaria organisms seem to be the most rigidly limited in their food requirements. Yet it has been recently shown (Rudzinska and Trager, 1959) that at least two species ingest their food, mainly hemoglobin, from the cytoplasm of their host red cell, much as do the free-living amoebae, although the ingested particles are either molecules or groups of molecules. Species normally occurring in the gut are most like the free-living forms in food habits, the intestinal amoebae, for example, ingesting bacteria and sometimes body cells much as their free-living relatives devour the microorganisms in their environment. In some parasites, however, parasitism has resulted in profound modification. *Opalina*, for example, an infusorian almost universally present in the colon of frogs, has no mouth and absorbs all its food through the cell surface, after the host has accommodatingly digested it. The exacting food requirements of most parasites can be adequately appreciated perhaps only by those experimenters who have tried to devise artificial media in which they will grow.

For the trypanosomes, most of which are easily grown in culture and have been much studied because some species produce serious diseases, it has been found that hematin, ascorbic acid (usually), and blood serum are required. Hematin is essential for normal respiration, and for one species of the family it has even been possible to determine that the per-parasite requirement is about 600,000 molecules. The role of ascorbic acid remains uncertain, but it is known that the parasites absorb it from the blood and concentrate it in their own cytoplasm. Blood serum is of course a highly complex substance; its contribution to the nutritional well-being of the parasites is no doubt also complex, but it likely functions in part as a supplier of the "B" group vitamins and of glucose (Lwoff, 1951).

Like the trypanosomes, the trichomonads are common parasites in many kinds of hosts and produce disease in some, and they are also easily cultured. Unlike the trypanosomes, however, they possess a cytostome and may ingest solid food such as erythrocytes, although it seems unlikely that much of their food requirement is met this way. Indeed it has also been stated that they do not possess proteolytic enzymes (Lwoff, 1951). Growth requirements differ for different species, but various sugars or starch serve as energy sources, and cholesterol, linoleic acid, ascorbic acid, and pantothenic acid may be essential.

Relatively less is known about the nutritional needs of the parasitic amoebae, of which there are also many species. *Entamoeba histolytica* has been studied most because of its great importance as a cause of human disease. Its normal diet in the colon seems to be erythrocytes, bacteria and doubtless other microorganisms, and dissolved substances derived from the tissues of the host, since it is an active tissue invader. But microorganisms, at least, are not essential for it may cause sterile abscesses in the liver and elsewhere. Determination of its exact nutritional requirements has proved very difficult, since it has not yet proved possible to grow it in axenic culture, but cholesterol, vitamins, especially of the "B" group, and amino acids seem to be necessary. Rice starch is greedily ingested by the amoebae and probably serves as the main energy source. Curiously enough, all these nutritional needs are satisfied by a diet of the easily cultivated trypanosome, *T. cruzi*, as shown by Phillips (1953). Of course this is a strictly acquired taste on the part of the amoeba, for it would never encounter these trypanosomes in its natural environment. A further remarkable observation was that only this species of trypanosome, of the many tested, would support growth, and that the organisms must be living when eaten.

Although numerous species of parasitic ciliates exist, few have been successfully cultivated, and more is known about *Balantidium coli* than any of the others. This ciliate grows well in a medium based on Ringer's solution and serum, especially if some finely divided starch is added. *Balantidium coli* is so fond of starch that it often fills itself with the granules to the point of utter satiety within five minutes, according to Nelson (1933). However it has catholic tastes, for red blood cells are also avidly ingested, and yeasts are eaten, though when there is a choice between yeast and starch the latter is always selected. All three foods are of course present in the colon where this parasite normally lives.

The Sporozoa are so large and varied a group that little can be said about their nutrition. The great majority are parasites of invertebrates, and since they have little importance they have been little studied. Because they lack a mouth and produce no pseudopods, and also because many of them are intracellular parasites during much of their life span, it is usually assumed that they absorb their food in dissolved form. Yet as we have just seen at least some species of malaria parasites actually ingest food, amoeba-fashion, from the substance of the host cell, though the resulting food vacuoles can only be seen with the electron microscope.

Digestion

Digestion among the Protozoa is believed to take place very much as it does in higher animals, the food vacuoles being comparable to the alimentary tract. In the amoebae they originate as a kind of "food cup" formed by the extension and joining of pseudopodia that capture the prey. The gastric vacuoles of ciliates form at the base of the gullet and then move gradually through the cytoplasm, often following a rather definite course about the nucleus. Careful study of the protozoan food vacuoles began almost two centuries ago, when Gleichen introduced the method of causing the organisms to ingest colored particles, thus making it possible to determine the course and fate of these tiny organelles. By using indicators in this fashion, changes in the pH within the vacuole can be followed. Usually the reaction is at first acid, sometimes reaching a pH as low as 4.0, becoming alkaline later; thus the action of the digestive enzymes is probably facilitated as in Metazoa. Proteolytic and carbohydrate-splitting enzymes have been demonstrated in some Protozoa, and it is clear that they must occur in many, for otherwise food organisms could not be digested. One of the most interesting cases is that of *Didinium nasutum*, which makes its captured paramecia supply not only nourishment but also the digestive enzyme dipeptidase required for their own digestion. Careful determinations of the amount of this essential protein-splitting ferment in didinia have shown that it is exactly equal to what was present in their ingested prey (Doyle and Patterson, 1942). Carbohydrate-splitting enzymes of at least two types are known to occur: those able to digest polymerized carbohydrates, such as starch and cellulose, and sugars. Some of the parasitic Protozoa producing cellulases are of great use to their hosts, which would otherwise be completely incapable of digesting their woody diet. Certain Protozoa are also believed to secrete lipolytic (fat-splitting) enzymes. Apparently digestion itself takes place rather rapidly. Only about a minute is required by *Didinium* to swallow a *Paramecium multimicronucleatum*, though its prey exceeds itself in size, and 20 minutes later little is left of the meal. Any indigestible or undigested food particles taken in by Protozoa are extruded from the body, and the ciliates often possess a special opening, the cytopyge or cytoproct, for this purpose.

It still seems at least possible, although doubted by most protozoologists, that in certain species of ciliates what appear to be gastric vacuoles are actually dilated portions of a minute tubule, the rest of which remains invisible because it is collapsed. Thus the vacuoles might be likened to the well-known oranges on their way down the ostrich's neck. Cosmovici (1933), by an ingenious method involving the immersion of colpidia and paramecia in the dextrin from saliva-dissolved starch and their subsequent treatment with iodine, demonstrated what he interpreted as a canal marking the course of gastric vacuoles through the cytoplasm.

Food Requirements and Cultivation

Although methods of cultivating Protozoa in the laboratory on chemically defined media, in the absence of other organisms (*axenic* culture), may not seem likely to shed much light on how these little animals live in nature, they may afford us much knowledge of living processes, and of how these correlate with events in the life cycle. They may also suggest evolutionary relationships, for it is probable that the biochemistry of an organism changes less than its morphology.

Actual determination of the cultural requirements of a given species of Protozoan is often difficult, because substances present in the environment in very low concentrations may nevertheless prove to be essential to the growth, multiplication, and even the continued existence of the organism.

A medium otherwise unsuitable has often proved successful for the growth of an organism because of the minute traces of some essential substance present in the cotton plugs of the culture tubes. It may likewise take only a trace of some toxic substance, perhaps derived from cultural utensils of glass, to inhibit multiplication. The precise composition of natural waters (or of the tissue fluids of a host) is always complex and difficult to determine. Yet if a species is to be successfully cultured, it is necessary to duplicate its natural living conditions as exactly as possible. If this is not done, even though the organism survives and multiplies, there is no assurance that its behavior and life cycle under artificial conditions will be the same as in nature.

If the protozoan being studied feeds on bacteria, as so many Protozoa do, the problem is further complicated because microorganisms of bacterial type are very often able to synthesize essential substances of a variety of kinds. Thus it may be quite difficult to discover whether a compound deemed to be a dietary requirement has for its ultimate source bacterial metabolism or the natural medium itself. Vitamin B_{12}, already mentioned as probably a nearly universal requirement of Protozoa (and of other animals as well), must apparently be elaborated in the first instance by bacteria or possibly some blue-green algae. Quite possibly this group of microorganisms serves as an important supplementary source of some of the other vitamins, particularly of the "B" complex.

Thus it becomes very desirable to be able to grow Protozoa under bacteria-free conditions. Apparently the first successful experiment of this kind was that of Zumstein, who grew *Euglena gracilis* this way in 1900. The bacteria-free culture of ciliates proved more difficult, largely because it was impossible to kill ingested bacterial spores without killing the eater, but it was accomplished by Oehler in 1919 after numerous earlier attempts had been made. He did not use a synthetic medium, but found that his organism (*Colpoda steinii*) required either peptone-glucose plus dead *Escherichia coli*, or particulate egg albumin. Since his time it has been possible to develop, for a number of species, media of more or less precisely known chemical composition. Once an approach to this ideal has been achieved, it may be possible to vary the concentrations of individual components of the medium, and thus eventually to blueprint quite exactly both the dietary preferences and the nutritional essentials of the organism. So far, however, the number of species of Protozoa so grown is relatively small.

Among the best known of this select group are members of the genus *Tetrahymena*, a group now comprising a number of species formerly assigned to the genus *Colpidium* and to several other genera. The species *T. pyriformis* (bearing such well-known aliases as *Colpidium campylum*, *C. striatum*, *Glaucoma pyriformis* or *piriformis*, and *T. geleii*) has been the ciliate most thoroughly studied from physiological and biochemical approaches to growth and metabolism. The synthetic medium* (largely the work of Kidder

*See Chap. 27 for detailed composition of this medium.

and his associates) used for its culture contains no less than 43 ingredients, comprising 17 amino acids, 9 vitamins, 9 mineral salts, and a few other miscellaneous items. The quantities needed of some of the items on this complicated bill of fare are so small that they must be measured in micrograms per liter, but they are none the less essential.

Since *Tetrahymena* in nature lives chiefly on minute microorganisms, such as bacteria, a good deal of adaptation must have been required of it before it could live successfully on a medium such as Kidder's, in which all the ingredients are in solution. That it is a highly versatile organism is shown by the success with which a number of different strains have been acclimated to the conditions of artificial culture, by numerous investigators in laboratories in widely separated parts of the world. Not the least significant of their findings is the discovery that strains from different sources may differ in their nutritional requirements, though morphologically indistinguishable. Remarkably enough, *T. pyriformis* appears to have nutritional requirements similar to those of mammals in many respects. The same ten amino acids are needed, though serine is also necessary for optimal growth. However Kidder noted that ammonium pyruvate could replace or supplement an otherwise deficient amino acid content in the medium, and suggested that this cheap source of nitrogen might be investigated further to determine its possible use in animal and even human diets.

As yet, though the number of protozoan species for which media of precisely known chemical composition have been devised is still relatively small, it is easier to grow the flagellates, particularly members of the Phytomastigina, in this way than Protozoa of other groups. Their cultural requirements seem to be simpler than those of the more animal-like types of Protozoa, and this is in accord with the theory that the chlorophyll-bearing forms are nearer the main line of protozoan evolutionary descent. As Lwoff (1951) has emphasized, colorless forms may arise from those with chlorophyll by the loss of plastids as in abnormal division, by the simple loss of chlorophyll because of factors in the environment, or even by a mutation involving a single gene. Mutations resulting in further loss of genes, or in their alteration, may account for the inability of many species of the more highly evolved Protozoa to perform certain steps previously possible in metabolic pathways, and it is not surprising to find that among such forms dietary requirements have become more exacting.

Thus if the plantlike flagellates are indeed more primitive than other Protozoa, they should be able to do with a diet simpler and more varied in composition than that needed, for example, by the relatively specialized ciliates. But even so the problem of devising suitable culture media for the Phytomastigina has proved far from easy. One of the pioneers in this field is Pringsheim, much of whose work was done in Prague. More recently, Hutner and Provasoli and their associates have been able to cultivate a number of additional species of chlorophyll-bearing flagellates, as well as some of their colorless relatives. The importance of cultivating the Protozoa can hardly be overestimated. It is the key to understanding their physiology and often their life history, and may help us understand their evolutionary relationships; for the disease-producing species cultivation is essential if we are to understand the mechanisms by which they injure their hosts.

The protozoan organism, despite its seeming simplicity, is much like other animal cells in its nutritional requirements, though individual species may differ widely. He who understands protozoan nutrition will come close to a knowledge of plant and animal nutrition generally. Since these problems are basally those of growth, abnormal as well as normal, greater knowledge of nutrition in the Protozoa is very likely to shed light on many unsolved problems of human physiology and medicine.

REFERENCES

Cosmovici, N. 1933. La nutrition et le rôle physiologique du vacuome chez les infusoires. La théorie canaliculaire du protoplasma. *Ann. Sci. Univ. Jassy*, 17: 294–336.

Doyle, W. L. and Patterson, E. K. 1942. Origin of dipeptidase in a protozoan. *Science*, 95: 206.

Hall, R. P. 1939. The trophic nature of the plant-like flagellates. *Quart. Rev. Biol.*, 14: 1–12.

Hutner, S. H. and Provasoli, L. 1951. *The Phytoflagellates*. In *Biochemistry and Physiology of Protozoa*, vol. 1, New York: Academic Press.

——, —— and Filfus, J. 1953. Nutrition of some phagotrophic fresh-water chrysomonads. Growth of Protozoa, *Ann. N. Y. Acad. Sci.*, 56: 852–62.

Jennings, H. S. 1931. *Behavior of the Lower Organisms*. New York: Columbia University Press.

Johnson, W. H. 1941. Nutrition in the Protozoa. *Quart. Rev. Biol.*, 16: 336–48.

—— 1956. Nutrition of Protozoa. *Ann. Rev. Microbiol.*, 10: 193–212.

Kidder, G. W. 1947. The nutrition of the monocellular organisms. *Ann. N. Y. Acad. Sci.*, 49: 99–110.

—— 1953. The nutrition of the ciliated Protozoa. Symposium on Nutrition and Growth Factors. Rome: Instituto Superiore di Sanita. Italy.

Leidy, J. 1879. *Fresh-water Rhizopods of North America*. U.S. Geol. Surv. Vol. 12.

Lilly, D. M. 1953. The nutrition of the carnivorous Protozoa. Growth of Protozoa. *Ann. N. Y. Acad. Sci.*, 56: 910–20.

Lund, E. E. 1941. The feeding mechanisms of various ciliated Protozoa. *J. Morphol.*, 69: 563–73.

Lwoff, A. 1951. *Introduction to Biochemistry of Protozoa. Biochemistry and Physiology of Protozoa*, vol. I. New York: Academic Press.

Lwoff, M. 1951. *The Nutrition of Parasitic Flagellates* ("*Trypanosomidae, Trichomonadinae*"). *Biochemistry and Physiology of Protozoa*, vol. I. New York: Academic Press.

——— 1951. *Nutrition of Parasitic Amebae. Biochemistry and Physiology of Protozoa.* vol. I. New York: Academic Press.

Nelson, E. C. 1933. The feeding reactions of *Balantidium coli* from the chimpanzee and pig. *Am. J. Hyg.*, 18: 185–201.

Parker, R. C. 1926. Symbiosis in *Paramecium bursaria. J. Exp. Zool.*, 46: 1–12.

Phillips, B. P. 1953. Studies on the cultivation of *Endamoeba histolytica* with *Trypanonsoma cruzi.* Growth of Protozoa. *Ann. N. Y. Acad. Sci.*, 56: 1028–32.

Rhumbler, L. 1910. Die verschiedenartigen Nahrungsaufnahmen bei Amoeben als Folge verschiedener Colloidalzustände ihrer Oberflächen. *Arch. Entw. Organ.*, 30: 194.

Rudzinska, M. A., and Trager, W. 1959. Phagotrophy and two new structures in the malaria parasite *Plasmodium berghei. J. Biophys. and Biochem. Cyt.*, 6: 103–12.

Sandon, H. 1932. The food of Protozoa. *Pub. Fac. Sci.*, Egyptian University (Cairo), No, 1, pp. 187.

Storm, J. and Hutner, S. H. 1953. Nutrition of *Peranema.* Growth of Protozoa. *Ann. N. Y. Acad. Sci.*, 56: 901–9.

Wilson, B. W. 1959. The growth and division of *Euglena gracilis* with ethanol as the sole carbon source. *J. Protozool.* 6(Supp.): 33.

Zumstein, H. 1900. Zur Morphologie und Physiologie der *Euglena gracilis* Klebs. *Jb. wiss. Bot.*, 34: 149.

ADDITIONAL REFERENCES

Holter, H. 1959. Problems of pinocytosis, with special regard to amoebae. *Ann. N. Y. Acad. Sci.*, 78: 524–37.

Lewis, W. H. 1931. Pinocytosis. *Bull. Johns Hopkins Hosp.*, 49: 17–27.

Mast, S. O. and Doyle, W. L. 1934. Ingestion of fluid by amoeba. *Protoplasma.* 20: 555–60.

7

Metabolism

"Amoeba is an entire organism in just the same sense that man is an entire organism. . . . From a physiological point of view, the Protista are very complex." Clifford Dobell, *Principles of Protistology*, 1911.

What differences there are between a protozoan cell (if we elect to call it that) and a metazoan cell result chiefly from the dependency of the metazoan cell on the rest of the organism for its well-being, and usually for its very existence. The protozoan secures its own food, by hook or often by crook, whereas the metazoan cell must have it supplied by its neighbors or relatives. Certain types of metazoan cells may, however, be grown in tissue culture, such as macrophages arising from marrow or spleen, or the well-known Hela cells originally derived from a human cancer and which have proved so useful in virus and other research. When grown under such conditions, or indeed while they are still a part of the intact body, they may be quite properly compared with Protozoa cultivated in chemically defined media such as Kidder's for *Tetrahymena*. The physiological needs and processes of protozoan and metazoan cells are then much the same, and both are highly complex. Nutritionally essential substances must be absorbed in dissolved form by each; physical conditions, such as temperature, pH, and tonicity, must be adjusted and maintained within relatively precise limits. Oxygen must be available for respiration, and CO_2 and other wastes must not be allowed to accumulate in the immediate environment. Within the protozoan and metazoan cell the chemical and physical changes, the sum of which we may call life, are certainly very nearly the same. Collectively they constitute metabolism, but for convenience food-getting and the broader aspects of nutrition have been discussed in the preceding chapter. Once within the metazoan cell or the protozoan organism, absorbed or ingested foodstuffs pass through the phases of metabolism.

Respiration

Living things are, of course, all alike in requiring oxygen. But they get it in different ways. Most of the free-living Protozoa secure their oxygen in dissolved form from the surrounding medium—just as metazoan cells get it from the tissue fluid or blood. However, some are either obligate or facultative anaerobes: that is, they are unable to utilize molecular oxygen (certain species are even poisoned by it) or can get on without it, and they then get it by intracellular degradation of oxygen-containing substances. Bottom-dwelling, free-living species, species living in water rich in decomposing organic matter (such as the fauna characteristic of the Imhoff tanks of sewage disposal plants), and numerous parasitic species belong in this category. The flagellates of termites and wood roaches and the ciliates of cattle are strict anaerobes.

Aerobic types absorb the oxygen they need by diffusion through the cell surface or membrane. Just how the rate is regulated we do not know; it cannot be explained simply as a function of the area of exposed surface available for gaseous exchange. If the relationship were this simple, larger species would be handicapped since their energy needs (other things being equal) would be proportional to their mass, whereas their ability to secure oxygen would depend on the total area of the cell surface; the smaller an object is, the more surface it will have in proportion to its volume. As far as we known, no organelles have been evolved among the Protozoa having for their specific use the facilitating of oxygen-carbon dioxide exchange, but obviously any structures increasing the cell surface would have this effect, as would also any departure from spherical form. It is also clear that the contractile vacuole, when present, must assist in the excretion of all soluble waste products, carbon dioxide included.

The oxygen requirements of Protozoa differ greatly. On the whole, the larger forms seem to use less per-unit volume than the smaller ones. This may be in part because of the difference just mentioned in surface-volume relationships; thus, although *Pelomyxa carolinensis*, a very large species, consumes more oxygen than *Amoeba proteus*, its consumption is actually only half as great per-unit volume. But there are numerous other governing factors; genetic (species), age (both of the individual and of the culture), and temperature, among others. Young individuals and young cultures use more oxygen than older ones. Growth itself seems to depend in part on the oxygen supply. Adolph (1931) exposed one lot of *Colpoda* cultures to ordinary air, and another to an atmosphere of one part air to three of hydrogen. After 24 hours he found individuals of the second lot to be considerably smaller than those in the first, and they had produced only half as many progeny. Thus it appears that young Colpodas—again like children—use much of their energy simply for growth, and stunting is the penalty when they cannot.

In an organism as small as most Protozoa it might seem that an ample supply of oxygen would be assured for all parts of the cell by diffusion alone. Calculations, however, have shown that this is not always so, particularly when there are no contractile vacuoles to speed the intracellular transport of water and dissolved substances. In forms such as these, *cyclosis* (protoplasmic streaming) apparently functions as the necessary mechanism.

Within limits, rise in temperature results in a higher respiratory rate (expressed as Q_{O_2}, or volume of oxygen/unit weight/unit of time). This last relationship may in part account for the relatively high oxygen requirements of the protozoan parasites of some warm-blooded animals, such as the malaria plasmodia and the trypanosomes. The availability of free oxygen is also a factor in many cases; so also is crowding, but the real reason here may be that proportionately less oxygen is available as population density increases. Or it may be that the organisms in a crowded culture tend to be smaller; Hartman (1927) showed this to be so in a population of avian malaria parasites as their numbers increased in the blood.

It seems quite certain that occupying a central place in the respiratory mechanism of the protozoan organism, as in others, are the pigments known as cytochromes. These are involved in oxygen reduction, together with the enzyme cytochrome oxidase. Cytochromes *a*, *b*, and *c* have been demonstrated in *Tetrahymena*, and one or more of them in a number of other Protozoa. It is of more than passing interest that cytochrome *c*, a pigment very similar if not identical with the heme of haemoglobin, should also occur in these so-called unicellular organisms.

Although measuring the oxygen consumed by an organism as small as a protozoan might seem a difficult undertaking, it is easier than it may sound, for the usual technique involves determining the amount consumed by a population of known density in unit time, and then calculating the consumption of a single individual. Or the calculation may be made on the basis of oxygen consumed per unit of dry weight of the organisms, or per-unit volume of cell substance. Although a number of species of Protozoa have been so studied, the oxygen use values obtained by different investigators are often not comparable, because of varying experimental conditions (e.g., temperature). Kalmus (1927) found a single *Paramecium caudatum* to consume 0.0052 mm.^{3} per hour at 21°C. and Adolph (1929) calculated the hourly consumption of *Colpoda steinii* to be 0.00055 mm.^{3} under essentially similar conditions. A very recent measurement of the hourly consumption rate of *Colpoda cucullus*, using the new and very delicate ultramicro diver method, gave a figure of 0.000121 mm.^{3} per individual at 24°C. (Pigon, 1959). The difference in the oxygen consumption in young and old cultures is illustrated by the figures for *Tetrahymena geleii* (now known as *T. pyriformis*) given by Pace and Lyman (1947). The amount used hourly per million organisms in a three-day culture at 15°C. was 271 mm.^{3}, but four days later it was only 70 mm.^{3}—a drop of almost 75 per cent.

Among parasitic Protozoa, trypanosomes are remarkable for their high respiratory activity. The blood forms of the very pathogenic species *Trypanosoma rhodesiense* consume oxygen at about 70 times the rate of a young rat, in terms of the dry weight of each (285 cc./hr./gm. as against 4.5 cc./hr./gm.,), according to von Brand (1951). Other pathogenic species, such as *T. cruzi*, live at a considerably lower rate, but it nevertheless seems probable that a high metabolic rate is in some way associated with ability to harm the host. The common rat trypanosome, *T. lewisi*, which (most regrettably) does its host no harm at all, has an oxygen consumption rate only about one-eighth that of *T. rhodesiense*. Perhaps also tied in with lack of pathogenicity is the low respiratory rate of the cultural stages of trypanosomes. These correspond to the stages normally occurring in insects serving as vectors, which seem to suffer no injury at all from their uninvited guests. Prolonged maintenance in laboratory animals may reduce the respiratory rate of some pathogenic trypanosomes, although the reason for the drop is entirely unknown.

Since the respiratory quotient ("R.Q."), or CO_2/O_2 ratio, varies with the nature of the fuel being consumed, it has been calculated for a number of protozoan species in the hope that light would thus be spread on their metabolic mechanisms. For *Balantidium coli*, a parasitic ciliate of the pig and man, Daniels (1931) obtained a value of 0.84, and for *Amoeba proteus* and *Blepharisma undulans* (the latter a free-living ciliate), Emerson (1929) reported a figure of approximately unity. R.Q. values approximating 1.0 indicate that carbohydrate serves as the main source of energy; if about 0.7 fat is the chief fuel, and if the value lies between these figures both carbohydrate and fat are being burned. Sometimes values considerably greater than 1.0 have been obtained, as with *Tetrahymena*. Pace and Lyman (1947) found young cultures of this ciliate to have an R.Q. of about 1.20 and old ones 2.8. This might be interpreted as an indication of a high rate of fat formation, and the organism does indeed produce fat, but the true explanation remains uncertain.

Energy Production

The Protozoa, despite their small size, require a considerable amount of energy for their manifold activities. The securing and digesting of food, its assimilation, growth and reproduction, and locomotion—all these involve the expenditure of significant amounts of energy. It is indeed likely that the individual protozoan lives more intensely than most metazoan cells of equal size, although there must be great differences in this respect among species, and even among individuals of the same species in different stages of the life cycle. Large organisms usually expend more energy than smaller ones, and the energy requirements of youth are greater than those of old age among the Protozoa as among the Metazoa. Active species need more energy than sedentary ones.

The mechanism of energy production in Protozoa appears to differ in no significant way from that in Metazoa. Despite the almost infinite variety in external appearance and even in internal organization and in life cycles, the cell seems to be much the same kind of energy machine whether it be animal or plant, protozoan or metazoan, or even a bacterium. Indeed very much of what we know about nature's machinery for energy production has been learned from intensive study of the microorganisms. In this the bacteria and yeasts have been most important, but Protozoa have played a part.

Carbohydrates of course are the main source of energy in Protozoa as in other living things, though proteins and fats are also used. For most Protozoa so far studied, various sugars, and particularly dextrose, are good fuels. Since starch is greedily ingested by some species, and paramylum and glycogen often occur as reserve foodstuffs, these too must function as energy sources. Curiously, the green flagellates are apparently unable to satisfy their energy requirements from glucose, at least when in culture, and this seems also to hold for the colorless members of the group, such as *Chilomonas*. It has been suggested that they may lack an essential enzyme for phosphorylation, or that glucose fails to get into the cell. Perhaps the chrysomonads are a partial exception; Hutner *et al.* (1953) found that when sucrose or glucose was added to cultures the growth of four species was accelerated. Glycerol had a like effect, which led to the suggestion that the flagellates may use sugars to produce glycerol, which in turn may function in the esterification of fatty acids.

The actual combustion of energy sources within the organism, whether they be carbohydrate, fat, or protein, is a complicated process, requiring the mediation of many enzymes. It varies somewhat according to the foodstuff being consumed, and whether metabolism is anaerobic or aerobic. Most Protozoa are aerobes, although some are facultative anaerobes as is *Tetrahymena* (Pace and Lyman, 1947). Strict anaerobes are common among parasitic species and some bottom-dwelling free-living species. However anaerobic metabolism is believed to be the more primitive, and it has been suggested that perhaps aerobic metabolism did not originate until chlorophyll-bearing plants arrived on the scene (van Niel, 1956). These may well have been forms from which the Phytomastigina of today evolved, with a resulting scheme of evolution like that shown in Figure 7.1 (Hutner and Provasoli, 1955). This scheme is notable both for its question marks and its omissions, but the paucity of knowledge about the biochemistry of many organisms makes such uncertainty inevitable.

The best known of the Protozoa as far as metabolism is concerned is probably *Tetrahymena*, because it can be grown on a chemically defined medium. Although strains differ somewhat in their nutritional requirements, indicating genetic individuality, the existence of the Krebs (tricarboxylic or citric acid) cycle has been virtually demonstrated, thus showing that its "metabolic mill," as the scheme has been aptly called, is like that of aerobic

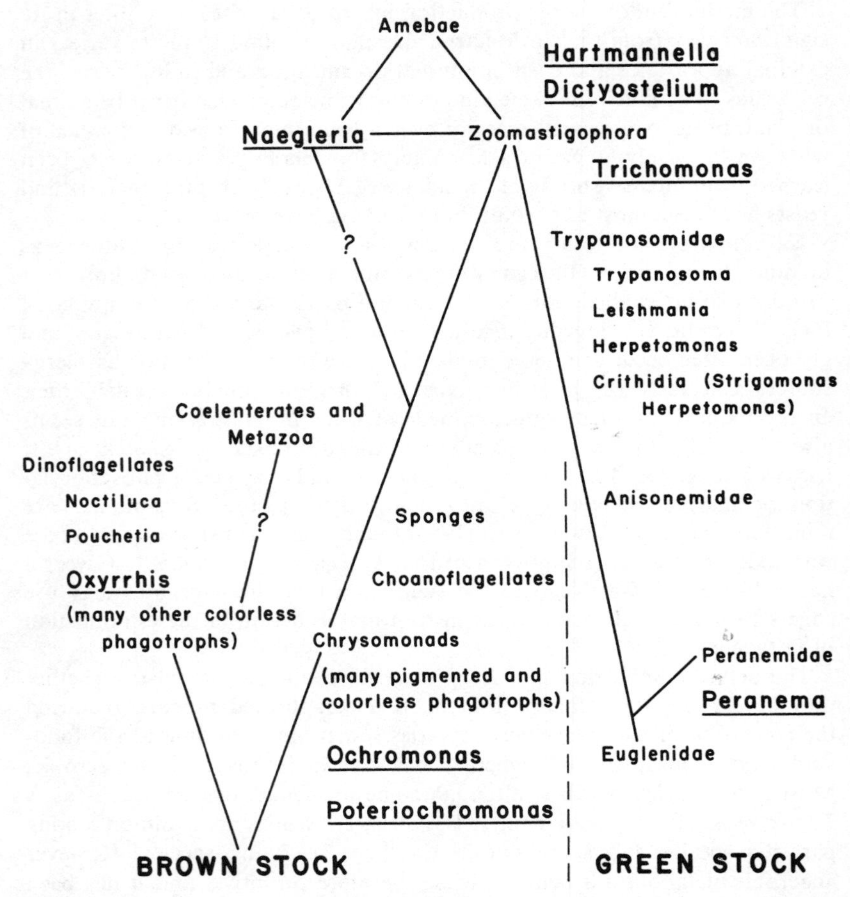

FIG. 7.1. Algal affinities of the flagellates. Phagotrophic forms in pure culture are underlined. "Brown stock" and "green stock" refer to the brown and green algae respectively. The table is of course incomplete. (Hutner and Provasoli, *Biochemistry and Physiology of the Protozoa.*)

organisms in general.* Many steps are involved, with the formation of carbon dioxide and water in the course of some of them. Anaerobic breakdown of glucose ("glycolysis") may also take place, with the formation of pyruvic acid, but this type of metabolism is less efficient in energy production and may also result in the accumulation of acids, especially lactic. Pyruvic acid may then be decarboxylated, CO_2 being given off, and the balance of the

*This cycle is discussed in detail in numerous texts; one of the better ones is Baldwin's *Dynamic Aspects of Biochemistry*, Cambridge University Press.

molecule in the form of acetyl-Co A (a compound with coenzyme A, or simply Co-A) then enters the Krebs cycle. The malaria parasites are good examples of forms normally anaerobic, the lactic acid they produce often accumulating to the detriment of the host (and even their own). There is now ample evidence that the Krebs cycle occurs in numerous species of Protozoa, including phytoflagellates such as *Euglena* and others able to use acetate (Hutner and Provasoli, 1951; 1955).

Tetrahymena may or may not be a "typical" ciliate; much less can it be regarded as a typical protozoan, if such exists. It is therefore unfortunate that so much of the nutritional research on Protozoa has been concentrated on it. *Holz (1966) remarks that in some respects its biochemical patterns resemble those of animals, in others green plants, and in still others, bacteria. Some of its biochemical characteristics, such as the composition of its DNA, are decidedly unusual. The intensive research on *Tetrahymena* has thus come closer to demonstrating that genera, and even species, differ in metabolic patterns than to showing that such patterns are applicable to ciliates or Protozoa generally. Some species of *Tetrahymena* (e.g., *T. pyriformis*) differ from others in ways as fundamental as the ability to synthesize lipids from simple or much more complex precursors.

In view of the complexity of the "mill" its over-all efficiency of about 60 per cent seems rather remarkable. But its complexity serves a useful purpose, for the organism could hardly make efficient use of the energy if it were not produced in a stepwise fashion, and this in turn is possible because much of it is stored in the form of high energy compounds such as ATP (adenosine triphosphate) and ADP (adenosine diphosphate), the function of which one might perhaps compare to that of a flywheel. Of course only a fraction of the total energy produced is used for movement, and in the conversion of chemical to mechanical energy there is inevitably some loss, with an efficiency of transformation of only about 20 per cent. It has been calculated that as a result only 1 per cent of the total energy output of an active protozoan such as *Paramecium* goes into ciliary movement.

Associated with activity of the protozoan cell there may also be electrical changes, just as there are in such metazoan cells as neurons. These phenomena have until recently been little studied in Protozoa, but *Eckert (1965) has been able to demonstrate characteristic alterations in potential associated with bursts of luminescence in an East Coast strain of *Noctiluca miliaris*. Similar changes could also be detected in a non-luminescent West Coast strain when the cells were stimulated and were found to be associated with tentacular activity, such as that occurring during feeding. Eckert remarked that electrical stimulation of the organisms resulted not only in light production by the luminescent strain, but in "several phenomena common to striated muscle twitches, namely threshold, summation, facilitation, and fatigue."

EXCRETION

The end products of metabolism in the protozoan consist of carbon dioxide, water, and nitrogen in the form of some relatively simple compound such as urea, uric acid, or ammonia—as in other animals. To these must be added oxygen in the case of chlorophyll-bearing forms. Water presents a special problem, since it also enters the body in excess whenever food is ingested, and in fresh-water forms osmotic flow through the cell wall into the cytoplasm tends to cause its accumulation in large amounts. The rate of input undoubtedly varies with circumstances and with the species; it has been estimated as between 2 and 4 per cent of body volume per hour from osmosis alone.

Carbon dioxide, because of its relatively high solubility, must rapidly diffuse through the cell and probably passes out of the organism through its exposed surface. Undoubtedly some may also get out in the contents of the contractile vacuoles (chiefly in the form of bicarbonate) when these are voided. Spallanzani suggested a respiratory function for these organelles as early as 1776, long before there was any real understanding of the respiratory process. For the chlorophyll-bearing Protozoa, and even for those with symbiotic algae, the excretion of endogenous carbon dioxide is no problem during daylight hours, for it is an essential raw material in photosynthesis. The excess oxygen produced as a by-product of photosynthesis diffuses through the cell wall into the surrounding water.

There is now little doubt that the contractile vacuoles and the structures often associated with them (such as the collecting canals and smaller contributory vacuoles) function chiefly to excrete excess water. The amount so voided may be quite considerable: *Paramecium caudatum*, with an apparently efficient system of feeder canals and two contractile vacuoles, may rid itself of water equivalent to its body volume in as little as a quarter of an hour, though its normal rate is slower. Most of this probably enters the organism through its surface, but other parts of the body may also be involved. Frisch (1937), as the result of intensive study of *Paramecium*, is convinced that in these ciliates most of the water enters through the esophageal surface, and that the rate of intake is largely under control of the ciliate, being perhaps determined in part by its varying need of oxygen. He regards the outer surface of these animals as impermeable to water and substances in solution. The food vacuoles of course also bring in much water, but they also void some when their indigestible food residues are defecated. Metabolic water probably accounts for a relatively small proportion of the total excreted.

A second function of the contractile vacuole in some Protozoa, and especially in the peritrichous ciliates and the suctoridans, is regulation of body volume. In the Suctorida it also helps condense the food, vacuolar output being greatly accelerated during feeding (Kitching, 1956).

Although the operation of the contractile vacuole attracted the attention of the earliest protozoologists, its actual emptying was not demonstrated until Jennings did so in 1904, and remarkably little is even yet known about the mechanisms involved. In the relaxation period ("diastole"), water may accumulate in the vacuole through the agency of phase separation, as by the squeezing out of molecular interspace water by the contraction of certain proteins. "Systole," or vacuolar contraction, occurs at a rate which tends to be constant for the species under given environmental conditions, and may be dependent on an intracellular timing mechanism. Such mechanisms are common in living things.

Contractile vacuoles, however, are absent in Sporozoa and other parasitic Protozoa, and in most marine forms; some ciliates and a few flagellates are exceptions in both cases. Most fresh-water dinoflagellates also lack them, although in this group the pusule (a vacuole of unknown function) may in part replace the contractile vacuole. Since marine and parasitic species live in a medium in which the concentration of salts and other dissolved substances is higher than it is within the organism, water tends to pass out of the cell because of osmotic flow and contractile vacuoles are presumably not needed. When certain of the more adaptable species are transferred to media of gradually increasing salinity, the rate of vacuolar pulsation diminishes and when a concentration equivalent to sea water has been reached the vacuole may even disappear. Similarly, marine species when transferred to fresh water sometimes develop a contractile vacuole spontaneously. It is rather difficult to understand why many parasitic and marine ciliates retain a contractile vacuole; apparently they find the maintenance of water balance an especially difficult problem. Perhaps the surfaces of these organisms are less permeable than in other groups of Protozoa. Some Protozoa have their own rather unique mechanisms of water balance regulation. Such a one is *Amoeba mira*, in which, according to Mast and Hopkins (1941), the food vacuoles perform this function. Other factors influencing water excretion include temperature, a drop in which, as might be expected, slows the rhythm of the contractile vacuole.

The protozoan also has the problem of excreting the end products of nitrogen or protein metabolism. What these are varies with the animal group; ammonia tends to be the main product in marine or fresh-water forms in which the covering membranes of the body are freely exposed to water, and through which this rather toxic substance can readily diffuse. Terrestrial animals are likely to excrete nitrogen-containing wastes in the form of the much less toxic urea or uric acid. Because of the small size of Protozoa it has been difficult to identify their nitrogenous excreta with exactness. However, Howland (1924) demonstrated urea, and Weatherby (1929) uric acid and ammonia, although it has not always been certain whether these substances originated with the Protozoa themselves, or possibly other microorganisms in the medium. Recent studies show that the ciliates, at least, form ammonia as an end product of nitrogen metabolism, but *Tetrahymena* may also produce some urea according to Seaman (1955). However, this is denied by *Dewey, Heinrich, and Kidder (1957).

It seems clear that there must also be other waste substances to be excreted. Whatever enters the cell must later leave it in one guise or another, and Protozoa require a number of mineral salts, as well as certain metal-containing vitamins. Unfortunately almost nothing is yet known about this phase of protozoan excretory activities. It is generally believed that the minute crystalline masses often seen in the cytoplasm of Protozoa represent waste products, and these have in many instances been thought to be phosphates of calcium. Crystals of other compounds, such as urates, oxalates, and carbonates, have also been reported. The frequency and abundance of such crystals appear to differ with the species, age of culture, stage in life cycle, and doubtless other factors. What the ultimate fate of these waste products may be, if such they really are, is uncertain. In some species they may be ultimately dissolved and voided with the contents of the contractile vacuole; in others they are apparently actually extruded through the cell surface.

Among other waste products known to be produced by certain Protozoa is the haemozoin or hematin, or simply heme (sometimes erroneously called melanin) of the malaria parasites, and their close relatives, the many species of *Haemoproteus*. It is an iron-containing pigment resulting from the breakdown by the parasites of haemoglobin, and is said to have a formula approximating $C_{33}H_{42}N_4O_4FeOH$. Although the role this compound many play in the symptomatology of malaria is uncertain (if indeed it plays any), it is likely that certain products of parasite metabolism are in many diseases in part responsible for the ills of the host. For example, it seems possible that *Trichomonas vaginalis* is pathogenic in part because of the amount of lactic acid resulting from its metabolism. In general, however, it has not been possible as yet to show that the metabolic waste products of parasitic Protozoa injure their hosts, or that toxins are produced.

Effects of Antibiotics

Antibiotics have proved so useful in the treatment of bacterial disease that equal success was hoped for in protozoan infections. It also seemed reasonable to expect that compounds which so strongly inhibited the growth of rickettsias and bacteria would exert a like effect on Protozoa, free-living as well as parasitic. In general antibiotics act by interfering at some point with the complicated enzymatic machinery of the organism, because of similarities in the enzymatic and antibiotic molecule, thus making it possible for the latter to combine with and inactivate the former. One might compare the antibiotic to a slightly off-size cog stopping the rotation of a wheel in a delicate machine, in this case, the metabolic mill. As the analogy suggests, antibiotics are more likely to be cytostatic than cytolethal. They inhibit cell reproduction and growth, and may also actually kill.

Despite their strong effects on many bacterial and some rickettsial infections (such as Rocky Mountain spotted fever), antibiotics have proved disappointing in the treatment of protozoan diseases, though there have been exceptions, of which the most outstanding is amoebiasis, or amoebic dysentery. Here a number of antibiotics have proved effective, especially terramycin, aureomycin, oxytetracycline, and fumagillin, although the benefits of such therapy are sometimes offset by alterations in the balance normally existing among the different organisms of the normal intestinal flora. Aureomycin has also been used with good results in the treatment of vaginal trichomoniasis, and in turkeys this drug, with penicillin and terramycin, has controlled the flagellate infection hexamitiasis. In general, however, Protozoa (including free-living species) have proved much more resistant to antibiotics than have bacteria.

Although the sulfonamides are not antibiotics their mechanism of action is similar, and they also have been tried in protozoan infections, as well as on some species of free-living Protozoa. Like the antibiotics, sulfonamides affect Protozoa only in relatively high concentrations, and to them paraminobenzoic acid is an effective blocking agent.

But for all the relative insusceptibility to these so-called miracle drugs exhibited by the Protozoa, these compounds have nevertheless been a great boon to protozoologists—and not simply to sick ones. For penicillin especially has greatly simplified the problem of obtaining Protozoa in axenic culture, and of keeping them free of subsequent bacterial contamination. Bacteria are also often secondary invaders in lesions primarily due to certain pathogenic Protozoa (e.g., amoebic ulcers of the colon), and here also antibiotics may be of value. Protozoologists owe a great debt to the discoverer of penicillin, Sir Alexander Fleming.

REFERENCES

Adolph, E. F. 1929. The regulation of adult body size in the protozoan *Colpoda*. *J. Exp. Zool.*, 53: 269–311.

——— 1931. *The Regulation of Size as Illustrated in Unicellular Organisms*. Springfield: Charles C Thomas.

von Brand, T. 1951. Metabolism of Trypanosomidae and Bodonidae. *Biochemistry and Physiology of Protozoa*, vol. I. New York: Academic Press.

Daniel, G. E. 1931. The respiratory quotient of *Balantidium coli*. *Am. J. Hyg.*, 14: 411–20.

Emerson, R. 1929. Measurements of the metabolism of two protozoans. *J. Gen. Physiol.*, 13: 153–58.

Frisch, J. A. 1937. Rate of pulsation and the function of the contractile vacuole in *Paramecium multimicronucleatum*. *Arch. f. Protistenk.*, 90: 123–61.

——— 1939. Experimental adaptation of *Paramecium* to sea water. *Arch. f. Protistenk.*, 93: 38–71.

Hartman, E. 1927. Certain interrelations between *Plasmodium praecox* and its host. *Am. J. Hyg.*, 7: 407–32.

Howland, R. B. 1924. On excretion of nitrogenous waste as a function of the contractile vacuole. *J. Exp. Zool.*, 40: 231–50.

Hutner, S. H. and Provasoli, L. 1955. Comparative biochemistry of flagellates. *Biochemistry and Physiology of Protozoa*, vol. 2. New York: Academic Press.

———, ———, and Filfus, J. 1953. Nutrition of some phagotrophic fresh-water chrysomonads, in *Growth of Protozoa. Ann. N. Y. Acad. Sci.*, 56: 852–62.

Jennings, H. S. 1904. A method of demonstrating the external discharge of the contractile vacuole. *Zool. Anz.*, 27: 656–58.

Kalmus, H. 1927. Das Kapillar-Respirometer: Eine neue Versuchsanordnung zur Messung des Gaswechsels von Microorganismen. Verläufige Mitteilung demonstriert und einem Beispiel: Die Atmung von *Paramecium caudatum. Biol. Zentralbl.*, 47: 595.

Kitching, J. A. 1938. Contractile vacuoles, *Biol. Rev.*, 13: 403–44.

——— 1956. Contractile vacuoles of Protozoa. *Protoplasmatologia, Handbuch der Protoplasmaforschung*, 3:.45.

Mast, S. O. and Hopkins, D. L. 1941. Regulation of the water content of *Amoeba mira* and adaptation to changes in the osmotic concentration of the surrounding medium. *J. Cell. and Comp. Physiol.*, 17: 31–48.

van Niel, C. B. 1956. Evolution as viewed by a microbiologist. In *The Microbe's Contribution to Microbiology*. Cambridge: Harvard University Press.

Pace, D. M. and Lyman, E. D. 1947. Oxygen consumption and carbon dioxide elimination in *Tetrahymena geleii* Furgason. *Biol. Bull.*, 92: 210–16.

Pigon, A. 1959. Respiration of *Colpoda cucullus* during active life and encyst-ment. *J. Protozool.*, 6: 303–08.

Seaman, G. R. 1955. Metabolism of free-living ciliates. *Biochemistry and Physiology of Protozoa*, vol. 2. New York: Academic Press.

Weatherby, J. H. 1929. Excretion of nitrogenous substances in Protozoa. *Physiol. Zool.*, 2: 375–94.

ADDITIONAL REFERENCES

Dewey, V. C., Heinrich, M. R., and Kidder, G. W. 1957. Evidence for the absence of the urea cycle in *Tetrahymena* *J. Protozool.*, 4: 211–19.

Eckert, R. 1965. Bioelectric control of bioluminescence in the dinoflagellate *Noctiluca. Science*, 147: 1140–42.

Holz, G. G., Jr. 1966. Is *Tetrahymena* a plant? *J. Protozool.*, 13: 2–4.

8

Locomotion

> "And the motion of most of these animalcules in the water was so swift, and so various, upwards, downwards, and round about that 'twas wonderful to see. . . ." Leeuwenhoek, Letter 6, September 7, 1674.

That animals move was probably the earliest characteristic man gave them, and to most people it remains the chief difference between animals and plants. But perhaps as great a marvel as any in nature is that animals as small as Protozoa, whose bodies comprise a single cell, can move, and even possess specialized organelles to do so. No wonder the motion of these tiny "animalcules" impressed Leeuwenhoek, for no eyes had ever seen it before.

To say that the motion of Protozoa is accomplished by flagella, cilia, or pseudopods is only part of the answer. There are also the challenges of the differences in these organelles, how they are produced, how their action is stimulated and coordinated, what their internal structure is, and how the mechanical energy they exert is produced from the prior chemical changes within the cell. We have too the problems of whether these organisms can adapt their motor responses to changing conditions, that is, learn, and what advantages motion brings to the organism. The evolutionary relationships of the various types of locomotor organelles is also an intriguing question; as is the manner in which they have been adapted to a variety of uses, some very different from their original functions.

How Protozoa move is of special concern because certain cells in the bodies of most Metazoa move in similar fashion. Phagocytic cells, like the leucocytes of our own bodies, are widely distributed in the different phyla of animals. The sperm cells of nemas are amoeboid, and those of most other animal types move with a long lashing tail, quite comparable to a protozoan flagellum. Ciliated cells are common; they are found on the gills of molluscs, in the nephridia of annelids, in the alimentary tract of amphibia such as the frog, and in the oviducts of man. Although their functions vary, the mechanism of flagellar and ciliary motion is probably similar wherever such organelles occur. Thus the results of research on the motor organelles of cells in any of these types of animals may throw light on the corresponding physiological problems of the rest.

In the Protozoa, motor activity serves ends different from those of similarly equipped cells of the Metazoa, for the protozoan is a living organism, not simply a part of one. Thus cilia are never adequate for locomotion in any but the smallest Metazoa, and their function is more likely to sweep clear some surface or to maintain a current in a minute duct. Coordinating the activities of cilia is necessary in a protozoan, where they may have many functions, but it may be less so in the metazoan ciliated cell, since the ciliated epithelia of Metazoa are usually concerned simply with the maintenance of a current in one direction over a surface. In the ciliates, therefore, a special mechanism has evolved for this function; it consists of a center and fibrils known as the "neuromotor system." However, there is still controversy about the existence and functioning of such a system, largely because differing methods of study seem to reveal different systems of fibrils, and their connections are not clear.

Types of Motor Organelles

Motor organelles among the Protozoa include not only pseudopodia, flagella, and cilia, but also contractile fibrils such as the myonemes, an example of which is the easily visible one in the stalk of *Vorticella.* Some Protozoa appear to lack any ability to move at all, as in some parasitic species. Others possess locomotor organelles only in certain stages of the life cycle, and still others may move in one fashion at one stage of the cycle and in another at a different stage. The amoeba ordinarily makes its way about the microscopic world with pseudopodia, but in some species flagellated forms, known as swarmers, are produced under certain conditions. These changes of character often hinder identifying a given protozoan, for one needs to know its past and perhaps a good deal about its future.

The origin of life occurred perhaps 1,000 million years ago—and any record of it has long since been hopelessly shrouded by the impenetrable mists of time; indeed, it may well have occurred more than once. Thus we do not know what the first Protozoa were like, nor anything about the nature of their ancestors. If we omit such groups as the Radiolarida and Foraminiferida, and the test-bearing flagellates—nearly all of which are marine Protozoa—the fossil record of the phylum tells us little. No known species of protozoan, fossil or living, can be regarded as primitive, and we therefore cannot be sure what type of motor organelle evolved first. The flagellates, however, are regarded as more primitive on the whole than either the Sarcodina or the Ciliophora, because of the close relationship of the green flagellates (Phytomastigina) to the green algae, evidenced by many resemblances both in morphology and life cycles. It is clear that chlorophyll-bearing organisms could subsist in a world in which there were no animals, but that the opposite would not be possible. Some bacteria also possess flagella, and since many of them are as nutritionally self-sufficient as chlorophyll-bearing organisms, this may be another reason for thinking that flagella were the first of nature's devices for locomotion. Simpler in both morphology and life history than either the green algae or the plantlike flagellates, and without locomotor organelles of any sort at any stage in their life cycles, are the blue-green algae. It has been suggested by some that both bacteria and the green algae may have evolved from forms very like the blue-green algae. From the green algae doubtless evolved certain groups of the chlorophyll-bearing flagellates, and from the latter the colorless flagellates. Some of the suggested relationships are shown in Figure 7.1. Axopodia may have been derived from flagella as there are similarities in the structure of both, although the former lack the internal fibrils of flagella, and may be tubular.

Little is definitely known about the actual mechanism by which chemical energy is converted into the mechanical energy of motion in the protozoan cell, but it is likely that the process is fundamentally similar to that which occurs in the muscle cells of the higher animals. Perhaps the current intensive studies of a newly discovered muscle protein may throw light on this problem. This substance, actomyosin, though not living, can be made to contract under suitable conditions. When a fiber consisting of linearly arranged actomyosin molecules is made, ATP added to the solution in which it is immersed causes it to contract. Thus the stored energy of the ATP molecule is converted to the mechanical energy of contraction.

FLAGELLA

Flagella (except bacterial) and cilia seem to be fundamentally similar in structure and manner of origin in the organism; differences between them are only in degree. One remarkable genus, *Ileonema*, possesses not only cilia but an organelle that has often passed for a flagellum; unfortunately nothing is known about its minute structure. But unluckily, the majority of flagellates are so small that their flagella are difficult to study as are the cilia of most forms. However, species with large and slow-moving flagella, such as *Peranema tricophorum*, have furnished suitable research material.

Most flagellates have not more than four flagella (a few forms, such as those found as symbionts in the gut of the termite and woodroach, may have many). Individual flagella are often specialized for different uses, especially when the number is small. Sometimes (as in certain soil-dwelling species) the flagellum (termed a "pulsellum") pushes the animal, but usually it has been thought to be a pulling organelle ("tractellum"). It may in some cases function as a kind of guide or rudder, or even as an anchor. Flagella, like cilia and pseudopodia, also often assist in procuring food. Perhaps some are tactile or sensory. In some species having two or more flagella, such as *Bodo caudatus*, one trails and gives the impression of serving as a runner, and often flagella differ in length; such characteristics aid in genus and species identification. Although flagella typically originate anteriorly there are numerous exceptions. Among free-living forms the dinoflagellates are unique in having laterally originating flagella, of which one of the two in a typical form is directed backward while the other courses in beltlike fashion around the body. The parasites also are often exceptional in their flagellar arrangement, perhaps because the nature of their environment poses special problems of locomotion. In trypanosomes the flagellum originates close to the posterior extremity, from whence it runs forward as the edge of a delicate undulating membrane, terminating as a short, freely moving fibril at the anterior end of the body; this is capable of quick reversal, and fits the parasite well for movement in the often swiftly moving blood of the host. A somewhat similar undulating membrane has been developed by the trichomonads, and this plus several flagella seems to function very efficiently in propelling them about in the vertebrate alimentary tract in which the majority of known species live. *Giardia*, another genus of vertebrate gut parasites, manages to achieve equally good results from the use of its eight flagella, symmetrically placed.

The mechanics of flagellar operation remains a problem, almost three centuries after the first observation of flagellated Protozoa and spermatozoa by Leeuwenhoek. Since a flagellate can vary its speed, turn, and even reverse itself, this minute filament must function in a highly complex fashion. *Jahn and colleagues have recently studied the action of flagella in a number of species, using high-speed cinematography, as Lowndes did. Variations were noted, but with a few exceptions (e.g., *Peranema*), there was a travelling planar sine wave originating at the base of the flagellum, although it could also start at the tip. In *Ochromonas*, a form with ciliary flagella, the mastigonemes—perhaps by giving surface roughness—seemed to cause the flagellum to pull the organism forward when theoretical considerations would have suggested a backward movement. But flagella have been a long time evolving, and just as a variety of morphological types exist (Úlehla in 1911 distinguished six), so, doubtless, do differences in operation.

Despite the great speed with which flagella and cilia seem to move, their motion is actually slow; it only seems fast because the distance moved is magnified while time is not. A further practical difficulty confronting the microscopist arises from the fact that different phases of the movement are seldom in the same plane, thus making it hard to follow the moving organelle through its entire cycle. In terms of distance, the rate of motion of the tip of even a rapidly oscillating cilium or flagellum during the work stroke is seldom greater than a few feet per hour, or generally less than a millimeter a second. Since watery solutions present a high frictional resistance to anything as small as these organelles, greater velocity would probably require an impossibly high rate of energy production. Relative size considered, flagellar movement is still respectably fast, for a flagellum may move through a distance of more than a thousand times its diameter in a single second. So likewise may be the movement of its owner. The tiny *Monas stigmatica*, measuring scarcely 6 micra, has been shown to cover a distance as great as 260 micra in a second, more than 40 times its length. As Lowndes (1945) remarks, this is a relative speed twice that of a "Spitfire" fighter plane and a thousand times that of a modern destroyer.

Earlier investigators believed that flagellar undulations could originate at either base or tip, but Lowndes (1944), who studied flagellar movement by high-speed cinematography (a technique almost necessary because of the apparent rapidity of such motion), has shown that they always start at the base. Furthermore, the flagellum, though anteriorly located, seems always

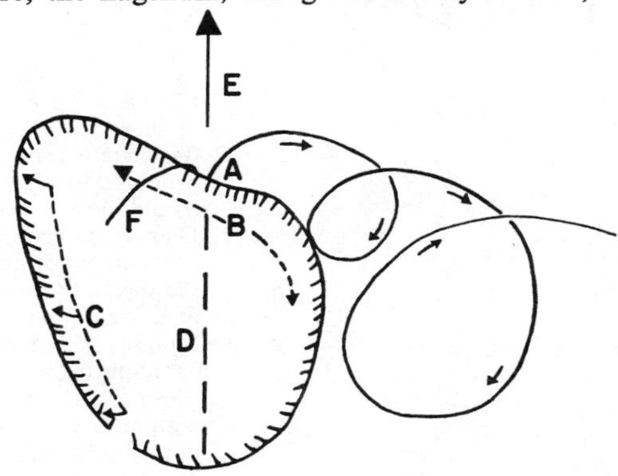

FIG. 8.1. Diagram of the small flagellate *Monas stigmatica* swimming. The organism is shown as a rotating inclined plane moving in the direction E. Waves moving from base to tip of the long flagellum furnish the motive power, causing the body to rotate and gyrate, edge C being raised above the plane represented by the page and edge B dropping below it. The very short flagellum F is thought to have a guiding or a sensory function. (After Lowndes, *Nature*.)

to *push* the organism, and he remarks, "The mechanical principle by which the organism is propelled is simply that of the inclined plane which is caused to rotate. In other words, it is that of the screw or propeller. Since the disturbances or waves pass down the flagellum in the form of a spiral they produce two distinct components. It is the resultant of these two components which causes the tip of the organism both to rotate and gyrate."

Lowndes' explanation becomes clearer when his figure of *Monas stigmatica* is consulted (Fig. 8.1). The longer of the two flagella furnishes the power and is directed backwards, its undulations proceeding from base to tip. Thus the organism is caused to rotate, its body serving as a propeller, as if edge B dropped below the surface of the page and edge C rose above it. Line D represents the axis of rotation and the arrow E the direction in which the organism moves. The very short flagellum F appears to play no part in propulsion but may be a guiding or sensory organelle. Locomotion in other flagellates seems similar, at least in *Euglena*, *Menoidium*, and *Peranema* (Fig. 8.2).

This explanation seems adequate for organisms which rotate as they travel, but it does not clearly account for the movement of a form such as *Peranema*, which (again according to Lowndes, 1936) does not rotate. There is no doubt,

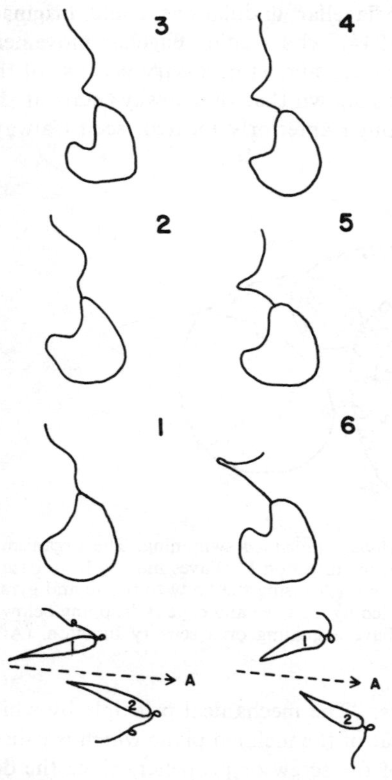

FIG. 8.2. *1–6*. Outline drawings show how flagellar contractions originate at base of flagellum of colorless flagellate *Peranema tricophorum* and progress to the tip, increasing in amplitude and velocity. *Bottom left*. Diagrammatic drawings of *Euglena* swimming normally. Proceeding in the direction indicated by the dotted line, it gyrates and rotates as it goes. *Bottom right*. *Menoidum incurvum*, a common colorless flagellate, whose action is similar to *Euglena*, although with more gyration. (Highspeed cinemicrography by Lowndes, *School Sci. Rev.*)

however, that here the flagellar waves also start at the base, increasing in amplitude and velocity as they go, also demonstrating that the flagellum "is an active unit and generates its own energy." Of course important problems are still unsolved, such as how the movement is triggered and controlled. And how, for example, does the organism initiate a halt or a change in direction? The most important problem of them all remains that of just how energy is generated in the flagellum. Some of the theories have been discussed in detail by Gray (1928) and Schmitt (1944). Probably the best guess is still that the basic mechanism of flagellar movement is essentially like that of muscle contraction. In both muscle fibrils and flagella (and even in axopodia and myonemes) there is reason to think that the structure is that of "oriented protein chains" (Schmitt, 1939). Brown (1945) makes the interesting suggestion that the impulse resulting in the flagellar wave of contraction may travel from base to tip along the coiled fibril constituting the sheath.

Cilia

Although the individual cilium has essentially the same structure as the flagellum, considerable differences in ciliary functioning arise from the need to coordinate the numerous cilia that an infusorian usually possesses. There are also differences in the type of movement characteristic of ciliates and flagellates, that of the former being much more uniform and less jerky.

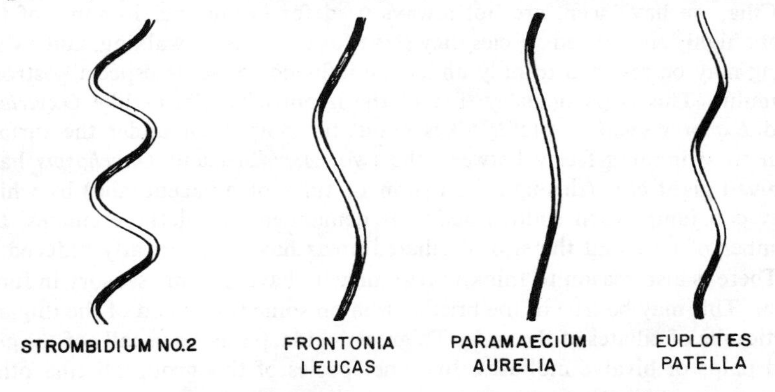

FIG. 8.3. Spiral paths of different types of ciliates. (After Bullington.)

Most ciliates appear to move in a spiral path, rotating on their axes as they go. Bullington (1925) found that this rotation was generally toward the left, particularly in marine types, which he interpreted as evidence that salt-water forms are probably older than species dwelling in fresh water. It is clear that spiral motion has the effect of bringing the organism into contact with a continually changing and relatively large area of its environment, and thus no doubt increases the likelihood of getting food or of getting away from unfavorable surroundings.

Bullington also concluded, from evidence furnished by observations on 164 species of ciliates, including the ciliated embryos of one species of suctoridan, that locomotion of this sort is probably characteristic of all the Infusoria, and that the particular type of spiral path was so characteristic of each species that it should have "pronounced taxonomic value" (Fig. 8.3).

Large ciliates (as perhaps one might have expected) were the swiftest swimmers, and the champion of them all he found to be *Paramecium caudatum,* which could dash through the water at the not quite breath-taking speed of 2,647 micra per second. The larger forms were also found to be more given to left-handed spiraling.

Jennings, who had made careful studies of motion in ciliates and other Protozoa twenty years earlier, believed that the direction and character of ciliate movement were largely dependent on body form and the stronger beating of the oral cilia, but Bullington found no such relation, nor did he find much variation in the direction of ciliary beat. Instead he concluded that spiral movement is due to "the oblique stroke of all the body cilia working together, and striking in the same direction." And he found the effective stroke in the left-spiraling species to be either obliquely backward or obliquely forward, according to whether the animal moved forward or backward. The maximum observed variation in the direction of ciliary beat was not quite 90°.

Cilia, we have seen, are not always used for swimming. In some of the more highly specialized species they serve for crawling or walking, and swimming may be resorted to only under the influence of some especially strong stimulus. This is particularly true of the hypotrichs. Forms like *Oxytricha* and *Euplotes* seem to prefer hikes about the bottom or under the surface film to swimming freely between the two. *Aspidisca* and *Onychaspis* have evolved giant cirri (through the fusion of tufts of adjacent cilia) by which they can jump when sudden action is demanded. In all these ciliates, the number of cilia and the size of ciliated areas have been greatly reduced.

There is also reason to think that some cilia have become sensory in function. This may be true of the bristles seen on some forms and of the thigmotactic cilia of ciliates of the order Thigmotrichida, parasites usually of the gills and palps of bivalve molluscs. In some species of this group all cilia other than those of the thigmotactic field have disappeared.

Then there are the sessile types—forms which prefer to stay put and lie in wait for their prey. Some of these, like the Suctorida, move about actively only in the younger or developing stages. Others, like *Vorticella* and its many relatives, may remain fixed for the time being but swim off when danger threatens, or perhaps when the hunting becomes poor in the neighborhood. Such ciliates usually develop a stalk by which they remain attached to the substrate. In the case of the vorticellids, this stalk is contractile by means of a central thread known as a myoneme (it would be most interesting to know just what the mechanism of contraction is; indirect evidence indicates that it may be similar to that of a muscle fibril). The myoneme is apparently a compound organelle in this case, since branches radiate from it into the base of the bell, which also contracts when the stalk does so. Myonemes occur in a number of ciliates; familiar examples are *Spirostomum* and *Stentor*.

It is now possible to study individual cilia under the electron microscope and it has been discovered that, minute as they are, their structure is nevertheless highly complex. Within each cilium are eleven fibrils, nine of which are peripheral and two central. Each of the former appears double. Flagella seem to have essentially the same structure, as do even the sperm tails of many of the higher animals. Underlying each cilium, as we have seen (Chapter 5), is a basal granule or kinetosome, and a short kinetodesmal fibril. The kinetodesmal fibrils, at least in *Tetrahymena*, overlap one another, and Metz and Westfall (1954) suggest that they may form a path for impulses causing ciliary movement, much as if they were nerve impulses. There is even reason to think that the same humoral mechanism, the acetyl choline-acetylcholine esterase system, may be involved as in nerve conduction across synapses and to muscle and gland cells (Seaman, 1951). Coordination of the often complex ciliary apparatus is generally believed to be through the neuromotor system, consisting of a center (the motorium) from which fibrils radiate to the various cilia and cilia-derived organelles (see Chapter 5). *Telotrochidium henneguyi*, a vorticellid ciliate, has evolved an especially complex system of this sort (Fig. 5.19). Although the uncertain relationship of the neuromotor system to other systems of fibrils in the ciliates has led some to question its coordinating function, the weight of evidence seems to make such a role probable, and such a system has been described from numerous species of ciliates.

The mechanism and type of motion of cilia and flagella are undoubtedly similar (*Jahn *et al.*, 1963), but there nevertheless seem to be differences in their operation, because all of the often numerous cilia must work together in propulsion and the relatively large, cumbrous organelles such as cirri and membranelles can hardly function as if they were single flagella. Three basal types of movement have been described by Gray (1928), although combinations of these are often seen. The simplest of the three resembles the swing of a pendulum, except that it is more rapid in one direction, the work stroke, than in the other. It would seem that the contractile or bending mechanism must lie at the base. This form of movement is often observed in large cirri, such as those of the hypotrichs. The second type involves a progressive flexing which begins at the tip while the cilium is still straight, and travels to the base so that the organelle becomes more and more hooked. This kind of movement is most often seen in large and rather slowly vibrating cilia. The third type consists of waves traveling from base to tip, and has been already described for flagella in which it is most common. Apparently the cilium is quite limp during its recovery cycle. As with the flagellum, ciliary movement seems to depend on the basal granule; if this remains intact the cilium may continue to beat for a time. It also appears that, as in the flagellum, the "mechanical energy of movement is generated along the path of the cilium" (Gray, 1928).

Some attempts have been made to determine the rate of beat of cilia, but without much success. Cilia seem to beat much more rapidly than most flagella, those of the gills of *Mytilus* (a mollusc) having a rate of approximately 720 per minute. Since the cilia of most Protozoa are so minute it is easy to see that such counting would entail difficulties, especially as "the rate of beat of a cilium is inversely proportional to its size" (Gray, 1928).

It is very hard to determine accurately the capacity of cilia for work, because their small size makes them in many respects not comparable to analogous larger mechanisms with which we are more familiar. In organisms as small as even the larger Protozoa, for example, momentum can hardly play a significant role because the cell surface (which determines the amount of frictional resistance with the surrounding water) is so great in relation to the mass. Most of the investigations in this field have been confined to the cilia of epithelium rather than those of Protozoa, and about all that can be certainly said is that muscle cells appear to be much more efficient devices for doing work than cilia.

PSEUDOPODIA

Except for axopodia, pseudopod formation and behavior are fundamentally different from motor phenomena exhibited by cilia and flagella. Pseudopodia are of course very temporary structures compared to the motor organelles of the Mastigophora and Infusoria. Yet despite their apparent simplicity, analyzing and understanding their behavior have been difficult and slow. Most of what we know is the result of microscopic observation, and it is still very limited. As De Bruyn (1947) remarks, "As to what causes such movement, we have only hypotheses," and he goes on to say, "Any theory of amoeboid movement will ultimately have to take into account the processes of energy production and metabolism. At present nothing is known about these processes insofar as they are involved in protoplasmic movement." There has been progress since this remark was made, but we are still in the realm of hypothesis.

To Ehrenberg, pseudopodia were extensions of the amoeba body forced out by musclelike contractions of other parts of the cell. But Ehrenberg was handicapped by his preconception of a protozoan as essentially metazoan in its organization, the various structures of the protozoan corresponding to metazoan organs. Dujardin, who took a more rational view of protozoan structure, thought such movement inherent in the properties of the protoplasm, which in amoebae he believed to be of an especially simple type, naming it "sarcode" (later made the basis of the class name Sarcodina by Bütschli). A few years later (1849), Ecker emphasized the ability of amoebic cytoplasm to contract, with a resulting expulsion of the noncontracting portion into the forming pseudopodium, thus contributing to the latter's growth.

Subsequent observations revealed that protoplasmic contractility was largely localized in a differentiated outer layer known as the cortex, but it also appeared that cortex (ectoplasm or ectosarc) and the passive inner cytoplasm (endoplasm or endosarc) were not permanently differentiated portions of the cell, since either might at times be converted into the other.

Of all the early studies, those of Schultze (1875) were probably the best. They were based on study of the large amoeba, *Pelomyxa palustris*, in which protoplasmic currents flowing from the posterior and central portions of the organism into the single, broad anterior pseudopodium were easily seen. Schulze concluded that the contraction of the posterior cortical layer pushed forward the protoplasm lying in front of it, and that this shift was balanced by a continuous transformation of contracted material into the passive central and posterior endoplasm. This explanation has proved essentially correct.

Toward the end of the nineteenth century the earlier ideas of pseudopod formation were largely discarded, and their place was taken by theories based on local changes in surface tension of the cell, or differences between the physical characteristics of cell surface and substrate. It is true that the physical traits of the medium may influence the kind of pseudopodium produced. Verworn showed many years ago that the blunt pseudopodia of the "limax amoebae" were transformed into the slender, pointed ones of the *Amoeba radiosa* type by the addition of potassium to the medium (Fig. 8.4). Mast more recently proved that changes in salt content would affect *Amoeba proteus* in similar fashion.

The surface tension theories of amoeboid movement were vigorously pushed by the pioneer protozoologists Rhumbler and Bütschli, but much of the recent work was done in America. Mast of Johns Hopkins in a series of brilliant studies showed that pseudopod formation in *Amoeba proteus* and *A. verrucosa* is associated with reversible sol-gel changes in the cytoplasm. Such changes he believed to be connected with changes in permeability of cell surface and resulting changes in water content, or tonicity. Local differences in surface tension, as he points out, could hardly be sufficient to cause the formation of pseudopodia in forms with a stiff pellicle, or to produce the force needed to cut organisms such as *Frontonia* or *Paramecium* in two, as some amoebae have been seen to do.

The protoplasm of the amoeba consists of three types, for which he coined the terms "plasmasol" for the central fluid portion, "plasmalemma" for the thin elastic surface layer, and "plasmagel" for the rigid layer between. His conception of amoeboid movement is often referred to as the "change of viscosity" theory, and was advocated earlier by several others, among them Hyman and Pantin.

Current theories are essentially refinements of Mast's view. Two have been seriously urged. One, the contraction-hydraulic hypothesis, assumes that pressure resulting from contraction of gel in the posterior region pushes the plasmasol into the advancing pseudopod. The other, that of "fountain-streaming," holds that there is a condensation at the anterior end (i.e., in the region of the postulated gel tube of the pseudopod), which pulls the plasmasol forward.

*Rinaldi and Jahn (1963) in some ingenious experiments in which motion pictures were made, using exposures of 5 or 10 seconds, and also shadow graphs, were able to follow the movement of particles in the advancing pseudopodia of *Amoeba proteus*. Streak photographs of particles in pseudopodia of the moving organisms were also made, and photographic evidence was checked with microscopic observations, using phase contrast, under both bright and dark field illumination. As a result, they concluded that only the contraction-hydraulic theory was tenable, since during pseudopod formation particles in every portion of the organelle moved forward and in withdrawal the reverse occurred.

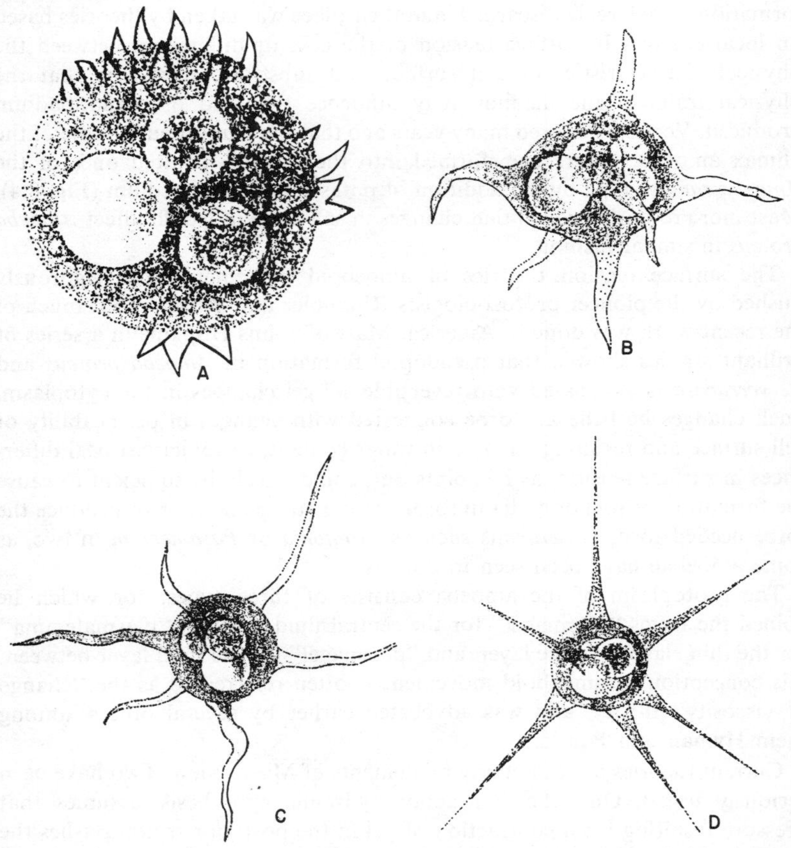

FIG. 8.4. *A. Amoeba radiosa*, a common fresh-water species, with its pseudopods just forming. *B, C, D*. Successive changes in appearance observed within a few minutes. *D* is the more typical condition, the radially extended pseudopodia, which may change very little in shape and size over a long time, considerably resembling axopodia. This amoeba varies considerably in size, but is seldom more than 50μ in diameter, and is usually much less. (After Leidy.)

It is now believed that changes in molecular structure may be at the bottom of the process. DeBruyn suggests that protoplasm can be thought of as "a three-dimensional network of protein chains, linked together by cross-linkages of side-chains. In such a structure contraction may occur either by a rearrangement of the side-chain connections, resulting in a narrowing of the meshes of the molecular reticulum, or by a folding of the protein chains themselves." More recently Goldacre and Lorch (1950), and Goldacre (1952), have adduced evidence that the latter mechanism is the one involved. As Goldacre remarks, "The amoeba is like a muscle cell in which the contraction is slow and confined to the rear end. Unlike in muscle, the contracted protein is soluble and is squeezed away to relax at the other end." As the amoeba progresses, the gelled cortical protoplasm at the posterior end contracts and is then liquefied, to be forced through the central portion of the animal where, by gelation, it forms the advancing pseudopod. This, according to Goldacre, involves the alternate folding (solation) and unfolding (gelation), respectively, of the protein molecules, which may be of so large a size that a single unfolded polypeptide chain with a molecular weight of 100 million could span the length of a 0.3 mm. amoeba. Figure 8.5, diagramming an amoeba, gives Goldacre's conception of what is going on in an amoeba on its way somewhere.

There is of course still the problem of where the amoeba gets its energy. Goldacre and Lorch believe that it comes from ATP, as in so many other biological processes, and they tested this hypothesis by injecting ATP into the tail of an *Amoeba discoides*. The resulting speed of streaming increased several times, indicating a speed-up in the processes of solation and gelation. The effect was transitory and could be observed as often as once a minute with repeated injections, indicating that no injury was done to the amoeba. This experiment seems to parallel *in vivo* the effects of ATP *in vitro* on actomyosin gel, which it causes to liquefy, and an actomyosin fiber which is thus made to contract. It is therefore further evidence that the mechanism of muscle contraction and pseudopod formation may be similar.

But, though this explanation may fit lobopodia it is difficult to apply it to axopodia and filopodia, as Noland (1957) remarks. Unfortunately the pseudopodia of Sarcodina other than the amoebae have received relatively little attention, doubtless because the larger species of amoebae have been so much easier to grow and study, but Jahn and Rinaldi (1959) have recently completed an interesting investigation of protoplasmic streaming in the filopodia of *Allogromia laticollaris*, one of the more primitive Foraminiferida. Constant movement of cytoplasm (plasmagel) in opposite directions on the two sides of the pseudopodial filaments is characteristic of these organisms and even though, to quote Jahn and Rinaldi, it may be "the simplest form of filament streaming known to exist," its explanation has proved a difficult problem. They suggest, however, a shearing mechanism, or "parallel displacement forces located between the adjacent surfaces of the two gel filaments, acting longitudinally and oppositely from one filament to the other so as to produce a two-way streaming." In simpler terms, the moving gel current on one side "crawls" on the oppositely moving surface of the other (Fig. 8.6). And this in turn they compare to the mechanism of contraction of muscle fibrils.

Thus it begins to appear that understanding molecular structure and behavior is as necessary to solving problems of how amoebae and other rhizopods function as it is in human physiology. Such behavior is orderly and always follows a definite pattern. Possibly then we should not be surprised at the observation of Schaeffer that not only is "the path of the

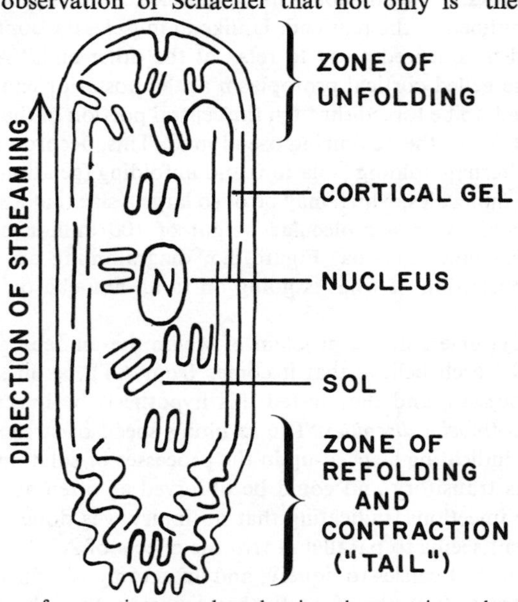

FIG. 8.5. Diagram of a moving amoeba, showing the protein molecules folding and unfolding, thus causing progressive solation and gelation of the cytoplasm, and the protoplasmic streaming which results in pseudopod formation. (After Goldacre and Lorch, *Nature*.)

amoeba . . . orderly" but that there is evidence that some type of mechanism coordinates the streaming of the amoebic cytoplasm. Random movement does not occur in the amoeba.

Axopodia, although they doubtless secure the energy they need for movement much as other types of pseudopodia, function rather differently and are unique in structure. They seem more like flagella, with an axial filament which may consist of minute tubular fibrils or plates (*Hovasse, 1965). Probably their chief function is the capture of food particles which easily adhere to their sticky surface, after which the axopodium can be quickly withdrawn. However, these versatile organelles may in some forms exercise a variety of other functions: they may be tactile, aid in keeping the organism suspended in the water or in swimming, and enable it to move on the substrate. Some Heliozoida do not have to rely wholly on axopodia to secure food or to make their way about, for they can also produce lobopodia or filopodia. There are even certain rare species of Radiolarida which produce pseudopods lacking an axial filament but able to function much like flagella.

According to Trégouboff, the role of axopodia in Radiolarida is not entirely certain, although they are believed to assist in keeping the organism floating and perhaps act as tactile organelles. But in the Heliozoida they are means of locomotion and operate in an interesting way. These Protozoa,

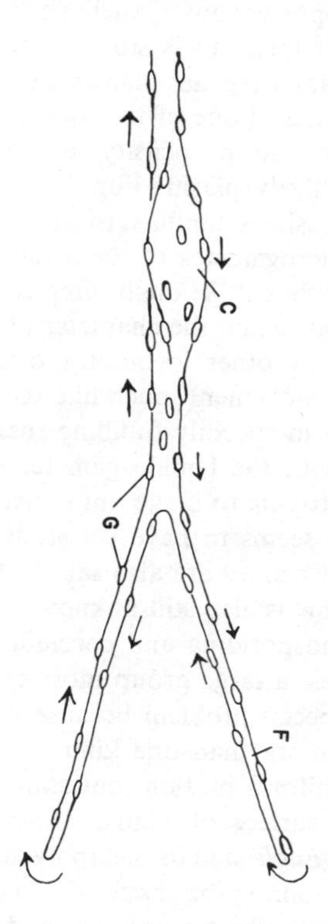

FIG. 8.6. Diagram of a pseudopodial branch of the foraminiferidan *Allogromia laticollaris*. The arrows show the direction of protoplasmic movement, which carries along the granules, G, and the small cytoplasmic mass, C. One of the two small branches is indicated by F. (After Jahn and Rinaldi, *Biol. Bull.*)

typically nearly spherical in shape, make their way along the bottom much like a rolling ball, usually quite slowly but sometimes with considerable speed. Penard, a French protozoologist who spent a lifetime of almost a century studying the "little animals" which so fascinated his great predecessor Leeuwenhoek, tells of the exploits of the fresh-water heliozoan *Acanthocystis ludibunda*, which "runs and flees away without ever getting tired, and in a single minute can cover a distance equal to twenty times its own diameter" (Penard, 1904). In terms of relative size, this would be equivalent only to a rather slow walk by a man, and it is no speed record even in the protozoan world. Yet it would be quite a feat if one were to do it as does the *Acanthocystis*, more or less by handsprings. It is described as being accomplished by a progressive shortening of the axopodia in the direction of movement, thus causing the animal to roll, the axopodia themselves being likened to stilts.

Locomotion in Sporozoa

Sporozoa possess no organelles of locomotion, yet some of them can move; how most of them do is still a puzzle. It is true that the microgametes of certain species have flagella, and certain other stages of a few species are amoeboid. Indeed one of the four species of human malaria parasites takes its name from its propensity for constant amoeboid activity: *Plasmodium vivax*, the "lively plasmodium." Yet we can hardly call this locomotion, since the parasite is confined to its tiny "lebensraum" within the host red cell. With the microgametes of the malaria parasites and other Haemosporidina, the case is different. Though often called "flagella" because of their filamentous appearance and the character of their movement, they nevertheless lack flagella or any other locomotor organelle. Yet they are extremely active, exhibiting a movement much like that of a vinegar eel, and apparently have no difficulty in speedily fulfilling their mission in life, which is to seek out and unite with the female gamete. So far their minuteness has foiled the researcher striving to make out structure with the ordinary light microscope, and no one seems to have yet studied them with the electron microscope. Malaria sporozoites are also said to be capable of some independent movement but how is also still unknown. This is true too of the motile stages of other Haemosporidina and coccidia.

Gregarines, a large group most species of which parasitize invertebrates, present a special problem because most of them are highly motile and are capable of more than one kind of movement. Typically they glide with a slow and uniform motion, but some of them also exhibit changes of body form of a variety of kinds. *Monocystis agilis*, common in earthworms, resembles a *Euglena* in its metaboly, dilating one part of its body after another and even assuming the shape of an hourglass. Other species bend themselves back and forth like vinegar eels, and still others glide slowly along leaving a trail of mucus behind them. There is also the problem of accounting for the movements of the attaching organ, termed an epimerite in some and a rostrum in others. It is said that the former may be capable of pseudopodial extension, and the latter is perhaps supplied with myonemes, contractile threads known to be abundant in the bodies of gregarines. Nevertheless, the manner in which the myonemes are used is still a puzzle, as is the role played by the trail of mucus. To quote Grassé, "In truth, in all this, a great many words and very few facts. The debate remains open."

BEHAVIOR

No account of the manifold chemical and energy transformations within the protozoan and their manner of getting about would be complete without some consideration of how all this enables the organism to relate itself to the outside world. Had Protozoa not learned, some 750 million years ago at least, to do this very successfully they would certainly never have survived until now from that distant day of the Archeozoic period when they first came into being. The early students of Protozoa were more impressed with similarities, as in behavior, these tiny organisms seemed to have to the higher animals than by any differences, other than size. Indeed Ehrenberg tried to homologize what he saw in the protozoan with superficially corresponding structures in the higher animals such as man. We know now that he was in grave error, for most of the structural peculiarities of Protozoa are different in kind rather than degree from those of the Metazoa.

But the reactions Protozoa exhibit to their environment are similar to those of other animals. That property of the cell (one might almost say "of protoplasm") enabling it to detect and react to environmental changes is known as irritability. It is customary to designate such a reaction ("taxis" or "tropism") as "positive" or "negative," depending on whether it involves a movement toward or away from the stimulating agent. Anyone who observes protozoan behavior with patience and meticulous attention to detail is sure to be struck with the variety of reactions they are capable of, and with their essential unpredictability. Under precisely controlled conditions a protozoan may seem to react in the same or nearly the same way to certain stimuli, but so do higher animals. In nature protozoan behavior is quite comparable to metazoan. This should not be surprising when the complex structure in many Protozoa is recalled. Even *Amoeba* is far from simple as we have seen, and as more refined techniques are developed for the study of minute structure its complexities seem steadily to increase.

Of course under natural conditions a protozoan is seldom or never subjected at a given time to a single stimulus. Nor is a given combination of stimuli ever likely to be precisely repeated. For this if for no other reason the behavior of the organism will appear to be constantly changing.

It is natural to ask whether a protozoan possesses "receptors," or specialized organelles to detect the various physical changes occurring in its environment. In general, the answer would have to be a qualified "no," or perhaps simply an admission of ignorance about the matter. But certain of the plant-like flagellates, such as the euglenids and some of the phytomonads, possess light-sensitive structures known as "stigmata" (singular, stigma), and the dinoflagellate family Pouchetiidae have an organelle of similar function but of different type known as an "ocellus." Likewise, the cilia or bristles of some ciliates seem to be sensory in nature. Certain regions of the organism in *Paramecium*, for instance, are more sensitive to contact stimuli than others. On the other hand, it is hard to see how there could be any localized areas of sensitivity in the naked rhizopods, which are no less aware of environmental changes than the apparently much more specialized ciliates.

The stimuli affecting protozoan behavior are as varied as those to which Metazoa react. Gravity, light, electricity, temperature, chemical composition and concentration, pH, osmotic pressure, contact with solid objects—all these, and probably others, influence the behavior of these minute organisms every instant of their brief life spans. But species and individuals react expectably in their own peculiar ways. For as each species differs genetically from others, so does each individual within a species from every other even when they are products of simple binary fission. Careful experiments have also revealed that previous experience may modify subsequent behavior, just as it does in the higher animals.

Probably the most thorough study of the relation between protozoan behavior and environment is Jennings', first published in 1906. Not only does it provide a detailed account of Jennings' own ingenious and thorough experiments, but it contains numerous references to often overlooked observations of such pioneer students of the Protozoa as Leidy, Rhumbler, Verworn, and others.

Even *Amoeba proteus* is shown by Jennings to exhibit relatively complex and highly adaptive behavior patterns. To simple stimuli, such as contact with a solid object (e.g., the experimenter's tiny glass rod), it reacts negatively or positively according to whether the agent is injurious or no. The variety of physical changes to which it is sensitive is surprising especially when one recalls its lack of specialized receptor organelles: light (except red), pH, chemical agents, electric currents, and temperature. A negative reaction is usually a change in direction just sufficient to avoid the objectionable agent, repeated until success is achieved, or "trial and error." But to an object of food, such as a *Euglena* cyst, it reacts much as a higher animal would, rolling it about when first attempts at ingestion fail (as they often do), until it is either successful or the baffling cyst is lost. According to Jennings, the chase may last for many minutes—for all the world like a tiny dog digging an elusive bone. And he remarks, "One who sees the behavior can hardly resist the conviction . . . that [it] is not purely reflex." Almost as remarkable as the behavior is that it can be achieved by an animal which seems so simple in structure. How, for example, is the necessary coordination of distant parts of the cell such as different pseudopodia accomplished?

The behavior of *Euglena viridis*, a common species of green flagellate, was also carefully studied by Jennings. Since *Euglena*, like green plants, makes its living by photosynthesis, its behavior patterns are apparently simpler than those of an organism like *Amoeba* that must hunt successfully or starve. *Euglena* seems to react chiefly and most strongly to changes in light intensity. Entry into a dark or dimly lighted area, where photosynthesis would obviously be hindered, results in an avoiding reaction or, if exposure to darkness is prolonged, in encystment. Direct sunlight also causes movement away from the area of intense illumination. The negative reaction consists of coming to a complete stop, followed by a sudden swerve toward the side bearing the eyespot (stigma), with continued rotation of the body on its long axis. The anterior end describes a circle, and when it reaches an area of more favorable light intensity the organism swims away in the new direction. Thus the animal returns to more nearly optimum light conditions. Jennings observed that the actions of *Euglena* show clearly that sensitivity to light is greatest at the anterior end where the stigma is located. Here the reaction is of a quite complex sort, and must require a high degree of coordination between different parts of the organism.

Since the ciliates include many of the highly specialized species of Protozoa, we might expect their behavior to be even more complex and adaptable than *Amoeba* and *Euglena* and their many close relatives. Jennings studied a number of species and he found *Stentor roeselii*, an attached tube-dwelling ciliate, especially interesting. This organism is sensitive to a variety of environmental agents. When a cloud of carmine grains, which it finds highly objectionable, is blown from a fine pipette into the water about an extended *Stentor*, it first bends to one side, and if this does not bring relief it contracts into its tube. If carmine remains suspended in its neighborhood, it contracts again and again after trial extensions, and finally leaves its tube completely and swims away. After swimming about for a time it stops, first at one place and then at another, each time seemingly to examine the prospective new home site by creeping over it, and finally selects what is presumably a suitable new building lot. It then makes itself a new tube in which to live, and proceeds to enjoy renewed peace and quiet.

Although Jennings cautioned against reading into this behavior evidence of capabilities comparable to those of the higher animals, in which both highly developed instincts and even reason determine so great a variety of action, he does say, "The reaction to any given stimulus is modified by . . . past experience, and the modifications are regulatory, not haphazard in character. The phenomena are thus similar to those shown in the 'learning' of higher organisms, save that the modifications depend on less complex relations and last a shorter time."

It thus appears that in the field of behavior, as in so many others in protozoology, there are still frontiers of fundamental knowledge to explore. Here we have, condensed into the tiny compass of a single minute unicellular (or acellular) organism, a mechanism complex enough to enable its possessor to meet a constantly changing environment with the same success achieved by the relatively huge forms of which we ourselves are one example.

REFERENCES

Brown, H. P. 1945. On the structure and mechanics of the protozoan flagellum. *Ohio J. Sci.*, 45: 247–301.

Bullington, E. E. 1925. A study of spiral movement in the ciliate Infusoria. *Arch. f. Protistenk.*, 50: 219–74.

DeBruyn, P. P. H. 1947. Theories of amoeboid movement. *Quart. Rev. Biol.*, 22: 1–24.

Dellinger, O. P. 1906. Locomotion of *Amoeba* and allied forms. *J. Exp. Biol.*, 3: 337–58.

Dujardin, F. 1841. *Histoire naturelle des Zoophytes. Infusoires.* Paris.

Ecker, A. 1849. Zur Lehre vom Bau und Leben der contraktilen Substanz der niedersten Thiere. *Z. wiss. Zool.*, 1: 218–45.

Ehrenberg, C. G. 1830. *Organisation, Systematik und geographisches Verhältniss der Infusionthierchen.* Berlin: F. Dummler.

Goldacre, R. J. 1952. The folding and unfolding of protein molecules as a basis of osmotic work. *Intern. Rev. Cytol.*, 1: 135–64.

——— and Lorch, I. J. 1950. Folding and unfolding of protein molecules in relation to protoplasmic streaming, amoeboid movement and osmotic work. *Nature*, 166: 497–500.

Grassé, P.-P. 1953. *Traité de Zoologie*, T. 1, Fasc. 2, Paris: Masson.

Gray, J. 1928.. *Ciliary Movement.* New York: Macmillan.

Hyman, L. H. 1917. Metabolic gradients in *Amoeba* and their relation to the mechanism of amoeboid movement. *J. Exp. Biol.*, 24: 55–99.

——— 1940. *Invertebrates: Protozoa through Ctenophora.* New York: McGraw-Hill.

Jahn, T. L. and Rinaldi, R. A. 1959. Protoplasmic movement in the Foraminiferan *Allogromia laticollaris;* and a theory of its mechanism. *Biol. Bull.*, 117: 100–18.

Jennings, H. S. 1931. *Behavior of the Lower Organisms.* New York: Columbia University Press.

Krijgsman, B. J. 1925. Beiträge zum Problem der Geisselbewegung. *Arch. f. Protistenk.*, 52: 478–88.

Lewin, R. A. 1955. Flagella–Variations and enigmas. *New Biology*, 19: 27-47.

Lowndes, A. G. 1936. Flagellar movement. *Nature*, 138: 210–11.

——— 1941. Mechanics of a flagellum. *Nature*, 148: 198–99.

——— 1943. The term *tractellum* in flagellate organisms. *Nature*, 152: 51.

——— 1945a. Swimming of *Monas stigmatica. Nature*, 155: 579.

——— 1945b. The swimming of *Euglena* and flagellar movement in general. *School Sci. Rev.*, No. 100, pp. 319–32.

Mast, S. O. 1926. Structure, movement, locomotion and stimulation in *Amoeba. J. Morphol. and Physiol.*, 41: 347–425.

——— 1931. Locomotion in *Amoeba proteus* Leidy. *Protoplasma*, 14: 321–30.

Metz, C. B. and Westfall, J. A. 1954. The fibrillar systems of ciliates as revealed by the electron microscope. *Biol. Bull.*, 107: 106–22.

Noland, L. E. 1957. Protoplasmic streaming: a perennial puzzle. *J. Protozool.*, 4: 1–6.

Oehler, R. 1919. Flagellaten und Ciliatenzucht auf reinem Boden. *Arch. f. Protistenk.*, 40: 16–26.

Pantin, C. F. A. 1923. On the physiology of amoeboid movement. I. *J. Marine Biol. Assn.*, 13: 24–69.

Penard, E. 1904. *Les Héliozoaires d'Eau Douce*. Geneva.

Pitelka, D. R. 1949. Observations on flagellar structure in Flagellata. *Univ. Calif. Pub. Zool.*, 53: 377–430.

Rhumbler, L. 1905. Zur Theorie der Oberflächenkräfte der Amoeben. *Z. wiss. Zool.*, 83: 152.

——— 1907. Review of Dellinger's paper on Locomotion of Amoebae and Allied Forms. *Zool. Zbl.*, 14: 614–17.

Schaeffer, A. A. 1920. *Amoeboid Movement*. Princeton: Princeton University Press.

Schmitt, F. O. 1939. The ultrastructure of protoplasmic constituents. *Physiol. Rev.*, 19: 270–302.

——— 1944. Structural proteins of cells and tissues. *Adv. in Prot. Chem.*, 1: 25–68.

———, Hall, C. E. and Jakus, M. A. 1943. The ultrastructure of protoplasmic fibrils. *Frontiers in Cytochemistry, Biol. Symp. X.* Jaques Cattell Press, pp. 261–76.

Schulze, F. E. 1875. Rhisopodenstudien. IV. *Arch. mikr. Anat.*, 11: 329–53.

Seaman, G. R. 1951. Localization of acetylcholinesterase activity in the protozoan *Tetrahymena geleii* S. *Proc. Soc. Exp. Biol. and Med.*, 76: 169–70.

Trégouboff, G. 1953. Classe des Acanthaires. In Grassé, P.-P., *Traité de Zoologie*, Paris: Masson. pp. 271–320.

Úlehla, V. 1911. Ultramikroskopische Studien über Geisselbewegung. *Biol. Centralbl.*, 31: 645-54; 657–76; 689–705; 721–31.

ADDITIONAL REFERENCES

Hovasse, R. 1965. Ultrastructure comparée des axopodes chez les héliozoaires des genres *Actinosphaerium*,*Actinophrys*, et *Raphidiophrys*. *Protistologica*, 1(1): 81–8.

Jahn, T. L. and Bovee, E. C. 1964. Protoplasmic movements and locomotion of Protozoa. In *Biochemistry and Physiology of the Protozoa* (S. H. Hutner, ed.). Vol. 3, pp. 61–129. New York: Academic Press.

———, Fonseca, J. R., and Landman, M. 1963. Mechanisms of locomotion of flagellates. III. *Peranema*, *Petalomonas*, and *Entosiphon*. *J. Protozool.*, *10* (Supplement): 11.

———, Landman, M., and Fonseca, J. R. 1964. The mechanism of locomotion of flagellates. II. Function of the mastigonemes of *Ochromonas*. *J. Protozool.*, 11: 291–6.

Rinaldi, R. A. and Jahn, T. L. 1963. On the mechanism of amoeboid movement, *Jour. Protozool.*, 10: 344–57.

9

Growth and Differentiation in the Protozoan Organism

"It is even extremely probable that, if in spite of their minuteness, one can succeed in cutting them into small pieces, each part will continue to live and become a complete Infusorian." Felix Dujardin, *Histoire Naturelle des Zoophytes. Infusoires.* 1841.

Of all the attributes of life, perhaps none is more typical than growth. To be sure, growth also occurs in such nonliving things as crystals. But a crystal grows by accretion; the growth of a living thing takes place from within. And the growth of a nonliving object does not involve reproduction in any sense, except the creation of crystalline form from the molecular arrangement characteristic of the substance. By contrast, the growth of a plant or animal involves primarily the reproduction within the cell of those chemical compounds unique to the species, even to the individual, and their use in a manner and pattern genetically determined. All that is as true of the Protozoa as of any of the higher animals and plants.

The mechanism of growth is still largely unknown; the way in which it is regulated is perhaps an even greater mystery. The secret of growth is, however, more than simply of biological interest; it is of great medical concern. Until it is understood, the genesis of abnormal growths, benign as well as malignant, will remain unexplained.

Growth in Protozoa

Protozoa present a real challenge to our comprehension of growth. They are not simple either in structure or in life history. We can grow some species, at least, in culture under precisely controlled conditions. The protozoan is an organism rather than the unit of a many-celled animal, and is thus not subject to control by other cells. It also multiplies more rapidly than Metazoa. Protozoa in their natural environment may be likened to metazoan cells in tissue culture, but protozoan cultivation is usually much easier. What laws govern growth in the protozoan organism? How is its maximum and characteristic size regulated? What factors determine differentiation? How does it achieve dedifferentiation (a thing impossible for most Metazoa) under certain circumstances? What is the relative importance of genetic and environmental conditions in the control of growth? None of these questions can yet be answered with finality, but we do know a good deal about these problems.

In actual growth, the protozoan organism passes through the stages of youth, maturity, and sometimes old age, much as do higher forms. This is true not only of the individual protozoan but, in many species, of the "clone," the population derived from a single ancestor. Indeed, such a population may be likened to the aggregate of cells comprising a metazoan, of which a single fertilized ovum was the ancestor. As far as the individual protozoan is concerned, growth usually continues until a certain size is reached. Then, after an interval of varying length, if conditions are favorable division takes place, and the daughter organisms in turn begin their life cycle. In one sense old age is not reached, and to the extent that continued division is possible the protozoan may be immortal. But the potential for indefinitely maintained reproduction varies with the species and, as we have seen, with environmental conditions.

Growth in the Individual

We can easily follow the growth stages of an individual protozoan, such as *Paramecium*, and make the measurements necessary for plotting increase in bodily dimensions. Such data are essentially of the same nature as for a multicellular animal—man, for instance. The rate of growth slows as age increases and the time for the next fission approaches, and it may reach a plateau. The bodily proportions of youth change with the onset of maturity, too, just as they do as the infant becomes the child and the child matures to adulthood. Young paramecia are relatively shorter and wider just after fission than they are after a period of growth. According to Wichterman (1953), Jennings found the ratio of breadth to length "considerably more than twice as great in the young as in the adults; in the former it was 63.136 per cent; in the latter 25.114 per cent." Of course, the time required for growth from earliest youth to maturity varies greatly with the species and also with the condition of the culture. In *Paramecium* or *Euplotes* it may be almost 24 hours or even longer in old cultures.

Unfortunately, few Protozoa are known as well as the various species of *Paramecium*, and most of the organisms in which growth has been carefully studied are ciliates. We have good reason, however, to think that basally growth in the different Protozoa is much the same. The growth curve of the soil amoeba *Hartmanella hyalina* is, for example, much like that of *Paramecium caudatum* and *Frontonia leucas* (Fig. 9.1). All these curves seem to be essentially sigmoid (i.e., more or less ς-shaped, resembling the Greek letter *sigma*).

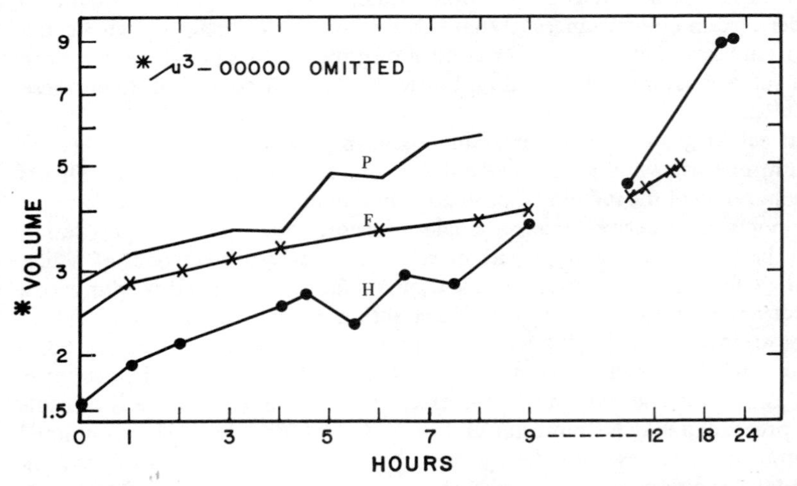

FIG. 9.1. Growth curves of *Paramecium caudatum* (P), *Frontonia leucas* (F), and *Hartmanella hyalina* (H), based on changes in volume. The two ciliates were studied by Popoff (1907, 1908) and the amoeba by Cutler and Crump (1927). (Data from Erdmann, 1920; curves from Richards, 1941, in Calkins and Summers, *Protozoa in Biological Research*.)

Size

The size range for the adults of any species of Protozoa varies within rather wide limits. Individuals differ genetically in most populations, and even when they do not, as in a laboratory culture derived from a single individual, a clone, there is still a rather substantial amount of variation. Jennings demonstrated this long ago in the "shelled amoeba" *Difflugia corona* (1920). Furthermore, in comparing any two populations variation resulting from different environments is inevitable. Considerable variability due to differences in physiological state is also to be expected. Starvation naturally results in a reduction in size, as in *Hartmanella*, though it rapidly recovers when food is again available (Fig. 9.2). The amount of food may also influence the life cycle, as it does that of *Enchelys* (Fig. 9.3). Woodruff noted almost half a century ago that individuals of *Oxytricha fallax* are smaller during periods of rapid division than when the rate is slower. In nature it is probable that most species of Protozoa are separable into races, just as the human species, and such races are prone to exhibit size differences. This is true, among others, of *Entamoeba histolytica*, the cause of

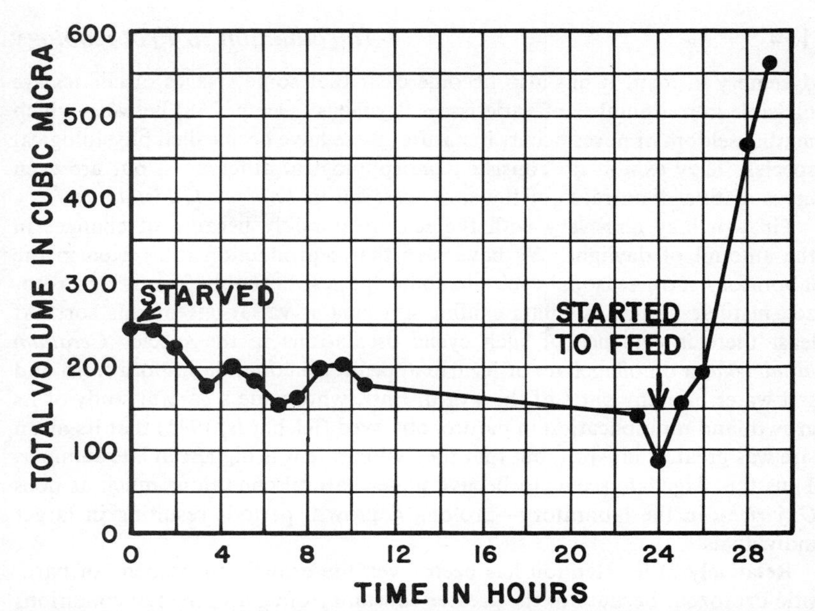

FIG. 9.2. Growth curve of the amoeba *Hartmanella*, showing the effect of starvation and subsequent feeding. The volume of the organism dropped by two thirds during slightly more than three weeks of deprivation of food. Temperature was maintained at 21°C. (Data of Cutler and Crump, 1927; graph from Adolph, 1931, in *Regulation of Size as Illustrated in Unicellular Organisms.*)

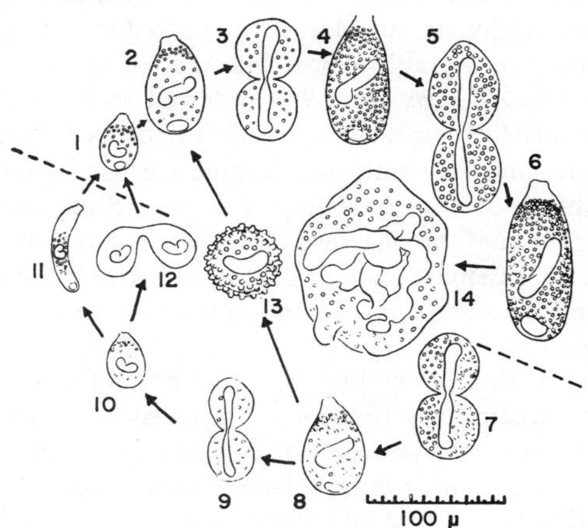

FIG. 9.3. The life cycle of *Enchelys mutans*, a holotrichous ciliate commonly encountered in cultures and showing marked dimorphism. There are the small forms (*11* and *1*), and the larger ones (*6*). Both are predators and live chiefly on *Colpidium* and *Glaucoma*, which are paralyzed instantly when they come in contact with the mouth area of *Enchelys*. *1–6*, growth and division in the presence of ample food; *7–11*, division and reduction in size when food is short; *12*, conjugation; *13*, cyst; *14*, an abnormal and gigantic form frequently developing in cultures with an oversupply of food for a prolonged period. The small forms (*11* and *1*) are very active hunters. (After Fauré-Fremiet.)

dysentery in man. It has also become clear that some species of ciliates are divisible into a number of varieties or "syngens" (see p. 238) between which mating seldom or never occurs in nature; these have been called physiological species. They exhibit no consistent morphological differences, but are even more distinct than races of the sort observed in *Entamoeba histolytica.*

Growth may also vary with the season, possibly because of changes in the amount of daylight. We have seen that reproduction rates even in the laboratory show seasonal cycles in some species. Probably free-living Protozoa in their natural habitats exhibit still greater variations of this sort. At least there is evidence of such cyclic oscillations in the species *Ceratium hirudinella*, a dinoflagellate of bizarre appearance common in both fresh and salt water in many parts of the world. Entz, who made a careful study of its growth and multiplication in nature, observed (Richards, 1941) that its mean size was greatest in April, but that the fission rate was highest in late summer. Thus this flagellate seems to behave under natural conditions much as does *Oxytricha* in the laboratory—prolonged growth periods resulting in larger individuals.

Relatively little attention has been given the growth phenomena of parasitic Protozoa. Because of the greater constancy of environmental conditions within a host, these organisms should furnish especially favorable material for the investigation of the laws of growth. Hartman (1927), however, made an intensive study of the growth of *Plasmodium cathemerium* (formerly known as *P. praecox*), a common malarial parasite of birds, and found its area to increase with maturation in an almost linear fashion (Fig. 9.4). Indeed, his measurements give a graph almost like that of the growth of the free-living ciliate *Frontonia leucas* (Fig. 9.1), referred to above. Not the least interesting aspect of Hartman's work was his discovery that the mean size of the parasites varies inversely with their numbers in the blood, a relationship he thought might stem from crowding. But the actual mechanism of this effect is obscure; it may be the result of excessive demands on the food supply, in this case chiefly the blood glucose. Probably the size of the host cell sometimes places a limit on the amount of growth possible for intracellular parasites.

Inheritance (partly through ploidy) plays an undoubted part in determining size, whether the organism is free-living or parasitic. Jennings early in his career undertook an extensive series of experiments to solve this problem, using two species of *Paramecium: P. caudatum* and *P. aurelia.* But instead of finding, as we might expect, that inheritance is important in governing the size of the individual ciliate, he found instead that it fixed in large degree the *mean size* of the members of a pure line. As he put it, "The *mean size* is therefore strictly hereditary . . ." but "In a given 'pure line' [progeny of a single individual] all detectable variations are due to growth and environmental action, and are not inherited." He continues, "In nature we find many pure lines differing in their characteristic mean dimensions" (1908).

The nucleocytoplasmic ratio is often suspected of being a major factor in triggering cell division and influencing cell size, but there is little reason to think that the relationship is simple. Division itself appears to depend in part on attaining a certain critical size (p. 188), but there comes a time when the machinery of fission is already in motion, so to speak, and reducing the size of the organism by amputating a bit of its cytoplasm no longer suffices to stop it; nuclear factors must be involved in this case. Nanney (1953) comments on the "complex interaction of nuclear and cytoplasmic factors"

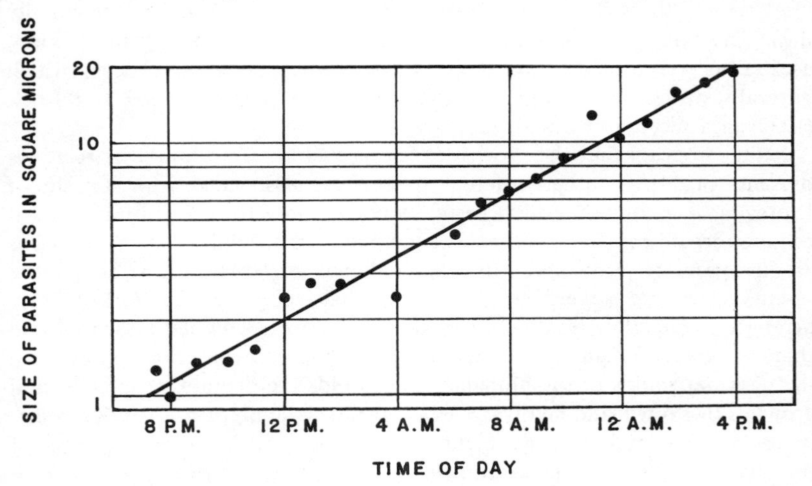

FIG. 9.4. Growth curve of *Plasmodium cathemerium*. The nature of the "curve" (it appears as a straight line, since it is plotted on semilogarithmic paper) indicates it to be exponential. The rate of growth appears to be constant during the period covered, which is the major part of the complete developmental cycle of 24 hours. (After Hartman, *Am. J. Hyg.*)

in determining differentiation in the ciliates. Chalkley (1931) concluded that for *Amoeba proteus* "nuclear division probably depends upon the volume of the cell" but "cytoplasmic division probably depends upon the volumetric ratio between cytoplasm and nucleus." Woodruff (1913), however, thought the "nucleocytoplasmic relation an incidental result rather than a cause of cell division" since "a wide variation in size of the cells and of the nuclei [of *Oxytricha*] occurs at all periods of the life of the race."

GROWTH IN COLONIAL TYPES

The development and maturation of the colonial Protozoa present another interesting aspect of the general problem of growth. The majority of these forms are members either of the phytomonad family Volvocidae or of the cilate order Peritrichida, the latter being therefore closely related to the familiar Vorticellas. *Volvox* has often been regarded as bridging the gap between unicellular types and the true Metazoa, since it is a hollow sphere of cells not unlike the blastula of higher forms. But its pattern of development does not support this view. Within the body of the parent the zygote undergoes a period of rapid cell division with little increase in size, culminating in the formation of a hollow sphaerule. A curious process known as "inversion" is then initiated, in which the sphere turns inside out. Sooner or later, the time depending largely on the temperature, the daughter colonies—which often number about eight but which vary with the temperature and the season—are liberated through a kind of birth pore, and the parent colony eventually dies. The young colonies grow for a time and then begin reproduction, asexual as well as sexual, on their own account. The whole process, which naturally varies in its details with the species, has been described by Pocock (1932) in a nicely illustrated monograph.

It thus appears that the growth of forms such as *Volvox* and its relatives presents problems rather different from those associated with unicellular Protozoa. A colony of *Volvox* certainly comes close to constituting a multicellular organism. By contrast, the colonial vorticellids such as *Epistylis* are in many respects more like the colonial coelenterates, although the resemblances are superficial. The individual zooids of an *Epistylis* are arranged symmetrically, division of the axial cells taking place unequally so that the smaller daughters are peripheral in position. From the latter develop the terminal zooids of the branches. The zooids are definite in number, and growth slows down and finally ceases as the normal limit of size for the species is reached. The entire colony might therefore be regarded as having some of the characteristics of a metazoan on the one hand and of a population of ciliates derived from a single parent on the other. In the latter case, as has already been seen, the normal course of events often involves division at a gradually diminishing rate until finally some form of internal reorganization, such as autogamy, endomixis, or conjugation supervenes.

To Fauré-Fremiet, to whom we owe much of our knowledge of these ciliates, the growth of *Epistylis* seemed to parallel the growth of metazoan organisms, since the ability of the constituent cells to divide and their degree of differentiation is determined by position in the colony, and not by age.

Intraorganismal Growth

The ability of most living things to regulate their size within the relatively narrow limits characteristic of the species is remarkable, but even more so is the mechanism by which nature stimulates and coordinates the growth of their component parts. The Protozoa are perhaps even more wonderful than their multicelled "superiors" in the animal kingdom, for in the compass of a single minute organism, its maximum dimensions often measured in micra, the complexity of pattern rivals anything else nature has created. Differentiation also includes the development of the various stages of life cycles, many of which are complicated. Not only must there be a mechanism governing the differing morphology of each stage, but a means must also exist by which the stages appear in their proper sequence.

In part, the nucleus is such a mechanism. Regeneration experiments have shown that without a nucleus neither regeneration nor long-continued life of the organism is possible (see Chap. 11). Among most ciliates, in which nuclear dimorphism is the rule, only the macronucleus is essential for repairing injuries and replacing lost parts; often a mere fraction of it is enough. It appears that nuclear genes are largely responsible for the development of species characteristics.

Yet experiments in nuclear transplantation on amoebae have shown that the cytoplasm also plays a significant genetic role. These experiments require skill and a micromanipulator of advanced design and even then they are not always successful. It has nevertheless been possible to remove and exchange the nuclei of such relatively large species of amoeba as *Amoeba proteus* and *discoides*, and in a few cases even to maintain cultures of the hybrid amoebae for several years. The amoebae exhibit characteristics of both species, proving that genes or their equivalent occur in cytoplasm as well as in the nucleus. Enucleated amoebae always die, though they may survive sometimes for several weeks; of course they never reproduce. Somewhat similar experiments were done by Tartar (1953) on two species of *Stentor*, *S. coeruleus*, and *S. polymorphus*. Macronuclear transplantation from one species to another was on the whole less successful than those on amoebae. Individuals with transplanted nuclei were able to survive and regenerate normally when a preponderance of their own cytoplasm accompanied the foreign nuclei; however, normal regeneration did not otherwise occur though there might be restoration of the shape characteristic of the species. In *Stentor* as in *Amoeba* the cytoplasm was found to play an important role in determining the expression of species, the cytoplasm of one species unable to substitute adequately for the other.

Morphogenesis in Ciliates

Unfortunately, however, experiments such as these on amoebae have not been widely attempted, and more is known about differentiation, or morphogenesis* as it is often called, in the ciliates than in other groups. Their obviously intricate organization and complex behavior have always attracted the attention of microscopists interested in the Protozoa, and the first careful studies of ciliate development were made almost a century ago.

It has now become clear that the mechanism of differentiation in ciliates

*"Morphogenesis" and "differentiation" are not, strictly speaking, synonymous. The former is the development or appearance of the form characteristic of the organism, whereas the latter is more accurately defined as an increase in complexity and organization. Yet the two kinds of development are really simply different aspects of the process of unfolding involved in the maturation of the organism, protozoan or otherwise.

is the same whether fission, encystment, conjugation, or the restoration of lost parts is involved. Fundamentally, it is dependent on the complex of ectoplasmic elements known as the "infraciliature" (Chap. 5).

As we have seen, the basal granules of this complex, like the genes, are self-duplicating units, but their potential is greater. From these granules or kinetosomes are developed cilia and cirri, their derived oral structures, trichocysts, trichites, organelles of attachment in sessile forms, and perhaps even the contractile vacuoles (Fig. 9.5). Thus these minute granules are indeed the structural (as distinct from the chemical) building blocks out of which the ciliate is made. The basal granules (blepharoplasts) of the flagellates seem to function in much the same way.

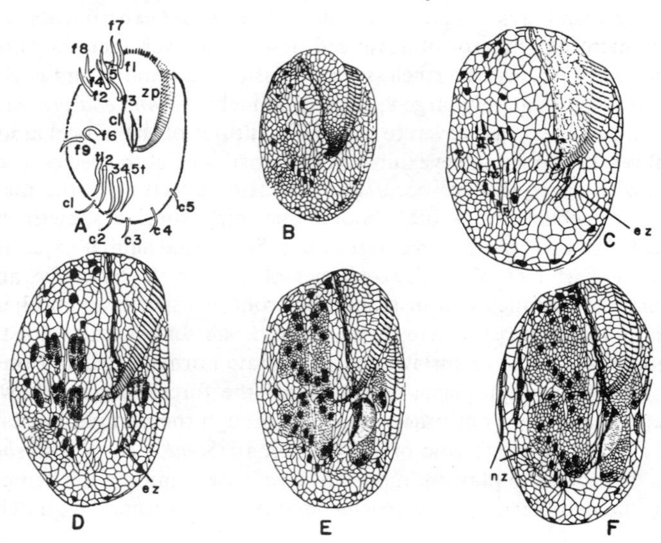

FIG. 9.5. The reorganization of *Euplotes patella* at division. The fibrillar network revealed by silver impregnation is shown in black in *B*, *C*, *D*, *E*, and *F*. *A* shows the cirri and adoral zone as seen in the living ciliate. c_1–c_5, posterior cirri; c.l., longitudinal crest; f_1–f_9, frontal cirri; l., undulating membrane; t_1–t_5, transverse cirri; z.p., preoral zone. *B*. Young ciliate. *C–F*. Prediviving form, showing the new cirri and new network. The peristome of the posterior daughter (opisthe) is formed from one kinetosome originating from the posterior sector of the parental peristome. *e.z.*, developing oral zone of the opisthe; *n.c.*, new cirri; *n.z.*, new network. (After Lwoff, 1950, *Problems of Morphogenesis in Ciliates.*)

It is significant that in the less primitive ciliates not all kinetosomes have equal potencies, although those within the same kinety appear to be equipotent. Some are more specialized than others, and without such specialized kinetosomes normal development of the young infusorian or the restoration of lost parts is not possible. For these purposes, the ectoplasm, of which all kinetosomes are a part, is more essential to the organism than is the endoplasm, any portion of which seems to be as easily replaceable as any other.

The specialized kinetosomes apparently play a role in the development of ciliates paralleling that of the embryonal centers known as "organizers" in the higher Metazoa.

Despite the remarkable powers of the kinetosomes, especially the more specialized ones, to determine the course of development, these minute bodies are not autonomous. Whatever structures ultimately form as the result of their influence appear in the species pattern. Evidently the functioning of kinetosomes and the genetic elements of the nuclei is reciprocal, and the endoplasm furnishes the milieu of both. Yet when we speak of nuclear influence in this case we have in mind only the macronucleus; the micronucleus (except as the macronucleus is originally derived from it) is not concerned. Because of its polyploid (or hyperpolyploid) nature even a small fraction of the macronucleus may be sufficient for normal differentiation of the ciliate organism. How? We do not know; perhaps some substance diffuses from it into the cytoplasm. Such a hypothetical substance, if it exists, requires continuous replenishment, for experiments have shown that removal of the macronucleus arrests all further development (Tartar, 1956). Evidently then, kinetosomes, endoplasm, and macronucleus are all essential and mutually dependent.

What stimulus triggers the reproductive process of the key kinetosomes? What initiates that train of events ending in the replacement of lost organelles? What sets in motion the rebuilding of normal body structure after the dedifferentiation of encystment? What induces the development of two new individuals, each fully equipped with the multiplicity of parts even a protozoan needs for normal living? We cannot answer these questions at present, and we know even less about what stops the process at the proper time once it is started. Weisz (1954), however, suggests that "morphogenesis in Protozoa may be viewed as a triphasic event," including a "preparatory phase" (dedifferentiation, followed by assembly of raw materials and building blocks), a "formative phase" (conversion of building blocks into new organelles), and a "regulatory phase" (reproportioning of parts, growth, etc.). The whole apparatus, he remarked, is "a servomechanism of exquisite complexity." Experimental work has shed at least a ray of light on its functioning: injury to a kinety, if specialized kinetosomes are involved, is itself a stimulus to the reproduction of neighboring kinetosomes. This observation does not, of course, explain how the underlying trigger mechanism operates. Perhaps it is basally little more enlightening than the familiar fact that in Metazoa the healing process requires initial injury to tissue to get started.

Following injury, or perhaps the environmental changes which in many species seem to initiate excystment, the specialized kinetosomes of certain kineties divide, forming an "anarchic field." Their daughter kinetosomes migrate, if necessary, to those areas of the organism where derivative organelles are destined to appear. Or the kinetosomes may thus respond to the as yet unknown premonitory stimuli which trigger fission and which originate within the cell. There are naturally marked differences among species in the details of what happens. Fauré-Fremiet, in discussing the diversity of form among ciliates, remarks that they are nevertheless remarkably homogeneous, and that even a slight difference in the initial blueprint or developmental pattern may modify profoundly the characteristics of the mature organism (1952). Here doubtless a complex interplay of kinetosomes and macronuclear genes is involved.

The blepharoplasts from which flagella arise are also capable of independent reproduction and on occasion even give rise to chains, just as kinetosomes form rows; they likewise persist during the nonmotile stages in the life history of their possessors.

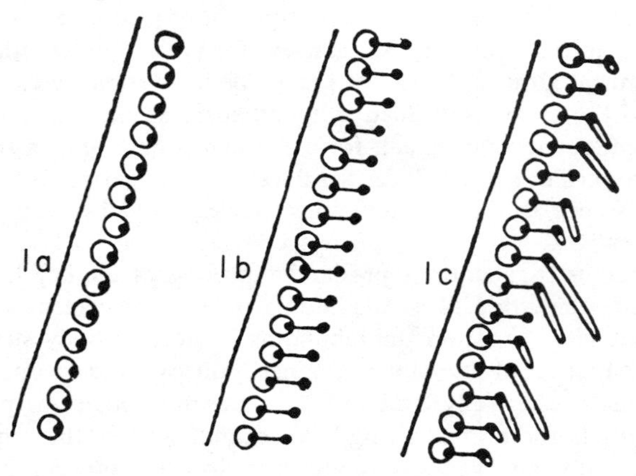

FIG. 9.6. Formation of trichocysts in the ciliate parasite *Gymnodinioides* of the hermit crab. *1a.* The normal ciliary row showing the kinetosomes. *1b.* Kinetosomes dividing, with the production of trichocystosomes. *1c.* Formation of trichocysts. (After Lwoff, 1950, *Problems of Morphogenesis in Ciliates.*)

Much of the research on the kinetosomes has been done on the apostomes, a group of curious parasitic ciliates occurring chiefly on crustacean hosts. These infusoria undergo remarkable changes in form in the various stages of their life cycles, which are in turn closely attuned to events in the life histories of their hosts. These transformations make the apostomes uniquely favorable material for the study of the potencies and fates of the kinetosomes.

Although there are numerous species differences in kinetosome behavior among these ciliates, in all species these tiny granules are extraordinarily versatile. Depending on their position in the organism, kinetosomes are capable of generating cilia, trichocysts, trichites, and other structures. Thus, since the kinetosomes persist even during periods of dedifferentiation and in the nonmotile stages of such forms as the Suctorida, the potential for the development of these varied organelles is preserved. Figure 9.6 shows a row of kinetosomes from which trichocysts are developing.

On what does the potency of a given kinetosome at a given time in the life cycle depend? Evidently the way in which a kinetosome will act at any time is a matter of its current environment, or as Weisz puts it, "the molecular species" in the neighborhood. The local presence or absence of certain enzymes, or of specific substances required for kinetosome growth and reproduction, possibly arising from the activities of the macronucleus, may affect kinetosome behavior. Conceivably the availability of essential nutrients in the diet may also influence it. Such variations might be correlated with stages

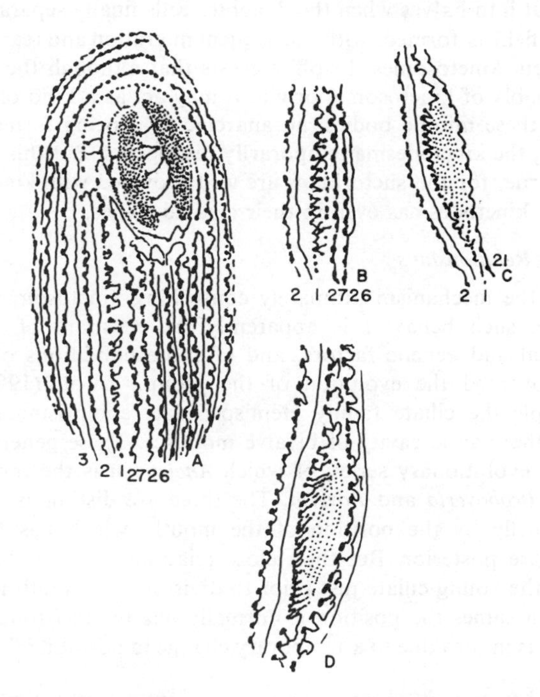

FIG. 9.7. Mouth formation in *Glaucoma scintillans*. *A*. The very beginning of the process; some kinetosomes have been produced at the left of the stomatogenic kinety 1. *B*, *C*, *D*. Later stages. In this ciliate the mouth is always produced by one kinety (kinety 1), and the behavior of the kinetosomes in this kinety may be followed in the figures. (After Lwoff, 1950, *Problems of Morphogenesis in Ciliates*.)

in the protozoan life cycle (as in the apostomes, in which encystment, for example, occurs only when the host moults), or with position in the organism. And it is essential to remember that enzyme production is gene-controlled.

Kinetosomes seem also to have other limitations. Some may be genetic, since in certain species the cytostome is always generated by a single kinety; such a kinety may be termed "stomatogenous." In other cases (e.g., the heterotrichous ciliates) a number of kineties may participate in the formation of the oral cilia. Mouth formation in *Glaucoma scintillans* is illustrated in Figure 9.7. A given kinetosome can produce only one type of organelle; when, for example, this is a trichocyst the power of further division is thereby lost. For unknown reasons, the conditions initiating division of some kinetosomes may not affect others, and some of the products of kinetosome fission often prove to have no future of their own, being simply absorbed into the cytoplasm. Nor, apparently, are all kinetosome divisions equivalent. Some kineties reproduce by elongation in ordinary cell division, the lengthened row being cut into halves when the daughter cells finally separate. In others, an anarchic field is formed, with subsequent migration and rearrangement of the constituent kinetosomes. Lwoff suggests that although the kinetodesma is itself probably of kinetosomal origin, it may act as a kind of guide to the behavior of these minute bodies, an anarchic field being formed when, for some reason, the kinetodesma temporarily disappears. But this is apparently not always true, for the suctoridans are well provided with kinetosomes but seem to lack kinetodesmas even in their ciliated stages.

Evolutionary Relationships

Although the mechanism ultimately controlling kinetosomal behavior is still obscure, such behavior is apparently the resultant of both present environmental and genetic factors, and of those conditions of the remote past that governed the evolution of the group. Lwoff (1950) mentions as an example the ciliate family Hemispeiridae, comprising three genera parasitic in the mantle cavity of bivalve molluscs. These genera apparently represent an evolutionary series, of which *Ancistrum* is the most primitive, followed by *Proboveria* and *Boveria*. The three are distinguished one from the other chiefly by the position of the mouth, which has become progressively more posterior. But their close relationship is confirmed by the fact that, in the young ciliate just prior to division, the mouth in *Proboveria* and *Boveria* assumes the position it normally has in *Ancistrum* throughout life. And this is in turn due to a temporary change in position of the formative kinetosomes.

In commenting on the variety of form in ciliates, Fauré-Fremiet (1952) remarks, "One can therefore conclude that the architecture of the ciliary apparatus governs in very large measure the body shape, and can be considered as the principal factor in morphogenesis." And the ciliary pattern, as we have seen, is determined by the kinetosomes. The similarity in submicroscopic structure of flagella and cilia, and the behavior of blepharoplasts and kinetosomes, suggests the probable origin of ciliates from flagellates, though no transitional forms between the two groups now exist.

Weisz (1954), following the lead of Fauré-Fremiet, divides ciliates into six groups according to the pattern of morphogenesis. Not all protozoologists concede that it is a good arrangement, but perhaps agreement at our present state of knowledge is too much to expect.

1. The *Prorodon* group. To this most holotrichs belong. Here the kineties are longitudinal and presumably equally specialized; within a kinety the kinetosomes are equipotential. Genera such as *Prorodon* and *Holophrya* (both members of the family Holophryidae) are generally thought to be relatively primitive.
2. The *Stentor* group. This includes some holotrichs and some heterotrichs. Members of this group possess longitudinal kineties as in Group 1, but one or several neighboring ventral kineties control both the formation of the cytostome and the polarity of the organism, as well as the contractile vacuole and the development of all surface organelles except body cilia. The kinetosomes within a kinety are equipotential.
3. The *Bursaria* group. This group also includes some holotrichs and some heterotrichs, and its characteristics are like those of Group 2, except that under certain circumstances (e.g., injury) all body structures other than the general ciliation dedifferentiate to be reformed after the development of one or (in fission) two anarchic fields of kinetosomes.
4. The *Euplotes* group. To this group belong the hypotrichs. Here the kinetosomes are not equipotential. In regeneration and fission there is initial total dedifferentiation of all body structures, with the subsequent development of several anarchic fields, the kinetosomes of which undergo extensive migration as their derivative structures appear.
5. The *Vorticella* group. In this group, which has received relatively little study, fall the peritrichs. The kineties are transversely arranged, but the mode of division is longitudinal. There is no dedifferentiation of the cytostomal apparatus, but the oral organelles of the young free-swimming ciliates are derived *in situ* from the same kinetosomes as those of the parent.
6. The *Podophrya* group, comprising the Suctorida. Kinetosomes are always present, but irregularly arranged in the adult, which is devoid of cilia. From these kinetosomes, after rearrangement in parallel rows, the cilia of the embryos are derived.

As Weisz further points out, more research will undoubtedly show that other groups of ciliates exist, and indeed some (such as those of the apostomes, discussed earlier) are already known.

Thus it appears that growth and differentiation in the ciliates, and to some degree in the flagellates, are the result of the interaction of three primary factors within the organism: the genes, the kinetosomes (both endowed with genetic continuity), and the cytoplasm (cortex and endoplasm) which furnishes the milieu in which the other two live and function.

Though it is possible to say, as Lwoff does, "The morphogenesis of a ciliate is essentially the multiplication, distribution, and organization of populations of kinetosomes and of the organelles which are the result of their activity," many problems in ciliate development are still unsolved. To the researcher they present an inviting and a continuing challenge.

REFERENCES

Adolph, E. F. 1931. *The Regulation of Size as Illustrated in Unicellular Organisms.* Springfield: Charles C Thomas.

Chatton, E. and Lwoff, A. 1935. Les Ciliés apostomes. Aperçu historique et général; étude monographique des genres et des espèces. *Arch. zool. exp. et gén.*, 77: 1–453.

Danielli, J. F. 1952. On transplanting nuclei. *Sci. Am.*, 186: 58–64.

Dujardin, M. F. 1841. *Histoire Naturelle des Zoophytes. Infusoires.* Paris.

Fauré-Fremiet, E. 1922. Le cycle de croissance des colonies de Vorticellides. *Bull. Sci. France et Belgique*, 56: 427.

——— 1930. Growth and differentiation of the colonies of *Zoothamnium alternans* (Clap. and Lachm.). *Biol. Bull.*, 58: 28–51.

——— 1948. Les mécanismes de la morphogenèse chez les Ciliés. Folia Biotheoretica, 3: 25–58.

——— 1952. La diversification structurale des Ciliés. *Bull. Soc. Zool. France*, 77: 274–81.

——— 1953. Morphology of Protozoa. *Ann. Rev. Microbiol.*, 7: 1–18.

——— 1954. Les problèmes de la différenciation chez les Protistes. *Bull. Soc. Zool.* 79: 311–29.

Hartman, E. 1927. Certain interrelationships between *Plasmodium praecox* and its host. *Am. J. Hyg.*, 7: 407–32.

Jennings, H. S. 1908. Heredity, variation, and evolution in Protozoa. II. Heredity and variation of size and form in *Paramecium*, with studies of growth, environmental action, and selection. *Proc. Am. Phil. Soc.*, 47: 393–546.

——— 1920. *Life and Death, Heredity and Evolution in Unicellular Organisms.* Boston: Badger.

Lwoff, A. 1950. *Problems of Morphogenesis in Ciliates.* New York: Wiley.

Nanney, D. L. 1953. Nucleo-cytoplasmic interaction during conjugation in *Tetrahymena. Biol. Bull.*, 105: 133–48.

Pocock, M. A. 1932. *Volvox* and associated algae from Kimberley. *Ann. So. African Museum*, 16: 473–521 (Part 3). Also *Volvox* in South Africa, pp. 523–635.

Richards, O. W. 1941. The growth of Protozoa. In *Protozoa in Biological Research*, G. N. Calkins and F. B. Summers (eds). New York: Columbia University Press.

Tartar, V. 1941. Intracellular Patterns: Facts and principles concerning patterns exhibited in the morphogenesis and regeneration of ciliate Protozoa. *Third Growth Symposium*, pp. 21–40.

——— 1953. Chimeras and nuclear transplantations in ciliates, *Stentor coeruleus* x *Stentor polymorphus. J. Exp. Zool.*, 124: 63–104.

——— 1956. IV. Pattern and substance in *Stentor*. In *Cellular Mechanisms in Differentiation and Growth.* Princeton: Princeton University Press.

Weisz, P. B. 1951. A general mechanism of differentiation based on morphogenetic studies in ciliates. *Am. Nat.*, 85: 293–311.

——— 1953. The embryologist and the protozoon. *Sci. Am.*, 188: 76–82.

——— 1954. Morphogenesis in the Protozoa. *Quart. Rev. Biol.*, 29: 207–29.

Wichterman, R. 1953. *The Biology of Paramecium.* Philadelphia: Blakiston.

Woodruff, L. 1913. Cell size, nuclear size, and the nucleo-cytoplasmic relation during the life of a pedigreed race of *Oxytricha fallax. J. Exp. Zool.*, 15: 1–22.

10

Encystment

"And not withstanding that most of these creatures are unable to stand the winter's cold, and so die, yet some of them survive, to propagate their kind: and this has been their lot from the very beginning of things." Leeuwenhoek, Letter 29, November 5, 1716.

That Protozoa occur so universally, at first thought seems in no way surprising. They are aquatic organisms for the most part, and so minute in size that water currents transport them hither and yon very easily. Yet they are also short-lived and among the most fragile of living things. Relatively slight changes in their environment are likely to be lethal. If the ability to form cysts were not so widespread, it is doubtful whether many of them would have the widespread distribution they do. Dispersal in the cyst stage has apparently been going on for many millions of years, for Deflandre (1951) has found fossilized cysts of Ebriedian flagellates in Miocene rocks. It is indeed probable that in no other phylum do we have so widespread a distribution. Cyst formation is probably of special importance for soil and fresh-water forms, and for parasites of the alimentary tract. Within the cyst, protected by its almost impermeable and dessication-resistant walls, the dormant animal may remain, like a Lilliputian Rip van Winkle, sometimes for many years, if conditions favorable for emergence are so long delayed.

Encystment

Protozoan cysts differ as much as the active stages of the species to which they belong. Generally they are smaller than the "trophozoite" (growth or vegetative stage), and the mass of protoplasm within the cyst wall has as little resemblance to the latter as the egg to the chicken. We know little about encystment in many species, but in those studied cyst formation often begins with a "rounding up" of the organism, followed by dedifferentiation and the loss of all its organelles. Cilia or flagella stop moving, pseudodopods are withdrawn, and soon a wall begins to be secreted about the animal. Often the cyst wall, which may consist of several layers, is elaborately marked in a way characteristic of the species. Yet there are few rules to which there are no exceptions. The ciliate *Frontonia depressa*, commonly found in the soil, produces cysts which differ little from the active form. Inactivity of cilia and the organelles about the mouth are the chief changes observed, aside from the secretion of a rather thin cyst wall. Even the contractile vacuole may pulsate slightly within the cyst.

Morphology

In shape, cysts are extremely variable (Fig. 10.1). In some species they may be spherical, in others oval, in still others typically irregular. *Entamoeba coli*, a parasite living in the human colon, builds itself a smooth spherical cyst. *Pleurotricha lanceolata*, a common free-living species, forms a spherical cyst but its surface is studded with minute, irregularly arranged knobs (Fig. 10.1, *B*, *E*). Whether *Paramecium* ever encysts is open to great doubt, although its ubiquitous occurrence in hay infusions and laboratory cultures is difficult to explain otherwise. It has been reported to produce cysts resembling sand grains, but attempts to confirm this observation have resulted in failure.

Occasionally cysts may be stalked. Such is the case with some Suctorida

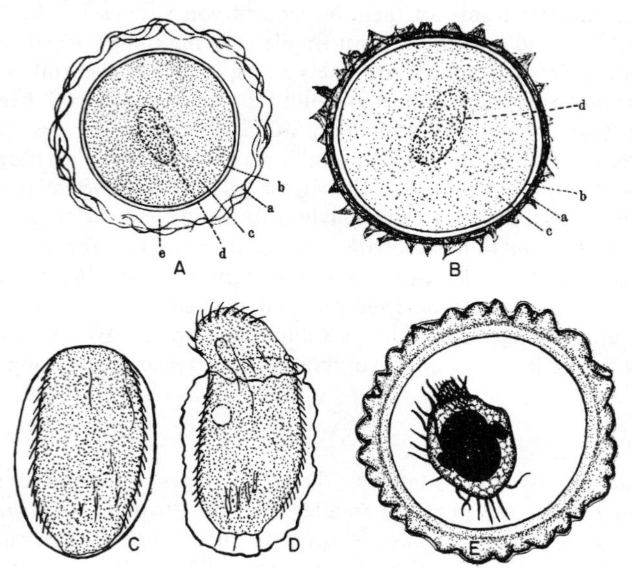

FIG. 10.1. Typical ciliate cysts. *A*. Cyst of *Stylonychia mytilus*. *B*. Cyst of *Pleurotricha lanceolata*. a, ectocyst; b, endocyst; c, plasma; d, nucleus; e, fluid-filled space between outer and inner walls. (Ilowaisky, 1926.) *C*, *D*. Excystation of *Stylonychia mytilus*. (Ilowaisky, 1926) *E*. *Pleurotricha lanceolata* apparently ready to excyst (original).

(Fig. 10.2) and certain species of *Euglena*. *Euglena pedunculata*, which received its species name from this peculiarity, forms a spherical cyst supported by a short thick stalk resting on a broad base. Other species seem less concerned about security, their cysts being unattached spheres free to move with the currents. Still others apparently encyst rarely or not at all. Marine Protozoa are said not to encyst, and the relative constancy of their environment no doubt lessens the advantages of encystment. But many

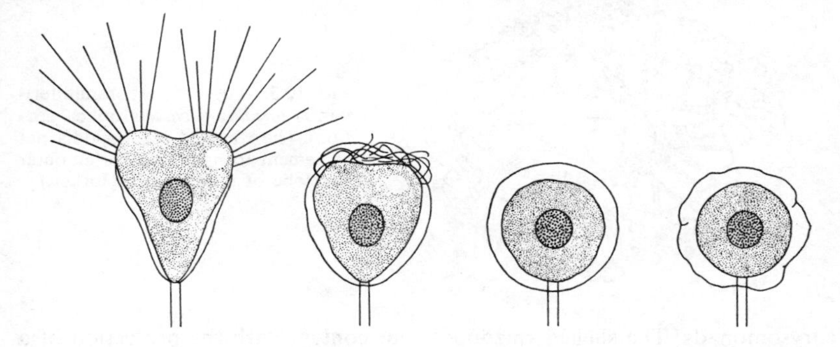

FIG. 10.2. Successive stages in the encystment of the suctoridan *Tokophrya lemnarum*. (After A. E. Noble.)

marine species undoubtedly do encyst; the cyst of a foraminiferidan is shown in Figure 10.3.

The cysts of Protozoa vary as much in size as their makers, and, as one might expect, in general the smallest cysts are constructed by the smallest species. Thus, the cysts of *Entamoeba coli*, the largest of the human intestinal amoeba, may measure as much as 30μ in diameter, while those of *Enteromonas hominis*, a minute flagellate also parasitic in the gut of man, may have dimensions as small as 6μ by 3μ.

Although an acquaintance with the morphology of the cysts of free-living species is of little significance to anyone but specialists, it is otherwise with the cysts of parasitic forms. Correct diagnosis of the intestinal Protozoa of man often depends on the ability of the parasitologist to identify cysts accurately. To a patient suffering from dysentery due to *Entamoeba histolytica* it is rather important that the laboratory distinguish this species of amoeba from *Entamoeba coli*, a harmless commensal.

Protozoan cysts probably vary less in composition than they do in shape and size. The cyst wall of the animal-like species has often been claimed to be of chitin, but this substance (a complex carbohydrate, often impregnated with mineral salts) is now said not to occur among the Protozoa. Kofoid *et al.* (1931) made a careful study of the chemical nature of the cyst wall of certain human intestinal Protozoa, and concluded that it had the properties of a keratin (a highly insoluble scleroprotein). It is probable that the cysts of most animal-like species have a similar composition, though often with the incorporation of mineral substances as reinforcement, and sometimes also of foreign bodies such as sand grains, debris, and the like. The plantlike flagellates in many cases construct their cysts of cellulose, added evidence of their close relationships to the green plants. In others, the cyst wall is of a gelatinlike material, as with many of the euglenids, or of silica, as in some

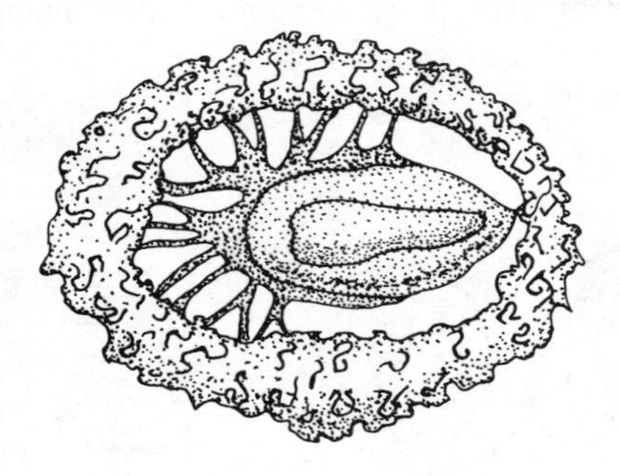

FIG. 10.3. A cyst of the foraminiferidan *Triloculina circularis* (optical section). The extended pseudopodia act as cement to hold together an outer envelope of mud. (After Hofker.)

chrysomonads. The shelled rhizopods, not content with the protection of a single or double-walled cyst like those of their close relatives, the amoebae, often retain the shell and form their cysts within it.

Advantages of Encystment

Why do Protozoa encyst? What advantages does it bring them? Both these questions have perplexed zoologists, and as so often is the case, the easy answer is not necessarily the correct one. Experiments show that the encysted animal can withstand adverse environmental influences that would kill the vegetative form. The cyst is far more resistant to drying, to chemical agents and to temperature extremes than the trophozoite. This is especially important for many species of parasites which must manage to survive in the outside world while awaiting a new host. Cysts of *Entamoeba histolytica* will even endure a concentration of chlorine in water five times that required to kill bacteria. The cyst is usually the infective stage of those parasitic Protozoa not requiring an intermediate host. Yet, remarkably enough, some species of intestinal Protozoa are apparently unable to form cysts. This is true of *Dientamoeba fragilis* and *Trichomonas* (all species, as far as is known), and also for *Entamoeba gingivalis* of the human mouth. But it seems likely that the first and the third are transmitted in ways which do not involve an intermediate sojourn in the hard outer world, while the trophozoites of *Trichomonas* are resistant enough so that the added protection of a cyst wall is not required.

Amazingly, Leeuwenhoek not only realized that his "little animals" must have some means of surviving such adverse influences as the winter's cold, but he may actually have seen the cysts of certain Infusoria. In Letter 29 (quoted at the beginning of the chapter), he tells us, "I can't forbear adding here, that I have allowed water animalcules, mixed with a little earthy matter, to lie dry in my closet for a whole winter: and when I put them again in water, I saw some of them unfold their limbs, which seemed to be wrapped up inside them, and swim about in the water." Unfortunately we don't know whether these were rotifers, which will withstand much drying, or ciliates, some of which form cysts that remain viable for years.

The actual longevity of protozoan cysts varies greatly. Cysts of some species of intestinal amoebae are apparently able to live only a few weeks outside the host. At the other extreme are cysts of certain free-living species. Goodey (1915), in some oft-quoted experiments, proved survival of cysts of the ciliate *Colpoda* from dried soil for 38 years, and those of certain small flagellates (e.g., *Bodo*) for 49 years. Hausman (1934) kept watering-trough sediment in an hermetically sealed glass tube for 20 years, and then found the flagellates *Mastigamoeba* and *Oikomonas* in cultures made from it.

Such longevity is truly remarkable in microscopic organisms as highly organized as even the simplest Protozoa are, and especially so when one remembers that it involves resisting the adverse effects of dessication, oxidation, and decay for so long. That protozoan cysts are at once so tenacious of life and so universally distributed is of course one of the reasons why the theory of spontaneous generation persisted until Pasteur's epochal experiments finally disproved it. It was difficult for the earlier biologists to conceive of continuing life under such unfavorable conditions, and when it appeared it seemed necessary to suppose that it originated *de novo*.

Laboratory experiments suggest, however, that the conditions under which encystment occurs may be very important in determining longevity. Wet cysts (cysts formed in cultures and not allowed to dry) probably do not live as long as dried cysts, as Bridgman (1957) points out. Although she was able to obtain excystment of the ciliate *Tillina magna* from cysts kept in dry soil 10 years, dry cysts produced experimentally on prepared sections of grass stems proved viable for no more than 16 weeks. Just what conditions favor the production of long-lived cysts is still very uncertain.

For the species, though not for the individual protozoan, encystment has the obvious advantage of facilitating dispersal. Because of the small size of cysts and their great resistance to adversity they may be carried long distances by wind and water, and in the guts of animals, to excyst perhaps years later when fortune again smiles.

Kinds of Cysts

But not all species of Protozoa form cysts, and when cysts are formed they may be of several kinds. It used to be said that cysts are of three types: resting or protective, digestive, and reproductive. There is in reality no clear distinction. The cysts just discussed were in general of the protective type, but within such cysts nuclear division often occurs in some species with concurrent or ultimate formation of a number of offspring. *Entamoeba coli*, a harmless intestinal parasite of man, normally produces an eight-nucleated cyst from which emerges an amoeba with a like number of nuclei; from it ultimately arise octuplet amoebulae.

Colpoda maupasi, a common, free-living ciliate, presents a most interesting contrast. Instead of producing a single type of cyst, this species may give rise

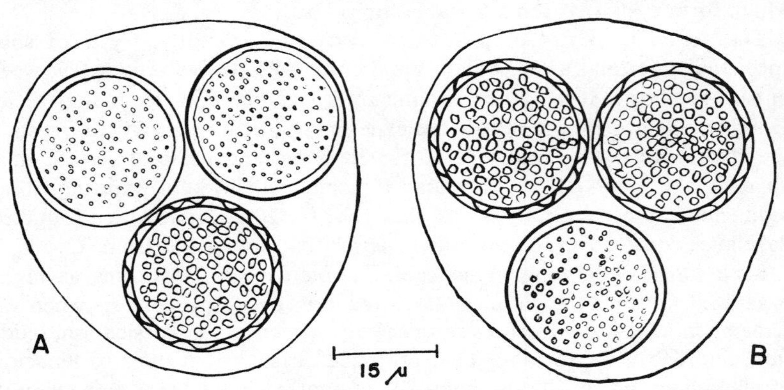

FIG. 10.4. *Colpoda maupasi* is a common ciliate which has the ability to form several different kinds of cysts, and may even excyst and re-encyst, all without leaving the original cyst. The figure shows two examples. *A*. One wrinkled and two smooth thick-walled monogenic resting cysts within the original quadrigenic parent cyst wall. *B*. A similar condition in which one smooth thick-walled and two wrinkled monogenic resting cysts are present. Both the characteristics of the medium and the stage of life cycle influence the production of cysts of specific types. (After Padnos, Jakowska, and Nigrelli, *J. Protozool.*, 1: 131.)

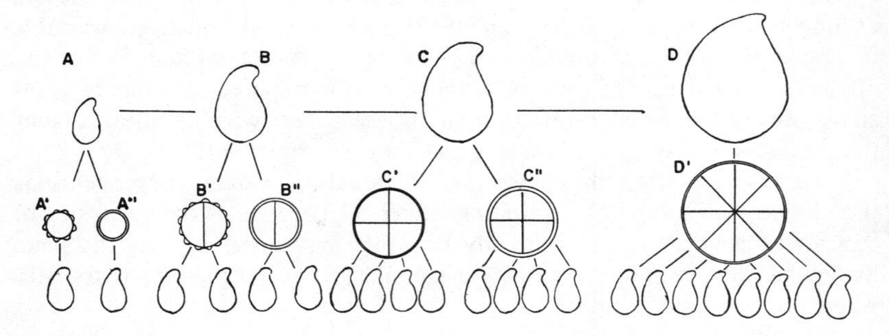

FIG. 10.5. The life cycle of *Colpoda maupasi*, a ciliate encountered in soil and in this case derived from hay. *A–D* represent growth phases of the trophic form with the resulting cysts; *A′*, monogenic wrinkled resting cyst; *A″*, monogenic smooth thick-walled resting cyst; *B′*, digenic wrinkled resting cyst; *B″*, digenic smooth thick-walled resting cyst; *C′*, quadrigenic thin-walled reproductive cyst; *C″*, quadrigenic smooth thick-walled resting cyst; *D′*, octogenic reproductive cyst. Line 3 shows, also in diagrammatic form, the individuals derived from the respective types of cyst. (After Padnos, Jakowska, and Nigrelli, *J. Protozool.*, 1: 131.)

to as many as seven, as Padnos, Jakowska, and Nigrelli (1954) have shown. These differ in morphology (their walls may be thick or thin, smooth or wrinkled), and in the number of offspring which may emerge: one, two, four, or eight. It is suggested that genetic factors may be concerned with the production of different cyst types. It also appears that there may be physiological differences since some cysts excyst spontaneously whereas others do so only under certain special conditions. Not surprisingly perhaps, the size of the family hatched from the cyst depends on the size of the parent, larger cysts and larger broods developing from larger individuals. Figures 10.4 and 10.5 show in diagrammatic form the cyst phase of the life cycle. Although this

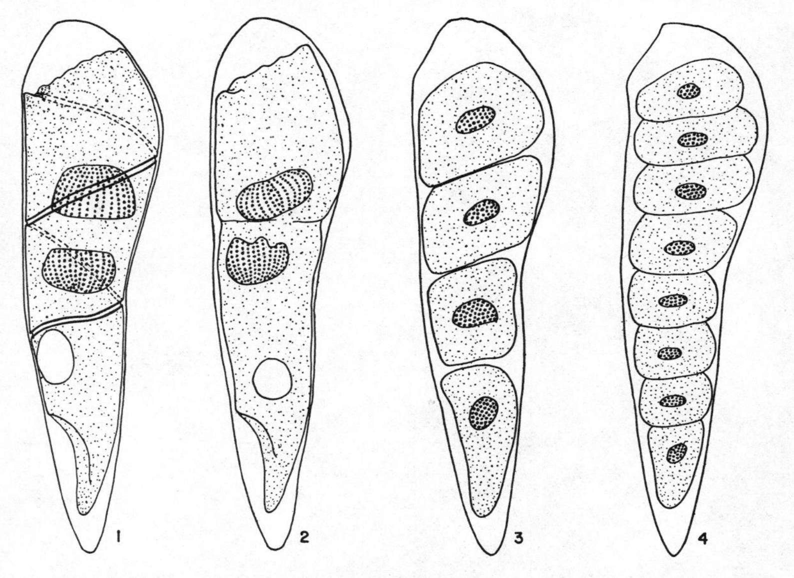

FIG. 10.6. Multiple fission within a dinoflagellate cyst. (After Kofoid and Swezy.)

looks complicated enough, a further complication may be introduced in certain cases, for occasionally daughter individuals within the cyst, apparently regretting a too hasty decision to re-enter the outside world, re-encyst.

Other species of ciliates, including some of *Colpoda*, produce what have been called "stable" and "unstable" cysts, though both may be of the protective type. These differ in the conditions required for excystment, and they may also differ in morphology; the stable type in *Woodruffia metabolica* has a much thicker wall than the unstable one, and excysts much more readily.

Reproductive cysts may also be of more than one type, for there are those in which multiplication occurs, as in some dinoflagellates (Fig. 10.6) and those in which a nuclear reorganization occurs without fission; an example of the latter is *Pleurotricha lanceolata*. Cysts of the second type are sometimes called *regeneration* cysts, since nuclear degeneration occurs initially with a subsequent rebuilding of the whole nuclear apparatus. For an extraordinary example of reproduction in a flagellate cyst, see Figure 10.7.

Digestive cysts are formed by some species, apparently to facilitate the digestion of a large meal—a kind of protozoan after-dinner siesta. For reasons unknown this habit has never developed among the amoebae, but

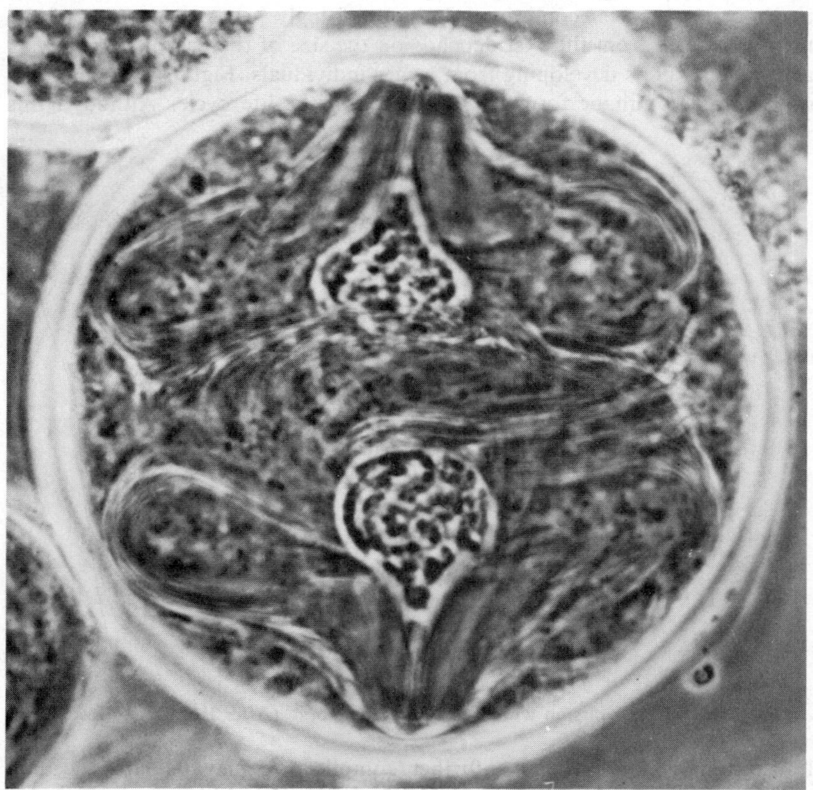

FIG. 10.7. A cyst of *Trichonympha* from the woodroach, ×1200. Two individuals are clearly visible within, the pair being formed by a precystic division of the encysting organism. (After Cleveland, *Proc. Am. Philos. Soc.*)

it occurs among the Vampyrellidae, proteomyxan organisms which are among the most beautiful denizens of the microscopic world. Among the ciliates, *Amphileptus* is an example of a form producing digestive cysts, sometimes dividing into a number of daughter cells afterward. *Polyspira*, an apostome ciliate parasitizing hermit crabs, does likewise.

The Sporozoa, which differ so much in so many other ways from the free-living Protozoa, are also quite different in the types of cysts they produce.

The majority are characterized by a sexual cycle culminating in the production of oocysts, which in turn often contain sporocysts. Within are the sporozoites, which are the infective stages; the formation and characteristics of such cysts will be discussed when the appropriate groups are considered. In still other cases, as in the blood-inhabiting species, the ability to form any resistant stage at all has been lost, probably because it is no longer needed, since the parasite is at no time outside a host.

Conditions Favoring Encystment

Just as there are many kinds of cysts it is likely that a variety of factors may induce encystment. In many species it seems to be a response to food deficiency. It was also noted long ago that certain free-living ciliates encysted when their cultures dried up. The increased concentration of dissolved salts may have been an indirect cause, especially as a loss of water by the encysting organism is known to occur. The absorption of water from the colon, with a resulting concentration of its contents, may similarly affect intestinal Protozoa, since cysts seldom occur in diarrheic stools though they may be very abundant in formed stools. Or perhaps in cases of diarrhea or dysentery trophozoites are swept out of the intestine too rapidly for encystment.

Although a lack of food and of moisture may be the most common causes of encystment there are other contributory ones, such as temperatures above or below optimum, a low pH, decreased oxygen supply, and the accumulation of waste products, all of which have been shown to be important under some conditions and for some species. Crowding is also often a factor, although it may exert its effect through the food deficiency and accumulation of metabolites which are likely to follow in its train. What the relative importance of these factors may be in inducing encystment in nature no one can say. The one thing which is certain is that they affect different species differently.

Intracystic Reorganization

In some ciliates and other Protozoa encystment may occur at certain periods in the life cycle, and more or less independently of environmental conditions. Such cysts are sometimes called "cyclic" cysts, and they are especially characteristic of parasites. Species such as the apostome ciliate *Gymnodinioides inkystans*, a parasite of the hermit crab, are good examples. Here encystment takes place on the exoskeleton or gills of the host after a brief free-swimming existence, and within the cyst then occurs a profound reorganization involving, among other things, an untwisting or detorsion of the organism so that it emerges as changed as an actor taking a second role in a play.

Reorganization within the cyst is not uncommon among ciliates, and the concurrent dedifferentiation followed (sometimes weeks or months later) by a redifferentiation is as complete as that of the developing butterfly chrysalis. There are few comparable instances among the Metazoa. Involved in the process in ciliates is often the disappearance and absorption of the macronucleus into the cytoplasm, while the micronucleus divides several times and most of the daughter nuclei also degenerate. Two may survive and unite, and from the resulting fertilization nucleus the normal nuclear complex is then built up. This process, known as "autogamy," was observed by the German biologist Fermor in the cyst of the hypotrichous ciliate *Stylonychia* as long ago as 1913. The absorption of macronuclear chromatin is believed to have the function of ridding the organism of the accumulated wastes of nuclear metabolism, and this process of purification is aided by the rejection of cytoplasmic excreta just prior to encystment.

While nothing comparable to this kind of intracyst reorganization is known to occur in flagellates and rhizopods, encystment among them may nevertheless be coupled with processes apparently sexual and which therefore involve nuclear change. The common shelled species *Euglypha scutigera*, studied by Penard, is an interesting example. Two individuals unite and give rise to a third very large binucleated one, which then encysts. Fusion of the two nuclei eventually occurs, and the resulting organism then forms a second cyst within the first. What happens then is not altogether certain. It is said that flagellated gametes arise from it and conjugate, with the production by the zygote of amoebulae. Unfortunately, although it is known that the formation of cysts (especially of the protective variety) is common among the Testacida, details of the process in the great majority of species remain unknown.

EXCYSTMENT

Whatever happens within the cyst, the individual leaving it is never the same as the one which entered it, sometimes, as we have seen, years before. Even though it may not have undergone nuclear reorganization it will have a brand new complement of organelles, and it may also be the product of several generations of reproduction. It is not too much to say that the metacystic organism has almost literally been "born again," and that it therefore has all the attributes of youth—renewed vigor, heightened metabolism, and usually a more rapid fission rate.

The actual process of excystment varies, but in only few cases has it been observed. Apparently the cyst wall sometimes has a minute opening through which the imprisoned organism can escape when conditions seem favorable, but this pore is usually invisible. Very likely it is produced by some enzyme secreted by the protozoan, and may therefore not be present at all until the very time of emergence. The approach of the time of release is heralded, particularly in the ciliates, by increased activity. The writer has often observed Pleurotrichas, imprisoned within their unyielding cyst walls, swimming actively about in their apparent attempt to escape.

Factors Governing Excystment

Just why excystment occurs when it does is still essentially a puzzle. It probably has little to do with the age of the cyst, except that enough time must elapse for whatever intracyst changes are typical of the species to occur, and the ability to excyst at all diminishes as the age of the cyst increases. In some cases cysts seem to be viable for only a few weeks, but in others, *Colpoda* for example, they may still be capable of excystment after many years, although the percentage viable lessens with the passage of time. This in turn certainly depends on environmental conditions—temperature, pH, probably the accompanying bacterial flora and its products, perhaps also on the fauna and its metabolic products. The concentration of various salts, largely determined by the degree of dessication, may also be important.

Low temperatures for most species are less damaging than higher ones. The writer has found many common free-living Protozoa in cultures made from ice collected outdoors in subzero weather, to which only distilled water was added. Some individuals, of course, may have survived freezing in the vegetative state, but most of them must either have emerged from cysts or were descended from such individuals.

The stimuli inducing excystment vary with the type of cyst and with the species. In general a favorable environment seems necessary. *Didinium nasutum* usually excysts readily in hay infusion, where under natural conditions its usual diet of paramecia is likely to be abundant. Jeffries (1956) found a very dilute lettuce extract the most effective excysting medium for *Pleurotricha lanceolata;* yeast extract and peptone in aqueous solution were less good. The amino acid glycine also worked well. Hydrogen ion concentration did not influence the percentage of encysting individuals particularly, but the higher the pH the slower was the process. Temperatures above or below about 20°C. also reduced the rate of excystment.

The ciliate *Woodruffia metabolica* produces both stable and unstable cysts, and occasionally intergrades. The former excyst readily in distilled water (in which the cysts of *Pleurotricha* refuse to emerge at all), while the latter come out freely in organic media, such as solutions of the dried water plant *Elodea* and alfalfa leaf meal.

Other species exhibit equal variety in the conditions they set for excystment. *Dileptus anser*, an interesting free-living ciliate easily identified by its long waving proboscis, excysts in 0.125 per cent wheat infusion. *Euplotes* excysts in tap water.

Little is known about the conditions required for excystment by parasitic Protozoa. *Gymnodinioides inkystans*, already mentioned, which parasitizes the hermit crab will only excyst when its host molts, which means that it must be sensitive to some substance produced by the host. Cysts of intestinal Protozoa are thought to excyst in the gut of the host, presumably due to the stimulation of digestive enzymes. For species living in birds and mammals, warmth is doubtless also a factor. The bacterial flora of the gut may also produce substances favoring excystment, as bacteria commonly used for food by some free-living ciliates are said to do.

The time required to emerge from the cyst of course varies with the species, and is known for only a few. Undoubtedly it also varies with the type of cyst. *Pleurotricha lanceolata* under the most favorable conditions requires five or six hours, as does *Entamoeba histolytica* when its cysts are placed in fresh culture medium at body temperature. Unstable cysts of *Woodruffia* will excyst under favorable conditions in little over an hour, or even less, but stable cysts may take as long as 12 or 15 hours.

Source of Energy

The question naturally arises "Where does the encysted animal get the considerable energy required for excystation?" The organism about to encyst may often be seen to have accumulated food reserves, such as amylaceous material, and masses of glycogen or other carbohydrate are also seen in cysts; such inclusions tend to disappear as the cyst ages, indicating utilization. The encysted organism is probably unable to absorb dissolved foodstuffs from its environment, since the material of which most cyst walls are made is nearly impermeable. Furthermore the cysts of many species live for months and even years under conditions where no absorption would be possible. *Oxytricha bifaria* conserves its stored food supply until the actual time of excystation, when it is rapidly consumed (Kay, 1945). Perhaps long-lived cysts use food more slowly than do those of the intestinal amoebae, in which the glycogen bodies rather rapidly disappear. There is little doubt that the encysted organism consumes its own substance if it remains within the cyst long enough, for excysted individuals may sometimes be induced to reencyst before they have time to ingest food; indeed this may happen with the same individual repeatedly. Under such circumstances the excysting organism is smaller each time it emerges.

It is also possible to measure the oxygen consumption of cysts by very delicate techniques recently developed. Such determinations show that respiration decreases with time in stepwise fashion in unstable cysts of *Colpoda cucullus*, remaining constant in the intervening intervals. For a cyst 26 days old, the oxygen used per hour amounted to 1 microliter $\times$ 10^5 (Pigon, 1959), although there was some variation; this is a rate many times lower than exhibited by vegetative individuals. It is therefore clear that even encysted organisms are burning an appreciable quantity of fuel, but the rate must approach zero in long-lived cysts with the passage of years.

In sum, although much has been learned about excystation in the Protozoa, many questions remain unanswered. Enough is known to make it certain that species differ greatly in the kinds of cysts they produce, and this may even be true of strains within a species, as in the case of *Colpoda maupasi*. Doubtless the explanation is a genetic one, yet this obviously does no more than state the nature of the problem. It is clear that conditions for both encystation and excystation also differ much for different species, but we can neither delineate them exactly nor explain just how they operate. Nor do we know just how commonly Protozoa encyst. Species within a genus may even show much variety in this regard, as in the genus *Euglena*. Marine Protozoa apparently seldom encyst (some species of Foraminiferida are apparently exceptions), perhaps because environmental conditions in the sea water in which they live change so little. Yet about this aspect of protozoology, as about so many others, our knowledge is still small. For the protozoologist with a mind for adventure into the microscopic world, much here remains to discover.

REFERENCES

Beers, C. D. 1927. Factors involved in the encystment of the ciliate *Didinium nasutum*. *J. Morphol. and Physiol.*, 43: 499–520.

——— 1927. The relation between hydrogen-ion concentration and encystment in *Didinium nasutum*. *J. Morphol. and Physiol.*, 44: 21–28.

——— 1930. Some effects of encystment in the ciliate *Didinium nasutum*. *J. Exp. Zool.*, 56: 193–208.

——— 1935. Structural changes during encystment and excystment in the ciliate *Didinum nasutum*. *Arch. f. Protistenk.*, 84: 133–55.

——— 1937. The viability of 10-year old cysts of *Didinium*. *Am. Nat.*, 71: 521–5.

Bridgman, A. J. 1957. Studies on dried cysts of *Tillina magna*. *J. Protozool.*, 4: 17–19.

Cleveland, L. R. and Sanders, E. P. 1930. Encystation, multiple fission without encystment, excystation, metacystic development, and variation in a pure line and nine strains of *Entamoeba histolytica*. *Arch. f. Protistenk.*, 70: 223-66.

Deflandre, G. 1951. Recherches sur les Ebriediens. Paleobiologie. Evolution. Systematique. *Bull. Biologique de la France et de la Belgique*, 85: 1–84, Fasc. 1.

Dobell, C. 1928. Researches on the intestinal Protozoa of monkeys and man. I: General Introduction; and II: Description of the whole life-history of *Entamoeba histolytica* in culture. *Parasitology*, 20: 357–412.

——— 1938. Researches on the intestinal Protozoa of monkeys and man. IX: The life-history of *Entamoeba coli*, with special reference to metacystic development. *Parasitology*, 30: 195–238.

Fermor, X. 1913. Die Bedeutung der Encystierung bei *Stylonychia pustulata*. *Ehrbg. Zool. Anz.*, 42: 380–84.

Garnjobst, L. 1937. A comparative study of protoplasmic reorganization in two hypotrichous ciliates, *Stylonethes sterki* and *Euplotes taylori*, with special reference to encystment. *Arch. f. Protistenk.*, 89: 317–81.

Gojdics, M. 1953. *The Genus Euglena*. Madison: University of Wisconsin Press.

Goodey, T. 1915. Remarkable retention of vitality of Protozoa in old stored soils. *Ann. Appl. Biol.*, 1: 395–99.

Grassé, P.-P. 1953. *Traité de Zoologie*. Vol. 1. Protozoaires; Rhizopodes, Actinopodes, Sporozoaires, Cnidosporidies. Paris: Masson.

Hausman, L. A. 1934. On the revivification of certain species of Protozoa after 20 years of encystment. *Am. Nat.*, 68: 456–62.

Ilowaisky, S. 1926. Material zum Studien der Cysten der Hypotrichen. *Arch. f. Protistenk.*, 54: 92–136.

Jeffries, W. B. 1956. Studies on excystment in the hypotrichous ciliate *Pleurotricha lanceolata*. *J. Protozool.*, 3: 136–44.

Johnson, W. H. and Evans, F. R. 1941. Dedifferentiation and redifferentiation of cilia in cysts of *Woodruffia metabolica*. *Trans. Am. Mic. Soc.*, 60: 7–16.

——— 1941. A further study of environmental factors affecting cystment in *Woodruffia metabolica*. *Physiol. Zool.*, 14: 227–37.

Jones, E. E. 1951. Encystment, excystment, and the nuclear cycle in the ciliate *Dileptus anser*. *J. Elisha Mitchell Sci. Soc.*, 67: 205–17.

Kay, M. W. 1945. Studies on *Oxytricha bifaria* Stokes. *Trans Am. Mic. Soc.*, 54: 267–82.

Kidder, G. W. and Stuart, C. A. 1939. Growth studies. II. The food factor in the growth, reproduction and encystment of *Colpoda*. *Physiol. Zool.*, 12: 341–47.

Kofoid, C. A., McNeil, E., and Kopac, M. J. 1931. Chemical nature of the cyst wall in certain human intestinal Protozoa. *Proc. Soc. Exp. Biol. and Med.*, 29: 100–02.

Lwoff, A. 1950. *Problems in the Morphogenesis of Ciliates*. New York: Wiley.

Manwell, R. D. 1928. Conjugation, division and encystment in *Pleurotricha lanceolata*. *Biol. Bull.*, 54: 417–63.

Mast, S. O. 1917. Conjugation and encystment in *Didinium nasutum* with special reference to their significance. *J. Exp. Zool.*, 23: 335–59.

Moore, E. L. 1924. Endomixis and encystment in *Spathidium spathula*. *J. Exp. Zool.*, 39: 317–37.

Padnos, M., Jakowska, S., and Nigrelli, R. F. 1954. Morphology and life history of *Colpoda maupasi*, Bensonhurst strain. *J. Protozool.*, 1: 131–39.

Penard, E. 1904. *Les Héliozoaires d'Eau Douce*. Geneva.

Penn, A. B. 1934. Factors which control encystment in *Pleurotricha lanceolata*. *Arch. f. Protistenk.*, 84: 1–32.

Pigon, A. 1959. Respiration of *Colpoda cucullus* during active life and encystment. *J. Protozool.*, 6: 303–08.

Strickland, A. G. R. and Haagen-Smit, A. J. 1948. Excystment of *Colpoda duodenaria*. *Science*, 107: 204–07.

Taylor, C. V. and Strickland, A. 1935. Some factors in the excystment of dried cysts of *Colpoda cucullus*. *Arch. f. Protistenk.*, 86: 181–90.

Tittler, I. A. 1935. Division, encystment, and endomixis in *Urostyla grandis*, with an account of an amicronucleate race. *La Cellule*, 44: 189–218.

11

Reproduction

"The observations caused me to view more nicely the lesser animalcules that were also swimming about in the water, and whereof the number was a good twenty times as many as that of the little animals aforesaid. Some of these animalcules were also coupled; and I saw that these likewise not only stayed long a-copulating, but also that one of the pair, whether it swam through the water or ran upon the glass, dragged the other forward, or trailed it after itself." Leeuwenhoek, Letter 71, March 7, 1692.

Reproduction is so universal an attribute of life that it is hard to imagine a time when almost nothing was known about its mechanism. Yet even today we have a great deal to learn not only about the mechanism of reproduction but about the forces that energize it and how they operate. If we completely understand reproduction in the Protozoa little of basic importance would remain for us to learn about the process in other animals, including man.

Essentially, reproduction is a process by which the cell produces a nearly exact replica of itself. Among metazoan organisms it is, therefore, the basis of growth as well as a method of multiplication. Sexual phenomena, although nearly universal in nature, are only a special phase of the reproductive process.

Reproduction is, of course, ultimately a process taking place within the protoplasm, involving the conversion of nonliving material into living matter of the unique sort of which a particular species is made. Individual genes act somehow as a template about which duplicate genes are created. So do other self-duplicating cell units, such as the mitochondria of all Protozoa, pyrenoids and chloroplasts when present, and the kinetosomes of ciliates and blepharoplasts of flagellates—the last two structures being probably homologous. Such replication, however, involves the formation of the ultimate building blocks of life, and while this is a matter of great interest to protozoologists, as to all biologists, it falls primarily in the realm of the biochemist.

These building blocks include desoxyribonucleic acid (DNA) and ribonucleic acid (RNA), and the proteins synthesized, as it is believed, with their aid. These substances are largely concentrated in the nuclei. Moses (1949) found the ratio of protein to RNA to DNA to be approximately 20: 2: 1 in both the micro- and macronuclei of *Paramecium*, and comments on the similarity of protozoan and metazoan nuclei in this regard. It is now thought that perhaps the DNA is the specific information-bearer for cellular syntheses, and that from it such information may be "translated into secondary templates of RNA" (Nanney, 1960). The RNA (most of it in ribosomes) seems to be concerned with protein synthesis.

Reproduction

Multiplication among the Protozoa occurs in a variety of ways, although they do not differ fundamentally. Most free-living Protozoa reproduce by simple binary fission, usually mitotic. Commonly recognized today, this fact was not always known. Leeuwenhoek observed the process but he did not appreciate its full significance, for he remarks only that one organism "dragged the other one forward, or trailed it after itself." The first fairly accurate account of division in the Protozoa was probably that of de Saussure, a physician of Geneva and a friend of Spallanzani. He is quoted by the latter (who had himself patiently watched the process of division in *Vorticella*) as saying in a letter: "You have good reason then, Monsieur, to think that the *Animalcules des Infusions* can, like polyps, multiply by continuous division and subdivision. You suggest this opinion only as a *doubt;* but the observations I have made on several species of these peculiar animals have convinced me that it can only be regarded as a *fact*. Such of these animals as are round or oval in form, without any proboscis or hook at the anterior end, divide in two transversely. They exhibit a constriction in the middle which increases little by little until the two halves are held together by a filament. Then the animal, or rather both animals, make great efforts to effect a separation and after their separation they remain several moments as if benumbed; but finally they begin running about here and there in the medium, just like the parent animal from which they were produced."*

Types of Division

The dividing forms de Saussure describes so accurately were certainly ciliates, since fission in this group is typically transverse, although the fission plane in *Vorticella* cuts the oral-aboral axis thus appearing to be longitudinal. *Cothurnia* belongs to this group (Fig. 11.1). Other exceptions also occur. In the Opalinas (although, as mentioned elsewhere, these may be more closely allied to the flagellates) division may be oblique. In still others, such as the stalked ectocommensals of the order Chonotrichida, budding may take place.

*Letter dated September 28, 1769.

In the flagellates division is usually longitudinal (Figs. 11.2 and 11.3) but there are exceptions to this rule too. Cleveland (1938) has described the life histories of two species of *Spirotrichonympha;* in one fission is longitudinal and in the other transverse, but here perhaps we should not be surprised, for the flagellate parasites of termites are almost unique in the protozoan world in many other ways. The dinoflagellates, perhaps because of the asymmetrical character of the thecae of many species, exhibit still another variant

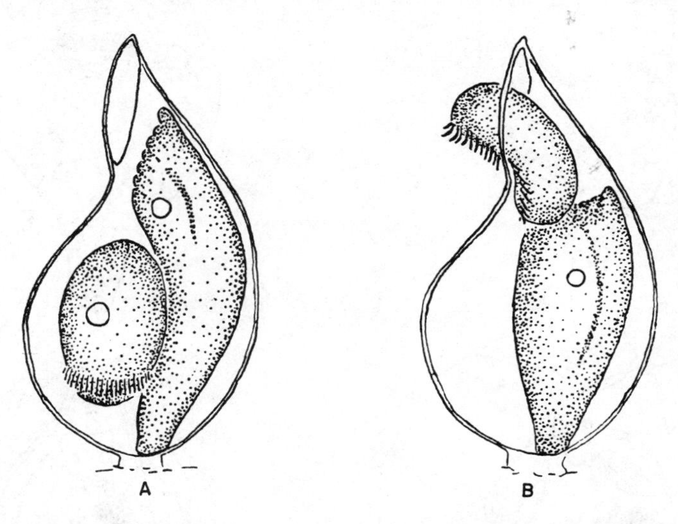

FIG. 11.1. *Cothurnia variabilis. A.* Division takes place within the lorica, and the daughter remains with the parent temporarily as shown. *B.* Here the daughter is leaving the parental lorica after its sibling (which inherits the lorica) contracts enough to permit egress. It then swims for 30 minutes or so, finally locating a suitable site on the crayfish gill and attaching itself. A stalk is secreted, then a lorica, and the cilia are lost. (After Hamilton, *Trans. Am. Micr. Soc.*)

of the process, for they often appear to divide obliquely (Fig. 11.4). Figure 11.5 shows the early stages of division in a *Noctiluca.* To add to the variety, something very close to multiple fission is frequently observed in trypanosomes (e.g., *Trypanosoma lewisi*, a species common in rats), where repeated nuclear division of the blood form occurs without the usual cytoplasmic fission, so that when the latter finally takes place a number of daughter cells may be formed simultaneously. During its intracellular phase in the gut of the rat flea, this trypanosome multiplies in yet another fashion, undergoing repeated division which culminates in the production of a number of small daughter cells, all remaining enclosed for a time in the covering membrane of the parent. Thus we have what amounts to a peculiar type of budding. The more usual type of division in trypanosomes is shown in Figure 11.6.

We cannot say in what plane division occurs among the naked Sarcodina

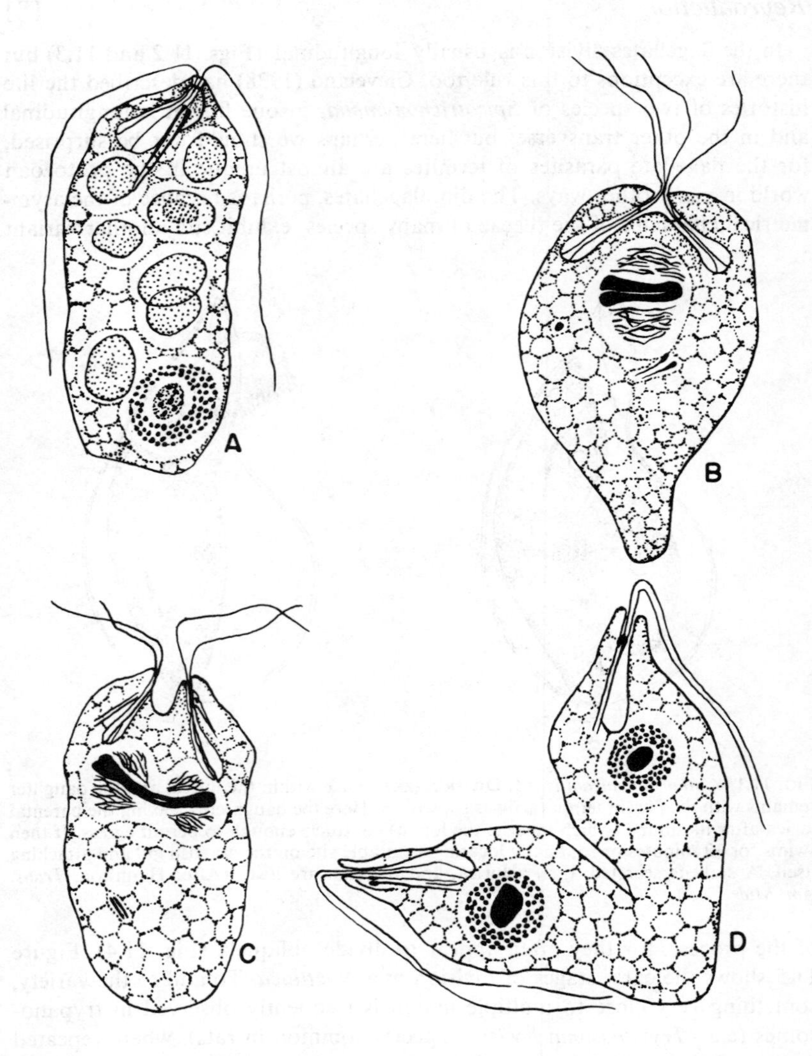

FIG. 11.2. Division in *Heteronema acus*, a euglenoid flagellate. *A*. Normal organism in the resting stage (interphase). There are numerous food vacuoles. *B*. The chromosomes are drawing apart and the elongated endosome has split in two. *C*. Organism in anaphase; cytoplasmic division started. Old pharyngeal apparatus nearly resorbed (it lies posterior to nucleus), and new one appearing close to neck of reservoir. *D*. Late telophase; daughter nuclei spherical and cytoplasmic division nearly complete. (After Loefer.)

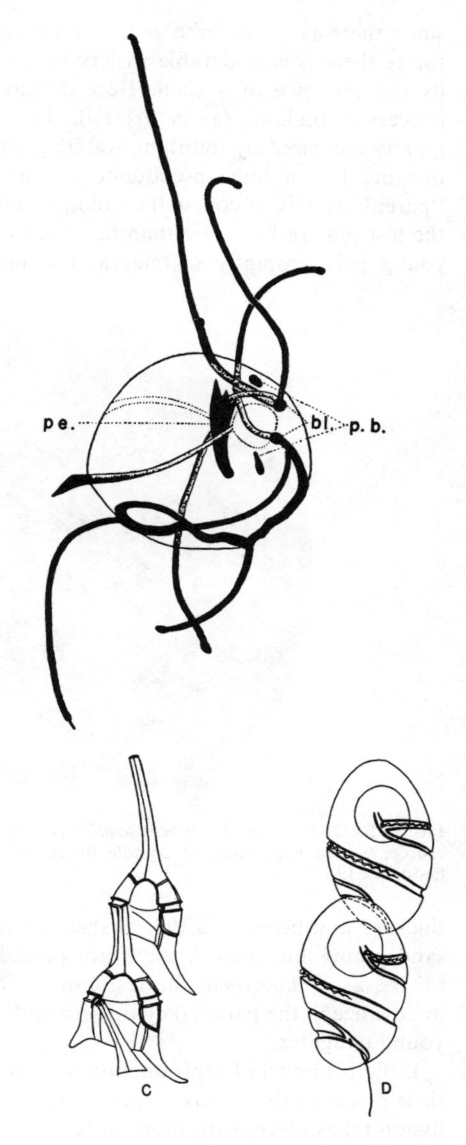

FIG. 11.3. An early prophase in the division of *Trichomonas hominis*. Two blepharoplast complexes are shown, from which arise two sets of mastigont structures; with each of the latter a parabasal body is associated. The original pelta is displaced and in disorganization; new ones have not yet appeared. bl., blepharoplast-complex; p.b., parabasal body; pe., pelta. (After Kirby, *J. Parasitol.*)

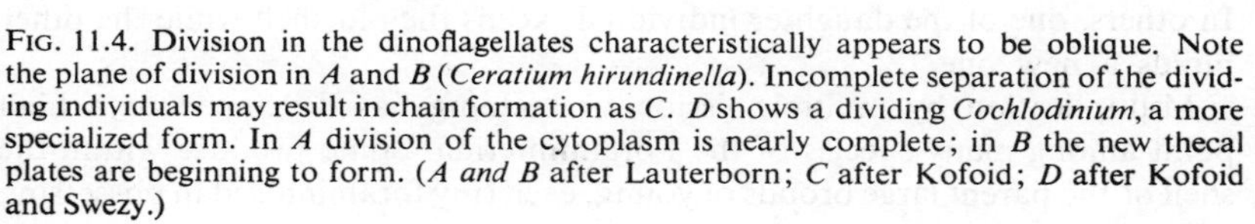

FIG. 11.4. Division in the dinoflagellates characteristically appears to be oblique. Note the plane of division in *A* and *B* (*Ceratium hirundinella*). Incomplete separation of the dividing individuals may result in chain formation as *C*. *D* shows a dividing *Cochlodinium*, a more specialized form. In *A* division of the cytoplasm is nearly complete; in *B* the new thecal plates are beginning to form. (*A and B* after Lauterborn; *C* after Kofoid; *D* after Kofoid and Swezy.)

since there is no definite point of reference, and even among the testate forms there is considerable variety because of the complications introduced by the presence of a shell. Here division often seems to be essentially a process of budding (as in *Arcella*), the protoplasm (whose volume may be greatly increased by imbibing water) gradually protruding through the shell opening like a huge pseudopodium, and eventually separating from the "parent" (really of course the sibling) after nuclear mitosis. The latter retains the test plus rather more than half the cytoplasm. The other member of the young pair promptly secretes a new membrane about itself which soon

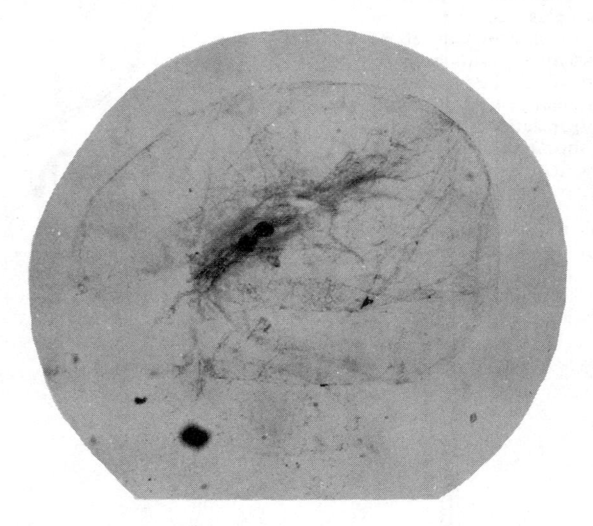

FIG. 11.5. Division in *Noctiluca scintillans*. What is apparently an anaphase is clearly visible, with a suggestion of spindle fibers. (This species also reproduces by multiple fission.) $\times 123$.

thickens and becomes a typical shell. In some species foreign bodies such as sand grains and diatom shells are picked up and fastened to the shell. In others, as in *Euglypha*, the test consists of minute siliceous plates secreted in advance in the parental cytoplasm and presented as a kind of dowry to the young daughter.

Little is known of reproduction among the Radiolarida, probably because their pelagic habitat makes them hard to obtain for study. In certain species fission takes place, with more or less equal division of the skeletal elements. In others, one of the daughter individuals keeps the old shell while the other builds a new one.

Multiplication by multiple fission or budding has developed to a high point among many species of the Foraminiferida. Some produce within the shell of the parent large broods of young, each tiny foraminiferidan possessing its own little shell of one or more chambers. Here the process of reproduction is essentially schizogony. At other stages of the life cycle gametes may be produced, sometimes in very large numbers.

Most Sporozoa, a group composed entirely of parasites, also reproduce by multiple fission, or schizogony. In this type of reproduction nuclear division occurs repeatedly until finally the cytoplasm follows suit, each bit of chromatin being, as it were, appropriated by a bit of cytoplasm (hence the name "schizogony," from the Greek *schizein* plus *gonos*, "to cut" and

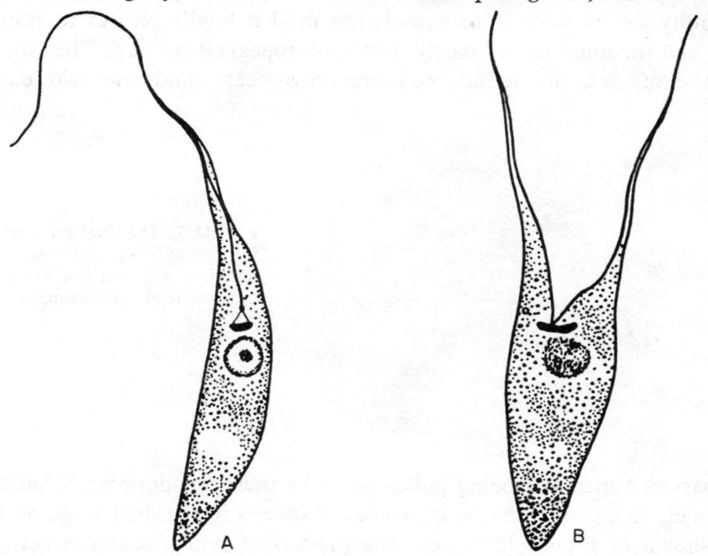

FIG. 11.6. Division in *Trypanosoma cruzi*. *A*. Elongated kinetoplast with its two connecting fibrils and the blepharoplast at the base of the flagellum. *B*. Elongation of the kinetoplast is evident, and the cytoplasm has started to divide rather precociously. (After E. R. Noble, *Quart. Rev. Biol.*)

"reproduction"). This type of multiplication may have become universal among the Sporozoa because the hazards of parasitism have placed a premium on numbers. Details of sporozoan life cycles vary greatly, but in most instances there are two distinct phases, one asexual and the other sexual. Reproduction in the latter is known as "gamogony" since it culminates in the production of gametes. The malaria organisms are good examples of this type of life cycle (Chap. 24). They develop asexually in the red cells of the host, certain of the young forms in every generation becoming sexual stages or "gametocytes." These mature in the mosquito, and the gametes of the two sexes fuse, the resulting zygote undergoing reproduction, or "sporogony," in the form of the oocyst. Sporozoites are finally formed, and these when injected at the time of the mosquito's bite multiply in the vertebrate host by schizogony.

Division in the Living Animal

In nonparasitic Protozoa, the first easily visible indication of division in the living organism is usually the appearance of a cytoplasmic constriction, longitudinal in flagellates and transverse in ciliates. *Amoeba proteus* withdraws its larger pseudopodia in preparation for division, and assumes a shape resembling a chestnut burr studded with very short, blunt spines—minute pseudopodia. As the time for actual fission approaches, the amoeba becomes steadily less responsive to stimulation until it finally refuses to respond at all, and the nucleus, ordinarily visible, disappears from view. Then the organism elongates, the telltale constriction appears, and the two daughters

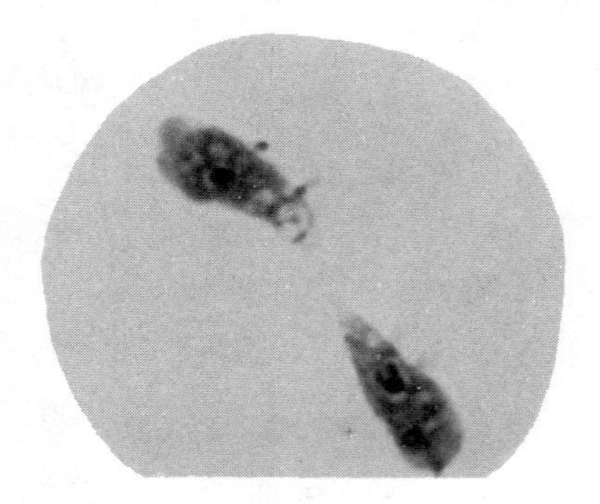

FIG. 11.7. The final stage of division in a free-living amoeba. The two daughter amoeba are still united by a filament of cytoplasm. $\times 884$.

separate, apparently being pulled apart by their pseudopodia (Chalkley and Daniel, 1933). An amoeba of a smaller species in the final stage of division is shown in Figure 11.7. But the process of reproduction is actually far advanced before all this happens, although the earliest premonitory changes are ordinarily detectable only in carefully stained preparations. Closer observation under the microscope may show a new flagellum developing for one of the daughter cells if the animal is a flagellate, or a new mouth for the posterior of two ciliate progeny. The appearance of such a cytostome with its associated structures is often an early sign of approaching division among the ciliates. The fate of the parental mouth and its organelles varies with the species; in some it appears to be retained by the anterior of the offspring, and in others it is resorbed while its successor is developing. Cirri are likewise resorbed during division of the hypotrichous ciliates (e.g., *Pleurotricha*, *Euplotes*), so that each daughter organism begins life with a new set. Whether this is also true of cilia generally is a moot question. It does seem to be true of those skeletal elements known as trichites, which in many ciliates constitute the so-called pharyngeal basket, as in *Chilodonella* (Fig. 11.8).

Nature, however, does not always endow the young protozoan with a completely new set of organelles. Although the old flagella may be resorbed and replaced, in many species they are inherited by one of the progeny. Or when there are an even number of flagella, they may be equally shared, each individual developing new ones to restore the count to normal. So too, the stigma may either divide or it may go to one of the daughter cells, a new one being formed by the other. Among the testate rhizopods, as in the case of *Arcella*, one of the offspring must often make do with the old shell. In forms with contractile vacuoles, the progeny may share them equally when more than one is present, or new ones may appear, as in *Amoeba proteus*, the old vacuole ceasing to beat with the onset of division and then vanishing. Trichocysts and food vacuoles may be shared, if not always equally.

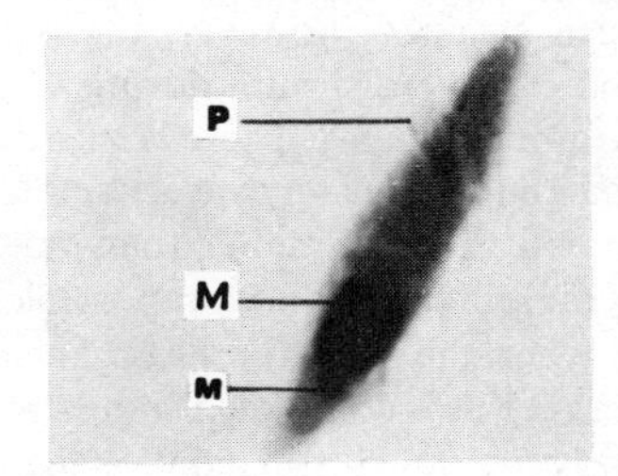

FIG. 11.8. Pharyngeal basket of *Chilodonella* sp. ×1040. P, pharyngeal basket consisting of trichites, the number of which varies with the species. The basket is apparently somewhat protrusible. M, macronucleus. m, micronucleus. The organism lies on its side.

Notwithstanding such exceptions, the young protozoan is on the whole a completely new creature, physiologically as well as morphologically, a point the great American protozoologist, the late Gary N. Calkins, was fond of emphasizing. That such is the case is attested by a higher metabolic and growth rate, just as in higher animals.

What Triggers Division?

What factors are responsible for initiating division? This is a difficult question to answer, although of course the influence of environmental conditions such as temperature, food availability, crowding, and the like is well known. Both in nature and in the laboratory, protozoan multiplication is often cyclical, apparently partly because of the cyclical recurrence of certain environmental factors such as light, temperature, etc. Thus, Beers (1948) observed that ciliates of the sea urchin exhibit "short periods of intense divisional activity alternate with long periods of nondivisional life" and the green species of euglenid flagellates have been found by Leedale (1959) to divide under natural conditions only in darkness. Doubtless they are too busy during the daylight hours in building up food reserves through photosynthesis to expend energy in reproduction. It has recently become possible to synchronize the division of *Tetrahymena* and *Amoeba proteus*, and such near-synchroneous multiplication of parasitic Protozoa within the host is often observed, as in the case of many species of malaria. Here the explanation seems to be a combination of genetic constitution of the parasite and cyclical physiological processes of the host. It has been suggested that fission is triggered by an imbalance between nuclear and cytoplasmic mass, or between volume of the organism and its surface, since the former is a measure of metabolism and the latter of the availability of food, oxygen, etc. Nanney (1953) remarks (though he had the nuclear changes of conjugation chiefly in mind), ". . . it is clear that the cytoplasm plays a critical role in directing the activities of the nuclei." More recently the possible role of a "critical size" has been urged, since it has been observed that a doubling of size always occurs between divisions. It was shown by Hartmann in 1928 that division in *Amoeba proteus* could be indefinitely delayed simply by periodically lopping off bits of cytoplasm. Regeneration occurred regularly, but as long as the critical size was not reached no division occurred. Even when this size has been attained, however, the onset of fission may occur only after a lag period of several hours during which the organism is doubtless busy with preparatory changes. There comes a time when these are so far advanced that the amoeba divides anyway, whether robbed of more cytoplasm or not. The daughters, though smaller than normal, grow to the usual size before themselves dividing. Still another possible trigger has been suggested: that the stage is set when the DNA of the organism has been finally doubled, but there is evidence that this is completed some time before division actually occurs.

Apparently we must still confess ignorance of the answer to this vital problem which concerns the human organism as much as the Protozoa. Adolph's comment in 1931 is still relevant, "On the other hand, reproduction is never regulated wholly by growth," even though for "*Stentor* and *Amoeba*, the conclusion may be drawn that the attainment of a certain size is necessary for reproduction."

Mitosis

The nuclear changes in protozoan mitosis are essentially like those seen in cells of higher animals and plants although there is much variety in details of the process. Among the higher animals and plants, an early indication of cell division is the migration of the extranuclear centrioles, or division centers, toward the poles of the cell, where they become centers for the asters of the developing spindle. A few Protozoa, such as certain Heliozoida and the remarkable symbiotic flagellates of the wood-feeding roach *Cryptocercus* and of termites, may show similar changes, but usually mitotic figures lacking asters are formed, although even here there may be a division center within the nucleus. The nuclear membrane in the dividing metazoan cell disappears, but in most Protozoa it persists during mitosis. Certain of the termite flagellates which have been studied most intensively by Cleveland (1949; 1953) are partial exceptions to the rule, since the nuclear membrane in them persists to a degree. What energizes the division center, or its equivalent in the cell, remains a mystery. Cleveland has shown that the whole process of mitosis, and even the behavior of the chromosomes, in the flagellates of *Cryptocercus* is under the direct control of hormones produced by the host, but the precise nature of such control is still unknown.

Some biologists formerly questioned whether organisms as low in the evolutionary scale as Protozoa were assumed to be could possess true chromosomes, comparable to those in Metazoa and in the higher plants. Such doubts are no longer entertained, for in all protozoan species thoroughly studied mitotic division has been found to occur, and chromosome numbers varying from two to several hundred have been counted. Nevertheless, mitotic division among the Protozoa is much more varied in form than in multicellular organisms. Attempts have been made to classify it, the simplest being called "promitosis," the more complex "mesomitosis," and the most advanced "metamitosis." Bělař (1926), who made a thorough study of the protozoan nucleus, suggested that two types only be recognized, the simpler to be called "paramitosis," and the more complex "eumitosis." A better scheme is that of Grassé (1952), who proposes two types also: "orthomitosis," in which there is a clearly marked metaphase, and "pleuromitosis," where the metaphase is lacking. There are differences, too, in the manner of insertion of the chromosomes.

The most thorough studies of mitosis in the Protozoa have been those of Cleveland (1949; 1953), who found the flagellate parasites of termites uniquely favorable material. Termites are widely distributed and thus easily available, and they are always infected, furnishing a kind of living culture tube. Nor is there a problem of host maintenance, since termites consider themselves well fed if only they are supplied with an ample diet of wood. Nuclear division here is pleuromitosis.

When the cytology of these remarkable Protozoa was studied, species of certain genera were found to have few and relatively large chromosomes. Chromosome behavior in division was therefore easy to follow, and—almost incredible good fortune—with proper illumination and suitable phase-contrast microscope equipment, chromosomes could be seen in the living organism. Several species of *Holomastigotoides* and *Barbulanympha* were chiefly used; *H. tusitala*, one race of which in the haploid state has only two chromosomes, proved to be the best. Cleveland did not rest with detailed and prolonged studies on these species; he states that he critically examined more than 500 species of flagellates, and that chromosome behavior is essentially alike in all.

Cleveland was able to observe chromosome pairing, segregation, crossing over, and the occurrence of spontaneous aberrations of various kinds, many of them lethal to the organisms. He also made interesting observations on the relation between chromosome behavior and that of the spiral flagellar bands which characterize these Protozoa. *Holmastigotoides tusitala* (Fig. 11.9) possesses five such bands, each spiraling about its body almost to the posterior end, and bearing flagella which seem to cover the body surface almost as if they were cilia. A further peculiarity of this species (and of some others) is its habit of transverse division, the posterior daughter being much the smaller, and receiving but one of the five flagellar bands; to insure this the band unwinds and moves posteriorly before division. The missing four bands are later regenerated, while the anterior daughter in like fashion makes good its donation to its sister. The centrioles play a prominent role in all this.

The various stages of mitotic division in this species are shown in Figure 11.10. The two chromosomes (the haploid condition is normal for this organism) appear as somewhat banana-shaped rods overshadowing the spindle itself with its centrioles and delicate fibrils. Although its general

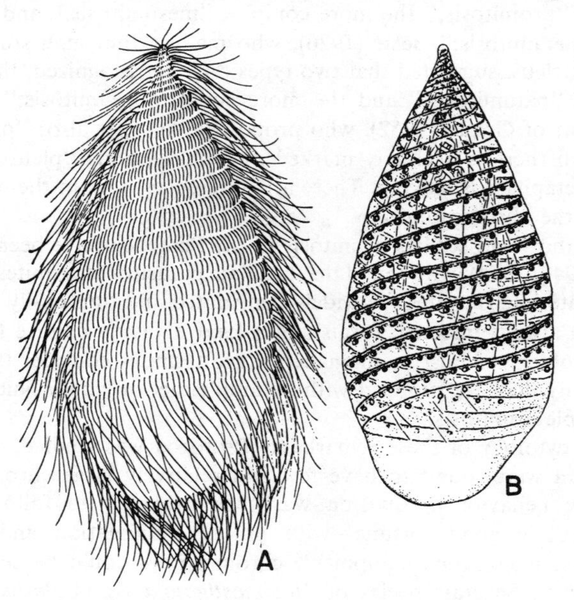

FIG. 11.9. *Holomastigotoides tusitala* of the termite *Prorhinotermes*. *A*. The living organism with its covering of flagella, those at the sides and posterior end being shown full length. *B*. The flagellar bands, with parabasal bodies regularly spaced along their courses, the long, thin axostyles extending anterio-posteriorly for nearly the whole body length. (After Cleveland, *Trans. Am. Philos. Soc.*)

appearance is strikingly different from the conventional schematic illustrations of mitotic spindles, the essential nuclear structures are all present. The linear replication of chromatinic material is clearly shown in Figure 11.10*i*. Although transverse duplication of chromosomes has been described in a few Protozoa it is no longer believed to occur. One of the ways in which chromosome behavior does differ, however, is the time in the life cycle of the organism when "rest" occurs. This Cleveland found to be ordinarily in the prophase, but in some genera it occurs in the telophase.

In commenting on the chromosomes themselves (Fig. 11.11), Cleveland remarks, "The basic pattern of chromosome morphology was laid down long ago in unicellular forms and there has not been much evolution in it as higher forms have developed. Rather, the evolution has been in the content—the chemistry—of the chromosomes." It is a rather intriguing thought that somewhere in our own 46 chromosomes there may still be a little DNA in the molecular image of our remote unicellular ancestors.

Mitosis in termite flagellates is of Grassé's pleuromitosis type, but the majority of Protozoa exhibit orthomitosis. The latter may, however, show many variations and Grassé lists nine major types, though he concedes

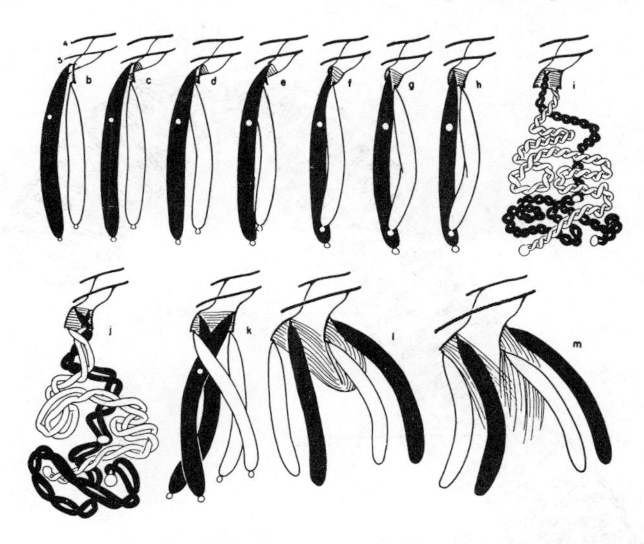

FIG. 11.10. Mitosis in the termite flagellate *Holomastigotoides tusitala*. *b–h*, telophase; *i–j*, prophase; *k*, metaphase; *l*, anaphase; *m*, telophase. Brief diagrammatic life cycle of centrioles, centromeres, chromosomes, and achromatic figure: *b–c*, new and old centrioles forming achromatic figure; *d*, one chromosome has shifted its connection from old to new centriole; *e–f*, flattening out of centrioles and achromatic figure; *g–h*, beginning of chromosomal twisting; *i*, chromosomes duplicated, producing many gyres (spiral circlets) of close-together relational coiling of chromatids. Centromeres duplicated; one in each group moves; other remains stationary. *j*, later stage in chromatids losing their relational coiling by unwinding. *k*, relational coiling practically gone, achromatic figure elongating and separating sister chromatids. *l*, central spindle bent, chromatids in two nonsister groups. *m*, central spindle pulled apart. (After Cleveland, *Trans. Am. Philos. Soc.*)

doubt about just where the Sporozoa should be placed. The matter is further complicated by the occurrence, in some species, of mitoses varying in a characteristic fashion at different stages in the life cycle. A few examples of species with different types of mitosis are discussed below.

On the whole, the mitotic process appears to be simplest in some amoebae and most highly developed in the ciliates and some Heliozoida. In the latter there may be extranuclear centrioles and aster formation just as in metazoan cells. Doubtless the mitotic machinery was evolved very early in the history of living things, since its occurrence is so nearly universal. The marvel is perhaps that it varies so little and not that it varies so much.

Some appreciation of the diversity in form which nevertheless exists in mitosis may be gained from illustrations showing the process in *Endolimax nana* and *Dientamoeba fragilis* (both small amoebae), *Euglena spirogyra* (a green flagellate), *Oxnerella* (a heliozoidan), and *Pleurotricha lanceolata* (a ciliate). Of the five examples, the two amoebae seem to exhibit a simpler form of division than the others. *Dientamoeba fragilis* (though its status as

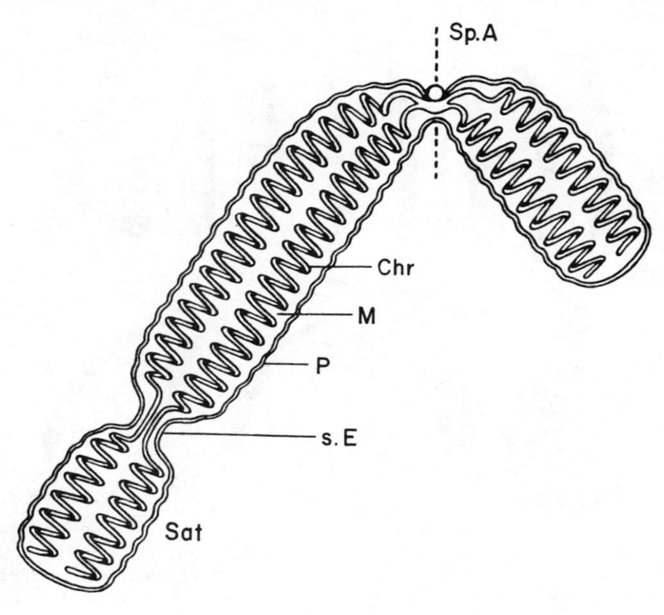

FIG. 11.11. The hypothetical structure of a chromosome. Sp.a., point of attachment of spindle ("kinetochor"); Chr., chromonema; M, matrix; P, pellicle; s. E., secondary constriction; Sat., satellite. (After Grell.)

an amoeba has been questioned by Dobell, 1940, who thought it perhaps a flagellate with an amoeboid phase) shows an especially simple spindle with centrioles and a connecting thread (centrodesmus), but without a clearly defined equatorial plate (Fig. 11.12). It apparently possesses six chromosomes, and these remain visible in the "resting nuclei," the binucleate condition peculiar to this species and probably representing an arrested division stage. The spindle developed by *Endolimax nana* is better defined, with a sharply marked equatorial plate and ten chromosomes (Fig. 11.13). As in the dividing nucleus of *Dientamoeba fragilis*, there are intranuclear centrioles and a centrodesmus. The entamoebae constitute one of Grassé's nine types of orthomitosis.

The division process in *Euglena spirogyra* is more complicated (Fig. 11.14). Here (Ratcliffe, 1927) the chromatin of the resting nucleus is arranged in paired strands of chromomeres. Within the nucleus is an endosome and also an "intranuclear body," the splitting of which is an early sign of fission. The endosome then divides, the nucleus constricts, but without forming any definite plate, and the two halves separate. Longitudinal fission of the body follows. Whether those euglenids exhibiting a sexual phase in their life

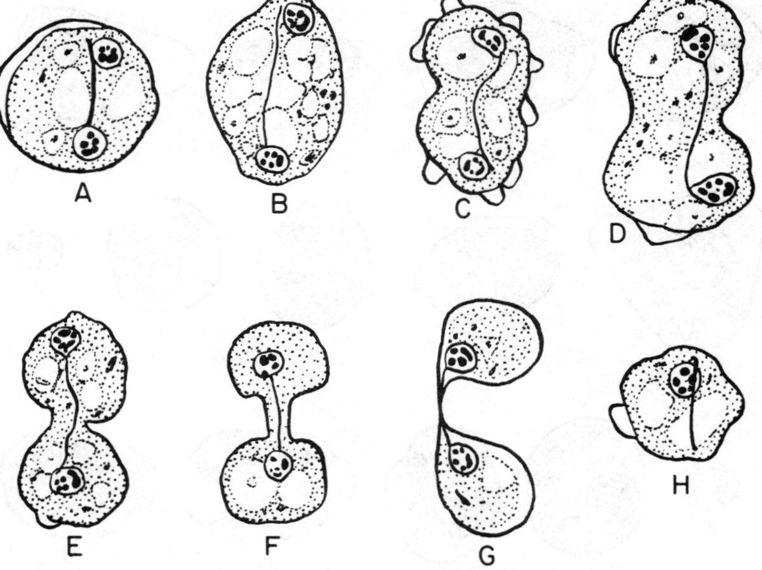

FIG. 11.12. Division in *Dientamoeba fragilis*. The normal binucleate forms of this species are considered to be in an arrested telophase stage, with chromosomes (six) still evident but no centrodesmus. When the division process is finally resumed a connecting thread, or centrodesmus, appears between the two nuclei and the stages shown are passed through. Uninucleate forms (*H*) constitute about 20 per cent of the normal population. (After Dobell, *Parasitol.*)

cycles resemble the phytomonads in being haploid while multiplying asexually is not known. Cytoplasmic elements such as the stigma, plastids, and kinetic apparatus also divide. New flagella are developed from the daughter blepharoplasts, according to the most recent accounts.

Oxnerella maritima (Fig. 11.15) is an example of a protozoan exhibiting what we may call typical mitosis ("metamitosis" according to the older usage). Dobell (1917), who first studied and described the species, called it "a minute and pretty" marine heliozoidan that crawls about the sea bottom or wanders around adhering to the surface film like a fly on a ceiling. When it is about to divide, the pseudopodia are withdrawn and they remain so until the process is complete. Mitosis in this and some other species of Protozoa (as pointed out by Schaudinn) involves as a preliminary step the division of the centriole, the halves of which then serve as focal points of the developing asters. Dobell commented on the similarity of behavior of the chromosomes (which numbered about 24) to those of metazoan cells, but he also remarked that the centriole, despite the apparent similarity of its role in division to that played by metazoan centrioles, could hardly be regarded as a homologue of the latter; rather, its primary function appeared to be serving as a center

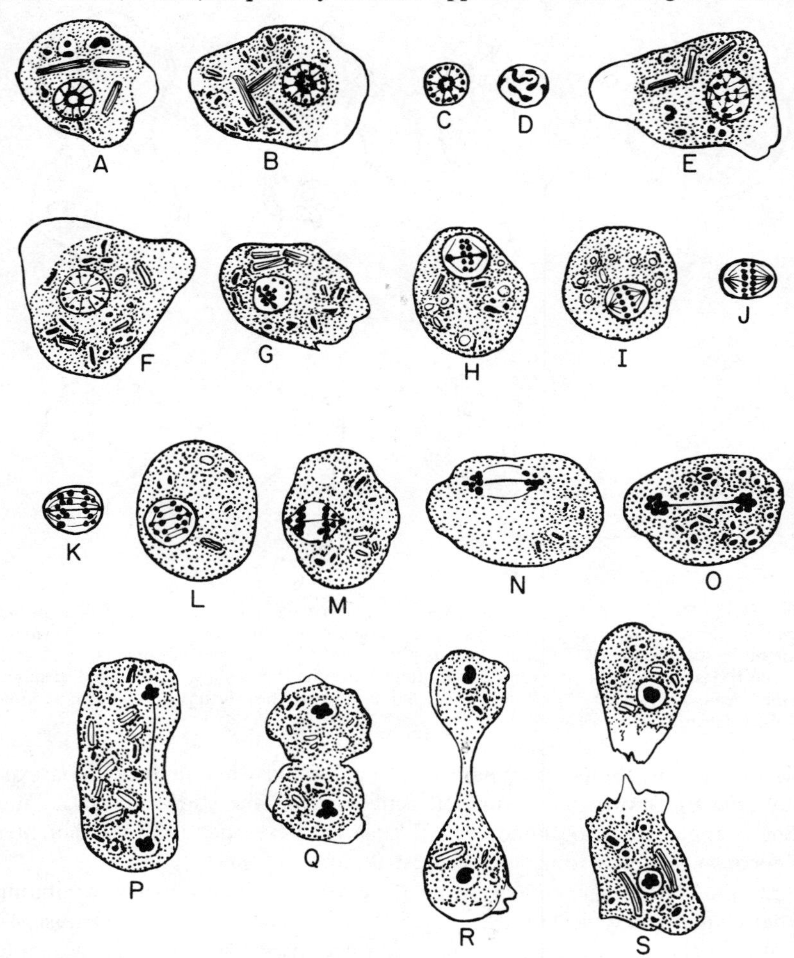

FIG. 11.13. Division of the human intestinal amoeba *Endolimax nana*. *A*, *B*, prophases; *C*, *D*, nuclei shown in prophase; *E–G*, later prophases; *H*, *I*, spindles showing equatorial plate; *J*, nucleus only, showing metaphase; *K*, nucleus only, showing early anaphase; *L–N*, various stages of anaphase; *O*, early telophase; *P–S*, various stages of telophase and final separation of daughter organisms. (After Dobell, *Parasitol.*)

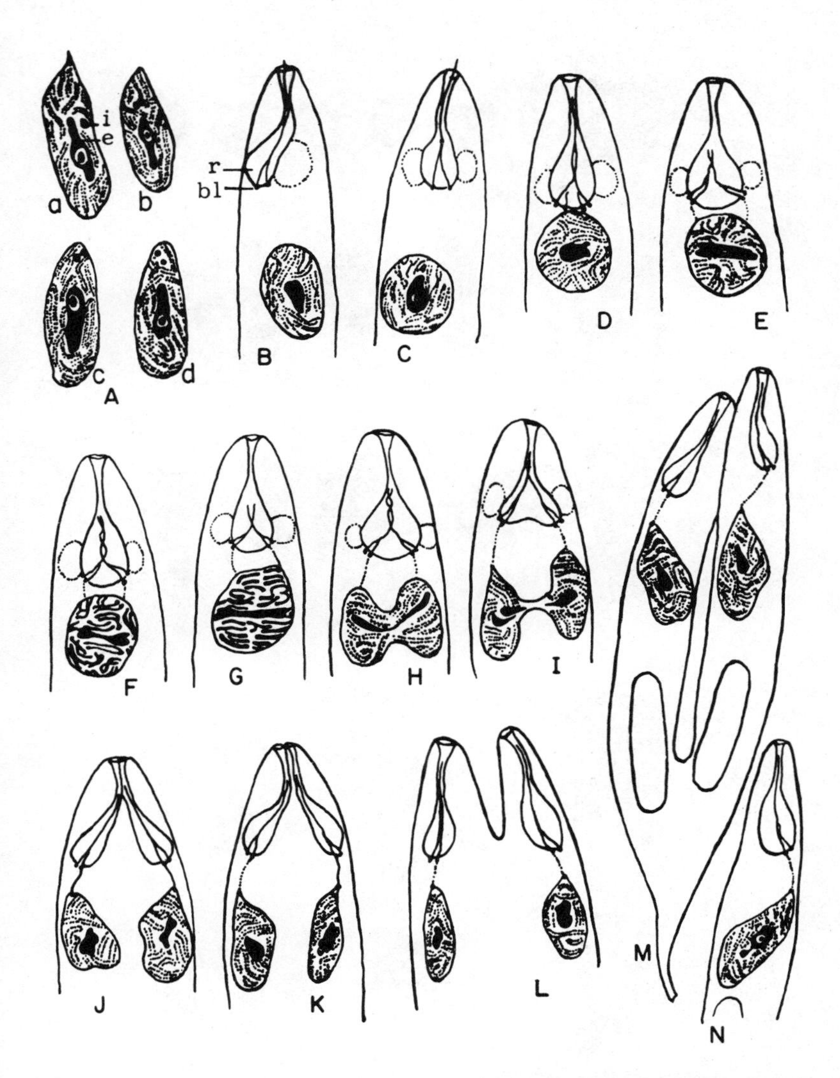

FIG. 11.14. Division in *Euglena spirogyra*. *A*. a, nucleus of vegetative form, with chromatin in form of paired rows of chromomeres; b, c, d, early prophase stages showing movement and division of intranuclear body (i). *B*, *C*, *D*. Chromatin threads shorten and thicken and the blepharoplasts (bl) divide; *E*, *F*. Endosome (e) elongates, and is composed of two parts; chromosomes lie in pairs; endosome continues to elongate and appears to consist of three parts. *G*. Metaphase; chromosome pairs separate. *H*. Anaphase; chromosomes divide longitudinally. *I*. Telophase; endosome completes division. *J*. Daughter nuclei separated; reservoir (r) completes division. *K–N*. Final stages of division, with formation of daughter organisms. (After Ratcliffe, *Biol. Bull.*)

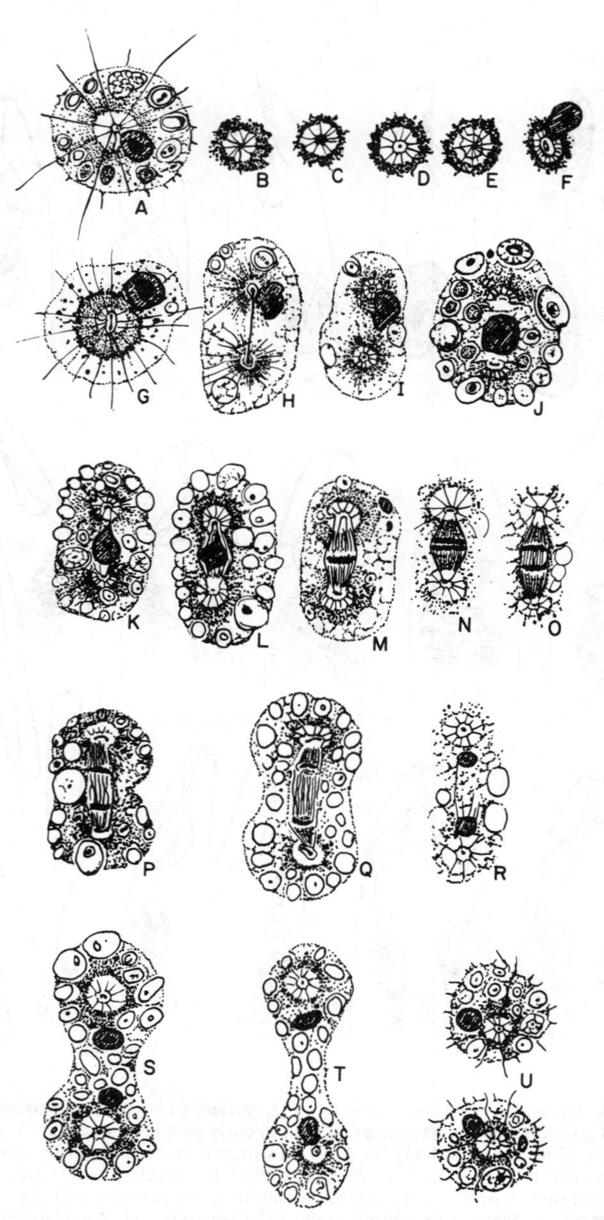

FIG. 11.15. Division in the marine heliozoan *Oxnerella maritima*. *A*, stained individual in resting phase; *B–E*, central area showing various aspects of the centroplast; *F*, central area and nucleus of an organism in an early stage of division, as indicated by centroplast. Note attachment of nucleus to it by three axopodial rays; *G*, *H*, division of centroplast; *I–L*, prophases; *M*, equatorial plate; *N*, metaphase; *O–Q*, anaphases; *R–T*, telophases; *U*, division completed. (After Dobell, *Quart. J. Micr. Soc.*)

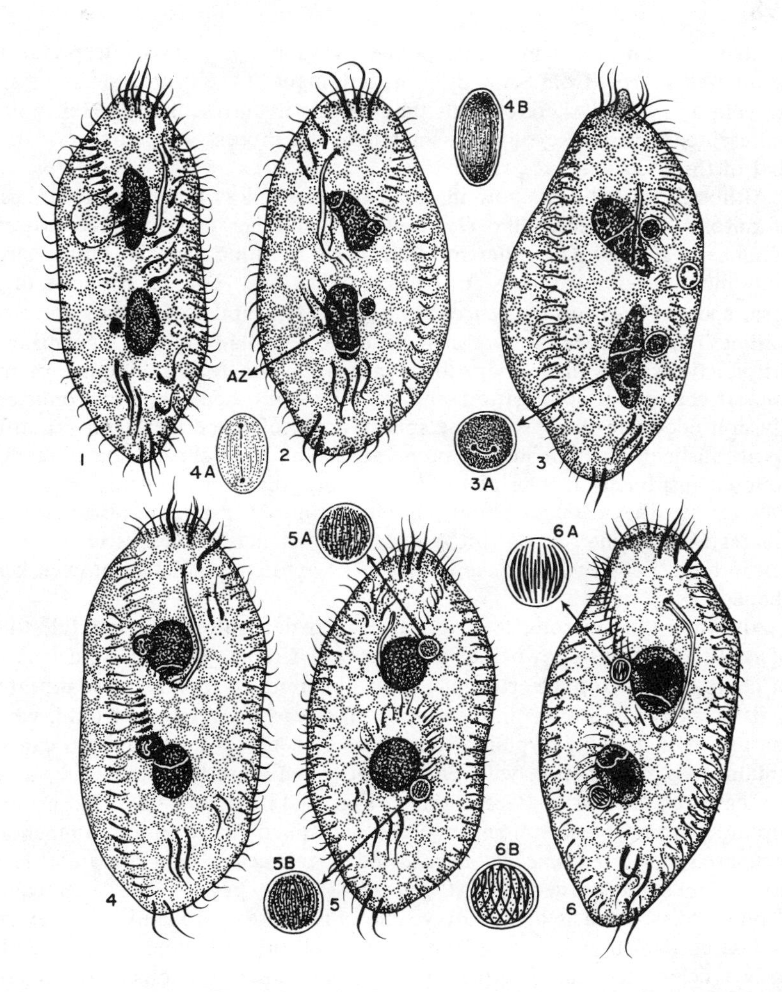

FIG. 11.16. Early stages of division in *Pleurotricha lanceolata*. *1.* A normal individual. *2.* Note the kernspalt (transverse split) in each macronucleus; a new adoral zone (AZ) is also beginning to appear in the region of the posterior macronucleus. *3.* The micronuclei show a division center, or endosome, in process of division, the halves being connected by a minute fibril; shown enlarged in 3A. *4.* A more advanced stage; kernspalts have moved nearly the length of the macronuclei, and the division centers in the micronuclei are more apparent; one of the micronuclei is shown enlarged in 4A; a somewhat later stage is shown in 4B. *5.* The chromatin in the micronuclei has condensed into definite threads; note the change in shape of the macronuclei; new cirri are beginning to appear in the neighborhood of the adoral zones; 5A and 5B show the micronuclei enlarged. *6.* A still later stage in division; note that in this individual the kernspalts are still visible; the events in micronuclei, macronuclei, and cytoplasm are not always synchronized in exactly the same fashion in different organisms; A definite spindle is now visible in the anterior micronucleus; both are shown enlarged (6A and 6B). (×275 in the larger illustrations; ×850 in the lettered illustrations. After Manwell, *Biol. Bull.*)

for the radiating axial filaments of the pseudopods. In this interpretation he differed sharply from Schaudinn, who thought of the Protozoa as "primitive animals," and therefore felt that mitosis in forms like the Heliozoida represented a further evolution, since here the process seemed to resemble that of the Metazoa.

Although the ciliates show in some respects a less highly evolved form of mitosis than species like *Oxnerella*, the number of guises the process assumes is greater. The differentiation of chromatinic material into macro- and micronuclei introduces further complications. The micronuclei of a form such as *Pleurotricha lanceolata* (Figs. 11.16 and 11.17) may develop spindles (if we include those of meiosis) of perhaps half a dozen recognizably distinct types: one in ordinary division, several others in the course of the nuclear changes of conjugation, and possibly still others when micronuclear division occurs in the cyst. These spindles may often be seen with beautiful clarity in haemotoxylin-stained preparations, and the chromosomes readily counted (especially in one of the two meiotic divisions); in this species of *Pleurotricha* they number about 40. The form of the spindle also varies in ciliates of different species: in *Spirostomum* it tends to be crescent-shaped, but in hypotrichs such as *Pleurotricha* it is typically spindlelike or even keg-shaped.

Although the macronucleus appears to divide amitotically, careful study of its behavior has shown that it is not that simple. The macronucleus is, of course, of micronuclear origin; what apparently happens is the repeated division of chromosomes in the course of its early development until, when mature, it is highly polyploid. Sometimes evidence of such division can be obtained by observation with phase-contrast microscopy during life, as in the case of *Tokophrya* (Grell, 1953) (Fig. 11.18.) Ultimately, the normal diploid chromosome complex may be multiplied many times; Sonneborn estimated that in *Paramecium aurelia* the macronucleus may contain at least 40 chromosome or gene sets ("genomes"). The process has been called "endomitosis," and is ordinarily visible only by refined staining techniques and after painstaking study. The concept of the polyploid nature of the macronucleus goes far to explain the ability of small fractions of the macronucleus to support the regeneration of fragmented organisms.

The actual behavior of the macronucleus in dividing ciliates varies considerably. Usually it simply elongates until a transverse constriction appears and then a complete separation of the daughter nuclei occurs. When it is beaded or elongate, as in *Stentor* and *Spirostomum*, it may condense into a somewhat rounded form as a first indicator of approaching fission. In forms normally possessing several macronuclei, each usually divides independently, but here too preliminary union of the nuclei may occur. In the macronuclei of the hypotrichs a reorganization band ("kernspalt"), appearing as a transverse cleft, may become visible at one end of the nucleus and move gradually to the other. Its function and significance remain unknown. Some ciliates are known in which the macronuclei never divide during vegetative life, but are instead formed by the micronuclei as needed.

Concurrently with nuclear division are other profound changes in progress in the organism; most of them have already been outlined. When final separation occurs one might expect that the two daughters (termed "proter" and "opisthe" for anterior and posterior respectively by protozoologists—

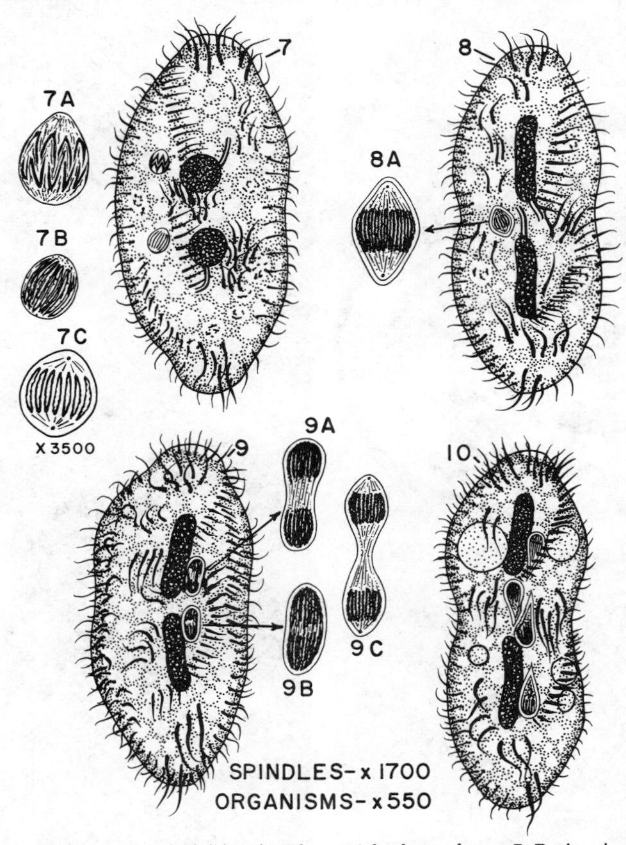

FIG. 11.17. Later stages of division in *Pleurotricha lanceolata*. *7*. Both micronuclei have formed spindles, which are shown enlarged in 7A and 7B; the cirri are larger and the macronuclei almost round. *7C*. A spindle, magnified about 1750 times, showing the appearance of the chromosomes and spindle fibers. *8*. An individual with a single micronucleus (such individuals are not uncommon), showing a nicely developed metaphase; The macronuclei have now elongated and the new cirri have become large and conspicuous. *9*. An individual with micronuclei in the anaphase. *9C*. A still later anaphase. *10*. An individual with both micronuclei in the telophase. Note that each micronucleus divides in such a manner that each of the daughter cells will receive one of the products, and also that division of the micronuclei is completed before that of the macronuclei. There is also a new contractile vacuole for each of the daughter cells to the right of each new adoral zone. (Organisms, $\times 275$; spindles, $\times 850$. After Manwell, *Biol. Bull.*).

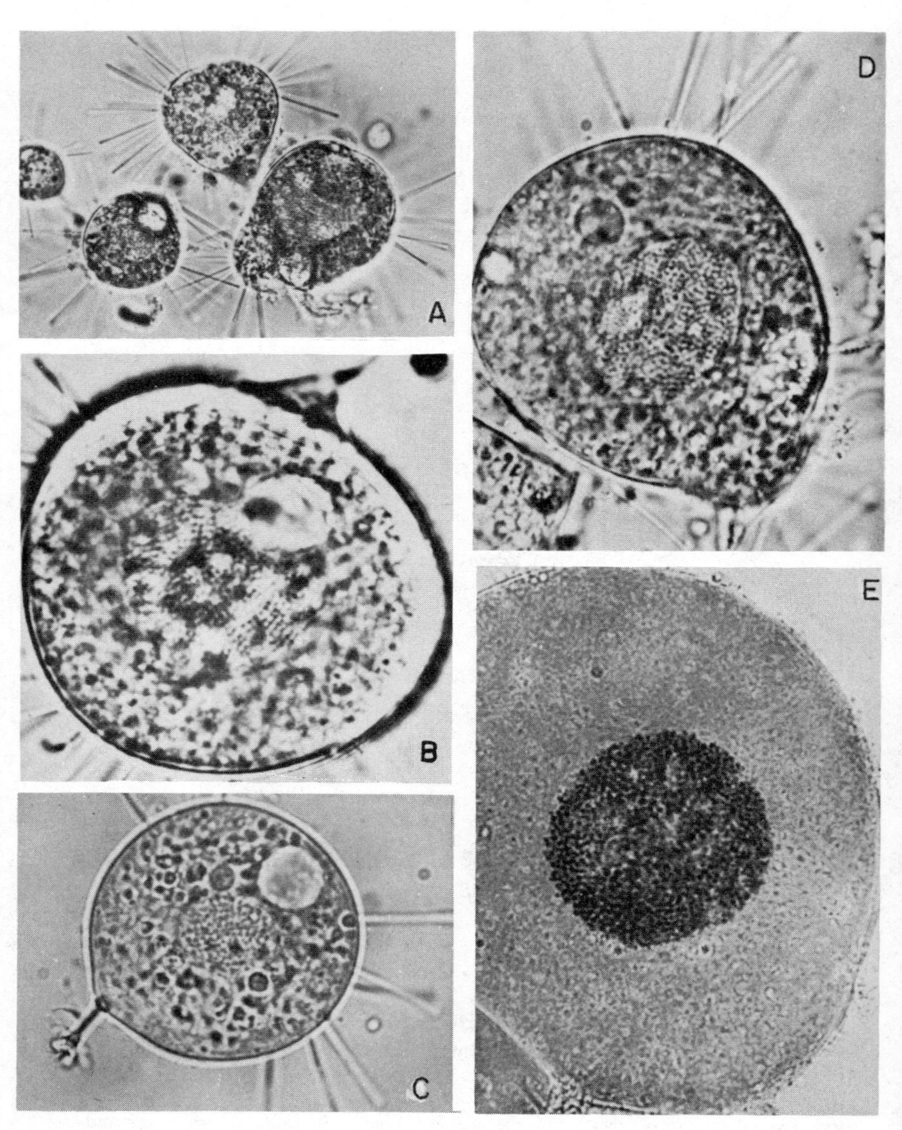

FIG. 11.18. Structure of the macronucleus in the suctoridan *Tokophrya*. The macronucleus is a large central mass filled with dumbbell-shaped bodies representing the chromosomes, which actually have a spiral, beaded threadlike structure. Since reproduction is by budding, the young forms (*C*) have smaller macronuclei (and therefore fewer chromosomes) than older ones (*D*). The paired chromosome arrangement represents the endomitotic chromatids. *A*. Group of four individuals in different stages of growth; ×400. *B*. One of the individuals shown in *A*, seen in life, ×900. *C*. A young individual, ×900. *D*. The largest form of the group depicted in *A*, seen in life, ×900. *E*. Section of one of the individuals shown in *A*, stained with acetic acid-carmine, ×400. (After Grell, *Arch. f. Protistenk.*)

perhaps to give their profession a greater aura of academic learning and scholarly respectability) would be exact duplicates, but this is not invariable. In some species of ciliates at least there may be more or less consistent size differences between anterior and posterior offspring, but this implies no genetic difference. Both of course have the same inheritance, and grow to the same size.

Meiosis

Like most living things, the life cycles of many Protozoa include sexual stages. Obviously a reduction of the chromosome count from the diploid to the haploid condition is necessary at some point; but when this occurs varies considerably. Among the plantlike flagellates and the Sporozoa it seems generally to be postzygotic, although numerous exceptions are known and many species have not been studied. Among the known exceptions are the Cnidosporidia and possibly *Monocystis* (a common parasite of earthworms). The Foraminiferida also present a curious situation: here there is an alternation of generations (p. 311), the microspheric (asexual) phase being diploid and the macrospheric (sexual) phase haploid (Le Calvez, 1950). Meiosis occurs in the agamonts. Thus, as Grell (1956) points out, these remarkable Protozoa have a heterophasic life cycle, otherwise known only in the plant kingdom. Until recently it has generally been thought that meiosis in the ciliates occurs in the second two of the pregamic, or "maturation," divisions of the micronuclei, occurring in conjugation or in autogamy and cytogamy. Most of the cytological studies of the life cycles of ciliates have agreed in this respect. Considerable doubt has been cast on this interpretation, however, and Sonneborn (1949) calls it "dubious." Catcheside (1949) goes further and remarks, "There is cytological evidence of the former (reduction in the first and second of the three pregamic divisions) in other Protozoa," although he also states that the nature of these divisions "has largely been resolved by genetic means." So far most of the genetic studies have been on *Paramecium*, which may turn out to be an exceptional species genetically, as it is in other respects. In any case, there are usually three pregamic or maturation divisions in ciliates undergoing conjugation or autogamy, and in *Paramecium* at least the first two of the three appear to be meiotic, the first reductional and the second equational. When more than one micronucleus is present all undergo maturation but only two of the final daughter nuclei survive, one to become the stationary and the other the wandering pronucleus. The actual form of the spindles is much like that in ordinary mitosis, but differences are sufficient in the stages so that each can usually be identified.

Budding

Although binary fission is probably the most common type of reproduction among the Protozoa, particularly among the free-living forms, multiplication

FIG. 11.19. *Right*. The life cycle of the suctoridan *Solenophrya micraster*. *1*. Free-swimming bud. *2*. Attachment to substrate; beginning of rounding; first appearance of tentacles. *3*. Rounding completed with an apparent increase in diameter. *4*. Lateral view of metamorphosing individual, showing the developing lorica. *5*. Further growth and elongation of tentacles. *6*. Cytoplasm begins to pull away from the growing lorica and more tentacles appear. *7*. Lateral view of an individual in stage shown in *6*. *8*. Cytoplasm pulled even more away from lorica. *9*. Top of lorica beginning to form; cytoplasm rounding up from the thin layer in the top of formed lorica with the formation of more tentacles. *10*. Top of lorica still growing and more tentacles forming and lengthening. *11*. Typical 24-hour-old adult; lorica completely formed over top of animal, with tentacles protruding through five equally spaced apertures.

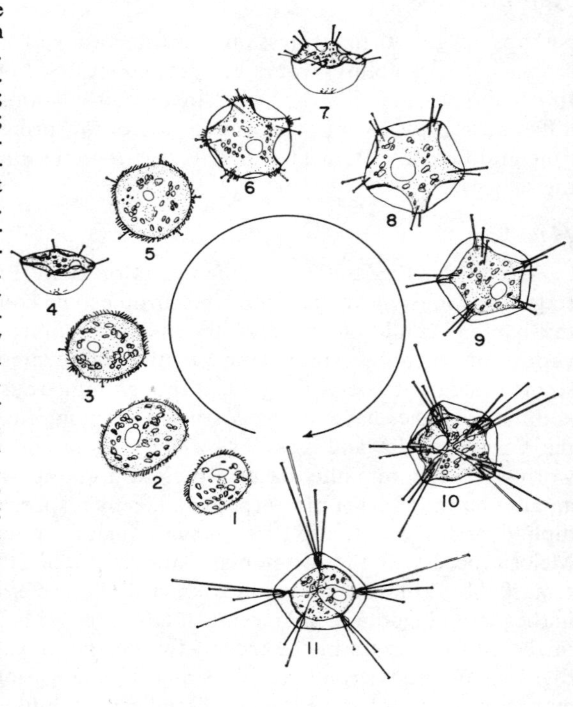

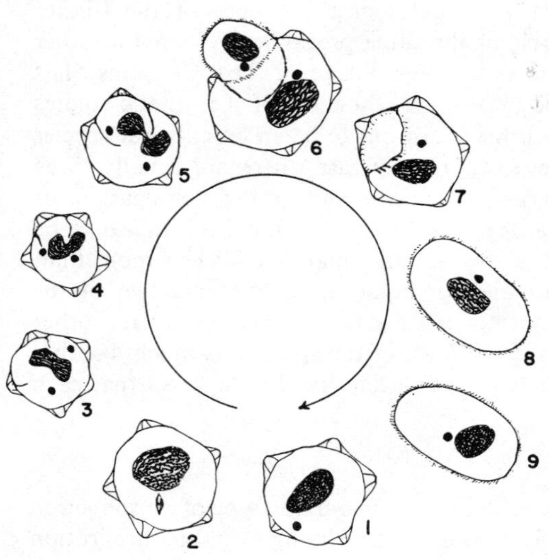

Left. Nuclear changes in bud formation in *Solenophrya micraster*. *1*. Adult initiating bud formation. *2*. Macronuclear "stranding" and "metaphase" of micronuclear division. *3*. Early stage of macronuclear constriction, with the start of surface invagination of body. *4*. Further macronuclear constriction and progress of surface invagination; cilia noticeable on exterior of bud. *5*. Bud is becoming apparent. *6*. Macronucleus has completed division and the budded daughter organism is making its escape. *7*. Parent after escape of daughter. *8*. Escaped daughter, showing cilia. *9*. A more advanced stage in the development of the daughter, showing the macronucleus restored to its normal granular condition. (After Hull, *J. Protozool.*, 1: 93.)

by budding (also called "gemmation") is very frequent. It may be either the usual mode of asexual increase, or limited to the sexual portion of the cycle. Since it is also mitotic, we might better speak of it as unequal division. The nuclear changes in a budding *Solenophrya*, a suctoridan, are shown in Figure 11.19 opposite.

Budding may be regarded as a form of fission in which the offspring are smaller than the parent (although minor size differences are often observed in the daughter organisms in ordinary division), and in which the parent in some species may continue to live and bud off additional progeny from time to time (Fig. 11.19). It occurs in all the larger taxonomic groups of Protozoa, although it seems to be less common among the ciliates. The Suctorida reproduce in this fashion, however, and so do such other groups as the chonotrichs, a group of sessile organisms living attached to invertebrates such as Crustacea, and the thigmotrichs, most of which are ectoparasites of molluscs.

Essentially, budding is of two types, although it shows many variations of detail. In the commoner, exogenous form, bits of cytoplasm, each with a little chromatin (but the characteristic number of chromosomes), are pinched off the periphery of the parent organism. Sometimes the chromatin is divided equally between parent and offspring; more often the young organism gets less. When equally shared, the process is obviously little different from ordinary fission. Exogenous budding is illustrated by the suctoridan *Ephelota*, which may form perhaps half a dozen offspring in this way at a time. Here the macronucleus branches in complex fashion, several branches being distributed to each of the progeny, while the micronuclei divide mitotically and also find their way to each daughter organism. A remarkable suctoridan parasite of *Ephelota*, *Tachyblaston*, lives as an adult attached to its stalk and from time to time produces buds which creep up the stalk into the body of the *Ephelota*, there to multiply by fission. The ciliated larvae so formed then make their way out of the host's body, and in turn take up an ectoparasitic existence on their stalks. Another interesting example of budding may be found in the holotrich genus *Radiophrya*, species of which are common parasites of oligochaete annelids. Here buds are formed at the posterior end, one after another, with the result that temporary chains are formed.

Endogenous budding is similar to exogenous budding except that the developing offspring start life within the body of the parent. It is said not to occur among the flagellates, but is otherwise rather common. The Suctorida again furnish particularly good examples, and in them this method of reproduction reaches perhaps its highest development. Sometimes only a single suctoridan is born at a time; in other cases they appear in litters, as in *Tokophrya* and *Acineta* respectively (Fig. 11.20). In any case the infant suctoridan is supplied with ciliary girdles by which it can make its way in the world; it does not settle down to the sedentary existence of its parent until it has satisfied the wanderlust of youth for a while, at the same time insuring

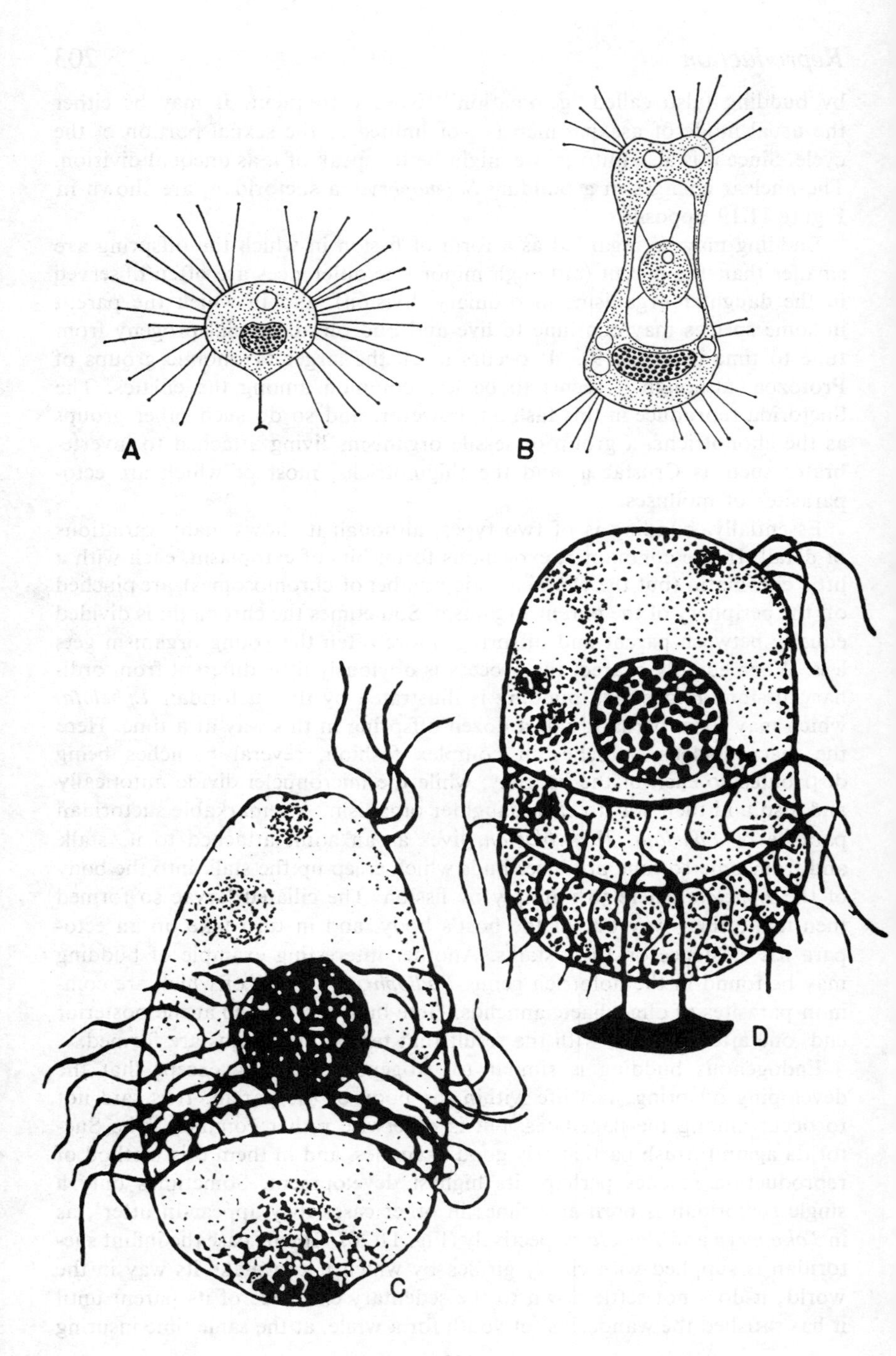
A
B
C
D

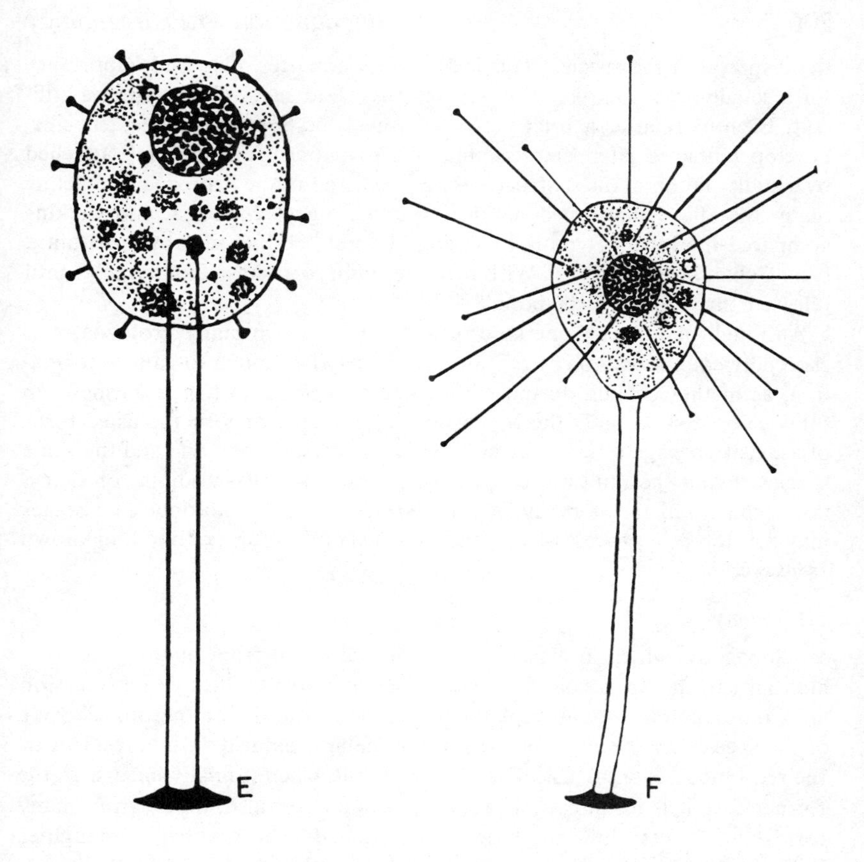

FIG. 11.20. Reproduction in the suctoridan *Tokophyra infusionum*. Ordinarily the young organism is ciliated, but in *B* the daughter is shown with tentacles. The parent is also somewhat abnormal, with its five contractile vacuoles; one is the usual number. A normal form is shown in *A*, with its stalk and attachment disc. (After Rudzinska and Chambers, *Biol. Bull.*) *C*. Free-swimming (larval) stage of the suctoridan *Tokophrya lemnarum*. A belt of four bands of cilia encircle the midregion of the body and a tuft of about six cilia is at the apical end. The broader end develops the stalk and pedicel while tentacles grow at the other end. Ordinarily the free-swimming stage lasts but a few minutes, and metamorphosis is complete in no more than an hour after attachment; however if attachment is delayed so also is metamorphosis. (A. E. Noble.) *D*. Young larva of the suctoridan *Tokophrya lemnarum* immediately after attachment to substrate. (A. E. Noble.) *E*. Early metamorphosis of the larva of *Tokophrya lemnarum*. Note the increase in stalk length and the disappearance of the cilia, together with the appearance of the tentacular buds. All this takes only a few minutes. (A. E. Noble.) *F*. Late stage in the larval development of *Tokophrya lemnarum*. Only separation of the tentacles into the two groups characteristic of the adult remains before maturity is reached. (A. E. Noble.)

the dispersal of the species. This is obviously a matter of special importance for a sessile organism which otherwise might find survival as a species difficult. Eventually after a brief period of travel the Suctorida lose their cilia, develop tentacles after the accepted species pattern, and become attached by a stalk and basal disc. Structures have evolved in the Protozoa paralleling many that have developed among even the higher Metazoa, and Calkins compared the birth of young following internal budding to growth within a brood chamber or uterus. Within it the embryo swims about freely until released through a "birth pore."

Although budding of one kind or another occurs in many Protozoa, it is especially common among the Sporozoa. It may be limited to gamete formation, as in the coccidia during microgametogenesis, and the gregarines. In other Sporozoa, notably the Myxosporidia, budding may be the usual form of asexual propagation. It may be of various types, even in one and the same species. From the multinucleated mass known as a plasmodium, or syncytium, characteristic of many of these species, smaller multinucleate stages may develop by a process of unequal fission, This is a form of budding known as *plasmotomy*.

Schizogony

Schizogony, which in essence is not very different from budding, is common among the Sporozoa. The nucleus divides repeatedly by binary fission until the maximum number of nuclei characteristic of the mature schizont of the species has been formed (as in the malaria parasites while resident in the red blood cells); or fission may be multiple when it finally occurs. In the former case it is mitotic, and probably in the latter also, although in many forms the process has not been much studied. The number of daughter individuals resulting from schizogony may reach hundreds or even thousands from a single parent cell, or it may be as few as four.

What might perhaps be called pseudo-multiple fission in organisms such as *Trypanosoma lewisi* has already been described; here a number of offspring may be formed simultaneously from the same parent both in the blood and in the flea. In the blood forms it is really precocious nuclear division (binary fission being the normal method), and in the latter case repeated division within the parent membrane. The group of daughter cells is collectively known as a "somatella" while separation is still incomplete. Thus this method of propagation is not greatly different from schizogony, but it exists in a somewhat different guise in genera such as *Trichomonas* (*Tritrichomonas*), a genus of parasites found in the gut, reproductive tract, and even the tissues of a number of vertebrates; it has also been described for certain free-living species. Here daughter individuals may be budded off from the somatella singly or in groups of two or three, and it is essentially exogenous budding.

Reproduction, Life Cycles, and Life Spans

No consideration of reproduction is complete without some knowledge of the life cycle of the species. Bound up with the life cycle is life span, which has an obvious relation to the number of individuals likely to be multiplying in any population at any one time. In addition, the mode of reproduction is apt to differ at various stages in the life cycle in Protozoa as well as in other living things.

It is also true, but often overlooked, that reproduction does not always result in increase in numbers. Literally, reproduction means no more than the creation of a new organism, and hence does not necessarily mean multiplication; thus the word should apply equally to the making over of an old one. In the ciliates conjugation, hemixis, autogamy, cytogamy, and sometimes even encystment (processes we shall discuss later) do not result in multiplication—they simply give rise to profound modifications of the organisms concerned.

The length of life of an average individual in any given species of protozoan varies with the environmental conditions, and is therefore virtually impossible to determine in nature. Abundance of food, optimum light, and especially optimum temperature, favor rapid multiplication. In the laboratory, division rate is affected by temperature in much the same way as is the velocity of a chemical reaction, doubtless because the process of fission is itself dependent on certain key reactions. The content in the medium of dissolved gases—notably oxygen and carbon dioxide but sometimes also hydrogen sulfide and methane—mineral salts, and pH, are important too.

Under the best conditions multiplication may be very rapid. The author has on numerous occasions counted 64 or more individuals in an isolation culture of *Colpoda* started from a single organism 24 hours earlier. At the end of the week, if such a rate were maintained, the potential progeny would number 2^{42}! And the total mass of such a population, if all survived and continued to reproduce, would soon exceed the mass of even the earth! The reproductive abilities of the free-living Protozoa generally, however, may not be quite equal to those of small ciliates such as the Colpodas. On the other hand, the reproductive potential of many parasitic species is even greater.

Life span is also limited by physiological state. The French pioneer protozoologist Émile Maupas postulated some seventy years ago periods of youth, maturity, and old age in the Protozoa, remarking, "My cultures have experimentally shown that these Protozoa do not escape the general law of senescence" (quoted from Calkins, 1923). The multiplication rate of the ciliates he worked with was at first rapid, slowing down with age until the organism died. He felt that senescence, otherwise inevitable, could be fended off by the onset of the sexual process we know as conjugation.

Although sexual phenomena occur widely among the Protozoa, conjugation is restricted to the ciliates. Since it is described in detail elsewhere (p. 229) we shall merely outline its cytological features here. Considerable differences in the Protozoa are exhibited by various species, but the nuclear changes in each member of a conjugating pair usually involve: (1) several micronuclear divisions, one being reductional, with (2) disappearance of all but two of the products, (3) degeneration of the macronucleus, (4) migration of one of the two surviving micronuclei to the other individual, (5) fusion to form a fertilization nucleus, and (6) several divisions of the latter, with reconstitution of the normal nuclear apparatus from its products. In the meantime, the organelles are resorbed and reformed, and finally the partners separate. Although conjugation is often considered a form of reproduction there is no multiplication. Its two significant features are genetic exchange and the extensive reorganization of each participant. Though Leeuwenhoek was the first to observe conjugation, Balbiani in 1858—more than a century and a half later—was the first to make a clear case for its sexual nature. It remained for Bütschli, the great German protozoologist, however, to show that the dividing micronucleus was in fact a nucleus and not a bundle of spermatozoa as Balbiani had thought.

After long controversy, Maupas' generalization as it relates to ciliates is now receiving increasing acceptance. Yet it is certain that species differ greatly in their abilities to fend off the onset of age. Environmental conditions play a large part; when they are sufficiently favorable it seems likely that a high rate of multiplication can continue indefinitely in some species. Woodruff's famous experiments with several species of ciliates, of which he successfully maintained actively reproducing cultures for many years without the intervention of conjugation, bear this out. Significant differences in the tendency to become senescent occur even among various strains of the same species, and species within the same genus, as Sonneborn has pointed out. Jennings (1939) stressed the importance of two factors affecting aging and death in the Protozoa: (1) "the substitution of reserve parts for those that become exhausted," remarking that in the ciliates the micronucleus may function "as a reserve" and (2) "genetic or hereditary . . . in all animals from the single celled Protozoa, through the invertebrates to man, the length of life is largely determined by inheritance." Fauré-Fremiet (1953) extends this idea when he remarks, "In sum, the notion of senescence among the ciliates suggests, though not necessarily, the possible evolution of the polyploid macronucleus toward functional impotence." The fragmentation and cytoplasmic absorption of the macronucleus which occurs in conjugation and other types of reorganization in the ciliates accord very well with this concept.

Autogamy and Cytogamy

In other experiments, Woodruff noted that even though division continued in his cultures of *Paramecium aurelia*, its rate fluctuated rather regularly. When individuals isolated at the low points of the curve were examined microscopically they were found to be undergoing extensive nuclear reorganization—fragmentation and disappearance of the macronucleus, several divisions of the micronucleus, and finally reconstruction of the former from one of the micronuclear products. The whole process seemed to be essentially like conjugation, except that it did not involve meiosis and only one individual was concerned; it was aptly compared to parthenogenetic

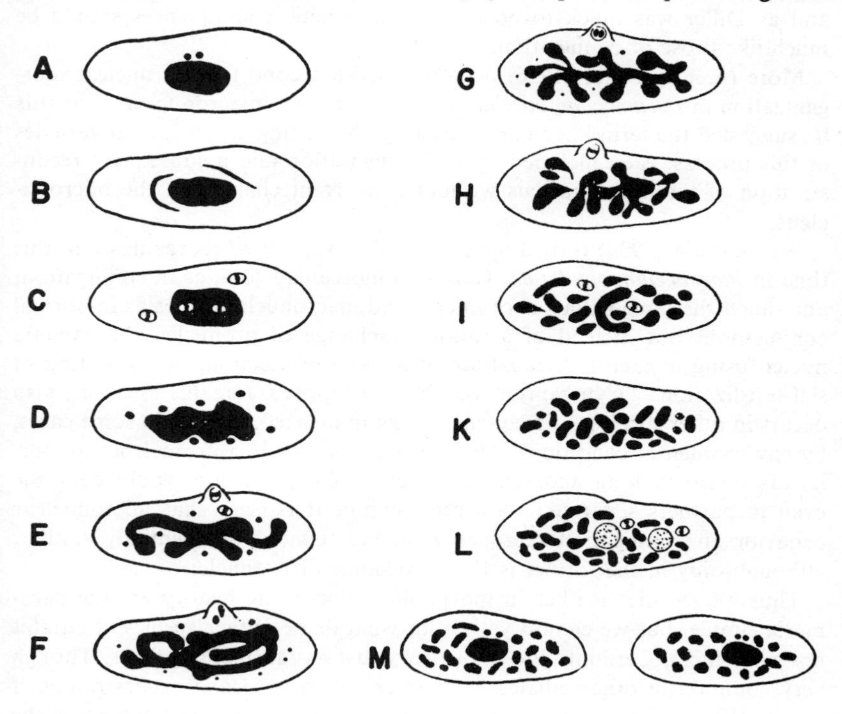

FIG. 11.21. Nuclear changes occurring in *Paramecium aurelia* during autogamy. *A*. Typical vegetative animal. *B*. Crescent micronuclei (shown as elongated streaks); characteristic of the prophase of the first prezygotic (maturation) division. *C*. Second prezygotic (maturation) division: four micronuclei, two of them lying in the concavity of the cup-shaped macronucleus. *D*. Eight micronuclear products of the second division; macronucleus preparing for skein formation. *E*. Variable numbers of nuclei continue to divide for a third time; here, two are functional; one lies in a bulge near the mouth ("paroral cone"). *F*. Potential gametic nuclei arise after the third division; degenerating nuclei from the second and third divisions may be present. *G*. Two gamete nuclei fuse in the paroral cone, with the resulting formation of a synkaryon. *H*. First (cleavage) division of the synkaryon. *I*. Second (cleavage) division of the synkaryon; macronucleus fragmenting. *J*. Four products of the second division; two transform into the new macronuclei and two remain micronuclei. *K*. Macronuclear anlagen (primordia) well developed; first cell division under way; micronuclei are dividing. *L*. Macronuclear anlagen segregated to the two daughter cells; old macronuclear fragments degenerating and about to be absorbed into the cytoplasm. (After Diller, *J. Morphol*., 59: 11.)

reproduction in the Metazoa. Later Diller (1934) showed that the process did involve meiosis after all, and that pronuclei were formed with subsequent fusion to produce a fertilization nucleus (synkaryon); reorganization similar to conjugation then ensued. What occurs then is "autogamy," rather than a more novel reorganization for which Woodruff and Erdmann (1914) had coined the name "endomixis." The sequence of nuclear events is shown in Figure 11.21: In this process both maturation and fertilization are involved, and as Diller was quick to point out, the genetic consequences should be much like those of conjugation.

More recently Diller (1936) demonstrated a second type of nuclear reorganization in *Paramecium aurelia*, involving only the macronucleus. For this he suggested the term "hemixis." Although he distinguished several varieties of this process, each included some fragmentation and a subsequent reconstitution of the macronucleus without concurrent changes in the micronucleus.

Wichterman (1939) turned up still another variant of reorganization, this time in *Paramecium caudatum*. Here two individuals join, as in conjugation, and much the same behavior of micro- and macronuclei ensues as in normal conjugation. But instead of a mutual exchange of pronuclei, the gametic nuclei fusing in each individual are of its own production, thus resulting in self-fertilization. "Cytogamy," the label proposed for this process, also occurs in other species, its frequency being influenced, at least in some cases, by environmental conditions. Though it is apparently a normal occurrence, it was observed long ago that the events of conjugation would continue even in partners separated soon after union. It is clear that micronuclear behavior in conjugation, autogamy, and cytogamy is essentially similar, although only in the former is there exchange of pronuclei.

Thus we see that neither in morphology nor in life history are the paramecia simple, but we cannot yet say to what degree other species of ciliates resemble them. Certainly life cycles vary just as morphology does. Though very common in other ciliates, encystment in *Paramecium* occurs rarely, if at all. Where encystment occurs it often involves a reorganization of the nuclear and organellar apparatus as profound as those just described. The cysts of many species also participate in the multiplicative process since intracystic division is a relatively common occurrence (p. 171).

Significance of Sexual Reorganization

What is the real significance of conjugation and its related intracellular transformation among the Protozoa? Woodruff sought the answer in a comprehensive series of experiments using *Spathidium spathula*, a ciliate selected largely because it does not exhibit autogamy in the vegetative state. He concluded that in this species conjugation results in an accelerated division rate, and that when conjugation is prevented in the cultures, although division may continue for a while, the organisms eventually die. Thus conjugation has a "direct survival value . . . [and produces] a profound physiological stimulation of the metabolic activities of the cell which is expressed in reproduction" (1925). When conjugation is prevented, some type of internal reorganization presumably takes its place. Such reorganization may take place even during ordinary division, as in ciliates of the family Loxodidae (Fauré-Fremiet, 1954). The same author suggests that comparable cycles of nuclear change may even occur among Sporozoa such as the gregarines and coccidia, and in the Foraminiferida.

Sexual phenomena such as conjugation play another essential role in assuring continued existence of the species. Like other forms of sexual reproduction, it results in a genetically diverse population, better able to survive the variations inevitable in any environment.

Fluctuation of Division Rates

Although the more marked fluctuations in division rates of laboratory cultures of the ciliates have often been correlated with profound internal changes like those described above, periodic pulses in rates of multiplication have been observed that could not be so explained. In certain instances at least they have apparently been seasonal. For example, Richards and Dawson (1927) followed cultures of *Paramecium aurelia*, *Blepharisma undulans*, and *Histrio complanatus* for three years and noted a "yearly cycle [of division rate] with a maximum during July." They regarded it as due to factors intrinsic in the organisms, since the tendency gradually diminished with prolonged cultivation under laboratory conditions.

Unfortunately, other groups of Protozoa have received less study than have the more readily cultivated ciliates. The research devoted to various species of *Paramecium* and *Stentor*, of *Euplotes* and *Tetrahymena*, has far exceeded that given to most other genera, and we cannot be sure just how broadly applicable are the results of this vast amount of investigation. But it is likely that such concepts as "youth, maturity, and old age" as applied to the Protozoa have numerous exceptions. The malaria organisms, for example, seem able to multiply indefinitely by ordinary multiple fission. On record are cases of human malaria where the infection has persisted for more than forty years, and strains of avian malaria have been maintained in the laboratory by blood transfer for many years. Sexual processes in malaria organisms can occur only in the mosquito, and there is no evidence of any other type of internal reorganization. Senescence, here at least, is no problem.

Probably protozoan parasites have, on the whole, a higher rate of multiplication than most free-living forms. This is to be expected, in view of the many hazards a parasitic mode of life imposes on the parasite. There is, for instance, almost always substantial resistance on the part of the host that the parasite must overcome. And there are the difficulties of getting from one host to another. During its residence in the blood, the malaria parasite in some instances reproduces every 24 hours, each segmenting form breaking up into an average of 16 or more offspring. During its tissue, or exoerythrocytic, stages, these forms may produce 100 or even 200 progeny each. But the protozoan "parent of the year" is probably *Eimeria bovis*, a coccidian parasite of cattle, where a single individual may bring into the internal bovine world not less than 100,000 young each two weeks! Other free-living and parasitic Protozoa may produce as many or more offspring in the same length of time, but over a number of generations, not in a single one.

Fortunately, the host possesses enough natural resistance, even against parasites that eventually prove fatal, to kill off a high proportion of the invaders. With the development of acquired immunity, resistance may increase to where it not only causes the death of many of the parasites but, as in malaria, also a decrease in their fecundity. Usually parasites do not long survive the death of the host, but some individuals of some species (e.g., *Toxoplasma*) may live for a time in the tissue if putrefaction is prevented by chilling, though they probably do not reproduce.

Regeneration

One final process, akin to those already discussed, remains to be mentioned. It is regeneration, or the ability of the protozoan organism to repair injuries. Not only the individual but the species benefits from this ability in the long course of evolution; but where populations are large, as is always the case with relatively small organisms like Protozoa, its value seems less clear. Animals on the higher rungs of the evolutionary ladder tend to lose this ability and it is therefore a little surprising that even the most highly evolved Protozoa under favorable conditions retain the ability to survive profound mutilation.

That large amoebae subjected to severe injury by microdissection can replace lost cytoplasmic structures has been known since the work of Gruber in 1886. As long as the nucleus remains complete, regeneration can occur even though 99 per cent of the cytoplasm is removed. Apparently, however, replacement of the shell by testate rhizopods is impossible, although the Foraminiferida, whose shells are constantly in process of formation and growth, can repair shell injuries.

The presence of the nucleus seems to be indispensable to regeneration in all Protozoa, although relatively little experimental work on this problem has been done on flagellates and Sporozoa. Among ciliates, however, only the macronucleus is usually necessary and often not all of this. Exceptions may exist though; Tittler (1938) observed that in *Uroleptus mobilis* "complete functional regeneration" after injury by electric current occurred only when both macronuclear and micronuclear material was present. A fraction of the total macronuclear substance may be enough for regeneration, and this is in accord with its apparently polyploid nature.

Among the ciliates, *Stentor* has been a favorite organism for the study of regeneration because of its relatively great size, though considerable research has also been done on *Paramecium*. Glass needles and the binocular microscope are usually employed for experimental surgery. The organisms are easily quieted by immersion in a viscous solution of methyl cellulose, an essentially nontoxic compound. Regeneration is most rapid in well-fed individuals, and small fragments, as long as they exceed a certain critical limit, regenerate as rapidly as larger ones. There are considerable differences, however, in regenerative ability and pattern among the various ciliate groups. *Paramecium*, a rather atypical infusorian in other respects, is peculiar in this also, for it can regenerate "only through feeding and subsequent growth," which led Tartar (1954a) to remark further that paramecia were "extraordinarily structurally conservative organisms." *Euplotes*, one of the most highly evolved genera of ciliates, forms two cirral fields and undergoes a nuclear reorganization involving both micro- and macronuclei, along with a resorption and complete replacement of all the old cirri, the latter happening through the agency of the anterior of the two fields of cirri (the posterior is resorbed); thus regeneration here is much like an abortive division (Yow, 1958). Regenerative ability is influenced not alone by the species (and therefore by the genetic constitution) but by other factors. In *Tetrahymena* larger fragments are more likely to regenerate than smaller ones, and those from the anterior regenerate more often and more rapidly than those from the posterior, thus indicating that the posterior portion is more quickly replaced (Albach and Corliss, 1959). Suzuki (1957) made the interesting observation that the old macronucleus in a conjugating individual of *Blepharisma* would not support regeneration, although the new one replacing it would. Pieces taken from the anterior and posterior thirds of *Stentor* regenerate faster than those from the middle third. It also appears that certain kinetosomes may be essential to regeneration of the oral structures, and hence of the complete animal. These kinetosomes are capable of migration from their original position to the normal location of organelles to which they give rise. The major steps in regeneration are shown in Figure 11.22.

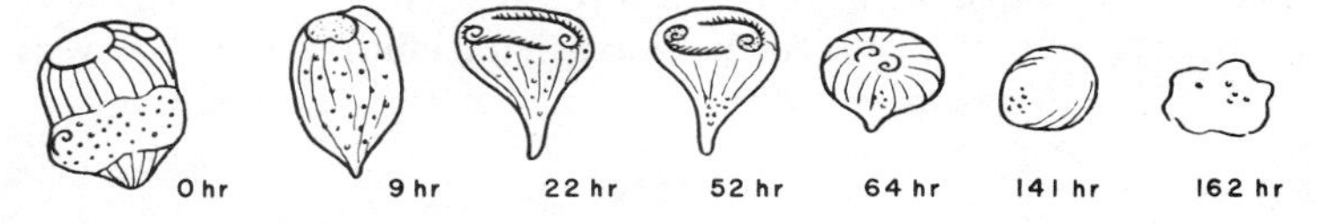

FIG. 11.22. Regeneration in a *Stentor* consisting of a *coeruleus* individual with oral area removed, and then fused with a half-cell of *S. polymorphus*, each fragment retaining its respective nuclei. Such individuals live for a time and then die. In this case, the regenerated double head had *coeruleus* characteristics (blue-green color, absence of symbiotic chlorellae). Successful (permanently viable) grafts require "the preponderance of the nucleus of one species in a preponderance of its own species cytoplasm." 0 hr.; operation. 9 hr.; diffusion of chlorellae throughout the mass, concentration of pigment in oral zone of *coeruleus* component. 22–52 hrs.; regeneration of two *coeruleus*-type heads with blue-green cytostomes, marked decrease in chlorellae. 64–141 hrs.; dedifferentiation of neoformations. 162 hrs.); death. (After Tartar, *J. Exp. Zool.*)

Closely related to the phenomena of regeneration are those following what is called homoplastic grafting (Fig. 11.22). This may be described as a way of producing Siamese twins of various kinds in the Protozoa—or, more accurately, producing aggregates of two or more individuals, the number being limited only by the size of the protozoan and the ingenuity of the experimenter. The technique was developed by Tartar for *Stentor*, and involves the apposition of freshly cut endoplasmic surfaces which readily fuse, even when individuals of different species are so joined. Such experiments have shown that masses of more than three grafted individuals tend to revert to triplets, doublets, or singletons, even though at first more than these numbers of oral structures may be produced. Thus there is a strong tendency to integration by "blending the constituent individuals into one" having the chacteristics of the species. Surprisingly, the Siamese twins and triplets were often capable of survival and reproduction in otherwise normal fashion, though forms with more than three heads were not. There was also physiological integration, since aggregates always contracted together. But it was apparently not possible for quadruplets or larger fused groups to feed normally, and death eventually supervened from starvation (Tartar, 1954b). Research such as this may throw much light on the biological basis of individuation.

In general, the process of repair and replacement of lost parts seems to be essentially like what goes on in normal fission and reorganization, even though different species exhibit many peculiarities. The hypotrichs, for example, are not satisfied with the simple replacement of a lost cirrus, but when one is lost resorb them all and proceed to form a complete new set (Balamuth, 1941). Yet the process is reproduction in its truest sense, since it is the literal re-creation of cytoplasmic structures (and even of nuclear ones, in the case of the macronucleus) in the species pattern.

It is entirely possible that if we understood just how the protozoan organism recovers from severe injuries, the secrets of growth, both normal and abnormal, might be revealed to be much the same in the higher animals and man. And it might then be possible for medical science to do much more to further healing and stimulate complete repair of injuries than it now is.

REFERENCES

Adolph, E. F. 1931. *The Regulation of Size as Illustrated in Unicellular Organisms*, Springfield: Charles C Thomas.

Albach, R. A. and Corliss, J. O. 1959. Regeneration in *Tetrahymena pyriformis*. *Trans. Am. Mic. Soc.*, 78: 276–84.

Balamuth, W. 1941. Contributions to the problem of regeneration in the Protozoa. In *Speciation. Defense Mechanisms in Plants and Animals. Biological Basis of Social Problems. Regeneration.* Jaques Cattell Press, pp. 257–70.

Balbiani, E. G. 1853. Sexuality of Infusoria. Comptes Rend., 46; *also* 1858, *J. de la Phys.*, I., and 1860, l.c., III.

——— 1891. Sur les regénérations successives du péristome comme caractère d'âge chez les Stentors et sur le rôle du noyau dans ce phénomène. *Zool. Anzeig.*, 14: 312.

Beers, C. D. 1948. The ciliates of *Strongylocentrotus dröbachiensis:* incidence, distribution in the host, and division. *Biol. Bull.*, 94: 99–112.

Bělař, K. 1926. Der Formwechsel der Protistenkerne. *Ergebn. u. Fortsch. Zool. Bd.* 6.

Bütschli, O. 1876. Studien über die ersten Entwicklungsvorgänge der Eizelle, die Zellteilung und die Conjugation der Infusorien, vol. 10. Frankfurt: Abh. Senck. Nat. Ges.

Catcheside, D. G. 1951. *Genetics of Micro-organisms.* London: Pitman.

Calkins, G. N. 1919. *Uroleptus mobilis.* II. Renewal of vitality through conjugation. *J. Exp. Zool.*, 29: 121–56.

——— 1923. "What did Maupas mean?" *Am. Nat.*, 57: 350–70.

Chalkley, H. W. and Daniel, G. E. 1933. The relation between the form of the living cell and nuclear phases of division in *Amoeba proteus* (Leidy). *Physiol. Zool.*, 6: 592–619.

Cleveland, L. R. 1938. Longitudinal and transverse division in two closely related flagellates. *Biol. Bull.*, 74: 1–40.

——— 1949a. Hormone-induced sexual cycles of flagellates. I. Gametogenesis, fertilization, and meiosis in *Trichonympha. J. Morphol.*, 85: 197–296.

——— 1949b. The whole life cycle of chromosomes and their coiling systems. *Trans. Am. Phil. Soc.*, 39 (New Series), Part I, pp. 1–100.

——— 1953. Studies on chromosomes and nuclear division. *Trans. Am. Phil. Soc.*, 43 (New Series), Part 3, pp. 809–69.

Cole, F. J. 1926. *The History of Protozoology.* London: University of London Press.

Diller, W. F. 1934. Autogamy in *Paramecium aurelia. Science*, 79: 57.

——— 1936. Nuclear reorganization processes in *Paramecium aurelia*, with descriptions of autogamy and "hemixis." *J. Morphol.*, 59: 11–67.

Dobell, C. 1917. On *Oxnerella maritima*, nov. gen., nov. sp., a new heliozoan, and its mode of division; with some remarks on the centroplast of the Heliozoa. *Quart. J. Mic. Sci.*, 62: 515–38.

——— 1940. Researches on the intestinal Protozoa of monkeys and man: X. The life history of *Dientamoeba fragilis:* observations, experiments, and speculations. *Parasitology*, 32: 417–61.

——— 1943. Researches on the intestinal Protozoa of monkeys and man: XI. The cytology and life history of *Endolimax nana. Parasitology*, 35 134–58.

Fauré-Fremiet, E. 1953. L'hypothèse de la sénescence et les cycles de réorganisations nucléaires chez les Ciliés. *Rev. Suisse Zool.*, 60: 426–38.

——— 1954. Réorganisation du type endomixique chez les Loxodidae et chez les Centrophorella. *J. Protozool.*, 1: 20–7.

Grassé, P.-P. 1952. Généralités. *Traité de Zoologie.* T. I, Fasc. 1. Paris: Masson

Grell, K. G. 1952. Der Stand unserer Kenntnisse über den Bau de Protisten kerne. *Verhandlungen der Deutschen Zool. Ges. in Freiburg*, pp. 212–51.

——— 1953. Die Struktur des Makronucleus von *Tokophrya. Arch. f. Protistenk.*, 98: 466–8, Heft 3/4.
——— 1956. Protozoa and Algae. *Ann. Rev. Microbiol.*, 10: 307–28.
Gruber, A. 1886. Beiträge zur Kenntnis der Physiologie und Biologie der Protozoen. *Ber. d. Naturf. Ges. Freiburg*, 1: 33–56.
Hartmann, M. 1928. Über experimental Unsterblichkeit von Protozoen Individuen. Ersatz der Fortpflanzung von *Amoeba proteus* durch fortgesetzte Regenerationen. *Zool. Jahrb., Abth. allg. Zool.*, 45.
——— 1952. Polyploide (polyenergide) Kerne bei Protozoen. *Arch. f. Protistenk.*, 98: 126–56.
Jennings, H. S. 1939. Senescence and death in Protozoa and invertebrates. In *Problems of Aging, Biological and Medical Aspects*, E. V. Cowdry (ed.). Baltimore: Williams and Wilkins.
——— 1940. Chromosomes and cytoplasm in Protozoa. A.A.A.S., *Pub. No.* 14, pp. 44–55.
Le Calvez, J. 1950. Recherches sur les Foraminifères. II. Place de la meiose et sexualité. *Arch. zool. exp. et gén.*, 87: 211.
Leedale, G. F. 1959. Periodicity of mitosis and cell division in the Euglenínae. *Biol. Bull.*, 116: 162–74.
Manwell, R. D. 1928. Conjugation, division and encystment in *Pleurotricha. Biol. Bull.*, 54: 417–63.
Maupas, E. 1888. Recherches expérimentales sur la multiplication des Infusoires ciliés. *Arch. zool. exp. et gén.*, (2) 6: 165–277.
——— 1889. La rajeunissement karyogamique chez les Ciliés. *Arch. zool. exp. et gén.*, 2ᵉ S., 7: 149–517.
Mazia, D. 1953. Cell Division. *Sci. Am.*, 189: 53.
——— 1956. The life history of the cell. *Am. Sci.*, 44: 1–32.
Moses, J. M. 1949. Nucleoproteins and the cytological chemistry of *Paramecium* nuclei. *Proc. Soc. Exp. Biol. and Med.*, 71: 537–9.
Nanney, D. L. 1953. Nucleocytoplasmic interaction during conjugation in *Tetrahymena. Biol. Bull.*, 105: 133–48.
——— 1960. Microbiology, developmental genetics and evolution. *Am. Nat.*, 94: 167–79.
Naville, A. 1931. Les Sporozoaires. *Mem. de la Soc. de Physique*, 41(Fasc. 1): 1–224.
Piekarski, G. 1941. Endomitose beim Grosskern der Ciliaten? Versucht einer Synthese. *Biol. Zbl.*, 61: 416–26.
Ratcliffe, H. L. 1927. Mitosis and cell division in *Euglena spirogyra* Ehrenberg. *Biol. Bull.*, 53: 109–22.
Richards, O. W. and Dawson, J. A. 1927. An analysis of the division rates of ciliates. *J. Gen. Physiol.*, 10: 853–8.
Sonneborn, T. M. 1957. Breeding systems, reproductive methods, and species problems in Protozoa. In *Species Problem*, Ernst Mayr (ed.). A.A.A.S. *Pub. No. 50*, pp. 155–324.
——— and Rafalko, M. 1957. Aging in the *Paramecium aurelia-multimicronucleatum* complex. *J. Protozool.*, 4(Suppl.): 21.
Spallanzani, L. 1777. Observations et Experiences Faites sur les Animalcules des Infusions. Paris: Gauthiers-Villers. (Reprint, 1920, of translation by Jean Senebier published in 1786 in Geneva.)

Suzuki, S. 1957. Morphogenesis in the regeneration of *Blepharisma undulans japonicus* Suzuki. Bull. Yamagata Univ. Nat. Sci., 4: 85–192.

Tartar, V. 1941. Intracellular patterns: Facts and principles concerning patterns exhibited in the morphogenesis and regeneration of ciliate Protozoa. *Third Growth Symposium*, pp. 21–40.

——— 1954a. Anomalies in regeneration of *Paramecium*. *J. Protozool.*, 1: 11–7.

——— 1954b. Reactions of *Stentor coeruleus* to homoplastic grafting. *J. Exp. Zool.*, 127: 511–75.

Tittler, I. A. 1938. Regeneration and reorganization in *Uroleptus mobilis* following injury by induced electric currents. *Biol. Bull.*, 75: 533–41.

Weisz, P. B. 1949. A cytochemical and cytological study of differentiation in normal and reorganizational stages of *Stentor coeruleus*. *J. Morphol.*, 84: 335–63.

——— 1955. Chemical inhibition of regeneration in *Stentor coeruleus*. *J. Cell. and Comp. Physiol.*, 46: 517–27.

——— 1956. Experiments on the initiation of division in *Stentor coeruleus*. *J. Exp. Zool.*, 131: 137–62.

Yow, F. W. 1958. A study of the regeneration pattern of *Euplotes eurystomus*. *J. Protozool.*, 5: 84–8.

12

Genetics

"... I have constantly observed that the parts of the divided animalcule become in a little while as large as the prevailing size in the population from which they have arisen; in sum, that one finds again in the progeny the same constancy and the same uniformity that one sees in the rest of Nature." Letter to Spallanzani from a Dr. de Saussure of Geneva, Switzerland, September 28, 1769.

Man has long been content to leave to folklore the solution of the origin of life, species patterns, and the fidelity with which these are perpetuated. Only within the last century has either problem been attacked with the tools of science. Gregor Mendel opened the door to understanding inheritance in multicellular organisms and thus indirectly to protozoan genetics. But Herbert S. Jennings of Johns Hopkins, whose monograph *The Genetics of the Protozoa* was published in 1929, made the first major breakthrough. His contribution was not only original research but the impetus to continued investigation through the training of students, who began their studies where his left off. Among Jennings' students, Tracy M. Sonneborn is especially well known because of his work on "mating types," a phase of ciliate genetics considered below.

Basically, inheritance in the Protozoa behaves in much the same way as in the Metazoa. There are chromosomes and genes in both, and in both the latter seem to consist of that most versatile of chemical compounds, deoxyribose nucleic acid (DNA), a substance which may predetermine the development of a cell into an amoeba or a man, depending on the particular assortment of molecular forms in which it appears. Likewise, partition and distribution to daughter cells in both Protozoa and Metazoa are almost universally by some type of mitosis.

DNA in the Protozoa

Nevertheless, though DNA is a nucleic acid it may not occur exclusively in the nucleus. Apparently, at least among the Protozoa, the cytoplasm may also possess genetic material, the "plasmagenes." The real nature of these bodies, and of other self-duplicating particles, such as the kinetosomes and their homologues the blepharoplasts, remains to be established, but there is reason to think that they may also contain DNA.

Chromosomes themselves contain chiefly protein, ribonucleic acid, and deoxyribose nucleic acid, but the first two of these seem to have little significance in determining the genetic constitution, or genotype, of the organism. Chemically, DNA consists of alternate sugar (desoxyribose) and phosphate groups arranged in long chains, probably in helical fashion. The almost infinite variety of form which such molecules may assume is due partly to variation in length and partly to groups such as the purines and pyrimidines, which may be attached as side-chains to the sugar components of the series. Thus a kind of chemical code has developed during the course of evolution by which genetic information is transmitted. Presumably it is the relatively minor changes produced in the molecular architecture of the DNA by radiation and other causes which have determined mutations, whether in the Protozoa or in multicellular organisms.

Kinds of Inheritance

Many careful studies of inheritance in the Protozoa have shown that it is, in general, of two kinds. During asexual reproduction, the daughter organisms might be expected to be identical. Certainly they share the nuclear components of the parent equally, but there is nevertheless a certain though very small variation. The second kind of inheritance occurs as the result of sexual multiplication, and here the genetic pattern may be quite complex, as with sexually reproducing organisms of other sorts. The offspring in this case show much greater variety, both of morphology and physiology.

Physiological differences among Protozoa of a given species in nature are of course very numerous, both in kind and origin. At bottom, doubtless all differences must be regarded as stemming from physiological traits of one kind or another. They include varying rates of reproduction, greater or lesser metabolic rates and variety in metabolic pathways, differing susceptibility to physical and chemical agents, and of course variety in stages of life history.

Some of these manifold characteristics are inherited, and some are not. All of them bear some relation to the environment. At any given time, the totality of characteristics of a protozoon, like those of any other living thing, is the product of its genetic constitution and of its environment. And the genetic constitution ("genotype") may be thought of, broadly speaking, as a resultant of the previous environment of the organism and its ancestors, experienced through vast expanses of time.

YOUTH, MATURITY, AND OLD AGE

Maupas, the French protozoologist of the last century whose work already has been mentioned, was probably the first to appreciate that Protozoa may go through a cycle comparable to youth, maturity, and old age. In studies of ciliates, he observed that as the age of his cultures increased, their vigor, as indicated by fission rate and the general appearance of the organisms, declined (Fig. 12.1). Clearly this is a matter of inheritance, since the offspring of senescent organisms also exhibit the characteristics associated with

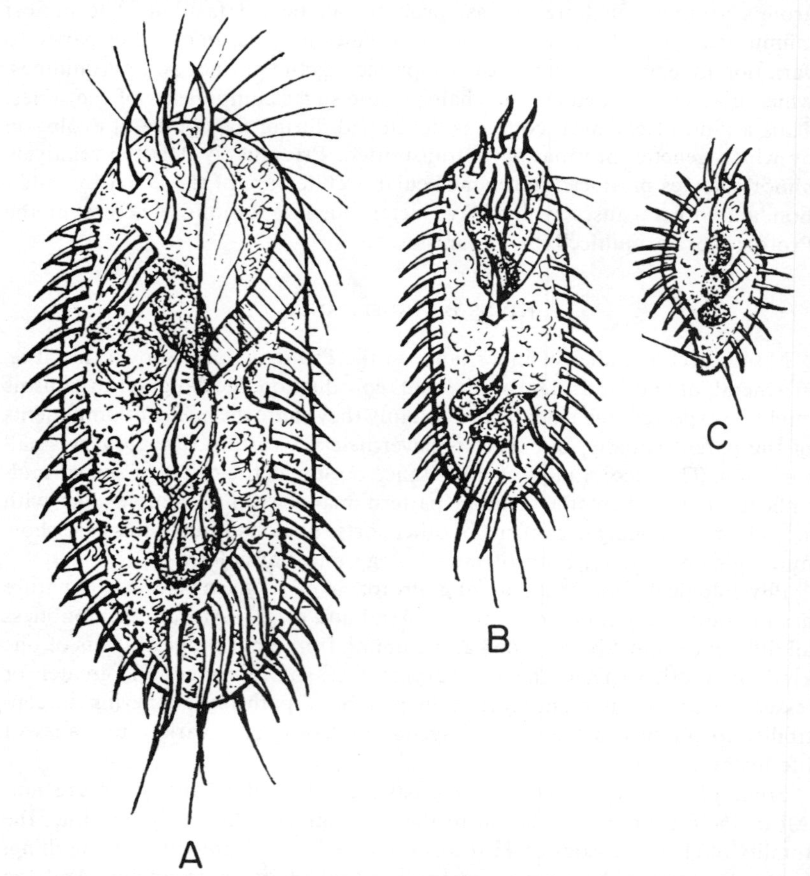

FIG. 12.1. Three successive stages in the decline and degeneration of the hypotrich, *Stylonychia pustulata*, as depicted by Maupas (1888). (After Jennings.)

age. Later work has shown that such changes are also a matter of environment, since with improvements in culture media and techniques they may often be postponed, sometimes almost indefinitely.

Perhaps because of Maupas' pioneering studies, the ciliates have remained favorites for this kind of research. One of the most studied has been what is often called the *Paramecium aurelia-multimicronucleatum* complex, used extensively by Sonneborn. He has noted that not only do different species vary in susceptibility to the effects of aging, but so do strains within a species. In one of his stocks of *Paramecium aurelia*, death always occurred after a maximum of 300 fissions in the absence of autogamy or conjugation, whereas in another the life span might extend to 350. The beneficial effects of autogamy varied according to when it occurred in the life cycle; if permitted to occur early and often senescence might be indefinitely postponed in vigorous stocks, but in lines already showing the effects of it might even cause premature death. Sonneborn remarked, "Yet, in spite of all the risks involved, the animals must undergo autogamy (or conjugation) or perish." Increasing age is likely to result in nuclear (especially macronuclear) changes of various kinds, inability of the organisms to form normal macronuclear anlagen during reorganization, and degeneration of the gullet. Even the micronucleus may ultimately be damaged as senescence advances, though selection of the more normal and vigorous individuals within a clone may postpone such injury for a long time.

Fauré-Fremiet has suggested that the degenerative changes of senescence may be the inevitable result of the evolution of the two types of nuclei among the ciliates, the micronucleus acting as a kind of genetic reserve, while the macronucleus governs the vegetative activities of the cell. With the acquisition of this function has come macronuclear development by endomitosis and polyploidization, a process carrying with it the "risk of irreversible impotence" due to chromosomal imbalance. Senescence may still be fended off for a time when environmental conditions are sufficiently favorable, or even indefinitely if cellular reorganization, such as that in conjugation and autogamy, occurs. Under such circumstances the macronucleus degenerates and is rebuilt through the agency of the micronucleus. Fauré-Fremiet further suggests that the greater susceptibility of some species to senescence may be a penalty of size and a high degree of differentiation.

Though written some years ago, the matter is well summarized in the words of Jennings (1939):

> Summarizing the relations of free-living cells to senescence and death, . . . we find many gradations, many diversities of constitution. Some free cells are so constituted that they are predestined to decline and death after a number of generations. Some are so constituted that decline occurs, but this is checked or reversed by substitution of reverse parts for those that are exhausted; they can live indefinitely, but are dependent on this substitution. In some the constitution is such that life and multiplication can continue indefinitely without visible substitution

of a reserve nucleus for an exhausted one; but whether this is due to the continued substitution, on a minute scale, of reserve parts for those that are outworn cannot now be positively stated. This perfected condition, in which living itself includes continuously the necessary processes of repair and elimination, is found in some free cells, but not in all.

Evidently longevity in the Protozoa, as in other species, man included, is a built-in characteristic—a matter largely of inheritance.

Very little is known about possible physiological changes associated with youth and age in the parasitic Protozoa. Certainly there is no evidence of anything of the sort in the malaria parasites, probably the most intensively studied parasitic organisms. These blood Protozoa continue to reproduce asexually at an undiminished rate in the vertebrate host for many years passing through thousands of generations, all without benefit of any kind of physiological reorganization. Yet throughout this period there is an inheritance of the morphology and type of reproduction appropriate to the vertebrate host (as distinguished from what would be observed in the mosquito), and clearcut evidence of a dependence of both on the environment furnished by the host. The length of the asexual cycle ("periodicity") is, in many cases, conditioned both by the genetic (species) constitution of the parasite and by the conditions imposed on it by the cyclic physiological changes in the host.

Inheritance in Parasites

Host-parasite specificity, that peculiar relationship of host and parasite which is sometimes so strict that only a particular host species will suffice for a given species of parasite, is also a genetic factor. Some species of avian malaria common in many passerine (perching) birds may be more or less readily adapted to the duckling, in which they apparently never occur in nature, but not to the chicken or parakeet; with other species of avian plasmodia, all attempts at adaptation have failed. Species which have become well adapted to life in the duckling may produce even heavier infections than they do in the canary, a more normal host since it is a passerine. But after being re-established in the canary for a time, it may be quite difficult to re-adapt the parasites to life in the duckling.

Genetic considerations also affect host susceptibility. For example, Huff (1931) has shown that in certain mosquitoes susceptibility or resistance to malarial infection is inherited in Mendelian fashion.

Uniparental inheritance in protozoan parasites has so far received relatively little study. This is largely because most hosts have sufficient resistance to make infections originating from single organisms both difficult and uncertain; indeed, more often than not they are quite impossible. But a beginning has been made. Several investigators have established malarial infections by injecting single intraerythrocytic parasites, and it was shown that the erythrocytic phase of the parasite could give rise to both exoerythrocytic forms and gametocytes.

Under some conditions environmental changes may induce the development of heritable alterations in parasites. One of the first such alterations to be demonstrated was the resistance that certain pathogenic trypanosomes acquire to subcurative doses of trypanocidal drugs, a phenomenon noted in 1909 by a great pioneer in immunology and biochemistry, Paul Ehrlich. In the course of long and painstaking experiments culminating in the discovery of arsphenamine ("606," the "magic bullet" which made possible for the first time the cure of syphilis), he found that arsenicals such as this, though initially capable even in small doses of causing a precipitous drop in the numbers of trypanosomes in the blood of experimentally infected animals, soon lost much of their efficacy. Exposure of the parasites to sublethal concentrations of the drug had resulted in an increasing resistance (often called "drug-fastness") to it, until finally even the maximum dosage tolerated by the unlucky rat proved insufficient to prevent death.

It was thus apparent that the chemical produced some inherent alteration in the trypanosomes. It persisted through many generations of asexual reproduction, sometimes disappearing spontaneously though gradually, and sometimes lost by a change in host species. In these parasites such drug-induced changes are believed to be in the cytoplasm rather than in the nucleus, and possibly to involve alterations in cellular permeability, though without much evidence other than their eventual disappearance. Actual alteration of genetic (nuclear) material would presumably be more permanent.

Drug-fastness, however, is not something peculiar to trypanosomes. The emergence of antibiotic-resistant strains of bacteria may well be of the same order, and the phenomenon in them has received much study. The most widely accepted explanation is that the resistance originates with mutant individuals, and that their progeny make up the resistant strain. It has been shown that mutations, though rare, may occur in some bacterial species with a frequency of about 10^{-7}, a far from negligible rate. But there is also the possibility that a direct adaptation of the bacteria to a changed environment (one containing the drug) may occur, and there is evidence for this view too. Both views raise many questions, for neither explains the precise nature of the changes occurring in the resistant cells. Sometimes the resistance may be due simply to the development of enhanced ability to use an alternative metabolic pathway.

The malaria parasites may also become refractory to such normally potent antimalarials as paludrine. It has also been observed that there are geographic differences in the response of malarial and leishmanial infections to therapy, curative doses of a drug ample for the disease in one area being insufficient in another. It thus appears that natural evolutionary processes have somehow led to the development of genetically and physiologically distinct strains in at least some species of parasitic Protozoa.

The mechanism of such acquired partial immunity to cellular poisons is not clear. The resistant organisms themselves possess the power of transforming the more injurious trivalent compounds of antimony and arsenic, to which they have been exposed, into the relatively nontoxic pentavalent salts. Evidently the phenomenon is much the same as the resistance insects develop to insecticides like DDT. But the origin of such resistant organisms is a matter of controversy. Some think it is essentially due to the effects of selection on a normally varying parasite population. Ordinarily in every population a very few individuals possess a tolerance much greater than the normal for such substances. The fortunate ones survive, while their more susceptible comrades are killed off, and give rise to a generation fitted to live in a previously lethal environment. But whether such individuals represent forms at the extreme end of the normal distribution curve, or mutants as in the case of antibiotic-resistant bacteria, we cannot say. Since gradual reversion to the normal condition is likely to occur eventually, mutant organisms in the parasitic Protozoa are probably not the usual explanation.

Quite different from such drug-induced changes, but of much interest because it is a temporary alteration affecting the reproductive mechanism, is the inhibition of division of the common rat trypanosome, *T. lewisi*, by some substance elaborated by the host. This antibody, which has been called "ablastin," apparently has no other harmful effect on the parasites. The mechanism by which reproduction is halted is not known. Presumably it operates through some cytological or other cellular change, since introduction of the parasites into a nonimmune host is followed by a prompt resumption of rapid multiplication. Strains of *Trypanosoma lewisi* resistant to ablastin apparently have never been observed.

Induced Heritable Changes

Many attempts have been made to produce genetically distinct races of Protozoa, sometimes because these organisms are assumed to be simpler in their organization than the multicellular forms and therefore should make favorable material for studying fundamental problems of inheritance. Probably the earliest experiment of this sort, and hardly matched since in thoroughness and meticulous execution, was Dallinger's (1887) (see Chap. 4). He undertook to acclimate several species of flagellates to higher temperatures, and succeeded after seven years of work in developing strains which would live and multiply at 158° F. (or 70° C.). But these strains lost their ability to live and reproduce at temperatures of 60° F. (15° C.), although the latter was close to their original optimum.

The German protozoologist Jollos (1913, *et seq.*) endeavored to produce lasting changes in *Paramecium* by exposure to heat and various chemicals such as certain arsenicals. He was apparently successful with wild populations because of the selection of normally occurring, genetically distinct stocks. Success with populations of uniparental origin was more difficult to achieve, but eventually a few individuals highly tolerant of the deleterious agents were observed and isolated. From them lines were built up in which the acquired resistance persisted through hundreds of asexual generations, although it was usually lost after conjugation. Because of their enduring nature, Jollos termed such alterations *Dauermodifikationen.*

In recent years a variety of mutagenic agents, especially radiation, have been employed in experiments on the genetic processes of the Protozoa. MacDougall (1929) was the first to achieve success in research of this sort. She repeatedly exposed *Chilodonella uncinatus* (then referred to as *Chilodon*) to strong ultraviolet light, with the result that in one of her cultures appeared a number of unusual individuals, some of them of relatively gigantic size, several with altered numbers of chromosomes (the normal being four),* and an occasional deformed organism. The radiation, if not too intense or prolonged, often resulted in ciliates of increased vigor.

She chose to regard as mutants only those ciliates in which the new characteristics persisted after encystment or conjugation or both. One such organism she described as "different from *Chilodon uncinatus* in practically every respect." It had a peculiar pharyngeal basket in which there were more trichites than usual, nuclei altered in position, five contractile vacuoles instead of two, and a changed pattern of ciliation. The progeny of this animal often encysted, though they did not conjugate, and were maintained in culture for sixteen months. Of the other variant organisms in the irradiated cultures, some lived and divided normally for a time, and one, a giant, did so for 68 generations.

Triploid and tetraploid individuals were also isolated from the irradiated cultures, and stocks of each were successfully established. They reproduced by fission and conjugated at intervals, the nuclei behaving in normal fashion. The increased number of chromosomes in each case was correlated with relatively minor modifications in morphology.

As with many other physical and chemical agents, the biological effects of ultraviolet light vary greatly with the dosage. In nature it is a significant source of the energy required for photosynthesis by the chlorophyll-bearing Protozoa. For others, slight exposure may be stimulating and result in an accelerated division rate, though more intense and prolonged radiation retards it. Curiously, subsequent exposure to visible light may result in partial recovery from the injury produced by excessive radiation. The mechanism by which two widely separated fractions of the radiant energy spectrum exert such antagonistic effects is not known. Ultraviolet radiation affects different parts of the cell differently; Shirley and Finley (1949) noted

*There are those who doubt that what appear to be chromosomes in the ciliates are in reality such. The uncertainty is based on an apparent disparity between the number visible in ordinary vegetative mitotic divisions and those seen in the first meiotic division; the latter seems larger than the former. Thus what seem to be chromosomes in ordinary mitoses may be in actuality chromosome aggregates.

that in *Spirostomum* excessive exposure resulted in fragmentation of the macronucleus.

Roentgen rays and radiation from radioactive isotopes have also been used extensively in experiments on the Protozoa. In general, the flagellates appear to be most sensitive and the ciliates suffer least. *Paramecium aurelia*, for example, can survive an exposure of 150,000 r. without death, and some ciliates are much more resistant. Mammals are killed by what are in comparison relatively minute doses; man, for example, has an LD50 (a dosage fatal to half those exposed) of only 450 r. Observed mutations following irradiation include changes in size and body shape, altered morphology and number of nuclei (even sometimes their loss), lessened swimming activities, and noticeably abnormal behavior.

Kimball and coworkers have shown that deleterious effects may follow radiation even in individuals of *Paramecium aurelia* which seem at first quite normal, although such effects might not become evident until after autogamy. Since autogamy results in completely homozygous individuals this is easily understandable. Radiation damage was evidenced by slower division rate of the progeny of injured organisms, and often by the early death of clones derived from them. The failure of radiation to produce immediate indications of damage was interpreted as due to the polyploid nature of the macronucleus. Since such a nucleus contains many sets of chromosomes the influence of occasional injured genes would be unlikely to be detected until autogamy rendered the animals homozygous. Kimball calculated that an exposure of 1000 r. would be expected to produce about two mutations per micronucleus.

Certain factors influence susceptibility or resistance to radiation injury. The first and apparently the most important is the physiological state of the organism. Protozoa in division or in some other form of reorganization seem most susceptible, and in this respect they are like tissue cells generally. Mutagenic agents of all kinds are most likely to exert their effects at such a time. Since almost all mutations are injurious to the progeny of irradiated organisms, the problem of protection from harmful amounts of radiation has recently received attention. Schoenborn (1956) tried the addition of a variety of substances important in metabolism to suspensions of ultraviolet-irradiated *Astasia longa*, and found that a certain amount of shielding was afforded by a number of amino acids. In most cases the effect was apparently due chiefly to a filtering action. Cysteine and adenine, however, exerted a greater protective effect than could be so accounted for, and this further protection might be a result of their ability to absorb some part of the injurious radiation even after they became incorporated into the substance of the cell. Fortunately for the Protozoa their natural resistance to radiation makes it unlikely that they will ever need such protection. In this age of artificially induced background radiation, some of it from such ubiquitously occurring elements as the long-lived strontium 90, we may well envy them.

Mutations of the sort described above involve considerable morphological changes, but this is not necessarily so. In nature physiological alterations unaccompanied by any visible differences in form may be the most frequent type. That physiologically distinct races exist in nature in some species of Protozoa is well known. But—probably because it is not yet possible to cultivate most Protozoa on chemically defined media—proved mutations involving physiological mechanisms, such as those produced by cytogeneticists in *Neurospora*, are as yet few.

Among the few seem to be those somewhat doubtfully claimed by Schoenborn (1954) as occurring in the colorless euglenid, *Astasia longa*. He exposed about 3,000 clones to various dosages of ultraviolet light and X-radiation, and in some of them injury sufficient to prevent or seriously hinder growth on the standard medium was observed. But to verify such changes as mutations, their behavior in sexual reproduction must be checked, and nature seems to have denied this species the gift of sex. So, Schoenborn remarks, "One can, therefore, [only] reach the tentative conclusion that gene mutations are responsible for the deficient strains of *A. longa*."

Sex in the Protozoa

Just as an understanding of sex is requisite for real comprehension of the processes of inheritance in the higher animals and plants, so it is with the Protozoa. Sexual phenomena among the latter are not universal (as we have just seen in *Astasia longa*), but they are very widespread. And they are so diverse that protozoan sex presents more problems than sex in the human species—acute as such difficulties may be in the latter at times. Nevertheless sex, wherever found in the world of living things, serves much the same biological purpose: a means by which an almost endless variety within the species may be achieved, thus facilitating survival and evolution. Its mechanisms differ only in detail.

Sexual phases characterize the life cycles of the Sporozoa, and of most of the green flagellates; they are also seen in the cycles of a few animal flagellates, such as those of the wood roach. They are a part of the life histories of many Sarcodina and most ciliates. Sex in the Protozoa may even be more widespread than now appears, for the life cycles of the great majority of species are even yet totally or partially unknown.

Most species (other than the ciliates) form true gametes which then fuse to form a zygote, just as in the multicellular organisms that reproduce sexually. These gametes often exhibit dimorphism; the male forms, the microgametes, are relatively small and provided with flagella to facilitate searching out the females, the macrogametes, which are larger and frequently provided with stored nutrients. But in many cases the gametes are alike in appearance —though of course always different physiologically. When different, they are "anisogamous" (from three Greek words meaning "not" (*an*), "equal" (*iso*), and "marriage" (*gamos*); when similar, "isogamous.") Anisogamy, which appears to be more common than isogamy, is well illustrated by the malaria parasites and other Haemosporidina. Here the male form, a minute filament produced in much greater numbers, actively seeks out the larger, immobile female cell.

Scytomonas subtilis, a free-living flagellate often found in feces, was thoroughly studied by the eminent English protozoologist Clifford Dobell, and is a good example of a species producing like sex cells. Although it belongs to the same family as *Astasia longa*, in which sexual phenomena are unknown, the life cycle of *Scytomonas* is rather complex, including not only longitudinal division with well-marked mitosis, but encystment and sexual fusion. In laboratory cultures, Dobell found the complete cycle required about ten days. At first the flagellates divide actively, but after approximately a week the sexual phase begins, the organisms pairing in permanent union. There is nothing in their appearance to differentiate them either from one another, or from ordinary vegetative reproducing forms. As Dobell remarked, "Every monad, apparently, is a potential gamete." Stained preparations showed that an equationa reductional division takes place after fusion of the organisms, followed by the extrusion of one or more chromatin granules; this last Dobell interpreted as perhaps a second heteropolar reductional division. Both the reduction nucleus proper and the granules degenerate, and the surviving nuclei fuse. The organism, which is in reality a zygote, then either divides for a time, with eventual encystment, or it may encyst immediately.

The life history of the malaria parasites is described elsewhere (Chap. 24) The gametocytes, which are intraerythrocytic forms and show little sexual differentiation, normally undergo maturation* in the gut of the mosquito, but the process also takes place readily in drawn blood and may be observed under the microscope. From the microgametocyte six or eight tiny but actively motile filaments (often called "flagella") are thrown off within a few minutes after the blood is taken, and proceed to seek out the larger and quiescent macrogametes, which have at the same time been formed by the maturation of the macrogametocytes. Details of the maturation process in the female cell are less well known, although it is said to involve the formation of polar bodies; this claim is very questionable.

Union of the micro- and macrogametes now ensues, and the zygote (known as an "ookinete" because it is capable of independent motion) makes its way through the stomach wall of the mosquito and undergoes nuclear division and eventual sporozoite formation. It is now believed that the diploid number of chromosomes occurs only in the zygote, not only in the malaria parasites but in gregarines and coccidia generally. Reduction to the haploid condition takes place in the first nuclear division after fertiliza-

*Maturation, as used here, does not imply chromosome reduction; it means simply maturing.

tion. According to Wolcott (1954, *et seq.*) the normal (haploid) number of chromosomes is two, species differences apparently being reflected only by peculiarities in chromosome morphology.

Yet another pattern of sexual expression occurs in the ciliates, in which gametes as such are not produced. Instead, the organisms themselves join in temporary union or conjugation (except among the vorticellids, where a specialized smaller form fuses with a larger individual in permanent union). Their micronuclei then undergo several, usually three, preparatory divisions, of which the first and second are generally meiotic. All but two of the daughter nuclei degenerate, irrespective of whether the normal micronuclear complement is one or more than one, and the macronucleus, having no further role to play, is likewise absorbed into the cytoplasm.

One of the two surviving nuclei, or "pronuclei," in each individual now migrates across the region of contact between the partners and fuses with the stationary nucleus. The product of this nuclear union, known in the older literature as the "amphinucleus" and now usually as the "synkaryon," then proceeds to divide, and from it are derived ultimately the micro- and macronuclear equipment typical of the species. Details of the pattern naturally vary considerably in different forms. For example, in *Paramecium* the prophase of the first of the maturation, or "pregamic," divisions of the micronucleus assumes a characteristic crescent shape, while in the hypotrich *Pleurotricha lanceolata*, this stage very much resembles a minute parachute. Figures 12.2 and 12.3 show details of the conjugation process in species of these two genera.

From this condensed account it may be seen that sexual reproduction in the ciliates is essentially like that in most Metazoa, although it does not result in any immediate increase in numbers. Likewise each exconjugant corresponds to a fertilized cell, or zygote, and its offspring constitute a clone. Individual organisms in such a clone bear the same relation to their single progenitor as do individual body cells to the ovum from which each was ultimately derived. More precisely stated, it is the migrating and stationary nuclei which play the parts of sperm and egg respectively, and the synkaryon corresponds to the zygotic nucleus. A difference may be noted, however, although the stationary nucleus may be compared to the egg, it really corresponds only to the egg nucleus, since there is nothing like yolk nor does it have its own complement of cytoplasm.

It is now evident that sexual reproduction in the Protozoa involves two kinds of inheritance patterns rather than the one seen in the great majority of Metazoa. In forms like the flagellates (particularly the Phytomastigina, the genetics of which are much better known than of the Zoomastigina), where the diploid condition is confined to the zygote and is therefore ephemeral, any characteristic will appear for which there is a gene in the particular chromosomal complex of the organism. There can be no question, within the clone, of the existence of individuals carrying recessive traits. But in the typically diploid ciliates, the mechanism exists for biparental inheritance exactly corresponding to that of the sexually reproducing multicellular organisms, including of course man. The reassortment of chromosomes and genes, however, that takes place in the reduction division of the flagellate zygote, and the occurrence within many ciliate species of diverse groups and mating types, makes the actual genetic picture somewhat complicated.

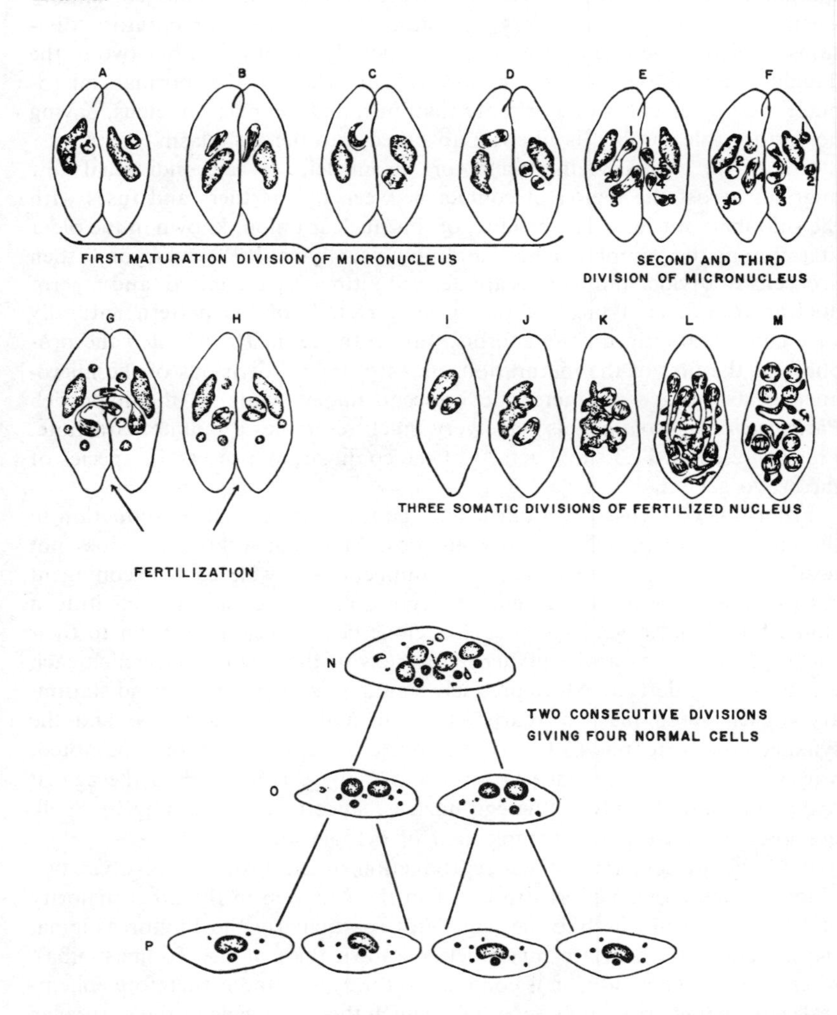

FIG. 12.2. A diagram of the fertilization process in *Paramecium caudatum*. Note the crescent-shaped spindles (c), and compare with the "parachutes" in *Pleurotricha*. (From Wichterman, *The Biology of Paramecium*; modified after Calkins.)

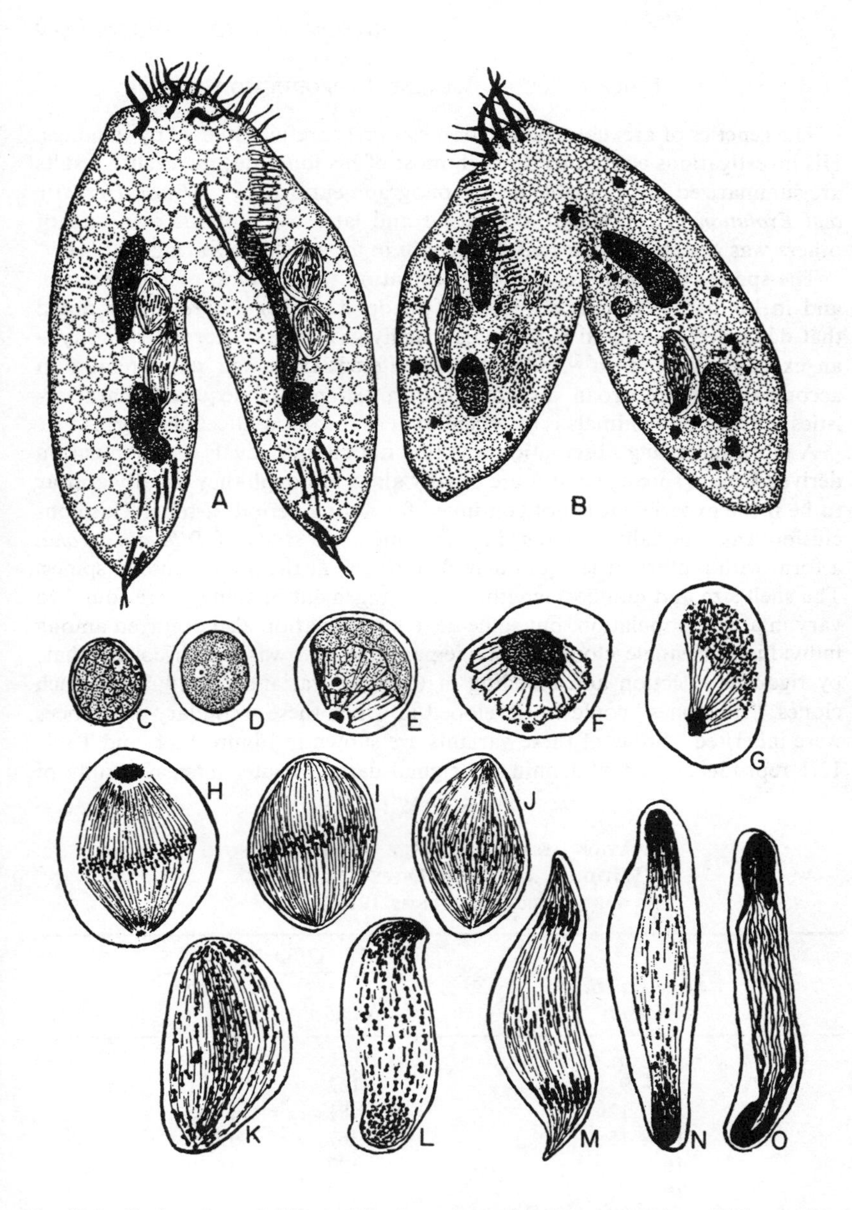

FIG. 12.3. Conjugation of *Pleurotricha lanceolata*: the first maturation division. The two pairs (*A* and *B*) show the micronuclei in the metaphase or later, although two of the micronuclei in *B* do not show clearly and one may be in a much earlier stage. *C–O* show the various stages arranged consecutively as worked out from a study of numerous pairs, beginning with the appearance of a division center and followed by the development of a typical "parachute." (After Manwell, *Biol. Bull.*)

Inheritance in Asexual Reproduction

The genetics of asexual reproduction was first carefully studied by Jennings. His investigations continued through most of his long life; the earlier results are summarized in a readable little monograph entitled *Life, Death, Heredity and Evolution in Unicellular Organisms* and later work of his own and of others was covered at considerable length in an article appearing in 1941.

The species to which he gave most attention were several of *Paramecium*, and in his earlier studies the shelled rhizopod *Difflugia corona*. He noted that deformed or injured individuals usually gave rise to normal progeny—an expected finding, in view of the dedifferentiation and redifferentiation accompanying protozoan division. The inheritance of acquired characteristics in these tiny animals is as exceptional as it is in multicellular forms.

A more surprising observation was that not all individuals in a population derived from a single ancestor are exactly alike, although they usually appear to be if the experiment is not continued for a long period of time. This conclusion was especially apparent in a very intensive study of *Difflugia corona*, a form with a more or less globular shell ringed at the aboral end by spines. The shell size and number, length, and arrangement of spines were found to vary in a wild population, but some degree of variation also occurred among individuals of single clones. More remarkable still was the discovery that, by rigorous selection and breeding of the most variant individuals in such clones, "subclones" could be developed in which these secondary differences were inherited. Some of these variants are shown in Figure 12.4, and Table 12.1 reproduces part of Jennings' original data. A later intensive study of

TABLE 12.1

Variation and Inheritance of Spine Length Within a Clone of Difflugia Corona (from Jennings 1916)

Spine Length of Parents	*Offspring*	
	Number	*Mean spine length*
4 - 6	21	10.38
7 - 9	162	11.01
10 - 12	481	11.85
13 - 15	367	12.90
16 - 18	129	14.39
19 - 21	26	14.34
22 - 24	15	16.34
25 - 31	18	17.06

Lengths given in units of 4 2/3 micra each; mean spine length for all, 12.54 micra.

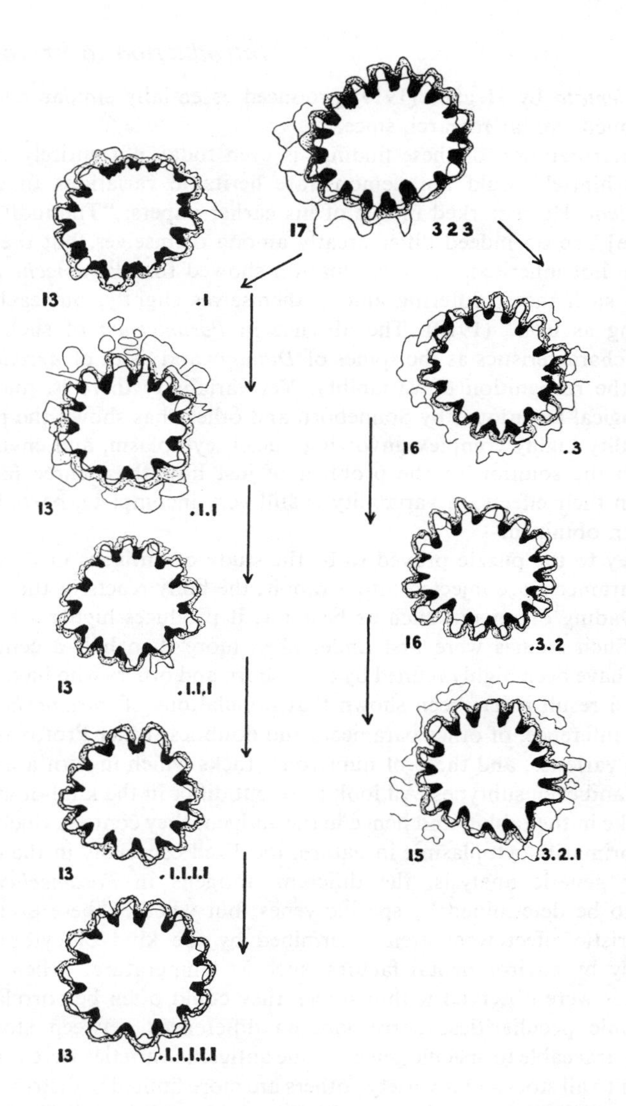

FIG. 12.4. Variable number of teeth about the mouth in different individuals of *Difflugia corona.* Here all the organisms had a like inheritance, no. 323 being the ancestor of the other eight. The first offspring (.1, at left) had 13 irregular teeth; from it came a line of six successive generations (five shown) each of which has 13 teeth. The third offspring of no. 323 (upper right, labeled .3) had 16 teeth; its offspring (.3.2) had 16; the offspring of the latter (.3.2.1) had 15. Numbers to the left of each individual refer to teeth; those to the right designate the individual. (From Jennings, *J. Exp. Zool.*)

Arcella dentata by Hegner (1919) produced essentially similar results, and so has much similar research since.

The interpretation of these findings is even today not entirely clear, and Jennings himself could not demonstrate heritable variations in clones of *Paramecium*. He remarked in one of his earlier papers, "The individuals of the [pure] line do indeed differ greatly among themselves, but these differences are not inherited. . . . Examination showed that *Paramecium* consists of many such races, differing among themselves slightly, but each race as unyielding as iron" (1910). The absence in *Paramecium* of such clearcut external characteristics as the spines of *Difflugia* and teeth of *Arcella dentata* hinders the recognition of variability. Yet variability there is, and the use of serological techniques by Sonneborn and others has shown the picture to be in reality highly complex, involving nuclei, cytoplasm, and environment. Although the solution to the problem of just how these three factors are related in their effects on variability is still very incomplete, partial answers have been obtained.

The key to the puzzle proved to be the study of antigens in *Paramecium*. When paramecia are injected into a rabbit, the body reacts in the same way as to invading organisms such as bacteria; it produces highly specific antibodies. Such studies were first undertaken more than half a century ago, but they have been highly refined by Sonneborn and others who have followed him. As a result it has been shown that populations of *Paramecium aurelia* (and, by inference, of other paramecia and doubtless other Protozoa) consist of many varieties, and these of numerous stocks which in turn are made up of types and even subtypes. All look alike but differ in the kind of antibodies they evoke in the rabbit, and hence in the antigens they contain. Such antigens appear primarily cytoplasmic in nature, localized especially in the cilia.

Under genetic analysis, the different antigens in *Paramecium aurelia* proved to be determined by specific genes, but whether these exerted their characteristic effect was often determined by the kind of cytoplasm and ultimately by environmental factors, such as temperature. When antigenic differences were observed within stocks they could often be correlated with cytoplasmic peculiarities; corresponding differences between stocks were generally traceable to specific genes. Some antigenic varieties were found to be common to all stocks in a variety; others are more limited in distribution. All told there is "an enormous variety of different antigenic types" (Beale, 1954).

To detect differences in antigens, a rabbit is first rendered immune by repeated injections of a centrifuged concentrate of paramecia injected into its blood circulation repeatedly for a week or two. By this time antibodies will have developed and the immune serum may be used to test other paramecia for specific antigens they may bear. To make the test, the serum is simply added, in various dilutions, to suspensions of the organisms. Those containing antigens will be immobilized and finally killed while others are unharmed by the antibodies in the serum, and the concentration of the antibody will be indicated by the degree of dilution at which it is still active (this is known as the "titre"). Since the first effect of the immune serum seems to be exerted on the cilia, which cease to beat, it is assumed that the antigens are found chiefly on their surfaces and perhaps also on the body surface. Tests such as these are astonishingly delicate and specific.

After finding that all populations of *Paramecium aurelia* contained organisms with a variety of antigens, and that a given organism was always characterized at a given time by a single antigen type, the next question was what determined such types. This was studied in paramecia reproducing asexually, in exconjugants, and in clones descended from individuals that had undergone autogamy (and were therefore homozygous). Individuals within clones were sometimes found to vary in antigenic type, but when these were separated and allowed to conjugate, and their offspring later rendered homozygous by autogamy, results like those shown in Figure 12.5 were

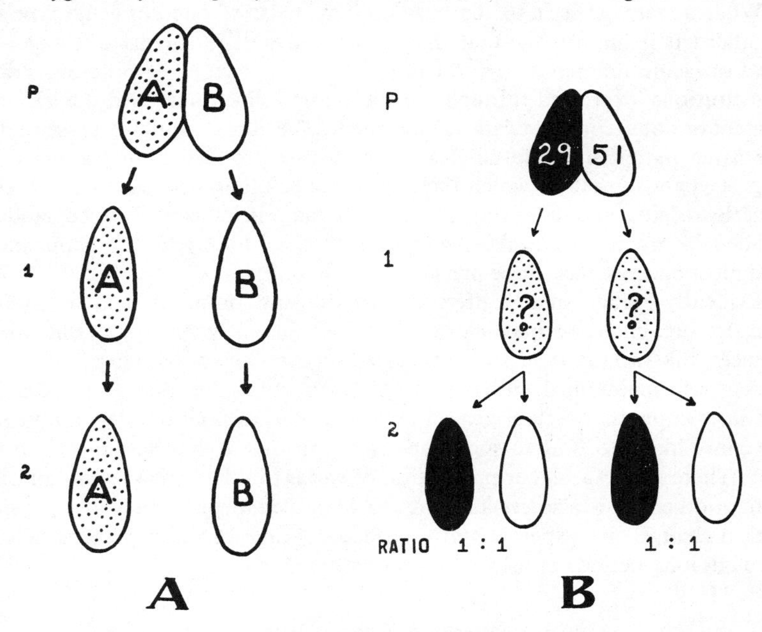

FIG. 12.5. The inheritance of antigenic differences in *Paramecium*. *A*. Inheritance of antigenic types A and B. P is the parental generation, and the next two generations are indicated by the numerals 1 and 2, the latter being obtained by self-fertilization. Each animal in generations 1 and 2 symbolizes an animal and its progeny obtained by repeated divisions. In this case difference in antigenic type appears to be due to some peculiarity of the cytoplasm. *B*. Here another antigenic type ("F") is involved. As in *A* above, the second generation is derived from the first after self-fertilization. Antigenic type in this case becomes apparent only after exposure to suitable conditions (i.e., the capacity to form the antigenic type rather than the type itself is inherited). For technical reasons the type cannot be determined in generation 1, but is revealed after self-fertilization in generation 2. (After Sonneborn, *Am. Sci.*)

obtained; thus it appeared that the two antigenic types in question were determined by a pair of alleles. However, it also appeared that when an exchange of cytoplasm occurred during conjugation, the type of antigenic reaction exhibited by the exconjugants might be altered. Thus it was evident that though ultimate control of antigenic type was nuclear (genic), immediate control was cytoplasmic; cytoplasm governed the action of the genes. It also found that nuclear control was not immediately exerted even through the cytoplasm, about five fissions being required before the antigenic type was finally determined. Further observation also showed that in some stocks the antigenic type might change with a relatively small change in temperature, or even with the amount of food available; however, it remained constant as long as such factors were maintained. Thus genes, cytoplasm, and environment combine in complex fashion to make the organism whatever it is at any given time.

What advantage is it to the paramecium to be of one antigenic type or another? It is improbable that there is any advantage (or disadvantage) at all. Antigenic differences are no more than a reflection of differing genic constitutions, expressed through the cytoplasm (which may also differ), and evident or not according to the environment. They are doubtless of essentially the same nature in these ciliates as in disease-producing microorganisms (e.g., trypanosomes), in which their existence has long been known. But the genetic origins and behavior of such characteristics can only be readily studied in organisms with sexual phases in their life cycles; trypanosomes and most bacteria lack these phases.

Evidently rather small differences in the environment, such as minor temperature changes, may cause variations in antigenic type. But what induced the variations in the Difflugias and Arcellas of Jennings' cultures, which were maintained under seemingly identical and constant conditions? We now know that certain genes in some species of multicellular organisms are more likely to mutate spontaneously than others, but we do not know why. There is also a higher proportion of variant individuals in some species than in others. It is also certain that considerable morphological and physiological changes are experimentally produced; some of these changes persist through long periods of asexual reproduction.

Mating Types and Sexual Reproduction

Though sex is nearly universal among living things, it is expressed in many different ways and such variations are even more marked among the Protozoa than in the Metazoa. When conjugation in the ciliates was first carefully studied its sexual nature seemed clear, but the existence of complications such as mutually infertile varieties and mating types in populations of apparently identical individuals was long unrecognized. The use of the term "mating type" is relatively new and should be restricted to the ciliates;

when applied to sexual stages in the flagellates, as it has been by some, it has a somewhat different meaning. Probably the original discovery of such types should be credited to Maupas, the French investigator whose work Cole (1926) said would "ever remain one of the most significant achievements in the whole range of zoological literature." Maupas noticed that conjugation did not ordinarily occur among closely related ciliates, but it was not long before many apparent exceptions were discovered, and his claims were therefore discounted. Now it is known that for conjugation to take place, the partners must indeed be physiologically different, which means that they are in general of different parentage. They are often found to differ morphologically as well. Such individuals are said to belong to different mating types. Thus a mating type may be thought of as the ciliate equivalent of sex. Much of what we know about the expression of sex among the ciliates we owe to Sonneborn's studies of *Paramecium aurelia*.

Species of ciliates, however, have their own distinctive sex patterns. For one thing, inability of closely related individuals to conjugate (at least in *Paramecium aurelia*) is not entirely a question of what we might call, for lack of anything better, "infusorian consanguinity." It depends rather on whether or not they possess macronuclei of like ancestry. Though members of a clone have, by definition, the same progenitor, some of them may at times undergo reorganization (e.g., autogamy), with concurrent reconstruction of the nuclear apparatus. An essentially similar reconstruction follows conjugation, but in the latter case biparental inheritance occurs and the synkaryon (fertilization nucleus) may therefore be heterozygous for various traits rather than invariably homozygous, as in postautogamous individuals. In the nuclear reorganization following conjugation each exconjugant emerges with two newly formed macronuclei, one of which each of its daughters by the first fission receives. Similarly, each of the two immediate descendants of a postautogamous individual receives a newly formed macronucleus. Since all descendants in each of these cases will have identical macronuclei as long as multiplication is by ordinary fission each line of descent is said to constitute a "karyonide" (also often spelled "caryonide"). A conjugating pair may therefore give rise to four karyonides, and an individual undergoing autogamy to two. Members of different karyonides may or may not be of the same mating type.

Some half dozen genera of ciliates have now been studied with reference to breeding patterns; of these certain species of *Paramecium* and *Euplotes* are the best known. Recently intensive study of *Tetrahymena pyriformis* has revealed that it, too, is well provided with mating types, a total of more than 40 now being known—"one of the largest breeding systems among the ciliates," as Elliott (1959) remarks. And there is no assurance that all of them are yet discovered. Painstaking collection and examination of samples of these ciliates from different localities, often widely separated, have shown that a number of groups or varieties exist, each consisting of two or more mating types; one variety of *Tetrahymena* contains at least nine. Conjugation normally occurs only between complementary mating types within a variety, or sometimes such mating types of what are presumably related varieties. When such relationships are less close, abortive mating reactions may occur but without actual conjugation. Table 12. 2, compiled by Sonneborn (1957), gives a partial picture of the situation as seen in *Paramecium aurelia*. Of the fifteen varieties presently recognized, only the eight between which some mutual conjugation is possible are included.

Clearly varieties of this kind constitute to all intents and purposes natural species, since interbreeding either does not occur or is rare. A variety is thus a natural gene pool, and yet morphological differences among varieties of this sort are slight or absent. There are therefore practical reasons for not calling these varieties species, and Sonneborn has coined the term "syngen" (or "whatever other new term meets with the approval of the experts") (Sonneborn, 1957). It seems likely that this new term will stick. When crossing does take place between varieties, the partners (progeny, in the genetic sense) are seldom viable. Such unions seem to be in defiance of natural law. But there seem to be exceptions to this rule, as to all others. Variety 16 of *Paramecium aurelia* (also known as *Paramecium multimicronucleatum*), for example, apparently has four mating types, each of which may mate quite promiscuously with any of the other three, and the case of *Tetrahymena pyriformis*, two varieties of which are known to contain eight or more mating types, has already been mentioned. On the other hand, change of mating type may occur occasionally within a clone, and then conjugation ("selfing") within the clone becomes possible, thus resulting in intensive inbreeding.

How did genetically independent units such as these varieties arise? Sonneborn suggests that geographic distribution has probably been the chief cause, just as with species of the conventional sort. Despite the wide distribution of most morphologically defined species of Protozoa, he believes that free-living forms among them have only a limited ability of dispersal. And he remarks, as others have done, that the generally accepted definition of species as a reproductively isolated group hardly applies to the Protozoa. It would be interesting to know whether mating types occur among the parasitic ciliates; geographic isolation, due to the propensity of many host species to travel, is less likely to occur, one would think.

The discovery of mating types in some ciliates enabled their genetics to be studied in much the same fashion as inheritance in the familiar laboratory strains of fruit flies and mice. The first important observation was that members of certain karyonides derived from wild individuals were always of like mating type, whereas in other karyonides they were of complementary types. It was also shown that in ordinary asexual reproduction the offspring were the same mating type as the parent. The explanation of this apparent contradiction is that the ability to produce one or two mating types is

THE SYSTEM OF SEXUAL REACTIONS AMONG MATING TYPES OF THOSE VARIETIES OF *P. aurelia* IN WHICH INTERVARIETAL REACTIONS OCCUR*
(From Sonneborn, 1957)

Variety	Mating type		1		3		4		5		7		8		10		14	
			I	II	V	VI	VII	VIII	IX	X	XIII	XIV	XV	XVI	XIX	XX	XXVII	XXVIII
1	I	..	—	+++	—	—	—	—	—	++	—	—	—	±	—	—	—	—
	II	..	C	—	+	—	—	—	++	—	++	—	±	—	—	—	—	—
3	V	..	0	C	—	+++	—	—	—	—	—	—	—	++	—	—	—	—
	VI	..	0	0	C	—	—	—	—	—	±	—	—	—	—	—	—	—
4	VII	..	0	0	0	0	—	+++	—	—	—	—	—	+++	—	—	—	—
	VIII	..	0	0	0	0	C	—	—	—	—	—	++	—	+	—	+	—
5	IX	..	0	C	0	0	0	0	—	+++	—	—	—	—	—	—	—	—
	X	..	0	0	0	0	0	0	C	—	±	—	—	—	—	—	—	—
7	XIII	..	0	C	0	0	0	0	0	0	—	+++	—	±	—	—	—	—
	XIV	..	0	0	0	0	0	0	0	0	C	—	—	—	—	—	—	—
8	XV	..	0	0	0	0	0	C	0	0	0	0	—	+++	—	—	—	—
	XVI	..	0	0	C	0	C	0	0	0	0	0	C	—	+	—	+	—
10	XIX	..	0	0	0	0	0	0	0	0	0	0	0	0	—	+++	—	?
	XX	..	0	0	0	0	0	0	0	0	0	0	0	0	C	—	+	—
14	XXVII	..	0	0	0	0	0	0	0	0	0	0	0	0	0	0	—	+++
	XXVIII	..	0	0	0	0	0	0	0	0	0	0	0	0	0	0	C	—

*Symbols above the diagonal refer only to occurrence of mating reactions (adhesion, agglutination); symbols below the diagonal refer to the occurrence of complete conjugation. Varieties 10 and 14 are newly discovered and still incompletely studied; their reactions may be more extensive than now known. +++ maximal reaction. ++ reduced mating reaction. + weak mating reaction. ± barely detectable reaction. —no mating reaction. C, conjugants formed. O, no conjugants formed.

inherited, depending on two allelic genes. The dominant gene of the pair determines the appearance of two mating types, whereas the recessive results in only one. The demonstration of heredity as a controlling factor in *Paramecium aurelia* required not only meticulously pedigreeing many individuals and studying of their progeny, both after fission and conjugation, but much time and patience.

As in other cases of inheritance where a single pair of alleles is concerned, paramecia may be purebred for either characteristic, or hybrid. If one considers a given cross (conjugating pair) involving individuals one of which is purebred for each of these characteristics, the exconjugants will all be hybrid, and exhibit the dominant trait, the ability to produce offspring of both mating types. Exchange of cytoplasm may also contribute to mating type change after reproduction (Fig. 12.6). Kimball felt that "some mechanism for self-reproduction of cytoplasmic materials" must exist. There is still the problem of how a single gene can produce two sorts of individuals, some of one type and some of another, but each type breeding true during ordinary asexual reproduction. Sonneborn found that this depended on the

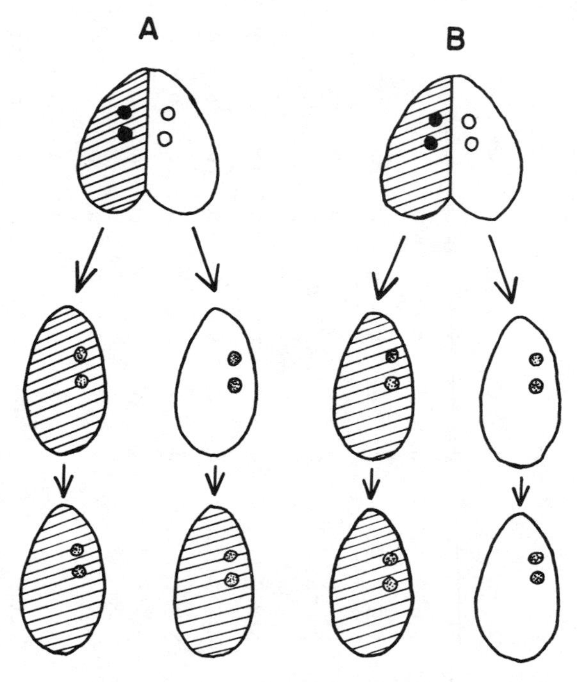

FIG. 12.6. Diagram showing differences between two groups, A and B, of varieties of *Paramecium aurelia* in inheritance at conjugation. In both the micronuclei (macronuclei are not shown) are identical, but in *A* the phenotype of the two members becomes identical while in *B* differences existing prior to conjugation persist. (After Kimball, *Am. Naturalist.*)

macronucleus, as indicated earlier, and that which of the two mating types a given macronucleus determines is, at least in part, a function of the environment. In his experience it was a matter of the temperature to which the exconjugants and their immediate progeny were exposed during the critical period of postmating nuclear reorganization. It might therefore be thought that the dominant gene operates by causing the development of a briefly more labile macronucleus; once the mating type has been "set" it is no longer susceptible of change. The problem, however, seems to have no such simple answer. Time of day as well as temperature plays a role, for Sonneborn observed that mating type in some strains exhibits a diurnal rhythm, changing at night and back again in the morning. Nanney (1957) observed that when cytoplasmic fusion was induced in conjugating pairs (type VII and type VIII), as by antiserum treatment, a change of mating type from type VII to VIII beyond the expected 50–50 ratio occurred. On the other hand a change in the reverse direction occurred in unstable karyonides, in which selfing occurred. In both cases the amount of change could be increased by exposure to higher than normal temperatures. Nanney suggests that two macronuclear functions are involved: one is control of the phenotype (mating type) and the other is control of the cytoplasmic conditions governing in part nuclear differentiation in the next generation. Evidently *Paramecium aurelia* is a highly complex animal.

Proof of the manner of inheritance of mating types, and also of the role of the macronucleus, was gained primarily from a study of individuals which had undergone autogamy. In this process (which, it will be recalled, is basically much like conjugation) there is a complete nuclear reorganization involving macronuclear degeneration and meiosis of the micronuclei. Two daughter pronuclei (derived from the same parental haploid nucleus, and hence genetically identical) fuse, thus restoring the diploid condition. It is evident that if the original hybrid condition is represented by the gene symbols "Aa," reduction will result in segregation to the "A" or "a" condition, and after fusion of the pronuclei it will be either "AA" or "aa." Careful observation of the mating behavior of the individuals in different clones will therefore reveal their true genetic pattern in mating types. Simply wait long enough for autogamy to occur, and then test karyonides derived from such individuals for ability to conjugate.

It was inferred that the gene responsible for the mating type operates through the macronucleus because the mating type is not fixed until the new macronuclear complex has been developed, whether after conjugation or after autogamy. Sonneborn states, "The clue to the process involved is given by those reorganized individuals from which two lines of descent arise at the first fission [following conjugation or autogamy]. At the same fission, each of the two resulting individuals receives one of the two new macronuclei normally formed in each reorganized individual" (Sonneborn, 1939). The crucial discovery was that postconjugation nuclear segregation, which occurred in some individuals at the first and in others at the second fission, coincided with the time at which mating type was also determined.

To determine how mating type was inherited, individuals of stocks in which only one mating type had been observed were crossed with individuals from stocks in which there were two. All such exconjugants gave rise to clones developing (after autogamy) two mating types. Thus it was clear that in the F1 generation the gene governing the appearance of two mating types was the dominant one.

The progeny of this cross were again crossed, and this time the exconjugants originated two-type and one-type karyonides in the ratio of approximately three to one. In the actual experiment, of 120 pairs of conjugants 88 bred karyonides showing the two-type condition and 32 the one-type condition. The results were typically Mendelian for the F2 generation. Half of these were then presumably hybrid, one-quarter purebred for the dominant (two-type) condition, and the other quarter purebred for the recessive (one-type) condition. The results of this and the following experiment are summarized in Table 12.3.

To clinch the matter, back crosses were tried of the two-type stocks, showing that they were indeed made up of two-thirds hybrid (heterozygous)

TABLE 12.3

INHERITANCE OF MATING TYPES I AND II IN PARAMECIUM AURELIA
(Based on Sonneborn, 1939)

Nature of Cross	*Number of Pairs*	*Progeny*[a] *(mating type)*	*Presumed Genotype* Parents	*Presumed Genotype* Progeny[b]
Purebred 2-type			AA	
by	149	all 2-type		Aa (all)
Purebred 1-type			aa	
F1 of above cross		2-type (81)	Aa	Aa ($\frac{1}{2}$)
by	158			
Purebred 1-type		1-type (77)	aa	aa ($\frac{1}{2}$)
F1 (hybrid)		2-type (88)	Aa	AA($\frac{1}{4}$)
by	120			Aa ($\frac{1}{2}$)
F1 (hybrid)		1-type (32)	Aa	aa ($\frac{1}{4}$)
F2 dominant		2-type (6)	AA or Aa	Aa (all)
by	19	1-type ($\frac{1}{2}$) } (13)		aa ($\frac{1}{2}$)
Purebred 1-type		2-type ($\frac{1}{2}$) }	aa	Aa ($\frac{1}{2}$)

[a]Numbers in parentheses refer to number of conjugating pairs of each cross giving 1-type or 2-type progeny. In the last cross *half* the progeny of 13 of the 19 pairs were of *each* type.

[b]Fractions in parentheses indicate the fractional number of each genotype in the offspring of a given type of cross.

and one-third purebred (homozygous) individuals for the two-type characteristic. Actual results of the back crosses were offspring developing both mating types from 81 of the 158 conjugating pairs, and one type only from the other 77 pairs—a ratio close to the theoretical 1: 1. In this way, Sonneborn remarked, was made "the first discovery of inheritance in Mendelian ratios in the ciliate Protozoa."

When mating types in *Paramecium aurelia* were first discovered it appeared that there might be only a few, but with the careful testing of paramecia collected from many sources the number of recognized types multiplied. There are now more than 30, divided into 16 varieties, each variety (with one exception) containing two mating types. The progenitors of these types were brought from localities throughout the world, and in nature some have a restricted distribution. Their mating behavior, doubtless in response to local conditions, has been found to vary somewhat with optima at different ranges of temperature, intensities of illumination, and other environmental factors.

Genetic study has also revealed that the varieties can be put into two groups, A and B. In Group A the cytoplasm seems to play no part in determining mating type, whereas its role in Group B is important. Some evidence indicates (Wood, quoted by Sonneborn, 1957) that a single gene governs the occurrence of cytoplasmic exchange during conjugation, but that the actual effect of this gene is strongly dependent on temperature and age of the clone, being greatest at low temperatures and in extreme youth. The way in which such differences work out in nature is obscure. Laboratory studies indicate that some varieties are extreme inbreeders, and others less so. Even conjugation, at least in some species (e.g., *Tetrahymena pyriformis*, according to Elliott, 1958) appears to be very often lethal. Yet there is no doubt that both genera have been highly successful in a biological sense, for both occur commonly throughout the world.

Inheritance of the Killer Trait

Although the discovery of mating types has furnished a highly useful tool for the study of ciliate genetics, so has Sonneborn's observation that some paramecia secrete a substance lethal to other paramecia, and the latter are remarkably susceptible to it. In this unique relation, two factors are involved. One is heritable, depending on a pair of allelic genes, and the other is a substance known as "kappa" occurring in the cytoplasm as very fine particles. "Kappa" is responsible for the production of a poison ("paramecin"), lethal to certain other paramecia, which are known as "sensitives." Fortunately for the species, members of what might be called the Kappa Fraternity (alias the "killers") are a small and select group, united by an actual kinship, since all possess a dominant gene. But to produce paramecin they must have "kappa" too. When homozygous for the recessive gene of the pair of alleles, any kappa present disappears and the organisms are then "sensitives." Figures 12.7 and 12.8 show diagrammatically what happens when killers are crossed with sensitives. It will be noticed that both genes and cytoplasm are concerned, as in the inheritance of mating types of Group B.

The actual nature of kappa remains in doubt, although its particles are known to consist at least in part of deoxyribose nucleic acid (DNA) and to be self-producing. A certain number of them must be present before the effects of kappa are evident. Actual transfer of kappa is effected across the cytoplasmic bridge which may unite paramecia during conjugation. Though only a few paramecia are killers, it is said that all except the killers are sensitive to paramecin. It is therefore evident that kappa has most of the characteristics of a virus, as even Sonneborn pointed out, though he originally

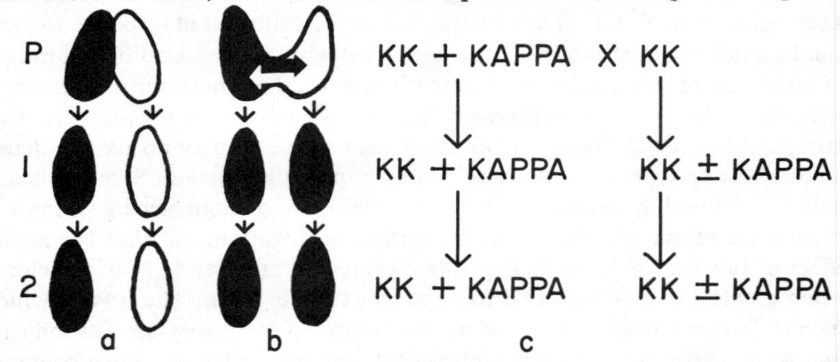

FIG. 12.7. Inheritance of killer and sensitive traits in *Paramecium*. Black symbolizes killers, white indicates sensitives. *P*, parents. *1*, first generation. *2*, second generation, obtained by self-fertilization in animals of first generation. Column *a* gives results when no cytoplasm is exchanged between parents, column *b* when the parents exchange cytoplasm. Column *c* represents the genes and cytoplasmic condition (presence or absence of *kappa*) in the various animals and cultures. (After Sonneborn, *Am. Sci.*)

preferred to call its particles "plasmagenes." The nature of paramecin has received some study, and this also seems to be a desoxyribonucleoprotein. It is a lethal poison indeed for sensitive paramecia, a single particle being sufficient to cause death. Very rarely, infected individuals become immune, and survive (*Mueller, 1965).

Later studies have shown that other kinds of killing agents and killers exist in the cloak and dagger microworld of paramecia. That the killing agents differ is evidenced by the various kinds of death they cause: the victims may develop humps on their bodies, their cytoplasm may be vacuolized, paralysis may develop, or they may be sent into attacks of spinning lasting hours or even days, from which only death brings relief. Most brutal of all perhaps are the "mate-killers." Death comes to the "sensitive" member of a pair of conjugants when it has been unlucky enough to choose a killer partner; here the lethal agent is not something in the culture fluid, but exerts its effects only during the period of contact. In all these cases the presence of kappa and degree of sensitivity to it are genetically determined, but the latter is influenced to a degree by the environment (e.g., starved individuals are more susceptible than well-nourished ones.)

These studies are significant not because they illumine the origin of characteristics in *Paramecium* relating to its normal living, but because they supply evidence for the occurrence of Mendelian inheritance in Protozoa. Indeed, in *Paramecium aurelia* some twenty traits dependent on single genes are now known, although no cases of linkage have been discovered. This is hardly surprising, as this species of *Paramecium* is said to have at least 30 to 40 chromosomes in the diploid condition. Still, the problem of making a chromosome map for *Paramecium aurelia* may prove less difficult than for

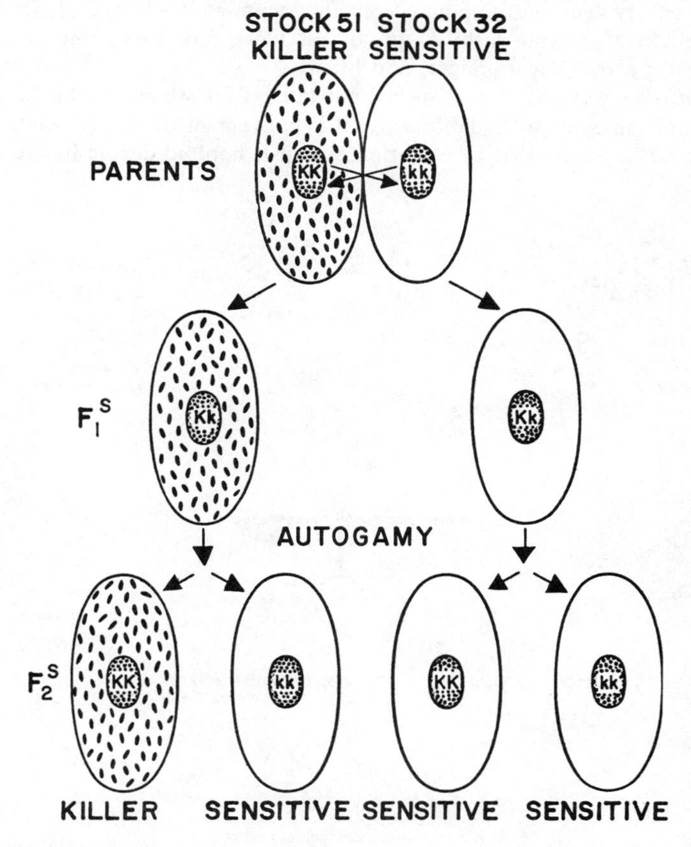

FIG. 12.8. The inheritance of the killer character in *Paramecium aurelia*. This character depends on two factors, the presence of *kappa* in the cytoplasm, and a dominant Mendelian gene in the nucleus. Without the latter (K) *kappa* disappears and will not reappear even with K, unless *kappa* is reintroduced from the cytoplasm of another killer organism. (After Beale.)

some other species of the genus, certain of which have many more chromosomes; *Paramecium calkensi* has been claimed to possess 150.

Inheritance in the Green Flagellates

Although much more has been learned about the genetics of ciliates, and especially *Paramecium*, than of any other Protozoa, inheritance in certain green flagellates is now being intensively studied, and they are proving equally good tools—even better from some points of view. Species of the alga-like genus *Chlamydomonas* have proved especially useful. The first researches on the genetics of this organism were made by Pascher, almost half a century ago, and he pointed out that some of its characteristics were inherited in Mendelian fashion. Important work has been done since by Ebersold, Levine, Lewin, Sager, and others.

Chlamydomonas has several special advantages for studies of this kind: it grows well on agar in the laboratory, it is convenient to handle, mutations seem to be easily induced by radiation, and it is haploid during its life cycle

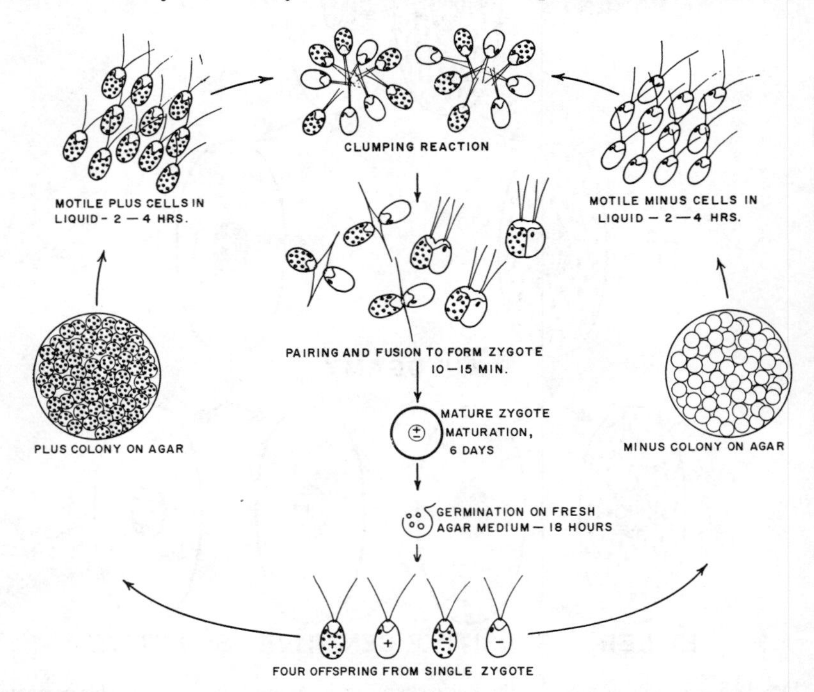

Fig. 12.9. Life cycle of *Chlamydomonas reinhardi*, a phytomonad useful in the study of protozoan (or algal) genetics. The "sex" is indicated by a plus or minus sign, and the presence or absence of a given inherited character (e.g., chlorophyll in organisms grown in darkness) by dots or their absence. (Modified slightly from Sager.)

except for a brief period in the zygote stage. The normal number of chromosomes in the haploid state is eight. The life cycle, first described in 1876, is shown in Figure 12.9, together with the duration of some of the stages under laboratory conditions. Gamete formation appears to be directly due to nitrogen depletion, and indirectly may be induced by exposure to light (Sager and Granick, 1954). The gametes themselves are isogamous, though by suitable techniques the sexes may be distinguished. Since they differ only physiologically, they may be referred to simply as "plus" or "minus" in mating type. After pairing and fusion, the zygote encysts; this is the only resistant stage in the life cycle, and in nature it doubtless aids in survival under adverse conditions. When maintained under suitable conditions in the laboratory the zygote germinates after about six days, with the emergence of from four to eight biflagellated offspring. These, which are motile in water, but not on agar, form colonies and may be examined for inherited characteristics. Such traits include biochemical deficiencies of the sort studied in the mold *Neurospora*, abnormalities in chlorophyll formation or functioning, paralysis of flagella, loss of eye-spots, and streptomycin resistance. Most of these mutant characters have been induced by exposure to radiation, particularly ultraviolet.

Different species of *Chlamydomonas* were crossed by Pascher (1916). Although certain parental characteristics showed Mendelian segregation in the offspring, others, such as body wall markings, exhibited apparent blending. Unfortunately Pascher did not continue research in this field. Recent work has involved the crossing of mutants of given species, particularly *Chlamydomonas reinhardi*. It has been found that most such characteristics segregate in Mendelian fashion, giving F1 ratios of 2:2 (there are no problems of dominance and recessiveness to complicate the picture, since the offspring are always haploid), and enough has now been learned about the genetics of *Chlamydomonas* to establish linkage of heritable characteristics into seven groups (Ebersold and Levine, 1959), presumably representing seven of the eight chromosomes. Certain characteristics, however, seem not to be transmitted in Mendelian fashion. One such is streptomycin resistance. Curiously, the determiner for this trait could apparently be passed along in the gametes of one mating type only (Sager, 1955). When present in the parent producing "plus" gametes it appeared in *all* the offspring, but did not appear at all when the streptomycin-resistant parent in the cross contributed the "minus" gametes.

Although a beginning has been made in our study of the genetics of Protozoa, it is still only a beginning. We know a good deal about inheritance in *Paramecium aurelia*, though less about other species in the genus. A start has been made in the study of *Chlamydomonas*. About inheritance in most of the other 50,000 species or so of the Protozoa, we are still profoundly ignorant. Probably it is safe to conclude that the Mendelian pattern is universal, with doubtless minor but interesting modifications here and there.

Protozoan genetics needs to be studied for its own sake, and not because, since Protozoa are assumed to be primitive, their genetic mechanism may prove to be especially simple. The author of a recent and otherwise excellent text in microbiology remarked "Protozoa are the most primitive members of the animal kingdom." The Protozoa may be minute, but they are not primitive.

REFERENCES

Beale, G. H. 1954a. *The Genetics of Paramecium Aurelia.* Cambridge University Press.

——— 1954b. Heredity in *Paramecium. Endeavour*, 13: 33–36.

Cole, F. J. 1926. *The History of Protozoology.* London: University of London Press.

Dallinger, W. H. 1887. The President's Address. *J. Roy. Mic. Soc.*, 7: 185–99.

Dobell, C. C. 1909. The structure and life history of *Copromonas subtilis*, nov. sp.: a contribution to our knowledge of the Flagellata. *Quart. J. Mic. Sci.*, 52: 75–120.

Downs, L. E. 1956. The breeding system in *Stylonychia putrina. Proc. Soc. Exp. Biol. and Med.*, 93: 586–7.

Ebersold, W. T. and Levine, R. P. 1959. A genetic analysis of linkage group 1 of *Chlamydomonas reinhardi. Zeitschr. f. Vererbungsl.*, 90: 74–82.

Elliott, A. M. 1959. A quarter century exploring *Tetrahymena. J. Protozool.*, 6: 1–7.

Gruchy, D. F. 1955. The breeding system and distribution of *Tetrahymena pyriformis. J. Protozool.*, 2: 178–85.

Hegner, R. W. 1919. Heredity, variation, and the appearance of diversities during the vegetative reproduction of *Arcella dentata. Genetics*, 4: 95–150.

Huff, C. G. 1931. The inheritance of natural immunity to *Plasmodium cathemerium* in two species of *Culex. J. Prev. Med.*, 5: 249–59.

Jennings, H. S. 1910. Experimental evidence on the effectiveness of selection. *Am. Nat.*, 44: 136–45.

——— 1916. Heredity, variation, and the results of selection in the uniparental reproduction of *Difflugia corona. Genetics*, 1: 407–534.

——— 1920. *Life and Death, Heredity and Evolution in the Unicellular Organisms.* Boston; Badger.

——— 1929. *Genetics of the Protozoa.* Bib. Gen. V., pp. 105–330. The Hague: Martinus Nijhoff.

——— 1940. Chromosomes and cytoplasm in Protozoa. A.A.A.S., *Pub. 14*, pp. 44–55.

——— 1941. Inheritance in the Protozoa. In *Protozoa in Biological Research*, G. N. Calkins and F. M. Summers (eds). New York: Columbia University Press.

Jollos, V. 1913. Experimentelle Untersuchungen an Infusorien. *Biol. Zentralbl.*, 33: 222–36.

——— 1921. Experimentelle Protistenstudien. I. Untersuchungen über Variabilitat und Vererbung bei Infusorien. *Arch. f. Protistenk.*, 43: 1–222.

——— 1934. Dauermodifikationen und Mutationen bei Protozoen. *Arch. f. Protistenk.*, 83: 197–219.

Kimball, R. F. 1950. The effect of radiation on the genetic mechanisms of *Paramecium aurelia. J. Cell. & Comp. Physiol.*, 35(Suppl. 1): 157–69.

Levine, R. P. and Ebersold, W. T. 1958. Gene recombination in *Chlamydomonas reinhardi. Cold Spring Harbor Symp. on Quant. Biol.*, 23: 101–9.

Lewin, R. A. 1953. The genetics of *Chlamydomonas moewusii* Gerloff. *J. Genetics*, 51: 543–60.

MacDougall, M. S. 1929. Modifications in *Chilodon uncinatus* produced by ultraviolet radiation. *J. Exp. Zool.*, 54: 95–109.

Manwell, R. D. 1928. Conjugation, division and encystment in *Pleurotricha lanceolata. Biol. Bull.*, 54: 417–63.

Maupas, E. 1889. Le rajeunissement karyogamique chez les Ciliés. *Arch. zool. exp. et gén.*, (2) 7: 149–517.

Nanney, D. L. 1957. Mating type inheritance at conjugation in Variety 4 of *Paramecium aurelia. J. Protozool.*, 4: 89–95.

Preer, Jr., J. R. 1957. Genetics of the Protozoa. *Ann. Rev. Microbiol.*, 11: 419–38.

Sager, R. A. 1955. Inheritance in the green alga *Chlamydomonas reinhardi. Genetics*, 40: 476–89.

——— 1960. Genetic systems in *Chlamydomonas. Science*, 132: 1459–65.

——— and Granick, S. 1954. Nutritional control of sexuality in *Chlamydomonas reinhardi. J. Gen. Physiol.*, 37: 729–42.

Schoenborn, H. W., 1954. Mutations in *Astasia longa* produced by radiation. *J. Protozool.*, 1: 170–73.

——— 1956. Protection against lethal damage induced by ultraviolet radiation. *J. Protozool.*, 3: 97–9.

Shirley, Jr., E. S. and Finley, H. E. 1949. The effects of ultraviolet radiation on *Spirostomum ambiguum. Trans. Am. Mic. Soc.*, 58: 136–53.

Sonneborn, T. M. 1939a. Sexuality and related problems in *Paramecium. Collecting Net*, 14: 1–6.

——— 1939b. *Paramecium aurelia:* Mating types and groups: lethal interactions; determination and inheritance. *Am. Nat.*, 73: 390–413.

——— 1942. Inheritance in ciliate Protozoa. *Am. Nat.*, 76: 46–62.

——— 1947. Recent advances in the genetics of *Paramecium* and *Euplotes. Advances in Genetics*, 1: 263–358.

——— 1949. Beyond the gene. *Am. Sci.*, 37: 33–59.

——— 1950. Partner of the gene. *Sci. Am.*, 183: 30–9.

——— 1951. Some current problems of genetics in the light of investigations on *Chlamydomonas* and *Paramecium. Cold Spring Harbor Symp. on Quant. Biol.*, 16: 483–503.

——— 1954. The relation of autogamy to senescence and rejuvenescence in *Paramecium aurelia. J. Protozool.*, 1: 38–53.

——— 1957. Breeding systems, reproductive methods, and species problems in *Protozoa.* A.A.A.S., *Pub. No. 50*, pp 155–324

——— 1959. Kappa and related particles in *Paramecium.* In *Advances in Virus Research.* New York: Academic Press.

——— 1960. The gene and cell differentiation. *Proc. Nat. Acad. Sci.*, 46: 149–65.

van Wagtendonk, W. J. 1946. The killing substance paramecin: chemical nature. *Am. Nat.*, 82: 60–8.

Wichterman, R. 1953. *The Biology of Paramecium*, Philadelphia, Blakiston.

——— 1957. Biological effects of radiations on the Protozoa. *Bios*, 28: 3–20.

Wolcott, G. B. 1954. Nuclear structure and division in the malaria parasite *Plasmodium vivax*. *J. Morphol.*, 94: 353–66.

——— 1955. Chromosomes of the four species of human malaria studied by phase microscopy. *J. Heredity*, 46: 53–7.

——— 1957. Chromosome studies in the genus *Plasmodium*. *J. Protozool.*, 4: 48–51.

ADDITIONAL REFERENCE

Mueller, J. A. 1965. Kappa-affected paramecia develop immunity. *J. Protozool.* 12: 278–81.

13

Principles and Problems of Classification

> "Hence, in determining whether a form should be ranked as a species or a variety, the opinion of naturalists having sound judgement and wide experience seems the only guide to follow." Charles Darwin, *The Origin of Species*, 1859.

Were this book written in the eighteenth century, the great pioneer in taxonomy Carolus Linnaeus might have contributed the chapter on classification. Our task would then have been easier. For Linnaeus recognized only three genera of Protozoa in the classical Tenth Edition of his *Systema Naturae*, published in 1758. Nor was the concept of species a problem for him: he believed that each kind of living thing was individually created, as told in the Biblical story in Genesis. If Linnaeus discerned few species of Protozoa, it was doubtless because he felt that the Lord, after creating the first ones, found them too insignificant to bother about further.

How Many Species?

Much has been added to our knowledge since Linnaeus did his great work in the middle 1700's. Although no protozoologist would risk his reputation by saying exactly how many species of Protozoa exist, the best estimates agree that even Weisz (1954), who commits himself no further than to suggest a figure of "50,000 plus," is an arch conservative on this matter. In implication at least, his is a gross understatement.

At present there are no reliable grounds on which to base a definite estimate of the number of protozoan species. On intensive study most genera of Protozoa have been found to contain many more species than were at first suspected. After more than twenty years of research Gojdics discovered that there are at least 155 species of *Euglena*, and it required a good-sized volume to list and describe them all. Protozoologists are increasingly interested in species of Protozoa parasitic in the lower animals and in the protozoan fauna of the oceans. Many ecological niches remain virtually unexplored. As the result of the operation of the laws of chance and mutation, new species may even be coming into being, at least occasionally. Since many species have also become extinct at various times in the past, and are doubtless still doing so, there is really no way in which we can ever be certain how many species of Protozoa exist.

The Concept of Species

We should remember at the outset that the very term "species" is a creation of man, not of nature. Like many concepts, it seems at first much more clear-cut than it is. We tend to overlook that a species is not a firmly stable group, but is labile and changes with time and even with geography. The members of each species are descended from ancestors which differed in numerous ways from the individuals now making up the group, and—if it survives long enough—genetic and environmental change will unite to make of the group in future millennia something quite unlike the assemblage of forms we see today.

Nor do we often pause, as we view the multiplicity of living types around us, to consider the possibly even greater number which have lived out their relatively brief span of existence in past ages, and are now represented, if at all, only by fossils or a few genes persisting in the gene pool of their descendants. For each surviving species two or three others may have dropped out of the ranks in the process of evolution, and are now beyond our ken. The Protozoa, minute in size and usually without hard parts, have left us fewer detectable remains even than other groups. How large the total of protozoan species now extinct may be is suggested by the number of known species of fossil Foraminiferida, a group of marine Protozoa whose calcareous skeleton lends itself easily to fossilization. Here the count already exceeds 26,000.

At first thought the concept of species as the basic biologic unit seems both simple and obvious. Yet in fact it is neither simple nor easy to apply. The conventional definition of a species is that it consists of a group of living things which (1) are more similar among themselves than to any outside the group, and (2) breed only within the group, or are infertile when cross-specific matings are attempted.

Many obstacles confront the biologist in applying this definition, even to the higher organisms. He is faced with intergrades, and careful observation often reveals exceptions even to the rule that members of a species are fertile only among themselves. What appear to be distinct species often overlap geographically, and interbreeding may occur. Furthermore, in describing a species there is always the fundamental problem of deciding just what characteristics are of taxonomic value. Differences in morphology were naturally the first to be used, and are certainly the most convenient for the purpose of species recognition. Yet physiological characteristics are none the less real, although they may be difficult to detect when unaccompanied by morphological differences.

Since the concept of species as a biological unit is artificial, and since no two individuals of any kind are wholly alike, deciding just how similar organisms should be to constitute a species is also a problem. And how much similarity must exist among species to justify their inclusion within the same genus? This dilemma, of course, confronts the taxonomist when he considers the creation of any taxonomic group, from subspecies to phylum. Unfortunately there are no universally recognized and easily applicable rules for its solution. Taxonomists, being human, naturally differ in their points of view. Those to whom any differences, however small, seem no less real in spite of their relative minuteness are known among biologists as "splitters." Those who consider a scheme of classification simply as a convenient tool to identify and group organisms are prone to regard details of morphology as trivial, and therefore prefer to think of them as taxonomically unimportant. Biologists call systematists of this persuasion "lumpers."

Any good scheme of classifying animals and plants should meet two conditions. The more important one is that the scheme be based on biological relationships, to the extent that these can be determined. It should also be convenient to use. The second of these conditions is much easier to achieve than the first. The determination of biological relationships is often difficult, even among animals we ordinarily consider the most highly evolved. Many of the types of evidence used with numerous species of Metazoa—structural resemblances based on comparative anatomy, and physiological, embryological, and fossil similarities—are either unsuitable or difficult to apply to Protozoa. In practice, evidence of relationships among the Protozoa is based chiefly on comparative morphology, together with resemblances (or differences) in life cycles. Recently, however, even the embryological concept has been applied to the developmental stages of certain ciliated Protozoa, and from their study conclusions have been drawn as to probable biological relationships (Lwoff, 1950).

Students, and even some contemporary biologists, often look upon taxonomy as insignificant. Fashions in biology, as in women's clothing and automobiles, change. The zoologist interested in physiology, genetics, and other experimental branches of biology frequently forgets that he must first know exactly what organism it is he wishes to work with. The amateur protozoologist developing an interest in Leeuwenhoek's "little animals" promptly discovers that none of them bears labels, and the initial task of recognition and identification seems formidable.

Fundamentally, the concept of species refers to a group of organisms endowed with a set of genes most of which are held in common. It is therefore a genetic concept, and it applies equally to the Protozoa and the Metazoa. This definition is difficult to use with organisms variously defined as unicellular or acellular. We can seldom test for intraspecific fertility, or for cross-specific sterility, even though the mechanism of inheritance in the Protozoa is essentially the same as in "higher" forms. Sexual processes are still unknown in many protozoan life cycles, and even when they occur, the details of the reproductive pattern are often such as to make attempts at cross-matings difficult among individuals presumed to be of different species, if indeed such experimental hybridization is possible at all.

Long ago the French amateur protozoologist Émile Maupas observed that not all ciliates of certain species would conjugate with one another. He accounted for this by supposing that closely related individuals are "infertile," but we now know that the explanation is more complicated. Although morphologically alike, individuals of some species have been shown to be physiologically different, and thus divisible into varieties. Each variety in turn consists of two (sometimes more) complementary mating types. In general conjugation or mating occurs freely between the mating types of a variety but not between these and individuals of other varieties (although there are some exceptions). Thus if inability to crossbreed with members outside the group is a criterion of species, each variety is equivalent to a species. Just how widely mating types occur among the ciliates is not yet known. Only a few species have been studied from this point of view, and all of them have been free living. Mating types are unknown among the flagellates and Sarcodina.

Thus, in practice, species of Protozoa are recognized and described on the basis of characteristic morphology, since no other method is really practical. Yet physiological peculiarities and distinctive life cycles are no less real, and can hardly be regarded as less important.

Parasitic species in particular frequently present special problems. Morphologically similar types sometimes occur in several or even in many host species. It was formerly assumed that physiological adaptation between host and parasite is always so complete as to justify the creation of a new species of parasite for each host species. Such "host-parasite specificity," as this strict relationship is called, may indeed exist, but in many cases it does not. The taxonomist must be sure of his evidence. To settle the question often involves a good deal of laboratory experimentation. The protozoologist's limbo is full of almost forgotten species of parasitic Protozoa that could not meet the test of strict host-parasite specificity.

A second difficulty making the life of a taxonomist hard is the frequent occurrence of intraspecific variation in many species of protozoan (and other) parasites. Races of *Entamoeba histolytica* (causative agent of amoebic dysentery in man) vary in pathogenicity and the size of cysts they produce. But should they be called "races," "varieties," or "subspecies"? There is no real solution to such problems, except eventual agreement based largely on convenience, or perhaps on a knowledge of the genetics of parasitic Protozoa we do not yet possess.

Many protozoologists, and probably most students, may not care to be concerned with such problems. But often they will have to identify species they plan to study. Then the question of just what critieria delimit protozoan species takes on real significance.

Using a Key

In practice, the first step in identifying an unknown species is usually resort to a key prepared by a specialist thoroughly familiar with the group of Protozoa to which the form is thought to belong. Such keys are generally constructed to permit an organism of unknown identity to be assigned to ever smaller groups, until the process of elimination places the form in the correct species. Keys are often made in the form of couplets, based on the presence or absence of some easily determined characteristic. Their use, therefore, involves successive choices based on prior determination of a series of characteristics.

Since *Paramecium* is a genus very commonly encountered, and embraces a number of species, a sample key which may be used for their identification follows. It has been made as simple as possible, in order better to illustrate the construction of a key and to facilitate its use.

Uniformly ciliated and somewhat cigar-shaped organisms, usually with trichocysts underlying the cilia, and with a prominent adoral groove leading to the mouth one-half or two-thirds of the body length back. . . . Genus *Paramecium* 1

1. Colorless forms. 2
   Green, with zoochlorellae; anterior end truncate. . . . . *P. bursaria*
2. Forms with two contractile vacuoles, fed by radiating canals. 3
   Forms with a single, centrally located contractile vacuole; no radiating canals and no trichocysts. . . . . *P. putrinum**
   Forms with contractile vacuoles, fed by vesicles rather than by radiating canals. . . . . *P. trichium*
3. Relatively long and cylindrical. 4
   Relatively blunt, somewhat dorsoventrally flattened, and rather truncate anteriorly. 6
4. Large (over 180 micra); pointed posteriorly. One micronucleus. . . . . *P. caudatum*
   Less pointed, but often even larger than above; 3 or 4, and sometimes up to 7 micronuclei . . . . *P. multimicronucleatum*
6. Length more than 120 micra. 7
   Seldom exceeds 115 micra; 4 (sometimes as few as 3 or as many as 8) micronuclei. . . . . *P. polycaryum*
7. Relatively large (120–210 micra); 3 or 4 (less commonly, up

P. putrinum is often regarded as a doubtful species.

to 8) micronuclei; brackish water. . . . . *P. woodruffi*
Smaller than above (110–40 micra); usually with only 2 (more rarely, up to 5) micronuclei; fresh or brackish water . . . . *P. calkinsi*

In constructing this key, we have utilized the more easily observed characteristics, such as shape, size, color, and number of contractile vacuoles and whether or not they are fed by radiating canals. Usually, identification can be made from these alone. If it is also necessary to know the nuclear characteristics, a little acidulated methyl green (see Chap. 27) may be run under the cover glass to make the nuclei visible, since in the living animal they are usually not perceptible.

Once the organisms have been identified as paramecia, to use the key, the color is first noted. If it is green, the species is *Paramecium bursaria*, and we need go no further. If the organism is colorless, note the contractile vacuoles (number 2 in key); these are always conspicuous in any paramecium. If there is but one vacuole and it has no radiating canals, we have *Paramecium putrinum;* if there are two, it is *P. trichium*. But if both contractile vacuoles are fed by radiating canals, we must then go further. Suppose our organism is *P. woodruffi*. This species has not only the two contractile vacuoles with their feeder canals, but is somewhat flattened dorsoventrally and of moderate size (e.g., 175 micra), and is somewhat truncated anteriorly. When treated with a little acidulated methyl green, several micronuclei can be counted. This takes us to number 3 in the key, then to number 6, and from there to number 7, and we have *Paramecium woodruffi*.

The identification of parasitic Protozoa presents different problems, although it may be easier, for the species of host, the habitat in the host or the kind of cell parasitized, and the character of infection may all be of assistance. A portion of a typical key for such identification follows. Let us assume that a protozoan parasite has been observed in the erythrocytes of a bird, and we want to ascertain the parasite species. We know it is a malaria parasite (and therefore a member of the genus *Plasmodium*) because (1) it produces pigment, and (2) reproductive stages have been seen in stained blood films.

Gametocytes round or irregular. 1
Gametocytes elongate. 6
1. Both asexual and sexual stages tend to be round, displacing nucleus of host cell. 2
Schizonts tend to encircle nucleus of host cell. 5
2. Canary not susceptible to experimental infections. 3
Canary susceptible to infection. 4
3. Nucleus of host cell displaced, but not very frequently expelled; pigment granules in gametocytes rather large and not very numerous; not known to occur in nature outside the Orient; 8–30 merozoites. . . . . *Plasmodium gallinaceum*

Nucleus of host cell displaced and often expelled; pigment granules in gametocytes numerous and fine; so far known to occur only in the padda bird of Europe, *Padda oryzivora*. . . *Plasmodium paddae*

4. Pigment of gametocytes relatively fine and dotlike; nucleus of host cell displaced, and often expelled by larger forms; number of merozoites varies according to strain, being usually 8–15; very common in passerine birds. . . . . *Plasmodium relictum*

Pigment of gametocytes coarse and often rodlike; nucleus of host cell displaced and frequently expelled, particularly by mature gametocytes; merozoites 6–24; marked quotidian periodicity; common in passerine birds. . . . . *Plasmodium cathemerium*

The next step in the solution of our problem is evidently to find and carefully examine the gametocytes (sexual stages), which are always present in the blood of malaria-infected birds. Because of their relatively large size, undivided chromatin, and numerous pigment granules (a degradation product of haemoglobin characteristic of malaria parasites) in their cytoplasm, they are easily recognizable. Since the gametocytes are seen to be "round or irregular," we pass to number 1 of the key. A second look shows that the larger parasites often displace the nucleus of the infected erythrocytes,* and therefore we go to number 2 in the key. In the example we are using it may appear that further progress is impossible until we have tried transmitting the infection to canaries. But if we read down the key we can eliminate the first of the four species listed because it is found in nature only in the Orient, and we know that our slide was made from a bird caught in the neighborhood. Similarly, the second species (*Plasmodium paddae*) occurs in the European "padda bird," and we are therefore justified in moving to number 4 of the key. A final look at the gametocytes discloses numerous granules of fine, dotlike pigment in the cytoplasm, and our task is completed: the species is *Plasmodium relictum*.

Keys, however, are not always available, and even when they are the student may not find their use easy. The professional protozoologist who undertakes the preparation of a key may find it difficult to put himself in the place of its prospective user, who may well have little experience in study of this sort. Devising a usable key is at best no simple task. Furthermore, natural variation within a species is often great enough so that fitting organisms into the limits set by a key is troublesome. Differing environments sometimes result in quite different size ranges for certain species. Even such relatively gross morphological traits as number of nuclei are not always constant (as a glance at the key for identification of species of *Paramecium* will show). The occurrence of genetically distinct races within many species may also complicate the picture.

*It will be remembered that only mammalian red cells lack nuclei.

Other Sources

Even with all the help the best key can give, recourse to the literature is often necessary. The older manuals and monographs, such as those of Stein, Kent, Leidy, and Stokes (to name only some of the better known) are still valuable. The zoologist of today, whether beginner or seasoned practitioner, must marvel that these pioneer protozoologists of a century or more ago, working with inadequate lighting and inferior microscopes, could produce descriptions and figures of such great accuracy. And they did so without benefit of the technicians and research grants many modern biologists consider indispensable.

But newer, more accurate, and more inclusive manuals are available. Among them are such comprehensive guides as *Fresh Water Biology*, originally edited by Ward and Whipple and recently brought up to date, and monographs on smaller groups of Protozoa such as *The Genus* "Euglena," by Gojdics. For the student interested in the ciliates, if he can read German there is nothing better than Kahl's *Urtiere oder Protozoa.* For these and others, the list at the end of this chapter may be consulted.

The final arbiter in the definitive recognition of a protozoan species is of course the original description, plus any later revisions and emendations. Unfortunately, these may not be readily obtainable, for they may be buried in older, perhaps obscure journals.

Nomenclature

Students are often confused when they discover that texts and articles in scientific journals frequently differ as to the correct name for an organism. Should the proper generic name of the organism responsible for amoebic dysentery in man be *Entamoeba* or *Endamoeba?* Some standard texts spell it one way and some the other. It is also possible to find the species referred to as "*histolytica*," "*dysenteriae*," "*tetragena*," or even "*dispar*." Which of these is correct? In theory, such differences are easily settled by reference to an internationally accepted code of rules known as the "International Code of Zoological Nomenclature." This code was promulgated by the International Zoological Congress meeting at Leyden, Holland, in 1895, and has been revised by other congresses since, the most recent in London in 1958. A permanent adjudicating body, known as the International Commission on Zoological Nomenclature, now sits as a kind of zoological supreme court to hear cases involving differences of opinion on the correct scientific names of various animals, and to make appropriate recommendations. The membership of the Commission includes recognized zoologists from different countries serving without compensation.

Despite the universal respect in which these men are held, their findings do not always stick, for biologists are all too human. For example, the correct genus name for the dysentery amoeba is now generally agreed to be *Entamoeba*. But though this usage is supported by Opinion 312 (issued December 17, 1954) of the Commission, the same body held *Endamoeba* to be the proper generic name in a decision made some years earlier (Opinion 99). Most American parasitologists and protozoologists accepted the latter opinion, but their opposite numbers abroad did not, and confusion resulted. Even now there are those who think that *Entamoeba histolytica* and *E. coli* should not be placed in the same genus, since the former causes dysentery in man whereas the latter is nonpathogenic, and opinion 312 of the Commission has especially permitted them to call the dysenteric amoeba *Poneramoeba* if they so desire.

Whenever doubt exists about the correct name of a species, the International Rules of Zoological Nomenclature stipulate that the name first given shall be the one used (the "law of priority") provided that it was accompanied by a recognizable description and "that the author has applied the rules of binary nomenclature." This last was, of course, the invention of Linnaeus and used in his famous Tenth Edition of the *Systema Naturae*. It is interesting that Linnaeus introduced this scheme of naming less because of its obvious convenience than because it promised to cut the publication costs of his catalogue of species.

The rules of nomenclature also require that the "scientific names of animals must be words which are either Latin or Latinized," and that these be "uninomial for subgenera and all higher groups, binomial for species and trinomial for subspecies." Thus the name for the parasite causing the most common kind of human malaria is *Plasmodium vivax*, and for the one frequently found in passerine birds, in which a subspecies (or variety) is recognized, *Plasmodium* (*relictum*) *matutinum*. In writing the scientific name of a species, the genus name should begin with a capital and the species name with a small letter unless the latter is based on the name of a person (e.g., *Euglena Hegneri*). The name of the author who originally described the species and the date the description was published are often given after the species name. Thus the common paramecium is referred to as *Paramecium caudatum* Ehrenberg, 1833.

Family names must end in "idae" and subfamily names in "inae." There is also a trend (but no rule) to use as suffixes to the names of orders and suborders "ida" and "ina" respectively. Beyond that, there is little uniformity in taxonomic usage.

Units of the same rank, whether species, genera, or even orders and classes, are by no means always comparable. On the contrary, they represent essentially value judgments of the taxonomist. The degree of biological relationship among organisms of a group which seems close enough for one protozoologist to include in the same genus, justifies to another protozoologist classification only into the same subfamily or even family.

Biological Relationships

Biologists have also differed, and are still not agreed, as to the overall position which Protozoa should occupy in the world of living things. It was natural for Leeuwenhoek to think of his "little animals" as essentially like the larger ones with which he was familiar, and more than a century later Ehrenberg still interpreted the structural details he saw in terms of organs and organ systems of man and other vertebrates. But his contemporary, the great French zoologist Dujardin, realized that in the Protozoa nature had created living beings of a more elemental and truly different sort. He termed them collectively *Infusoria*, and remarked that most of them were quite transparent, evidenced little structure (except for vacuoles), and multiplied usually by "spontaneous division." Yet with prophetic vision he also recognized that mere inability to distinguish minute structure under the microscope is not proof that no such structure exists. Even today most zoologists regard the Protozoa as a rather primitive group of organisms, sharing a cellular organization with the "higher" animals and plants, and constituting one relatively small phylum in a total of perhaps thirty making up the animal kingdom. There is a growing tendency, however, to put Protozoa into a subkingdom or even a kingdom by themselves, to recognize that they are unique enough in structure to be thought of as acellular or noncellular.

A knowledge of biological relationships should underlie any scheme of classification. Unfortunately, we do not yet have this knowledge for the Protozoa, either as a group or as an assemblage of smaller groups. Yet evidence throws light on certain probable kinships. Because of their apparently simple structure, it was thought originally that the amoeboid rhizopods were the most primitive. Now it is almost agreed that the green flagellates came first, largely because chlorophyll-bearing organisms obviously could have existed in a world devoid of animals, and because of certain similarities in the life cycles of these flagellates and algae. Some species of the order Chrysomonadida (see p. 284) may closely resemble these hypothetical primitive types. By the loss of chlorophyll and adoption of predatory habits the colorless flagellates (Zoomastigina) probably arose from the plantlike forms. With the appearance of multicellular animals many of these flagellates must have been ingested, and as the eons passed some of the more adaptable ones were probably able to continue life within the bodies of their captors even more successfully than Jonah is said to have done in the whale. Thus may have arisen that great gathering of polyglot Protozoa we call the Sporozoa. There is no actual evidence for this descent except certain aspects of the flagellate and sporozoan life cycle, particularly the common occurrence of sexual stages and the postzygotic meiosis characteristic of the green flagellates (ancestral to the colorless forms) and many of the Sporozoa. The rhizopod amoebae (and, ultimately, doubtless other Sarcodina) are thought to have originated from the colorless flagellates. The evidence for this last lies in the rather common occurrence of life cycles including both flagellated and amoeboid stages. This fact, indeed, makes the proper classification of some species difficult. Is such an organism a flagellate or an amoeba? Often one cannot be sure, and a decision can be made only by looking at the life cycle as a whole.

In determining the degree or nature of the relationships of individual groups of Protozoa, we often know so little that we can only speculate. When there is evidence for the origin or relationships of a group, however, we shall discuss it in the appropriate place in the text.

Identification

In practice, identification of any species of protozoan requires, first of all, careful and patient observation under conditions as nearly normal for it as possible. Structural details, seemingly unimportant because of their minuteness, should never be overlooked. Whatever the type of organism, we must determine accurately its size, its shape (and whether constant or changeable), its color, and the type of movement if mobile.

Locomotor organelles, once identified as flagella, pseudopodia, or cilia, should be studied carefully. Know the number and relative lengths of the flagella, the type and manner of extrusion of pseudopodia, the ciliary pattern, and the relative size and lengths of cilia and of the organelles derived from them.

When a mouth, or cytostome, is present, as it is in many flagellates and most ciliates, its position and the presence and type of associated organelles (e.g., the undulating membrane of some ciliates) is useful in species determination. The cytostome is also the point of reference in describing the ciliary pattern, and in many free-living flagellates it is near the place of origin of the flagella.

Chlorophyll, to which plantlike flagellates owe much of their color, occurs in the form of discrete bodies (called "chromatophores" or "chloroplasts") of a characteristic form, number, and arrangement. These characteristics should be carefully noted, because they are often useful in identification of species. Although we usually think of chlorophyll as green, its actual color varies or may be masked by other pigments, so that chlorophyll-bearing Protozoa vary greatly in hue. For the same reasons, the numerous species of free-living Protozoa whose cytoplasm contains symbiotic algae are not necessarily green. Color in Protozoa also extends to discrete pigmented bodies, such as the orange-red eyespots, or stigmas, of many of the plantlike flagellates. Specialized living cytoplasmic bodies, colored or otherwise, are often called "plastids."

Masses of starch or starchlike material, known as paramylum or paramylon, distinctive in shape, size, number, and position, may also be frequently noted in these flagellates. In Protozoa lacking chlorophyll, a related carbohydrate known as glycogen (animal starch) may similarly occur.

Observe carefully such external characteristics as the presence and character (shape, size, color, composition) of tests and loricae, if any, and look for surface markings, such as the striations often observed on the plantlike flagellates, and the wrinkled appearance of certain free-living amoebae.

Aside from the tests or houses characteristic of many Protozoa, some species (e.g., certain ciliates) may possess internal skeletal elements, particularly the trichites frequently occurring about the cytostome. Other internal skeletal structures may also be present, as in some of the cattle ciliates, a fascinating and highly specialized group of parasites. Such skeletal organelles may usually be seen in the living animal without great difficulty.

In fresh-water forms, contractile vacuoles are the rule, and their position and number are usually constant for a given species. Feeder vacuoles, or vesicles, and canals should also be sought.

Nuclei are of course always present, but they are seldom easily discerned in life. Sometimes they may be seen as discrete bodies, lacking the inclusions usually occurring in the cytoplasm, in organisms which have been kept for some time under the cover glass. The size, shape, number, and position of the nucleus or nuclei, if discernable, are of considerable use in species identification.

Nor are these morphological characteristics the only ones. Others, such as trichocysts, cytopyge (cell anus, or cytoproct), and karyosomes within the nucleus should be noted. As we discuss specific groups, later in the book, we shall consider these characteristics.

The determination of parasitic species may present added problems. The host species, the parasite's habitat in the host, and the nature of the life cycle, are all important. The stages of the parasite occurring in the host, and where and under what conditions they are seen, as in the blood Protozoa, may be indispensable for species recognition. Here a word of caution must be added. Species of parasites can seldom be considered apart from their hosts. The same species of protozoan may exhibit different morphology in different host species. Sometimes only carefully controlled laboratory experimentation will decide whether two seemingly different species are indeed distinct. Conversely, apparently identical species on careful experimental test occasionally turn out to be physiologically distinct.

Accurate species identification must rest on the study of numerous individuals. We have already learned that variation in any natural population is considerable, and no single individual can safely be judged typical. But this variation, though always a problem, lends fascination to the game. When played according to the rules, as Leeuwenhoek and his many successors discovered, the study of the Protozoa is not only interesting to the scientist but fun for the naturalist, be he a beginner or a confirmed addict.

Broader Aspects of Protozoan Classification

Classification at the higher levels presents as many problems as correct genus and species identification, and is in some respects even more difficult, since actual biological relationships can never be certainly known; they must instead be inferred on morphological and developmental grounds. However, the electron microscope has added much to what we know of morphology, and better culturing methods have often revealed important new facts about development. At the same time, multiplication of known species has added to the need for a correct grouping into families, orders, and classes.

The most recent attempt to construct a biologically defensible taxonomic scheme was made by the *Committee on Taxonomy and Taxonomic Problems of Protozoologists (1964). As the Committee admits, their scheme must be regarded as still tentative, though they think it a "marked improvement" over previous schemes.

Among the more important changes they suggest are union of flagellates (p. 291) and Sarcodina (p. 322) into a single subphylum ("Sarcomastigophora"), splitting the Sporozoa (pp. 364–5) into two subphyla ("Sporozoa" and "Cnidospora"), placing the ciliated frog and toad parasites collectively called opalinids (p. 443) into a separate superclass ("Opalinata") of the Sarcomastigophora, and putting the babesias (pp. 362 and 554) and *Toxoplasma* (p. 559) each into a distinct class ("Piroplasmea" and "Toxoplasmea") respectively. The former would be moved to the Sarcodina.

Their scheme at the lower levels does not differ greatly from the older ones; whether it will receive general acceptance remains to be seen, but as they themselves say, it will not be the last word.

REFERENCES

Becker, E. R. 1933. Host-specificity and specificity of animal parasites. *Am. J. Trop. Med.*, 13: 505–23.

Bütschli, O. 1880–89. "Protozoa" in *Bronn's Klassen und Ordnungen des Tierreichs*. Leipzig.

Calkins, G. N. 1933. Derived organization. Taxonomic structures. In *Biology of the Protozoa*. Philadelphia: Lea & Febiger.

Copeland, H. F. 1956. *The Classification of the Lower Organisms*. Palo Alto, California: Pacific Books.

Corliss, J. O. 1959. Comments on the systematics and phylogeny of Protozoa. *Syst. Zool.*, 8: 169–90.

——— 1960. The problem of homonyms among generic names of ciliated Protozoa, with proposal of several new names. *J. Protozool.* 7: 269–78.

Dobell, C. 1912. Some recent work on mutation in microorganisms. *J. Genetics*, 2: 201–20.

Edmondson, W. T. 1959. *Ward and Whipple's Fresh-water Biology*, 2d ed. New York: Wiley.

Gojdics, M. 1953. *The Genus Euglena*. Madison: University of Wisconsin Press.

Grassé, P.-P. 1953. *Traité de Zoologie*. Anatomie, Systématique, Biologie. T. 1, Protozoaires. Fasc. 1 et 2. Paris: Masson.

Hall, R. P. 1953. The classification of the Protozoa. In *Protozoology*. New York: Prentice-Hall.

Hatch, M. 1941. The logical basis of the species concept. *Biological Symposia*, vol. 4, pp. 223–42. Jaques Cattell Press.

Hoare, C. A. 1955. Intraspecific biological groups in pathogenic Protozoa. *Refuah Veterinarith*, 12: 258–63.

Hyman, L. H. 1940. Classification. The acellular animals. Phylum Protozoa. In *The Invertebrates. Protozoa Through Ctenophora*. New York: McGraw-Hill.

——— 1959. Retrospect. In *The Invertebrates. Smaller Coelomate Groups*. New York: McGraw-Hill.

Kahl, A. 1930–35. *Urtiere oder Protozoa*. 1. Wimpertiere oder Ciliata (Infusoria). Jena: Verlag Gustav Fischer.

Kent, W. S. 1880–81. *A Manual of the Infusoria*. London: David Bogue.

Kudo, R. R. 1954. *Protozoology*, 4th ed. Part II. Taxonomy and special biology. Springfield: Charles C Thomas.

Leidy, J. 1879. *Fresh-water Rhizopods of North America*, U.S. Geol. Surv., Vol. 12.

Lwoff, A. 1950. *Problems of Morphogenesis in Ciliates*. New York: Wiley.

MacDougall, M. S. 1929. Modifications of *Chilodon uncinatus* produced by ultraviolet radiation. *J. Exp. Zool.*, 53: 95–109.

——— 1931. Another mutation of *Chilodon uncinatus* produced by ultraviolet radiation, with a description of its maturation processes. *J. Exp. Zool.*, 58: 229–36.

Manwell, R. D. 1957. The problem of intraspecific variation in parasitic Protozoa. *Systematic Zoology*, 6: 1–6.

Pearse, A. S. (ed.) 1936. Zoological names. A list of phyla, classes, and orders. *A.A.A.S.*, p. 24 (Duke University Press).

Pénard, E. 1904. *Les Héliozoaires d'Eau Douce*. Geneva.

Sand, R. 1901. *Etude Monographique sur le Groupe des Infusoires Tentaculifères*. Bruxelles.

Stein, F. R. 1859–83. *Der Organismus der Infusionsthiere*. Leipzig.

Stokes, A. C. 1888. A preliminary contribution toward a history of the fresh-water Infusoria of the United States. *J. Trenton Nat. Hist. Soc.*, 1: 71–344.

Van Cleave, H. J. 1943. An index to the opinions rendered by the International Commission on Zoological Nomenclature. *Am. Midland Naturalist*, 30: 223–40.

Weisz, P. B. 1954. *Biology*. New York: McGraw-Hill.

[ADDITIONAL REFERENCES FOR THIS CHAPTER WILL BE FOUND ON PAGE 296.]

14

Mastigophora

"The Monads are also among the simplest of all the Infusoria, they multiply in almost all infusions, and have no other organs visible than their whiplike filaments which have only been perceived in these modern times, and which can be clearly made out only with the finest microscopes, and with the greatest precautions." Felix Dujardin, *Histoire Naturelle des Zoophytes. Infusoires.* 1841.

Of all Protozoa, the flagellates, or Mastigophora, are perhaps the most varied in their characteristics. This is true of their morphology, their life cycles, their habits, and even their habitats. They range from forms almost as minute as the larger bacteria to organisms like *Volvox*—large enough to be perceived by the naked eye (Fig. 14.1). In shape they exhibit almost endless variety. They may have a single minute flagellum, a single large one, or many flagella. Their life cycles may involve little more than reproduction by simple fission, with mitosis of an apparently primitive sort, plus (usually) a cyst stage, or they may have sexual stages with gametes corresponding to those generated by the higher animals; sometimes there is even multiplication by multiple fission. In some species reproduction may occur within the cyst. The behavior of the flagellates, though relatively little studied, is certainly no less complex than that of other Protozoa.

Despite these and other complexities, the flagellates are usually regarded as the most primitive group in the phylum, although as one would expect of a group which has certainly existed on earth for many hundreds of millions of years, they are in no true sense primitive. Of the two large subdivisions into which the class is split, the Phytomastigina and the Zoomastigina, we shall consider the former first, since the latter are probably derived from them. The essential difference between the Phytomastigina (plantlike forms) and the Zoomastigina (animal-like flagellates) is of course their manner of nutrition. In the former group starch or paramylum is the chief anabolic product and there is often a cellulose jacket or cyst wall. Among the animal-like types these substances are not found, and glycogen occurs instead, though lipids (fats and oils) may supplement it as a reserve foodstuff.

A second difference separating the two great groups of flagellates is the apparent absence, except in organisms such as the flagellate symbionts of wood roaches, of sexual stages in the life cycles of the Zoomastigina. It will be recalled that such stages also characterize the higher algae, and are important evidence for the belief that these organisms and the Phytomastigina

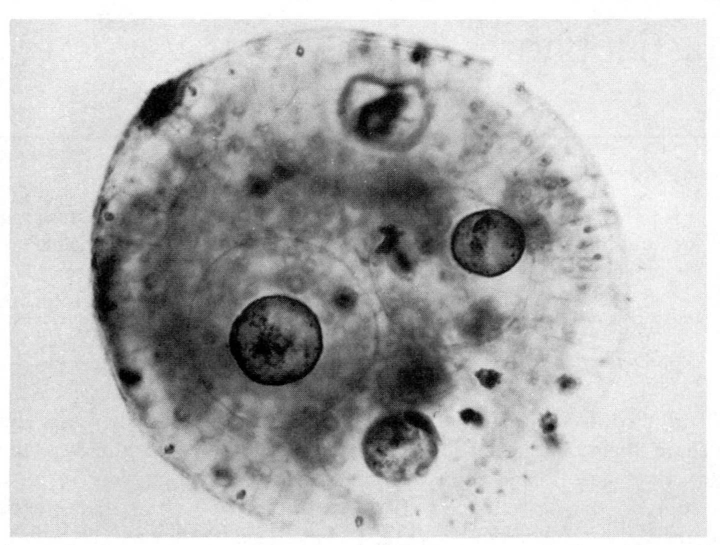

FIG. 14.1. *Volvox weismanni* containing several encysted zygotes and other minute unknown parasitic amoebae, ×288. The zygotes are distinguished by the cyst wall; a group of four amoebae appears in the lower right quadrant.

are closely related. The absence of such stages in the cycles of the animal-like flagellates poses a problem, however. If these Protozoa evolved from the plantlike types, and if sexual reproduction is a mechanism facilitating survival and evolutionary change, as most biologists believe, one wonders how the Zoomastigina have done so well without the boon of sex.

Flagellates differ much in their manner of making a living, the places they make it, and in their modes of life. Most are solitary, swimming freely about in their tiny world. Some species, however, are sessile, being attached to the substrate by a stalk. Others, more gregarious, live in communities which might be likened to the population of an apartment house; such is the remarkable species *Rhipidodendron splendidum*. Still others must be regarded as communist in their philosophy of life, since they live in colonies in which there is nearly complete egalitarianism, each member making his own living (by photosynthesis), but sharing all else with his comrades. Yet there may be a trace of class distinction even here, for there are likely to be some cells of the colony (as in the fresh-water group of Volvocidae) which have taken over the reproductive functions, at least to the extent that these are sexual. Nor are the colonies themselves all alike in some species; some may be sexual and others parthenogenetic.

Although very many species of flagellates are free-living some are parasitic. As one might expect, parasitism is exceptional among the Phytomastigina, yet—surprisingly enough—it is not unknown. Among the Zoomastigina it

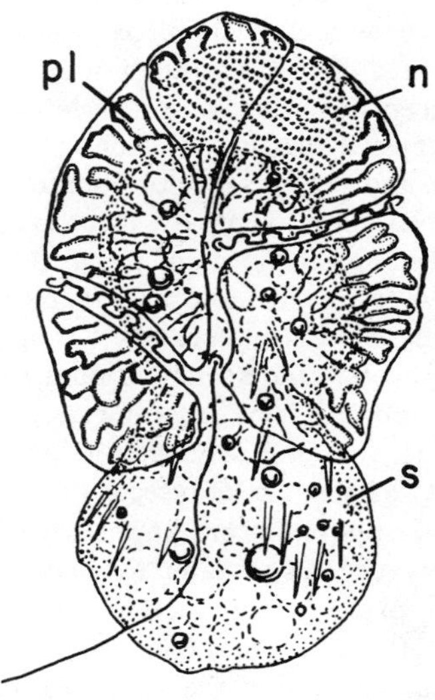

FIG. 14.2. *Gyrodinium pavillardi* ingesting the ciliate *Strombidium*. S, *Strombidium*; n, nucleus; pl, plastid. (After Grassé's reproduction of an unpublished figure by Biecheler, *Traité de Zoologie*.)

is common, and some of the most dangerous parasites both of man and other animals are to be found in this group (see Chaps. 19 and 20), yet even among the animal-like flagellates free-living species probably make up the majority; these are common in stagnant waters the world over. Most free-living and parasitic flagellates probably take their food in dissolved form. Some, though, may ingest particles, if minute enough, and individual trichomonads (parasites of a variety of hosts) have occasionally been seen to take in red cells. Certain dinoflagellates may even ingest other Protozoa (Fig. 14.2).

THE PHYTOMASTIGINA

The plantlike flagellates are regarded as probably the most primitive of Protozoa because, since they possess chlorophyll, they are nutritionally independent of other living things, except perhaps for their vitamin B_{12} requirements; this last the bacteria and blue-green algae may furnish. Thus they are in no way dependent on green plants, and could easily survive in a world almost devoid of other forms of life, at least as long as physical conditions were favorable and an adequate supply of carbon dioxide was available. A second reason for thinking the green flagellates primitive is the resemblance of the life cycles of many of them to those of the green algae. Though certainly not the first living things in the world they, or forms like them, must have appeared very early.

The green flagellates are typically planktonic organisms, floating and swimming about in ponds, lakes, and seas, but they are also numerous in the still waters of swamps and bogs. The green flagellates of today seem well adapted to their environment, for like most other Protozoa, they are found in nearly every environment where life is possible. Of course some species have a much more limited distribution than others. What they chiefly require is moisture, light (except for a few very adaptable species, which are capable of living in the darkness if that becomes necessary, taking on an animal-like nutritional pattern), and the presence of organic and inorganic substances of the proper kinds in solution. They are therefore obviously incapable of an aerial existence, although the cysts of some species are resistant enough to desiccation to survive for a time, and may therefore be dispersed with wind-blown dust. Dispersal, as with numerous other species of Protozoa, may also be accomplished through the agency of animals, in the guts of which cysts swallowed with the food or drinking water may be transported, unaffected by the digestive juices, for long distances. Doubtless they are also often the recipients of charity on the part of insects, on whose feet they ride.

The Mastigophora are important to us for a number of reasons. Some of these tiny organisms, and especially those with chlorophyll, have a very essential place in the food chains of the larger animals, man included. Without them the more minute crustacea, often collectively called Entomostraca, would find their food supply sharply curtailed. Insect larvae and young fish of many species, in whose diets these crustacea are important, would then feel the pinch, and soon man himself in many places would go hungry. Nor would the lack be simply in quantity: the algae, and their close relatives the green flagellates, are also to a large degree the ultimate source of some of the vitamins, such as vitamins A and D.

The life cycles of the plantlike flagellates are in many species still unknown; in others they often bear a strong resemblance to those of the typical green algae, to which they may be closely related. Still others of these flagellates, such as the vast assemblage of marine forms known as the dinoflagellates, the mostly minute and ubiquitous chrysomonads, and the collared types (choanoflagellates) are thought to be more closely akin to the brown algae. But neither the green nor the brown algae should be thought of as really primitive; the blue-green algae are believed to be closer to the primordial stock from which both plants and animals doubtless descended, for their life histories and structure are both simple. From them, in the opinion of some, not only the higher algae but even the bacteria may have been derived. However, it is not necessary to suppose that all these groups had a common origin; botanists today are inclined to think that they may have arisen independently.

Included in the life cycles of many species of green flagellates, as of many green algae, are sexual stages. The gametes often seem to be alike, but staining may reveal differences between the sexes. In *Volvox* the female gamete is relatively large and immobile and the male much smaller, and flagellated. In algae, too, the gametes may be alike (isogamous) or unlike (anisogamous), as also in *Volvox*. It is an interesting fact that in most of the chlorophyll-bearing flagellates (and also the green algae) only the zygote is diploid, a reduction division occurring soon after sexual union, so that the organisms commonly encountered are haploid. This condition also seems to hold among many of the Sporozoa, and by some it has been interpreted as having phylogenetic significance.

Curiously enough, the Euglenidae appear to reproduce only asexually, though there are unconfirmed reports of sexuality in one or two species. Perhaps partly for this reason, the botanists, who still claim these organisms, have put them in a separate phylum, the Euglenophyta, although it is only fair to add that more is then included in the group than the organisms recognized as euglenids by protozoologists. As botanically constituted, the phylum is said to embrace a total of 25 genera and 235 species. Botanists also concede that these organisms may properly be regarded as either plants or animals, and perhaps the best solution to the problem of where to place them is simply to give them dual citizenship in both animal and plant kingdoms.

Euglena spirogyra

Flagellates differ so much among themselves that it is hard to pick out any form for special description as typical. But the common and very beautiful species *Euglena spirogyra* will serve as an example of the Phytomastigina.

Euglena spirogyra is a rather large species, and is shown in Figure 14.3. A number of varieties exist, differing in size and other ways, but individuals are seldom less than 100 micra long and perhaps 10 micra wide when extended (Euglenae are highly changeable, or "plastic," in shape). Because of its size and cosmopolitan distribution this species was one of the earlier ones to be discovered; the German pioneer in the study of protozoology, Ehrenberg, gave it its name in 1838.

When swimming freely, *Euglena spirogyra* is spindle-shaped and somewhat pointed at the posterior end. At the anterior end projects a flagellum, which may be a quarter as long as the body, and it propels the animal through the water with a gentle gliding motion. The motion and the frequent changes in shape often cause the animal to be mistaken for a small worm when first seen. The flagellum, though relatively large, may not be easily visible because of its constant oscillation.

Also at the anterior end is the cell mouth, or cytostome, with a gullet

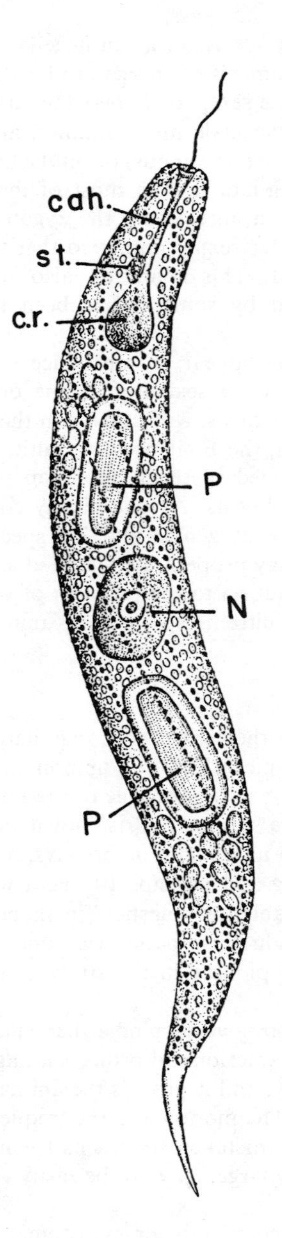

FIG. 14.3. *Euglena spirogyra.* The canal (can.) is also referred to as esophagus, but apparently does not function in food ingestion; st., stigma; c.r., reservoir; P, paramylum bodies; N, nucleus. (After Stein.)

or canal leading from it to a cavity of uncertain function, known as the reservoir. Since Euglenidae apparently do not take in particulate food, both mouth and gullet are misnomers. The flagellum has its origin in the posterior portion of the reservoir wall, and near its base is the contractile vacuole.

Externally, the body is covered by tiny knobs arranged in closely spaced longitudinal rows which twist about the body in spiral fashion—hence the species name. Similar spiral striations occur in many species of Euglenidae, but they are usually finer; the structure is very complex (* Mignot, 1965).

The animal (or plant) is a bright green, due to the presence of chlorophyll bodies known usually as chromatophores, although they may also be referred to as chloroplasts. These have an arrangement characteristic of the species; in *Euglena spirogyra* they consist of numerous small discs, very closely spaced. Anteriorly and adjacent to the canal is a bright orange eyespot, or stigma. This organelle is very often present in the chlorophyll-bearing flagellates, although its structure varies. Back of it, in *Euglena*, lies a body known as a flagellar swelling because of its intimate association with the base of the flagellum; it is light-sensitive and thus analogous in its function to a retina. The stigma itself seems to act as a kind of light filter, absorbing certain wave lengths. The first careful study of the function of the flagellate eyespot was done as early as 1882 by Engelmann, a German physiologist who is best remembered for his study of the fine structure of striated muscle. He showed that the stigma of *Euglena* was light-sensitive, and that response to changes in intensity of illumination occurred only when the incident beam was focused directly on this organelle. Even Ehrenberg realized this, for his name for the genus is derived from the Greek roots *eu* for "good" and *glene* for "eye."

Internally, the euglenids have a very complex structure indeed (* Mignot, 1966). In the midregion lies the nucleus, which in the living animal is usually visible only as a roundish area, devoid of chlorophyll or other inclusions. Sometimes the endosome can also be seen, but staining is likely to be required to reveal its presence. Near each end of the nucleus is a large, link-shaped body of paramylum, and sometimes there may be a third somewhere in the cytoplasm. The storing of reserve foodstuffs in the form of paramylum, rather than as true starch, is one of the ways in which the euglenids differ from what we might call plants proper. Paramylum (derived from two Greek terms meaning "related to starch") is a polysaccharide, but does not give the intense blue-black reaction of starch treated with iodine. Many green flagellates possess pyrenoids, protein bodies which in most species of *Euglena* appear as transparent round structures within the chromatophores, where they are thought to function in the deposition of paramylum, but these are absent in *Euglena spirogyra*.

The flagellum, as in most Euglenidae, may be seen even in the living animal to emerge from the cytostome. Stained specimens reveal that it originates on the posterior wall of the reservoir, where it appears to represent the union of two delicate threads, or axonemes, each of which seems to terminate in a tiny body called a blepharoplast lying on the floor of the reservoir. Electron microscopy, however, has shown each fibril to have the same internal structure as the flagellum itself, indicating that the euglenids now having a single flagellum evolved from stock possessing two; the blepharoplasts, too, are revealed as simply extensions of the internal fibrillar structure of the axonemes. From one of the blepharoplasts it is said that a tiny fibril, or rhizoplast, extends to the nuclear membrane, thus suggesting that the flagellum is under the ultimate control of the nucleus. The flagellar swelling associated with the eyespot also suggests a possible mechanism by which *Euglena* reacts to changes in illumination.

The final structure of the flagellum has received much study. In many flagellates it is so small that the ordinary light microscope reveals little. Electron microscopy, however, has shown that flagella and cilia have essentially the same makeup. Each consists of a peripheral sheath enclosing eleven longitudinal fibrils, of which two are in the center and the others about the periphery (Fig. 5.9). In the flagellum the latter two appear to be continuous with the above-mentioned pair of axonemes.

The similarity of flagella and cilia also extends to the basal bodies of both. Though given different names, the blepharoplasts of flagella and the kinetosomes of cilia are thought to be homologous. Axopodia, too, are in certain respects like flagella, since they likewise consist of an inner core (the axial filament) and a surrounding sheath, and may in some cases gently oscillate, like a flagellum in slow motion. But there are important differences, since the sheath of the axopodium consists of a layer of fluid, freely flowing cytoplasm (as may be easily seen from the visible movement of granules within it); electron microscopy has also shown that the axopodium lacks the typical "9+2" pattern of internal fibrils of the flagellum and cilium. Thus the theory formerly rather widely held that the axopodium might have evolved from the flagellum seems no longer tenable.

In only one phylum of animals (the Nemathelminthes, or unsegmented round worms) are flagellated or ciliated cells apparently absent. In all the rest the sperm actively search out the egg by the use of flagella to propel themselves about, epithelial membranes may make use of ciliated cells to keep in motion the film of covering moisture, and flame cells in some aid in the excretion of fluid wastes. Even among the plants, notably mosses and liverworts, the male reproductive cells may be flagellated. Thus it appears that nature's basic patent on the flagellum, taken out when life was very young—perhaps more than two billion years ago—proved eminently practical, and has been little improved on since.

Other Euglenids

So far, *Euglena spirogyra* may be considered a fairly typical member of the Phytomastigina. But it is nevertheless exceptional in certain respects.

It lacks the more or less rigid cellulose jacket with which many of the plant-like flagellates are clothed, and (unlike numerous species of *Euglena*) it is not known to encyst. But this seems to be no great handicap, for it is said that ordinary motile forms may nevertheless survive for months, or even years, under unfavorable conditions, without very noticeable morphological change and (as one might expect) without division. Some species of *Euglena* (as well as of many other groups) may under such conditions go into what is known as the *palmella* stage. These are aggregates of cells which have lost their flagella and surrounded themselves with jelly. They differ from true colonies in that each palmella contains a variable number of cells.

When the population of palmella-producing species becomes dense enough, the surface water of a pond may become matlike, and this is especially evident in a breeze. If in addition the Euglenae happen to be red species (e.g., *E. sanguinea*, *haematodes*, *demulcens*), in many of which a marked change in light intensity results in a change in color, the effect may be spectacular as well as surprising. Gojdics tells of ponds red at sunset with *Euglena* bloom and green the following morning at dawn.

It is possible to grow easily in cultures only a few species of Euglenidae. One is *Euglena gracilis*, a very common form. It is only about half the length of *E. spirogyra*, with (usually) seven to ten circular chromatophores, a number of minute ovoidal paramylum bodies, a rather long flagellum, and it is very faintly striated. Both axonemes are easily visible in the living organism. Because of the relative ease with which this species can be cultured, much physiological work has been done on it. It apparently required only the presence of certain mineral salts, vitamin B_{12}, and enough light for growth and multiplication, and is thus an example of a completely "photoautotrophic" form. Although formulating media of precisely known composition suitable for the cultivation of most species of *Euglena* is difficult, the majority grow well in suitable mixtures of soil and water.

Among the close relatives of *Euglena* are a number of common species, including some of the most beautiful of the plantlike denizens of the microscopic world. The different varieties of *Phacus* (Fig. 14.4), a genus of flattened forms often twisted into bizarre shapes, are good examples. They have a rigid pellicle with closely spaced longitudinal striations, and a conspicuous orange-red stigma. The posterior end is usually drawn out into a point.

There are also numerous colorless species in the order. Some are even placed in the genus *Euglena*, and under artificial conditions of darkness certain species (e.g., *E. gracilis*) lose their chlorophyll and take on an animal pattern of nutrition. That species without chlorophyll should belong in the subclass Phytomastigina is a biological paradox, but comparative morphology and the nature of reserve food stuffs show clearly that colorless species and those with chlorophyll are closely kin. Versatile species such as *Euglena gracilis*, when compelled by darkness to live as animals do, then require certain organic extracts containing nitrogen, such as peptones.

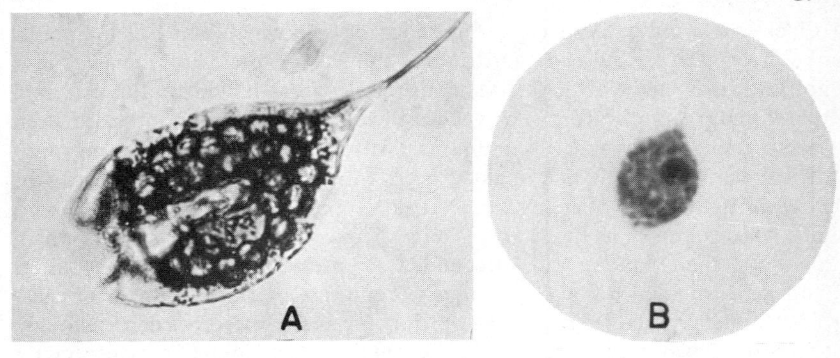

FIG. 14.4. *A. Phacus pleuronectes*, a common euglenid flagellate. The flagellum is not visible, nor is the circular paramylum body usually seen near the center, ×676. *B. Phacus acuminata*, also common. The single flagellum is not visible. The short, sharp posterior spike is characteristic. ×676.

In view of the great diversity of types such as those just discussed, we may ask what characteristics euglenids possess in common. They are (1) the presence of paramylum as a metabolic product, (2) the possession of one, or in some species, two flagella, (3) in most cases, a cytostome and canal (gullet) leading from it. The great majority possess green chromatophores, and most species live in fresh water.

Volvox, a Colonial Form

The order Phytomonadida, to which *Volvox* belongs, is a group in which the ultimate product of photosynthesis is starch, rather than the paramylum of the euglenids. Some species also produce oil. As in the euglenids, some species are red and a few are even colorless. Most of the phytomonads are rather small, roundish forms, with one, two, or occasionally even four flagella. The great majority share the preference of the euglenids for fresh water.

Volvox, though not a typical phytomonad, is an interesting type. (Fig. 14.5). Unlike most members of the order, it is colonial and often rather large; some colonies reach a diameter of three millimeters, and contain as many as 50,000 cells, though other species are much smaller. They are always green. It is hardly surprising that they attracted Leeuwenhoek's attention, and he has left us an easily recognizable drawing of a colony, reproduced in Figure 14.6. They were also studied by Ehrenberg, to whom we owe the family name. The generic name is one of the few of Protozoa which we can credit to the great systematist Linnaeus.

Although certain cells in a colony may be sexually differentiated, the great majority are alike, each with its own chromatophores, a pair of equally long flagella, and a stigma. Yet they share whatever protection numbers may give, and all contribute to locomotion, the flagellar movement being coordinated. This sharing apparently reminded someone of life in a monastery, for the name "coenobium," also applied to a monastery or convent, is often used instead of "colony" for these communities of cells.

The life cycle of a colony begins when the zygote is formed by the fusion of a flagellated microgamete, or sperm, with the larger, immobile macrogamete, or egg. Actual fertilization has not yet been observed, but the zygote

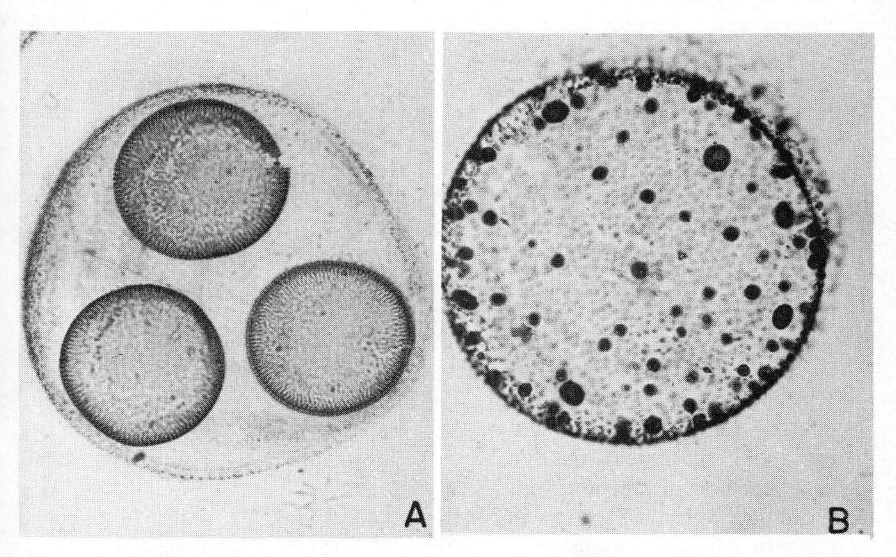

FIG. 14.5. Colonial flagellate *Volvox*, ×145. *A. Volvox* sp. containing three maturing embryos. *B.* Small asexual colonies developing inside parent (in various stages of development).

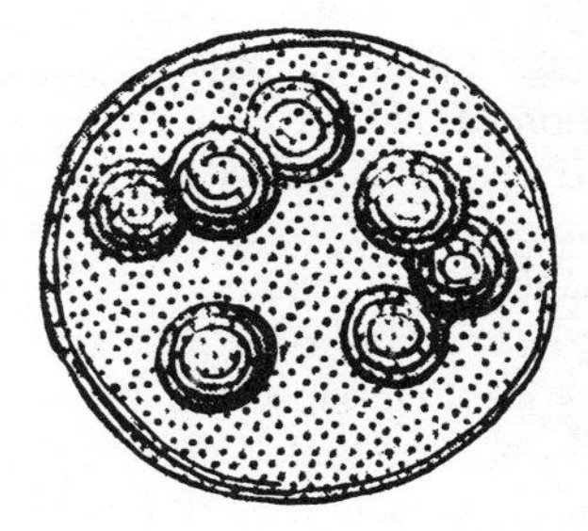

FIG. 14.6. Leeuwenhoek's drawing of *Volvox*. (From Dobell, *Antony van Leeuwenhoek and his Little Animals*.)

is known to be relatively resistant, with a tough, spiny jacket. This is the guise the organism assumes when it must cope with adverse conditions, such as drought and cold. In this form it settles to the pond bottom, there to await the moisture and warmth it needs for growth. When conditions are favorable, as they usually are in the spring, the resistant envelope is shed and division soon produces a typical sphere of cells. At first, however, their flagellar ends are directed inward, and inversion, a process similar to turning the finger of a glove inside out, is necessary to orient them properly. This is accomplished through a pore (the "phialopore").

At other times reproduction may be sexual, a few cells near the lower pole of the colony (which has a definite axial differentiation) dividing until eventually they form a small daughter colony within the body of the parent. Like the sexually produced colony, this colony has its constituent cells facing inward, and inversion is necessary before the young organism is ready for life on its own. Sometimes granddaughter colonies may even be observed developing within the daughters, as they were by Leeuwenhoek. He remarked in a letter dated January 2, 1700, that on August 30, 1698, he saw in ditch water "a great many green round particles, of the bigness of sand-grains," and goes on to say, "In this water were included two of the foresaid round bodies, and these were of the biggest sort; and contained in each of them were five little round particles, which inclosed particles were pretty well grown in size: and in a third big body there lay seven round lesser particles. These last were uncommon small. . . . Furthermore I perceived, after the space of five days, that the small particles inclosed in the third large body were not only grown in bigness, but I could also then discern that from inside these small particles other less round particles were to come forth." The "green round particles" were *Volvox* colonies, and the particles within them were developing daughters and granddaughters. With the birth of daughters, the parent dies. Probably in most species of *Volvox* this is the usual method of reproduction.

Gamete formation occurs in somewhat similar fashion. Mary A. Pocock, a South African biologist, has worked out some details of the life cycle of several species of *Volvox*. Sperm are produced by the separation and dispersion of small flagellated cells which at first constitute a "male" colony. Macrogametes, or eggs, develop from cells which have grown, but without division and formed at first a "female" colony. These cells, when ready for fertilization (at which time they are known as "oospheres"), are relatively large and well provided with reserve foodstuffs in the form of starch and oils. Since the actual act of fertilization has not yet been seen, their immediate subsequent history remains unknown.

Doubtless details of the process vary in different species, just as there is variety in other aspects of the life cycle. For example, *Volvox capensis* is monoecious, the same individual producing male and female gametes, whereas *V. rousseletii* is dioecious, some colonies being male and others female. Although asexual reproduction seems to be the rule in most species of *Volvox*, at least the greater part of the time, individual cells of a fragmented colony are incapable of regenerating the rest of the colony. Cellular differentiation seems to occur rather early in the development of a *Volvox* colony, and to be irreversible.

Organisms like *Volvox* have long held a special fascination for biologists interested in evolution. Although we do not believe they actually represent ancestral forms of the higher Metazoa, their striking resemblance to the blastula stage of a cleaving egg suggested that here, at least, was a living group transitional between unicellular and multicellular animals. To pioneer evolutionists, such as the German zoologist Haeckel, the existence of colonial organisms such as these seemed almost like proof of the law of recapitulation, "Ontogeny repeats phylogeny." The process of inversion occurring in the development of *Volvox* has a certain resemblance to the gastrulation of the metazoan embryo. Nevertheless, the apparent evolutionary significance of forms like *Volvox* has tended to diminish as knowledge of metazoan development has increased. Still, *Volvox* continues to fill a unique position. As the contemporary French biologist, Jules Pavillard, puts it, "... In the world of the Protista, it [*Volvox*] has reached an advanced state of development of which one can scarcely find the equal among organisms belonging to the higher levels in the taxonomic scheme."*

The Dinoflagellates

Another interesting group of plantlike flagellates, although their appeal is of a different sort, is the Dinoflagellida. As in the euglenids, chromatophores are not invariably present, but the dinoflagellate affinities of the colorless forms are evident from their morphology. In structure the dinoflagellates differ from anything else in the protozoan world, though one can hardly pick out a "typical" species for description.

Their most unusual characteristic is the possession of two flagella, usually originating at different points and in different planes. One of them encircles the body in belt fashion, sometimes winding about twice or even three times. It lies in a transverse groove known as a "girdle" (Fig. 14.7). The other is directed backward, and is situated in a longitudinal groove, the "sulcus." It appears to be largely responsible for locomotion, pushing the organism forward, and perhaps functioning somewhat like the tail of a spermatozoon. The spiral path in which the animal moves is doubtless due to their combined action plus the asymmetry of body form. Possibly the ribbonlike shape assumed by the transverse flagellum is also a factor in its movement.

In appearance the dinoflagellates exhibit the greatest variety. Some are naked, while others wear a coat of elaborately embossed armor, often called a "theca." This consists of cellulose, to which added toughness may be given by impregnation with mineral salts. It may be constructed of plates, or as two valves, or it may be one piece. Often there are projections, and these with the asymmetry and the sculpturing may combine to give the organisms a very bizarre form.

The unarmored dinoflagellates are often brightly colored, displaying in the ocean, the natural habitat of most of them, pastel hues of exquisite beauty. They have been aptly compared to the butterflies and orchids of the

*Grassé, P. – P., 1952. *Traité de Zoologie*. Tome 1, Fasc. 1, p. 187. Paris: Masson.

terrestrial world, though reduced in size a thousand times. Many, too, are bioluminescent. Of the latter, *Noctiluca* (Fig. 14.8) is best known, but its light is shed chiefly on waters inshore. *Pyrocystis* lives on the high seas. However, bioluminescence is not limited to the dinoflagellates; it occurs in other Protozoa, and in a variety of multicellular animals, as well as in some bacteria and fungi (*Seliger *et al.*, 1962).

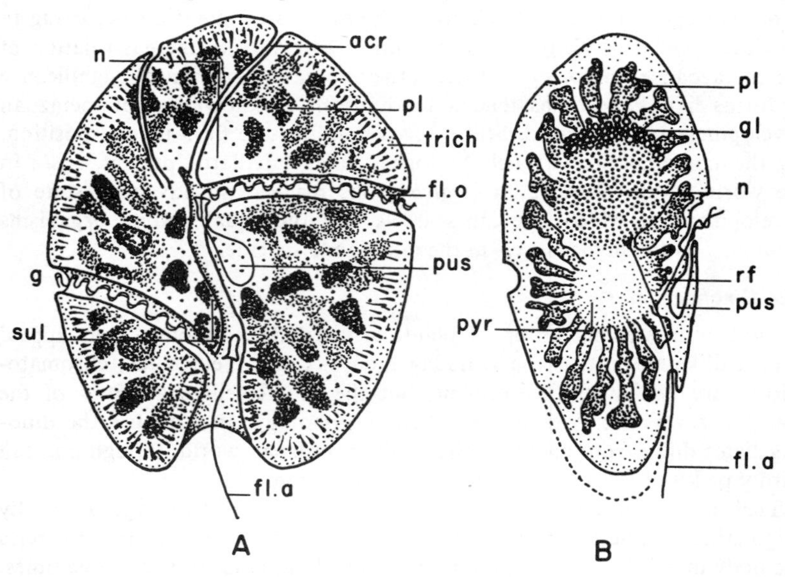

FIG. 14.7. A dinoflagellate, *Gyrodinium pavillardi* Biecheler. *A*. Face and profile in optical section: g, girdle; sul, sulcus, or longitudinal furrow; acr, a narrowed extension of the longitudinal furrow (acrobase); pus, pusule; fl.o, undulating flagellum; fl.a, axial flagellum; pl, one of the branches of a plastid; trich, trichocysts; n, nucleus; gl, lipid globules; pyr, pyrenoid; rf, probably a flagellar root. *B*. Cross section. (After Biecheler, *Traité de Zoologie*.)

Dinoflagellates make up a large and significant part of the plankton especially that of the ocean. The late Charles A. Kofoid, a distinguished American protozoologist for whom the dinoflagellates held a special fascination, wrote, "The dinoflagellates form an exceedingly important part of the ocean meadows, the source of the primitive food supply of the sea, both in the number of individuals and the total mass of living substance produced." There are also a very large number of species—just how large, we cannot yet say. Kofoid was himself the discoverer of many of them. Since the open ocean is probably the preferred habitat of most members of the group, any estimate of the total species count is very hard to come by. Furthermore, even if they are captured, many forms are so delicate that they perish before anything more than the most cursory laboratory examination can be made.

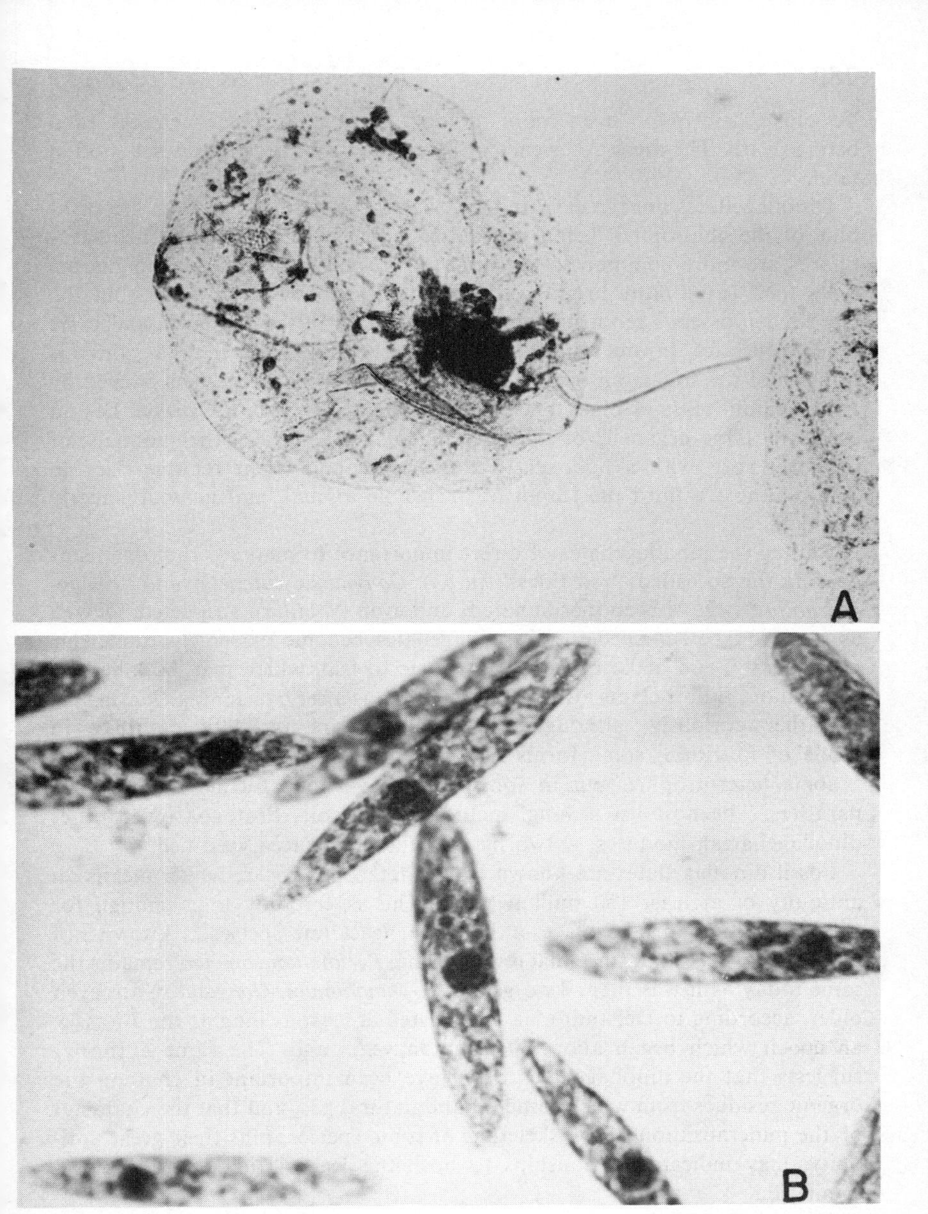

Fig. 14.8. *A. Noctiluca scintillans* (also known as *N. miliaris*). Extending to the right is the tentacle, capable of active movement. The flagellum is much smaller and does not show. The tentacle may function in locomotion, and the flagellum in food-getting. The heavily staining mass below the tentacle represents the nucleus, food body, and fat droplets which are not clearly differentiated from one another. ×145. *B. Euglena* sp., showing the spindle shape typical of *Euglena* and the central nucleus. Faintly visible in the upper righthand corner is the single flagellum, and in several of the individuals are suggestions of the spiral striations seen in many of the species. ×1150.

To make description even more difficult, their appearance changes even before death. The mere experience of being caught in a plankton net is often fatal.

Dinoflagellates are typically red or brown, the natural green or greenish-blue of the chlorophyll being masked by some other pigment. Chromatophores are more common in the armored forms. For some reason marine types tend to be more brightly arrayed than those of fresh water, but the latter may possess a reddish stigma similar to that of the euglenids and some other flagellates. In one family of ocean dwellers, the Pouchetiidae, there is in place of the stigma an ocellus, consisting of a transparent lens backed by an amoeboid mass of black or reddish pigment. *Erythropsis cornuta* has an especially large organelle of this sort. There must be few other animals, of whatever sort, with a light-sensitive device of equivalent relative size—in length almost a third the length of the whole animal, and in width nearly as great.

Among the dinoflagellates of direct importance to man are the organisms causing the so-called "red tides" (p. 53). *Gonyaulax catenella* and *Gymnodinium brevis* have been incriminated, and even *Noctiluca* suspected, as well as numerous other species. These flagellates become extremely numerous occasionally, and make the sea water toxic to fish, which may be killed by the million. Edible clams also become poisonous after consuming *Gonyaulax*.

Rather curiously, although dinoflagellates are typically plantlike in mode of nutrition, some forms have not only lost their chlorophyll and become heterotrophic, but in some cases have gone further and become parasites. The host list is long, including numerous Protozoa (even other dinoflagellates), molluscs, flatworms, annelids, crustacea, and fish.

Fossil dinoflagellates are known from Cretaceous rocks, which means an antiquity of at least 150 million years. This is certainly long enough for much evolution to take place, although at least one species is known not to have changed at all since that remote time. *Peridinium conicum* remains the same today as it was then. Two genera (*Gymnodinium*, *Gonyaulax*) are even older, according to Deflandre, having existed at least as long as the Jurassic, an epoch which began about 170 million years ago. The same authority suggests that the dinoflagellates may have been important in creating the organic residues from which came oil and natural gas, and that the character of the mineralization of the skeletons of some species, plus their great antiquity, may indicate relationships to both the Radiolarida and the Foraminiferida.

The color which is so conspicuous in many dinoflagellates may be due to several different pigments, notably chlorophyll *a* and *c*, and certain xanthophylls. These may be aggregated into chromatophores, usually associated with pyrenoids, or they may be diffuse. Variety of coloration is especially great in the unarmored species.

Reserve foodstuffs appear in the form of fats and oils, and starch;

one or both may be present. The former are likely to be concentrated in colorless globules about the nucleus, which itself is rather peculiar and therefore called a "dinokaryon." The nucleus ordinarily appears as a mass of many small chromatin particles, arranged in straight or curved rows, the whole surrounded by the nuclear membrane. Each row is said to be a chromosome, and in some forms to be visible even in the living animal.

Leading from one of the flagellar orifices, generally the posterior, is a canal ending in a reservoir called the "pusule"; two such organelles may be present, and the canal may connect with both flagellar openings. The pusules are filled with a delicately rose-tinted liquid, and in some species they are so large as to be very conspicuous. Their function remains unknown, although they have been sometimes suspected of being concerned with the absorption of food.

As in other flagellates, there are associated with the flagella fibrils and blepharoplasts, although in this group they make up a rather more complex system than usual. An intricate pattern of argentophil lines and minute bodies has been revealed by the use of the Klein silver impregnation method. These seem analogous to the "silver-line system" of the ciliates, but their function, if any, remains unknown.

Perhaps the most unexpected aspect of dinoflagellate structure is the occurrence, in some species (e.g., *Polykrikos*, Fig. 14.9), of nematocysts. These are similar enough to coelenterate nematocysts to cause some observers to suggest that they may not be normally present, but may represent something ingested. But it is now almost agreed that they are actually organelles, and Kofoid comments, "In the nematocysts of the Coelenterates the anatomical features show only a slight advance over those found in the dinoflagellates" (Kofoid and Swezy, 1921). Trichocysts are also commonly observed, and they exhibit considerable variety of form. Their real function in these flagellates is as obscure as it is in most ciliates. Because trichocysts are organelles of so unique a sort, there have been numerous speculations about their evolutionary origin, but it seems certain that they must have evolved quite independently in dinoflagellates and ciliates.

The problem of food-getting has been solved by the dinoflagellates in various ways. Presumably the great majority of chlorophyll-bearing species depend entirely on photosynthesis for their carbohydrates, and, less directly, for their fats and proteins. Some of them may also ingest solid food, and this may even be accomplished by armored forms. Biecheler has given an account, quoted by Chatton, of the way in which *Gyrodinium pavillardi* captures and then devours the ciliate *Strombidium*, itself an active swimmer and presumably protected by trichocysts. But simple contact with a sticky substance secreted by the flagellate is sufficient to trap and paralyze the luckless *Strombidium*, and only a few minutes is required for its complete ingestion. This is achieved by separation of the apical lips of the longitudinal furrow, thus creating the equivalent of a mouth. Apparently the dinoflagellates also often entrap other organisms in a net of fine, branching pseudopodia, which they extrude through pores of the theca.

Since most dinoflagellates are pelagic, and thus conditioned to an environment which is a complex of many chemical, physical, and biotic factors, it is scarcely surprising that cultivating them in the laboratory has proved difficult. However this has been done with a few species, among them a *Peridinium* and a *Gyrodinium*. Their growth requirements include vitamin B_{12} and such trace elements as zinc, manganese, and copper. Maintenance of various dinoflagellates is also often possible in a balanced aquarium for a time.

The life cycles of the dinoflagellates are both varied and complex. Multiplication is ordinarily by binary fission, but mutiple fission is also encountered.

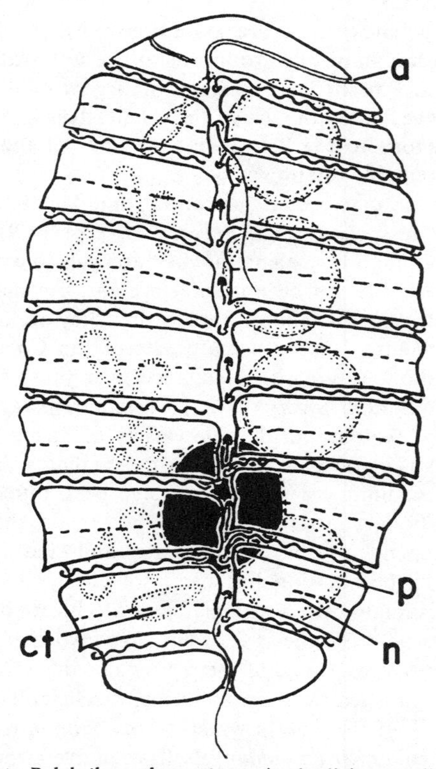

FIG. 14.9. Dinoflagellate *Polykrikos schwartzi* seen in the living condition (ventral view). *a*, acrobase; *ct*, cnidocyst; *n*, nucleus; *p*, ingested prey. It is in effect a chain of undivided individuals, nuclear division having taken place without cytoplasmic division (hence the numerous nuclei and the large number of flagella, although the latter are also doubled in number because of a double centro-flagellar apparatus annexed to each nucleus). Note presence of cnidocysts, as in Sporozoa of the cnidosporidian group. (After Grassé, *Traité de Zoologie*.)

Probably mitosis is the rule in nuclear division, but in many species the process has not yet been studied. Similarity of the reproductive process to that of other flagellates, however, ends here, for the plane of division is usually oblique. Multiple fission also occurs in some species. Not uncommonly, separation of the daughter individuals may be delayed, with resulting chain formation.

For the armored forms disposal of the old theca and equipment of the daughter cells with new ones presents a problem which is solved in many ways. Sometimes one of the daughters receives the entire parental estate, while the other is left only the privilege of building a new one. Or the theca may be abandoned, each daughter starting life naked. In still other cases it may be divided, half being bequeathed to each of the offspring along with the obligation to match it.

Cyst formation is common among the dinoflagellates, and is often carried to an extreme, secondary cysts being formed (usually after prior division of the encysted individual) within the primary one. When intracystic multiplication takes place, two or many offspring may result. Cysts are of several kinds. Among holozoic species, of which there are many among the dinoflagellates, encystment seems frequently to be the equivalent of an after-dinner siesta, the sated individual remaining within the cyst until digestion is more or less complete. Or the cysts may be thick-walled and so resistant to adversity that they remain viable for many months or even years. Kofoid noticed that encystment was relatively uncommon among the chlorophyll-bearing species, and that binary fission in forms lacking chromatophores apparently occurred most often while in the cyst.

One of the most remarkable characteristics of many members of this order is their polymorphism. In some cases several stages in the same life cycle have been thought to belong to different species. This sort of life history is shown in Figure 14.10, although the question mark at the top indicates that knowledge of it is still incomplete. It is said that the so-called "*Pyrocystis* stages," involving the production of small cysts and vegetative forms, may repeat themselves many times before the eventual appearance of the relatively gigantic spherical stages and the globular cysts, with their broods of daughter crescentic cysts within. From each of the latter emerge, in turn, from two to eight granddaughter "gymnodinia."

What factors determine the change from one of these stages to the next remains uncertain, and it is equally uncertain what advantages so complex a life cycle confer on the species. But that there must be advantages is indicated by its widespread occurrence among the dinoflagellates. The adoption of parasitism as a mode of life by some dinoflagellates has made an already complex cycle even more so. It may include, in addition to stages of the *Gymnodinium* type and cysts of a variety of kinds, amoeboid forms.

Sexual stages also have been reported for a few species, among them one of *Noctiluca*, but their general occurrence is still doubtful. As a group, the

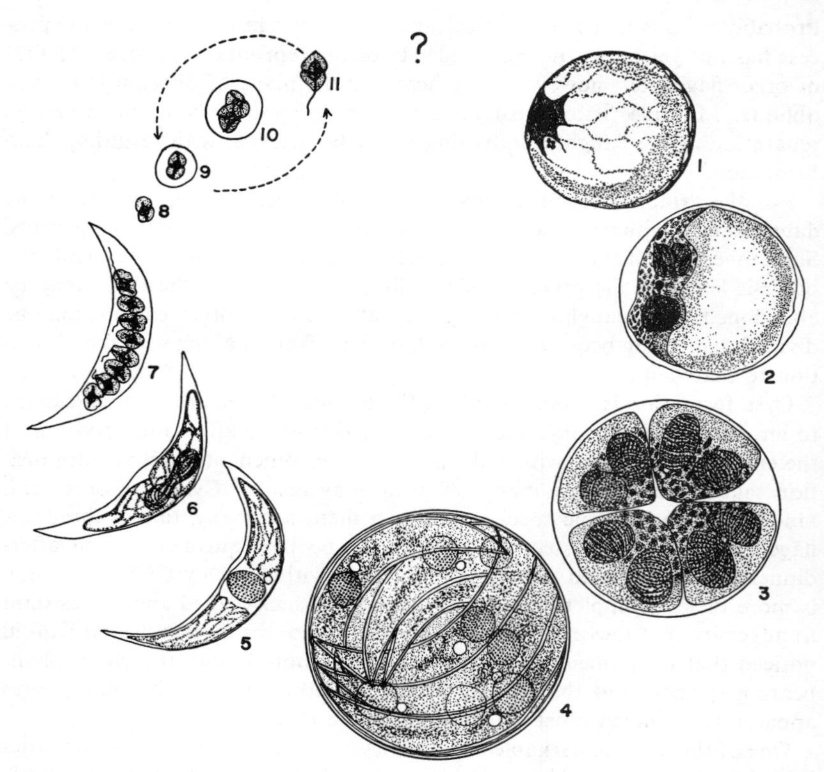

FIG. 14.10. Life cycle of *Gymnodinium lunula*, an unarmored dinoflagellate. *1–7*, *Pyrocystis* stage; *8–11*, *Gymnodinium* stage. *1*, large globular form; *2*, formation of first cleavage nuclei, primary cyst stage; *3*, second cleavage with fourth division of nuclei completed; *4*, formation of crescent-shaped spores, secondary cysts; *5*, single spore released from the cyst; *6*, beginning of division of the spore; *7*, completion of spore divisions with the formation of eight *Gymnodinium* individuals; *8*, *G. lunula* escaped from the cyst; *9*, formation of tertiary cyst; *10*, division of encysted individual; *11*, individual escaped from cyst. At this point further encystment may take place, and stages *9–11* may be repeated many times before the beginning of the next stage. Stages *1–4* are said to require about six hours. (After Dogiel, modified by Kofoid and Swezy.)

dinoflagellates are badly in need of intensive study. Their antiquity and the great diversity of form, habit, and habitat suggest that any researcher with patience and skill will be amply rewarded.

The Chrysomonads

Variety, however, is not limited to the three orders of plantlike flagellates already briefly discussed. It also occurs in the remaining three orders, the Chrysomonadida, Cryptomonadida, and Chloromonadida. The chrysomonads are undoubtedly closely related to the brown algae, and there is reason to think that some of the animal flagellates and the rhizopods may have originated from them.

The chrysomonads are relatively small and usually lack a cell wall. Many of them are prone to form pseudopodia, which they use for food capture. Like the dinoflagellates, some chrysomonads have become holozoic or saprophytic, but most of them possess chromatophores (generally of some shade of yellow or brown), and are therefore autotrophic. Fat and leucosin (a carbohydrate of uncertain nature, said to be characteristic of the group) are metabolic products. Near the anterior end of the organism is usually a stigma, and in the same neighborhood one or two flagella take their origin. The habitat may be either fresh or salt water.

Colony formation and attachment by a stalk are not uncommon. One frequently encountered genus of the former sort is *Synura*, members of which are said sometimes to give drinking water a cucumberlike odor when present in large numbers. *Stylochromonas* is an example of the second type. It is a marine resident, with a curious cytoplasmic collar surrounding its single flagellum; in this respect it bears a striking resemblance to the collared flagellated cells of sponges, known as "choanocytes." Except for the presence of two golden-brown chromatophores and its salt water habitat it also differs little from the collared animal-like flagellates of the Codosigidae, a family of the Protomonadida (to be considered below).

The life cycles of the chrysomonads typically include multiplication by longitudinal binary fission, the formation of siliceous cysts, and often the development of palmella stages. Rather surprisingly, in view of the undoubted algal affinities of the group, sexual stages seem conspicuously absent.

Sometimes placed in a class by themselves, but usually considered a family of chrysomonads, are the Coccolithina. These flagellates are almost exclusively marine or brackish water forms, although a few species are known from fresh water. The family name is derived from the peculiar skeletal elements known as coccoliths that cover their bodies. These consist of minute perforate or imperforate plates, secreted either at the surface or in the surface membrane and often of very bizarre appearance. Because of their calcareous nature they apparently lend themselves readily to fossilization, and as a result the group is known to have existed at least since Jurassic time, some 170 million years ago, and it is probably much more ancient than this. Coccoliths, as the plates are often called, are often abundant in chalks and the group apparently played a very important role in the genesis of many limestones. It has also been suggested that some kinds of fossil coccoliths, like certain genera of fossil Foraminiferida, may indicate the prevalence of milder climates at some periods in the past. To Ehrenberg belongs the credit for the first recognition of these peculiar protozoan fossils, more than a century ago.

The Coccolithina are organisms with one or two flagella, and two to four yellowish or yellowish-brown chromatophores. This latter characteristic has caused many botanists to believe that these flagellates must be closely related to the Chrysophyta (yellow-brown algae), but their actual origin is uncertain. The presence of chromatophores implies photosynthesis, yet deep-sea dredging has shown that these organisms often occur at depths of several thousand feet. Evidently an animal-like mode of subsistence is possible for such forms, for at these depths there can hardly be enough light for autotrophic nutrition.

Although the life cycles of the Coccolithina are very incompletely known, reproduction appears to be usually by longitudinal binary fission, with some type of mitosis. The coccoliths are shared equally by the two daughter cells. Sometimes there may also be multiple fission, with the formation of rather numerous offspring. Cysts may be produced, but these are calcareous rather than siliceous as in the more typical chrysomonads. For at least one species (*Coccolithus fragilis*), a life cycle has been suggested that includes two quite different phases, depending on whether the environment is coastal or in deeper waters. The organisms may take on no less than five different guises, each however with two flagella, as well as assuming the palmella form, encysting, and giving rise to numerous spores. With some of these transformations go also striking changes in color.

One other group of Chrysomonadida, widely distributed as fossils, deserves mention. They are the Silicoflagellidae, a family known from Cretaceous rocks, and thus at least 140 million years old. They persist in the oceans today, but apparently are seldom present in great numbers. Their shells, as the family name suggests, are of silica and superficially resemble those of the Radiolarida, for which their fossil remains have sometimes been mistaken. Nevertheless the starlike shape of the shells of these flagellates, especially when considered together with other characteristics, seemed unique enough to Deflandre to justify placing them in a class by themselves, the Silicoflagellida.

The Cryptomonads

The Cryptomonadida, another order of the Phytomastigina, are mostly small flagellates with two flagella, generally slightly unequal in length. Their minuteness may have suggested their name, since it is derived from the Greek words *kryptos*, meaning "hidden," plus *monas*, "alone." One or two, rarely three, chromatophores are typically present. These are usually brown, but may be some shade of green or even red. In a few species they may even be entirely absent. The pigmented types may also contain a stigma. Some of these forms can change color according to the nature of the substances dissolved in the water. Thus, Hollande states, individuals of one and the same species may appear varying tints of red, greenish, or even blue.

The cryptomonads are not a large group, but some of them are very common. One of the species most often seen in fresh water is *Chilomonas paramecium*, which is a colorless form but otherwise fairly typical. The body shape and "cytopharynx" (really a longitudinal groove) of the organism apparently reminded Ehrenberg, who coined both generic and specific names, of the ciliate *Paramecium*, and the liplike process at the anterior extremity suggested the name *Chilomonas* (from the Greek *cheilos*, "lip").

At the base of the cytopharynx lie a group of granules which electron microscopy has shown to be trichocysts. Hollande states that they may even assist in locomotion when suddenly discharged, thus furnishing a kind of rocket-aided takeoff. The two somewhat unequal flagella originate close to the anterior end, adjacent to the cytopharynx, and the single contractile vacuole is located in the same neighborhood. Conspicuous are the numerous starch granules with which much of the endoplasm is filled. When viewed with the ordinary light microscope they often appear to be very light blue, but this is an illusion since the organisms entirely lack pigment. Just back of the midregion lies the single nucleus, nearly spherical and containing a karyosome.

Like other cryptomonads, *Chilomonas* puts out no pseudopodia. This may be in part a consequence of the relatively thick cuticle.

In view of the ubiquitousness of the species, it would seem easy to cultivate, and for a colorless form its requirements are relatively simple. Yet it nevertheless needs a medium of rather complex composition. Salts of calcium, magnesium, ammonium, and iron are among the ingredients required, plus some simple organic carbon source and the vitamin thiamin. The organism grows well on this medium and has been widely used as a research tool in biochemical and other investigations.

The Chloromonads

Though several other orders of plantlike flagellates are sometimes recognized, we shall discuss only the Chloromonadida. It is a small and poorly known group of fresh- and brackish-water forms, most of which are only rarely encountered. They possess grass-green chromatophores, but lack pyrenoids, and appear to produce only oils as a product of anabolism. There is no stigma.

The cuticle, though thin and fragile, seems to be thick enough to prevent much change in shape and the extrusion of pseudopodia. Two flagella, one trailing, are present and are associated with the usual blepharoplasts and fibrils. There is, in most forms, a dorsoventral differentiation, the dorsal surface being convex and the ventral one bearing a longitudinal groove of varying length. The centrally located nucleus bears a nuclear cap and is of considerable size. All the chloromonads so far studied possess trichocysts, mucus-producing organelles, or both. A contractile vacuole is present, situated generally in the anterior third of the body above the nucleus.

The life cycles include encystment, at least in some species, and reproduction by longitudinal binary fission, apparently with mitosis. The relation of the Chloromonadida to other groups, either of the plantlike flagellates or the algae, remains obscure, although attempts have been made to show kinship to the dinoflagellates.

The Ebriedians

The Ebriedians, recognized as a class by Deflandre, are best known from their fossilized siliceous skeletons. These consist of a system of tiny rods built around an internal central spicule, and bear a certain resemblance to the skeletons of Radiolarida. Four families and more than 75 genera are known from their fossil remains, but only two genera survive in the oceans today, and the group is unknown from fresh water. By geological standards the Ebriedians are not old, going back only to Eocene times, some 60 million years ago.

The morphology of these organisms, except as it relates to their skeletons, can be judged only from the two modern representatives. There are two flagella, and a nucleus of the dinokaryon type. The nucleus is reminiscent of the dinoflagellate nucleus, but it is doubtful that these flagellates have any relationship to the dinoflagellates. Chromatophores are absent, but the cytoplasm may be tinted a light brown or rose. Nutrition is animal-like, and the anabolic product consists of minute fat droplets.

Since the Ebriedians are pelagic organisms not commonly encountered, little is known about their life cycles. Reproduction is said to be by simple fission, but it has been little studied The critical study of fossil Ebriedians has revealed more about their life cycles than we know of those now living. Fossilized cysts have been found, and different stages in life histories are suggested by fossil skeletons of varying degrees of complexity.

THE ZOOMASTIGINA

The animal-like flagellates, or Zoomastigina, constitute a much smaller assemblage than do the plantlike types, and one with fewer subdivisions. Still, they present much variety, and include many species of parasites, some of which are of economic and medical importance. Though the free-living species probably are not important in the economics of nature, they do have much biological interest and many of them are commonly found in both fresh and salt water and in the soil.

Since one must suppose that they arose from the plantlike flagellates, the evolution of the Zoomastigina presents a challenging problem. A change in nutrition, involving loss of chlorophyll, may not be difficult and is certainly not rare in nature; colorless types occur frequently even among the Phytomastigina, and at least a few normally green species get along successfully in darkness when compelled to do so. They then lose their chlorophyll, and revert to a saprophytic or holozoic mode of existence.

On the whole, the free-living Zoomastigina are small forms and the majority, like most Phytomastigina, possess only one or two flagella. Some are very minute indeed. Many individuals of the relatively common species *Oikomonas termo*, for instance, are little larger than a mammalian red cell. The species *Caviomonas mobilis*, parasitizing the guinea pig, does not crowd the intestine of its host with its diameter of three or four micra.

These organisms show little observable detail of structure, and they seem to justify Dujardin's description of them as "among the simplest of all Infusoria." Yet refined techniques of study, such as examination with electron microscopy, belie the apparent simplicity. *In vitro* cultivation, where

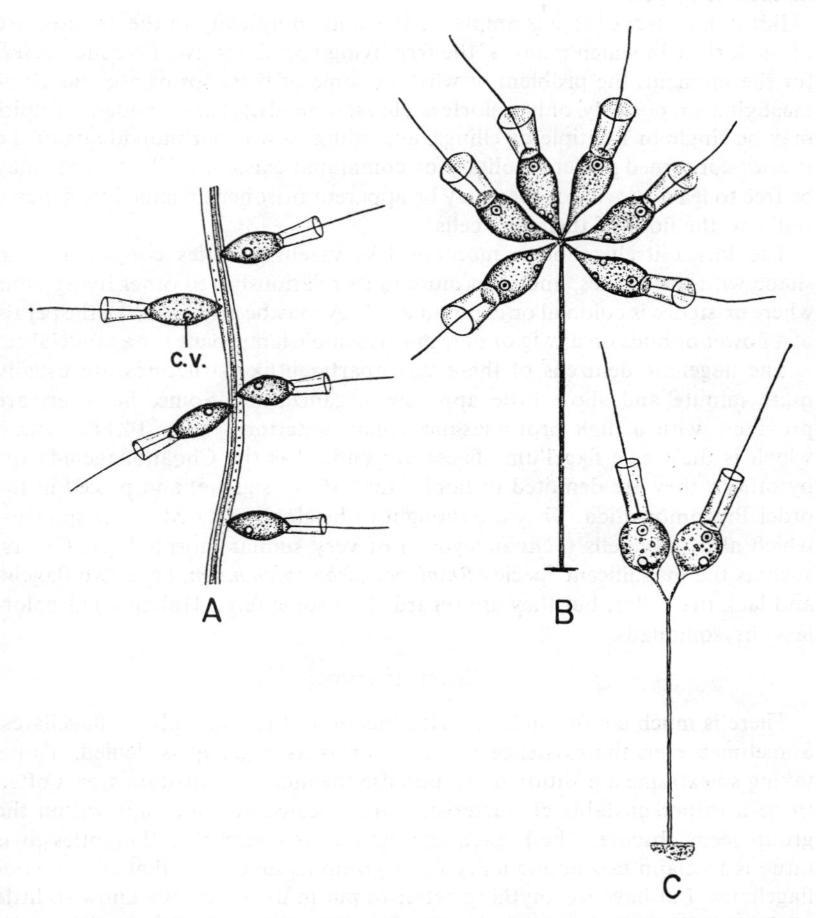

FIG. 14.11. *Codosiga botrytis*, a collared flagellate. *A*. Young individuals attached to a *Vorticella* stalk. *B*. Colony of six individuals. *C*. Two individuals resulting from recent fission and still on the parental stalk. c.v. Contractile vacuole. The collars are retractile, and the flagellates swim backwards (i.e., the flagellum trails) when free. (After Stein, from Minchin.)

it has been successfully accomplished, as for the haemoflagellates, demonstrates that their nutritional requirements are often highly complex.

For the larger species of animal flagellates, ordinary methods of microscopic examination reveal a complex organization rivalling anything seen elsewhere in the protozoan world (see Fig. 18.1, which depicts a species of *Trichonympha* from the woodroach.) Curiously, the parasitic species of animal flagellates show a greater complexity of visible organization than do most of their free-living cousins, and such complications are also usually evident in their life cycles.

But in any event there is ample variety and complexity in the architecture of the loricae in which many of the free-living flagellates live. (We may defer, for the moment, the problem of whether some of these forms are true Zoomastigina or possibly only colorless chrysomonads.) Their residential units may be single or multiple dwellings, according to whether individuals of the species concerned prefer a solitary or communal existence. The tenants may be free to leave at will, or they may be apparent prisoners, chained by delicate stalks to the floor of their tiny cells.

The lorica itself, though more or less vaselike, varies considerably in shape with the species, and even more in its relationship to other living units where existence is colonial or communal. They may be arranged like the petals of a flower or buds on a twig or they may resemble lampshades on a candelabra.

The flagellate denizens of these tiny apartmentlike structures are usually quite minute and show little apparent organization. Some, however, are provided with a high protoplasmic collar anteriorly (Fig. 14.11), within which is the single flagellum. These are known as the Choanoflagellida or, by others, they are demoted to family rank (Codosigidae) and placed in the order Protomonadida. They are thought to be closely related to the sponges, which also have cells ("choanocytes") of very similar morphology. Others, such as the magnificent species *Rhipidodendron splendidum*, have two flagella and lack the collar, but they are regarded by some (e.g., Hollande) as colorless chrysomonads.

CLASSIFICATION

There is much confusion in the classification of the animal-like flagellates. Sometimes even the existence of these forms as a group is denied. Those taking so extreme a position do so because the mode of nutrition seems often to be a rather unstable characteristic, and because relationships within the group seem obscure. The French zoologist Grassé remarks, "Doubtless it is fatuous to claim taxonomic unity for a group as diverse as that of the Zooflagellates. But have we anything better to put in its place? We know so little of the constituent groups it embraces as to lack knowledge of the differential characteristics which would allow us to separate it into natural phyla and determine their real relationships."

Grassé recognizes three superorders in the class Zoomastigina. In the first of these, the Protomonadina, he places five orders, and in the second, the Metamonadina, eight more, with a tentative ninth. The third superorder, the Opalinina, comprises only the opalinid parasites of frogs and toads. These have traditionally been regarded as ciliates, and have been grouped together as the Protociliata.

A system of this sort has much to commend it, but it seems premature in view of our ignorance of much of the basic biology of so many of these flagellates. The scheme of classification used here, both for plant- and animal-like forms, is essentially that which has been employed by most American protozoologists, differing little from Hall's (1953) and Kudo's (1954). It is largely based on mode of nutrition and nature of reserve foodstuffs, ability to form pseudopodia (when present), and sometimes on the type of skeletal elements and on body form. The number of flagella, though an artificial criterion, is also a useful characteristic, because it can be more easily determined than can the structure of the entire locomotor apparatus (or "mastigont," more strictly defined as all structures developing from the blepharoplasts). The characteristics of the mastigont are a better index of degree of biological relationships. Life cycles, when known, are taken into account, as is sometimes habitat. Yet there is little doubt that all that is presently known in many cases fails to indicate biological relationships, and as knowledge increases the scheme will be extensively revised. In outline form it is as follows:

Subphylum PLASMODROMA Doflein. Protozoa with either flagella or pseudopodia (sometimes with both), or in the case of the Sporozoa (all of which are parasitic), with no locomotor organelles, although often capable of independent movement nevertheless. One type of nucleus only, but frequently multinucleate.

Class MASTIGOPHORA Diesing. Organisms using flagella for locomotion, although pseudopodia may be produced by some.

Subclass 1. PHYTOMASTIGINA Doflein. A very varied assemblage of forms, although the majority possess chromatophores of chlorophyll and carry on photosynthesis. (For this reason, botanists often claim them as algae, to which they are undoubtedly closely related.) Amoeboid, flagellate, and palmella stages often characterize the life cycles, which may also include a sexual phase.

Order 1. CHRYSOMONADIDA Stein. A large group of mostly minute forms, with two (sometimes one or three) flagella. Usually no body wall, with the result that the body form is quite changeable ("plastic" or "metabolic"). The chromatophores are discoid, and some shade of brown, yellow, or yellowish-green. Starch is always absent, but leucosin and fats may be present. There is often a stigma. The habitat is fresh or salt water; also soil.

Suborder 1. EUCHRYSOMONADINA Pascher. Chrysomonads in which the flagellate stage is dominant. A variable number of families and, often, subfamilies, are recognized, differentiation being chiefly on flagellar characteristics and body form.

Suborder 2. SILICOFLAGELLINA Borgert. Marine flagellates with a siliceous skeleton, a single flagellum, greenish-brown chromatophores, and reserve foodstuffs in the form of leucosin. Extensively represented as fossils.

Suborder 3. COCCOLITHINA Lohmann. Almost always marine; rarely in fresh water. Usually two equal flagella. Yellowish-green, or brownish chromatophores. Usually very small forms, with a skeleton of calcareous platelets of varied shapes. Many fossil types known.

Suborder 4. RHIZOCHRYSODINA Pascher. Amoeboid phase dominant (it is said that the flagellate phase is unknown). Aggregates or even true plasmodia may be formed by the amoeboid stages. Chromatophores and test may or may not be present. The identity of this group is often doubted.

Suborder 5. CHRYSOCAPSINA. Pascher. These flagellates have the palmella stage dominant; in some it may be quite absent, or is at least unknown.

Order 2. HETEROCHLORIDA Pascher. Dorso-ventrally differentiated, usually naked forms with two unequal flagella and (sometimes) pseudopodia also. One or more yellowish-green, or yellow chromatophores; reserve foodstuffs in the form of leucosin and lipids, but no starch. Encystment endogenous; cysts with double wall, no pore, and of silica.

Suborder 1. EUHETEROCHLORINA Pascher. Flagellate stage dominant, although sometimes with pseudopodia.

Suborder 2. RHIZOCHLORIDINA Pascher. Plasmodial (nonflagellated) stage dominant; pseudopodia lobose or filamentous.

Suborder 3. HETEROCAPSINA Pascher. Palmella stage dominant.

Order 3. CRYPTOMONADIDA Ehrenberg. Forms with two ribbonlike flagella, subequal; usually a rigid pellicle, often with dorso-ventral differentiation. Ventral groove ("pharynx") frequently present. Both colorless and chlorophyll-bearing forms occur; in the latter, one or several chromatophores (brown, some shade of green, even red). Reserve foodstuffs: starch or fatlike substances. A rather small, but well-defined group.

Order 4. DINOFLAGELLIDA Bütschli. A large and very important, varied, and predominantly marine group. There are two flagella, arising independently: one, directed backward, lies in a longitudinal groove ("sulcus"), while the second, ribbonlike in form, winds around the body, belt-fashion, in a left-handed spiral (usually posteriorly), following a transverse groove ("girdle"). Cellulose envelope ("theca") in the majority of species, often elaborately sculptured. Nucleus of a rather peculiar type (known as "dinokaryon"). Mode of nutrition varied: typically autotrophic. with brownish-green chromatophores. Usually two vacuoles ("pusules"), filled with rose-colored fluid.

Suborder 1. PROROCENTRINA Poche. Bivalve theca, without distinct grooves. Chromatophores always present.

Suborder 2. GYMNODININA Kofoid. Naked, but with grooves, and sometimes with internal skeleton. Mostly marine. (*Noctiluca*, remarkable both for its phosphorescence and its size—it may reach a diameter of 2,000 micra—belongs to this suborder.)

Suborder 3. PERIDININA Schütt. In these forms the theca consists of discrete plates, and is rather thick. The number and arrangement of these plates furnishes the basis for further subdivision.

Suborder 4. DINOCAPSINA Pascher. Palmella stage dominant; flagellate stage *Gymnodinium*-like.

Suborder 5. DINOCOCCINA Pascher. Here the so-called "protococcus" or "pyrocystis" stage is dominant (thin-walled and without flagella).

Order 5. PHYTOMONADIDA Blochmann. A large group of predominantly small, mostly fresh-water forms. Typically with green chromatophores (though some species are colorless); starch always present, and often oils also. Usually a cellulose envelope. Unicellular, or in colonies ("coenobia"), in which the number of cells is definite. Flagella usually 2 or 4, but sometimes up to 8. Sexual reproduction common; encystment and palmella formation also occur. Diploid only in zygote.

Order 6. EUGLENOIDIDA Bütschli. Rather large organisms, typically with bright green chromatophores, and one or two flagella. But numerous colorless forms exist, and some species may lose their chlorophyll in darkness, then adopting a saprozoic mode of nutrition. (A few may also appear red, due to an accumulation of haematochrome.) Paramylum is the reserve foodstuff (characteristic of this order only), supplemented by fats and oils. At the anterior end of the body is an opening leading into a canal ("gullet"), ending in a cavity ("reservoir"); the flagella protrude from this. Mostly fresh-water forms. Hollande (1952) divides the order into five large families.

Order 7. CHLOROMONADIDA Klebs. A small group of relatively little-known organisms, with (usually) grass-green chromatophores (hence the name, from the Greek "chloros," for light green), and one or, more commonly, two flagella (of which one trails). The reserve foodstuff is oil. Body is in most species dorsally convex, with a ventral groove. Trichocysts and/or mucus-producing organelles present. No stigma.

Subclass 2. ZOOMASTIGINA Doflein. These flagellates differ from the colorless members of the Phytomastigina chiefly in the nature of the stored carbohydrate, which is glycogen rather than starch or paramylum; fats and oils may also be present. Cellulose body coverings do not occur; except in the cyst stages there is usually no rigid body envelope. Most forms are rather small, and (except in some species of parasites or symbionts) have only one or a few flagella. Sexual stages are, except in one group, unknown.

Order 1. RHIZOMASTIGIDA Bütschli. A small group of flagellates (or of Sarcodina), with from one to four flagella, and the ability to put out pseudopods. The correct taxonomic position for these organisms is uncertain.

Order 2. PROTOMONADIDA Blochmann. Mostly small organisms with one or two flagella, prone to change shape (plastic); some are colonial and many are parasitic. Holozoic or saprozoic nutrition. Life cycle usually simple, except in the single family Trypanosomatidae. Includes the collared forms of the family Codosigidae, regarded as closely related to the sponges, and often put into a separate order, the Choanoflagellida or Craspedomonadida.

Order 3. POLYMASTIGIDA Blochmann. Flagellates with from three to eight flagella, uninucleate or sometimes multinucleate. Usually very small, and often provided with a cytostome. Includes many parasitic species.

Order 4. TRICHOMONADIDA Kirby. Uninucleate or multinucleate forms (not binucleate), with from three to six flagella, one of which is typically trailing, an axostyle, and a well-developed parabasal apparatus. The trailing flagellum often forms part of an undulating membrane. Parasites in many host species.

Order 5. HYPERMASTIGIDA Grassi and Foa. Uninucleate flagellates with numerous flagella (hence the name); gut parasites of roaches (especially wood roaches), and termites.

REFERENCES

Allegre, C. F. and Jahn, T. L. 1943. A survey of the genus *Phacus* Dujardin (Protozoa; Euglenoidina). *Trans. Am. Mic. Soc.*, 62: 233–44.

Bonner, J. T. 1950. *Volvox. Sci. Am.*, 182: 52.

Bütschli, O. 1883–7. Mastigophora. In *Bronn's Klassen und Ordnungen des Tierreichs.* Bd. I, Abt. II. Leipzig.

Chatton, E. and Hovasse, R. 1934. L'existence d'un réseau ectoplasmique chez les *Polykrikos* et les précisions qu'il fournit à la morphologie peridinienne. *C. r. Soc. Biol.*, 115: 1039–41.

Cleveland, L. R. 1956. Brief accounts of the sexual cycles of the flagellates of *Cryptocercus. J. Protozool.*, 3: 161–80.

Conrad, W. and Van Meel, L. 1952. Matériaux pour une monographie de *Trachélomonas* Ehrenberg, C., 1834, *Strombomonas* Deflandre, G., 1930, et *Euglena* Ehrenberg, C., 1832, Genres d'Euglenacées. Inst. Roy. Sci. Nat. Belgique.

Corliss, J. O. 1955. The opalinid infusorians: Flagellates or ciliates? *J. Protozool.*, 2: 107–14.

Deflandre, G. 1936. Les Flagellés fossiles. Aperçu biologique et paléontologique. Rôle géologique. Actualités scient. et industr. *Expos. de géol.*, vol. 335. III. Paris.

——— 1951. Recherches sur les Ebriédiens. Paléobiologie. Evolution. Systématique. *Bull. Biol. France et Belgique*, 88: 1–84.

——— and Cookson, I. C. 1954. Micropaléontologie. Sur le microplancton fossile conservé dans diverses roches sédimentaires australiennes s'étageant du Crétace inférieur au Miocène supérieur. *C. Acad. Sci.*, 239: 1235–8.

Doflein, F. and Reichenow, E. 1952. *Lehrbuch der Protozoenkunde.* Teil 2. 1. Hälfte: Mastigophora und Rhizopoden. Jena: Verlag Gustav Fischer.

Fuller, H. J. and Tippo, O. 1955. Thallophyta: Algae. In *College Botany.* New York: Holt.

Gojdics, M. 1953. *The Genus Euglena.* Madison: University of Wisconsin Press.

Grassé, P.-P. 1952. *Traité de Zoologie.* T. 1, Fasc. 1. Paris: Masson.

Hall, R. P. 1939. The trophic nature of the plant-like flagellates. *Quart. Rev. Biol.*, 14: 1–12.

——— 1953. The Mastigophora. In *Protozoology.* New York: Prentice-Hall.

Hartog, M. 1906. Flagellata. In *Cambridge Natural History*, vol. 1. London: Macmillan.

Hutner, S. H. and McLaughlin, J. J. A. 1958. Poisonous tides. *Sci. Am.*, 199: 92.

———, Provasoli, L., McLaughlin, J. J. A. and Pintner, I. J. 1956. Biochemical geography: some aspects of recent vitamin research. *Geog. Rev.*, 46: 404–7.

Hyman, L. H. 1940. *The Invertebrates: Protozoa Through Ctenophora.* New York: McGraw-Hill.

Jahn, T. 1946. The euglenoid flagellates. *Quart. Rev. Biol.*, 21: 246–74.

——— 1955. Adventures among the flagellates, and other matters. *J. Protozool.*, 2: 1–5.

Jane, F. W. 1955. Famous plant-animal: *Euglena. New Biology*, 19: 114–25.

Kent, S. 1880–1. *A Manual of the Infusoria.* London: David Bogue.

Kirby, H. 1944. Some observations on cytology and morphogenesis in flagellate Protozoa. *J. Morphol.*, 75: 361–421.

——— 1947. Flagellate and host relationships of trichomonad flagellates. *J. Parasitol.*, 33: 214–28.

Kofoid, C. A. and Swezy, O. 1921. *The Free-living Unarmored Dinoflagellata.* University of California Press.

Kuhn, A. 1921. *Morphologie der Tiere in Bildern.* I. Heft., 1. Teil. Flagellaten. Berlin: Bornträger.

Pitelka, D. R. 1945. Morphology and taxonomy of flagellates of the genus *Peranema* Dujardin. *J. Morphol.*, 76: 179–89.

Pascher, A. 1914. *Flagellatae. Die Süsswasserflora Deutschlands.* Heft 1 und 2. Jena.

——— 1927. *Die Süsswasserflora Deutschlands, Österreichs und der Schweiz.* 4. Volvocales. Jena.

Pocock, M. A. 1932. *Volvox* and associated algae from Kimberley. *Ann. So. Afr. Mus.*, 16: 473–521, Part 3 and *Volvox* in South Africa, 16: 523–646, Part 5.

Pringsheim, E. G. 1941. The interrelationships of pigmented and colourless Flagellata. *Biol. Rev.*, 16: 191–204.

——— 1955. The genus *Polystomella. J. Protozool.*, 2: 137–45.

Provasoli, L. and Pintner, I. J. 1953. Ecological implications of *in vitro* nutritional requirements of algal flagellates. *Ann. N.Y. Acad. Sci.*, 56: 839–51.

Reichenow, E. 1931. Parasitische Flagellata. In Grimpe, *Die Tierwelt der Nord- und Ost see*. Teil 2.

Schiller, J. 1935. Dinoflagellatae. In Rabenhorst's *Kryptogamen-Flora*, vol. 10, Heft 3.

Smith, G. M. 1950. *The Fresh-water Algae of the United States*. New York: McGraw-Hill.

Stein, F. 1878 and 1883. Der Organismus der Infusionsthiere. III. Abt., Der Organismus der Flagellaten oder Geisselinfusorien. Hälfte 1 und 2. Leipzig.

West, G. 1916. *Algae*, vol. 1. Cambridge University Press.

ADDITIONAL REFERENCES

Mignot, J.-P. 1965. Ultrastructure des Eugléniens. I. Comparaison de la cuticule chez différentes espèces. Protistologica, 1(1): 5–16 (4 plates).

——— 1966. Structure et ultrastructure de quelques Euglénomadines. *Protistologica*, 2(3): 51–117 (23 plates).

Seliger, H. H., Fastie, W. G., Taylor, W. R., and McElroy, W. D. 1962. Bioluminescence of marine dinoflagellates. *J. Gen. Physiol.*, 45: 1003–17.

ADDITIONAL REFERENCES FOR CHAPTER 13

Committee on Taxonomy and Taxonomic Problems of the Phylum Protozoa (B. M. Honigberg, Chairman). 1964. A revised classification of the phylum Protozoa. *J. Protozool.*, 11: 7–20.

International Code of Zoological Nomenclature. 1964. XV International Congress of Zoology. London: International Trust for Zoological Nomenclature.

15

Sarcodina

> "Without trace of nerve elements, and without definite, fixed organs of any kind, internal or external, the Rhizopod,—simplest of all animals, a mere jelly-speck—moves about with the apparent purposes of more complex creatures. It selects and swallows its appropriate food, digests it, and rejects the insoluble remains. It grows and reproduces its kind. It evolves a wonderful variety of distinctive forms, often of the utmost beauty, and, indeed, it altogether exhibits such marvellous attributes, that one is led to ask the question in what constitutes the superiority of animals usually regarded as much higher in the scale of life." Joseph Leidy, *Fresh-water Rhizopods of North America*, 1879.

Most people, even many biologists, still think first of *Amoeba* when Protozoa are mentioned (Fig. 15.1). *Amoeba*, they suppose, stands at one end of the evolutionary ladder and man at the other. The Sarcodina, of which the amoeba has become the most familiar example (although by no means the most typical), are often presented as interesting chiefly because of their assumed primitiveness. Yet the amoeba, seemingly so simple in its organization, is highly complex. It has its nucleus and constituent chromosomes, its contractile vacuole (except when marine or parasitic), its mitochondria, and frequently a rather complicated life history.

There are better reasons for studying the Sarcodina, ancient though they are, than any light they may shed on our very remote ancestors. Sarcodina are common in almost all environments where life is possible and are often therefore our close neighbors. Some have contributed largely to the formation of sedimentary rocks on which our buildings stand, and of which they may be built. Others have become parasites, and a few cause serious diseases of man and animals. Some, too, have proved excellent research material for the study of such fundamental biological problems as

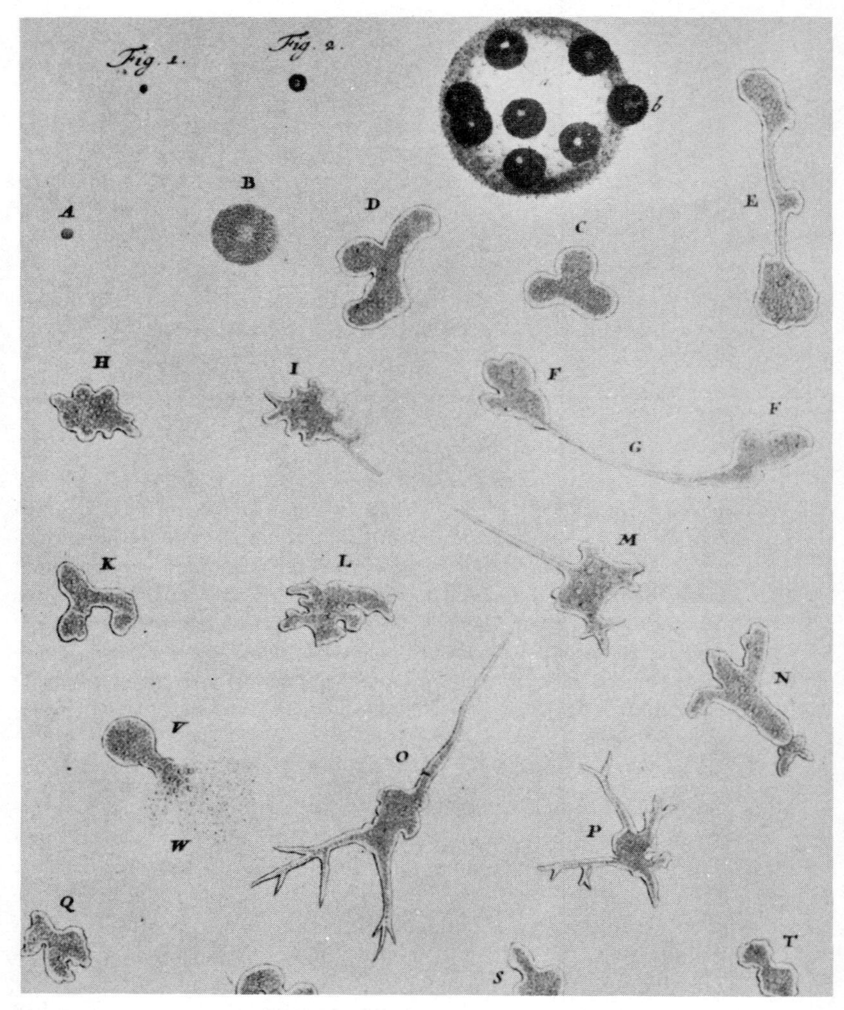

FIG. 15.1. A plate originally published in 1755 by Rösel, in which he described "Der kleine Proteus," the amoeba shown in the drawings A-T. *G* is a dividing form. Figs. *1–3* are various stages of *Volvox*, the first perhaps natural size. (Reproduction from Mast.)

the mechanism of protoplasmic contraction and the interrelationships of nucleus and cytoplasm.

CHARACTERISTICS OF THE CLASS

The class Sarcodina includes organisms sharing the ability to form temporary cytoplasmic extensions called pseudopodia, with which they move about and capture food, but they have little else in common. Besides, there are

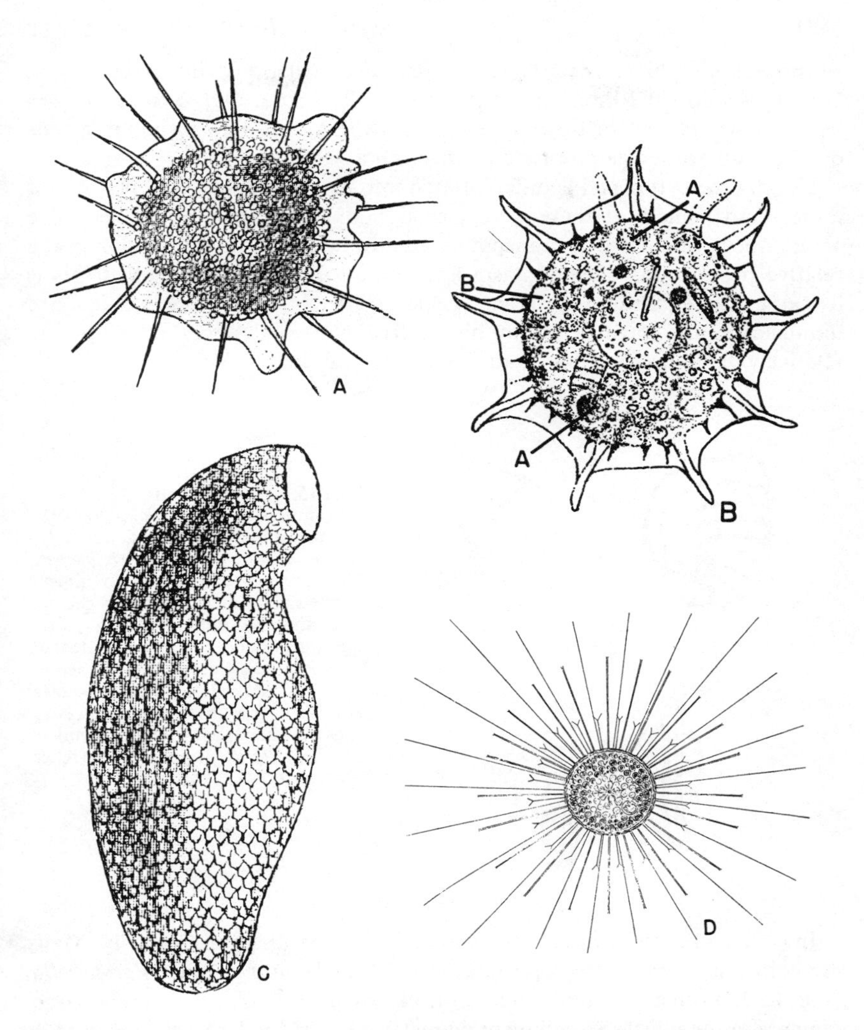

FIG. 15.2. *A. Vampyrella lateritia*, a heliozoan-like proteomyxan. Diameter 30–40μ. (After Conn.) *B. Arcella vulgaris*, a common fresh-water shelled rhizopod. View of living individual, seen from the oral side. a, nucleus (there are characteristically two); b, contractile vacuole. The circle in the center shows the shell opening, through which pseudopods are extruded. The boat-shaped object shown in the cytoplasm is an ingested diatom; others are shown on the opposite side to one of the nuclei, although these are of a different sort. (After Leidy.) *C. Cyphoderia ampulla*, a testate rhizopod making a retort-shaped shell of chitinlike material, to which is affixed a coating of scales laid on like shingles. The shell may be colorless or some shade of yellow. Length 60–200μ. (After Conn.) *D. Acanthocystis turfacea*, a common fresh-water heliozoan, characterized by numerous long pseudopodia and many radially arranged tubular needles, forked at their ends. These needles are of two sizes, the shorter ones having the larger forks. The cytoplasm is green. Covering the animal is a thin skeleton of siliceous plates, and underlying this is a second layer made up of the expanded bases of the needles. Size variable, but about 50–60μ. (After Penard.)

Mastigophora which, as we have seen, capture their prey with pseudopodia, Nor is pseudopod formation limited to the Protozoa. Our own leucocytes and phagocytic cells of other types possess them, as do corresponding cells of numerous other vertebrate and invertebrate Metazoa.

Usually the cytoplasm is differentiated into ectoplasm and endoplasm, but there is no pellicle. If there were, pseudopodia could not be formed. The ectoplasm is often clearly transparent or hyaline. Pseudopodia differ in the relative proportions of endoplasm and ectoplasm they contain, and this is useful in distinguishing different species. The majority of Sarcodina are uninucleate, and most divide by binary fission. Sexual stages are known in relatively few species other than the Foraminiferida.

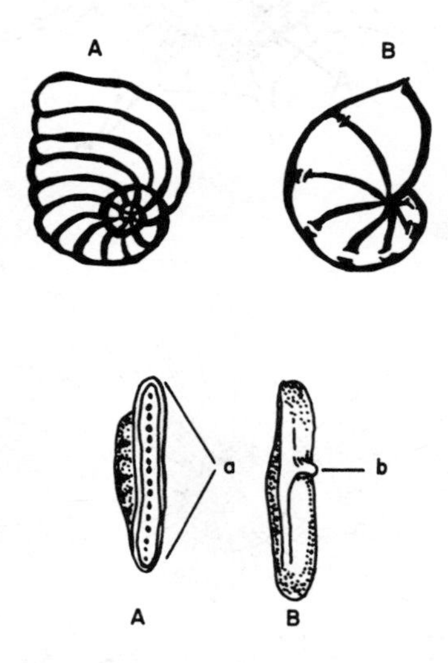

FIG. 15.3. The shell structure of the foraminiferidan *Peneroplis pertusus*, (*A* above and below), *Cristellaria*, (*B* above), and *Anomalina ariminensis*, (*B* below). The first has an imperforate shell wall, the second is perforate, and the third shows a single aperture. *A* and *B* above are median sections; below are shown the septal faces. "a" in the lower left (*A*) shows a row of apertures. Note the openings (foramina) between the chambers in *Cristellaria* (upper B). (After Chapman.)

In color, size, and general appearance Sarcodina differ enormously. Many are colorless; some are colored, like the delicately orange-hued *Vampyrella*, (Fig. 15.2*A*) others are bioluminescent, like many Radiolarida. Some Sarcodina are minute, little exceeding in diameter some of the larger bacteria, while others may be measured in millimeters, or even inches.

Though many are naked others are clad in shells, frequently of great beauty. Through openings in the shells pseudopodia may be put out or withdrawn. In the fresh-water shelled forms (the Testacida) there is a single such opening as in *Arcella* and *Cyphoderia*, but one fresh-water heliozoidan is characterized by siliceous plates and spicules (Fig. 15.2), and there may be many minute pores in the shells of the marine types known as Foraminiferida (though the name, from the Latin "window-bearing," refers to openings between the shell chambers, rather than to the perforations leading to the exterior) (Fig. 15.3).

Most Sarcodina prefer a solitary existence, but some of the Mycetozoida (also often called Myxomycetes, because of a long-assumed kinship to the fungi) include in their life history an amoeboid phase in which they are very social, even marching in armies to whatever destination may seem most desirable to amoebulae. But they are apparently guided by a substance called acrasin rather than by the leadership of some one of their own number.

There are also a few sessile forms, for example, the foraminiferidan *Haliphysema*. Still others, like the marine heliozoidan *Actinomonas*, may anchor themselves at times only to swim off later (in this case, with flagella, which it is also able to produce).

That the Sarcodina, despite the slowness with which they move, may be great travelers is indicated by the occurrence of many species throughout the world, or at least in widely separated regions. This is especially true of the marine types, such as the Radiolarida and some naked forms, which are normally pelagic, drifting about with wave and current.

Although all Sarcodina produce pseudopods in at least some stage of their life cycle, they very often enclose themselves within cysts, emerging weeks, months, occasionally even years later. This is a trait especially characteristic of fresh-water and parasitic forms, and one they of course have in common with many other kinds of Protozoa, and even with animals of other phyla. Some species produce flagella at certain stages in their life cycle, and if these were all one could observe, their true place among the Protozoa might be very hard to determine.

ANTIQUITY

Of all the Protozoa, the Sarcodina are of the greatest known antiquity. Fossil foraminiferidan and radiolaridan shells have been found in Cambrian and pre-Cambrian rocks—the oldest in which recognizable fossils of any sort are known to occur.* Some of the older works may have incorrectly identified certain of these supposed foraminiferidan fossils, there is no doubt of their occurrence in Ordovician rocks, and thus of their existence at least 425 million years ago. Fossil Radiolarida of Cambrian age have been identified, and therefore we know that they have existed for more than half a billion years. Some of these extremely ancient species were very similar to those still living. In places radiolaridan fossils are almost unbelievably abundant. It is said that in Barbados in the West Indies they constitute the bulk of strata 1,100 feet thick, and similar rock layers in the Nicobar Islands of the Indian Ocean exceed 2,000 feet in thickness. Just as the Foraminiferida and Radiolarida of the past have contributed greatly to the

*There are one or two recent reports in the literature of the finding of recognizable algal and protozoan fossils of considerably greater age, but these need confirmation. That of Tyler and Barghoorn (*Science*, 119: 606, 1954), for example, reports the occurrence of several types of algae and fungi, and what was thought to be a flagellate, in pre-Cambrian rocks of the Canadian Shield, of a probable age of about 2 billion years.

limestone and silica comprising much of the earth's sedimentary cover rock, those living in the oceans today will fill a similar role for the future. Fossil Foraminiferida are useful as stratigraphic indices for the oil driller, and they are said to have been important in the postwar development of the oil industry in Japan.

Fossil Foraminiferida in the accumulated sediments of the sea bottom are sensitive indicators of fluctuations in climates of the past, since some species are highly temperature-sensitive. With climatic changes go gradual changes in ocean temperatures. Thus, dredgings from the Atlantic and Caribbean sea bottoms reveal stratified deposits of certain species of *Globigerina*, indicating that temperatures began to decline about 80,000 years ago, until they reached a minimum about 65,000 years later with a gradual rise since. This is evidence that the last recession of the great continental ice sheets probably occurred at about the latter time, and is apparently still continuing.

Habitats

The majority of Sarcodina are ocean dwellers, and of these, most are members of the two great orders Radiolarida and Foraminiferida. The latter group is the larger, and includes, according to Cushman, at least 50 families

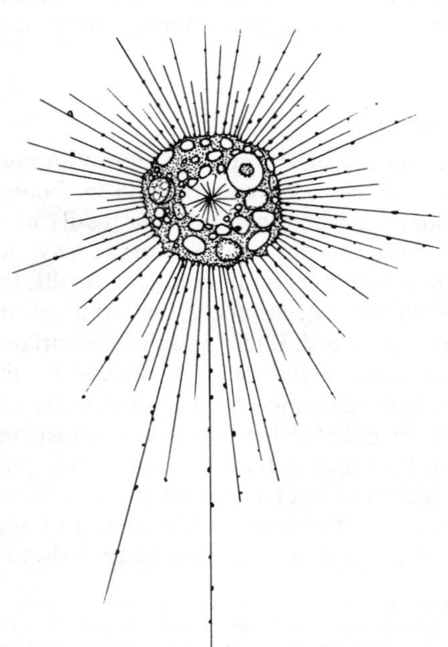

Fig. 15.4. *Oxnerella maritima*, a marine heliozoidan. The rounded inclusions in the cytoplasm are ingested swarm spores of green algae; drawings from life. (After Dobell, *Quart. J. Mic. Sci.*)

with possibly more than 30,000 species. The Radiolarida are less well known, but Haeckel, who made the most extensive study of this group, described some hundreds of species almost a century ago. At present, more than 5,000 species have been described, and doubtless many more exist, among them naked marine heliozoidans, such as *Oxnerella maritima* (Fig. 15.4).

The Foraminiferida are usually equipped with elaborately chambered shells of pseudochitin heavily impregnated with calcium carbonate. They are preferentially bottom or beach dwellers, perhaps because of the weight of their shells, and creep about the substrate. A few are naked, a few live in fresh water, and a few are sessile. These last are often found attached to marine vegetation. Those without shells are, in general, believed to be more primitive than the rest.

The Radiolarida are chiefly pelagic. Unable to swim, they drift with the ocean currents, and are especially numerous in warmer waters. But not all Radiolarida are surface dwellers. Some live at depths of 1,200 or 1,500 feet, and a few species have been dredged from abyssal deeps of several miles, where night is eternal. There all organisms must depend for food ultimately on the bodies of other organisms which drop from the surface.

Amoebae have adapted themselves to almost every conceivable niche: some are marine, like the Radiolarida and all but a few Foraminiferida, but others are restricted to fresh water. Here they may be found in ephemeral puddles, the bottom ooze of ponds and the silt of lakes and streams, swamps and bogs from the Arctic to the equator and from mountain meadows to the reedy seaside, and the surface layers of soil everywhere. A few species are found even in the highly saline lakes of arid inland regions. Moss and sphagnum, especially where the water is soft, are likely to contain them in abundance. Apparently they are less prone to occur in limestone pools and springs, as Leidy observed almost a century ago. Testacida (e.g., *Centropyxis aculeata*, Fig. 15.5) are also numerous in similar situations, but they appear to be especially fond of life in the acid waters of peaty bogs.

Although it is true that many rhizopods are more likely to be found in some environments than in others, some species seem to be nearly cosmopolitan, occurring in both salt and fresh water and in surface soils as well as silt. Individuals of such species may sometimes even be transferred from one habitat to another of very different sort, especially if the change is gradual. On the other hand, some species are very particular about their environment, and fail to survive for long a change of any magnitude. Parasitic species, doubtless because of the constancy of the environment provided by their hosts, to which they have become accustomed over a long time, are the least adaptable of all.

By some authorities, the Testacida are regarded as representative in fresh water of the environmental position occupied by Foraminiferida in the oceans. But there is little actual resemblance between the two groups. Not only is the composition of the testacidan and foraminiferidan shell quite different, but so also is its architecture. Only in a very few species of Testacida is there even a suggestion of a chambered structure. The life cycles typical of the two orders differ although the cycles are known for only very few.

Apparently the possession of a shell, though it confers obvious protective advantages, may be a handicap to adaptability, for the naked Sarcodina are found in many more habitats than are their armored brethren. Perhaps we should expect this, for shells imply a high degree of specialization.

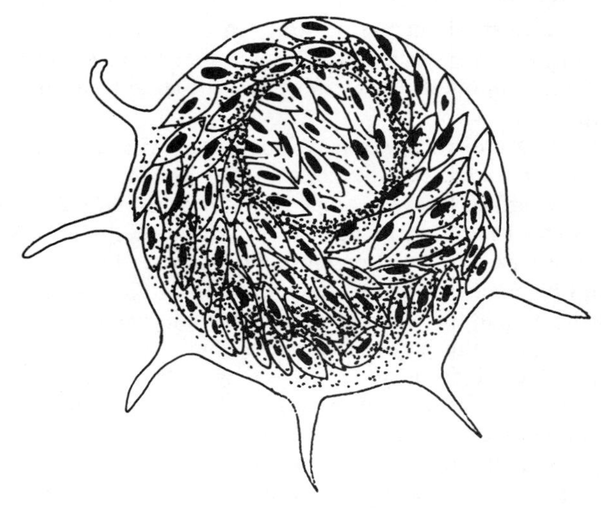

FIG. 15.5. *Centropyxis aculeata*, a testate rhizopod with a shell, usually yellowish or some shade of brown, and often covered (as in the specimen shown) with diatom shells. The shell mouth is somewhat eccentric. Size 100–150μ. (After Leidy.)

Among the more cosmopolitan of the Sarcodina are many of the soil dwellers. These tend to be species which also occur commonly in fresh water, and even in the feces of animals—more particularly, in those of herbivores. Sandon, an English biologist who studied Protozoa of the soil extensively, could only find eleven species of naked rhizopods (of which eight were amoebae) and two of testacidous types which seemed to be entirely restricted to this habitat. But he found in soil numerous species of Sarcodina known also to occur in aqueous environments of various kinds, both fresh water and salt and even in sewage. The list included, too, species normally preferring water plants and mosses. These Protozoa are minute enough to find ample room in the films of moisture adherent to soil particles and such an environment is not really very different from that supplied by natural bodies of water, except, perhaps, for the character of food it normally furnishes and its relative abundance. Periods of drought are doubtless met by ready encystment.

Beach sands constitute a special niche in environments of this kind, and it is one about which we know little; indeed, as far as Protozoa are concerned, next to nothing. Yet they are numerous in beach sands. One of the few studies of this kind yet made showed a census of about 800 Protozoa to the cubic centimeter. Another similar study, done on Trout Lake in Wisconsin, gave much larger figures, the count per cubic centimeter of sand in some cases reaching 50,000. Although the bacterial population was much higher, the protozoan density very much exceeded that of other animal organisms, such as rotifers, water bears, and copepods. Among the Protozoa the Sarcodina made up a minority.

But findings such as these raise more questions than they answer. What are the interrelationships of individuals, species, and larger taxonomic groups among such populations? How do we explain the oscillations in population size that occur from time to time? What are the effects of periodic changes in the physical characteristics of the environment? Some species appear to be restricted to beach sands, and to have developed in the course of their evolution certain special adaptive features. Such species are said to be "psammophilous." Yet beaches, particularly those near fresh water, have a relatively short life, geologically speaking. Thus we may suspect that the evolution of these psammophilous species has perhaps been relatively rapid. The limnologist Pennak remarked, in the course of a symposium on lake biology held some years ago, "We believe that the Protozoa of beaches present a wide opportunity for the protozoologist from the points of view of physiology and ecology as well as taxonomy."

Life Cycles

At present what we know of the life cycles of Sarcodina is enough to show the great variety within the group. Sexual stages may or may not be included and encystment may or may not occur. Though ordinary reproduction is usually by fission, the process differs greatly in details. During the interkinetic, or resting, phase the nucleus in most species appears to be of the vesicular type, often with a conspicuous karyosome, and may even be Feulgen-negative, though becoming positive with the onset of division. (In many species, more than one nucleus is present.) In some, a part of the nuclear substance may be of plastin (a hypothetical substance based on ill-defined staining differences, and of uncertain nature). The chromatin is often condensed into granules at the nuclear periphery, in a layer just within the nuclear membrane. Although plastin and chromatin both stain heavily with hematoxylin, traditionally one of the best nuclear stains, differentiation is easy by the use of the Feulgen stain, which is a delicate test for the presence of desoxyribonucleic acid (DNA). This substance, of course, is the raw material of which the genes of all living things are made.

Unfortunately, the nuclear behavior of many species of Sarcodina has been inadequately studied, or not at all. But fission is accompanied by true mitosis, as in other Protozoa, with the appearance of spindles and chromosomes, and often of centrioles or division centers. The nuclear membrane is likely to persist, as in most other Protozoa (although it does not do so in dividing metazoan cells, as a rule). If centrioles are present there may be a suggestion of astral rays.

If chromosomes in Sarcodina, as in most other living things, are the bearers of genes, and if the number of chromosomes is a rough measure of the total number of genes (and hence of the number of genetically transmitted traits), at least some species of Sarcodina must be highly complex, despite their apparently simple structure. What a dissertation might result from an intensive study of the genetics of *Amoeba proteus*, reported by Jepps to possess "500 or more definite chromosomes"! Although chromosome number in the relatively small number of species of Sarcodina in which it has been determined varies from a very few to the hundreds in *Amoeba proteus*, there is almost equal variety in other aspects of nuclear division. The multinucleate condition characteristic of some genera (such as *Pelomyxa*, some species of which may have as many as 1,000) adds to the complexities of reproduction.

Division

Division appears to be a somewhat simpler process for naked Sarcodina than for shelled types, in which at least one of the daughters must be supplied with a new test. But even among amoebae there is much variety. Sometimes, as in certain of the smaller species, evidence of definite chromosomes is lacking even though a typical spindle is formed. Or the chromosomes may be clearly visible, as in the division of the common small intestinal amoeba of man, *Endolimax nana*. The English protozoologist Dobell, who extensively studied this species, reported its chromosome number to be ten. One wonders that so tiny an animal—only a little larger than a red cell—should have genes enough to fill so many chromosomes, for its total of heritable traits must be small. In still other amoebae chromosomes may be so numerous as to defy the researcher wishing to count them: such as Dawson and his coworkers found to be the case with *Amoeba dubia*.

The division process is influenced by a number of factors. Temperature, of course, is one of them. The rate is slowed by cold and accelerated, up to point, by heat. Since mitosis is in reality a composite of many steps, environmental changes affect them unequally, as Chalkley and Daniel observed in the case of *Amoeba proteus* subjected to varying temperatures. For each species there is undoubtedly an optimum, or at least an optimum range.

Nutrition, too, is naturally a factor of importance in the division process. Some kinds of food, as in other animals, man included, suffice for continued growth and multiplication, whereas others are inadequate. *Amoeba proteus*, for example, seems to require a varied diet. When fed the small ciliate *Colpidium*, unsupplemented by anything else, the amoebae become sluggish and corpulent. Division is still possible, although its rate is much slowed. A diet of the colorless flagellate *Chilomonas* is even worse, for growth and multiplication continue for only a few days, with gradual death of the amoebae thereafter.

The time required for mitosis, like the frequency of its occurrence, is variable, and the process may be slowed or even interrupted by environmental changes, perhaps to resume its normal course later. Rafalko found that in the amoebo-flagellate *Naegleria gruberi* mitosis normally requires only 15 to 18 minutes. This species—a protozoan Dr. Jekyll and Mr. Hyde—may appear either in the guise of a typical small amoeba or an equally typical flagellate, according to circumstances. However, environmental factors, even changes as slight as varying currents in the medium, may cause a temporary halt in the process. Another interesting peculiarity of this species is that the flagellate phase seems to be a response to unfavorable conditions which, if continued, result in encystment. Mitosis requires a favorable environment, and is thus characteristic of the amoeboid phase.

An unexpected and rather surprising discovery was that division, in at least some species of amoeba, is a cyclical matter and may be related to the time of day (or night). In the citrus-waste species *Flamella citrensis* we have a good example. Here Bovee (1956) found fission to be nocturnal and apparently to occur every 24 hours. For a long time it was widely believed that such cyclical reproduction was more or less limited to certain parasitic Protozoa, such as some of the species of malaria plasmodia, and it had seldom been observed in free-living Protozoa.

As in many other kinds of living things, both manner and rate of multiplication are also influenced by stage in the life cycle. Among the entamoebae (intestinal parasites, of which the majority of known species are from vertebrates), although the cyst is produced by a single individual, nuclear division in many species occurs within it, with the eventual formation of at least as many amoebulae as nuclei. In the species most thoroughly studied, *Entamoeba histolytica* and *E. coli* of man and other primates, a single multinucleate organism emerges from the cyst, and proceeds to divide until a brood of uninucleate metacystic amoebulae result. Details of the process naturally vary with the species and remain unknown for most of them. Some of the nuclei in the newly hatched amoeba may even divide, while others pass to the daughter amoebulae without further division. Energy for the series of mitoses taking place within the cyst apparently comes mainly from stored glycogen, which gradually disappears as the cysts mature.

Certain larger amoebae, such as the familiar species *Amoeba proteus*, are said to form from time to time large broods of minute cysts or spores, from which tiny amoebulae eventually emerge. Taylor states, in speaking of *Amoeba kerrii*, a species she described, "Hundreds of encysted amoebulae may be formed from one mature amoeba" (1947; 1962). This has been questioned by some and needs confirmation; much research is still needed to elucidate completely the biology of the amoebae. The process initially involves breaking up the nucleus, and forming from it chromatin blocks which, by further fragmentation, give rise to the nuclei with one of which each amoebula is supplied. But all this nuclear multiplication should not be regarded as simply fragmentation, or as a form of nuclear multiplication by repeated amitosis. According to the English protozoologist Jepps, it is an orderly process that should be thought of as "functional mitosis."

The type of asexual reproduction known as "plasmotomy" also seems frequent in some genera. Plasmotomy involves the fission of multinucleated individuals into two or more smaller daughters, each also with numerous nuclei, but without concurrent nuclear division. *Pelomyxa*, a fairly common genus of fresh-water forms, exemplifies this type of reproduction. These organisms are of more than ordinary interest for several reasons.

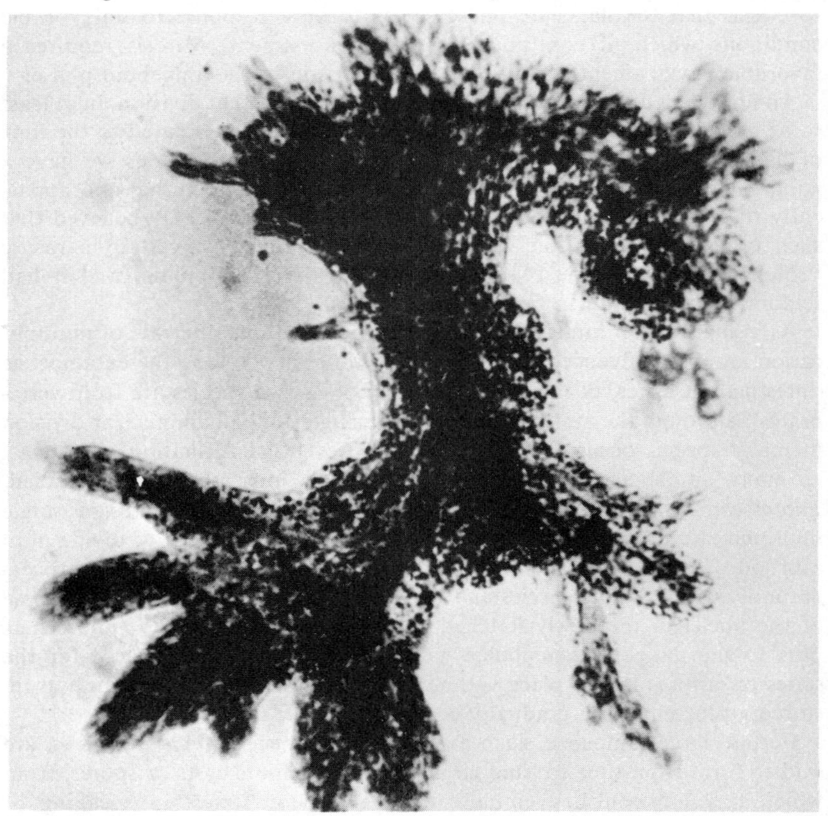

FIG. 15.6. *Pelomyxa illinoisensis*. Amoebae of this species often measure 500μ or even 1000μ in length when extended, and characteristically extend many pseudopods, although they may extend but one (the "clavate" forms), and there is a distinct pellicle. The cytoplasm is full of minute granules. (After Kudo, *J. Morphol.*, 88: 145.)

Some individuals of certain species contain extraordinary numbers of nuclei (sometimes more than 1,000), and they are relatively gigantic. An extended *Pelomyxa* may be a fifth of an inch (5,000 micra) long (Fig. 15.6). Of course, even though cytoplasmic fission may be unaccompanied by nuclear division, the latter must occur at some time, and such division may be nearly simultaneous and synchronous, at least in the species studied by Kudo.

Mitosis in the shelled Sarcodina is similar to that in the naked forms, although the presence of the test makes for complications. *Arcella vulgaris* one of the commonest species, initiates the process of division by extruding through the shell aperture a mass of cytoplasm destined to be the dowry of the daughter-to-be. Then the two nuclei undergo mitosis, the infant Arcellas receiving in each case one of the products. At about the same time secretion of a new shell begins, and when it is complete the two youthful organisms

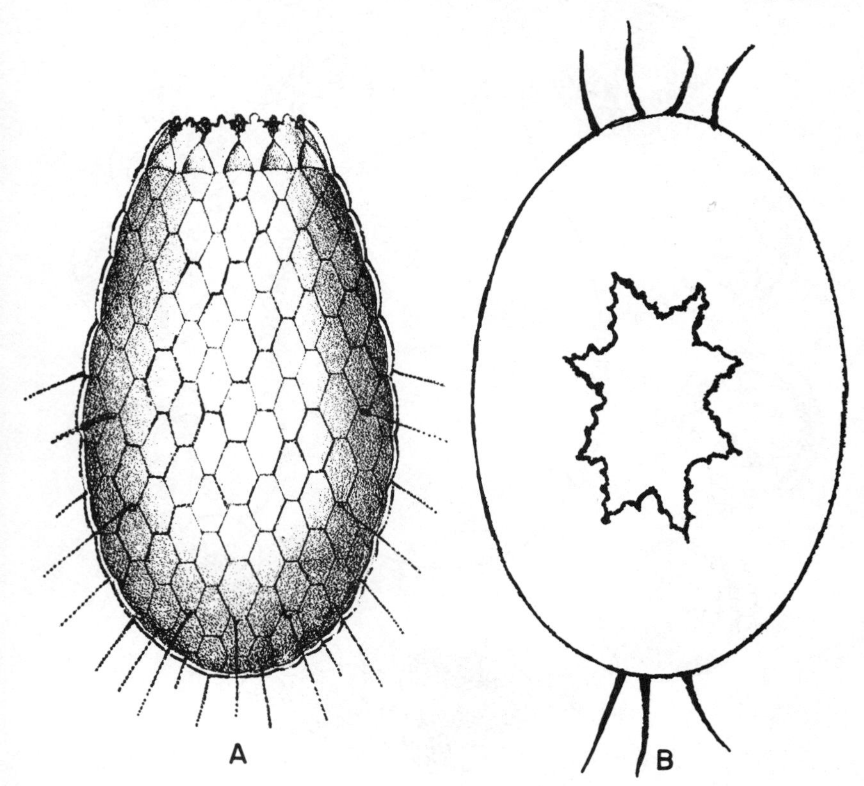

FIG. 15.7. *A. Euglypha ciliata*, a frequently encountered testate rhizopod, about 100μ. The shell appears to consist of transparent hexagonal plates, usually provided with a variable number of hairlike processes. This species is common in sphagnous swamps. *B.* Shell of *Euglypha ciliata* shown in cross section. (After Leidy.)

separate. By now any distinction between them has disappeared, except that one retains the old shell, which is recognizable because of its darker color. The two are in every other respect, as far as one can tell, identical twins.

Another frequently encountered testate form is *Euglypha* (Fig. 15.7), which has a shell composed of siliceous platelets. An individual about to reproduce prudently accumulates a reserve supply of these platelets in its cytoplasm. Early in the process of parturition (an apt term in this case), the platelets destined for the daughter are extruded along with her complement of cytoplasm, and are soon arranged in proper fashion on the surface. When this process is nearly complete, the single nucleus divides mitotically as each of the two nuclei of *Arcella* do, and the daughter receives one of

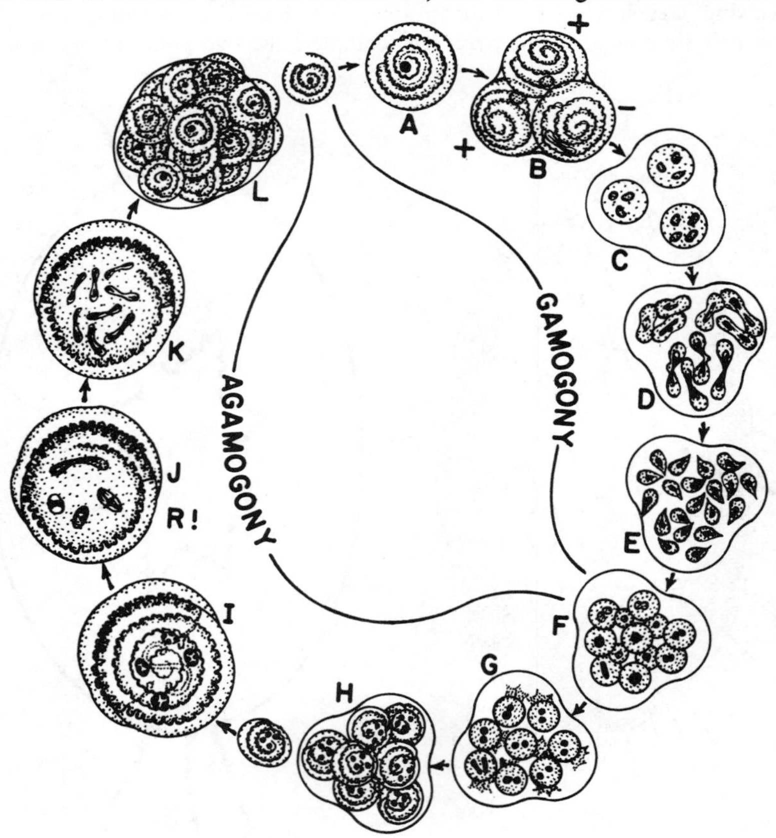

FIG. 15.8. Life cycle of the foraminiferidan *Patellina corrugata*. *A*. Association of two gamonts. *B–C*. Formation of gametes. *D*. Fertilization. *E*. Zygotes. *F*. Four-nucleated agamonts. *G*. Release of the agamonts. *H*. Growing agamont. *I–K*. Meiosis. *L*. Release of the agametes or gamonts. R!, chromosome reduction in stage *J*. (After Grell, originally from data of Myers and Le Calvez.)

the products. In this protozoan the food being digested is also said to be shared, so that both young Euglyphas start life on their own with nutrient reserves.

Mitosis in another testate rhizopod, *Lesquereusia spiralis*, is heralded by a change in appearance several hours before the actual start of the process and a cessation of pseudopodial activity. The entire process is said by Stump to require about 45 minutes, but the metaphase takes only five. For a form of its minute size (only about 22 micra in diameter) it is remarkably well supplied with chromosomes; Stump counted between 175 and 200.

The Foraminiferida, most of which have rather large chambered shells, have been much studied in recent years but the life cycles are as yet known for only a few species. Here there are two phases, one asexual and the other sexual, the former involving reproduction by multiple fission. Related to these two phases are also two forms within each species, the "megalospheric" and the "microspheric," so named because the initial chamber, the "proloculum," is larger in one than in the other. There is also a difference in the number of nuclei, the megalospheric phase being uninucleate while the microspheric often has more nuclei than there are chambers in the shell, with an even further increase when adulthood has been reached (Fig. 15.8).

Eventually the microspheric phase, which is the result of "conjugation or sexual process" (Cushman, 1955), liberates a swarm of minute amoebalike young (Fig. 15.9) by multiple fission, each organism representing one of the nuclei with its appropriated bit of cytoplasm. The infant Foraminiferidan then proceeds to the secretion of the proloculum of a new test, destined to be of the megalospheric phase, and gradually other chambers are added until whatever number may characterize the species have been formed. In the meantime the amount of cytoplasm increases and numerous nucleoli appear, although there is no actual nuclear multiplication.

Finally the nucleus fragments, and the individual bits of chromatin divide mitotically, with the eventual liberation of a multitude of flagellated gametes, although this interpretation has been questioned by some, including Kofoid. He and others believed that these minute organisms were flagellate parasites rather than gametes, and that the actual gametes are spherical amoebulae, each with a haploid number of chromosomes. According to Cushman (1955), however, the flagellated forms pair, and each resulting zygote secretes a proloculum of the microspheric phase. Although smaller to begin with, growth results in a larger adult form than that developing from the megalospheric proloculum.

The Foraminiferida thus present an interesting contrast to the Testacida, although even those species which have been adequately studied present considerable variety. In each stage of the life cycle of the former group a new shell is produced, and a sexual stage seems to be the rule. Instead of binary fission with mitosis, as in the Testacida, here there is multiple fission. This very complex life cycle apparently originated among the Foraminiferida

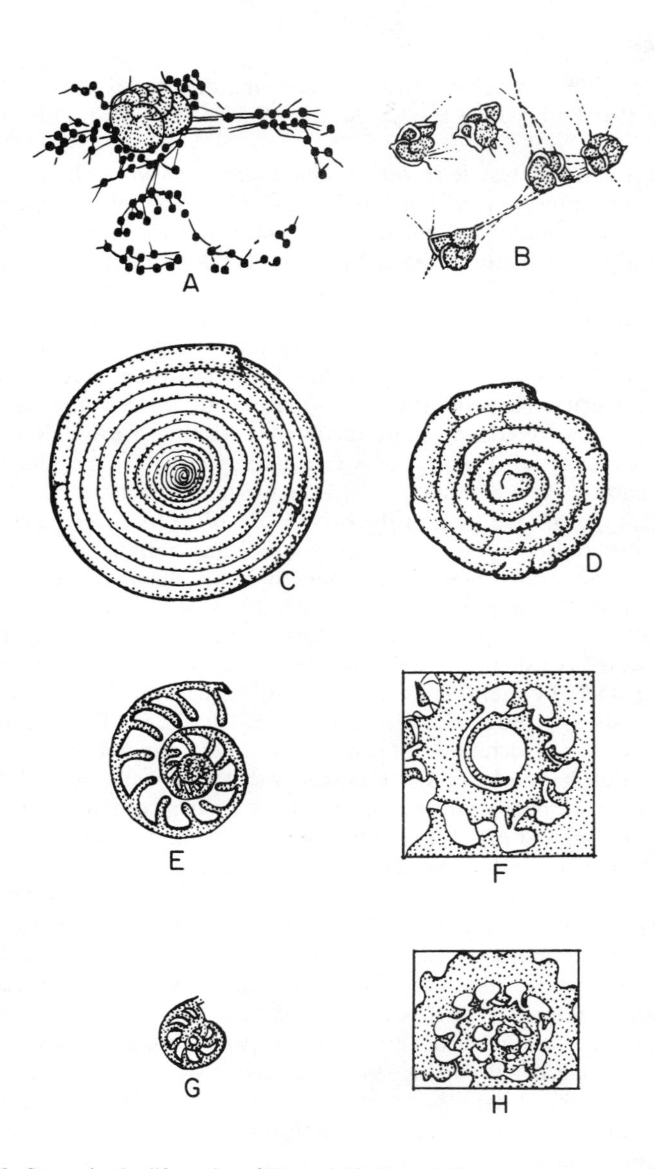

FIG. 15.9. Stages in the life cycles of Foraminiferida. *A*. Parent test with megalosphaeric young. *B*. Later four-chambered stage of the same generation. *C*. Microsphaeric adult of *Ammodiscus*. *D*. Megalosphaeric adult of the same. *E*, *G*. Microsphaeric and megalosphaeric sections of *Operculina*. *F*, *H*. Megalosphaeric and microsphaeric sections of *Peneroplis*. (After Cushman; *A*, *B*, *E*, *F*, *G*, *H*, originals after Lister.)

early in their evolution, for some of the genera in which it occurs are known from very ancient rocks. Evidently there has been ample time for differences in life cycles to develop, and it appears that those who have believed the gametes to be amoeboid were right with regard to some species, and the distinction between megalospheric and microspheric forms is not always easy to make, nor is their significance entirely clear. Cushman, probably the greatest authority on the group, suggested that the latter should be regarded as "conservative," exhibiting evolutionary recapitulation; Jepps is less sure. In any case, there is opportunity and much need for more research on the Foraminiferida.

Radiolarida and Acantharida

About the life cycles of the two other important marine groups, the Radiolarida and Acantharida, much less is known. For many years they were placed together in the single group Radiolarida. Of the two, the Radiolarida are a much larger assemblage, comprising thousands of species. Some authorities think them enough different to call each a class.

Most Radiolarida exhibit homaxonic (spherical) symmetry, though radially and bilaterally symmetrical forms are known. Their skeletons are usually of silica, and often of surpassing beauty. Within is a central capsule of pseudochitin. The Acantharida have a much simpler skeleton consisting of 20 spicules, radially arranged according to "Müller's Law" (i.e., along circles corresponding to the climatic circles of the globe—equatorial, tropical, circumpolar). The spicules are said to be of strontium sulfate, though the evidence is equivocal. A central capsule is often absent. The pseudopodia of Acantharida are axopodia; Radiolarida form very delicate reticulopodia.

The life cycles of Acantharida include, in at least some of the few species yet studied, both asexual and sexual reproduction. The former has rarely been seen, but involves binary fission, with doubling of the spicules and their equal sharing by the two daughters. Sexual reproduction includes the formation of isogametes, each having two flagella of unequal length.

Asexual reproduction is said to be more common among the Radiolarida than among the Acantharida, and may be binary or multiple fission, but in few (if any) species has the entire cycle been worked out. Binary fission involves a true mitosis with spindle formation; as many as 1500 or even more chromosomes have been counted. (Fortunately the nuclei of Radiolarida are relatively large—otherwise counting so many would be impossible.) The central capsule divides first, followed by the extracapsular portion. Where the skeleton is of the compact type, one of the daughters receives the entire skeleton, and the other must make a new one. When the skeleton is in two parts ("valves"), each daughter gets one and must then match it. In still other cases, when the skeleton is largely of spicules, partition is also equal; here, again, each daughter must make the other half.

Little is known about sexual reproduction in the Radiolarida. However, at times multitudes of biflagellated cells are formed which have been regarded as isogametes, but their subsequent development has not been followed. There may be released also minute flagellated cells of two sizes, long interpreted as anisogametes, but now known to be in reality parasitic (and probably symbiotic) dinoflagellates. These are lacking in the Acantharida, though zooxanthellae, doubtless also symbionts, are present.

With more knowledge of life cycles we might at least make intelligent guesses about evolutionary relationships; it is only possible at present to say that Acantharida are probably related to the Heliozoida.

Heliozoida

The life cycles of the Heliozoida, which constitute the other great division of the Actinopoda, are better known than those of the Radiolarida, but much still remains to be learned. Some of the pioneer work on this group was done by a distinguished German physician-turned-zoologist, Richard Hertwig. Although he is better remembered for his ideas on the significance of some nuclear changes occurring in ciliates during conjugation, he also wrote a famous paper on the fresh-water heliozoidan *Actinosphaerium* in 1899. Here also the nuclear phenomena interested him particularly, and he described mitosis and figured chromosomes only ten years after they had been first definitely seen and so designated. *Actinosphaerium* is normally provided with several hundred nuclei. As the animal grows they divide mitotically until finally plasmotomy takes place with the formation of two new organisms. It may be that observation of *Actinosphaerium* with its precocious nuclear mitoses and deferred cytoplasmic splitting inspired in part Hertwig's famous theory that a disproportion in the nucleo-cytoplasmic ratio triggered cell division.

Hertwig was also able to see and describe the sexual phase of the life cycle of *Actinosphaerium*. He noted that in the fall, as nutrition began to suffer, most of the nuclei disintegrated and were absorbed into the cytoplasm, and the organisms then encysted. Within the cyst each remaining nucleus then appropriated a bit of cytoplasm, and after separation into an equal number of uninucleate cells, secondary cyst walls were secreted about each. Then the organism within (known now as a "primary cytospore") divides once, and each daughter nucleus divides twice more, the first of the two nuclear fissions being meiotic. The products are the gametes, and these then unite to form the zygote. (The whole process is a beautiful example of complete inbreeding.) The next event in this complex life story involves the secretion by the zygote of a cyst wall about itself, so that it is now very well protected with its own wall, the enveloping wall of the cytospore and, outside both, the wall of the mother cyst. In this condition it may lie dormant for some weeks, at least.

Within the cyst the nucleus of the zygote divides once or twice, and—since the young forms seen in culture are uninucleate—it seems likely that cytoplasmic division occurs soon after the organisms excyst.

Whether life cycles as complex as this are characteristic of the Heliozoida generally is not known. However, a rather similar sequence of events was shown by the zoologist Bělař to occur in the common fresh-water "sun animal" *Actinophrys sol*—not unexpectedly, since *Actinosphaerium* and *Actinophrys* are presumably close kin (Fig. 15.10). But, in any case, there are many variations. The mere possession of skeletal elements, or the stalk required for a sedentary existence, would require modifications in the mode of cell division. One or both characterize some species.

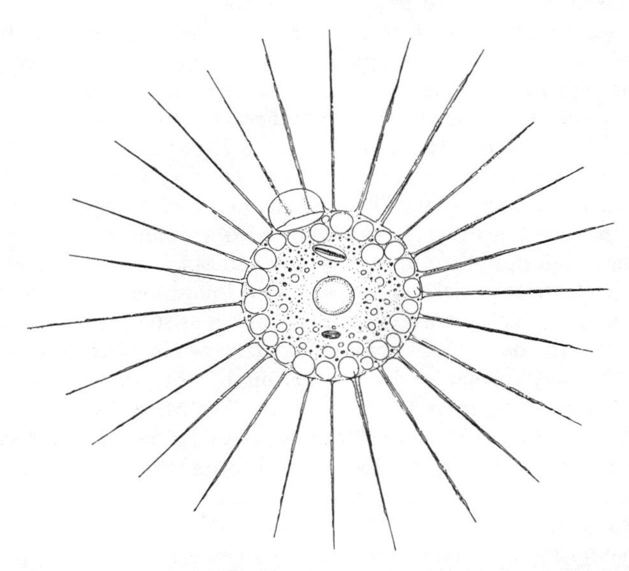

FIG. 15.10. *Actinophrys sol*, a common fresh-water heliozoan. The ectoplasm is typically highly vacuolated, but usually contains only one contractile vacuole. The axopodia are said to be tubular. It is usually colorless, but there is a yellow variety. Size about 50μ. (After Penard.)

An especially interesting aspect of the biology of *Actinophrys* is its ability as shown by Bělař, to maintain itself by asexual reproduction for many hundred generations when environmental conditions are sufficiently favorable. Similar observations have been made on other Protozoa. Woodruff made them in his long-continued study of *Paramecium* (and also of *Spathidium* and *Blepharisma*), and Hegner called attention to the persistence, sometimes for many years, of the malaria parasites in the vertebrate host, where they multiply only by asexual reproduction.

There are thus two especially interesting aspects of the life cycles of these organisms: the ability of some to reproduce almost indefinitely without the intervention of a sexual phase, and the occurrence in others of a sexual cycle in which there is complete inbreeding. Apparently a periodic reassortment of genes is by no means always a biological necessity for the survival of a species.

"*Slime Molds*"

The life histories of the Sarcodina known as "slime molds" are also interesting, although the group is not a large one (fewer than 500 species are known) and is something of a misfit among the Protozoa. Some aspects of the life cycles of these organisms suggest relationships to the fungi, and these peculiarities have so impressed the botanists that the slime molds have often been claimed by them as plants of a sort. They resemble true plants in their use of cellulose (a typically plant product), which they employ as a building material for their cyst walls, and are like the Myxobacteriales in their propensity to produce slime. Yet there are amoeboid stages in the life cycles of the slime molds which move and ingest solid food, as do the true amoebae.

These remarkable organisms (still generally assumed to be Protozoa, and which Bonner, 1959, suggests may have evolved from common soil amoebae), form at certain periods in their life cycles large protoplasmic masses, which are either multinucleate (and hence properly called "plasmodia" or "syncytia"), or appear to be so. These masses have a slimy consistency, and from time to time spores are developed from them. The whole process is strongly reminiscent of spore formation among the fungi. It has thus seemed logical enough to call these organisms "slime molds," or Myxomycetes, from the Greek words for slime and fungus. Since use of this term seemed to concede that they were in fact fungi, zoologists have often preferred to call them Mycetozoida, meaning "fungus animals," though this hardly seems to solve the problem.

Unfortunately for everyone concerned, the slime molds apparently constitute not one group, but four—all of probably different ancestry. The resemblances between them seem rather superficial. Bonner, an American zoologist especially interested in slime molds, includes them all as Mycetozoida, but remarks that "they represent four anomalous groups of unknown origin."

The first of the four groups, variously known as Myxomycetales, Myxogastrales, or Myxomycetes, is by far the largest, comprising some 400 species. Here the spores germinate minute young forms called "swarmers," each of which speedily develops two flagella of unequal length. These are capable of quick metamorphosis, losing the flagella and becoming amoebulae. In either stage they may fuse in pairs, thus revealing that they are in reality gametes. The resulting zygote develops into a plasmodium as nuclear divisions succeed one another, without corresponding cytoplasmic division. The needed food is secured by ingesting bacteria, or even cannibalizing comrade swarmers.

If conditions continue favorable, the plasmodium grows until it may reach a diameter of several inches—a gigantic size for a protozoan. Finally the surface becomes studded with small projections, each of which develops into a small stalk with a knob at the top—a kind of spore case, the whole structure being known as a "sporangium." Within it the nuclei undergo meiosis, and each haploid daughter nucleus then secretes a cellulose envelope. These structures are the fruiting bodies, or spores.

The cycle just given (and shown somewhat diagrammatically in Fig. 15.11), is generalized, and its details vary with the species. For the other three groups the pattern is rather different, and since they are relatively small

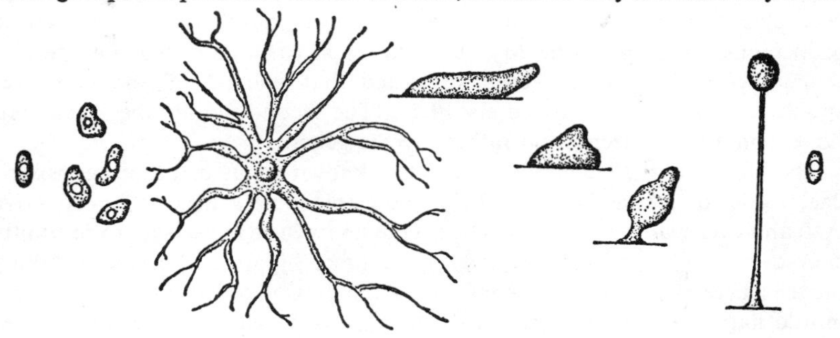

FIG. 15.11. Stages in the life cycle of *Dictyostelium*, a slime mold. At the left is an elliptical spore, about 8 by 3 micra. From this emerges an amoebula (four are shown immediately to the right of the spore). The amoebulae feed chiefly on bacteria but sometimes on each other, and go through a number of divisions. They then aggregate, forming the branched "pseudoplasmodium" (shown to the right of the amoebulae) and from this develop the "migrating" and "culmination stages" (the three bodies to the right of the pseudoplasmodium). The final stage is the fruiting body at the top of the stalk, within which are formed the spores (one appears at the extreme right). (After Bonner, *The Cellular Slime Molds*.)

examples will not be given. But this is not to say that these groups are not important. The Plasmodiophorales include some species of plant parasites, such as *Plasmodiophora brassica*, the cause of clubroot disease in cabbage, which do much damage to crops or other plants of value. And the Labyrinthulales, though an even smaller group (it has but a single genus), includes a species destructive to eel grass, a water plant that furnishes food and shelter, directly or indirectly, to numerous marine animals, and is thus important to man. Numerous species of algae are also attacked. A widespread epidemic of eel grass infestation with this parasite occurred some years ago, with profound consequences to marine coast life (Renn, 1936).

The Acrasiales, last of the four groups, include some dozen species, all of them soil dwellers. Their life cycle lacks a flagellated stage. The amoebulae which are liberated from the spores (the spore case is of cellulose here also) remain uninucleate, growing and multiplying mitotically as do true amoebae.

But quite unlike the latter, they possess a gregarious instinct which causes them eventually to begin a migration to certain trysting points. There they form a mass known as a "pseudoplasmodium," since they retain their individuality as cells. For this reason Bonner has dubbed them the "social amoebae."

Proteomyxida

The Proteomyxida (or Proteomyxa) is so ill defined and so incompletely known that little that is definite can be said of life cycles of its members. *Labyrinthula*, referred to above as an important parasite of eel grass, has been placed in this order by some protozoologists, but it seems likely that its proper position is with the Mycetozoida, as already indicated. Of the two remaining families, it has been suggested that one (the Pseudosporidae) should also perhaps be moved elsewhere. The second family, the Vampyrellidae, contains a dozen or so rather diverse genera.

The life cycle of *Pseudospora*, the better known genus of the Pseudosporidae, is said to involve the production by the amoeboid stage of a reproductive cyst, and eventually of gametes. The zygote, in its turn, is also said to promptly encyst. The organisms themselves appear to be highly polymorphic, taking on the guise of a minute amoeba, or of a heliozoidanlike form, or a highly motile flagellate with two flagella—all, it is said, within a few minutes of observation.

The Vampyrellidae, many species of which are notable for their beautiful orange or reddish coloration, are said to have similar life cycles. These include small uninucleate stages, larger multinucleate forms (plasmodia), and cysts, which may also be multinucleate.

Behavior and Feeding Habits

Of all the Sarcodina, the behavior of *Amoeba* is probably the most studied; Jennings (1906) devoted a chapter to this protozoan in a pioneering monograph entitled *Behavior of the Lower Organisms*, which is still a classic, and several of his observations are mentioned below.

The feeding habits of *Amoeba proteus* have come in for much attention. Leidy noted that this amoeba is largely vegetarian. But he also commented on its ability to capture rotifers, and described the capture and ingestion of a *Urocentrum*—a very active ciliate indeed. Evidently the race is not always to the swiftest, even in the protozoan world.

Schaeffer, an American protozoologist for whom amoebae had a compelling fascination, made an extensive survey of food choice in several species of the genus and remarked in a paper published in 1917, "The problem of choice of food has turned out to be very intricate and difficult; much more so than was at first suspected." But he did find that "*Amoeba* is capable of exercising very nice discrimination in feeding between two particles of different composition—one digestible, the other not—lying very close together." Yet in some cases indigestible particles were eaten, and selection did not seem to be based on chemical properties "in any known case." A moving particle was more likely to be eaten than one standing still. And he concluded, "Many individual acts of [food] selection cannot therefore be explained when standing alone, but only when the amoeba's past, especially its immediate past, is known." Jennings had reached somewhat similar conclusions much earlier, remarking in 1906, "The Amoeba conducts itself in its efforts to obtain food in much the same way as animals far higher in the scale," and "The [feeding] reaction is thus complex; at times, as we have seen, extremely so." Like Schaeffer he noted that species differed in their manner of food taking.

To say that a living organism exhibits instinctive or complex behavior of any sort is of course only labeling such behavior—not explaining it. To explain this behavior some very ingenious experiments have been devised, for example Bonner's (1949). He placed some of his "social amoebae" on glass shelves immersed in water and observed, when these were separated, that the amoebae attempted to span the gap to reach the mass of previously aggregated amoebae toward which they were attracted. As a result of experiments such as this he concluded that these amoebae secrete a substance called "acrasin" to which they are highly sensitive. This substance, apparently formed first by a "leader," is the guide for the others. Toward this point they all start their trek, and the migration is further accelerated by more acrasin which they, in their turn, are stimulated to secrete. When the behavior of this amoebic army was further followed, it became evident that its mobilization was the first step in a differentiation culminating in spore formation. Certain amoebae were found destined to be stalk formers and others for the formation of spores, whereas still others dropped back to make up a rear guard whose duty it would be to generate the basal disc.

The study of problems such as these is made much easier by cultivation techniques. Numerous species of Sarcodina can be grown in the laboratory, but usually only in association with other organisms, although amoebae of the so-called *Hartmanella* group may be grown axenically. Only very recently has it become possible to cultivate an amoeba in a chemically defined medium. This has been achieved by Adam (1959), an English protozoologist, for a species of soil amoeba of the genus *Acanthamoeba* (= *Hartmanella*). The composition of the medium is highly complex, and reminiscent of that developed for the ciliate *Tetrahymena*. The only vitamins required are B_{12} and thiamine. We may expect continued progress in this direction. Kopac has recently predicted that "research on *Amoeba* in 2158 A.D." will involve the extended use of such cultivation, and that it will in turn make possible the experimental production of new varieties by transplantation of various cellular elements, the more precise study of growth and division, and a better understanding of what he calls subcellular ecology. Likewise, it should shed more light on symbiotic relationships. Extended to parasitic species, it should also help illumine the relationships between amoebae and their hosts, and perhaps even the evolution of these relationships.

Classification

On the whole, division of the Sarcodina into classes, orders, and sometimes smaller taxonomic groups, is based on peculiarities of the pseudopodia. But pseudopodial morphology (as well as other structural characteristics) is influenced by physical conditions, such as temperature and salt content of the medium. Contractile vacuoles are, in general, confined to fresh-water species. The characteristics of the shell, when there is one, are likewise important in classification.

Rather than using pseudopodial morphology, *Bovee and Jahn (1964) believe the mechanism of pseudopodial movement to be a better basis for the classification of Sarcodina above family level. They would recognize two classes (with appropriate subclasses and orders): (1) Autotractea, for forms possessing axopodia, filopodia, or myxopodia, and (2) Hydraulea, for "amoebae" of various kinds, including those with shells (e.g., Testacida). Such a scheme they think more accurately reflects biological kinships. But it seems to ignore the problem of what to do with other Protozoa producing pseudopodia (such as some Mastigophora and even certain Sporozoa).

Some Sarcodina possess, sometimes at the same and sometimes at different times, axopodia and flagella. Certain protozoologists therefore place

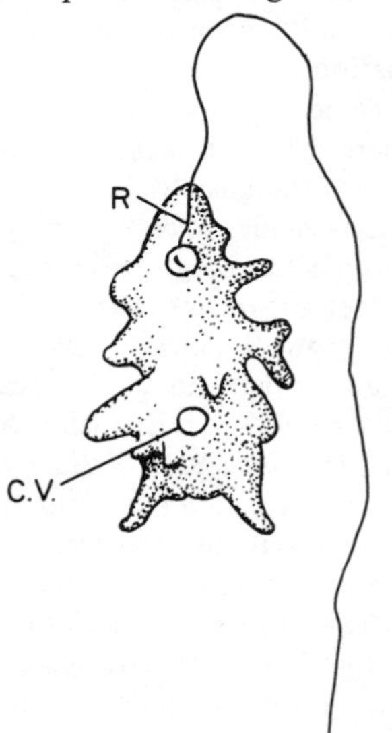

FIG. 15.12. *Mastigamoeba* from life. C.V., contractile vacuole; R, rhizoplast. (After Klug.)

these organisms in a separate order (as is done in the skeleton scheme used below). The ordinal name Helioflagellida is based on the Greek *helios* ("sun") and the Latin *flagellum* ("whip"). The possession of both axopodia and flagella might suggest descent from either the Mastigophora or the Heliozoida, or even conceivably from both. But on the whole little is known about most of these forms, and it is not possible to say from what they were derived.

There are also true Heliozoida which have flagellated stages in their life cycles; whether such stages are simply nature's attempt to facilitate migration (after all, axopodia are poor means of getting about the world), or are gametes, it is in most cases impossible to know.

We must not forget the organisms with both lobose pseudopodia and flagella placed in the order Rhizomastigina (or Rhizomastigida). This small assemblage, comprising only a few genera, is variously regarded as belonging with the Sarcodina or the Mastigophora, although in the classification adopted in this book it is placed with the latter. The difficulty is, of course, to decide which means of locomotion is the dominant one, but it also involves a knowledge of the life cycles and their interpretation. Complex cycles, including sexual stages, have been claimed for some. Flagella are retained throughout, and the organisms are restricted to fresh water. *Mastigamoeba* (Fig. 15.12) is one of the commoner genera.

It is thus clear that we cannot always know whether we are dealing with Sarcodina or Mastigophora. Nature has been a long time developing the biological relationships which biologists have only lately begun to study.

Of the groups so far discussed, three (Helioflagellida, Heliozoida, Radiolarida) fall into the subclass Actinopoda, while all the others are lumped together as Rhizopoda by the taxonomist. However, the term "rhizopod" is still often used for any organism producing pseudopods, just as ciliates are still sometimes referred to as "infusoria."

Although the skeleton classification given below may seem quite simple and straightforward, it is by no means acceptable to all protozoologists. Disagreement becomes greater as one goes down the taxonomic scale, through suborders, families, genera, and species. Essentially this disagreement is due to lack of knowledge. The English protozoologist Jepps remarks of the Testacida, "It is doubtful whether the whole life history is known for a single member of the group . . ." and some of the other groups of Sarcodina are little better known. This is especially true of the great marine orders Foraminiferida and Radiolarida. These groups are difficult to study in the laboratory, because it is hard to know and duplicate their normal environmental conditions. Similar difficulties may confront the protozoologist interested in the study of many of the soil-dwelling Sarcodina. Numerous uncertainties surround even the much studied *Amoeba*. Kudo remarks in a recent paper, "While most of us . . . seem to agree that the genus *Amoeba* includes a great variety of amoeboid organisms and should be divided into a number of genera, there is at present no clear-cut criterion, agreeable to all concerned, on which genera can be founded."

The student should, therefore, not be misled by the discovery that protozoologists and protozoology texts disagree. Almost any scheme of classification in current use is defensible, at least to a point, and it is probably impossible to find any two texts in complete agreement. But with the discovery of new facts disagreement should diminish.

The following classification is essentially Hall's (1953) and not greatly different from Kudo's (1954).

Class SARCODINA Hertwig and Lesser. Protozoa using pseudopodia for locomotion and food-getting, and lacking a definite pellicle (though in some cases movement seems to be by a protoplasmic flow not involving the formation of true pseudopodia). Flagella may characterize gametes, when such are a part of the life cycle, or may act simply as an alternative method of locomotion. Some species are sessile and many produce shells.

Subclass 1. ACTINOPODA Calkins. Here the pseudopodia are typically axopodia (although lobopodia may also be produced in some cases), and the organisms are usually floating in habit, though shells and other skeletal elements may often be present. Flagellate stages are known for a few.

Order 1. HELIOFLAGELLIDA Doflein. A relatively small group in which both axopodia and flagella are present, and which, in at least some cases, also possess a central granule noticeable especially in division.

Order 2. HELIOZOIDA Haeckel. These are largely fresh water in habitat, spherical (homaxonic) in form, and provided with many radiating axopodia. The cytoplasm is highly alveolar. Many are naked, but some produce shells or are clad in a gelatinous envelope. Some of the latter have spicules.

Order 3. ACANTHARIDA Haeckel. Marine organisms with a skeleton of 20 spicules (strontium sulfate?), arranged radially according to "Müller's Law." A central capsule is often lacking, especially in the more primitive types; when present it is imperforate (except to allow passage of spicules). Pseudopodia of axopodial type. Zooxanthellae (rather than parasitic dinoflagellates) present. Binary fission, at least in simpler forms; sexual reproduction involves isogametes. Fossil remains rare.

Order 4. RADIOLARIDA Haeckel. A very large group of strictly marine and usually pelagic forms. Siliceous skeleton; often very complex. Symmetry usually homaxonic (spherical) or radial; size up to 2,000 micra (especially in warm waters). Central capsule with openings ("pylea"), permitting cytoplasmic connection between endoplasm and ectoplasm. Produce reticulopodia. Life cycles very poorly known; believed to include sexual phase. Fossil remains abundant (e.g., in Barbados). See p. 27 *et seq.*

Subclass 2. RHIZOPODA von Siebold. Amoeboid Protozoa which do not develop axopodia or possess alveolar cytoplasm. Formation of shells, often impregnated with mineral salts and/or covered with foreign bodies (sand grains, etc.) is common, and multinuclearity not infrequent.

Order 1. PROTEOMYXIDA Lankester. Here are placed a number of naked forms producing filopodia, and (often) flagellate swarmers. Many species are parasitic on algae or the related chlorophyll-bearing flagellates (e.g., *Volvox*).

Order 2. MYCETOZOIDA de Bary. These are the slime molds, organisms which have life cycles sufficiently like those of fungi to be so classed by many botanists (and perhaps by some protozoologists). They may form true plasmodia, or aggregates which resemble true plasmodia (and are hence called "pseudoplasmodia"), though in the latter case the constituent organisms do not actually fuse. Included in the life cycles are amoeboid and flagellate stages (though the latter are lacking in the Acrasiales), but fungus-like spores are also usually produced. Some are stalked at times, and some are parasitic.

Order 3. AMOEBIDA Ehrenberg. In this order are placed the typical amoebae, with which almost everyone is familiar, at least by name. Movement is by lobopodia or by a streaming of the entire body of protoplasm, with a resulting sluglike progress (the small "limax" amoebae are so called, from the Latin *limax*, "slug"). Pseudopods also function in food-getting, or may be formed without evident use. Ectoplasm and endoplasm are usually easily distinguishable. In a few cases flagellate stages are known to be part of the life cycle, and in others many amoebulae are formed from time to time from the parent organism. Amoebae may be parasitic or free living, fresh water, or marine.

Order 4. TESTACIDA Schultze. Here are placed Sarcodina with shells, almost invariably of a single chamber, of pseudochitin, often impregnated with mineral salts; foreign bodies (sand grains, diatom shells, detritus) may also be incorporated. The shell is usually a single piece, but may consist of secreted scales. Reproduction is ordinarily by binary fission, but the life cycles are very incompletely known. Habitat: fresh water.

Order 5. FORAMINIFERIDA d'Orbigny. A very large group, mostly of multichambered (polythalamous) shelled forms; shell heavily impregnated with calcium carbonate in a foundation material of pseudochitin. The shell is usually pierced by many openings making extrusion of pseudopodia possible, and the chambers are connected with openings (the "foramina," from the Latin for "windows") by which cytoplasmic communication is maintained; it may have imbedded in it sponge spicules or other foreign materials. But there are a few forms, all belonging to a single family (the Allogromiidae), which are naked or have a nonporous chitinous shell. In this order are the largest Protozoa known (if one excepts the plasmodia of some Mycetozoida); indeed, certain fossil types reached a diameter of several inches, and one (genus *Neusina*) of a family now living may measure as much as 8 inches. Few foraminiferidan life cycles are completely known, but those that are include both sexual and asexual stages.

REFERENCES

Adam, K. M. G. 1959. The growth of *Acanthamoeba* sp. in a chemically defined medium. *J. Gen. Microbiol.*, 21: 519–29.

Bělař, K. 1923. Untersuchungen an *Actinophrys sol* Ehrenberg. I. Die Morphologie des Formwechsels. *Arch. f. Protistenk.*, 46: 1.

——— 1924. Untersuchungen an *Actinophrys sol* Ehrenberg. II. Beiträge zur Physiologie des Formwechsels. *Arch. f. Protistenk.*, 48: 371.

Bonner, J. T. 1949. The social amoebae. *Sci. Am.*, 180: 44–7.

——— 1959. *The Cellular Slime Molds.* Princeton: Princeton University Press.

Bovee, E. C. 1956. Some observations on the morphology and activities of a new amoeba from citrus wastes, *Flamella citrensis* n. sp. *J. Protozool.*, 3: 151–5.

Calkins, G. N. 1933. Special morphology and taxonomy of the *Sarcodina.* In *Biology of the Protozoa.* Philadelphia: Lea & Febiger.

Cash, J. 1905. Rhizopoda. In *British Fresh-water Rhizopoda and Heliozoa,* vol. 1. London: Ray Society.

Chalkley, H. W. and Daniel, G. E. 1933. The relation between the form of the living cell and the nuclear phases of division in *Amoeba proteus* (Leidy). *Physiol. Zool.*, 6: 592–619.

Crowder, W. 1926. Marvels of Mycetozoa. *Nat. Geog. Mag.*, 49: 421–43.

Cushman, J. A. 1955. *Foraminifera.* Cambridge: Harvard University Press.

Dawson, J. A., Kessler, W. R. and Silberstein, J. K. 1935. Mitosis in *Amoeba dubia. Biol. Bull.*, 69: 447–61.

Deflandre, G. 1928. Le genre *Arcella* Ehrenberg. *Arch. f. Protistenk.*, 64: 152.

Dobell, C. 1943. Researches on the intestinal Protozoa of monkeys and man. XI. The cytology and life history of *Endolimax nana. Parasitology,* 35: 134–58.

Doflein, F. and Reichenow, E. 1952. *Lehrbuch der Protozoenkunde.* Teil 2, 1. Hälfte. Jena: Verlag Gustav Fischer.

Grassé, P.-P. 1953. *Traité de Zoologie.* T. 1, Fasc. 2. Paris: Masson.

Hall, R. P. 1953. *Protozoology.* New York: Prentice-Hall.

Halsey, H. R. 1936. The life cycle of *Amoeba proteus* (Pallas, Leidy) and of *Amoeba dubia* (Schaeffer). *J. Exp. Zool.*, 74: 167–203.

Hertwig, R. 1899. Über Kerntheilung, Richtungskörperbildung und Befruchtung von *Actinosphaerium eichhorni. Abh. bayer. Akad. Wiss.*, 19: 631.

Hirschfield, H. L. (ed). 1959. Biology of the *Amoeba.* (Includes Research on the *Amoeba* in 2158 A.D., by M. J. Kopac) *Ann. N.Y. Acad. Sci.*, 28: 403–704.

Hofker, J. 1950. Wonderful animals of the sea: Foraminifera. *Amsterdam Nat.*, 1: 60–79.

Jennings, H. S. 1906. The behavior of *Amoeba.* In *Behavior of the Lower Organisms,* New York: Columbia University Press.

Jepps, M. W. 1956. *The Protozoa, Sarcodina.* Edinburgh: Oliver and Boyd.

Kofoid, C. A. 1934. An interpretation of the conflicting views as to the life cycle of the Foraminifera. *Science,* 79: 436–7.

Kudo, R. R. 1952. The genus *Pelomyxa. Trans. Am. Mic. Soc.*, 71: 108–13.

——— 1954. *Protozoology.* Springfield: Charles C Thomas.

Lankester, S. 1909. *A Treatise on Zoology.* Part 1, Fasc. 2. Introduction and Protozoa. London: Adam and Charles Black.

Le Calvez, J. 1938. Recherches sur les Foraminifères. I. Développement et reproduction. *Arch. zool. exp. et gén.*, 80: 163.
Leidy, J. 1879. *Fresh-water Rhizopods of North America.* U.S. Geol. Surv., vol. 12.
Mast, S. O. 1943. *Amoeba* and *Pelomyxa* vs. *Chaos. Turtox News*, 16: (No. 3).
Milne, L. J. and Margery J. 1951. The eel-grass catastrophe. *Sci. Am.*, 184: 52.
Minchin, E. A. 1912. The Sarcodina. In *An Introduction to the Study of the Protozoa.* London: Edward Arnold.
Moulton, F. R. 1939. Problems in Lake Biology. A.A.A.S., *Pub. No. 10* (includes article by R. W. Pennak).
Penard, E. 1902. *Faune Rhizopodique du Bassin du Léman.* Geneva.
——— 1904. *Les Héliozoaires d'Eau Douce.* Geneva.
——— 1908. *Catalogue des Invertébrés de la Suisse.* Fasc. 1, Sarcodinés. Geneva: Museum d' Histoire Naturelle de Genève.
Rafalko, J. S. 1947. Cytological observations on the amoebo-flagellate *Naegleria gruberi. J. Morphol.*, 81: 1–44.
Renn, C. E. 1936. The wasting disease of *Zostera marina.* I. A phytological investigation of the diseased plant. *Biol. Bull.*, 70: 148–58.
Sandon, H. 1927. *The Composition and Distribution of the Protozoan Fauna of the Soil.* Edinburgh: Oliver and Boyd.
——— 1957. Neglected animals—Foraminifera. *New Biology*, 24: 7–32.
Schaeffer, A. A. 1917. Choice of food in *Amoeba. J. Animal Behavior*, 7: 220–58.
——— 1926. Recent discoveries in the biology of *Amoeba. Quart. Rev. Biol.*, 1: 95–118.
——— 1926. Taxonomy of *Amoeba.* Carnegie Inst. Wash., *Pub. 345.*
Schewiakoff, W. 1926. Die Acantharia. Fauna u. Flora Neapel, No. 37.
Stump, A. B. 1959. Mitosis in the rhizopod *Lesquereusia spiralis. J. Protozool.*, 6: 185–9.
Suess, H. E. 1956. Absolute chronology of the last glaciation. *Science*, 123: 355–7.
Sussman, M. 1956. The biology of cellular slime molds. *Ann. Rev. Microbiol.*, 10; 21–50.
Taylor, M. 1923. Nuclear divisions in *Amoeba proteus. Quart. J. Mic. Sci.*, 67: 39–46.
——— 1924. *Amoeba proteus:* some new observations on its nucleus, life history and culture. *Quart. J. Mic. Sci.*, 69: 110–50.
——— 1947. *Amoeba kerrii* (n. sp.): morphology, cytology and life history. *Quart. J. Mic. Sci.*, 88: 99–113.
——— and Hayes, C. 1944. *Amoeba lescheri* (n. sp.): its morphology, cytology and life history. *Quart. J. Mic. Sci.*, 84: 295–328.
Wailes, G. H. 1921. Heliozoa. In *British Fresh-water Rhizopoda and Heliozoa*, vol. 5. London: Ray Society.

ADDITIONAL REFERENCES

Bovee, E. C. and Jahn, T. L. 1964. Movement mechanisms in sarcodinan taxonomy: Subclasses and orders in the classes Autotractea and Hydraulea. *J. Protozool.*, *11* (Supplement): 29.
Taylor, M. 1962. Ways and waywardnesses of *Amoeba. Amer. Sci.*, 50: 576–96.

16

Ciliophora

"Having one Evening been examining a little of this slime . . . I was diverted with the sudden Appearance of a little Creature whose Figure was intirely new to me, moving about with great Agility, and having so much seeming Intention in all its Motions, that my Eyes were immediately fixed upon it with Admiration. Its Body in Substance and Colour resembled a Snail's: the shape thereof was somewhat elliptical, but pointed at one End, whilst from the other a long, slender and finely proportioned Neck stretched itself out, and was terminated with what I judged to be an Head, of a Size perfectly suitable to the other Parts of the Animal. In short, without the least of Fancy, which is ever carefully to be guarded against in the use of the Microscope, the Head and Neck and indeed the whole Appearance of the Animal had no little resemblance to that of a Swan . . ." Description of a common ciliate, *Lacrymaria olor*, in Henry Baker, *Employment for the Microscope*, 1753.

The ciliates were undoubtedly the first Protozoa to capture the interest of the microscopist: some of them are giants in their microcosmos; they are continually active; and they are astonishingly intricate in bodily organization.

Despite their ubiquity, the Ciliophora may contain fewer species than the other major groups of Protozoa. A special problem exists with respect to the characteristics defining a ciliate species, because some genera include reproductively isolated groups or varieties made up of mating types. These have been appropriately termed "genetic species" and given the name "syngens" by Sonneborn (Chap. 12). Furthermore, new species, including parasitic ones, are still being discovered and many environmental niches are still unexplored.

Many species which have been described are probably not valid, but only time and labor can establish whether any given grouping is in fact a species.

Natural variation among ciliates is considerable, and in numerous instances variant strains have been accorded species status.

To catalogue the species of ciliates described in the zoological literature would be a monumental task. But we may get some idea of the number of ciliate species from careful studies such as those of Kahl, the German investigator who made a most painstaking survey, excluding only the tintinnids, a very large group almost wholly confined to the ocean, and the parasites (although he did include the ectocommensals). He listed about 2,400 distinct species and subspecies, as well as a large number of species names which he regarded as synonyms. Most of these species are undoubtedly cosmopolitan. There are, in addition, many parasitic species. Possibly 6,000 ciliate species, in all, have been described.

Definition

Except for the protociliates (parasites of frogs and toads) whose cilia should perhaps rightfully be regarded as flagella, all Ciliophora possess cilia at some stage in their life cycle. They are unique among the Protozoa in possessing two kinds of nuclei, known as macronuclei and micronuclei because of size differences. Macronuclei are involved only with the more mundane activities of life—control of nutrition, metabolism, regeneration, and probably of growth. The micronucleus seems to be concerned only with inheritance. Although amicronucleate strains of ciliates are occasionally encountered, individuals without macronuclei never are. If Protozoa are regarded as cells, this nuclear division of labor makes the ciliates cells of a unique kind.

External Characteristics

Size, Shape, Color

As in other groups of Protozoa, the range of size among the ciliates is great. The smallest measures 10 or 12 micra long—little larger than a human red blood cell, while the largest (such as the gigantic *Spirostomum ambiguum*) may reach a length of 3 millimeters. Thus the difference between the two extremes is some 300-fold, and of course a much greater one in volume.

Most species are colorless, but some are brightly colored. *Mesodinium rubrum* contains chromatophores of haematochrome in its ectoplasm giving it, and even the sea water in which it lives, a blood-red tint (Fig. 16.1). Others exhibit beautiful pastel colors. *Blepharisma undulans* when abundant in laboratory cultures lends them a delicate lavender hue, but no genus surpasses *Stentor* in the colors it displays. At one end of the spectrum is the rose-tinted *Stentor igneus;* at the other end is *S. multiformis*, garbed in dark blue or green (Fig. 16.2). Lying between are species tinted blue, amethyst, yellow, and brown. Some of these, like the familiar *Paramecium bursaria*, owe their brilliance not only to natural pigments but to symbiotic algae; still others are entirely colorless, although they may contain symbiotic bacteria, as do several common species of *Euplotes*. On the whole, symbiosis, or mutualism, is less common among the ciliates than in other major groups of Protozoa.

In form there is equal diversity within the class. Although the more highly evolved types approach bilateral symmetry, with well-marked dorso-ventral and antero-posterior differentiation, there are numerous examples of radial symmetry, or something very close to it. Such are the Peritrichida (e.g., *Vorticella*) and some of the Suctorida, as well as *Didinium*. The majority

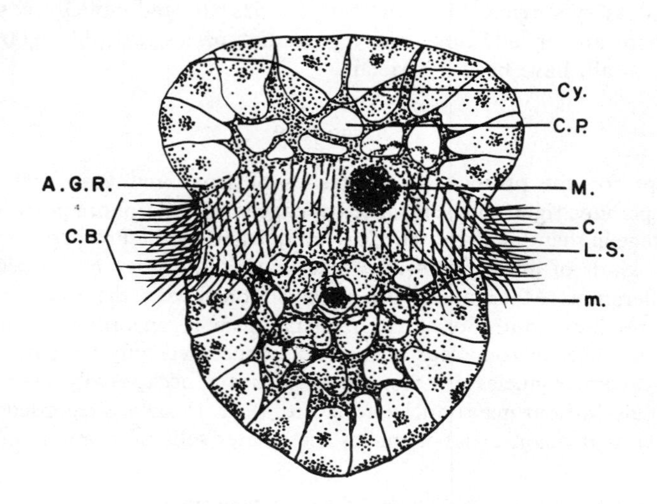

FIG. 16.1. *Mesodinium rubrum*, a marine ciliate which may cause red water (not to be confused with the dinoflagellates causing the "red tides" poisonous to fish). A.G.R., anterior granular ring; C., cilia; C.B., ciliary band; C.P., chromatophore platelets; Cy., cytostome; L.S., longitudinal striations; M., macronucleus; m., micronucleus. (After Powers, *Biol. Bull.*)

of such forms are sessile during most of their existence, being attached to the substrate by a stalk. Some forms have not only a stalk, but also a test or house, which they seldom leave, e.g. *Folliculinopsis* and *Cothurnia* (Fig. 16.3). Certain species have gone a step farther, providing their homes with a lid which they close behind them when in danger. *Caulicola*, a fresh- or brackish-water ciliate closely related to the Vorticellas, is a good example.

Though attached most of the time, some of this group of ciliates, such as the many species of *Vorticella*, do move about occasionally. They rapidly grow an accessory ring of cilia at the base of the bell (the "telotroch" stage), and take off for foreign parts, leaving the stalk behind. In Suctorida, however, only the embryos have cilia. The adults are sessile, having lost all trace of locomotor organelles (e.g., *Tokophrya*, Fig. 16.4). All they do is await,

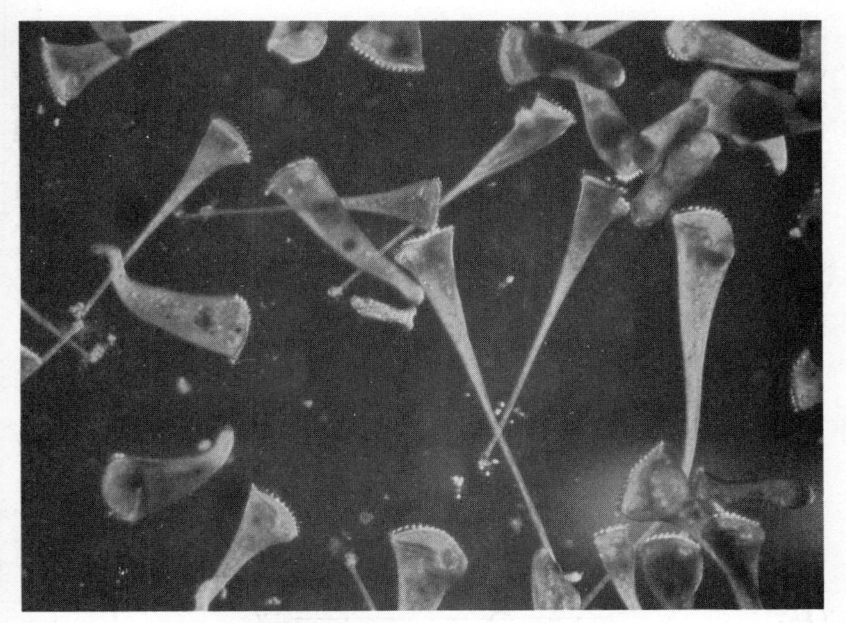

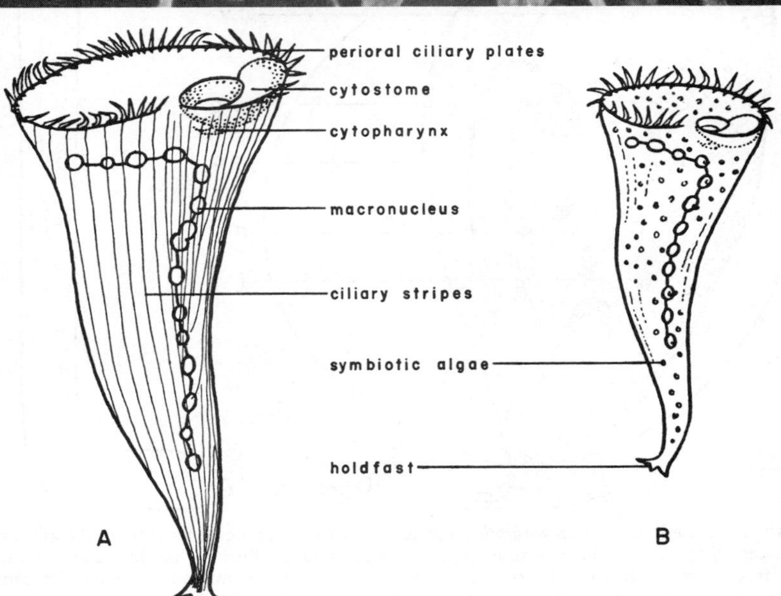

FIG. 16.2. *Above. Stentor* sp. with a much smaller rotifer also shown almost exactly in the center. Most of the ciliates are fully extended but a few are partially or wholly contracted. (General Biological Supply House.) *Below. A. S. coeruleus* a common blue-green species, sometimes reaching a length of 2 mm. *B. S. polymorphus* is usually grass-green, due to the presence of symbiotic algae (*Chlorella*) and considerably smaller. (After Tartar, *J. Exp. Zool.*)

with tentacles outstretched, the arrival of prey. Some species can apparently exercise a certain degree of selection, for some Protozoa swim about unharmed among the tentacles, while others are caught and paralyzed the instant they are touched.

Although most ciliates are solitary, some species are gregarious and

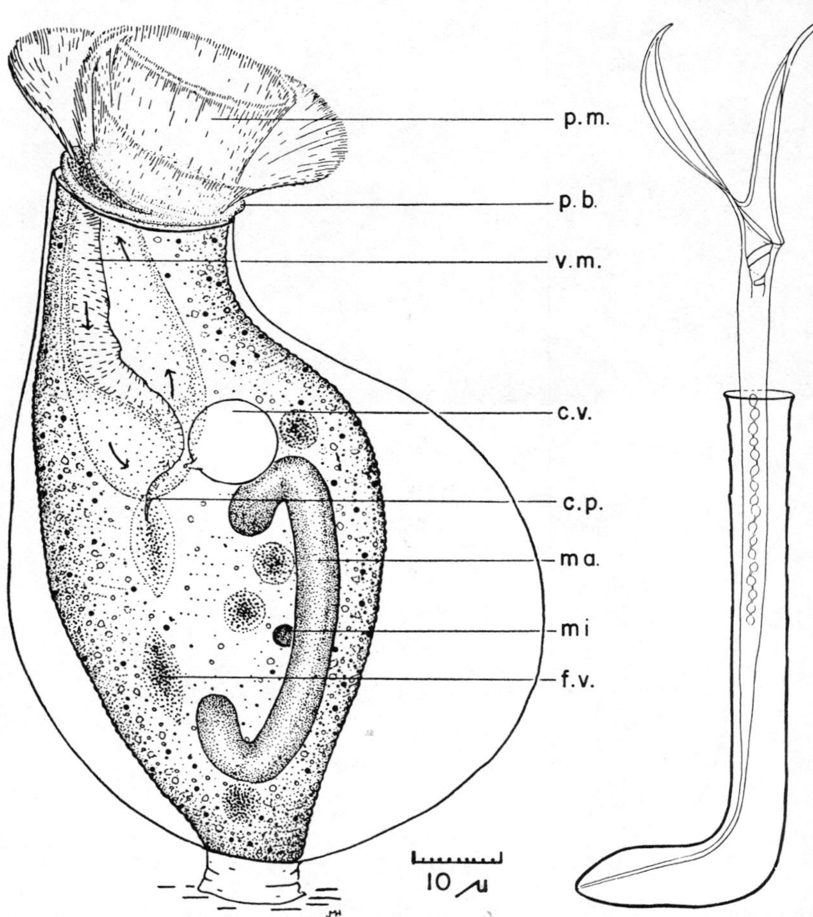

FIG. 16.3. *Left. Cothurnia variabilis*, a loricate peritrich, is an ectoparasite on the gills and neighboring parts of the common crayfish. As in other Cothurnias, the lorica stands on a stalk. c.p., cytopharynx; c.v., contractile vacuole; f.v., food vacuole, spindle-shaped because it is in the "pharyngeal tube"; ma., macronucleus; mi., micronucleus; p.b., peristome border; p.m., peristomal membranes; v.m., vestibular membrane. (After Hamilton, *Trans. Am. Mic. Soc.*) *Right. Folliculinopsis producta*, a sessile marine ciliate living in a tube or lorica. When extended the organism reaches a millimeter in length. The macronucleus is beaded and may be seen in the central portion of the body. The long anterior processes are the peristomal lobes. The tube or lorica is yellowish brown while the ciliate is a dark green-blue. (After Fauré-Fremiet.)

welcome others of their own kind, if only for selfish reasons, since cannibalism is far from unknown. Species of *Vorticella* often live in communities occasionally numbering many individuals. There are other genera of the Peritrichida in which treelike colonies are formed, the individual members living at the end of twiglike branches. In some (*Epistylis*) only the individual bells contract. In others (*Carchesium*, Fig. 16.5) the stalks of the different animalcules contract with the bell, just as in *Vorticella*, while in the genus *Zoothamnium* the contractile elements of all members of the colony are joined, with the result that the entire colony contracts at once. For most observers this is a dramatic spectacle.

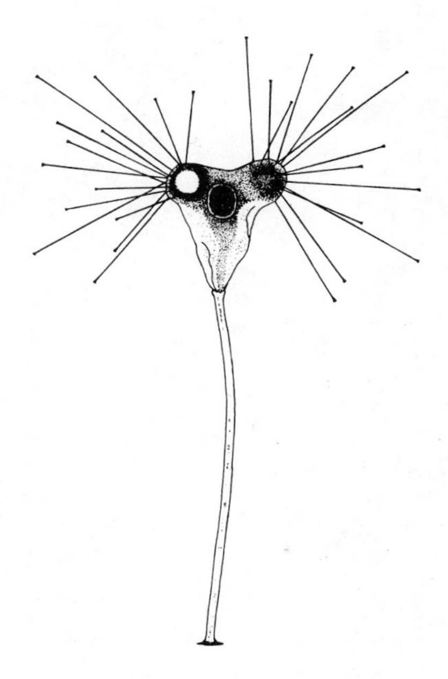

FIG. 16.4. *Tokophrya lemnarum*, a typical suctoridan. Ordinarily the tentacles are arranged in two groups as shown, each tentacle appearing knobbed (though really cupped) at its distal end. As usually seen the organism is somewhat heart-shaped, but viewed from the side it looks more like a club. The size is very variable, but the length of the body (less the stalk) averages about 53μ and the width 38μ. (After A. E. Noble.)

Free-swimming ciliates are usually naked except for cilia, but a few are armored to a moderate degree. The various members of the genus *Coleps* possess cortical skeletons of calcareous plates, between which cilia protrude. Of what advantage these skeletal elements may be, one can only guess. There is nothing to indicate that *Coleps* is thus rendered any the less palatable to its enemies.

Another kind of armor is exhibited by the ciliates of the great group comprising the order Tintinnida. The great majority of the many hundred species known are marine, although about a dozen occur in fresh water. They are free-swimming and pelagic, being carried aimlessly hither and yon by the ocean currents. Yet they bear loricae of vaselike form, within which they live and which one might think would be a burden. The architecture of the loricae, which show the greatest variety of shapes (Fig. 16.6), accurately reflects, as Kofoid and Campbell (1929) remark, ". . . the movements of the body, of the adoral membranelles, and of the ciliary lines on the body, in conjunction with the outpouring of the stored-up [bodily] secretion." And they further note that this secretion may itself contain coccoliths, skeletal remains of the coccolithid flagellates used as food.

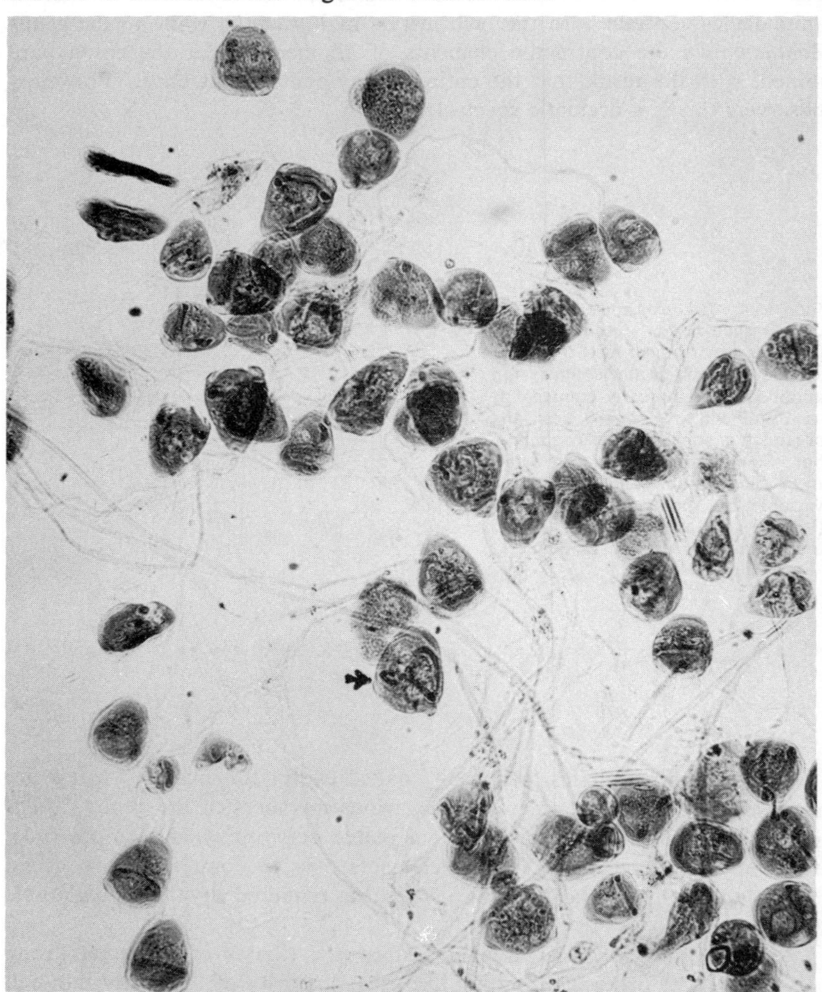

FIG. 16.5. A portion of the colony of the colonial ciliate *Carchesium*. In this genus, the individual zooids contract independently, since the myonemes are not interconnected. Hence they resemble so many *Vorticellas*, having only a part of the stalks in common. A few individuals show the characteristic band-shaped macronucleus (one such is indicated by the arrow). ×116.

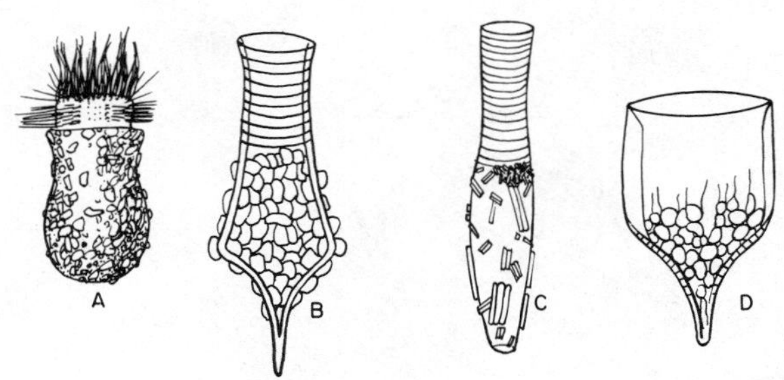

FIG. 16.6. Tintinnid loricae: *A. Codonella cratera*, a rare fresh-water species (after Fauré-Fremiet). *B. Codonellopsis gaussi* from the Antarctic (after Laackmann). *C. Laackmanniella naviculaefera*, another planktonic form from the Antarctic (after Laackmann). The adherent objects are diatoms. *D. Epiplocylisundella* from the Red Sea (after Ostenfeld and Schmidt).

The Mouth

In most ciliates, though in not quite all, the mouth (or cytostome) is a conspicuous structure, but in some forms (e.g., the orders Astomatida and Apostomatida) it is very small or even absent. Primitively, no doubt, the cytostome was located at the anterior end, where it has remained in such holotrichous forms as *Holophrya* (Fig. 16.7). In some species the mouth is always open and probably always in use, receiving whatever edible particles may be swept into it. In other ciliates it remains normally closed, opening only when needed to take in food. The size of a good mouthful for some of

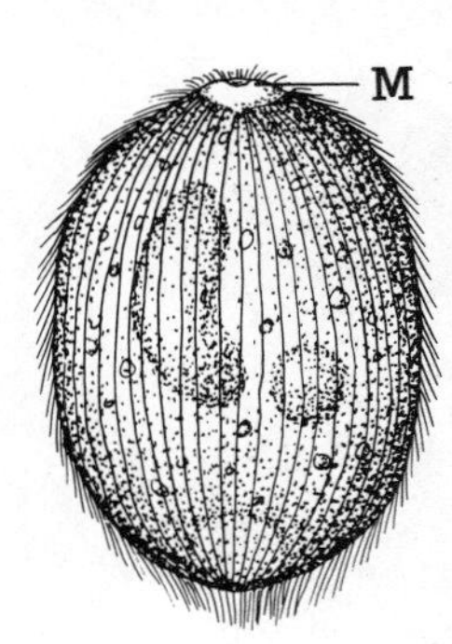

FIG. 16.7. *Holophrya discolor*, a holotrichous ciliate, probably a primitive sort. The mouth (M) is at the anterior end, and the ciliary meridians radiate from it. (After Bütschli.)

these organisms is amazing. Occasionally the captor's cytoplasm seems to amount to little more than a thin envelope stretched around the still struggling prisoner within. Sometimes the mouth is more or less circular, but in numerous others it is slitlike.

Associated with the mouth in many species are highly developed organelles, such as membranelles and undulating membranes (Fig. 16.8). Both are derived from cilia, as shown by the underlying basal granules. The membranelles are rectangular or ribbon-like in form, each consisting of fused

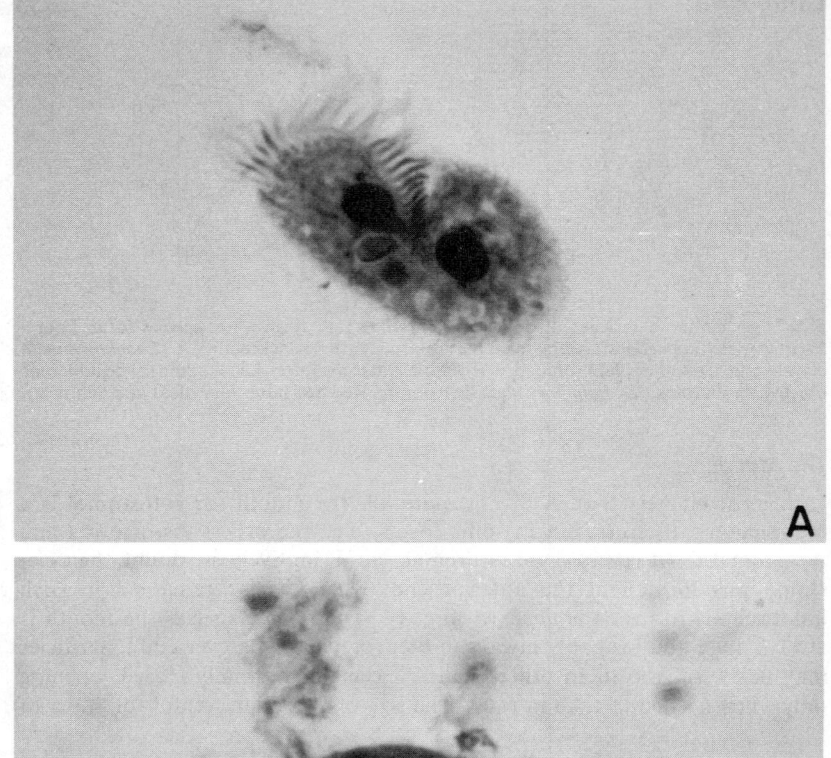

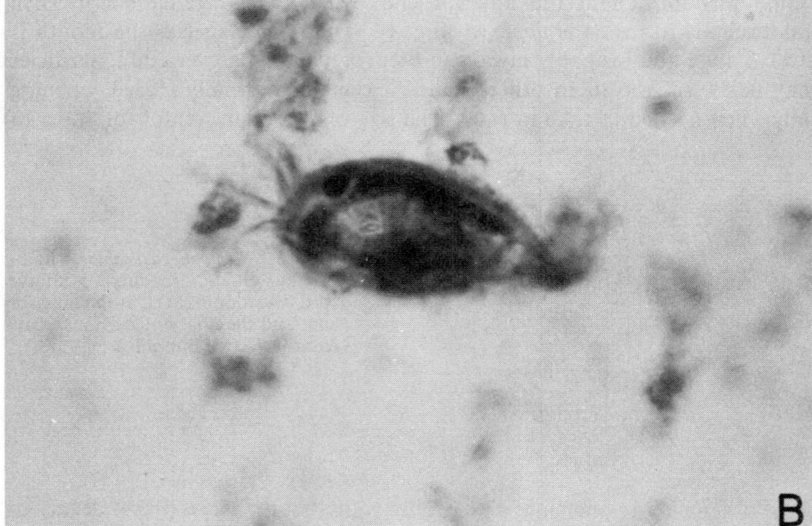

FIG. 16.8. Hypotrichous ciliates, ×676. *A. Oxytricha* sp. (ventral view) showing especially well the adoral zone and the membranelles of which it largely consists. The two macronuclei and, slightly anteriorly and to the left, the micronuclei are clearly visible. The marginal cilia are also easily seen, although the ventral and anal cirri do not show. *B. Euplotes* sp. (ventral view), one of the most highly specialized among the free-living Ciliophora, but the cirri of which this ciliate makes such remarkable use are only visible at the anterior end. Note the characteristic horse-shoe-shaped macronucleus, with the micronucleus set slightly into its anterior outer edge.

cilia, the whole being aptly likened to a flat paintbrush. Measured in terms of the basal granules underlying them, they are from two to four such granules across. The "adoral zone" associated with the mouth in so many ciliates, particularly those of the subclass Spirotricha, consists of a row of such membranelles on the left side of the mouth cavity or peristomial field. The undulating membranes—often of relatively enormous size as in *Pleuronema* and *Cyclidium*—are diaphanous, consisting of a fused row of long cilia. The area immediately about the mouth, or perhaps more exactly, the buccal cavity, is known as the "peristome." The adoral zone is to be distinguished from the peristome; the former leads to the mouth ("adoral" being derived from the Latin *ad*, meaning "to," and *os*, meaning "mouth;"

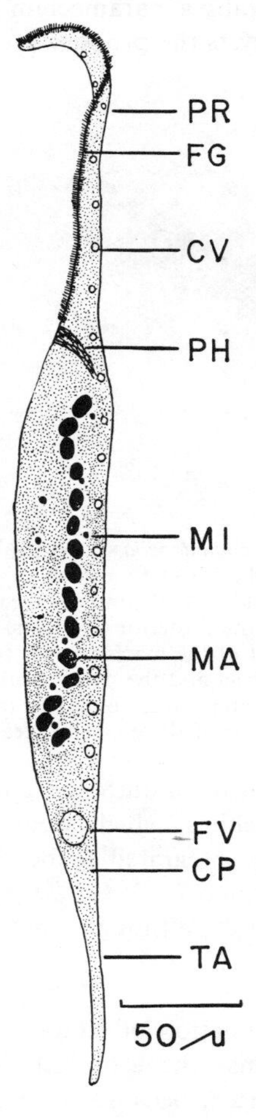

FIG. 16.9. *Dileptus monilatus*, a common holotrichous ciliate. CP, position of cytopyge; CV, contractile vacuole (there are about 20 in all); FG, feeding groove; FV, food vacuole; MA, macronucleus; MI, micronucleus; PH, pharyngeal basket; PR, proboscis; TA, caudal process. (After Jones and Beers, *J. Elisha Mitchell Sci. Soc.*)

"peristome" from the Greek *peri*, meaning "about," and *stoma*, also meaning "mouth").

From the mouth there is usually an esophagus or gullet leading to the cytoplasm, the anterior portion of which may be called the "cytopharynx." Often this is itself fitted out with rows of cilia, as well as with a complex system of cytoplasmic encircling fibrils (myonemes) to facilitate swallowing.

In some species the oral region also exhibits other modifications. A *Dileptus* (Fig. 16.9) waving its proboscis about as it moves through the water impresses most students. The graceful *Lacrymaria olor* constantly stretches its swanlike neck, in this direction and that, apparently in search of food. More impressive still is a *Didinium nasutum* stalking paramecia. At intervals it suddenly stabs a paramecium with its proboscis, paralyzes it by means of the trichocysts the proboscis is armed with, and then swallows the victim whole.

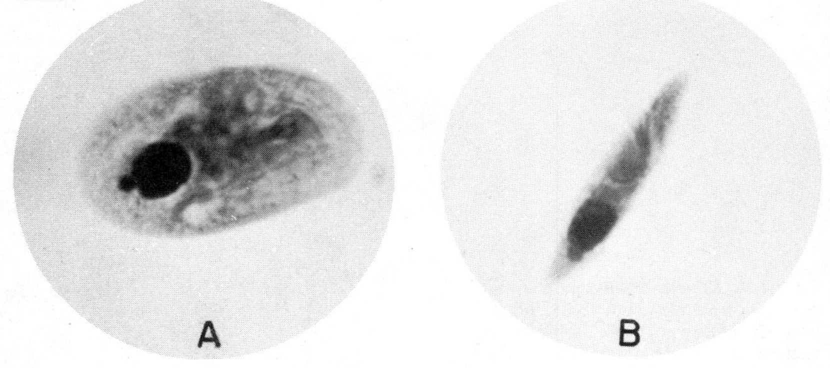

FIG. 16.10. *A. Chilodonella* sp. Ventral view showing the macro- and micronuclei and the pharyngeal basket; also four rows of cilia may be quite clearly seen on the left (more convex) side, and several rows less clearly on the right. The micronucleus lies against the macronucleus, and posterior to it. The pharyngeal basket has a position anteriorly corresponding to that of the macronucleus posteriorly. ×832. *B.* Side view of *Chilodonella*, showing the nuclei and the pharyngeal basket, which typically projects slightly from the ventral surface, and makes about one and one-half turns internally. It is said to contain ten trichites, and four of these can be clearly distinguished. The rest would be out of focus. ×832.

Lying about the mouth and gullet, in some ciliates, are skeletal structures known as trichites, which seem to serve a stiffening function. These may be sufficiently integrated to be collectively termed a "pharyngeal basket," as in the common *Chilodonella* (Fig. 16.10). In this ciliate the basket is said to be protrusible, although what advantage this may be we do not know.

Body Regions

Close examination of ciliates shows the body to consist of two roughly defined regions—the ectoplasm and the endoplasm. The former is often called the cortex, because it covers the animal as bark covers a tree. The cortex gives rise to the covering membrane, known variously as the "pellicle" or "periplast," (e.g., the conspicuous ridges on the carapace of *Euplotes*). From it also stem superficial structures, such as cilia and their derived organelles (cirri, membranelles, undulating membranes), trichocysts, skeletal plates, and the complex system revealed by silver impregnation (Fig. 16.11). Closely associated with the cortex, if not actually derived from it, are the contractile threads or "myonemes" seen in many species of ciliates, such as

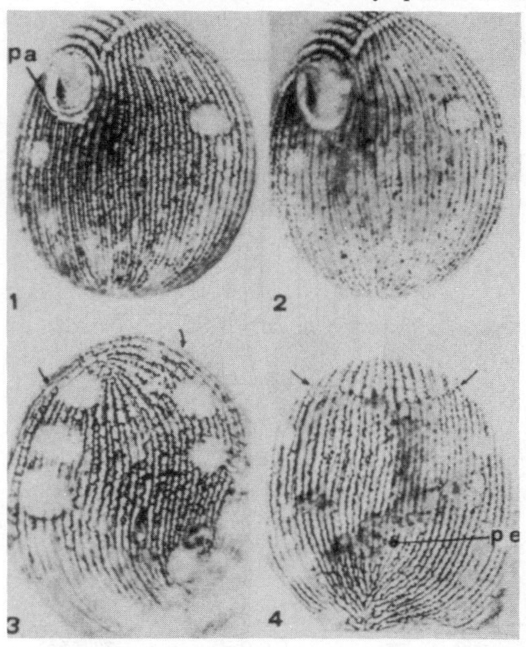

FIG. 16.11. The "silver-line system" of the tetrahymenid ciliate *Glaucoma scintillans* seen in organisms treated with Klein's silver impregnation method. (from Gelei, 1935). *1* and *2* show individuals, each of which was stained a little differently. *1.* The parallel rows of cilia and mouth are especially well shown; pa, basal section of the paroral membrane; near it, on the left, may be seen the circumoral fibril. *2.* Interciliary fibrils visible between the ciliary rows. *3.* An organism dorsal side uppermost. The arrows indicate the interruption of the meridian by the polar "suture line" in this and the following figure. *4.* Similar to, but slightly differently oriented from the organism shown in the preceding figure; pe, excretory pore.

Vorticella and *Spirostomum*. In the blue sessile ciliate *Stentor coeruleus* there are not only longitudinal myonemes but spirally wound myonemes surrounding the gullet, and associated with each a delicate fibril which has been called a neuroid because of its presumed nervelike function.

Ciliation

The organelles concerned with movement are more highly developed in the ciliates than in other Protozoa; the same evolutionary tendencies at work in the development of the higher animals have apparently operated with the ciliates too. Just as the numbers of vertebrae, ribs, appendages, teeth, and the like have been sharply reduced with the climb up the evolutionary ladder, so have the number and distribution of cilia. Presumably ciliation covered the entire body in the primitive ciliates, as it still does in many holotrichs. The cilia themselves are usually arranged in parallel longitudinal rows, originating in the neighborhood of the mouth. But in the Oligotrichida,

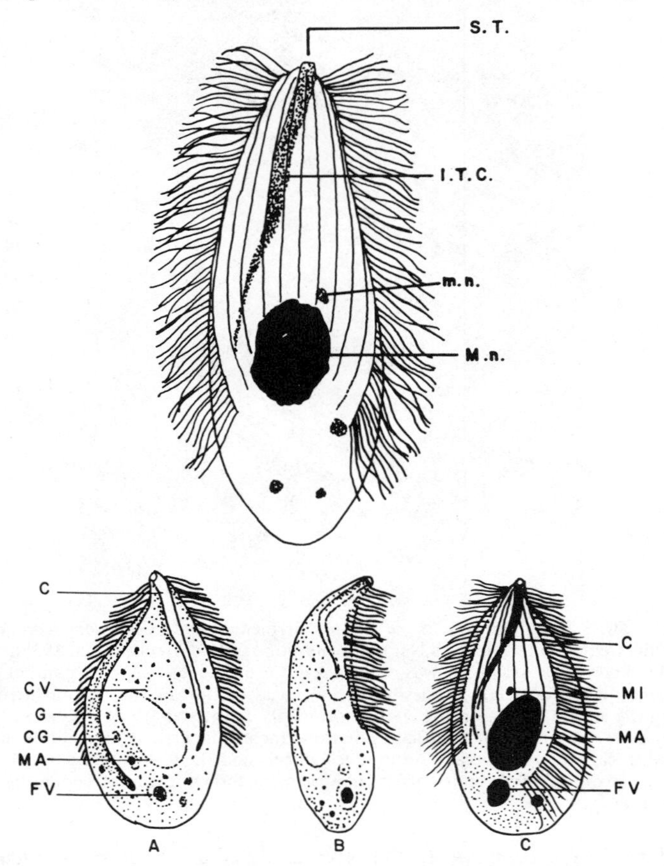

FIG. 16.12. Thigmotrich ciliates. *Above. Ancistrocoma dissimilis*, parasitic on the gills and palps of the rock-boring piddock (a mollusc). I.T.C., internal tubular canal; M.n., macronucleus; m.n., micronucleus; S.T. contractile suctorial tentacle. *Below. Heterocineta phoronopsidis*, parasitizing the tentacles of *Phoronopsis viridis*. *A*. Dorsal aspect. *B*. Lateral view from right side. *C*. Ventral view (stained preparation). c, internal tubular canal; cg, cytoplasmic granule; cv, contractile vacuole; fv, food vacuole; g, dorso-lateral groove; ma, macronucleus; mi, micronucleus. The tentacle, which is contractile and continuous with the internal tubular canal, is used for attachment to the epithelial cells of the host's tentacles. (After Kosloff, *Biol. Bull.*)

Hypotrichida, and Peritrichida, all three highly evolved groups, much of the body is now naked of cilia, and those remaining, especially in the Hypotrichida (e.g., *Euplotes*), are often fused to form the relatively enormous appendages known as cirri. The uses of these organelles have also changed, so that in some genera (e.g., *Aspidisca*, *Onychaspis*) they are employed not only for swimming but for creeping, walking, and even jumping.

It has been claimed by some that there are sensory organelles in some ciliates, such as the dorsal bristles of *Euplotes*. There is no doubt that these organisms are highly sensitive to environmental stimuli of a variety of kinds, but the function of these structures is very difficult to prove. Certainly many species, particularly pigmented ones, are highly sensitive to light, and yet they lack any such photoreceptors as the eye spots of the green flagellates.

The thigmotactic cilia of the ciliates of the order Thigmotrichida (e.g., *Heterocineta phoropsidis*, Fig. 16.12) may also be in part sensory in function. These organisms attach themselves to the gills and palps or mantle cavities of their molluscan hosts, using the thigmotactic cilia for this purpose. These cilia are usually especially long, while the body cilia are reduced and even absent; so also may be the cytostome, a suctorial tentacle and internal canal leading from it taking its place.

Other structures of ciliary origin occur on numerous species of ciliates. Such are the caudal bristles seen on species such as *Uronema* and *Pleuronema*. Often there is only one, but there may be several or many. Their function is doubtful, though it has been said to be sensory in some forms. In *Urocentrum* so often observed rotating madly about in infusions, they have fused to form a cirruslike organelle which apparently has a threefold function: it secretes a filament of mucus and is thigmotactic, adhering to the mucus and thus serving as a temporary anchor. Held fast by this, the organism whirls and swings this way and that.

INTERNAL STRUCTURE

Fibrillar Systems

With the specialization of the locomotor organelles has apparently come the development of what is generally known as the neuromotor apparatus. (This has been more fully discussed in Chapter 5.) It consists of a center, the motorium, located usually anteriorly and often near the mouth, from which fibrils radiate to the membranelles and other oral motor structures, and to some at least of the cirri and cilia. Whether this fibrillar apparatus actually has the coordinating function suggested by its name is not wholly certain, but the results of microdissection as well as the arrangement of the fibrils make it probable. We do not yet know how widely this system occurs in the ciliates. Special staining techniques and great skill in interpreting the results of these studies are required of the investigator. Other systems of fibrils have also been described in ciliates, but their functions are even less clear, and it is impossible to say whether or not the various systems correspond. The use of silver staining techniques has shown in many species what appears to be a system of cortical fibrils, though in part these may only be surface markings, despite the name "silver-line system" often applied to it.

Connecting the basal granules are fibrils which appear to be arranged in parallel fashion among the rows of cilia. These have recently been studied with the electron microscope and interesting details regarding their structure and arrangement have been revealed. Unfortunately, the techniques required for electron microscopy preclude examination of the living animal, and it is usually necessary to examine fragments and piece together the results. One of the most complete studies so far made was of several species of *Paramecium* by Metz, Pitelka, and Westfall (1953). They disintegrated the organisms by "sonic dissection" and observed the pellicular bits with both phase and electron microscopy. The pellicle itself is three-layered (Pitelka, 1959) (Fig. 16.13). Through the pellicle, the surface of which consists of closely packed polygons, pass the cilia. Each cilium originates in a basal granule (the kinetosome) and leaves the pellicle through a ring lying in the center of one of the polygons. Also originating from each basal body on its right is a tapering fibril which joins other similar fibrils to make up an anteriorly directed compound fiber, or bundle (the kinetodesma); each fibril being short, they overlap one another like shingles. Individual fibrils appear to consist of a larger axial fibril with a very minute one wound about it in a close helix. The fibrillar bundles (kinetodesmas) are supposed to constitute the neuromotor fibrils revealed by staining techniques and ordinary light microscopy, although if this is so one can only speculate about how an impulse travels along a bundle. It must have to jump from one fibril to the next, somewhat as a nerve impulse crosses a synapse. A later study of *Tetrahymena* by Metz and Westfall (1954) showed the kineties of this organism to have a structure similar to those of *Paramecium*. Other ciliates have also been investigated with the electron microscope, among them *Nyctotherus*. Here King and Beams (1958) were able to show that the cilia had the usual nine peripheral fibrils (each double) and two central. The latter ended in a bulb at the level of the pellicular membrane, while the nine peripheral fibrils within the cilium continued to form the basal granule. The dorsal bristles of *Euplotes* likewise, according to Roth (1956), have a structure typical of cilia. The motorium, visible in a few preparations, seems to consist of fibrils individually like those of the kinetodesmas.

Trichocysts

Many ciliates are provided with trichocysts (see Chap. 5). These are easily seen even in the living animal, where in forms like *Paramecium* they occur in long rows just underlying the cilia. They may be of several types; there are those so well known in *Paramecium* and the "toxicysts" of *Dileptus* and *Actinobolina*. In some ciliates, such as *Paramecium*, trichocysts are widely

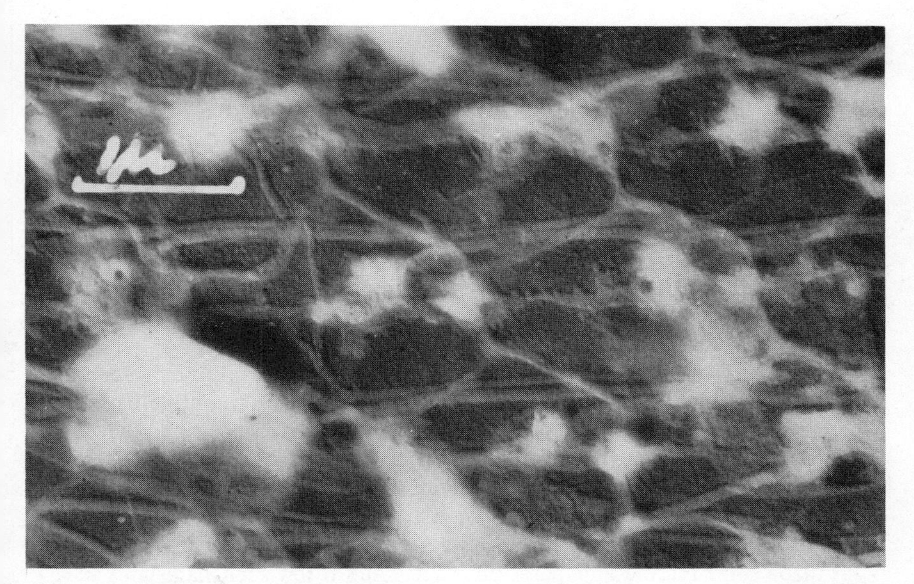

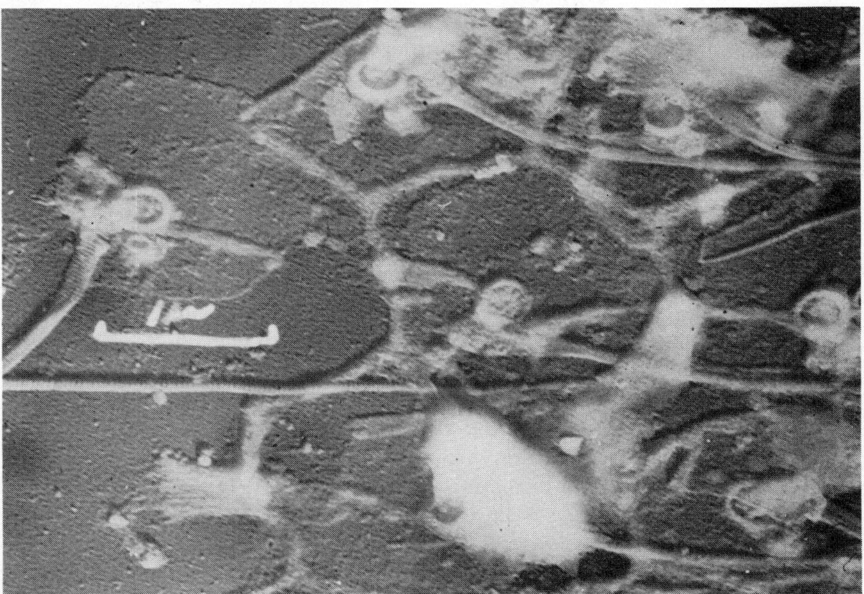

FIG. 16.13. Structure of the pellicle of *Paramecium aurelia* as revealed by electron microscopy. *Above*. A fragment viewed from the outside. The polygon latticework and subpellicular bundles of kinetodesmal fibrils are both shown. ×21,600. *Below*. A similar fragment viewed from the inside. Remnants of trichocysts are attached to the polygon cross bars (lower left). Single and double (right center) ciliary rings and accessory rings (upper left) appear in the centers of the polygons. Kinetodesmal fibrils are also to be seen. ×24,000. (After Metz, Pitelka, and Westfall, *Biol. Bull.*)

and quite uniformly distributed over the body, but in others (e.g., *Dileptus*) they may be limited to the area about the mouth, or to tentacles, as in *Actinobolina*, or to papillae. Trichocysts such as those of *Paramecium* are thought to be defensive, though without much proof. When discharged, it is much as if the whole organelle were turned inside out. Toxicysts are presumably used for offense as well as defense, and doubtless also aid in the capture of prey. Trichocysts seem to originate by division of the kinetosomes, as do cilia.

Contractile Vacuoles

At or near the aboral or posterior end of the more primitive or generalized ciliates, such as many of the holotrichs, is typically a contractile vacuole. When the mouth has shifted its original anterior position, the contractile vacuole has often shifted too, so that in many cases it is laterally placed. In many species (e.g., *Paramecium*) there are two, or even many, as in *Amphileptus*. These vacuoles are often filled from a number of radially arranged canals or vesicles, as in *Paramecium* and *Euplotes*. Unlike other Protozoa some marine and parasitic ciliates and a few such flagellates have retained their contractile vacuoles. We do not understand why excreting water in a medium with the osmotic properties of sea water or tissue fluids requires special provision in these organisms.

Food Vacuoles

Food vacuoles, like contractile vacuoles, are a constant feature of ciliates, except of course in starved organisms. Since they are temporary organelles, their number varies from time to time. Their formation and behavior have been most studied in *Paramecium*, and both Wichterman (1952) and Kitching (1956) give good summaries of what is known about them. On the whole, their size tends to be rather constant in a given species of ciliate. They begin as a dilatation of the esophagus (the "esophageal sac") which receives food particles from the anterior portion of the gullet, or "cytopharynx." When a certain size has been reached the sac is pinched off, and begins its course through the cytoplasm as a gastric vacuole; apparently the course is determined by cytoplasmic streaming or cyclosis, and is therefore a fairly regular one about the nuclei. The formation of the sac seems to be initiated by the entry of food particles. A system of encircling fibrils (myonemes) about the cytopharynx is present in many ciliates to facilitate swallowing. The process of digestion within the vacuole is discussed in Chapter 6. Some ciliates also possess a cell anus ("cytopyge" or "cytoproct") through which any indigestible residue is defecated to the exterior.

The Nuclei

The nuclei are seldom visible in life and staining is usually required for their study. The nuclear apparatus varies greatly in different species. There may be one or several nuclei of each kind, and occasionally many, as in *Stentor*, which is said to possess sometimes as many as eighty or ninety micronuclei. Usually, however, there is one micronucleus and it is often embedded in the macronucleus, or lies so close that it is difficult to discern. Although the micronucleus is generally round or oval, the macronucleus varies greatly in form, spherical or oval in some species and ribbonlike, beaded, reniform, or horseshoe-shaped in others. The macronucleus may occasionally be seen, even in living ciliates, though usually only when they are moribund under the coverslip. In *Dileptus* it is present as many discrete granules, dispersed throughout the cytoplasm and not surrounded by nuclear membranes. Within a species, however, the nuclear characteristics are constant with little variation, though, as noted above, amicronucleate individuals are occasionally seen, and in certain species the micronuclear number is somewhat variable.

Other Cytoplasmic Inclusions

Aside from very minute bodies, such as the mitochondria present in the great majority if not in all cells, the Golgi apparatus, and the like, the cytoplasm of ciliates may contain symbiotic algae and bacteria, globules of one sort or another, and pigment spots such as those seen in certain species of *Ophryoglena* (e.g., *O. macrostoma* and *O. flavicans*). Occasionally one may also see in them parasitic organisms, even other Protozoa. The larger species of *Ophryoglena*, for example, are often invaded by a smaller species named, appropriately enough, *O. maligna*.

Skeletal Elements

The trichites which may occur about the mouth, forming what is often called a "pharyngeal basket," and the tests or loricae in which some ciliates live, have already been mentioned. But there may also be internal skeletal elements as in certain of the cattle ciliates. *Polyplastron* (the very name means "many plates") has an internal armature of five plates. We are not sure what use such stiffening structures have; some protozoologists have suggested that they do not constitute an internal skeleton at all, but may be bodies of paraglycogen or hemicellulose which serve as food reserves. Or it may be that the viscosity of the medium, consisting as it does of digestive juices and shreds of vegetation, makes some reinforcement of the protoplasm necessary.

Life Cycles

Division

Although ciliates differ greatly in morphology, they exhibit life cycles which have much in common. Division in the great majority of species is transverse, and in this they are unique among the Protozoa. Only the protociliates and the peritrichs differ from the usual pattern; in them division is longitudinal or nearly so. The first of these two groups may not be ciliates at all for they lack a cytostome and have nuclei of only one type, so that they seem to have more in common with the flagellates, and the second are in most cases sedentary and have a symmetry approaching the radial type; thus neither group is at all typical.

The micro- and macronuclei behave quite differently in division. The former divides first, exhibiting a typical mitotic spindle, with what appear to be true chromosomes, characteristic of the species in both number and shape. Yet these "chromosomes" may apparently be chromosome aggregates instead, at least in some species, and the micronuclei themselves seem to be polyploid instead of diploid on occasion. Even in meiosis it is said that chromosomes may be aggregated to some extent (Sonneborn, 1949). Even so the apparent chromosome number may be quite large, reaching in some species several hundred or more.

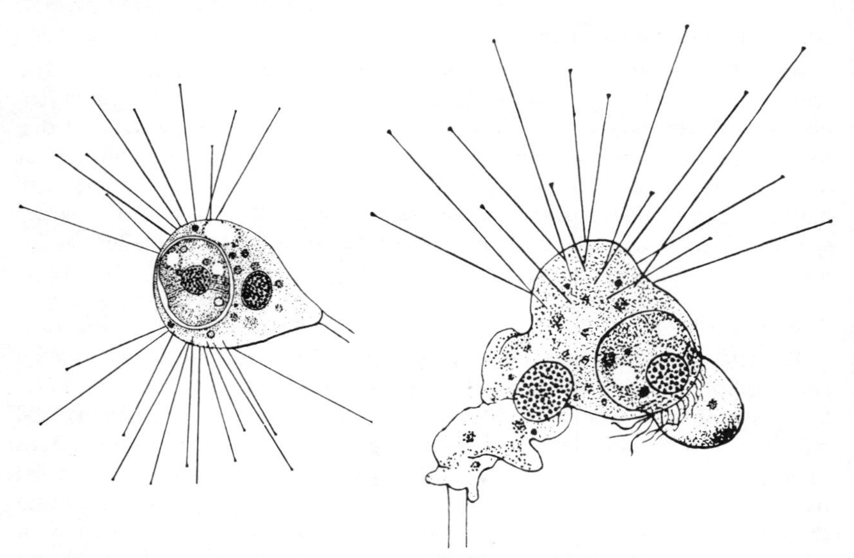

FIG. 16.14. *Left. Tokophrya lemnarum* with brood pouch. The ciliated bands of the embryo may be seen encircling it in belt fashion. *Right.* "Birth" in *Tokophrya lemnarum.* The ciliated embryo is being discharged from the body of the parent by a process involving the forcible contraction of the maternal body, thus suggestive of mammalian labor. (After A. E. Noble.)

When division of the micronucleus is nearly complete, the macronucleus divides amitotically, elongating and then constricting in the center, with the ultimate formation of two daughter nuclei. When more than one nucleus of either or both kinds is present, the division process in nuclei of each kind is nearly synchronous. There is no suggestion of mitosis or of chromosomes in the fission of the macronucleus, but other evidence, particularly the fact that a very small fragment of it may on occasion regenerate the whole, shows it to be probably polyploid. And the macronucleus is, of course, a product of the micronucleus in the first place, since it always arises at the time of conjugation (and after certain other types of cellular reorganization, notably autogamy) from a product of micronuclear division, the fusion nucleus (synkaryon). Details of nuclear behavior, of course, vary with the species of ciliate and the stage of the life cycle.

In some ciliates budding occurs. Even among the peritrichs it has sometimes been suggested that what appears to be longitudinal division could be a form of budding, since one of the two daughters is often smaller than the other, and the larger one retains the stalk while the other swims away after developing a ring of aboral cilia and must establish itself elsewhere. In the Chonotrichida, which are ectoparasites of Crustacea, the buds are generally external and lateral in position, though there may be a brood pouch in which the young ciliate develops. Among the Suctorida, budding is in many cases internal and there is apparently a true brood pouch (a "marsupium"), from which the young ciliated embryos escape when development is complete (Fig. 16.14). One could, accurately enough, call it birth.

Sex in the Ciliates

Sexual phenomena are widespread and probably universal among the ciliates. Whether such processes are a form of reproduction is debatable, since sexual union among the ciliates typically does not involve gametes and permanent union, with subsequent fission of the resulting zygote, but is instead a temporary joining of two individuals (Chap. 12). Thus there is no multiplication, although the division rate of the exconjugants often shows an increase compared to the rate before conjugation. There are, however, exceptions to the rule that sexual union is temporary. With the vorticellids it is permanent. Among them differentiation of the participating organisms takes place, the smaller (presumably the "male") seeking out and uniting with the larger cell, considered essentially an egg. The resulting form behaves much as an exconjugant in other species of ciliates, the process of sexual union among the ciliates being known as conjugation, as we have seen.

One of the most remarkable aspects of sex among the ciliates is the occurrence of mating types (Chap. 12). There may be within a species many varieties, in each of which there are two or more mating types, although we do not yet know just how widespread this phenomenon is. Since, except between complementary types in presumably related varieties, there is no sexual attraction, such varieties are essentially physiological species ("syngens") as we have seen. Thus the inability of outbreeding to occur between varieties seems to circumscribe the value of the sex process to the species as a group. And even though conjugation (or autogamy) seems necessary to ward off senescence and eventual death of the clone it may itself be a highly lethal process. Elliott (1959), in speaking of experiences with *Tetrahymena*, remarks, "Another difficulty, and one that was the most provoking of all, was the high lethality following conjugation. . . . One is hard pressed to explain how, in natural populations, a process so lethal as this can possibly be tolerated by a species."

Other processes, such as autogamy, cytogamy, and even cellular reorganization of the sort so long termed endomixis, occur among ciliates and may be conceived of as related to the profound changes typical of conjugation. Their occurrence varies with the species of ciliate, and they are described elsewhere (Chap. 11). But in these processes only a single individual is involved. Hence such a reorganization has often been likened to parthenogenesis, although the comparison is certainly inexact, since no multiplication results.

Encystment

Encystment seems to be nearly universal in the ciliates, as in most other Protozoa (Chap. 10). Little is known about the process in most species, but it involves a far-reaching reorganization of both nuclear and cytoplasmic structures, sometimes with reproduction in the cyst and sometimes not. The cysts themselves show the widest variety of form and architecture. Some are spherical and smooth, such as those of *Nyctotherus cordiformis*, a parasite commonly found in frogs and toads. Or they may be spherical and studded with knobs, like the cysts of *Pleurotricha lanceolata*. Another species of *Nyctotherus*, *N. ovalis*, frequently seen in the cosmopolitan cockroach, produces beautifully symmetrical pear-shaped cysts with a knob where the stem of the pear might be expected to be. *Paramecium caudatum*, despite its widespread occurrence, apparently produces no cysts at all. Ciliate cysts vary so much among species that they would ordinarily be unidentifiable except to a specialist, even as ciliate cysts. The material the cyst walls are made of has often been said to be chitin, by analogy with the composition of the supposedly similar horny coverings of Crustacea and insects. But it is now said to consist of some polysaccharide. It is, in any case, tough and inert chemically, and apparently relatively impermeable.

Habitats

The ciliates are highly adaptable to various habitats. That they are a highly adaptable lot is evident from the ubiquity with which they occur. There are ciliates in almost any cup of water we quaff from a roadside brook or spring, or from the wooded lake which supplies the city in which we live. Many species (often in a rather high population density) occur in shallow ponds and bogs that support vegetation. Highly polluted waters are favored habitats for many species. Leeuwenhoek found his ciliates "in rain, which had stood but a few days in a new tub, that was painted blue within."*

*Letter No. 18, October 9, 1676.

They are abundant in brackish and salt water and in tide pools. Some species are tolerant enough of salinity to occur in both salt and fresh water, though it is doubtful whether many of them could survive a sudden change, especially when in the vegetative state. Cysts, however, are much more resistant.

Species of ciliates have been found in relic lakes, where the rate of evaporation has exceeded inflow over a long period of time. Edmondson, for example, identified 76 species in the Devil's Lake complex of North Dakota, where the salinity amounts to about 1 per cent, and Pack found two species in the Great Salt Lake, once the great glacial Lake Bonneville, whose salt concentration far exceeds the ocean's. In both cases, the ancestors of these forms must have been resident in fresh water.

Some species of ciliates are often abundant in the damp surface layers of the soil, and they are said by Sandon to occur in similar environments all over the world. Whether soil is the normal habitat of all of them is questionable, since aquatic forms are washed into and away from the soil during floods, and ciliate cysts may be dispersed by wind and animals as well as water. Naturally the ciliate population of soils varies greatly with physical conditions, but Sandon found no surface soils collected from many parts of the world wholly lacking in ciliates. Some species, such as *Colpoda*, *Pleurotricha*, and many of *Vorticella*, are especially common in soil; others are found only rarely. Sandon lists 267 species of ciliates, characteristic of other habitats, which have also been found in the soil. These, however, were not all different, since certain of the species occurred in a variety of situations. In all, approximately a hundred distinct species seem to have been found by Sandon himself in the total of soil samples examined, although he is not entirely clear on this, and he lists no ciliate species as restricted to a soil habitat. The apparent failure of any ciliate species to adapt itself exclusively to a soil existence is interesting, if true, since a number of such species of flagellates and rhizopods are known.

There are, however, certain species of ciliates characteristic of littoral or beach sands, some being modified in structure and locomotor organelles for their habitat. These forms are sometimes referred to as psammophilous.

Of all ciliates, only the tintinnids seem to be almost exclusively resident in the ocean. Kofoid, who studied them exhaustively, stated that of the 752 species known at the time (about 1930), only 12 are fresh-water types. It is worth noting that Kofoid himself, with the assistance of a colleague, Arthur Campbell, described 276 of these. Though most marine species are found in tropical waters, the group is cosmopolitan and even polar waters have their tintinnid fauna. Although Kofoid could state that these ciliates outnumbered those of any other major group, known species of gymnostomes and peritrichs now surpass them, but one wonders whether this would be true if the tintinnids had had protozoologists since Kofoid who were equally interested in them. It seems almost certain that numerous species remain to be discovered.

There are also many parasitic species among the ciliates. The host list includes, perhaps, most of the free-living species of the animal kingdom. The protociliates (although they may in fact be more closely allied to the flagellates) parasitize the tailless amphibia (Anura), and few frogs escape infection. Leeuwenhoek was the first to see these ciliates, and he has left us a figure of one (Fig. 16.15). His drawing is crude but remarkably good considering that he saw the animal through a single-lens microscope.

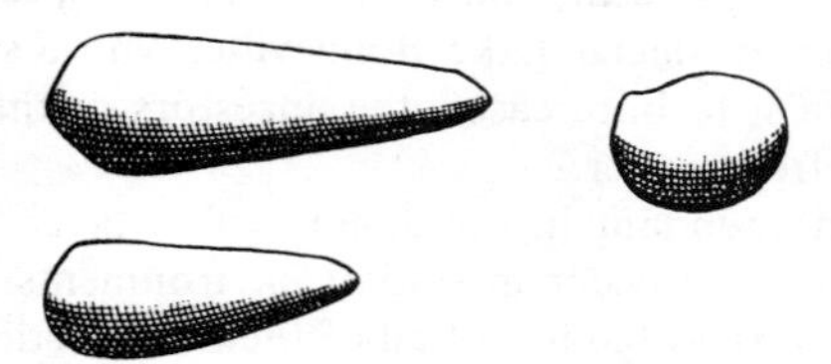

FIG. 16.15. *Left.* The parasitic ciliate *Opalina dimidiata* from the colon of frogs. *Right. Nyctotherus cordiformis*, another gut ciliate of the frog. Drawings accompanying Leeuwenhoek's letter of July 16, 1683.

The parasitic ciliates known as the Astomatida completely lack a mouth. Their life cycles, which are tied in with those of their host in remarkable fashion, have been much studied by the French zoologists, Chatton and Lwoff. The sexual activities and time of molting (ecdysis) of the hosts frequently determine when the parasites divide and encyst, their size and shape, and the arrangement of their cilia.

Ciliates also occur commonly as parasites of vertebrates, and are discussed at greater length in Chapter 22. Among them, for example, are the important and deadly fish parasites belonging to the genus *Icthyophthirius*. These occasionally become a serious problem in hatcheries. Nor are man and his domestic animals immune. *Balantidium coli* is a species shared by man and the pig, and a very similar species is found in guinea pigs. In pigs little harm seems to result, but balantidiasis in man may be a serious or even fatal infection.

Protozoa are sometimes the victims of parasitic ciliates. Penard, the French protozoologist whose interest in the Protozoa extended for most of his ninety-odd years, mentioned a species of *Enchelys* which robs a *Difflugia* of its patiently stored reserve food stuffs, and the case of *Ophryoglena maligna*, an internal parasite of another species of *Ophryoglena*, has already been mentioned.

Many species of ciliates have taken up existence as commensals. These include the hitchhikers of the protozoan world. Many of them ride about as passengers on Crustacea on which they are often abundant, perhaps because these hosts are so regularly present in pools and bogs, as well as on snails, fresh-water annelid worms, and even rotifers. Among fresh-water invertebrates, shrimps of the genus *Gammarus* are popular with these ciliates. Penard studied the infusoria of the fresh waters of France and named 14 species belonging to 11 genera as commensals of the shrimp, and also suggested that his list of ciliates commensal on this and other invertebrates was incomplete. Some species of ciliate commensals occur rather commonly on more than one species of host, but the majority appear to be quite host-specific. This is a little surprising, since most of these commensals are simply affixed to these hosts, though some, Penard further remarked, just "follow the 'host' about everywhere."

It would be interesting to know what physiological mechanisms are involved in associations like these between a ciliate and its larger partner. What is the attraction which a *Gammarus* has for its numerous ciliate passengers? What is there to cause certain Protozoa to follow some other invertebrate species about everywhere? The explanation is doubtless tied in with the physiology of both "host" and parasite. Perhaps food is more easily had for the latter under conditions of such intimate association, but this explanation can only be partial, and simply opens up more questions.

These problems are probably like those involved in attempting to explain why some ciliate species (or other Protozoa) "prefer" certain environments to others. Hegner, for example, found an assortment of Protozoa in the bladders of bladderworts and pitcher plants, together with a variety of living and dead organisms of other kinds. These plants are normally insectivorous, and possibly their protozoan fauna may have originated from cysts introduced by trapped insects, since the cysts of some Protozoa are dispersed in this way. Evidently some species survived digestion by the enzymes secreted by the plants while others were destroyed, and a series of experiments in which Protozoa were experimentally introduced into the pitchers and bladders showed that this was in fact the case. Paramecia, as a rule, were killed by the bladderwort within an hour or two (though a few lived much longer), while Euglenas were not much injured. Yet in the pitcher plant the experimentally introduced paramecia not only lived, but even multiplied, and certain ciliates (e.g., forms provisionally identified as belonging to the genera *Holosticha* and *Prorodon*) were found to be more or less constantly present. Observations like these may have a bearing on the way parasitism originated.

Evolution

Thus, like the other major groups of Protozoa, ciliates have successfully adapted themselves to almost every kind of environment in which protozoan life is possible. In bodily organization, however, the ciliates certainly stand near the top of the evolutionary ladder. Though they have come a long way their kinship to the flagellates, from which they are believed to have originated at some remote time, is evidenced by the remarkable similarities electron microscopy has revealed in the structure of flagella and cilia, the apparent homology of the basal granules of these organelles, and even in some aspects of their biochemistry (*Holz, 1964). Evolutionary development has led to much variety in living ciliates, with losses in some forms so that few cilia remain and gains in others; *Paramecium*, for example, is said to have more than 3,000 kinetosomes and associated fibrils in its cortex (Sonneborn, 1955).

What has happened in the evolution of the Protociliata is not so clear. Has parasitism cost them the cytostome their ancestors may once have had? Or did they never evolve to that point? Did they once have the dimorphic nuclei typical of other ciliates? Or is the absence of such specialization evidence of a closer relationship to the flagellates? Corliss (1955, 1956) and Grassé (1952) have urged the exclusion of the opalinids from the Cilophora, but to the author this seems premature and they are accordingly here retained. Their evolution, however, poses fascinating problems.

Classification

As our knowledge of the ciliates has gradually increased, since their discovery in 1675, they have been grouped in a variety of ways. For a time they remained the "Animalculae," and they and microscopic life in general were lumped together as Infusoria because of their universal occurrence in water containing vegetable matter, or "infusions." Ledermüller introduced this designation in 1760. A quarter of a century later O. F. Müller proposed a classification in which he termed the Protozoa "Animalcula Infusoria," though he restricted the name "Infusoria" to organisms without visible motile organs. After the lapse of slightly more than a half century Dujardin (1841) came close to the modern scheme when he recognized three groups of Protozoa: rhizopods, flagellates, and ciliates, though all were embraced under the blanket term "Zoophytes" ("plant-animals"). Apparently, however, he realized that this was something of a catchall, for he remarked:

> The setting apart, the distinction of these beings, could only have taken place slowly, and little by little by little. First insects and their larvae were separated out, then the gill-bearing crustacea and entomostraca; still later, worms and Zoophytes were recognized, forms which had been confused with microscopic organisms. Lately a variety of other objects have also been taken out, such as fragments of mollusc gills; but on the other hand, quite inappropriately, have been reunited the rest, now the Zoosperms and again whole families of microscopic Algae, Desmids, and Diatoms.

And he then makes a comment which has the air of truly modern biological skepticism: "A more rigorous definition of the true Infusoria should certainly be established; but however much care one may take in establishing it, this class still remains an aggregation of very different types, having only negative characteristics in common." Four years after the publication of Dujardin's classic *Histoire Naturelle des Infusoires*, in which these remarks were made, von Siebold took a step in this direction when he divided all Protozoa into two classes, Rhizopoda and Infusoria, though he omitted the flagellates, since he thought them plants. Here the term "Infusoria" was used in approximately its modern meaning, although the name "Ciliophora" has largely superseded it. The first to use the word "ciliate" in its present sense was Stein in 1857. Bütschli contributed to a clear understanding of the place of the ciliates among the Protozoa in the 1880's.

Classification of the ciliates, however, is still in a state of flux. The scheme which follows is that of Corliss (1956), who based it largely on previous work of Fauré-Fremiet (1950). The only change the author has made is to retain the Protociliata as the first of the three subclasses. As any good taxonomic

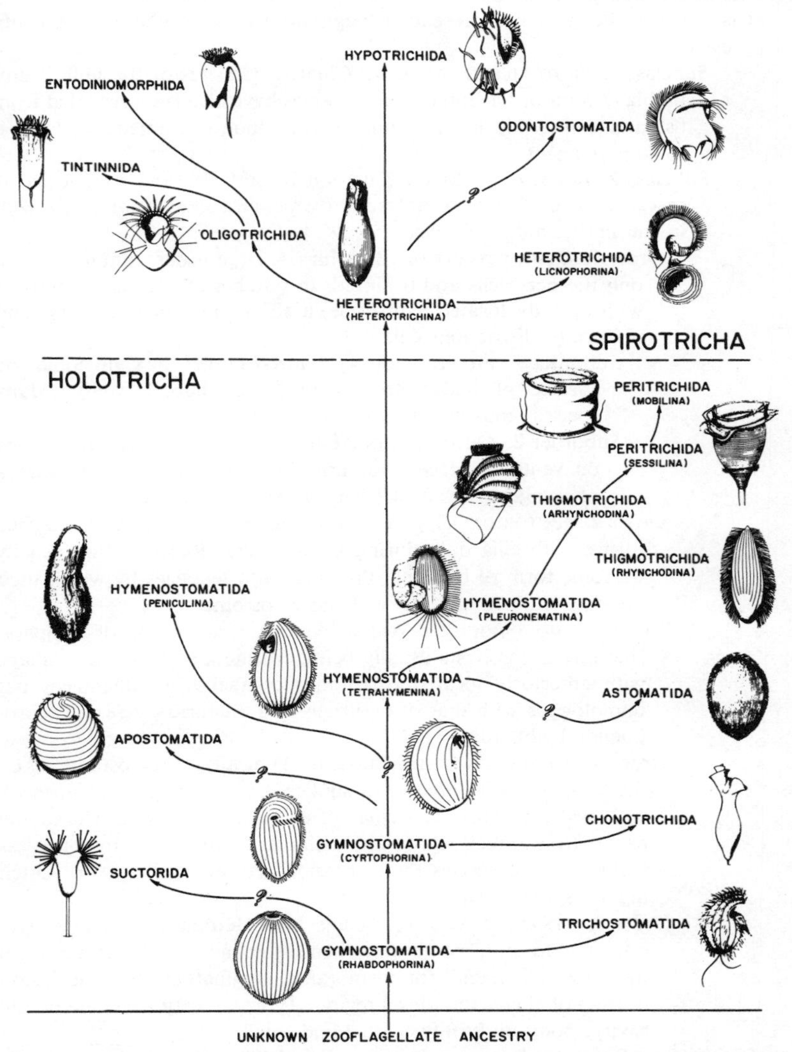

FIG. 16.16. Genealogical tree showing possible ancestry and probable relationships of the ciliates. (After Fauré-Fremiet and Corliss.)

scheme should, this one follows the probable pattern of evolution within the group, as reference to Figure 16.16 will show. Division into the various taxonomic groups is based chiefly on ciliation and differences in the mouth and its associated organelles.

Subphylum CILIOPHORA Doflein. Protozoa possessing cilia or cilia-derived organelles at some stage in the life cycle.

Class CILIATA Perty. Cilia present throughout the active phases of the life cycle.

Subclass 1. PROTOCILIATA Metcalf. Ciliated parasites of the tailless amphibia (Anura or Salientia); a few species have also been reported from fish and snakes. Nuclei all of one type; no mouth. Gametic fusion said to be permanent.

Subclass 2. HOLOTRICHA Stein. Ciliation primitively (and still, in many cases) quite uniform. Adoral zone of membranelles primitive, except in the Peritrichida. No cirri.

Order 1. GYMNOSTOMATIDA Bütschli. A large order, ranking behind only the peritrichs and tintinnids in number of species. Primitively with apically located cytostome, a simple axis of symmetry, and uniformly distributed cilia.

Suborder 1. RHABDOPHORINA Fauré-Fremiet. Cytostome on surface of body; no vestibulum or buccal cavity. Many species; mostly carnivorous.

Suborder 2. CYRTOPHORINA Fauré-Fremiet. Cytostome located on ventral surface in anterior half of body. Usually with a pharyngeal basket of trichites. All herbivorous.

Order 2. SUCTORIDA Claparède and Lachmann. Typically attached forms, with cilia only during larval stages. Reproduction usually by some form of budding. Provided with tentacles by which their prey are trapped and ingested; no cytostome.

Order 3. CHONOTRICHIDA Wallengren. A small group of ectoparasitic forms, the hosts usually being crustaceans. Vaselike in shape with anteriorly located peristome; organelle of attachment not homologous with that of peritrichs and Suctorida. Asexual reproduction by budding.

Order 4. TRICHOSTOMATIDA Bütschli. Typically, the cytostome lies on the floor of the *vestibulum* (a sunken area of the ventral surface), with walls bearing free cilia, often densely packed. Hence the name: from Greek words for "hair" and "mouth". In some less highly evolved species the cytostome may be nearly polar. Often highly asymmetrical.

Order 5. HYMENOSTOMATIDA Delage and Hérouard. Ciliates with one or more oral membranes (hence the name of the group, from the Greek "hymen," for membrane), originating from the fusion of rows of cilia in the adoral region. Typically with a ciliated buccal cavity; body uniformly ciliated and small.

Suborder 1. TETRAHYMENINA Fauré-Fremiet. A large group, typified by the much-studied genus *Tetrahymena*. Mostly small, and little differentiated forms, with a cytostome lying on the floor of the buccal cavity, and the latter fitted with three membranelles (usually on left) and a single undulating membrane on right. (But there is variety in the suborder, and one mouthless form has recently been referred to it on the basis of its infraciliature).

Suborder 2. PENICULINA Fauré-Fremiet. In this group the organelles associated with the mouth are more highly evolved than in the preceding one: there are typically *peniculi* (a *peniculus* is an organelle regarded as a homologue of an adoral zone membranelle, with a base four granules in width and many long); there may also be a vestibulum.

Suborder 3. PLEURONEMATINA Fauré-Fremiet. Cytostome centrally located, and usually with a conspicuous undulating membrane; long dorsal cilia thought to be of thigmotactic function.

Order 6. ASTOMATIDA Schewiakoff. Forms lacking a mouth (hence the name), but otherwise so varied as to suggest a polyphyletic origin (Corliss, 1956). About 200 species, parasitizing a great variety of hosts. Budding frequent, with chain formation.

Order 7. APOSTOMATIDA Chatton and Lwoff. A very small group, with only about three dozen species. The mouth is ventral and very small, appearing within a peculiar structure known as the "rosette." The apostomes are parasitic, mostly on crustaceans, though they may have two hosts.

Order 8. THIGMOTRICHIDA Chatton and Lwoff. Mostly molluscan parasites and provided with an anteriorly situated area of thigmotactic cilia (hence the name) by which they attach themselves to their hosts.

Suborder 1. STOMATINA Chatton and Lwoff. Cytostome subequatorially or terminally placed.

Suborder 2. RHYNCHODINA Chatton and Lwoff. In this group the cytostome has been replaced by a sucker, which may be at the distal end of a tentacle.

Order 9. PERITRICHIDA Stein. A very large order, with more than a thousand known species, many of which are sessile and stalked; even the free-swimming forms may possess a basal disk for attachment. At the anterior end is a conspicuously ciliated disk, the ciliary ring winding counterclockwise (viewed orally).

Suborder 1. SESSILINA Kahl. Attached by a stalk, except in the motile (telotroch) stage, in which there is a ring of cilia encircling the aboral end; this stage may originate by transformation of the sessile form, or by fission (which is nearly longitudinal).

Suborder 2. MOBILINA Kahl. Unattached and free-swimming throughout life cycle (except for temporary attachment). Often commensal or parasitic on a variety of hosts.

Subclass 3. SPIROTRICHA Bütschli. In this rather large group the adoral zone membranelles are well developed, and wind clockwise (or to the right) from distal end to mouth. Peristome not drawn out beyond general body surface.

Order 1. HETEROTRICHIDA Stein. A large order. Ciliation (except that of adoral zone, which may also have an undulating membrane) is typically essentially uniform. Cirri and sensory bristles sometimes present. (In two families, the Peritromidae and the Licnophoridae, body ciliation is much restricted or lacking.)

Order 2. OLIGOTRICHIDA Bütschli. A small order, in which body cilia are few or absent, although there are cirri or bristles in some. The membranelles of the adoral zone are large and may be used for locomotion; there may be a short band of relatively weak membranelles used for food-getting. Fewer than a hundred known species.

Order 3. TINTINNIDA Kofoid and Campbell. A remarkable group of almost exclusively marine forms, comprising more than 900 species. Adoral zone prominent; little body ciliation. Taxonomy largely on basis of the lorica.

Order 4. ENTODINIOMORPHIDA Reichenow. A small order with but two families (though rather numerous species), all parasites of herbivorous mammals. Little body ciliature, but strongly developed adoral zone membranelles. Often bizarre in form. Commensal or perhaps symbiotic to some degree.

Order 5. CTENOSTOMIDA Kahl. A very small group with less than three dozen species. Fitted with a carapace; mouth with only eight membranelles resembling cirri.

Order 6. HYPOTRICHIDA Stein. Dorsoventrally flattened, with a stiff pellicle. No cilia on dorsal surface (though there may be bristles); cirri have replaced cilia on ventral surface. Adoral zone membranelles highly developed, and there may be an undulating membrane. Kernspalt commonly observed in macronucleus during division.

REFERENCES

Calkins, G. N. 1933. Special morphology and taxonomy of the Infusoria. In *The Biology of the Protozoa.* Philadelphia: Lea & Febiger.

Chatton, E. and Lwoff, A. 1935. Les ciliés apostomes. I. Aperçu historique et général; étude monographique des genres et des espèces. *Arch. zool. exp. et gén.*, 77: 1–453.

Corliss, J. O. 1955. The opalinid infusorians; flagellates or ciliates? *J. Protozool.*, 2: 107–14.

——— 1956. On the evolution and systematics of ciliated Protozoa. *Syst. Zool.*, 5: 68–91, 121–40

——— 1957. Nomenclatural history of the Higher Taxa in the subphylum Ciliophora. *Arch. f. Protistenk.*, 102: 113–46.

——— 1958. The phylogenetic significance of the genus *Pseudomicrothorax* in the evolution of the holotrichous ciliates. *Acta Biol.*, 8: 367–88.

——— 1959. An illustrated key to the higher groups of the ciliated Protozoa, with definition of terms. *J. Protozool.*, 6: 265–84.

Doflein, F. and Reichenow, E. 1953. *Lehrbuch der Protozoenkunde. Ciliophora.* Jena: Verlag Gustav Fischer.

Dujardin, F. 1841. *Histoire Naturelle des Zoophytes. Infusoires.* Paris.

Edmondson, C. H. 1920. Protozoa of the Devil's Lake complex, North Dakota. *Trans. Am. Mic. Soc.*, 39: 167–98.

Elliott, A. M. 1959a. A quarter century exploring *Tetrahymena. J. Protozool.*, 6: 1–7.

——— 1959b. Biology of *Tetrahymena. Ann. Rev. Microbiol.*, 13: 76–96.

Fauré-Fremiet, E. 1936. La famille des Folliculinidae (Infusoria, Heterotricha). *Mem. Roy. d'Hist. Nat. Belgique*, Fasc. 3, pp. 1129–75.

——— 1950a. Morphologie comparée et systématique des ciliés. *Bull. Soc. Zool. France*, 75: 109–22.

——— 1950b. Morphologie comparée des ciliés holotriches trichostomes. *Anais Acad. Brasil. Ciencias*, 22: 257–61.

——— 1951. La diversification structurale des ciliés. *Bull. Soc. Zool. France*, 77: 274–81.

——— 1951. The marine sand-dwelling ciliates of Cape Cod. *Biol. Bull.* 100: 59–70.

——— 1952. Symbiontes bactériens des ciliés du genre *Euplotes. C. r. Acad. Sci.*, 235: 402–3.

——— 1955. La position systématique du genre *Balantidium. J. Protozool.*, 2: 54–8.

——— Rouiller, C. and Gauchery, M. 1956. La structure fine des ciliés. Présentation de micrographies électroniques. *Bull. Soc. Zool. France*, 81: 168–70.

Furgason, W. H. 1940. The significant cytostomal pattern of the "*Glaucoma-Colpidium* group," and a proposed new genus and species, *Tetrahymena geleii. Arch. f. Protistenk.*, 94: 225–66.

Hall, R. P. 1953. Ciliophora. In *Protozoology.* New York: Prentice-Hall.

Hegner, R. W. 1926. The Protozoa of the pitcher plant. *Biol. Bull.*, 50: 271–6.

Kahl, A. 1932–5. *Urtiere oder Protozoa.* I. Wimpertiere oder Ciliata. (Infusoria). Parts 18, 21, 25, 30. Jena: Verlag Gustav Fischer.

Kent, W. S. 1880–81. *A Manual of the Infusoria.* London: David Bogue.

King, R. L. and Beams, H. W. 1958. The cilia of *Nyctotherus ovalis* Leidy. *J. Protozool.*, 5 (Suppl.): 11.

Kitching, J. A. 1956. Food Vacuoles. In *Protoplasmatologia, Handbuch der Protoplasmaforschung.* III. Cytoplasma-Organellen.

Kofoid, C. A. and Campbell, A. S. 1929. A conspectus of the marine and fresh-water Ciliata belonging to the suborder Tintinnoinea, with descriptions of new species principally from the Agassiz Expedition to the Eastern Tropical Pacific, 1904–5. *Univ. Calif. Pub. Zool.*, 34: 1–403.

Kudo, R. R. 1966. 5th edition. *Protozoology.* Springfield: Charles C.Thomas.

Maupas, E. 1883. Contribution à l'étude morphologique et anatomique des infusoires ciliés. *Arch. zool. exp. et gén.*, (Ser. 2) 1: 427–664.

Metz, C. B., Pitelka, D. R. and Westfall, J. A. 1953. The fibrillar systems of ciliates as revealed by the electron microscope. 1. *Paramecium. Biol. Bull.*, 104: 408–25.

Metz, C. B. and Westfall, J. A. 1954. The fibrillar systems of ciliates as revealed by the electron microscope. II. *Tetrahymena. Biol. Bull.*, 107: 106–22.

Noland, L. E. and Finley, H. E. 1931. Studies on the taxonomy of the genus *Vorticella. Trans. Am. Mic. Soc.*, 50: 81–123.

Penard, E. 1922. *Etudes sur les Infusoires d'Eau Douce.* Geneva.

Pitelka, D. R. 1959. Ultrastructure of the silver-line system in three tetrahymenid ciliates. *J. Protozool.*, 6 (Suppl.): 22.

Powers, P. B. A. 1932. *Cyclotrichium meunieri* sp. nov. (Protozoa, Ciliata); a cause of red water off the coast of Maine. *Biol. Bull.*, 63: 74–80.

Roth, L. E. 1956. Further electron microscope studies of *Euplotes patella. J. Protozool.*, 3 (Suppl.): 5.

Sand, R. 1901. Etude Monographique sur le groupe Infusoires. Tentaculifères. *Ann. Soc. Belge Microscopique*, vol. 24, 25. 26.

Sonneborn, T. M. 1949. Ciliated Protozoa: cytogenetics, genetics, and evolution. *Ann. Rev. Microbiol.*, pp. 55–80.

Stokes, A. 1886–8. A preliminary contribution toward a history of the freshwater Infusoria of the United States. *J. Trenton Nat. Hist. Soc.*, 1: 71–344.

Wichterman, R. 1953. *Biology of Paramecium.* Philadelphia: Blakiston.

Yusa, A. 1957. The morphology and morphogenesis of the buccal organelles in *Paramecium* with particular reference to their systematic significance. *J. Protozool.*, 4: 128–41.

ADDITIONAL REFERENCE

Holz, G. G., Jr. 1964. Nutrition and metabolism of ciliates. In *Biochemistry and Physiology of the Protozoa.* (S. H. Hutner, ed.). Vol. 3, pp. 199–242. New York: Academic Press.

17

Sporozoa

"In the past, the evil [pébrine] had been sought in the worm and even in the seeds [eggs], but my observations prove that it develops chiefly in the chrysalis, especially in the mature chrysalis, at the moment of the moth's reproduction. The microscope then detects its presence with certainty . . . ." Louis Pasteur in 1866. Quoted from Dubos, *Louis Pasteur*, 1950.

The Sporozoa constitute a large group. Formerly regarded as a subphylum, they are now more commonly treated as a class in the subphylum Plasmodroma. Grassé treats the group almost as a subkingdom, and splits it into three classes. Collectively they make a polyglot assortment of organisms which, Calkins once remarked, have "ever been a puzzle to systematists." All are parasitic and all exhibit highly complex life cycles, but some are more parasitic and more complex in their life cycles than others. They may or may not have an alternation of hosts, and in choice of hosts they are limited only by the extent of the animal kingdom. In morphology, too, there is the greatest diversity in form and size.

There is a good deal of confusion about the classification of the Sporozoa, and the scheme at the end of this chapter has the unqualified approval of probably no protozoologist; indeed, it is hard to find even two texts that agree. Nevertheless, although opinions differ on the taxonomic rank of some groups, there is fair agreement about naming the larger and more common ones. In general, spore characteristics form the basis for group separation.

History

The term "Sporozoa" was coined by Leuckart, a famous German pioneer in parasitology, in 1879. Leuckart disapproved of zoological names based on purely negative characteristics, and felt that spore formation was about the only trait possessed in common by all members of the group. Yet the name was an unfortunate one (although perhaps no better could have been done) for the term "Sporozoa" ("spore animals") implies something unique about the ability to form spores. But spores are what the biologist says they are. When a bacteriologist speaks of a spore, he means a resistant structure formed by a bacterium under adverse environmental conditions. To a mycologist, a spore is the fruiting body of a fungus. The botanist may mean the reproductive bodies of the fern when he talks of spores. Applied to the Sporozoa, "spore" usually refers to the daughter cells resulting from multiple fission after fertilization, but even here it is rather loosely used. It can at least be said, however, that sexual phenomena and multiple fission (or "sporulation") are universal among the Sporozoa.

How many species of Sporozoa there are is unknown, but the number is certainly large. Labbé, at the turn of the century, included in his monograph what he believed to be a complete list of all species described up to that time, a total of "239 certain, 259 uncertain, 18 subspecies and 15 varieties"; little more than a decade later Minchin wrote that the Sporozoa constitute a group of "a great number of parasitic organisms" (1912). It is of course essential to remember that differences of opinion still persist as to just what organisms should be classed as Sporozoa. Wenyon (1926), for example, placed the Cnidosporidia in a class by themselves. But on the whole the types of organisms regarded as Sporozoa in Labbé's day are still so regarded, and this large and heterogeneous assemblage of parasites continues to grow.

Although the first observation of a sporozoan was apparently made by Leeuwenhoek, when he saw "oval corpuscles" (coccidia) in the bile of a rabbit, the history of the class for practical purposes covers little more than a century and a quarter.

The first sporozoan to be described in detail was a gregarine parasitic in beetles, though Dufour, who published an account of it in 1826, mistook it for a minute fluke. Its real protozoan affinities were first recognized in 1848 by the Swiss biologist Kölliker, who taught for many years at the German university of Würzberg. The peculiar gregarine habit of associating end to end in temporary chains of two or more individuals was first described in 1837 by the German zoologist von Siebold.

The involved life cycles characteristic of many Sporozoa were not discovered until much later; probably most of them are still unknown. The idea that there might be an alternation of generations was suggested in 1892 by Pfeiffer, but it was not demonstrated until the mosquito–bird cycle of avian malaria was revealed in 1898 by the Englishman Sir Ronald Ross,* who was knighted for this discovery. The famous English physician Sir

*The tick transmission of *Babesia bigemina*, the causative organism of Texas cattle fever, was demonstrated five years earlier by the American investigators Smith and Kilbourne, but here only the means of transmission was involved. Interestingly, a recurrent suspicion that mosquitoes might be the transmitters of malaria existed in the minds of some long before the theories of Manson and the epochal discoveries of Ross and Grassi.

Patrick Manson had shown a quarter century earlier that the microfilarian parasite of man, *Wuchereria bancrofti*, was similarly transmitted. The Italian Grassi just managed to beat Ross in the race to prove that anopheline mosquitoes transmit human malaria. And a young American medical student, W. G. MacCallum, filled in the final gap in the cycle by his discovery of the sexual nature of the gametocytes. At about the same time Siedlicki published his classic study of the sexual cycle of a coccidian (*Aggregata eberthi*, parasitic in the cuttle fish and the crab) and of a gregarine.

In the years since much has been added to our knowledge of the Sporozoa, and particularly of such important members of the group as the malaria parasites. We realize that there are many more species of Sporozoa than was once thought, and new ones are being constantly added. The work of Cleveland has given striking hints of the great age and probable evolutionary origin of these parasites, which may have arisen from the flagellates. Chatton and Biecheler's discovery of the coccidia-like life cycle of certain parasitic dinoflagellates even suggests a relationship of dinoflagellates and Sporozoa.

Importance

Some Sporozoa are of the greatest importance to man. The malaria parasites cause disability or fatal disease in an estimated one-eighth of the world's population at any one time. The coccidia, although they give rise to few human illnesses, often cause much mortality among domestic animals. Other Sporozoa, such as the Cnidosporidia, cause serious fish diseases whose prevention requires constant vigilance. Nor are the invertebrates immune to sporozoan attack. Some Cnidosporidia are highly pathogenic to bees and silkworms. One of Pasteur's monumental researches dealt with the cnidosporidian silkworm disease pébrine, at that time so serious that it threatened economic paralysis of whole provinces in France. Because of the magnitude of their injury, some of the more important Sporozoa are considered at greater length in chapters to follow.

The More Important Groups

The gregarines, though not of immediate significance to man, are nevertheless a numerous tribe. Despite complexities of structure and life history, they may well be the most primitive of the Sporozoa. The body plan of adults is unusual, divided as most of them are into two major parts, the anterior of which is further divided. They often associate themselves end to end in groups of two or more—a relationship called "syzygy." Their gregarious habits are responsible for the name Gregarinida. Their hosts are almost invariably invertebrates.

The Coccidiomorphida, a much more important group than the gregarines, are mostly intracellular parasites of the gut epithelium or of the blood.

Though they occur in some invertebrates, they are best known from vertebrate hosts. Usually there is one host, although some (e.g., the blood parasites known as Haemosporidina) have two. The order Coccidiomorphida is divided into two suborders, the Coccidina and the Haemosporidina, of which the latter are blood parasites. The first of the two is much the larger, and includes numerous species causing heavy losses to the stockman and poultry raiser; the second is chiefly important because to it belong the malaria plasmodia. The name "coccidia" is borrowed from the Greek *kokkus*, meaning "seed," which the oocysts produced by these parasites are thought to resemble.

The Myxosporida and Microsporida are chiefly parasites, sometimes highly injurious, of fish. Other microorganisms usually placed with the Sporozoa are the muscle-dwelling parasites called Sarcosporidia, the much studied genus *Toxoplasma*, and the Haplosporidia, species of which have been described from a great variety of hosts. Few solid facts are known about the last two groups.

Characteristics of the Sporozoa

It is difficult to define the Sporozoa in terms of morphology. Not only are differences in form and size great, but locomotor organelles in what may be called the adult stages are always lacking. Labbé, who wrote a still useful monograph on the Sporozoa over a half century ago, described the group in this way:

> Uninucleate or multinucleate Protozoa always exhibiting an external protoplasmic covering [the ectoplasm or pellicle]; never possessing cilia or flagella in the adult state; having endosmotic nutrition; invariably parasites within cells or tissues or cavities of other animals; reproducing by division, and especially by sporulation. Nucleated sporozoites, amoeboid or nonamoeboid, may be the direct or indirect [spore] product of sporulation. (1899)

Labbé's definition is still essentially accurate, although certain parts of it cannot be accepted without qualification.

Almost every conceivable shape is to be seen among the Sporozoa. Not only does comparison of species reveal extreme variety, but different stages of the life cycle of the same species usually show little resemblance. In the coccidia, the intracellular vegetative forms, or trophozoites, are large and the sicklelike sporozoites are minute. The contrast between adult gregarines (often called "sporadins") and the sporozoites from which they develop is even more striking. The former are large vermicular organisms, provided at the anterior end with a remarkably specialized organelle of attachment, the "epimerite," while the latter are very small, delicately drawn-out filaments.

Size variation is no less extreme. The merozoite of a species such as *Plasmodium rouxi* (a malaria parasite of birds) is one or two micra in diameter. The relatively colossal *Porospora gigantea*, parasitic in certain crustacea and molluscs, has a diameter of half an inch or more.

Locomotor organelles are absent among adult Sporozoa, yet many species are capable of independent movement at certain periods in their life history. The microgametes, for example, are often flagellated and active. Similarly, the zygote may be quite mobile (in which case it is known as an "ookinete"), but it has no demonstrable organelles to help it move, and we do not understand how it makes its way about its microworld. Still others, such as the gregarines, exhibit through most of their life cycle the ability to glide about with a peculiar sinuous motion much like that of the ookinetes. The mechanics of gregariniform movement, as it is often called, are obscure. It has been laid to rhythmical contraction of cytoplasmic fibrils or "myonemes," and also to a combination of this contraction with the secretion of mucus. In some, progression is much like that of a slug, without change in body shape. In others, it has been compared to the behavior of a *Euglena*, with its frequent changes in form, or to the nervous agitation of a vinegar eel. The variety of behavior among gregarines suggests that highly interesting results might come from intensive study.

Sometimes amoeboid movement may be observed among certain of the Sporozoa. The life cycle of the myxosporidan parasites of fish includes a minute and active stage known as an "amoebula," capable of doing a good deal of traveling with the help of its pseudopods. In the common malaria parasite of man, *Plasmodium vivax*, there is intense amoeboid activity even though the parasite has only the very limited "lebensraum" provided by its host erythrocyte. On the whole, it is not very clear what advantage accrues to those Sporozoa with motility, since it is of no assistance in getting from one host to another, and apparently does not even make possible any extensive reconnoitering about the bodies of their hosts. The intense protoplasmic activity sometimes seen in forms such as *Plasmodium vivax* may be only a physical reflection of the high metabolic rate associated with the rapid growth and frequent fission entailed by parasitism. Yet it is characteristic of only a few species.

Rapid growth and fission of course require an efficient mechanism for food absorption and excretion of wastes. It is generally supposed that this is accomplished in all Sporozoa by osmosis and diffusion, as stated by Labbé, but possibly this is not always so. Rudzinska and Trager (1957) recently showed that certain species of malaria plasmodia actually ingest hemoglobin in amoeboid fashion, even though they are of course intracellular. Electron microscopic studies of other Sporozoa may show, as in this case, that residence within the cell is no barrier to taking solid food, though the particles ingested may be extremely small.

Excretion is mainly by osmotic exchange, and contractile vacuoles are never seen among the Sporozoa. Yet some waste materials may accumulate as insoluble masses (e.g., the pigment, or hemozoin, of malaria parasites) later to be discarded at the time of fission.

Life Cycles

Diversity of life cycles equals the variety in form and size of the Sporozoa. There is always a sexual and an asexual phase, and often multiplication occurs in both. In some, however, reproduction is limited to the former, as with the majority of the gregarines. Here the sporozoite grows and eventually initiates the gamete-producing cycle ("gamogony"). When there is an alternation of hosts, commonly the sexual stages occur in one and the asexual stages of the cycle in another host. The former is therefore known as the "definitive" and the latter as the "intermediate" host (the terms "primary" and "secondary" are also used). Thus the mosquito is the definitive host of the malaria parasites and the vertebrate (e.g., man) the intermediate.

Careful study of the behavior of the nucleus during fission of the Sporozoa has almost always found it to be mitotic, with the appearance of true chromosomes. Sometimes the asexual phase results in the production of almost incredibly large numbers of progeny, as in the case of the cattle coccidian *Eimeria bovis*, with its brood of about 100,000 merozoites from each parental schizont (see Chap. 25). In other species the corresponding numbers may be very small; the avian malaria parasite *Plasmodium rouxi* begets a family of only four. But where fecundity is small, the short duration and frequent repetition of the reproductive cycle may effectively balance limited increase in numbers per generation. Thus asexual reproduction of the malaria parasites may continue for many months or years, and perhaps for the lifetime of the host.

Details of the sporulation, or multiple fission process, vary greatly. Sometimes, as with the erythrocytic parasites known as babesias or piroplasms, it appears to be a simple budding. *Babesia bigemina*, a cattle parasite, takes its name from the formation of two daughters in this fashion from the parent. The process as observed in ordinary blood films greatly resembles binary fission. Budding, or "gemmation," either external or internal is of common occurrence among the Sporozoa. It may take place, with reference to the host, extracellularly or intracellularly, and one or more divisions often occur within the oocyst and sporocyst.

Ultimately, and sometimes soon after the onset of the infection, sexual stages appear. These may mark the culmination of the asexual cycle (except in the majority of the gregarines), or they may be produced in greater or lesser numbers throughout the course of the infection, as in the malaria parasites. Maturation of the gametocytes (immature sexual stages) into the ripe sex cells (gametes) corresponding to eggs and sperm may take place either in the original host (as in most coccidia) or in a second host, when the cycle involves host alternation.

Normally, fertilization promptly follows gametogenesis. The resulting zygote initiates a period of reproduction with the formation of infective stages known as sporozoites. If the cycle includes a single host, the sporozoites are usually developed within a resistant envelope or envelopes, constituting the walls of the oocyst and the sporocysts it contains. Thus protection is assured until entry into a new host becomes possible. Otherwise, the sporozoites remain naked, as do those of the malaria plasmodia, where the shelter of the mosquito's body seems all that is necessary.

Transmission from one host to another is accomplished in many ways, but like most parasites the Sporozoa assume a passive role and allow the host to encompass his own infection. Infection generally results from ingesting ripe oocysts, which are somehow stimulated to germinate and liberate their loads of sporozoites in the host's gut. But apparently some species of Sporozoa have found this method too chancey, and have evolved a cycle in which each of the two hosts infects the other. Thus the Haemosporidina, to which the malaria parasites and their relatives belong, require two hosts. One is usually a biting fly and the other a terrestrial vertebrate. The former infects itself when it feeds on the blood of the parasitized vertebrate. It passes the microorganisms to the next victim when enough time has elapsed for completing the sexual cycle with its resulting myriad sporozoites.

The parasites may also be transferred when one host eats another, as in the species *Hepatozoon canis*, a parasite of dogs and ticks. The tick, *Rhipicephalus sanguinis*, becomes infected when it feeds on the blood of infected dogs, and they in turn acquire the parasites when the irritation of the bites causes them to lick their wounds and inadvertently swallow the parasites.

A surer mode of transfer occurs in some species where the parasites invade the eggs, as they often do in cycles involving ticks. Although this kind of transmission is convenient, it is relatively rare. An example, however, is the microsporidian *Perezia fumiferanae*, a parasite important in the natural control of the spruce budworm, an insect responsible for heavy damage to spruce in our northern forests. Its life cycle has recently been thoroughly worked out by Thompson (1957). In mammals the parasites sometimes pass the placental barrier, but such transmission is even less common than the ovarian. The ubiquitous species *Toxoplasma gondii* makes regular use of it, although this parasite also succeeds in getting from one host to another in a variety of other ways, some of them still undiscovered.

Once within the host, the parasite promptly begins the appropriate phase of its life cycle, usually doing little injury. Doubtless in the course of evolution both parasites and hosts that could not make satisfactory mutual adjustments became extinct. Intracellular forms kill the cells in which they live, but host regenerative capacity compensates for the loss. Immunity soon develops and puts an end to further harm to the host from the parasite. There are of course exceptions. Man suffers severely from some kinds of malaria, and many of the coccidia are highly pathogenic.

Host Specificity

For reasons little understood, most parasites exhibit marked host-specificity. This is fortunate, since otherwise we would all be susceptible to attack by a multitude of parasitic species. Such specificity, however, varies greatly with the species of parasite, but with the great majority of Sporozoa it is strict. Furthermore, certain groups of parasites tend to occur only in certain host groups. Malaria, for example, occurs commonly in birds (particularly the perching, or passerine, forms), much less commonly in reptiles and mammals, and hardly at all among the amphibia. And though there are numerous species of malaria in reptiles, most of them seem restricted to lizards. Among mammals, rodents and primates seem to be the most favored hosts. Coccidial infection is seen frequently in the terrestrial vertebrates, and particularly in birds and mammals, yet invertebrates seem much less prone to infection. Gregarines, though a parasite group with all the earmarks of great antiquity, have apparently never been able to extend their host range much beyond insects and a few other invertebrates. The Myxosporida are seldom found in any hosts except fish. The explanation of this spotty host distribution presents a difficult but intriguing problem.

Classification

The scheme of classification given in outline form below was formulated for the Sporozoa in 1936 by a committee working under the auspices of the American Association for the Advancement of Science. It is part of a larger scheme constructed to apply to the entire Animal Kingdom. The taxonomy of the Protozoa was done under the chairmanship of L. E. Noland of the University of Wisconsin, and represented the majority opinion of some forty American protozoologists.

Class. SPOROZOA Leuckart (1879). Entoparasitic Protozoa usually reproducing by multiple fission, with life cycles including both sexual and asexual phases; locomotor organelles absent in adult stages.

Subclass 1. TELOSPORIDIA Schaudinn (1900). Sporozoites elongate, lacking polar capsules. (The name, from the Greek *telikos*, "end," implies that spore formation occurs only at the end of the trophic or growth period, but it is not entirely appropriate for forms other than the Gregarinida).

Order 1. GREGARINIDA Lankester (1866). Parasitic in early stages on or in epithelial cells, but free in body cavities later; gamonts typically associate in groups of two or more ("syzygy"); hosts almost invariably invertebrates, especially annelids and insects.

Order 2. COCCIDIOMORPHIDA Doflein (1901). Typically intracellular in all stages; gamonts dimorphic ("anisogamous").

Suborder 1. COCCIDINA Leuckart (1879). Usually intracellular parasites of epithelium, generally of digestive tract; reproduction in both sexual and asexual phases; sporozoites generally contained within sporocyst, lying in oocyst with resistant walls; wide host range.

Suborder 2. HAEMOSPORIDINA Danielewski (1886). Life cycle involves two hosts, a vertebrate and invertebrate, the sexual phase occurring in the latter; sporozoites naked in body of invertebrate.

Subclass 2. CNIDOSPORIDIA Doflein (1901). Sporozoites with polar capsules containing coiled thread, somewhat after the fashion of the nematocysts of coelenterates; no alternation of hosts. Majority of species are parasites of fish, but occur also in annelids, arthropods, and others.

Order 1. MYXOSPORIDA Bütschli (1880). Spores with bivalve chitinous covering membrane; organism within (known as "sporoplasm") with one to four capsules variously arranged; sporoplasm becomes "amoebula" upon emergence; nearly always tissue or body cavity parasites of fish.

Order 2. ACTINOMYXIDA Stolç (1899). Spores within trivalved membranous covering, each with three polar capsules and one to many sporoplasms; mostly parasites of annelids and the related marine burrowing worms known as sipunculids.

Order 3. MICROSPORIDA Balbiani (1883). Very minute intracellular parasites, producing extremely small spores contained within a one-piece envelope, and provided with one or two polar filaments; wide host range, but most often parasites of arthropods and fish.

Subclass 3. SARCOSPORIDIA Balbiani (1882). Parasites producing large tubules (Miescher's tubes) filled with vast numbers of minute, crescentic spores. Striated muscle attacked; occur in a wide range of vertebrates. (There is now some reason to think that these organisms may not be Protozoa, but perhaps fungi.)

Order 1. SARCOSPORIDA Balbiani (1882). With characteristics of subclass.

Subclass 4. HAPLOSPORIDIA Caullery and Mesnil (1899). Sporozoa producing spores without polar capsules; spores germinate into amoebulae, which undergo nuclear fission during growth, becoming multinucleate. Parasitize a great variety of hosts, but especially annelids and fish.

REFERENCES

Bütschli, O. 1880–2. Sporozoa. In *Bronn's Klassen und Ordnungen des Tierreichs*. Bd. 1.

Calkins, G. N. 1901. The Sporozoa. In *The Protozoa*. London: Macmillan.

——— 1933. Special morphology and taxonomy of the Sporozoa. In *The Biology of the Protozoa*. Philadelphia: Lea & Febiger.

Cole, F. J. 1926. *The History of Protozoology*. London: University of London Press.

Doflein, F. and Reichenow, E. 1953. Sporozoa und Ciliophora. In *Lehrbuch der Protozoenkunde*, Teil 2, Hälfte 2. Jena: Verlag Gustav Fischer.

Grassé, P.-P. 1953. *Traité de Zoologie*. Tome 1, Fasc. 2. Paris: Masson.

Hartmann, M. 1923–5. Sporozoa. In Kükenthal-Krumbach, *Handb. d. Zoologie*, Bd. 1.

Hartog, M. 1906. Sporozoa. In *Cambridge Natural History*, vol. 1. London: Macmillan.

Hyman, L. M. 1940. *The Invertebrates: Protozoa Through Ctenophora.* New York: McGraw-Hill.

Labbé, A. 1899. Sporozoa. In Bütschli's *Das Tierreich.* Berlin: Verlag R. Friedländer.

Minchin, E. A. 1912. *An Introduction to the Study of the Protozoa, with Special Reference to the Parasitic Forms.* London: Arnold.

Naville, A. 1931. Les Sporozoaires. *Mem. Soc. Phys. et Hist. Nat., Genève*, 41: 1–223.

Pearse, A. S. (ed). 1936. *Zoological Names: A List of Phyla, Classes, and Orders.* A.A.A.S., Duke University Press.

Rudzinska, M. A. and Trager, W. 1957. Intracellular phagotrophy by malaria parasites: an electron microscope study of *Plasmodium lophurae. J. Protozool.*, 4: 190–9.

Thomson, J. G. and Robertson, A. 1929. *Protozoology. A Manual for Medical Men.* New York: Wood.

Thomson, H. M. 1957. *Perezia fumiferanae* Thom, a protozoan parasite of the spruce budworm, *Choristoneura fumiferana* (Clem.). Ph. D. thesis, McGill University, Montreal.

Wasielewski, T. v. 1896. *Sporozoenkunde. Ein Leitfaden für Aerzte, Tierärzte und Zoologen.* pp. 162. Jena: Verlag Gustav Fischer.

Wenyon, C. M. 1926. *Protozoology.* New York: Wood.

18

Parasitism and Symbiosis

"A parasite, for instance, is a shocking and a baneful monster, yet still Nature has infused into his blandishments a not unpolished charm."
Socrates in Plato's *Phaedrus*.

Parasites, like neighbors, are talked about but not easily described. And like neighbors they may be welcome, unwelcome, or ignored by us. We all have our parasites, although we may never have given them a thought, for most of them are either harmless or injurious only under exceptional conditions. Even Robinson Crusoe on his lonely island was not free of them—of this we can be quite certain. Had he been when he landed there, probably he would have soon acquired parasites from the wild animals thereabouts.

Perhaps as good a definition as any of the term "parasite" is to call it an uninvited guest, for the word derives from the two Greek words *para* and *sitos*, meaning literally "eating beside one another." Among the Protozoa, parasitism is extremely common. Hosts include plants and animals of all the phyla. There are parasitic species of Mastigophora, Sarcodina, and Ciliophora, and of course all the members of the class Sporozoa are parasites.

Types of Parasitism

Unlike guests in general it is uncommon for parasites to be enthusiastically welcomed by their hosts. A few associations of host and parasite have become so close that both profit by it, and sometimes neither could exist without the other: these are the "symbionts," and the relationship is called "symbiosis."* But more frequently the parasite is simply an uninvited guest

*Unfortunately the use of the terms "parasite," "symbiont," and "commensal" is surrounded by some confusion. Parasites are sometimes regarded as organisms which necessarily harm their hosts, while "symbiosis" (which, by derivation, means only "life together") may be used for any association of organisms in which no mutual harm is done. Some biologists prefer "mutualism" as an alternative term for symbiosis, especially when both partners benefit. "Commensalism" is used to describe a relation in which neither partner is harmed, and one is benefited; the word is derived from two Latin roots meaning "together at the table," so that its original meaning is strikingly like that of parasite.

for whom the host involuntarily and unconsciously pays the board bill. Such a relationship is known as "commensalism," and only the parasite is benefited. Exceptionally the host finds that the coming of the parasite has cost him his most precious possession, his health, and then the parasite is said to be a "pathogen." When the injury is extreme, the parasite may itself be done out of home and subsistence by the death of its host. Since this eventuality is equally bad for both, most parasitologists think that such a relationship is relatively recent: with the passage of the millennia either a condition of mutual toleration between host and parasite results, or both host and parasite become extinct.

Those who disagree with this view point out that when the free-living ancestors of parasitic species first turned to parasitism, they presumably had not yet evolved a mechanism that could harm their hosts (for example, enzymes, able to dissolve the host's tissues). Or perhaps evolution had not yet resulted in the loss of certain enzymes by the parasite that made it absolutely dependent on its host for nutriment. As Ball remarks (1943), supporting the idea that pathogenicity on the part of a parasite may be the end result of long evolution rather than its beginning: "But in the present stage of our knowledge of the evolutionary past of parasites, there are few parasitologists who would hazard an opinion that parasite A is relatively harmless because, from the standpoint of evolution, it is just embarking on its nefarious career and hasn't learned perhaps how to be a successful burglar or pickpocket; and that parasite B is, on the other hand, relatively harmless, because it is an old hand at the game and has learned that if it doesn't steal too much, the loss will hardly be noticed, while grand larceny or murder may quickly cause the destruction of the malefactor." Ball has put it well. Unfortunately, the problem does not lend itself to solution by experiment. Frequently an unaccustomed host may be successfully infected with a given parasite, but the resulting disease may be mild, moderately severe, or fatal. One cannot predict the outcome.

Of course the host-parasite relationship is always one of degree. Beneficial associations vary from complete interdependence, as in the case of the famous flagellates of wood roaches and termites, to slight or occasional benefit (certain bacterial inhabitants of the colon, for example, supplement their host's vitamin requirements) to a partnership (commensalism) in which there seems to be neither significant benefit nor harm to the host. Similarly, pathogenicity may be exceptional, occasional and usually slight, or almost invariable and severe. Both the inherited genetic constitution of the host and of the parasite are among the variants determining what happens in specific cases.

Sometimes parasites are themselves parasitized; indeed it is said that parasites of parasites of parasites exist. A form which parasitizes a parasite is said to be a "hyperparasite." An example is *Nucleophaga*, said to be a "vegetable organism," about which little is known. *Sphaerita* is a somewhat similar form. Both parasitize certain species of human intestinal amoebae.

Origin of Parasitism

In considering the origin of parasitism, we assume that parasitic species evolved from free-living ancestors. At first thought, the change in habitat seems a long step, and we wonder about the actual nature of the evolutionary process experienced by the prospective parasite (and perhaps the potential host). Presumably it ordinarily involved more than natural selection from a normally variant population of individuals nutritionally and environmentally adaptable enough to reproduce within the host-to-be. We may hazard the guess that perhaps the first protozoan parasites of Metazoa, at any rate, established themselves in the gut. Here both food and living conditions were probably less novel than in a tissue habitat. It seems likely that only the occasional mutant individual had the degree of adaptability requisite for life within some other animal. Perhaps a fortuitous combination of mutations in both the potential parasite and its prospective host made the establishment of the host-parasite relationship possible.

In any case, a very considerable change in diet must have been required of the ancestral parasitic forms. We know from experimental work that mutations often involve such alterations (studies of the fungus *Neurospora* and of certain Protozoa, e.g., Schoenborn, 1954, on the common flagellate *Astasia*). The initial modifications required of the parasite, therefore, were probably biochemical. The profound transformations in morphology in the evolution of many parasitic species, and the development of very complex life cycles, doubtless occurred more gradually. In evaluating the evolutionary significance of such changes, it is well to remember that not all of them are necessarily advantageous to the parasite. Some may have been essentially neutral. Others that had survival value generally cost the parasite something; for example, the remarkably close physiological adaptation exhibited by the human malaria plasmodia to life in man has at the same time made them incapable of survival and multiplication in any other vertebrate host. Experimentally induced mutations, like those occurring naturally, usually result in certain losses to the organism, so that the survival requirements become more rigid.

If mutations occurring in nature at random seem not to account for the remarkable degree of adaptation to their hosts exhibited by many parasites, we must consider two conditions favoring such evolution among the Protozoa. One applies to evolution generally: the total amount of time available for changes in living things certainly exceeds half a billion years—a figure quite beyond comprehension. The other concerns the size of the population from which mutant individuals can be selected by nature. Although the proportion of mutating individuals in any population is always extremely low, in a very large population there will always be at least a few. Among minute and rapidly reproducing organisms such as the Protozoa, the number of individuals of a given species, even during a short time, approaches infinity; thus the number of mutants would, in an absolute sense, be large.

Since encystment is very common in the Protozoa, parasitism may first have arisen as the result of the ingestion of cysts. Parasite cysts are present everywhere in nature, and the prospective host no doubt swallowed many of them daily in food and drink. Although the vegetative forms are ingested even oftener, unlike the cysts, they are usually quickly destroyed by digestive juices. But suppose that an occasional encysted organism, perhaps mistaking its new and temporary environment for the depths of a warm pond, emerged from its cyst. Rarely—very rarely—it may have managed to endure the disillusion of finding itself in a foreign world, and not only survived but reproduced. There have been few laboratory experiments to determine the ability of free-living protozoan species to cope with conditions such as those encountered by our hypothetical pioneer in the parasitic world, but one was reported a few years ago by Robert Hegner. He introduced euglenas and paramecia into the alimentary tract of cockroaches and found some of these Protozoa still viable after 5 and 24 hours respectively. For most of them, however, the experience was fatal. Possibly ectoparasites (of which there are some even among the Protozoa) adopted their mode of existence earlier than those living within their hosts, but they are of course restricted to water-dwellers.*

From parasitism of the alimentary tract to parasitism of the tissues and internal organs may not have been a big jump. Some intestinal Protozoa, such as *Entamoeba histolytica*, may invade the mucosa and submucosa; this particular amoeba also frequently manages to make its way to the liver via the portal circulation, and there it produces dangerous abscesses. Others, like the coccidia, regularly reside in the mucosal lining and are occasionally seen in other tissues, including erythrocytes.

When picked up by blood-sucking invertebrates in the course of a meal, a parasite may conceivably have been adaptable enough, during that vast span of time since evolution began, sometime to survive and multiply in its new environment, and thus a vertebrate-invertebrate host cycle may have originated. More likely, in many cases, a change may have developed in the reverse direction. Insects, antedating in their evolution many terrestrial vertebrates, perhaps passed their own parasitic Protozoa to the insectivorous mammals and birds which ate them. Usually both insect host and parasites would be digested thereafter, but occasionally some of the parasites may have resisted dissolution by enzymes of the alimentary tract and adapted themselves to their new and sheltered world. Of course, no transfer to a second host could occur unless such organisms managed somehow to colonize the blood or tissue fluids, and this must have been a rather difficult evolutionary step. Yet that it occurred is suggested by the life cycles of certain

*It is of interest that ectoparasitic Protozoa were among the first to be observed. Trembley (according to Dujardin, 1841) saw them on *Hydra* in 1744.

trypanosomes (*T. lewisi* of the rat, or *T. cruzi* of man and other mammals). These are regularly transmitted from insect to vertebrate when the latter, irritated by the bite of the insect, licks the point of attack and swallows its small adversary, or perhaps scratches into the tiny wound a little of the insect's feces, in either case acquiring its intestinal parasites. Many insects thus cooperate with their parasites' need of new hosts by defecating on the prospective host as they feed. Trypanosomes and their close relatives the leishmanias could thus have become parasites of vertebrates, at least when the latter were terrestrial in habitat. And as we might expect, many flagellates have never adapted themselves to hosts other than insects, and are still passed from one host to another in ancestral fashion when the cyst stage is ingested. Yet McGhee has recently shown that a flagellate of this group (*Crithidia euryophthalmi*) must also be considered a potential parasite of vertebrates. Of course, pharyngeal secretions of insects transmit many trypanosomes and leishmanias, and doubtless this mechanism represents a more recently evolved pattern of transfer.

Parasitic Flagellates

Among protozoan parasites of the class Mastigophora, the trypanosomes and leishmanias constitute one of the largest and best-known groups. They have no known close free-living relatives, although the crithidial and herpetomonad stages occurring in the invertebrate host (and also in cultures) resemble typical free-living flagellates in appearance. On the whole, the evidence suggests that their parasitic habit originated far back in the remote past.

Although the family Trypanosomatidae, to which both trypanosomes and leishmanias belong (as well as certain closely related parasites of insects and plants), includes several species pathogenic to man or some animals, pathogenicity is the exception rather than the rule. Since apparently the vertebrate rather than the invertebrate host is harmed, we have another reason for thinking that the family originated as a group of parasites largely restricted to insects.

Many species of flagellates have adapted themselves to life in the alimentary tract of vertebrates as well as invertebrates. Among such parasites are those belonging to the genera *Trichomonas*, *Giardia*, and *Chilomastix*. All three of these genera are large, and their hosts include man, numerous other vertebrates, and occasionally invertebrates. Like the flagellates of blood and tissues, most of those living in the alimentary tract of vertebrates cause little harm to their hosts. But here again certain species, such as several of the trichomonads, may do serious injury. *Trichomonas vaginalis*, probably once a migrant from the colon to the human vagina, is a common cause of severe vaginal irritation. *Trichomonas gallinae*, often found in pharyngeal secretions of the pigeon family, is a frequent cause of death in squabs;

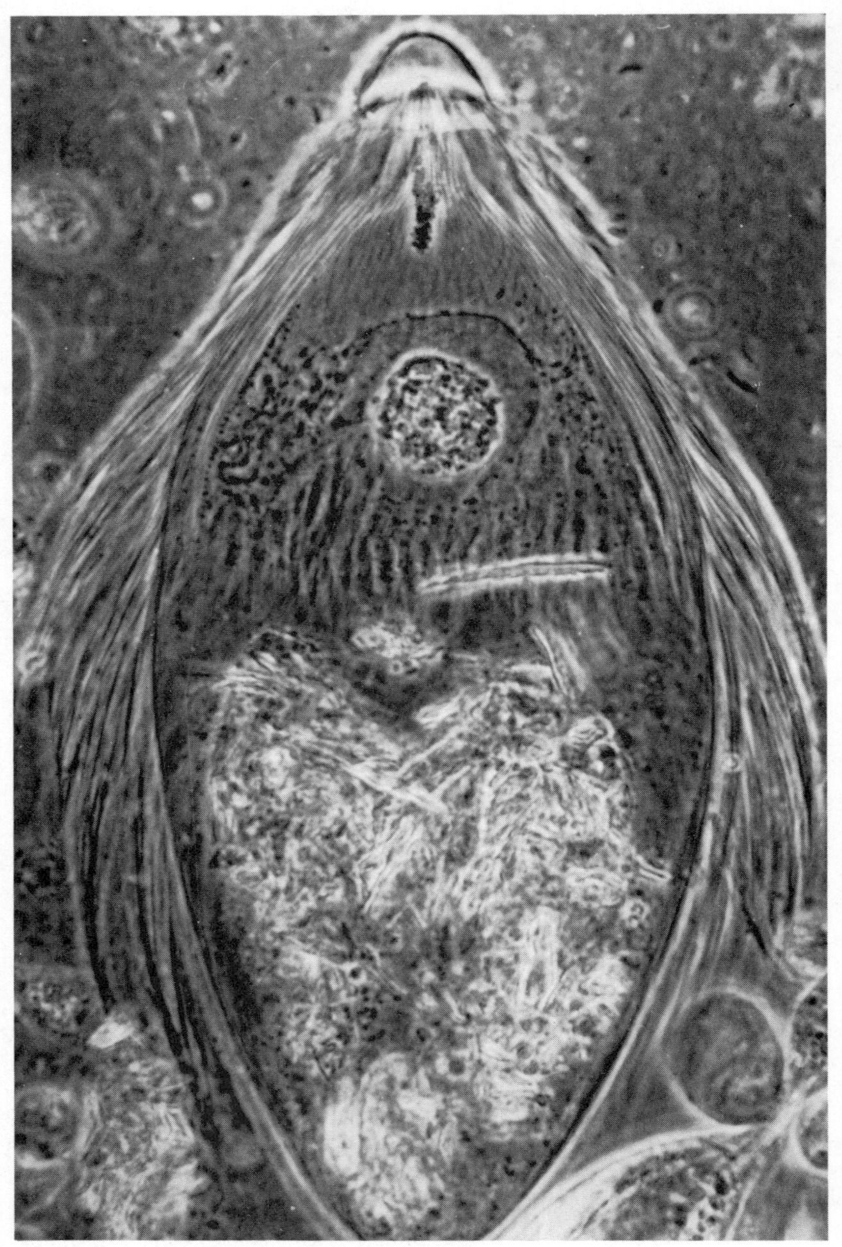

FIG. 18.1. A trophozoite of *Trichonympha* sp. of the woodroach, ×1500. Note outer, unflagellated cap at the anterior end, the flagella extending posteriorly, the nucleus (about a quarter of the way back), and numerous wood particles in the posterior half of the animal. (After Cleveland.)

some strains are highly fatal to adult birds, where the parasites may invade such organs as the liver. Although many parasitic species of Protozoa have become so completely host-adapted that it is difficult to cultivate them *in vitro*, and sometimes even harder to transfer them to other host species, trichomonads are exceptions to the rule: they are usually easily grown on relatively simple host media, and some of them are versatile enough to infect hosts far removed, biologically speaking, from their normal ones. For example, *T. gallinae*, when experimentally introduced into the mouse, is capable of causing a severe infection.

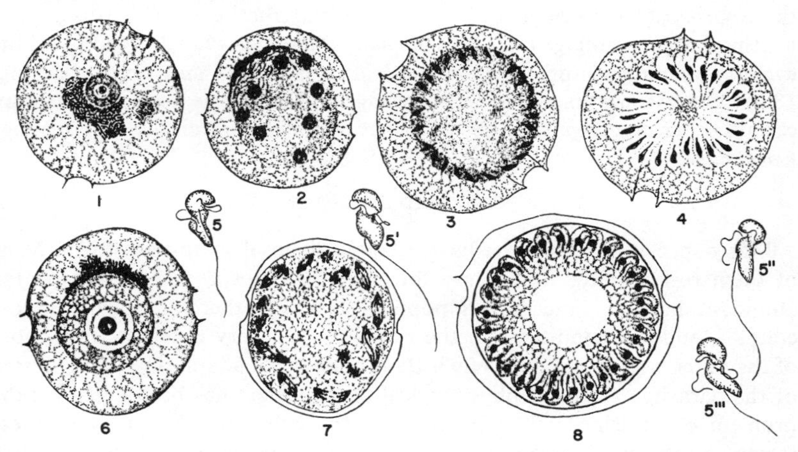

FIG. 18.2. Dinoflagellate *Coccidinium Duboscqui* parasitic on dinoflagellate genera *Glenodinium* and *Peridinium*. *1.* A young parasite in contact with host-cell nucleus, which it gradually destroys. *2, 3, 4.* Stages involving repeated mitoses with eventual formation of flagellated dinospores, believed by Chatton and Biecheler to be microgametes. They have a typical dinoflagellate form, and are shown in *5, 5′, 5″*, and *5‴*. The nuclei from which these dinospores arise are shown arranged in a circle about the host cell periphery *2, 3, 4. 6, 7, 8.* Stages similar to those above, but taking place within a cyst which is formed within the host dinoflagellate after its death and subsequently drops to the pond bottom. From this series of stages macrogametes are believed to be formed, although they have not been actually observed. In *7* the spindles with their four to five V-shaped chromosomes may be seen; this is the haploid number, since only the zygote is diploid among the dinoflagellates. (After Chatton and Biecheler, *C. R. Acad. Sci.* 199: 252.)

Numerous flagellate parasites of importance to man belong to other genera. Some cause diseases in domestic animals, and an interesting example is *Histomonas meleagridis*, the organism giving rise to the serious infection of turkeys known as "blackhead." This species hit on the bright idea of using the egg of a nematode parasite of chickens and turkeys, *Heterakis gallinae*, as a vehicle to carry it from one host to another.

The remarkable flagellates of termites and woodroaches were studied intensively in recent years by L. R. Cleveland at Harvard University. Certain species of these Protozoa have taken over completely the task of digesting the wood on which some castes of their hosts feed, with the result that when the latter are artificially deprived of their parasites they speedily die of starvation (Fig. 18.1). Unlike other animal flagellates, these forms in the woodroach exhibit complex sexual phenomena in their life cycles, and this and other peculiarities convinced Cleveland that their parasitism originated in the very remote past, even before that of the Sporozoa. He goes so far as to suggest that the life cycles of these unique flagellates indicate a common ancestry with the Sporozoa, a group already briefly discussed.

Another interesting suggestion concerning a possible flagellate origin of the Sporozoa has been made by Chatton and Biecheler. They discovered a remarkable dinoflagellate, *Coccidinium Duboscqui*, which exhibits a life cycle combining features seen in both dinoflagellates and Sporozoa (Fig. 18.2). Since some stages of this dinoflagellate, which is itself a parasite of certain other dinoflagellates, resemble coccidia it was given the generic name *Coccidinium*.

Amoeboid Parasites

Many species of Sarcodina have adapted themselves to parasitism. Most of them reside in the alimentary tracts of their hosts, and subsist on the abundant bacterial and fungal population found there, especially in the colon. Doubtless conditions in the colon are not very different from those of the outer world, and relatively little physiological adaptation was required of the parasite. Yet what physiological change there has been enhances the problem of devising artificial media for the cultivation of many of these forms. As far as we know, none of them can do more than survive for a relatively brief interval outside their former host while awaiting a chance to invade a new one; multiplication in pond and soil, where so many of their free-living relatives reside comfortably, is no longer possible.

Host species of parasitic amoebae are, as perhaps one should expect, extremely varied. They range all the way from other Protozoa, such as the common colonial flagellate *Volvox*, to man, who harbors at least six amoebic species belonging to four genera. Most species of parasitic amoebae are not tissue invaders, but a few may be pathogenic. *Entamoeba histolytica*, as we have seen, may invade the intestinal mucosa and submucosa, and sometimes even produce skin ulceration and abscesses in the liver, brain, and lung; amoebic appendicitis also occurs. *Entamoeba invadens*, occurring in the colon of reptiles, is apparently even more injurious to its hosts. Although closely related species of amoebae are also known to parasitize insects (and doubtless other arthropods), they apparently do no harm. Perhaps here the association of host and parasite is a more ancient one.

Parasitic Ciliates

Ciliates, too, are well represented among parasitic Protozoa. One of the larger and more common genera is *Balantidium*, which has species occurring in hosts as far separated in the zoological spectrum as man and the cockroach. Frogs and salamanders are especially well provided with ciliate parasites of this genus. There is little reason to think that most such ciliates harm their hosts, although balantidiasis in man is likely to be serious.

Other species of parasitic ciliates are numerous, though none of them occurs in man. The family Opalinidae, with, it is said, at least 160 species found mostly in frogs and toads (Anura), is wholly parasitic. These forms inhabit the lower portion of the colon, and the incidence of infection in amphibia of this order is extremely high. Like the flagellates of roaches and termites (to which parasites Corliss, 1955, suggests they may be more closely related than to the ciliates *sensu strictu**), the Opalinidae seem to have no free-living relatives. According to Metcalf (1923; 1929; 1940), who studied them thoroughly, they appear to live as commensals rather than as symbionts. Metcalf has constructed a genealogical tree for the Anura based on certain peculiarities of distribution of both host and parasite species, and shows that these peculiarities furnish additional evidence for the existence in earlier geological periods of land bridges between continents, including a "tropical trans-Pacific route between South America and Malaysia or Australasia." Although Metcalf's conclusions have been severely questioned, studies such as these add evidence, if any were needed, that all knowledge is indeed one, and that a specialized field such as protozoology is in fact one aspect of a great whole.

The ciliates found in the stomach of cattle and other ruminants constitute an especially striking group, although the characteristic that first arrests the attention of the zoologist is their remarkably complex structure. Indeed the organization of some of these forms seems more elaborate than that of simpler Metazoa such as *Hydra*. One needs only to compare figures of this possibly degenerate coelenterate with those of any species of *Polyplastron*, one of the most highly evolved genera of this large group of ciliates. Dogiel (1946) suggested that these forms, all of which are members of the family Ophryoscolecidae, became parasites at least as far back as the Eocene period some 60 million years ago. Certain genera, such as *Troglodytella* of apes and chimpanzees, are no longer parasitic in ruminants, but Dogiel remarks that they may well have evolved from ophryoscolecids originally harbored by forest-dwelling hoofed mammals.

Another widespread group of ciliates, although its species apparently have much less in common than do the numerous species found in the ruminants, is the assemblage of forms occurring in the gut of sea urchins. Most of these echinoderms seem to play host to a very varied ciliate fauna, although there is no evidence that they are thereby much inconvenienced. Unfortunately, these infusoria have so far been little studied. In some cases

**Sensu strictu*, Latin for "in the strict sense," a phrase often used in biology, especially by taxonomists.

the association of host and parasite seems not to be very close; for example, one of these species, *Entodiscus borealis*, survives very well for extended periods in sea water, but reproduces only in the body of its host (Powers, 1933).

In some respects, the apostomatous ciliates of Crustacea are among the most extraordinary infusoria. The life histories of a number of them have been worked out by the well-known French protozoologists E. Chatton and A. Lwoff, and are described in a monograph published in 1935. These ciliates have formed the basis for fundamental researches into the factors controlling protozoan development and differentiation, which are, as Lwoff (1950) remarks, ". . . under the triple control of genes, of self-reproducing cytoplasmic particles, and the environment." The life cycles of the apostomatous ciliates, as in the case of termite and roach flagellates, seem to be closely coupled with corresponding changes undergone by the host. For example, excystation may occur only at the molt, doubtless stimulated by the host's hormones, as indicated by recent work of Trager (1957).

Sporozoa

Of all groups of protozoan parasites, the largest and perhaps the most important is that great assembly of diverse forms embraced in the class Sporozoa. These organisms have little in common except that all lack flagella, pseudopodia, and cilia (although some are nevertheless capable of independent movement), and that they typically reproduce by a process of multiple fission often called "sporulation," producing daughter organisms known as "spores" (whence the name of the class). There are no free-living Sporozoa, and sporozoan morphology and life histories furnish no clues to their evolutionary origin, unless we accept Cleveland's recent suggestion of a possible relationship between this class and the flagellates of roaches. Even Cleveland's hypothesis only puts the problem a little further back in time: the divergence occurred at the "beginning of Tertiary, if not much earlier,"* or at least 70 million years ago. Relationships among the various groups within the class Sporozoa are no clearer.

Included among the Sporozoa are some of the most important parasites of man and the lower animals. We need only be reminded of the genus *Plasmodium*, with its 50-odd species producing malaria in man and animals. Human malaria has influenced the history of civilization much more profoundly than have most battles, and even today it affects several hundred

*Cleveland's statement is worth quoting in full: "This ancient primitive roach, *Cryptocercus*, has preserved types of protozoan sexual behavior which are probably older than the Sporozoa. The Mastigophora and Sporozoa probably existed as a single group a long time ago because the sexual cycles in both groups are too closely related at the present time to have been produced by parallel evolution. If these highly developed types of sexual cycles and behavior existed, much as they do today, prior to the divergence of Mastigophora and Sporozoa—and everything indicates that they did—we have very conclusive evidence that sex in Protozoa has existed for a very long time" (1949).

million people each year. The related coccidia affect many kinds of vertebrates and invertebrates, as do the rather less closely related babesias; Protozoa of both these groups cause serious diseases in numerous species of domestic and wild animals. And the Cnidosporidia and Microsporida, largely represented in both invertebrates and vertebrates, often cause considerable mortality in fish and honeybees, to name but two commercially important examples. Still, as with most other kinds of parasitic Protozoa, injury to their hosts by Sporozoa is probably the exception rather than the rule.

Parasites of Protozoa

We have already seen that certain Protozoa have become quite indispensable to their hosts. Similarly, many Protozoa themselves have parasitic organisms which are almost equally indispensable. Among such symbionts are both algae and bacteria. The cytoplasm of numerous Foraminiferida and Radiolarida is filled with algae, without doubt making the problem of their subsistence easier. Among fresh-water Protozoa, *Paramecium bursaria* is a familiar example. Some species of *Euplotes* regularly contain numerous bacteria in their cytoplasm which are presumed to be symbionts. Careful study of such partnerships between Protozoa and other microorganisms would doubtless uncover evidence of relationships as interesting as those already mentioned between protozoan symbionts and their metazoan hosts.

That some Protozoa are parasitized by other Protozoa we have already seen, but the author knows of no case where such an association is beneficial to both. On the contrary it is often harmful, as in the case of the amoeba which parasitizes the colonial flagellate *Volvox*. There are also Sporozoa which have as their hosts other Protozoa, and there is little doubt that many of them are injurious.

Mutual Adaptability

Perhaps one of the most fascinating aspects of parasitism is the remarkable capacity for adaptation that both host and parasite have been compelled to exhibit over the course of time. On the part of the host, physiological changes have occurred with evolution and continue to occur in response to the actual presence of the parasite. These changes both limit the harm the parasite may do, and perhaps govern its reproductive behavior as well as influencing its life cycle. On the part of the parasite, profound modifications in its nutritional pattern have developed (doubtless as the result of mutations), and the requirement that new hosts be found from time to time has resulted in the acquisition of various new abilities, such as increased reproductive capacity, adaptation to vector hosts, and the like. The discovery of the real nature of such host-parasite relationships is still one of the great challenges of the biologist.

REFERENCES

Ball, G. H. 1943. Parasitism and evolution. *Am. Nat.*, 77: 345–64.

Becker, E. R., Schulz, J. A. and Emmerson, M. A. 1930. Experiments on the physiological relationships between the stomach Infusoria of Ruminants, and their hosts, with a bibliography. *Iowa State J. Sci.*, 4: 215–51.

Chatton, E. and Lwoff, A. 1935. Les Ciliés apostomes. 1. Aperçu historique et générale; étude monographique des genres et des espèces. *Arch. zool. exp. et gén.*, 77: 1–453.

Cleveland, L. R. 1949. Hormone-induced sexual cycles of flagellates. I. Gametogenesis, fertilization, and meiosis in *Trichonympha. J. Morphol.*, 85: 197–296.

Corliss, J. O. 1955. The opalinid infusorians: flagellates or ciliates? *J. Protozool.*, 2: 107–14.

Dogiel, V. 1946. Phylogeny of the family Ophryoscolecidae from a palaeontological and parasitological standpoint. *Zoologicheskii Zhurnal*, 25: 395–402.

Dujardin, M. F. 1841. *Histoire Naturelle des Zoophytes. Infusoires.* Paris.

Hegner, R. W. 1924. Parasitism among the Protozoa. *Sci. Monthly*, 19: 1–16.

——— 1926. Homologies and analogies between free-living and parasitic Protozoa. *Am. Nat.*, 60: 516–25.

——— 1929. The viability of paramecia and *Euglena* in the digestive tract of cockroaches. *J. Parasitol.* 15: 272–5.

——— 1938. *Big Fleas Have Little Fleas.* Baltimore: Williams & Wilkins.

Huff, C. G. 1938. Studies on the evolution of some disease-producing organisms. *Quart. Rev. Biol.*, 13: 196–206.

Hungate, R. E. 1950. Mutualisms in Protozoa. Ann. Rev. Microbiol., 4: 53–66.

Lwoff, A. 1950. *Problems of Morphogenesis in Ciliates.* New York: Wiley.

McGhee, R. B. 1957. Infection of chick embryos by Crithidia from a phytophagous Hemipteron. *Science*, 125: 157–8.

Manwell, R. D. 1955. Some evolutionary possibilities in the evolution of the malaria parasites. *Ind. J. Malariology*, 9: 247–53.

Metcalf, M. M. 1923. The opalinid ciliate infusorians. Smithsonian Institution, U.S. National Museum, *Bull. 120.*

——— 1929. The Opalinidae and their significance. *Proc. Nat. Acad. Sci.*, 15: 448–52.

——— 1940. Further studies on the opalinid ciliate infusorians and their hosts. No. 3077, *Proc. U.S. Nat. Museum*, 87: 465–634.

Powers, P. B. A. 1933. Studies on the ciliates from sea-urchins. II. *Entodiscus borealis* (Hentschel) (Protozoa, Ciliata). Behavior and physiology. *Biol. Bull.*, 65: 122–36.

Schoenborn, H. W. 1954. Mutations in *Astasia longa* induced by radiation. *J. Protozool.*, 1: 170–8.

Taliaferro, W. H. 1955. Specificity in the relationship between host and animal parasites. In *Biological Specificity and Growth*, E. Butler (ed.), Princeton: Princeton Univ. Press.

Trager, W. 1957. Excystation of apostome ciliates in relation to molting of their crustacean hosts. *Biol. Bull.*, 112: 132–6.

——— 1960. Intracellular parasitism and symbiosis. In *The Cell*, J. Brachet and A. E. Mirsky (eds.), New York: Academic Press.

19

Flagellates of the Alimentary and Reproductive Tracts

> "All the particles aforesaid lay in a clear transparent medium, wherein I have sometimes also seen animalcules* a-moving very prettily; some of 'em a bit bigger, others a bit less, than a blood globule, and all of one and the same make. Their bodies were somewhat longer than broad, and their belly, which was flat-like, furnisht with sundry little paws, wherewith they made such a stir in the clear medium and among the globules, that you might e'en fancy you saw a pissabed running up against a wall; and albeit they made a quick motion with their paws, yet for all that they made but slow progress . . ."   Leeuwenhoek, Letter 34, November 4, 1681.

In a group as ancient as the Mastigophora and comprising so many species, we can expect many of them to have discovered the great convenience of a parasitic mode of life. Very few species of Metazoa are without their flagellate parasites, and plants also harbor them, though for some unexplained reason much less commonly. Some flagellates even parasitize other Protozoa.

Most protozoan parasites of the vertebrate alimentary tract probably adopted the parasitic mode of life long ago, but few belong to genera with free-living species nor do they resemble any free-living forms. Nevertheless, they must have originated from free-living ancestors. Like most Mastigophora, the intestinal flagellates produce cysts and these are the infective stages. However, to the protective function is often added that of reproduction, for nuclear division takes place within the cysts of some species. The vegetative forms multiply by longitudinal mitotic fission, as do other Mastigophora, but details of the process vary greatly. Occasionally a life cycle may even include multiple fission, as in *Trichomonas fecalis*.

*Dobell considers that the "little animalcules" seen by Leeuwenhoek in his own excrement were undoubtedly the common intestinal flagellate *Giardia lamblia*.

Parasitic species occur in the gut of so many host species that merely to list known forms would require many pages, and we can be sure that many of them have not yet been discovered. We shall consider some of the more important of these species, particularly the half dozen or so found in man and a few that have economic significance because of the diseases they cause in domestic animals. The characteristics of the human species of flagellates are summarized in greater detail in Table 19.1.

Parasites of Man

Trichomonas

Of the human parasites, trichomonads are undoubtedly the most common. They number three species: *Trichomonas tenax* (also known as *T. buccalis*) of the mouth, *T. hominis* of the colon, and *T. vaginalis* of the vagina. It seems entirely possible that the first of the three was also the first to parasitize man, and that invasion of the colon occurred when some of the more adaptable of the organisms were swallowed. *Trichomonas vaginalis*, despite its name, is also often an inhabitant of the male urinogenital tract and prostate, and may well have originated as a wanderer from the colon. Morphological differences separating the three species are so slight that it has often been thought that such classification was unjustified, but Wenrich (1944; 1947), who has probably studied these parasites more thoroughly than anyone else, has found morphological differences which seem constant and definite. There also appear to be physiological differences, for *Trichomonas vaginalis* seems to need a high pH (four to five) for optimum growth.

A rather high proportion of the world's population harbors one or more of these species of parasites, although accurate figures are not available. Most surveys have been done on atypical samples, such as hospital or clinic patients, prisoners, and inmates of mental institutions; some have also been done on college students. But no such institutional population can be regarded as a fair sampling of the population at large. Then, too, the methods and competence of different investigators vary, and the rate of infection differs according to geographic area and economic and cultural levels. Thus the figures on relative incidence of these and other parasites in Table 19.2 must be taken as only rough, if conservative, approximations. These flagellates are common in monkeys, although with a different relative frequency. Hegner (1929) found *Trichomonas hominis* in nearly 100 per cent of the animals he examined, and *T. tenax* was common, but *T. vaginalis* was seen only in a single individual. Infection with *Trichomonas hominis* is also common in chimpanzees. It seems likely that both simian and human hosts have inherited these parasites, so to speak, from their common ancestral stock. There seem to be no significant geographic differences in the incidence of human trichomoniasis of any of the three types; whether there are such differences in monkey and chimpanzee infection is entirely unknown.

TABLE 19.1

CHARACTERISTICS OF FLAGELLATES OF HUMAN ALIMENTARY AND REPRODUCTIVE TRACTS

A. TROPHOZOITES

MORPHOLOGY OF TROPHOZOITES

Species	*Habitat*	*Cyst*	*Dimensions* Length Mean (micra)	Length Range (micra)	Width Mean (micra)	Width Range (micra)	*Form*	*Flagella* Number and Position
Trichomonas tenax[a] (O. F. Mueller, 1773) Dobell, 1939	Mouth	None	6.5	5–12	8	7–10	Variable	4; anterior (& undulating mb.)
Trichomonas hominis[a] (Davaine, 1860) Leuckart, 1879	Intestines (esp. cecum)	None	7.5	5–14	8.5	7–10	Variable	3–5; anterior (& undulating mb. with free flag'm.)
Trichomonas vaginalis[a] Donné, 1837	Vagina & male urethra	None	13	7–23	10	5–15	Variable	4; anterior (& undulating mb.)
Chilomastix mesnili (Wenyon, 1910) Alexieff, 1912	Intestines (esp. cecum)	Yes	14	6–20	6	3–10	Pear-shaped	3; anterior (posterior—in cytostome)
Giardia lamblia Stiles, 1915	Small int. (esp. duod'm.)	Yes	14	9.5–21	7	5–15	like half-pear	8; posterior
Embadomonas intestinalis Wenyon & O'Connor, 1917	Intestines (region?)	Yes	6.5	4–9	3.5	3–4	elongate oval	2; anterior (1 of above directed posteriorly)
Enteromonas hominis da Fonseca, 1915	Intestines (region?)	Yes	7	4–10	4.5	3–6	elongate oval	3; anterior 1; posterior

[a]Wenrich (1944) thinks there are slight, but fairly constant, differences in these three species of *Trichmonas*, but others believe that in the same environment (e.g. in culture) the three are identical. Though Wenrich states that even "cultural experiments show distinctive physiological differences for the three species of *Trichomonas*," elsewhere, he concedes "that in the living condition, especially in culture material, it is much more difficult to make out the distinctions . . ." (Wenrich, 1947).

(TABLE 19.1, cont.)

B. CYSTS[b]

Species	*Form*	*Size* (range & mean)				*Nuclei*
		Length (micra)		*Width* (micra)		
Chilomastix mesnili	oval; button-like anterior knob	7–10	8	4.5–6	5	One (central or eccentric karyosome)
Giardia lamblia	elliptical; double outline	8–12	10.5	7–10	7.4	2–4 (sometimes more); central karysome
Embadomonas intestinalis	lemon-shaped; double outline	4.5–7	5.5	3–4.5	4	One (central karyosome)
Enteromonas hominis	elliptical; double outline	6–8	7	3–4	3.5	2 (1 at each pole); large karyosome

[b]On the whole, the cyst stages are much less variable in shape than are the trophozoites. Diagnosis based on cyst morphology is therefore often easier, particularly as cysts are more likely to be seen. The internal structure of *Chilomastix* and *Giardia* is also complex, and diagnostic.

TABLE 19.2

THE RELATIVE INCIDENCE OF FLAGELLATES OF THE HUMAN ALIMENTARY AND REPRODUCTIVE TRACTS*

Species of Parasite	*Incidence of Infection*
Trichomonas vaginalis	25% (women); 4% (men)
Trichomonas tenax (*T. buccalis*)	10%
Trichomonas hominis	2%
Giardia lamblia	15%
Chilomastix mesnili	4%
or	
Embadomonas (*Retortamonas*) *intestinalis*	rare
Enteromonas hominis	rare

*Compiled and averaged from various sources.

In morphology all three species are so similar as to make differentiation difficult for anyone not a specialist, but since each has its own localization in the host this is of no practical importance. Flagellates of the genus *Trichomonas* possess an undulating membrane edged by a flagellum, as in the trypanosomes. In *Trichomonas hominis* the free end of the flagellum may, according to Wenrich, extend some distance beyond the membrane. Because of the undulating membrane, trichomonads have occasionally been mistaken for trypanosomes by inexperienced investigators. The anterior flagella vary in number from three to five, leading some workers to suggest dividing the genus into three genera: *Tritrichomonas*, *Tetratrichomonas*, and *Pentatrichomonas*. At the base of the flagellum is a complex structure known as the parabasal apparatus, from which extends posteriorly the longitudinal axostyle, said to be fibrillar in nature, and which frequently appears to project as a sort of spike or tail (Fig. 19. 1). The finer makeup of the organism is best seen in stained specimens, though such preparations are often disappointing to the inexperienced. Many of the parasites exhibit distortion of body shape because of failure of the fixing agent to preserve normal body form. The nucleus is, of course, visible only when stained, and lies just back of the parabasal apparatus. Dividing forms exhibit from three to six chromosomes, according to the species.

Living trichomonads are best identified from their peculiar jerky motion and their undulating membrane, which is generally easily seen, especially in the more slowly moving individuals. Since no known species of *Trichomonas* produces cysts, the actively motile forms are of course the only stages to be sought.

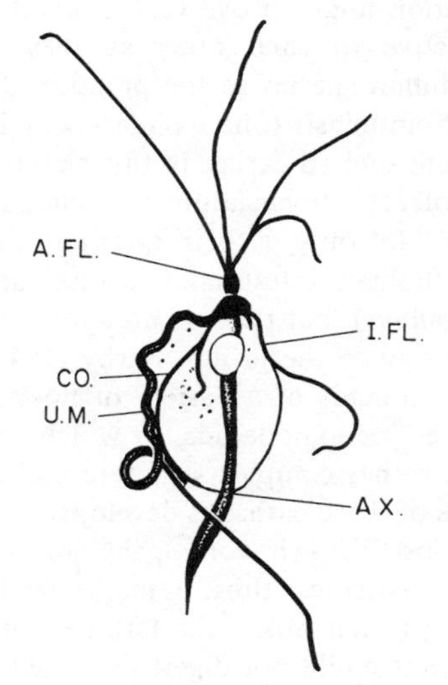

FIG. 19.1. *Trichomonas hominis*. A. FL., anterior flagella. AX., axostyle. CO., costa. I. FL., independent flagellum. U. M., undulating membrane. (After Kirby.)

Whether the trichomonads of man are pathogenic or not has been much debated. At least there seems to be little doubt that *Trichomonas vaginalis* causes an annoying and persistent vaginitis in women, although infection in the male seldom produces symptoms. A variety of drugs and methods of treatment have been employed, among them stovarsol, carbasone, diodoquin, and polyoxyethylene nonyl phenol plus a chelating agent. The results of treatment, however, are often disappointing since apparent cure may be followed by relapse, though in some cases this may actually be reinfection from the husband who is likely to feel no need of treatment. Probably trichomoniasis due to this species of flagellate may be regarded as a venereal infection.

Luckily *Trichomonas hominis* is seldom pathogenic, and any trouble it may cause is limited to a mild diarrhea. How it is transmitted remains a problem, since there is no cyst stage. Apparently the lack of cysts is no handicap, for the vegetative forms tolerate considerable changes in environmental conditions although they do not seem to live long in highly diluted sewage. Hegner (1929a) found these flagellates surviving at least 24 hours in feces-contaminated milk, and he remarked: "These (and other) experiments . . . render practically certain the transmission of human intestinal flagellates by flies and unsanitary conditions to human beings as a result of the contamination of milk and probably other types of food and drink." The rat, experimentally susceptible to infection with *Trichomonas hominis*, as well as to several species of its own, may also serve as a reservoir host (Wenrich and Yanoff, 1927; Hegner and Eskridge, 1937).

It may be fortunate for public health officers as well as for the rest of us that *Trichomonas tenax* is also nonpathogenic, for the problem of control of the infection might prove very difficult as long as men and women remain as attractive to each other as they have always been. The addiction of the human species to the practice of kissing insures a high incidence of oral trichomoniasis (one wonders why it is not much higher than the mere 10 per cent quoted earlier in this chapter).

The correct nomenclature and classification of the trichomonads is still disputed. Not only is there question about the species name (some believe that the human intestinal trichomonad should be known as *Pentatrichomonas hominis*), but there is no agreement about the taxonomic rank which should be given the group. Kirby (1947), who made an extensive study of the trichomonads of a variety of hosts, thought they should constitute an order, the Trichomonadida, in which he also placed a number of flagellates having a rather complex structure and found in termites. From the remote ancestors of these parasites developed, as the result of "a flood tide of evolution" (to use his own words), the great number of species collectively known as hypermastigotes, those remarkable Protozoa which have become indispensable to their hosts, the termites and woodroaches, since without them these insects could not digest their diet of wood.

Giardia

More common as an intestinal parasite than *Trichomonas* is *Giardia lamblia* (still sometimes known as *Lamblia intestinalis*). Indeed some surveys suggest that of all the Protozoa of the human alimentary tract, this species falls behind only *Entamoeba coli*, *E. histolytica*, and *Endolimax nana* in incidence. Certainly it is equally cosmopolitan. Rather wisely, in view of the lack of protozoan competition there, *Giardia lamblia* has selected the duodenum and jejunum as its place of residence.

Giardia lamblia is perhaps the most interesting of all flagellate intestinal parasites in its morphology. The trophozoites, particularly when stained, resemble a caricature of a tiny human face. Lying where the eyes should be are two nuclei. Between them and extending a little below are fibrillar structures like the nose in a cartoon and an obliquely placed, heavily staining

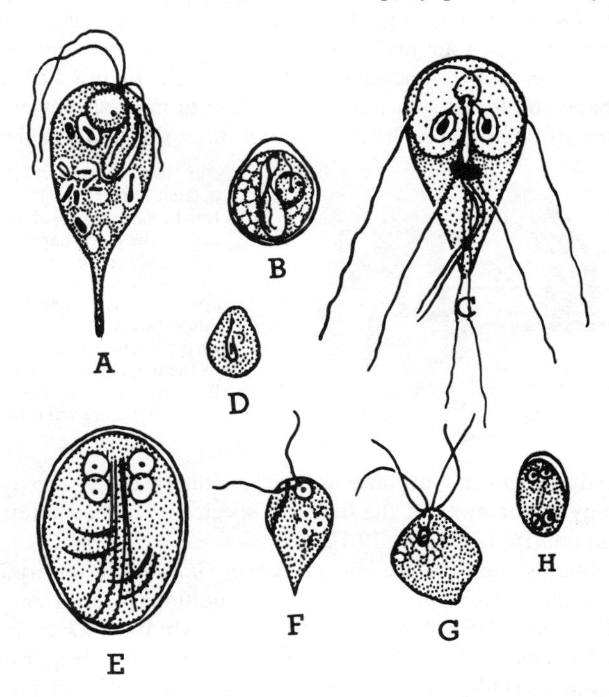

FIG. 19.2. Intestinal flagellates of man. *A. Chilomastix mesnili* (trophozoite). *B. Chilomastix mesnili* (cyst). *C. Giardia lamblia* (trophozoite). *D. Embadomonas intestinalis*, also known as *Retortomonas intestinalis* (cyst). *E. Giardia lamblia* (cyst). *F. Embadomonas intestinalis* (trophozoite). *G. Enteromonas hominis* (trophozoite). *H. E. hominis* (cyst). (After Hegner, Cort, and Root.)

structure looking somewhat like a contorted mouth. There are even four pairs of posteriorly directed flagella to answer for whiskers (see Fig. 19.2). To the zoologist however, *Giardia* is particularly interesting because it displays an almost perfect bilateral symmetry rare among the Protozoa.

In shape *Giardia* resembles a pear cut in half, with the core and adjacent pulp removed. This cup-shaped depression seems to function as a sucking disc by which the organism adheres to the intestinal mucosa. It is said that occasionally the parasites are so numerous that they cover the inner surface of the intestine almost as thickly as the pile on a carpet. The gastrointestinal symptoms frequently seen in giardial infection may result from local irritation or interference by the parasites with the proper absorption of digested food, since *Giardia* seems not to be a tissue invader or to cause demonstrable lesions. Yet there is a record of a laboratory infection in which moderately severe gastrointestinal symptoms resulted (Tsuchiya and Andrews, 1930).

Unlike the trichomonads, all known species of *Giardia* produce cysts. Those of *Giardia lamblia* (Fig. 19.3) are oval in shape, relatively large, and possess from two to four nuclei, although supernumerary forms may have up to 16. A normal four-nucleated cyst is shown in Figure 19.2, *E*. Except for the shape and the greater number of nuclei in most specimens, as well as the absence of free flagella, the cysts do not differ greatly from the vegetative forms, and recognition of both stages is usually easy. The comparative morphology of the cysts of the different species of human intestinal flagellates is summarized in Table 19.1B.

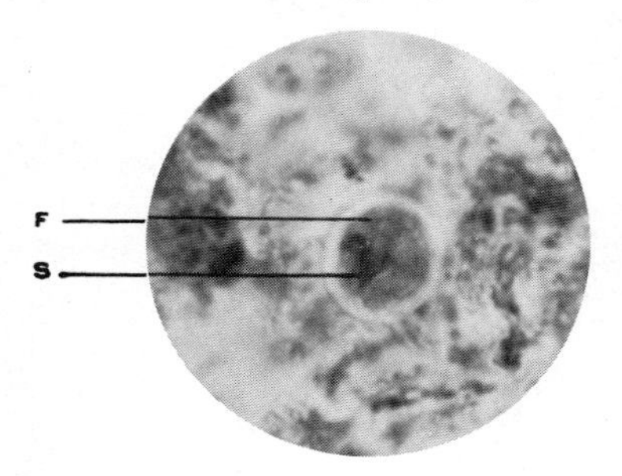

FIG. 19.3. Cyst of *Giardia lamblia* in a fecal smear, ×1180. The characteristic features, visible in the figure, are (1) the oval shape, (2) clear space surrounding the organism within the cyst wall, and (3) the "shield" (S), appearing as a dark V-shaped body. Also shown, though faintly, is a part of the fibrillar network (F), which is, in the trophozoite, connected with the flagella. Not visible are the nuclei, of which there are from two to four.

What circumstances induce encystation in *Giardia* are uncertain. To say, as some authors do, that "unfavorable conditions" result in encystment helps little. Filice (1952) in a comprehensive study of *Giardia duodenalis* in the rat concluded that a lowered pH (about 6.7) was responsible for the cyst formation usually occurring in the caecum and studied the process in detail. Apparently dilution of the feces and a short exposure to normal body temperature are sufficient to induce excystation. Probably it is safe to assume that the same factors will induce like effects on the human species of *Giardia*. Of all the human intestinal flagellates, *Giardia lamblia* is probably the easiest to diagnose. The odd pear-shaped trophozoites and the oval cysts are unlike those of any other species.

Chilomastix

Chilomastix mesnili infections are relatively rare and of no known clinical significance. Like most Protozoa of the alimentary tract, it resides in the colon.

In shape it resembles a pear and is usually smaller than *Giardia*. The posterior end of the body projects as a somewhat asymmetrical spinous process. This and the three anteriorly directed flagella give the animal a superficial resemblance to *Trichomonas*, and the French zoologist Davaine, who seems to have been the first to see both *Trichomonas hominis* and *Chilomastix mesnili*, just about a century ago, regarded the two species as slightly differing varieties of a protozoan he called *Cercomonas hominis*. However, *Chilomastix* moves in a less wobbly fashion and lacks the relatively conspicuous undulating membrane of *Trichomonas*. It also possesses a large oval mouth or cytostome, with a small fourth flagellum within (Fig. 19.2, *A*). The cysts of *Chilomastix mesnili* are lemon-shaped, with a buttonlike knob at the anterior end, and contain an easily visible nucleus and cytostome (Fig. 19.2, *B*). Except for their relative minuteness, recognition is not difficult.

Embadomonas (*Retortomonas*)

Embadomonas intestinalis (also known as *Retortomonas intestinalis*) is a very rare human parasite. Its habitat in the body is not entirely certain, but is believed to be the colon where it apparently does no harm. A minute organism rather variable in shape, it has two anterior flagella, a nucleus lying close to their point of origin, and a "well-developed cytostome" (Brumpt) (Fig. 19.2, *F*). Cysts resemble those of *Chilomastix*, but are only about half as large and usually appear to have a simpler internal structure (Fig. 19.2, *D*). Chandler (1955) suggests that this parasite may in fact be an insect boarder gone astray, perhaps because of the accidental swallowing of its normal host by man. The type species of the genus is indeed a parasite of crane flies. Unfortunately *Embadomonas intestinalis* has not been much studied because of its rarity and harmlessness.

Enteromonas

Like *Chilomastix mesnili*, *Enteromonas hominis* has three anteriorly directed flagella, and by some accounts a fourth in the opposite direction, but the latter is attached for most of its length to the edge of the body, with only a short free end (Fig. 19.2, *G*). In size the human species of *Enteromonas* and *Embadomonas* are much the same. The cysts of *Enteromonas* are elliptical, and contain from two to four nuclei, equally divided between the poles (Fig. 19.2, *H*). Although even less is known about *Enteromonas hominis* than about *Embadomonas intestinalis*, there is no reason to think that it is pathogenic; probably it also resides in the colon.

Doubtful Species

Other species of flagellates have been described from the human intestinal tract, but their validity is doubtful. Probably some represent coprozoic forms (normally free-living species which find diluted feces a suitable habitat) derived from cysts ingested with food or water. Others probably belong to one or another of the intestinal species already listed, and were described in the mistaken belief that they were new. Or they may just possibly have been flagellates normally parasitic in the lower animals, and adaptable enough to live occasionally in man.

Parasites of Animals

Trichomonas

Trichomonas is a large genus with a host range including both vertebrates and invertebrates. Only one free-living species has been described, and there is some doubt whether this was in fact a trichomonad.

One of the most important of the trichomonads occurring in the lower animals is *Trichomonas foetus** of cattle. This is in many respects similar to *Trichomonas vaginalis*. The parasite apparently persists in the preputial cavity of the bull for the life of the animal and does little harm. In infected cows, however, conception is likely to be delayed and abortion commonly occurs. Thus, as in man, the male may act as a symptom-free carrier, while the female is seriously affected. Trichomoniasis in cattle occurs throughout the United States, and is indeed probably world-wide; there is no doubt that it is a disease of considerable economic importance.

Another species of *Trichomonas* which may be highly pathogenic is *T. gallinae*, a common parasite of doves and pigeons and probably the same as *T. diversa*, a cause of disease in turkeys (Stabler, 1938). Inhabiting the pharynx and esophagus of its hosts, it is probably most often transferred to young birds by the parents in the course of feeding. Contaminated water, however, may be a vehicle of infection. The disease is frequently fatal to squabs, and some strains of the parasite are highly virulent even for adult birds. Mourning doves are often killed by it, and epizootics resulting in the deaths of thousands of these birds have occurred in Alabama. Stabler (1954) made the interesting suggestion that the final extinction of the passenger pigeon may have been due to *Trichomonas gallinae*, although of course it can never be proved.

Almost as common as *Trichomonas gallinae* in pigeons is *Trichomonas augusta*, occurring in frogs, toads, and salamanders; its incidence in frogs is said to be at least 90 per cent. Because the infection rate is so high it makes excellent material for laboratory study, but it has no known effects on its hosts.

Another species able to produce infection in frogs is *Trichomonas* (or *Tritrichomonas*) *fecalis* isolated by Cleveland (1928) from diluted human feces and extensively studied by him. That this is a valid species seems quite certain, but that it is in fact a human parasite is at least doubtful.

Trichomonas fecalis was, however, of more than ordinary interest for several other reasons. It "grew well in hay infusions" and would "grow

*A species of *Trichomonas* (*T. suis*) also occurs in pigs, and there is some reason to think it may be the same as *T. foetus*.

in practically any fluid which supports bacterial growth," and was thus considerably less fastidious than most other species of *Trichomonas*. Though normally reproducing by binary longitudinal fission as do other flagellates, it also often underwent multiple fission. Cleveland remarked that "Syncytia or plasmodia containing at least one hundred individuals were observed many times," such forms sometimes being even visible with the naked eye. Similar forms have been seen in the case of *Trichomonas augusta*. However multiple fission of this sort must not be confused with that normally occurring in the Sporozoa. It is still essentially binary fission, with division of the nucleus and some other cellular components out of phase with cytoplasmic partition.

Giardia and Chilomastix

Species of *Giardia* have been described from many vertebrates, and one has even been described from a parasitic nematode (Wenyon, 1926). The host range includes frogs, tadpoles, rabbits, numerous species of rodents, primates other than man, and other vertebrates. As often with parasitic organisms, species of *Giardia* have frequently been named on the basis of an assumed host specificity alone; this however is a notoriously untrustworthy criterion for specific taxonomic recognition.

The morphology of *Giardia* (and morphology should, at least in the present state of knowledge, always be the basis for the creation of species) is known to vary somewhat with ecological factors, such as diet, and Filice (1952) believes that the actual number of species in the genus may be relatively small. Many of the species described in the past he prefers to regard as races; thus he divides *Giardia duodenalis* into ten races, and designates the human parasite as "Race *lamblia*," that of the rabbit "Race *duodenalis*," etc. Since the very concept of species differs with circumstances and usage (and even with the biologist), it seems likely that custom will dictate the use of many species names, even when strict application of logic and taxonomic law indicates otherwise. Furthermore, the problem is in essence genetic, and knowledge of the genetics of most Protozoa, and of almost all the parasitic species, hardly yet exists.

Much the same taxonomic problems exist with *Chilomastix*. The genus is moderately common in vertebrates, but the distinctiveness of many of the species which have been described is still uncertain. Most such species, as in so many other cases, have been based on an assumed host specificity. The host list includes in part guinea pigs, rats, goats, fowls, and even fish.

Histomonas

Among flagellates which cause serious injury to domestic animals, perhaps none is more interesting than *Histomonas meleagridis*, etiological agent of "blackhead" in turkeys (Fig. 19.4). Its special interest arises because of its remarkable life history, which includes both amoeboid and flagellated stages, and its epidemiology. Infection may arise from the ingestion of organisms found in the droppings of diseased birds, possible only for a brief time, since the parasites soon die after leaving the host, or more usually from picking up the infected eggs of the cecal nematode *Heterakis gallinae*, common in chickens. Such eggs are very resistant and may survive in the soil for many months. Thus an infected poultry range may remain dangerous for turkeys for a long time. The parasitized eggs seem in no way injured, and when picked up by healthy fowls a double infection by worm and flagellate follows, but the latter, for the turkeys especially, is much more pathogenic. Chickens usually survive the infection and become carriers, but among young chicks the mortality may be considerable. The parasite lives and multiplies in the caeca and, in turkeys especially, may spread to the liver.

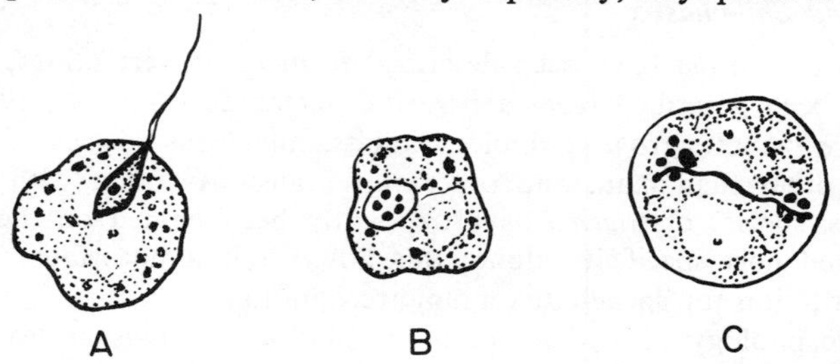

FIG. 19.4. *Histomonas meleagridis*. *A*. Normal form. *B*. Non flagellate or amoeboid form. *C*. Late anaphase of nuclear division. Note the centrodesmus and blepharoplasts, and the absence of flagella. (After Dobell, *Parasitol.*)

There is little local tissue reaction to the parasites in the turkey, which may in part explain the greater injury that results. Although turkeys of all ages are susceptible to infection, younger birds are more often affected, and the mortality among them sometimes approaches 100 per cent. It is clear that without *Histomonas* our Thanksgiving dinners would be considerably less expensive, for no method of treating blackhead is very effective, and the cost of protecting birds from infection is not small. It is never safe to allow them on a range recently used for chickens, in which the incidence of infection is often extremely high, or even to raise the two kinds of fowl in close association.

The amoeboid stages of *Histomonas* are relatively minute (from 8μ to 15μ in diameter) and occur in cecal and hepatic tissues, where they reside between rather than within the cells. They are apparently capable of actual migration from one area to another. The flagellate forms usually possess a single flagellum, but occasionally as many as four are developed. Even these forms, however, are capable of active amoeboid movement and may ingest bacteria, cell fragments, and starch grains. The flagellates tend to be slightly larger than the amoeboid stages, although the smallest of the former may be as little as 4μ in diameter. The nuclear changes in division are strikingly similar to those seen in *Trichomonas* (Tyzzer, 1919; Bishop, 1938). Sometimes nuclear division occurs without cytoplasmic division, and in this binucleate and even quadrinucleate individuals are formed. It appears that the old flagellum is discarded and each daughter must provide itself with a new one. There are from six to eight chromosomes.

The diphasic life cycle of *Histomonas* is reminiscent of such coprozoic forms as *Tetramitus rostratus*, a species which has been studied intensively by Bunting (1926) and Bunting and Wenrich (1929). Here also there are both amoeboid and flagellated stages, as well as a cyst stage lacking in both *Histomonas* and *Trichomonas*. This species, which finds conditions in diluted feces much to its liking, could doubtless adapt itself to existence within the colon. With the development of an alternative method of transmission to new hosts, such as the acquisition of a vector (or its equivalent, the egg of an intestinal nematode like *Heterakis*) the cyst stage might be dropped from its life cycle. Thus it may be that parasitic species such as *Histomonas meleagridis* evolved from forms like *Tetramitus*, although the resemblance noted above in the patterns of nuclear division remains unexplained.

Hexamita

Another important turkey disease is caused by a flagellate known as *Hexamita meleagridis*. It is a parasite of rather remarkable appearance somewhat resembling *Giardia*, and it is bilaterally symmetrical with two anteriorly placed nuclei. The organism causes a severe enteritis, particularly in poults, in which it resides in the intestine, less commonly in the caeca, and more rarely reaches the abdominal cavity and liver. About one-third of turkeys are infected, and outbreaks of the disease have sometimes resulted in a mortality among poults of from 70 to 80 per cent. Other species of fowl do not seem susceptible. Apparently surviving poults become carriers of the parasite, and transmission occurs through contaminated food and water. Presumably this takes place when trophozoites are ingested, since cysts have not been seen (although they have been described for *Hexamita intestinalis* of the frog).

As with most of the genera already discussed, *Hexamita* occurs in a wide variety of hosts. McNeil, Hinshaw and Kofoid (1941) prepared a host-parasite check list, hosts including such diverse types as leeches, trematodes, insects (both larvae and adults), oysters, fish, amphibia, tortoises, birds, and mammals (notably rodents and monkeys). The number of species of *Hexamita* is uncertain, since little is known about the degree of host specificity exhibited by various strains of the parasite, or other phases of the biology of *Hexamita*.

Other Flagellate Parasites of the Lower Animals

The parasitic flagellates so far considered are, of course, only a minute sampling of the many species of such organisms. Such forms swarm in the gut of most insects, for example, although little definite knowledge exists about most of them. Flagellate species chosen for special mention directly affect our well-being or our pocket books, and often both since they add to the cost of food and leather. But in considering parasitic organisms we should beware of thinking that parasitism and disease are always to be equated; most species of parasites appear to cause no harm to their hosts at least most of the time. On the other hand, parasitic disease is extremely important in maintaining the balance in nature, and thus disease in any living thing and the organism causing it is significant to man. Paraphrasing John Donne, "no living thing is an island" and the balance of nature is infinitely delicate.

REFERENCES

Becker, E. R. 1943. Protozoa. In *Diseases of Poultry*. Biester & Devries.

Bishop, A. 1935. Observations upon a *Trichomonas* from pond water. *Parasitology*, 27: 246–56.

——— 1936. Further observations on a *Trichomonas* from pond water. *Parasitology*, 28: 443–5.

——— 1938. *Histomonas meleagridis* in domestic fowls (*Gallus gallus*), cultivation and experimental infection. *Parasitology*, 30: 181–94.

Bunting, M. 1926. Studies on the life-cycle of *Tetramitus rostratus* Perty. *J. Morphol. and Physiol.*, 42: 23–71.

——— and Wenrich, D. H. 1929. Binary fission in the amoeboid and flagellate phases of *Tetramitus rostratus* (Protozoa). *J. Morphol. and Physiol.*, 47: 37–87.

Chandler, A. C. 1955. *Introduction to Parasitology*, 9th ed. New York: Wiley.

Cleveland, L. R. 1928. *Tritrichomonas fecalis*, nov. sp. of man; its ability to grow and multiply indefinitely in faeces diluted with tap water and in frogs and tadpoles. *Am. J. Hyg.*, 8: 232–78.

Craig, C. F. and Faust, E. C. 1951. *Clinical Parasitology*. Philadelphia: Lea & Febiger.

Filice, F. P. 1952. Studies on the cytology and life-history of a *Giardia* from the laboratory rat. *Univ. Calif. Pub. Zool.*, 57: 53–146.

Hegner, R. 1928. Experimental studies on the viability and transmission of *Trichomonas hominis*. *Am. J. Hyg.*, 8: 16–34.

——— 1929a. The viability of trichomonad flagellates in milk. *J. Parasitol.*, 16: 47–8.

——— 1929b. The Protozoa of wild monkeys. *Science*, 70: 539–40.

——— and Eskridge, L. 1937. Persistence in rats of human intestinal trichomonad flagellates. *Am. J. Hyg.*, 26: 124–6.

Hinshaw, W. R. and McNeil, E. 1941. Carriers of *Hexamita meleagridis*. *Am. J. Vet. Res.*, 2: 452–8.

Jira. J. 1958. Zur Kenntnis der männlichen Trichomoniase. *Zentralbl. Bakt., Parasit., Infectionsk. u. Hyg.*, 1 Orig., 172: 310–29.

Kirby, H. 1945. The structure of the common intestinal trichomonad of man. *J. Parasitol.*, 31: 163–75.

——— 1947. Flagellate and host relationships of the trichomonad flagellates. *J. Parasitol.*, 33: 214–28.

McNeil, E., Hinshaw, W. R., and Kofoid, C. A. 1941. *Hexamita meleagridis*, sp. nov. from the turkey. *Am. J. Hyg.*, 34 (Sect. C.): 71–82.

Morgan, B. B. 1944. Host list of the genus *Trichomonas* (Protozoa: Flagellata). *Trans. Wis. Acad. Sci.*, 35: 235–45.

——— and Hawking, P. A. 1948. *Veterinary Protozoology*. Minneapolis: Burgess.

Rees, C. W. 1938. Observations on bovine venereal trichomoniasis. *Vet. Med.*, 33: 1–16.

Stabler, R. M. 1938. The similarity between the flagellate of turkey trichomoniasis and *T. columbae* in the pigeon. *J. A. V. M. A.*, 93: 33–4.

——— 1954. Parasitological reviews: *Trichomonas gallinae:* a review. *Exp. Parasitol.*, 3: 368–402.

Trussell, R. E. (introduction by E. D. Plass). 1947. *Trichomonas Vaginalis and Trichomoniasis*. Springfield: Charles C Thomas.

Tsuchiya, H. and Andrews, J. 1930. A report on a case of giardiasis. *Am. J. Hyg.*, 12: 297–98.

Tyzzer, E. E. 1919. Developmental phases of the protozoon of "blackhead" in turkeys. *J. Med. Res.*, 40: 1.

Wenrich, D. H. 1944. The comparative morphology of the trichomonad flagellates of man. *Am. J. Trop. Med.*, 24: 39–51.

——— 1947. The species of *Trichomonas* in man. *J. Parasitol.*, 33: 177–88.

——— and Yanoff, J. 1927. Results of feeding active trichomonad flagellates to rats. *Am. J. Hyg.*, 7: 119–24.

Wenyon, C. M. 1926. *Protozoology*, vol. 1. New York: Wood.

20

Trypanosomes and Their Relatives

> "His end was to be overtaken by the sleeping sickness (*'illat an-nawm*) which is a disease that frequently befalls the inhabitants of those countries, and especially their chieftains. Sleep overtakes one of them in such a manner that it is hardly possible to awake him. He (the King) remained in this condition during two years until he died in the year 775 A.H. (A.D. 1373–74)." A fourteenth-century Arab writer quoted in H. H. Scott, *A History of Tropical Medicine*, 1939.

Of the infinitely complex factors determining human history, one of the most crucial is physical health and stamina. Though seldom explored in history textbooks, freedom from debilitating diseases can make all the difference in the rise and fall of nations, in peace as well as war. The growth of any civilization, indigenous or imported, has been obstructed in many parts of Africa by such diseases as African sleeping sickness. On the other hand, freedom from its ravages has contributed greatly to the strength of the nations of the New World. How can we account for our good fortune?

Perhaps we should thank that Providence which ordained the Ice Ages, for the widespread colder climate may have exterminated the tsetse fly in North America, where fossil remains show it was at one time abundant. Many fossils have been collected from the famous Later Oligocene insect-bearing shales of Florissant in Colorado, deposits at least 40 million years old, dating from a period when climatic conditions were much milder than they are now. Since tsetse flies transmit the trypanosomes of sleeping sickness, the disease cannot exist in their absence. To be sure, the flies may occur in an area, and not the disease, but in such a case the arrival of infected individuals would probably result in its speedy establishment—just as malaria took root in North America, carried by African Negroes brought in by the slave traders. *Anopheles* mosquitoes were already abundant and the stage was set for the villain of the piece, the malaria *Plasmodium*. Despite the best efforts of the slavetraders to bring in healthy Negroes (for the early signs of sleeping sickness were even then well known, and it was recognized that Negroes afflicted with it would eventually die, whereas those with malaria usually recovered), mistakes were often made. Had the tsetse fly been indigenous in this country then it is hard to imagine what tragedy might have resulted.

FAMILY TRYPANOSOMATIDAE

The family Trypanosomatidae is a moderately large one with many species but only six genera. Of these genera, the genus *Trypanosoma* is by far the most important. Relationships of the mammalian species (some of which cause serious illness in man and animals) are shown graphically in Fig. 20.1, and are further considered in a paper by * Hoare (1964), who suggests the creation of a number of subgenera. Hosts also include many of the lower vertebrates, and numerous invertebrates as well. Of the various genera (Fig. 20.2), *Trypanosoma* must be considered the most highly evolved, although this is reflected more in the complexity of the life history of many species than in their morphology.

Next to the trypanosomes in importance as agents of human disease are members of the genus *Leishmania.* Man is susceptible to infection by three species of the genus, and all of them are pathogenic. *Leishmania* may also infect other mammals, and the disease in dogs is sometimes a source of human infection, since the insect vector probably feeds at times on dogs.

In the opinion of Grassé (1952) the two genera *Leptomonas* and *Herpetomonas* should be combined as *Leptomonas.* But little is known about the life histories of the great majority of species or about their host-parasite relationships, and the problem of synonomy of the two genera cannot be settled. The difference between the two genera, if there is one, lies in the life cycles. *Crithidia* forms are said to appear in the life cycle of herpetomonads but not in that of leptomonads (Fig. 20.2). Most of the species are parasitic in insects, but they have also been reported from nematodes, rotifers, and molluscs. Apparently few (if any) of them are pathogenic.

Crithidia is a genus of invertebrate parasites (the hosts are chiefly insects) about which little more is known than of the two preceding genera. One of the best known species is *Crithidia gerridis* of the common water bug (*Gerris fossarum*). Until the life cycles of more of the species are worked out, even the validity of the genus will remain uncertain, as Grassé remarks. The chart shows that members of this genus are separated from others in the family by the occurrence in their life histories of crithidial, leptomonad, and leishmanial stages.

The phytomonads are plant parasites, probably best known from the euphorbias, in the acrid, milky juice, or latex, in which they live (Fig. 20.3), but they have also been found in numerous other kinds of plants, including a number of the Compositae (e.g., dandelions). Except in their life history,

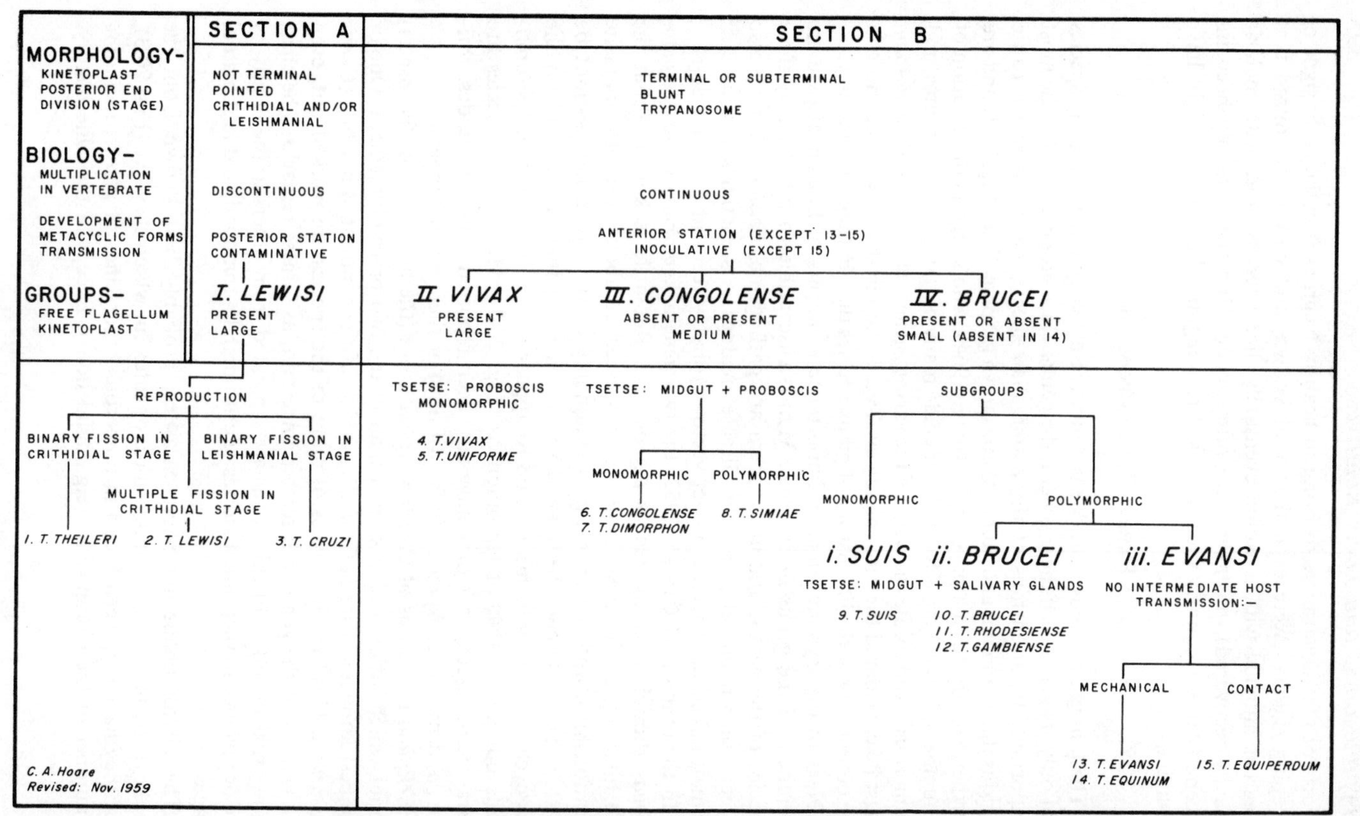

FIG. 20.1. Classification of important mammalian trypanosomes, (After Hoare, 1959.)

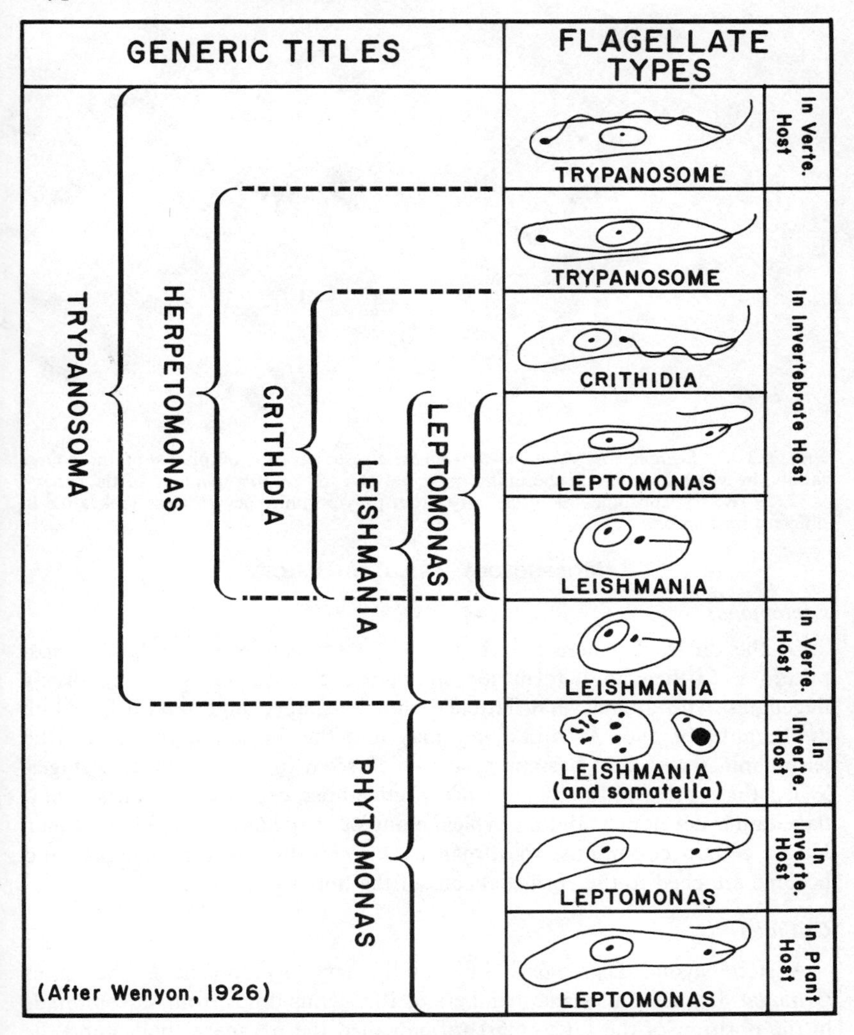

FIG. 20.2. Genera of Trypanosomatidae. (Modified from Wenyon, *Protozoology*.)

which involves a plant and an insect host, these parasites are essentially like leptomonads and are often placed with them. They have a very cosmopolitan distribution. Perhaps because of a widely held belief that the phytomonads do not harm their hosts, they have received little careful study. Yet these parasites may be pathogenic to the plants harboring them; *Leptomonas leptovasorum* often kills coffee trees in Dutch Guiana, and some species parasitizing euphorbias are harmful.

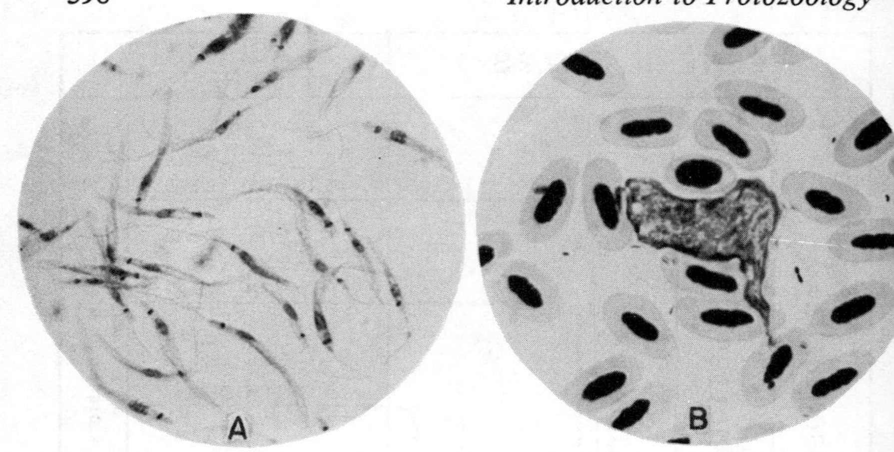

FIG. 20.3. *A. Leptomonas* (*Phytomonas*) *elmassiani*, a parasite of plants of the spurge family, in which it occurs in the milky juice. ×1232. *B. Trypanosoma* sp. of the canary, ×1232. The trypanosomes of birds, very often polymorphic, nevertheless look alike in different host species.

MORPHOLOGY AND LIFE HISTORY

Leptomonas

Of the different genera in the group, *Leptomonas* is probably the most primitive. In their adult form, leptomonads are much like some free-living flagellates, with a somewhat flattened, spindle-shaped body and a flagellum at the anterior end. At times they may lose the flagellum, round up (the leishmania stage), and become cysts, which seem to be the infective stages. When these are swallowed by a prospective host, excystment occurs, a new flagellum is developed, and the typical elongate shape of the adult is assumed. Under certain conditions, the organism may again lose the flagellum and become attached to the epithelial cells of the host's gut.

Crithidia

A step beyond *Leptomonas* in evolutionary development is the genus *Crithidia*. In the adult stage members of this genus differ from leptomonads in the position of the kinetoplast (often called the parabasal body) and the point of flagellar origin, both of which have moved posteriorly until they are just in front of the nucleus. The flagellum constitutes the edge of a short undulating membrane which terminates at the anterior end of the organism. As with the leptomonads there is a leishmaniaform stage within a minute spherical cyst, and ingestion of this by a susceptible insect may result in new infection. The parasite then emerges, develops a new flagellum (passing through a leptomonas-like stage as it does so) and develops finally into a typical crithidia. Attached forms with flagella also occur, as in the genus *Leptomonas*.

Herpetomonas

The life cycles of the herpetomonads, also little-known insect parasites, appear to include all the stages characterizing the crithidias, plus that of the trypanosome, achieved simply by the further migration of the kinetoplast and flagellar insertion to a position posterior to the nucleus. The flagellum, however, extends through the body to its point of emergence at the anterior end, rather than constituting the edge of an undulating membrane. Furthermore, whereas the true trypanosomes have both an invertebrate and a vertebrate host, the herpetomonads are restricted to the former, infection developing when the encysted stages are ingested. The herpetomonads seem not to be pathogenic to their insect host.

Trypanosoma

Trypanosomes are minute blood-inhabiting Protozoa, thin and rather leaflike in cross section, usually somewhat crescent-shaped, with a central nucleus and a rather heavily staining posteriorly placed kinetoplast. In stained specimens a single flagellum appears to originate from this body, but in suitable preparations a minute body (the blepharoplast) may also be seen adjacent to the kinetoplast, and a short fibril connecting the former to the base of the flagellum. The flagellum itself continues laterally, enclosed by a fold of cuticle, until it projects a short distance at the anterior end. The flagellum and cuticular fold constitute an undulating membrane, a very effective propelling device in the blood plasma or tissue fluid in which the parasite lives. The morphology, however, varies according to the stage in the life cycle and whether the host is vertebrate or invertebrate. Differences in

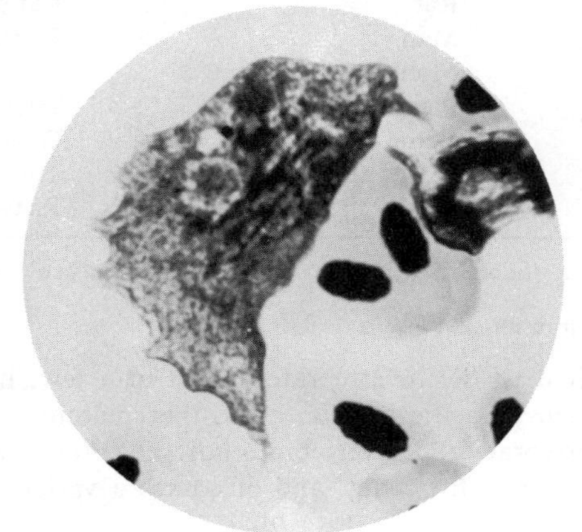

FIG. 20.4. *Trypanosoma rotatorium* of the frog, ×1078. Note its relatively enormous size, as compared with trypanosomes of mammals. It is also much more sluggish. The kinetoplast is visible as a small black dot not far from the left upper edge, about one-third of the way back from the anterior end. The undulating membrane (left edge) and short terminal flagellum are also easily seen.

the morphology of trypanosomes seem to depend also on the class of host parasitized. Thus, mammalian trypanosomes tend to be relatively small and without the longitudinal striations which often characterize the larger forms occurring in birds (Fig. 20.3*B*). But these are not typical; trypanosomes are notoriously variable in morphology, even within a single host, and are accordingly often said to be "polymorphic." Figure 20.4 shows a species of trypanosome common in frogs.

Since trypanosomes spend none of their time outside a host they do not go through a cyst stage, but at some time in the life cycle they round up and lose their undulating membrane and flagellum, becoming at the same time intracellular. This is known as the leishmania stage (Fig. 20.5), and in most

TABLE 20.1

Species	*Vertebrate Host*	*Invertebrate Host*
T. congolense	Horses, cattle, sheep	Tsetse flies
T. cruzi	Man, armadillos, rodents, opossums, and perhaps pigs	Triatomid bugs
T. dimorphon	Horses	Tsetse flies
T. equiperdum	Horses	None*
T. evansi	Horses, mules, donkeys, cattle, camels, elephants, dogs	Tabanids and other bloodsucking flies†
T. gambiense	Man	Tsetse flies
T. lewisi	Rat	Rat fleas
T. rhodesiense	Man	Tsetse flies
T. simiae	Monkeys	Tsetse flies
T. suis	Pigs	Tsetse flies
T. theileri	Cattle	Tabanid flies
T. vivax	Cattle, sheep, goats, horses	Tsetse flies
T. uniforme	Cattle, sheep, goats	Tsetse flies

*Though there was doubtless an invertebrate host once it has now been lost, and transmission is by coitus.

†Not true hosts; transmission mechanical.

species it is limited to the invertebrate host. In a few, however, notably *Trypanosoma cruzi*, parasitic in man and other mammals, this stage may occur in both vertebrate and invertebrate hosts. Leptomonad and crithidial forms likewise occur in the latter, and of course a variety of intermediate forms as well.

The relationships of the more important species of mammalian trypanosomes, in terms of morphology and life history, are shown in Figure 20.1. Certain of these species are considered in some detail in a later section of this chapter. The majority have their home in Africa, where they constitute a

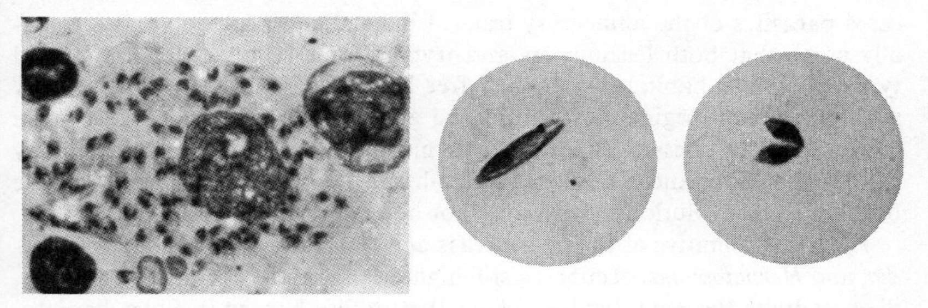

FIG. 20.5. *Left. Leishmania donovani*, in a contact preparation of hamster spleen, ×952. Transmitted by a midge of the genus *Phlebotomus. Center* and *Right. Leishmania donovani* in culture, ×1078. Here the parasites assume the leptomonad form, typical of the insect host.

serious threat to man and domestic animals. The species listed on the chart appear in the table above, together with their main vertebrate and invertebrate hosts.

Leishmania

The genus *Leishmania* comprises a relatively small group of flagellates having a life cycle with only two stages: the leptomonad forms in the midgut of the insect host, and the intracellular forms (considered the adult stage) found in the vertebrate. The morphology of the parasite in its intracellular stage differs in no way from that of the corresponding forms occurring in the genera already discussed. Indeed all the species of *Leishmania* known seem alike morphologically, and can only be distinguished by their predilection for certain hosts and their behavior in the host.

Phytomonas

The phytomonads retain the leptomonad form in their plant host, but in the transmitting insect (apparently some species of sap-sucking bug) leishmania form stages, or even aggregations of such stages known as a "somatella," may occur. Chains may also occur. It has been suggested that some of the flagellates normally living in the gut of plant-feeding insects may be adaptable enough to continue existence in the plants preyed upon, but *Phytomonas* is another genus about which relatively little is known.

EVOLUTION

Although four genera—*Leptomonas*, *Herpetomonas*, *Crithidia*, *Phytomonas*—of the six in the family are of no medical or veterinary importance, a study of their morphology and life histories throws much light on the probable evolution of the trypanosomes. The first three of the four are in most cases parasites of the alimentary tract of insects, and protozoologists generally agree that both leishmanias and trypanosomes evolved from ancestral types of similar habitat. What their free-living ancestors were like, or how such parasitism originated, can only be guessed, but at least we can make some plausible guesses. Since there is much similarity in the morphology of these parasites and a number of free-living species who possess a single flagellum, the evolutionary gap may not be really very wide.

The most primitive of the six genera is *Leptomonas*. A step beyond is *Crithidia*, and *Herpetomonas* occupies a still higher rung on the evolutionary ladder, if one admits the now disputed claim that its life history includes herpetomonad stages. The kinetoplast has gradually migrated posteriorly, with the concurrent development of an undulating membrane, until the trypanosome form was finally evolved. But the life cycle continued to include leptomonad and crithidial stages.

The genus *Leishmania* is generally thought to represent a group in which the trypanosome and crithidial stages have been dropped from the life history. However, there are those who think the genus may have arisen independently. Indeed, species of *Leishmania* are known, e.g. *L. chameleonis* of chameleons, in which the leptomonads seem to be the only stages occurring in the life cycle. Their habitat is the intestine. Other species of saurian Leishmanias may occasionally leave the intestine for the blood stream, and still others apparently have found the blood so good a medium that they no longer live anywhere else in their lizard hosts. Little is known about the mode of transmission of any of these forms, but it is hard to see how an invertebrate could serve as a vector for parasites restricted to the intestine of the vertebrate.

It is possible that flagellates such as those of the lizard should not be included in the genus *Leishmania*, or it may be that they have become secondarily so well adapted to life in the vertebrate that the insect host has been dropped from the life cycle. In any case, one may assume that these parasites are still primitive, and that adoption of an intracellular existence, with concurrent loss of the flagellum, occurred later in evolution.

Presumably the phytomonads became plant parasites in much the same way as did trypanosomes and leishmanias in vertebrates. Some insect leptomonads doubtless proved adaptable enough to survive and multiply in plants after being injected into their juices by their sap-sucking hosts. Certainly these flagellates are a highly adaptable lot. McGhee (1957) was recently able to establish in chick embryos crithidia of an undetermined species from a plant-feeding bug, *Euryophthalmus davisi*. These organisms were then successfully carried through a number of transfers from embryo to embryo.

Within the genus *Trypanosoma* Lavier (1942–43) has suggested that the long, slender types occurring in fish, salamanders and newts, and tortoises and turtles may be the oldest. These trypanosomes were perhaps originally parasitic in aquatic blood-sucking invertebrates such as leeches. Then come the broader forms with the kinetoplast displaced anteriorly, and the nucleus posteriorly; trypanosomes of this type are often found in frogs and toads, reptiles other than those with shells, birds, and some mammals. Finally, there are thinner forms with a more anteriorly situated nucleus; these last occur in mammals. Lavier also recognized certain transitional types.

REPRODUCTION AND CULTIVATION

Reproduction in all genera of the family Trypanosomatidae is by simple binary fission (longitudinal, in the flagellated stages), and it occurs in each of the hosts involved in the life cycle. As the chart in Figure 20.1 shows, however, there are significant differences among species in this regard, and the same species vary in behavior in vertebrate and invertebrate hosts. *Trypanosoma cruzi*, for example, multiplies in its vertebrate host only by binary fission of the leishmania stages, whereas most other members of the genus do so in the trypanosome stage. The process of division may be splendidly observed in the organisms of a young culture, although here the stages seen are those normally occurring in the invertebrate.

At various time sexual stages have been claimed to occur in the life histories of several species, but the evidence has never been strong enough to carry conviction. Failure to find sexual phenomena in these organisms, however, certainly does not prove that they do not occur. The truth is that trypanosomes are sufficiently polymorphic, and variable enough in their behavior, both in their living hosts and in culture, to make difficult the interpretation of unusual forms which may simulate pairing or even gamete formation.

Cultivation of many of these species is relatively easy, but for some obscure reason only the forms characteristic of the insect host usually develop *in vitro*. The medium first devised, and still very commonly used, consists of agar, sea salt, distilled water, and mammalian (usually rabbit) blood. This is known as N.N.N. (Novy-MacNeal-Nicolle's) Medium (see p. 569). It works well for numerous species of trypanosomes, though not for all, and is particularly well suited to the leishmanias. Organisms often lose much of their virulence when grown for some time in culture, however, probably because even the most ingeniously devised culture media fail to duplicate exactly the physiological conditions of the normal vertebrate host. The stages occurring in the invertebrate host may have somewhat less exacting nutritional requirements, and this is the reason why the parasites revert to this status in culture; it does not, however, explain why they at the same time may lose virulence.

MAN AS HOST

Man is host to three species of trypanosomes: *Trypanosoma gambiense*, *T. rhodesiense*, and *T. cruzi*. The first of these is the cause of the chronic but nevertheless fatal form of African sleeping sickness,* a disease widespread in west Africa. The second of the three species gives rise to a much more rapidly fatal type of the same disease, but is fortunately limited to a much smaller area of southeastern Africa, although there is some overlapping of the ranges of the two. Interestingly enough, the morphology of both species of trypanosomes is almost identical even though they differ sharply in virulence. There is reason to think that both *Trypanosoma rhodesiense* and *gambiense* arose, by physiological differentiation, from *T. brucei*, a species producing a benign and common infection in bush bucks, though highly virulent for laboratory animals such as the white rat. These two species of trypanosome are morphologically indistinguishable. All three of the species are transmitted by tsetse flies (genus *Glossina*). *Trypanosoma gambiense* has as its most common vector *Glossina palpalis;* for *T. rhodesiense* it is *G. morsitans.*

In man the parasites live and reproduce chiefly in the blood stream. Some manage to reach the lymph nodes, and perhaps also cause injury through the elaboration of toxic substances; they are never tissue parasites, though they invade the central nervous system late in the course of the disease, and may be demonstrable in the cerebrospinal fluid. The irritation they cause in the brain and cord is responsible for the motor disturbance and stupor characteristic of the later stage of the malady.

In the fly the flagellates are found chiefly in the intestine, but some migrate to the buccal cavity and salivary glands through which they reach the body of the new host during the insect's bite. Luckily they are not passed from one generation of flies to another.

Control of sleeping sickness by measures directed at reducing the tsetse fly population is relatively difficult because the flies carry their larvae until the pupal stage is reached and the larval stages are thus protected from attack. The density of the fly population has been sharply cut in many areas, however, by clearing the underbrush in which they find shelter and by insecticides such as DDT and dieldrin. The number of new cases of sleeping sickness has greatly diminished in recent years, not only because of such measures but probably also because of the widespread use of tryparsamide and certain newer drugs to treat the disease, thus reducing opportunities for infection of the vector.

The third of the three species of trypanosomes living in man is *Trypanosoma cruzi* (Fig. 20.6) occurring chiefly in South and Central America although it is often found in rodents in our own Southwest. It has even been isolated and identified recently from Maryland raccoons (Walton *et al.*, 1958). Unlike the other two, this species is not at all choosy about its vertebrate and invertebrate hosts. Several species of armadillos, monkeys, cats, dogs, bats, foxes, and other mammals have been found infected. Dogs and cats and

*It is important to remember that sleeping sickness, like many other diseases, may have different causes. What occurs in the United States is properly known as encephalitis and is a virus infection.

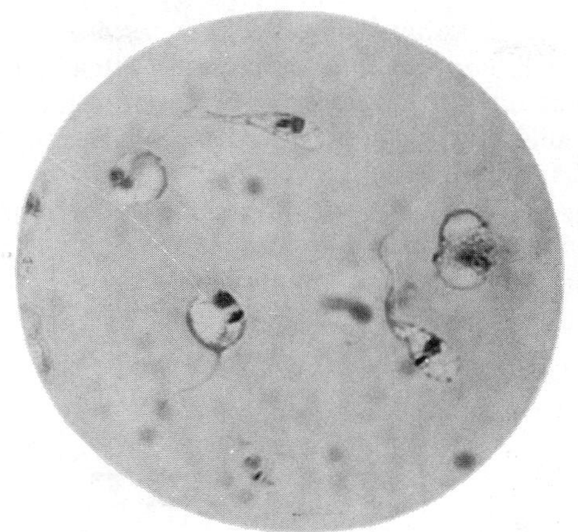

FIG. 20.6. *Trypanosoma cruzi* in culture, ×1078. Trypanosomes in culture assume the stages normally found in the insect vector. Note the altered position of the kinetoplast. In the forms shown it lies either anterior to or beside the nucleus.

possibly some other mammalian hosts may serve as reservoirs of human infection. A variety of arthropod species serve as vectors, although triatomid bugs seem to be favorites. In Brazil *Panstrongylus megistus*, a large and vicious blood sucker of nocturnal habits, is the most important transmitting agent. Even bedbugs have been experimentally infected.

Trypanosoma cruzi is as unorthodox in its life cycle as it is versatile in its choice of hosts (Fig. 20.7). Unlike most other species of mammalian trypanosomes, it develops leishmania stages in the internal organs, especially the cells of the reticulo-endothelial system and the heart muscle. It is these forms which reproduce, as noted above. When some of them "spill over" into the blood stream, they pass through the leptomonad and crithidial stages, becoming the typical trypanosomes with the characteristic "C" shape and the conspicuous terminal kinetoplast of this species (Fig. 20.8). Peculiarly slender forms and forms intermediate between this and the more typical short broad "C" have also been observed. Crithidial stages also develop in the gut of the transmitting bug, becoming small trypanosomes (the infective stage) within a week or so. Some of these are both very slender and very active, and Chagas, the discoverer of *T. cruzi*, interpreted them as "male," but without much reason. The trypanosomes, completing their cycle in the insect, pass out with its feces, and if deposited on a human during the biting process, may be scratched into the wound causing Chagas' disease. Although it is not as rapidly fatal as either of the other forms of trypanosomiasis, Chagas' disease is nevertheless serious, especially in children, and as yet has not been effectively treated.

The genus *Trypanosoma* is, however, not the only one in the family to cause serious illness in man. Three species of *Leishmania* are important as human parasites: *L. donovani*, *L. tropica*, and *L. braziliensis*. The first of these causes kala-azar (visceral leishmaniasis), an often fatal disease, moderately common in Mediterranean countries and parts of the Far East. Oriental sore, a relatively mild skin disease which also goes under several other names, is caused by *Leishmania tropica*. It has much the same geographical distribution as *Leishmania donovani*, but where one is common frequently the other is not, though for what reason is not known. The third species, *Leishmania braziliensis*, occurs in South and Central America, where it

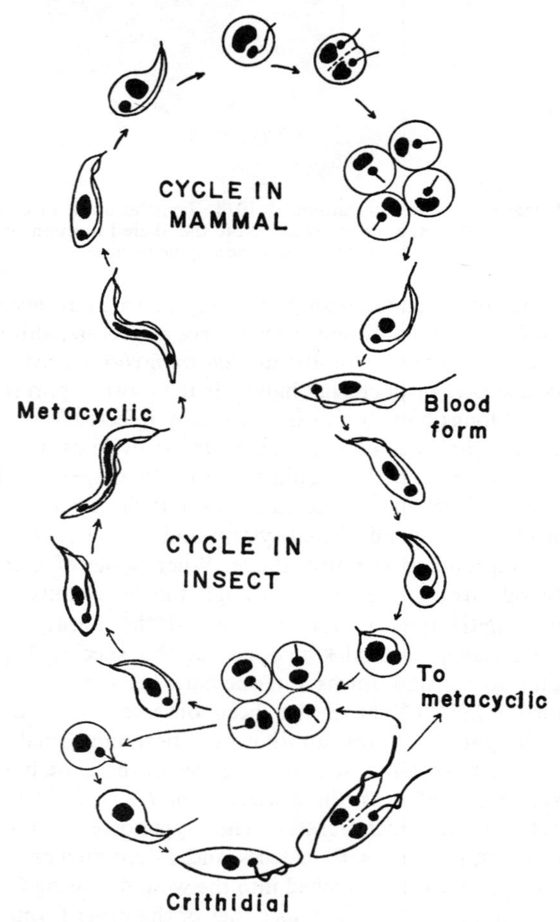

FIG. 20.7. Diagrammatic schema of the life cycle of *Trypanosoma cruzi*. Note that leishmaniaform stages occur in both the insect and vertebrate hosts. Reproduction may take place in both leishmania and trypanosome stages in the former (or in culture) but only in the leishmania stage in the vertebrate; hence multiplication in the latter occurs only in the tissues. (After E. R. Noble, *Quart. Rev. Biol.*)

gives rise to a serious and often mutilating infection of the nasopharynx, and to obstinate and progressive lesions of the skin.

The parasites in each case are so nearly alike that they can be distinguished only by their varying effects on the host: a different disease is produced by each. The differing clinical effects must reflect physiological peculiarities of each species of *Leishmania*, but nothing is yet known of the real nature of these peculiarities. Since there are also apparently variant strains within each species, it is evident that here lie opportunities for detailed genetic studies of the parasites, and possibly of genetic host differences as well.

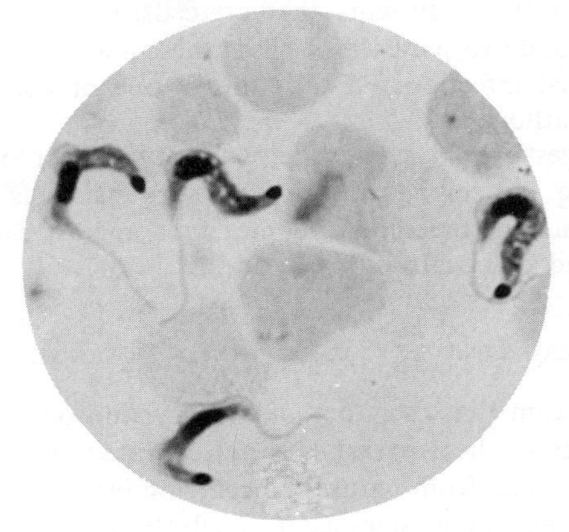

FIG. 20.8. *Trypanosoma cruzi* in the blood of a nurseling rat, $\times 1540$. Note the characteristic "C" shape of several of the parasites, and the blunt posterior end, with the large, almost terminal kinetoplast (parabasal body).

Unfortunately, man is, as we have seen, not the only susceptible host in nature, and hamsters have proved susceptible in the laboratory, particularly to *Leishmania donovani*. Because of the occurrence of natural infections of this species new human infections might arise from time to time even though all existing cases were cured. Dogs function thus in China and in Mediterranean countries.

Transmission of all the leishmanias is apparently the work of small midges belonging to the genus *Phlebotomus*, and thereby hangs a fascinating tale of research. Early studies had shown that the organisms producing leishmaniasis readily assumed the flagellate form in bedbugs, but despite many and repeated attempts transmission by bedbugs could never be proved. Then it was noted that kala-azar and *Phlebotomus argentipes* in India had a similar geographic distribution, and it was not difficult to show that these insects became heavily infected with the flagellated forms of *Leishmania* when fed on a patient ill with leishmaniasis. But here again the complete cycle of parasite transmission was very hard to demonstrate until it was discovered that the substitution of a diet of raisin juice or glucose solution for blood, after the initial blood meal, made the insects highly infective for volunteers. Apparently the parasites do not survive in the midge in nature, at least in sufficient numbers to produce infection in an ordinarily susceptible new host, unless the vector follows a blood meal with a drink of some fruit juice, but the preferred fruit remains unknown.

For treatment of kala-azar and mucocutaneous leishmaniasis antimony salts (such as neostibosan) are chiefly relied on, but a complete course of therapy requires considerable time, and some strains of Leishmanias are more resistant to drugs than others. Oddly enough, it is the macrophages of the reticulo-endothelial system—the very cells mainly responsible for the resistance of the body to parasites such as these—which seem most susceptible to leishmanial infection. Perhaps this is one of the reasons why the parasites are so pathogenic.

Since the midges are so small, they cannot be kept out of houses by ordinary screening, but control with residual sprays such as DDT has proved effective. Prevention of breeding and measures against the larvae have so far not been found practical.

Animal Diseases Caused by Haemoflagellates

Here it might seem that the story of the Trypanosomatidae should end, but the importance of these parasites is not confined to their role as agents of human illness. Aside from naturally occurring infections of leishmaniasis in animals other than man, significant chiefly because they may indirectly originate human infection, a number of species of trypanosomes also give rise to serious diseases of domestic animals, and others are of interest because they may be used as tools for research into fundamental problems of medicine.

Trypanosoma equiperdum causes a serious infection of horses known as dourine or "horse syphilis," since it is transmitted by coitus (Fig. 20.9). In all probability it once had an insect vector, as do most other mammalian trypanosomes, but when transfer by direct contact proved so much more

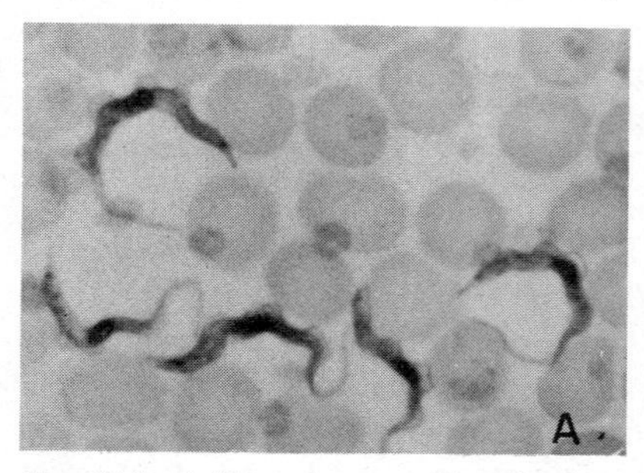

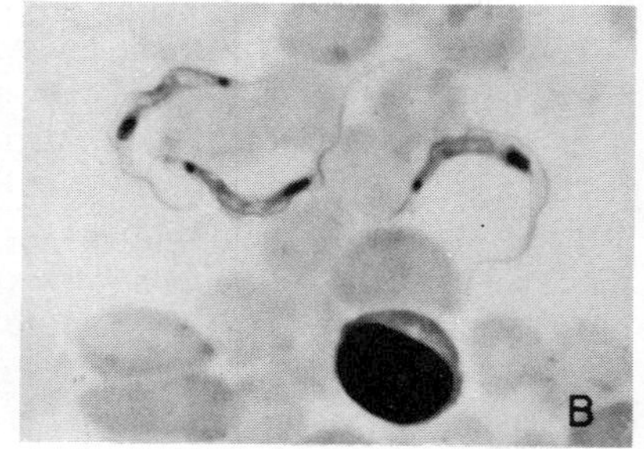

FIG. 20.9. *A. Trypanosoma equiperdum* in the blood of a rat. Lethal to horses, it is transmitted by coitus. *B. Trypanosoma lewisi.* ×1020.

convenient for the parasite, the need for the insect vanished. The contrasting effects which the same parasite often has on different host species is well illustrated by the behavior of *T. equiperdum* in horses and rats. In the former it produces a chronic disease, with eventual involvement of the nervous system and usually a fatal ending. Rats are never naturally infected, but they are highly susceptible to experimental inoculation with the parasites, and death invariably follows within a week or less.

No other trypanosomes of domestic animals are important in the United States, but elsewhere in the world the story is different. Among pathogenic species is *Trypanosoma evansi* causing the disease surra, often fatal to horses and only slightly less serious to cattle, camels, and elephants, as well as some other mammals common in tropical and subtropical regions. This trypanosome appears to be closely related to a South American species, *T. equinum*, that causes a disease of horses known as mal-de-caderas and that differs from almost all other species of trypanosomes by having no kinetoplast.* Since its absence seems to cause the parasite no inconvenience, one wonders what the real function of the kinetoplast can be.

In Africa *Trypanosoma brucei*, of which *T. gambiense* and *rhodesiense* are quite certainly variants, produces a severe and often fatal infection of pigs, dogs, cattle, and even camels, despite its apparent lack of pathogenicity for its normal host, the bush buck. (Fig. 20.10.) Sheep and goats also suffer from its effects, though they are less susceptible than most other domestic animals.

One other trypanosome of animals must be mentioned: *T. lewisi*, an extremely common parasite of rats (Fig. 20.9). It has no practical importance since infected animals are no more inconvenienced by the parasite than most children are by German measles, but it has proved excellent research material in investigating certain basal problems in the biology of trypanosomiasis. The trypanosome is easily transmissible in the laboratory by the intraperitoneal inoculation of a drop or two of infected blood (p. 575), although in nature and also in the laboratory (if desired), the rat flea serves as the vector, infection occurring either when the dejecta of the insect are scratched into the bite by the irritated host, or when the flea is itself swallowed as the rat licks its wounds.

Trypanosoma lewisi is an example of that group of trypanosomes which develop on the posterior portion of the gut of their insect hosts; this is termed development in the "posterior station." Since this locale is believed to have been the ancestral habitat, these trypanosomes are considered more primitive than species causing African sleeping sickness, which are restricted to the "anterior station," or those parts of the alimentary apparatus in the

*Although *Trypanosoma equinum* is the only species in which the absence of a kinetoplast is the rule, in several other species a certain proportion of the parasites may lack it, both in naturally acquired and experimental infections. Hoare (1954) remarks that the loss of the structure probably originated as a mutation due to irregularity in division, and that such individuals may then have been able to outbreed the normal ones in the case of the ancestral forms of *T. equinum*.

immediate neighborhood of the buccal cavity. Infection of a new vertebrate host, in the latter case, occurs only when the insect injects parasites in the actual process of sucking blood.

Blood-sucking insects and other invertebrates may transmit trypanosomes by still a third mechanism. The proboscis of the insect functions exactly as if it were a contaminated hypodermic syringe passed around a group of drug addicts, with the resulting spread of disease. An outbreak of malaria some years ago in a New York prison was actually traced to such a cause.

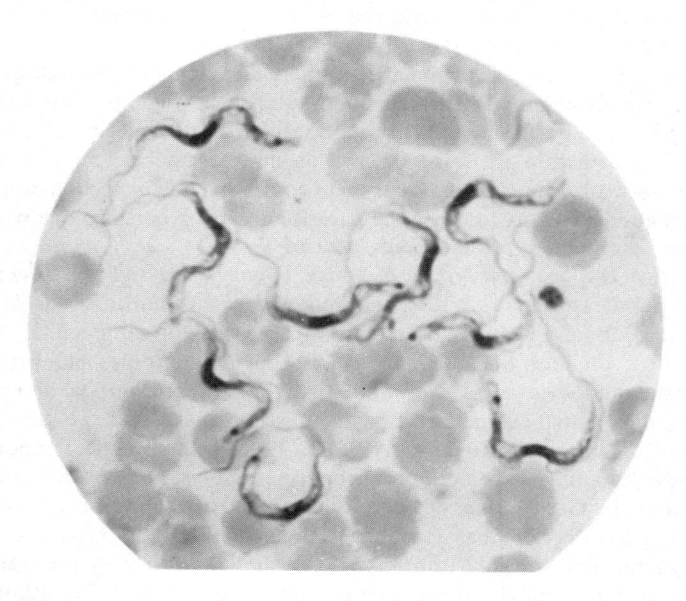

FIG. 20.10. *Trypanosoma brucei*, a species almost identical in morphology with *T. rhodesiense* but occurring normally in various kinds of game animals in Africa. *T. rhodesiense* may have originated as a variant strain of *T. brucei*. *Trypanosoma brucei* also causes the serious cattle disease "*nagana*." ×1540.

A hungry horse fly or stable fly, interrupted in its feeding by the flick of its victim's tail, may attack another animal while its mouth parts are still moist with trypanosome-containing blood. In such transfer parasite morphology of course does not change, since the insect is not really a host.

One of the most interesting discoveries made on *Trypanosoma lewisi* concerned the mechanism by which the rat copes with the infection. Instead of the more usual destruction of the parasites by phagocytosis, the rat elaborates a reproduction-inhibiting substance known as ablastin. Thus no more young forms are produced. By the time all the presently existing trypanosomes have reached maturity, other antibodies develop to kill them off by stages, leaving the rat with a permanent immunity. Somewhat paradoxically, this trypanosome also may benefit its host, apparently by reducing the latter's thiamin requirements. As a result, young infected rats gain weight more rapidly than normal ones (*Lincicome *et al.*, 1963; 1965). It may thus be a symbiont.

Why are some trypanosomes virtually harmless to their hosts while others may be lethal? Why does *Trypanosoma lewisi* have no visible effect on the rat, when *Trypanosoma equiperdum* is able to reproduce in the rat's blood stream without any check? It is not yet possible to give a definite answer to either of these questions, but probably the differences in pathogenicity depend, at least in part, on metabolic differences in parasite species. These must in turn arise from genetic differences. Further, the pathogenic trypanosomes are on the whole greedier carbohydrate consumers, failing to oxidize it completely. A heavy burden is thus put on the carbohydrate reserves of the host, possibly (because of toxic by-products) interfering with the host's carbohydrate metabolism. Although not yet demonstrated, toxins may originate from the protein metabolism of the parasite.

Trypanosomes have a remarkable ability to adapt themselves to both therapeutic drugs and antibodies, apparently the result of a change in the surface properties of the parasite by which less drug is absorbed or the lethal effects of antibodies are resisted. Once such changes occur they are passed on to all later generations of the trypanosomes, although they are likely to be lost in passing through the insect vector. It is therefore very important to administer a sufficiently large dose of a curative drug to eradicate all the parasites. And the host, if it fails to extinguish the infection by the first burst of antibody production, must begin all over again and produce a completely different antibody the second time and thereafter until either the parasites have all been killed or it succumbs to the invaders.

Thus we find that the trypanosomes and leishmanias present us with many problems, not the least of which is their successful control in domestic animals. In Africa the diseases they cause are especially important, and may have been more significant in retarding the advance of civilization in this largest of continents than all other obstacles combined. Many other problems concern parasite physiology and its relation to the host's. The number of unanswered questions may seem remarkable when we remember that the first trypanosome, *T. rotatorium* of frogs, was discovered in 1842, more than a century ago. Since then much has been learned, but not enough. We know that this family of flagellates has many species and is one of the most important to man, though the majority of its members do not appear to harm their hosts. But we have little idea of how many species actually exist, and it might even be said that we are not sure just what a species of trypanosome really is. Doubtless some of the more pathogenic among them play a significant part in the maintenance of biological balances, yet we have little real information on the matter. And more study of the group might also reveal a far better blueprint of the evolutionary relationships of its members than we now possess.

REFERENCES

Becker, E. R. 1923. Observations on the morphology and life-history of *Herpetomonas muscae-domesticae* in North American muscoid flies. *J. Parasitol.*, 9: 199–213.

Faust, E. C. 1949. The etiologic agent of Chagas' Disease in the United States. *Bol. Oficina Sanit. Panamer.*, 28: 455–61.

Hoare, C. A. 1936. Morphological and taxonomic studies on mammalian trypanosomes. I. The method of reproduction in its bearing on classification, with special reference to the *lewisi* group. *Parasitology*, 28: 98–109.

——— 1948. The relationship of the haemoflagellates. *Proc. Fourth Internat. Cong. Trop. Med. and Hyg.*, pp. 1110–6. Washington D.C.

——— 1954. The loss of the kinetoplast in trypanosomes with special reference to *Trypanosoma cruzi. J. Protozool.*, 1: 28–33.

——— 1957. Classification of trypanosomes of veterinary and medical importance. *Vet. Med. and Ann.*, 3: 1–13.

Holmes, F. O. 1925. The relation of *Herpetomonas elmassiani* to its plant and insect hosts. *Biol. Bull.*, 49: 323–37.

Kraneved, F. C., Houwink, A. L., and Keidel, H. J. W. 1951. Electron microscopical investigations on trypanosomes. I. Some preliminary data regarding the structure of *Trypanosoma evansi.* Koninkl. Nederl. Akademie van Wetenschappen, Amsterdam. *Proceedings*, Series C., 54: 393–9.

Laveran, A., and Mesnil, F. 1912. *Trypanosomes et Trypanosomiases*, 2d. ed. Paris.

Lavier, G. 1942–3. L'évolution de la morphologie dans le genre *Trypanosoma. Ann. Parasit. hum. comp.*, 19: 168–200.

Manwell, R. D. 1955. An insect Pompeii. *Sci. Month.*, 80: 356–61.

McGhee, R. B. 1957. Infection of chick embryos by *Crithidia* from a phytophagous Hemipteron. *Science*, 125: 157–58.

Noble, E. R. 1955. The morphology and life-cycles of trypanosomes. *Quart. Rev. Biol.*, 30: 1–28.

Taliaferro, W. H. 1924. A reaction product in infections with *Trypanosoma lewisi* which inhibits the reproduction of the trypanosomes. *J. Exp. Med.*, 39: 171–90.

Walton, B. C., Bauman, P. M., Diamond, L. S., and Herman, C. M. 1958. The isolation and identification of *Trypanosoma cruzi* from raccoons in Maryland. *Am. J. Trop. Med. and Hyg.*, 7: 603–10.

Wenyon, C. M. 1926. *Protozoology*, vol. 1. New York; Wood.

Willett, K. G. 1961. Recent Developments in research on the African trypanosomiases. *Trans. N. Y. Acad. Sci.*, Ser. II, 23: 233–36.

Wolcott, G. B. 1952. Mitosis in *Trypanosoma lewisi. J. Morphol.*, 90: 189–99.

[ADDITIONAL REFERENCES FOR THIS CHAPTER WILL BE FOUND ON PAGE 457.]

21

Amoebae of the Alimentary Tract

"My excrement being so thin, I was at divers times persuaded to examine it; and each time I kept in mind what food I had eaten, and what drink I had drunk, and what I found afterwards: but to tell all my observations here would make all too long a story." Leeuwenhoek, Letter 34, November 4, 1681.

Except for certain species of bacteria, man probably harbors no parasites as commonly as he does the amoebae of the mouth and intestines. The incidence of infection by these organisms has never been accurately determined, and it varies with different classes of the population and in different parts of the globe. Nevertheless, if all amoebic infections could be identified, they would undoubtedly be found to exist in over half the people in the world.

Nor is man especially favored by nature in this respect. Many of the lower animals, both vertebrate and invertebrate, harbor these tiny stowaways. Parasitism by amoebae is not only very common, but the number of such species of amoebae must be very great; they must constitute a particularly adaptable tribe of Protozoa. Frequently the amoebae of the lower animals closely resemble species occurring in man, and often they even belong to the same genera. In a few cases, the species seem identical.

Most parasitic amoebae are found in the alimentary tract of their hosts; relatively few are tissue invaders. They are usually dwellers in the colon, where the environment is rich in digested or partly digested foodstuffs, swarming with microorganisms of many kinds. They seem equally fond of both, and no doubt to their free-living amoebic ancestors such an environment must have seemed a real Garden of Eden when first they entered it. Most parasitic amoebae are content to make their hosts foot the board bill, for most do their hosts no harm.

Life Cycles

The life cycles of the great majority of these amoebae are still unknown, but the cycles of several human species are most likely typical of the group. These have been worked out in detail by Dobell, and also by Cleveland. Usually there are trophozoite, precystic, cystic, and metacystic forms. Sexual stages appear to be wholly absent. The trophozoites feed, grow, and eventually divide. In some species true mitosis has been observed. Although Kofoid and Swezy (1925), for example, state that *Endamoeba dysenteriae* (=*Entamoeba histolytica*) undergoes true mitosis, and that there are six chromosomes, other authors (e.g., Dobell) question this.

At intervals, in response to stimuli which are still largely unknown, the trophozoites become precystic forms which in turn develop into cysts. In the case of *Entamoeba histolytica*, encystment in cultures seems to depend largely on the character of the food present, as for example, rice starch and some kinds of bacteria. The cyst may remain viable outside the body of the host for a few weeks or less, although no one knows just how long. Probably environmental conditions have much to do with it. If fortune is kind, they may be ingested with food or water by a suitable potential host, and infection results. The trophozoites are incapable of setting up an infection except in a few species, such as *Entamoeba gingivalis* and *Dientamoeba fragilis*, which seem unable to produce cysts.

The cyst stage is essential in the life cycles of parasites such as the intestinal amoebae, which lack vectors. Not only does it bridge the gap between hosts, but it is highly resistant to adverse environmental influences, such as drying and harmful chemical agents. It also frequently serves for reproduction, its nucleus dividing repeatedly until the number of nuclei within the cyst characteristic of the species results. Cysts may also contain structures such as the chromatoidal bars (whose function is unknown) and stored food in the form of glycogen masses; both chromatoids and glycogen tend to disappear with age.

When conditions are favorable for excystment, the imprisoned amoeba becomes active and finally gains freedom through a pore in the cyst wall. In the species in which it has been observed, the process takes only a few hours. After liberation the young or metacystic amoeba feeds actively, and soon divides. Details of the process of fission undoubtedly vary greatly in the different species. If the encysted organism contains more than one nucleus, subsequent development may be decidedly complicated until the normal uninucleate condition is regained.

The nature of the stimuli required for excystment is still unknown. For species parasitizing warm-blooded animals, warmth, moisture, and presumably the presence of suitable food (such as certain bacteria) are necessary.

THE AMOEBAE OF MAN

Although parasitic amoebae occur in other families, most of the known species are sufficiently alike to be placed in a single family, the Endamoebidae, containing (according to Kudo) six genera. The amoebae of man fall into four of these, and number six species. Some protozoologists and parasitologists still believe that one or more additional genera, and perhaps several other species of human amoebae, should be recognized. Most of them, however, follow Dobell (1919), as we do also. It is likely that errors in

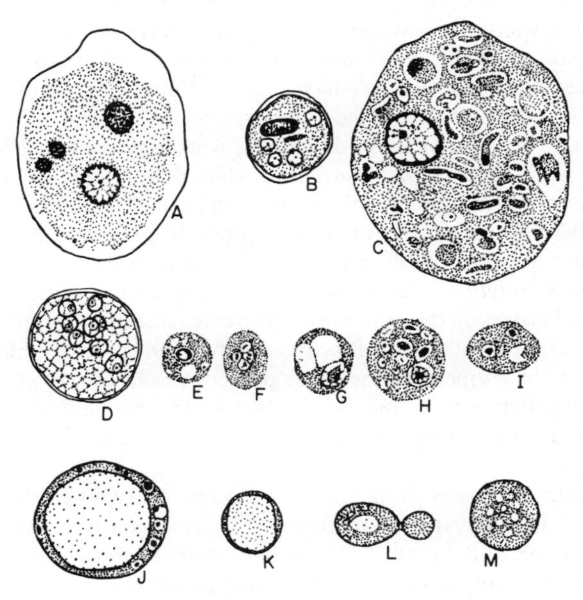

FIG. 21.1. Intestinal amoebae and other organisms of man, ×1000. *A. Entamoeba histolytica* (trophozoite). *B. E. histolytica* (cyst). *C. Entamoeba coli* (trophozoite). *D. E. coli* (cyst). *E. Endolimax nana* (trophozoite). *F. E. nana* (cyst). *G. Iodamoeba bütschlii* (cyst). *H. I. bütschlii* (trophozoite). *I. Dientamoeba fragilis*. *J., K.* Large and small forms of *Blastocystis hominis*, a very common fecal organism of vegetable nature. *L.* Budding yeast. *M.* Intestinal mold. (After Hegner, Cort, and Root.)

technique, and inability to recognize artifacts or to distinguish between protozoan cysts and other objects, such as the yeast and vegetable cells always present in feces, have led to the description of many new "species" of amoebae.

Of the six recognized species of human amoeba (Fig. 21.1), one, *Entamoeba gingivalis*, lives in the mouth between the gums and teeth. The others inhabit the intestine, although the preferred site of *Iodamoeba bütschlii* and

Dientamoeba fragilis is not definitely known. *Entamoeba histolytica, E. coli,* and *Endolimax nana* reside in the colon, or caecum, and *Dientamoeba fragilis* has also been found in the latter situation on occasion. Only the first of these is certainly pathogenic, although *Dientamoeba fragilis* may cause mild gastrointestinal symptoms in some cases. That it is at times a pathogen is supported by its occurrence sometimes in diseased appendices, where it has been found filled with ingested red cells (Burrows *et al.*, 1954), although even under these circumstances it is apparently not a tissue invader.

Entamoeba histolytica

The first observation of amoebae in man was made by Loesch in 1875. He discovered what we now call *Entamoeba histolytica* in the stools of a dysenteric patient in St. Petersburg and succeeded in transferring it to puppies. The studies of Loesch have been followed by many others on *Entamoeba histolytica* and on the other species of human amoebae. The most important have been those of Dobell, who first brought order out of chaos in his work entitled *The Amoebae of Man*, published in 1919, and later did a number of other equally authoritative studies. He not only made a very careful study of which species of those claimed to parasitize man are valid and which are not, but (with Jepps) added several to the list.

Entamoeba histolytica is universally recognized as the cause of amoebic dysentery, and occurs all over the world. The true incidence of amoebiasis (the term usually used for infection with this amoeba) varies greatly, higher in the tropics than in the temperate zones, and greater in some classes of population than in others. Between 10 and 20 per cent of the population are believed to be infected, even in the United States, where sanitary standards are relatively high.

Even among university students 4.1 per cent of 1,060 freshmen at the University of Pennsylvania were found in a comparatively recent survey (Wenrich, Stabler, and Arnett, 1935) to be carriers of *Entamoeba histolytica.* Nevertheless, they appeared to be "at least as healthy" as the other 95.9 per cent, showing the varying reaction of different individuals to such parasitism, and also perhaps the occurrence of relatively avirulent strains of the amoeba. Still, it is probable that any *histolytica* infection can cause trouble sooner or later.

Control of human amoebiasis is primarily a matter of proper sewage disposal and well-sanitated food and water supplies. Failure to protect the latter resulted in the widespread epidemic of amoebic dysentery occurring during the Century of Progress Exposition in 1933. The outbreak, involving more than a thousand detected cases and 58 deaths, was traced to makeshift plumbing connections from which sewage leaked into an ice-water tank of one of the hotels. From this hotel the water was piped to the upper floors of the other hotel across the street. Other investigations made as a direct result of this epidemic revealed potentially dangerous makeshift plumbing in a number of American cities, and there is little doubt that they exist even today in most communities.

Flies have also been incriminated in the spread of intestinal amoebae. Cysts may survive for a time in the gut of the fly, to be later regurgitated, or deposited with the dejecta on human food. (House flies are greedy creatures, and rather than stop feeding when satiated, they often simply regurgitate their already swallowed meal, and begin all over again.) Control of the house fly is probably one of the most important elements in restricting the spread of human amoebiasis.

Treatment of amoebiasis is fortunately not difficult, although a permanent cure (in the sense that no relapses occur) is sometimes hard to achieve, and since little residual immunity results from amoebic infection, reinfection is always possible. Emetine usually clears up the clinical indications of amoebiasis, but recurrences are common. For more lasting results, some of the newer iodine compounds (such as diodoquin, chiniofon, and vioform) are much more effective. Carbarsone (an arsenical) and the antibiotics terramycin and aureomycin are also employed.

It is still uncertain just how *Entamoeba histolytica* injures its host, nor is it known why some individuals suffer seriously while others seem unaffected. Although some of the organisms live in the lumen of the colon there is no reason to think that they do any damage. The ulcers in the intestinal wall characteristic of amoebic dysentery are generally thought to result from the direct invasion of the mucosa by the parasites, and this they are believed to accomplish with the aid of a tissue-dissolving substance known as a cytolysin. Yet the existence of such a substance has never been demonstrated. In any case, the amoebae are found in large numbers nested in the ulcerated intestinal lining, together with bacteria which may play an important role as secondary invaders (Fig. 21.2). The effectiveness of antibiotics may be due more to their action on the bacteria than on the amoebae.

Entamoeba histolytica is a rather small amoeba, usually measuring from 18μ to 20μ in diameter, though the cysts are often considerably smaller than this. Ectoplasm and endoplasm are generally sharply differentiated. Pseudopods, which have a characteristic glassy appearance because they are formed from ectoplasm, are rapidly extruded and often as rapidly withdrawn. The nucleus is seldom visible in the living organism, but ingested red cells are commonly seen in the cytoplasm, and are valuable in diagnosis.

In stained preparations the nuclear structure is easily distinguished and characteristic of the species. The nuclear membrane is extremely delicate and thin, with minute chromatin granules arranged in a row around its inner edges. In the center (although it may sometimes be eccentric) is a small karyosome measuring perhaps 0.5μ in diameter, often surrounded by a clear area, or halo. The entire nucleus in the trophozoite is generally 4μ or 5μ in diameter.

Stained cysts possess from one to eight nuclei, but the usual number in mature cysts is four. Nuclear structure is like that of the trophozoite, although the more nuclei there are the smaller their size. Young cysts may contain one or more heavily staining chromatoidal bars, usually with blunt ends, and occasionally one or two glycogen masses resembling vacuoles. Other inclusions, such as food particles, disappear when the cyst is formed. Most cysts are spherical in shape, although oval ones are sometimes seen. The

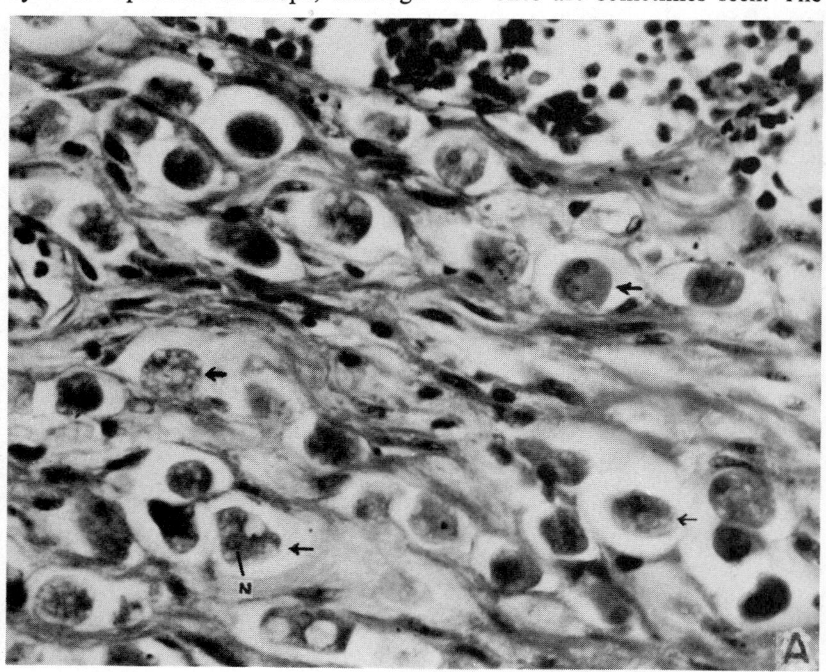

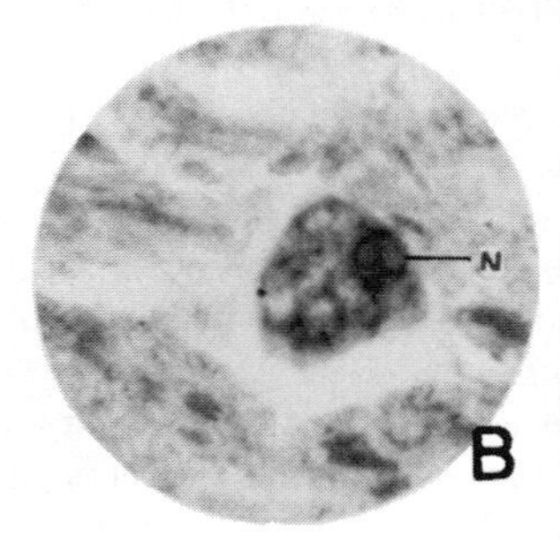

FIG. 21.2. *A. Entamoeba histolytica* in a cross section of the colon wall, ×570. Arrows indicate several of the amoebae, and numerous others may be readily seen, some with the characteristic nucleus (n) with its central (or near central) karyosome (e.g., amoeba indicated by arrow at lower left). *B. Entamoeba histolytica* in the intestinal wall from a human case, ×1095. N, nucleus.

cyst wall seems to consist of a single layer which, (according to Kofoid, McNeil and Kopac, 1931) is flexible, tough, and highly resistant to most chemical agents. Chemical tests suggested that it consists of keratins (albuminoid proteins which seem to make up the bulk of such animal structures as hair, nails, and feathers).

In size, the cysts of *Entamoeba histolytica* differ considerably, but generally range between 18μ and 25μ in diameter. Large and small races exist, and those that produce the smaller cysts are often less pathogenic than others. Some studies indicate that, rather than two, there are five such races all distinguishable by cyst size. Presumably with such differences go heritable characteristics of a variety of other kinds, and if our techniques for the study of amoebic genetics were as highly refined as they are for the fruit fly *Drosophila*, it is intriguing to think of what a monumental monograph some scholar could write on the subject.

The relationships of the different stages within the life cycle are shown in Figure 21.3. The quadrinucleate amoeba which emerges from the cyst feeds voraciously, and very soon begins a series of rapid divisions resulting

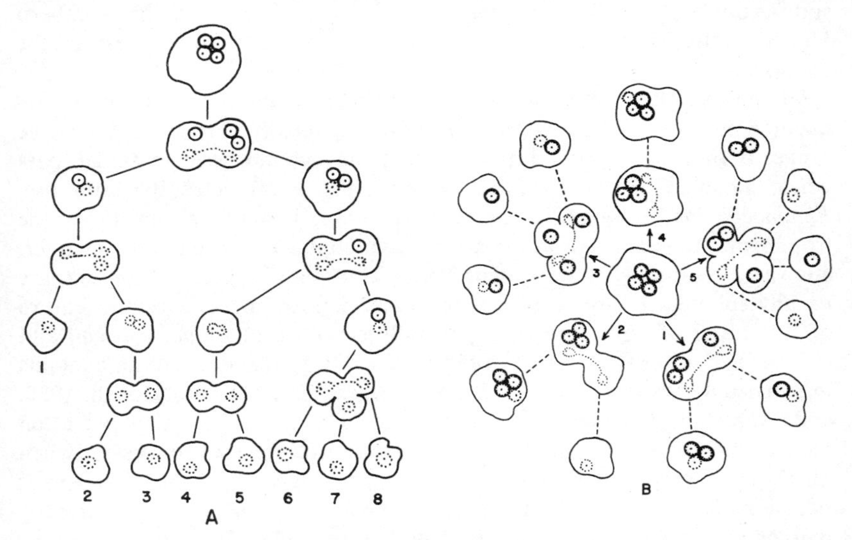

FIG. 21.3. *A*. The commonest pattern of metacystic development of *Entamoeba histolytica*. Eight small amoebae eventually develop from the excysted quadrinucleate organism. The nuclei originally present in the cyst are represented by solid circles, each containing its karyosome, and the daughter nuclei resulting from successive divisions are shown with dotted lines. The amoebae in the final brood of eight do not all originate in exactly the same way: no. 1, for example, is a product of two nuclear divisions, and 8 of three. *B*. The five possible patterns of metacystic development of *Entamoeba histolytica* shown diagrammatically as they begin. 1 is the commonest, and is followed through in *A*. The subsequent history of the metacystic amoebae from eight-nucleated cysts, of which there are always a small proportion, is not known. The time required for completing metacystic development varies, but is normally about 12 hours. (After Dobell, *Parasitol.*)

in the eventual formation of the typical uninucleate trophozoite. Prior to each division, one of the original four nuclei divides, and then (except in the first such division) a small uninucleate amoeba is pinched off, until eight such amoebulae are produced. In the meantime various combinations of cystic and metacystic nuclei are possible in the daughter organisms, the two types of nuclei distinguishable by their differing size, so that (according to Cleveland and Sanders, 1930) at this stage in the life history as many as 24 kinds of amoebae may exist.

Entamoeba coli

Although *Entamoeba histolytica* is the only species of amoeba known to cause serious disease in man, the other species are equally interesting to the protozoologist who, like Leeuwenhoek, finds a fascination in any of these "little animals"; the practical importance of being able to distinguish pathogenic and harmless types is obvious. Of the latter, *Entamoeba coli* is probably the most common. The incidence of infection by *Entamoeba coli* varies considerably in different regions and among people of differing habits and status, but at least 40 to 50 per cent of the population in the southern United States is infected, and perhaps half as many in the rest of the country.

Of the intestinal amoebae of man, *Entamoeba coli* is the one most frequently mistaken for *E. histolytica*. Both organisms overlap in their size range, both produce spherical cysts that may contain chromatoidal bars when immature, and the nucleus in both has a rather similar structure. *Entamoeba coli*, however, is usually considerably larger in all its stages, the trophozoite is less active and extrudes pseudopodia in a different manner, and the cysts usually contain eight nuclei. Chromatoidal bars, if present, are usually splintery in appearance. The minute karyosome, a constant feature of the nuclei of both species, is generally eccentric in *Entamoeba coli* and central in *E. histolytica*. The cytoplasm of *coli* trophozoites seldom contains ingested red cells (it is often said that it never does, but both Dobell, 1936, and Tyzzer and Geiman, 1938, have observed them), though they are often seen in the corresponding stage of *E. histolytica*. Nevertheless, even when all these differences are taken into account, diagnosis of the organism is occasionally doubtful. The life cycle of *Entamoeba coli* has been carefully worked out by Dobell, and closely resembles that of *E. histolytica*. From the multinucleated, mature cyst, emerges an amoeba with an equal number of nuclei. Cytoplasmic division follows, without the prior nuclear divisions characteristic of *E. histolytica*, finally forming sixteen amoebulae.

Endolimax nana

Almost as common as *Entamoeba coli* is *Endolimax nana* (Fig. 21.4), some surveys showing an incidence of nearly 30 per cent. Like *Entamoeba histolytica*, it produces cysts with four nuclei (Fig. 21.5), but the resemblance ends here, for species belonging to the genus *Endolimax* have nuclei with a very heavy, massive karyosome.

The life cycle of *Endolimax nana* resembles in broad outline those of *Entamoeba histolytica* and *E. coli* (Dobell, 1943). Under favorable conditions (such as the presence of a suitable bacterial flora) cysts are formed from the trophozoites after two preparatory divisions. Within the cyst, the single nucleus divides mitotically, eventually producing four or occasionally as many as eight small daughter nuclei. Excystment takes place through an

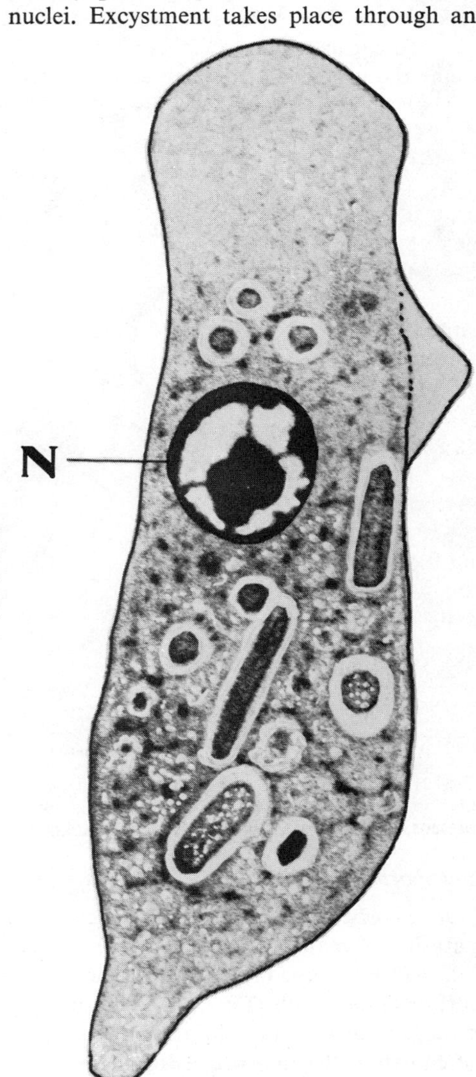

FIG. 21.4. Trophozoite of *Endolimax nana*. Note the nucleus (N) with the characteristic large karyosome. The other inclusions are ingested food particles, probably bacteria. (After Dobell, *Parasitol.*)

extremely minute pore in the cyst wall, and the hatchling amoeba then undergoes a series of binary fissions (without division of the nuclei) by each of which a single uninucleate amoeba is pinched off, until there are as many such metacystic amoebae as there were nuclei. These tiny organisms are said by Dobell to measure in some cases only 2μ in diameter, making them probably the smallest amoebae ever observed. Feeding is active, however, and growth rapid, and multiplication (always mitotic) by binary fission soon takes place. Somewhat surprisingly in view of its small size, *Endolimax nana* has ten chromosomes. One wonders how many genes are required to determine the genetically transmitted characteristics of so minute a living thing.

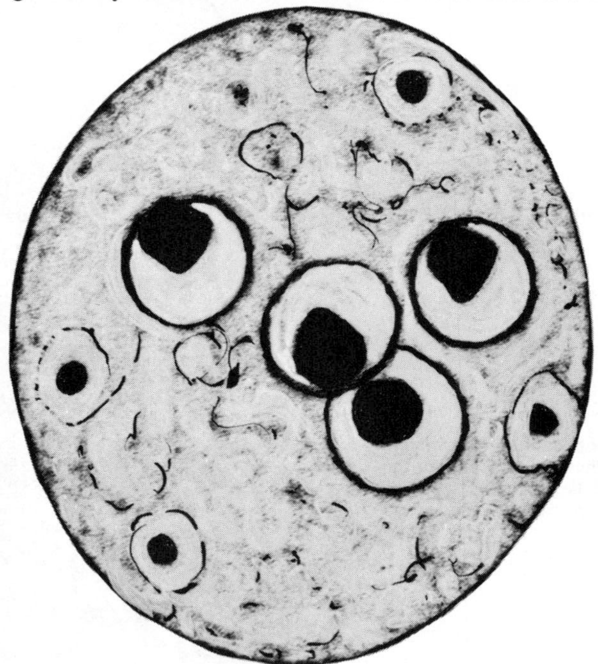

FIG. 21.5. Cyst of *Endolimax nana*. Cysts are usually more or less oval, but differ considerably in shape and less in size. Four nuclei, each with the typical large karyosome, are present, but other inclusions such as chromatoidal bars, are absent. (After Dobell, *Parasitol.*)

Iodamoeba bütschlii

Iodamoeba bütschlii (still sometimes called *I. williamsi*) is a relatively rare parasite of man, although some 5 per cent of the general population is infected. For some mysterious reason, trophozoites are seldom seen, even in soft or liquid stools (Fig. 21.6). On the other hand, cysts with the very large glycogen vacuole and the irregular shape characteristic of the species are often extremely numerous (Fig. 21.7).

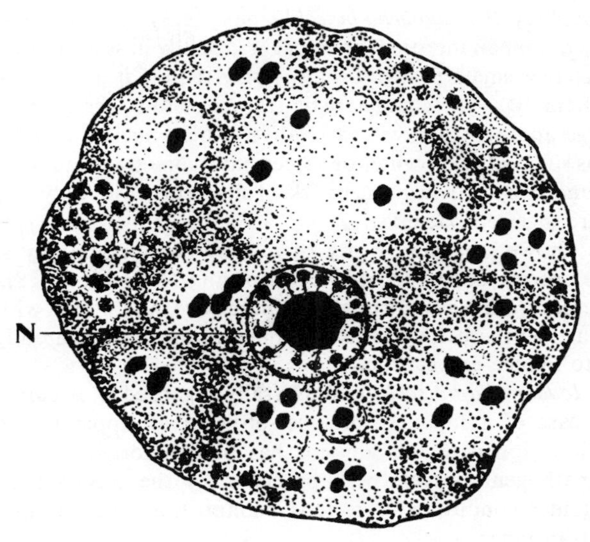

FIG. 21.6. *Iodamoeba bütschlii* (trophozoite). Note the nuclear structure: a large, heavily staining karyosome surrounded with delicate radiating fibrils, each with a peripherally placed granule. (After Wenrich.)

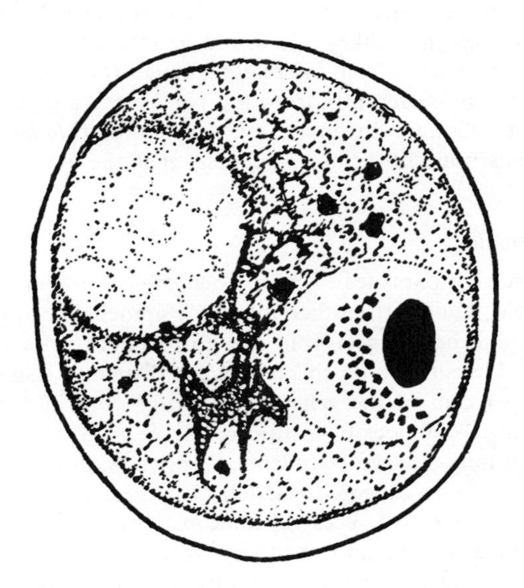

FIG. 21.7. Cyst of *Iodamoeba bütschlii*. Note the characteristic nuclear structure and the glycogen vacuole. The latter appears empty in stained preparations, and stains a deep brown in fecal smears treated with iodine. From a human host. (After Wenrich, *Am. Philos. Soc.*)

The morphology of *Iodamoeba bütschlii* makes easy its differentiation from other species of human intestinal amoebae, especially in stained preparations. The trophozoite is small, though usually larger than that of *Endolimax nana*, averaging about 10μ. The characteristic feature of this species is its nucleus, which is large and provided with a relatively thick nuclear membrane. The karyosome is situated at its center or somewhat to one side, and often appears to be surrounded by a delicate network of filaments. The only cytoplasmic inclusions are ingested bacteria.

As far as known, the life history of *Iodamoeba* is essentially similar to those of the species described above. According to Smith (1928), a single amoeba emerges from a cyst through a minute pore, and moves with unusual activity for a time thereafter. Moisture and warmth are apparently all that is required to induce excystation.

Although *Iodamoeba bütschlii* is universally regarded as a harmless commensal, the case of a Japanese prisoner-of-war who apparently died of a generalized infection with this amoeba has been reported (Derrick, 1948). Occasional pathogenic strains perhaps occur, or the species may become pathogenic under conditions of serious malnutrition, such as those which existed in this instance.

An apparently identical amoeba is a very common parasite of pigs, and *Iodamoeba bütschlii* is believed to be primarily a porcine rather than a human intestinal guest. In some areas one pig out of five is infected, and this amoeba is also very common in monkeys. It would be interesting to know whether man acquired the parasite from his remote simian (or presimian) ancestors, or has acquired it more recently from the pig, or the pig perhaps acquired its *Iodamoeba* from man. In any case, the fact that *Iodamoeba bütschlii* apparently harms none of its hosts suggests that the association is a very ancient one.

Dientamoeba fragilis

Of all the amoebic parasites of man, *Dientamoeba fragilis* is perhaps the most remarkable. Unlike the other intestinal amoeba, it apparently never produces cysts, and how it gets from one host to another is a mystery. The true incidence of infection with this amoeba is disputed. Usually it is found in less than 1 per cent of the general population, but in some surveys 30 to 50 per cent of the inmates of institutions have been found to harbor the parasite. Even among university students, Wenrich, Stabler and Arnett (1935) found 4.3 per cent to be *D. fragilis* carriers. The mechanism of transmission, therefore, must be effective. Because some patients harboring the parasite exhibit eosinophilia, and because the trophozoites bear a strong resemblance to *Histomonas* (a flagellate parasite of turkeys usually transmitted in the eggs of a nematode worm, *Heterakis gallinae*), Dobell has suggested that some round worm of man may be involved in a similar way, though this is so far without proof. Burrows and Swerdlow (1956), however, have recently presented circumstantial evidence that the pinworm *Enterobius vermicularis* may play such a role. They found the worm and *Dientamoeba* associated in twenty times as many diseased appendices as chance would indicate, and they also saw in *Enterobius* eggs in these cases "small, ameboid, uninucleate organisms" whose nuclei greatly resembled those of *Dientamoeba.* Since *Enterobius* is an extremely common parasite of man, parasitism of the eggs of this worm by *Dientamoeba* would make transmission very easy; its inability to encyst is thus perhaps no handicap at all.

There is evidence that *Dientamoeba fragilis* may be mildly pathogenic. No one has yet demonstrated any specific lesions due to the parasite, but patients with mild gastrointestinal symptoms otherwise unaccounted for have been relieved when the amoebae disappeared from the stools. Indeed the amoeba has recently been found in the appendix in a number of cases of appendicitis, and Swerdlow and Burrows (1955) remark, "The frequency with which *Dientamoeba fragilis* is found in the appendix, its preference for red blood cells when they are available, and the concomitant occurrence of fibrosis in the appendiceal wall in all cases of *D. fragilis* infections of the appendix incriminate it as a pathogen."

Diagnosis of infection with *Dientamoeba fragilis* rests on recognizing the

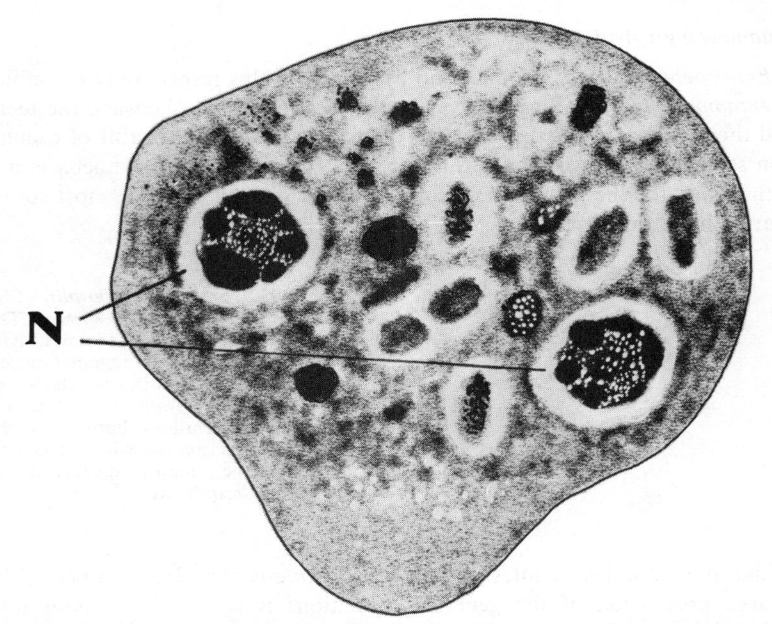

FIG. 21.8. Trophozoite of *Dientamoeba fragilis*. About 80 per cent of the trophozoites of this species exhibit two nuclei (N). These have a characteristic grapelike structure, the chromatin granules (representing the "grapes") probably corresponding to chromosomes, of which there seem to be six. The binucleate condition is the result of arrested mitosis. (After Dobell, *Parasitol.*)

trophozoites in the feces. It was formerly believed that they lived for only a short time outside the body (hence the species name "*fragilis*"), but they apparently survive for a considerable period. The writer has seen active specimens in stools kept in the laboratory for at least eight hours after passage. The amoebae were prone to round up at first, but soon began putting out short stumpy pseudopods, though usually without much progressive motion. Adding a little water to the stool causes the prompt dissolution of the amoebae in what has been termed a characteristically "explosive" manner. Staining with haematoxylin reveals the presence of two nuclei in about 80 per cent of the organisms, each nucleus possessing a karyosome of grape-like appearance (Fig. 21.8). The amoebae usually measure from 5μ to 12μ, though larger specimens are sometimes seen. Cytoplasmic inclusions consist of bacteria and other debris; red cells do not occur.

Division in the amoeba has been thoroughly studied by Dobell (1940) and Wenrich (1944). Cytoplasmic fission occurs first, producing two uninucleate offspring. The nucleus of each new individual then divides by mitosis, restoring the binucleate condition typical of the majority of individuals in most *Dientamoeba* populations. Dobell states that six is the usual number of chromosomes but Wenrich could find only four.

Entamoeba gingivalis

Entamoeba gingivalis produces no cysts, and in this respect only resembles *Dientamoeba fragilis*. Since it lives in the mouth (usually between the teeth and the gums) transmission is easy; the widespread human habit of osculation seems made to order for this parasite. Fortunately this amoeba is not pathogenic, for the problems of control would dismay even the most stout-hearted health officer!

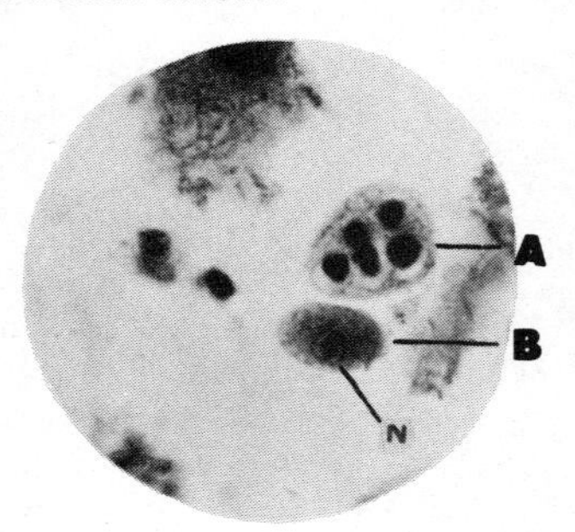

FIG. 21.9. *Entamoeba gingivalis*, × 570. Two individuals are shown. The upper one (a) contains several heavily staining remnants of ingested nuclei; inclusions such as this are characteristic of *E. gingivalis*. The lower one (b) lack inclusions but shows the typical nucleus (n) with its central karyosome, hardly distinguishable from *E. histolytica*.

Like most species of intestinal amoebae, *Entamoeba gingivalis* occurs in a large proportion of the general population; it is found in about one healthy mouth in ten. Although it obviously contributes nothing to the well-being of an infected individual, there is no evidence that it causes injury. Persons suffering from pyorrhea, a very common condition in the middle-aged and older, however, exhibit up to 95 per cent incidence; and as many as 50 per cent of children having carious teeth with much accumulated tartar but otherwise healthy mouths may be similarly infected.

Entamoeba gingivalis closely resembles *E. histolytica* in morphology (Fig. 21.9). There is the same central karyosome in the nucleus of both, but the clear halo seen about the karyosome of the latter species is replaced in *E. gingivalis* by a cloudy or shaded area in stained specimens. Like *histolytica, Entamoeba gingivalis* has a predilection for the cells of its host (in this case, the leucocytes, the disintegrating nuclei of which often crowd the amoeba's cytoplasm), but it is not a tissue invader. Its frequent occurrence in pyorrheic lesions and about carious or tartarous teeth probably indicates only that the amoeba finds such conditions favorable to its existence. Yet its ability to ingest and digest leucocytes (and, in culture, erythrocytes), and its occurrence in nearly all the gingival pus pockets of pyorrhea should excite suspicion until its harmlessness has been wholly proved.

Recognizing *Entamoeba gingivalis* presents no difficulty, since it is the only amoeba known to occur in the human mouth. The trophozoites are found in pus pockets about accumulated calculus below the gum line, and are numerous about the filamentous bacteria involved in tartar formation. In stained organisms the remains of ingested leucocytes are especially conspicuous. The amoebae measure from 12μ to 20μ.

Other Species of Human Amoebae

Numerous other species of amoebae have been described from the human alimentary tract, and even from various tissues and lesions, but those described above are the only ones generally accepted by parasitologists. Recently, however, evidence shows that two other species, *Entamoeba hartmanni* and *E. polecki,* both first described by Prowazek in 1912, may be valid and may occur in man. According to Burrows (1959), they are distinguishable from *Entamoeba histolytica* on grounds of size and morphology, and he thinks that what has usually been called "the small race of *Entamoeba histolytica*" is in reality *E. hartmanni.* If such is the case, *Entamoeba hartmanni* is indeed common, for about half the reported infections *E. histolytica* would be in this group. *Entamoeba polecki* was seen by Prowazek in pigs and in a child in Saipan, and several human infections with it have been described by other investigators since. Burrows and Klink (1955) believe that the species may be much commoner in man than is generally thought. Since both *Entamoeba hartmanni* and *E. polecki* closely resemble *E. histolytica,* their validity must for the present remain uncertain.

Other species whose validity is doubtful but which are often mentioned in the literature include:

Councilmania lafleuri Kofoid and Swezy, 1921
Councilmania tenuis Kofoid, 1928

Councilmania dissimilis Kofoid, 1928
Entamoeba dispar Brumpt, 1925
Entamoeba tetragena Viereck, 1907
Karyamoebina falcata Kofoid and Swezy, 1924, 1925

Diagnosis

Since the occurrence of parasitic amoebae of man is worldwide, accurate recognition is necessary. Diagnosis of human amoebiasis can only be made with certainty by finding cysts or trophozoites in the feces, since many persons are carriers and show no symptoms. Obviously, it is important to be able to distinguish *Entamoeba histolytica* from the nonpathogenic species, and there is no doubt that mistakes have often been made, with unfortunate results for both patient and doctor. Craig and Faust (5th ed. 1951, p. 109) remark, "In fact, the differential diagnosis of parasitic amoebae constitutes the most difficult task in clinical parasitology." Negative findings do not always mean freedom from infection, for cysts may be few in number; special methods (p. 590) and repeated examinations may be required to find them. It is likely, too, that the abundance of cysts in the feces varies considerably from time to time. It is worth remembering that the presence of *any* species of intestinal protozoan means that circumstances have at some time been favorable for infection with others, for all are acquired in the same way, by fecal contamination of food and water.

For the convenience both of the student and of those who may be entrusted with the laboratory examination of stool specimens, the characteristics of the various species have been summarized in Table 21.1.

Amoebae of the Lower Animals

An adequate picture of the problem of human amoebiasis is impossible without some knowledge of the amoebae of the lower animals. Amoebiasis in these hosts is a source of information about differing life cycles, and may give us a better understanding of the mechanism of pathological changes in human infections; it may also shed light on other factors in the host-parasite relationship.

Species of *Entamoeba* have been described from a great variety of animals: the guinea pig, domestic pig, mouse, rat, goat, horse, rabbit, cattle, birds, and even fish. Frogs are often parasitized by a species known as *Entamoeba ranarum*, not only almost indistinguishable from *E. histolytica*, but capable of causing liver abscess just as the latter species often does in man. Nor are reptiles immune to amoebiasis. Ratcliffe and Geiman (1938) studied both spontaneous and experimentally induced cases of infection with *Entamoeba invadens*, using as hosts lizards, alligators, and snakes, and concluded that "we cannot be certain that reptiles ever recover spontaneously." Not only

TABLE 21.1

MORPHOLOGY OF AMOEBAE OF MAN

TROPHOZOITES (*in vivo*)[a]

Characteristics	*Entamoeba gingivalis* (Gros, 1849)	*Entamoeba histolytica* (Schaudinn, 1903)	*Entamoeba coli* (Grassi, 1909)	*Endolimax nana* (Wenyon and O'Connor, 1917)	*Iodamoeba bütschlii* (*williamsi*) v. Prowazek, 1912)	*Dientamoeba fragilis* Jepps and Dobell, 1918
Habitat	mouth (especially about gums and teeth)	usually lining of colon	lumen of colon	lumen of colon	lumen of colon (probably)	colon (probably)
Size:[b]						
range	5–60μ	15–60μ	15–50μ	6–15μ	5–20μ	3–22μ
usual	(12–25μ)	(18–25μ)	(20–30μ)	(8–10μ)	(9–13μ)	(5–12μ)
Mobility	moderately active	marked, progressive	sluggish, not progressive	active (in fresh stools)	sluggish (soon lost)	active (in freshly passed stools)
Pseudopodia	hyaline, ectoplasmic (often several)	clear, glassy (finger-like, rapidly put forth)	grayish, not refractive	hyaline (finger-like in fresh stools)	hyaline, broad	hyaline, broad, blunt.
Nucleus	seldom visible	seldom visible	usually visible (ring of refractive granules)	seldom visible	usually invisible	usually invisible
Cytoplasm	ectoplasm and endoplasmic differentiation sharp	ectoplasm and endoplasmic differentiation sharp	grayish; cytoplasmic differentiation indistinct	granular	granular, vacuolar	granular, vacuolar; ecto-endoplasmic differentiation fairly sharp
Cytoplasmic inclusions	leucocyte nuclei and bacteria	often red cells	bacteria and debris	bacteria and debris	bacteria and debris	bacteria and debris

[a]Morphology of precystic and metacystic forms may differ somewhat from that of ordinary trophozoites, and is not included in this table.

[b]Races of differing size have been shown to exist for *Entamoeba histolytica*, and may also exist in some or all of the other species.

TABLE 21.2

MORPHOLOGY OF THE INTESTINAL AMOEBAE OF MAN

Characteristics	*Entamoeba histolytica*	*Entamoeba coli*	*Endolimax nana*	*Iodamoeba bütschlii*
CYSTS (unstained)				
Size[a]	5–20μ	12–35μ (12–22μ usually)	5–12μ in length (mean 7–8μ)	6.4–16.6μ (mean 9.1μ–highly variable)
Shape[b]	round	round	oval	irregular
Nuclei (number)	1–8 (usually 4)	1–32 (usually 8)	1–4 (usually 4)	1
Appearance of nuclei	refractile bodies (difficult to see)	refractile rings	usually not seen	usually not seen
Chromatoidal bars (young cysts)	blunt; refractile	splintery; refractile	absent?	absent
Glycogen masses	often 1 or 2 seen as vacuolar masses in young cysts	often a large vacuolar mass in young cyst	absent	very large
CYSTS (stained)				
Nuclei				
Nuclear membrane	delicate	relatively heavy	thin; no chromatin granules	relatively heavy
Karyosome	small; usually central	easily seen; usually eccentric	large; heavy	large, heavy; usually eccentric; granular
Chromatoidal bars	blunt (intensely black with haematoxylin)	splintery (intensely black with haematoxylin)	none	none
Glycogen	unstained with haematoxylin; deep brown with iodine			

[a]Races differing in size exist in *E. histolytica* and *E. coli* and perhaps for the other two species also.

[b]Shape of cyst in all four species somewhat variable.

did the amoebae produce lesions of the colon in these hosts, but liver abscesses and even gastric ulcers.

Even Protozoa may be infected by species of these ubiquitous parasitic amoebae. *Volvox* harbors one and so does *Zelleriella*, an opalinid ciliate from frogs and toads (Stabler and Chen, 1936). In some cases certain of these amoebae were found to be themselves parasitized by a "*Sphaerita*-like" organism (believed to be a fungus). Thus we have a case of a hyper-hyper-parasite (a parasite which parasitizes a parasite which parasitizes a parasite)!

Mouth amoebae are also ubiquitous. *Entamoeba gingivalis* was found by Hegner and Chu (1930) in nearly half of the Philippine monkeys they examined, and similar forms have been seen in the mouths of dogs, cats, and horses. Dogs may even harbor oral amoebae of human origin, possibly as the result of too intimate relationships with their over-affectionate owners. One wonders how natural infections are acquired, since these animals have never been observed to kiss, but since the trophozoites of *Entamoeba gingivalis* are highly resistant to changes in pH, chilling, and even drying, infection may result from the ingestion of contaminated food and water.

Among the lower animals, amoebae of the *Endolimax* type have been reported from a variety of mammals, birds, reptiles, and even amphibia. *Iodamoeba* has not only been found in monkeys, gorillas, and pigs, but even in South African turtles. Only *Dientamoeba* so far appears to be unrepresented among vertebrates below the primate level.

Invertebrates and perhaps even plants have their share of parasitic amoebae. Endamoebae are especially widely distributed in nature, and have been found in hosts as diverse as the box elder bug, cockroach, grasshopper, larvae of the crane fly, and bee; even the *Hydra* may be infected. Species of both *Endamoeba* and *Endolimax* have been reported from leeches, and amoebae belonging to the latter have also been seen in fleas and box elder bugs. Amoebae of other genera have been described from bees and grasshoppers, but *Iodamoeba* and *Dientamoeba* are so far not known to occur in invertebrate hosts. Intensive search would probably result in the discovery of many additional species.

Origin of Parasitism by Amoebae

The origin of such widespread parasitism must have occurred in the distant past. The frequency of coprozoic types, or free-living forms which find animal excreta a favorable environment, suggests the first step in such evolution. The cysts of such forms are ubiquitous and can pass through the alimentary tract of their "hosts" without injury, as is evidenced by their frequent occurrence in stools which are not immediately examined after passage. The literature of parasitology contains many descriptions of new "species" of intestinal amoebae that were really coprozoic in origin. Such errors are easy to make since, except for nuclear structure, parasitic amoebae do not appear to differ in morphology very much from many of the more minute free-living species.

Despite the intensive study which the amoebae of man have received, our knowledge is still fragmentary. Although we can easily cultivate many parasitic amoebae (p. 569) we still know little of their physiology. Careful search of animals and plants for parasitic Protozoa is likely to reveal many new species, knowledge of which might help us to form some conception of how human parasitism by these rhizopods developed in the first place. It is unlikely that such species will all prove to be commensals. Some may be pathogens important to the farmer or to the conservationist, and others may be found beneficial from a human point of view because they harm injurious hosts.

REFERENCES

Boeck, W. C. and Drbohlav, J. 1925. The cultivation of *Endamoeba histolytica. Am. J. Hyg.*, 5: 371–407.

Bundesen, H. 1935. Plumbing in relation to infectious disease. *Am. J. Trop. Med.*, 15: 455–66.

Burrows, R. B. 1959. Morphological differentiation of *Entamoeba hartmanni* and *E. polecki* from *E. histolytica. Am. J. Trop. Med. and Hyg.*, 8: 583–9.

——— and Klink, G. E. 1955. *Entamoeba polecki* infections in man. *Am. J. Hyg.*, 62: 156–67.

——— and Swerdlow, M. A. 1956. *Enterobius vermicularis* as a probable vector of *Dientamoeba fragilis. Am. J. Trop. Med. and Hyg.*, 5: 258–65.

———, ———, Frost, J. K. and Leeper, C. K. 1954. Pathology of *Dientamoeba fragilis* infections in the appendix. *Am. J. Trop. Med. and Hyg.*, 3: 1033–9.

Cleveland, L. R. and Sanders, E. P. 1930. Encystation, multiple fission without encystment, excystation, metacystic development, and variation in a pure line and nine strains of *Entamoeba hystolytica. Arch. f. Protistenk.*, 70: 223–66.

Craig, C. F. 1934. *Amebiasis and Amebic Dysentery.* Springfield: Charles C Thomas.

Derrick, E. H. 1948. A fatal case of generalized amebiasis due to a protozoon closely resembling, if not identical with, *Iodamoeba bütschlii. Trans. Roy. Soc. Trop. Med. and Hyg.*, 42: 191–8.

Dobell, C. 1920. *The Amoebae Living in Man.* New York: Wood.

——— 1928. Researches on the intestinal Protozoa of monkeys and man. I. General introduction., and II. Descriptions of the whole life-history of *Entamoeba histolytica* in cultures. *Parasitology*, 20: 357–412.

——— 1938. Researches on the intestinal Protozoa of monkeys and man. IX. The life-history of *Entamoeba coli* with special reference to metacystic development. *Parasitology*, 30: 195–238.

——— 1940. Researches on the intestinal Protozoa of monkeys and man. X. The life-history of *Dientamoeba fragilis:* observations, experiments, and speculations. *Parasitology*, 32: 417–61.

——— 1943. Researches on the intestinal Protozoa of monkeys and man. XI. The cytology and life-history of *Endolimax nana. Parasitology*, 35: 134–58.

Faust, E. C. 1954. *Amebiasis*. Springfield: Charles C Thomas.

Hegner, R. W. 1928. The evolutionary significance of the protozoan parasites of monkeys and man. *Quart. Rev. Biol.*, 3: 225–44.

——— and Chu, H. J. 1930. A comparative study of the intestinal Protozoa of wild monkeys and man. *Am. J. Hyg.*, 12: 62–108.

Kofoid, C. A., Hinshaw, H. C., and Johnstone, H. G. 1929. Animal parasites of the mouth and their relation to dental disease. *J. Am. Dent. Assn.*, August, p. 20.

———, McNeil, E., and Kopac, M. J. 1931. The chemical nature of the cyst wall in human intestinal Protozoa. *Proc. Soc. Exp. Biol. and Med.*, 29: 100–2.

——— and Swezy, O. 1925. On the number of chromosomes and the type of mitosis in *Endamoeba dysenteriae*. *Univ. Calif. Pub. Zool.*, 26: 331.

Kudo, R. R. 1944. *Manual of Human Protozoa*. Springfield: Charles C Thomas.

Leidy, J. 1879. On *Amoeba blattae*. *Proc. Acad. Nat. Sci. Phila.*, 31: 204–5.

Loesch, F. 1875. Massenhafte Entwickelung von Amöben im Dickdarm. *Arch. f. Path. Anat.*, 65: 196–211.

Morris, S. 1936. Studies of *Endamoeba blattae* (Bütschli). *J. Morphol.*, 59: 225–62.

Phillips, B. 1950. Cultivation of *Endamoeba histolytica* with *Trypanosoma cruzi*. *Science*, 111: 8–9.

Ratcliffe, H. L. and Geiman, Q. M. 1938. Spontaneous and experimental amebic infection in reptiles. *Arch. Path.*, 25: 160–84.

Smith, S. C. 1928. Host-parasite relations between *Iodamoeba williamsi* and certain mammalian hosts (guinea pigs and rats). *Am. J. Hyg.*, 8: 1–15.

Stabler, R. M. and Chen, T.-T. 1936. Observations on an *Endamoeba* parasitizing opalinid ciliates. *Biol. Bull.*, 70: 56–71.

Swerdlow, M. A. and Burrows, R. B. 1955. *Dientamoeba fragilis*, an intestinal pathogen. *J.A.M.A.*, 158: 176–8.

Tyzzer, E. E. and Geiman, Q. M. 1938. The ingestion of red cells by *Endamoeba coli* and its significance in diagnosis. *Am. J. Hyg.*, 28: 271–87.

Wenrich, D. H. 1940. Studies on the biology of *Dientamoeba fragilis*. *Proc. 3rd Int. Cong. Microbiol.*, pp. 408–10.

——— 1944a. Nuclear structure and nuclear division in *Dientamoeba fragilis* (Protozoa). *J. Morphol.*, 74: 467–91.

——— 1944b. Studies on *Dientamoeba fragilis* (Protozoa). IV. Further observations, with an outline of present-day knowledge of this species. *J. Parasitol.*, 30: 322–38.

———, Stabler, R. M., and Arnett, J. H. 1935. *Endamoeba histolytica* and other intestinal Protozoa in 1,060 college freshmen. *Am. J. Trop. Med.*, 15: 331–45.

Wenyon, C. M. 1926. *Protozoology*. New York: Wood.

22

The Parasitic Ciliates and the Opalinidae

"Because I couldn't satisfy myself about the animalcules . . . I continued my observations, and at last I came across some frogs in whose dirt, which I took out of the guts, I beheld an inconceivably great company of living animalcules, and these of divers sorts and sizes. . . ."* Leeuwenhoek, Letter 38, July 16, 1683.

Like the parasitic flagellates and amoebae, the majority of ciliates that have adopted parasitism as a mode of life have selected (if we may use the word) the alimentary tract of their hosts as their home. It is easy to reach and it is presumably also easy in which to make (or take) a living. In it there are no enemies except occasional hyperparasitic microorganisms, and of course the leucocytes of the host itself. But the latter are stirred into activity only when the parasite injures its host sufficiently to excite inflammation, and as we have seen, most parasitic organisms do little injury.

Although little fossil evidence supports the assertion, we are certain that the Ciliophora are a very ancient group, if less ancient than the flagellates and the amoeboid Protozoa. Perhaps for this reason fewer of them have become parasites. Possibly, too, the greater specialization of structure the ciliates have evolved implies refinements in their physiological requirements that make adaptation to life within other organisms difficult.

Nevertheless, many species of parasitic ciliates are known, and surely many remain to be discovered. Some, like those of the genus *Balantidium*, are much like their free-living brethren in structure and life cycle; others, such as the Opalinidae,† do not resemble any other known Protozoa.

*Dobell considers that the "animalcules" were undoubtedly *Opalina* and *Nyctotherus*.

†Although the Opalinidae have usually been regarded as ciliates there is an increasing body of opinion that they may be more closely related to the flagellates. The established usage is however followed here.

BALANTIDIUM COLI

Of the parasitic ciliates (some are shown in Fig. 22.1), *Balantidium coli* is the only species certainly occurring in man. It inhabits the colon, where its effects are similar to those of *Entamoeba histolytica*. Fortunately, balantidial infection (or balantidiasis) is far less frequent in most parts of the world than amoebiasis, but it is very common in pigs, in which it seems usually to do little harm. Man is relatively resistant to infection, and perhaps for this reason he is often a carrier. Nevertheless, although the parasites may be transmitted from person to person, most human cases appear to be of porcine origin.

Doubtless *Balantidium coli* was originally a parasite of swine, as it still is primarily. When man domesticated the pig he often lived under the same roof with it, as he does even now in many places, and there was soon a mutual exchange of parasites. From his valued four-footed house guests man picked up *Balantidium coli* and reciprocated by presenting them with one of his own amoebae, *Iodamoeba bütschlii*, or so it is generally believed. The pig and man also share the intestinal round worm, *Ascaris lumbricoides*, but which of the two host species has the other to thank for its original infection remains uncertain.

Laboratory infections in several species of experimental animals, including man and the monkey, have been achieved, but the degree to which natural infection exists in these and other species of lower animals has not been settled. Scott (1927) believed that the *Balantidium* of the guinea pig belonged to the same species as that of man and the pig, and worked out the life cycle in some detail, finding it essentially like that of the free-living ciliates. Recently Westphal (1957) has shown that guinea pigs are easily infected with the pig balantidium, although they seem quite uninjured by it. Additional evidence of the ubiquity of this ciliate is its discovery in rats by Netik *et al.* (1958), although cross-infection experiments should be performed before one can assume that the rat balantidium is actually the same as the form occurring in pigs and man. Quite possibly physiologically distinct races of *Balantidium* exist; Young (1950) found it difficult to infect human volunteers even when the parasites are derived from another human host. There is still some controversy as to whether the pig may not harbor a species distinct from *Balantidium coli*. McDonald (1922) described a ciliate he called *Balantidium suis* on the basis of rather slight morphological differences, but Levine (1940) found that these differences often disappeared after cultivation. Most parasitologists believe that the pig and man harbor the same species.

Balantidial infection is common in numerous species of vertebrates, among them frogs and toads (in which the incidence is very high), fish, tortoises, birds, and cattle. In most instances we are fairly sure that different

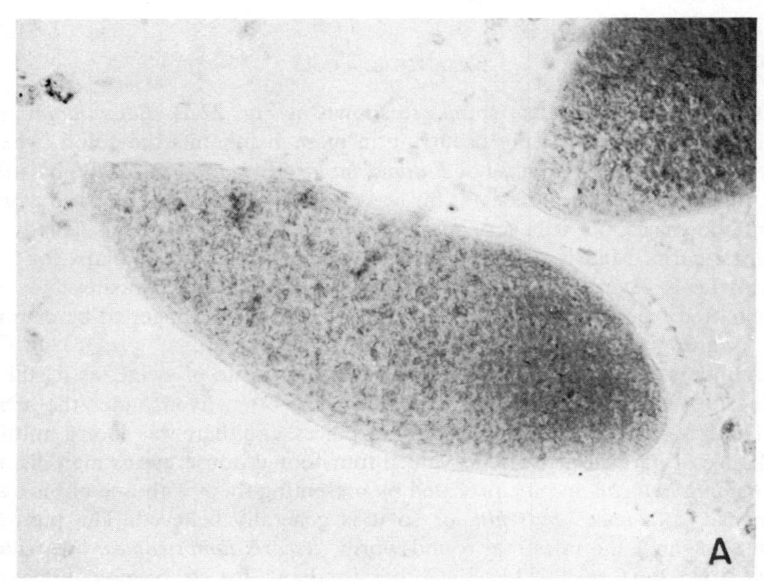

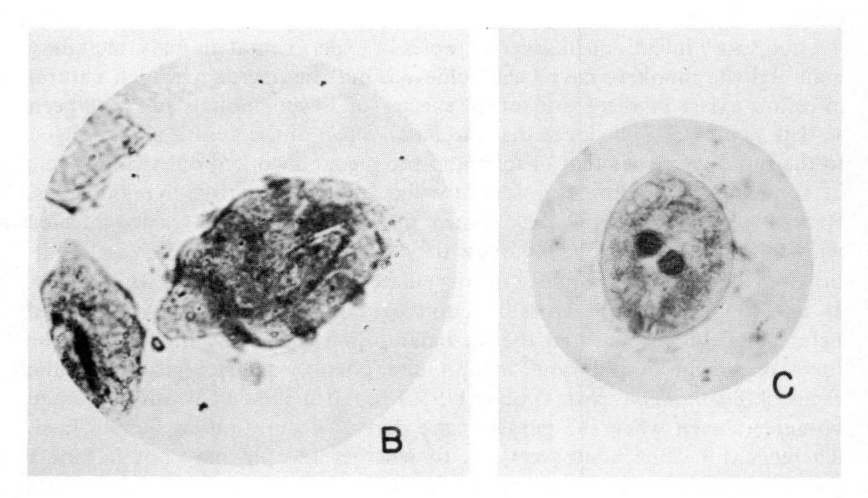

FIG. 22.1. Parasitic ciliates. *A. Opalina ranarum*, consisting of a flattened plate of protoplasm with many nuclei, all of which are of a single type. These may be faintly seen as rings of chromatin in the cytoplasm. $\times 385$. *B. Troglodytella abrassarti*, a common ciliate in the colon of chimpanzees. The three ciliary zones may be clearly seen, the two central belts being discontinuous (i.e. forming incomplete circlets). Also visible is the L-shaped macronucleus, the lower bar of which lies just under the most posterior circlet, or that nearest the pointed end. $\times 195$. *C. Balantidium coli*. The peristomal opening is visible near the anterior tip, and the cytopyge at the other end. Though the figured specimen seems to have two macronuclei, a single kidney-shaped macronucleus is the rule. $\times 385$.

species of *Balantidium* are involved, but there is ample opportunity for much research on this genus, especially as it is well represented in the gut population of a variety of invertebrates. Thorough and detailed studies of the nutritional requirements and host-parasite relationships of the balantidia of different hosts might shed much light on the probable evolutionary history of the genus. The name *Balantidium* was originated by Claparède and Lachmann for a gut ciliate found in frogs, and previously placed in the genus *Bursaria* by Ehrenberg. Fauré-Fremiet (1955) has shown that the real relationships of these parasites are with the Plagiopylidae, free-living ciliates of the Trichostomatida, on the grounds of resemblances in infraciliature and division.

Scott found that the *Balantidium* of the guinea pig reproduces by mitotic binary fission, conjugates similarly to *Paramecium caudatum*, and forms cysts in which, however, no reproduction occurs. The cyst is, as with other intestinal Protozoa, the infective stage.

Balantidium coli is a relatively large ciliate and has a superficial resemblance to *Paramecium*, although the position of the mouth is entirely different. Indeed Malmsten, who discovered the organism in 1857, called it *Paramecium coli.* It ranges in length between 50 and 100μ, and has a width about three-quarters as great. Its shape is more or less oval, with a circular cross section. At the anterior end, slightly eccentric in position, is a conical peristome with a mouth or cytostome at its base. Posteriorly and almost opposite lies the cytopyge or cell anus. The ciliation is nearly uniform, only the peristomial cilia being longer than the rest. Stained specimens show a large kidney-shaped macronucleus, in the concavity of which rests a small micronucleus. Unlike most parasitic Protozoa, such ciliates may have a contractile vacuole and *Balantidium coli* has two. The larger one lies anteriorly, while the other is close to the cytopyge.

The cyst is spherical or oval, varying in length between 45 and 65 micra. The cyst wall appears double, and in stained specimens the nuclei are conspicuous. The cytoplasm in stained preparations often seems to have drawn away somewhat from the cyst wall, leaving a considerable space. Organisms about to encyst round up; and with the secretion of a cyst wall activity within gradually ceases. Excystment may occur in cultures, though under natural conditions it probably takes place only within the colon of the prospective new host. Scott found the process requiring about half an hour. Little is known about the longevity of cysts in nature, although Scott remarked, "The cyst apparently retains its viability indefinitely if kept from complete drying." She also commented, however, on the greater length of time required by older cysts for excystation, which probably indicates a reduction of vitality with age.

Human infection with *Balantidium coli* apparently occurs throughout much of the world, although clinical cases appear to be uncommon in most areas. Since the majority of cases in man are thought to originate from contact with pigs, it would seem that the incidence of balantidiasis in countries such as India, where because of religious scruples little pork is consumed, would be markedly lower than elsewhere, but critical studies of this kind have not been made.

Treatment of balantidial infection is often followed by relapse, as in many other protozoan diseases. Nevertheless, the clinical manifestations can usually be cleared up by drugs such as carbasone (an arsenical) and the antibiotics aureomycin and terramycin. Extended post-treatment observation is necessary to determine whether an apparent cure is permanent.

NYCTOTHERUS

Several species of ciliates, other than *Balantidium coli*, have also been claimed as human parasites, but their validity is questionable. Among them are *Balantidium minutum* and *Nyctotherus faba*, both described by Schaudinn in 1899. Schaudinn was a justly famous microbiologist, but since his discoveries of these parasites have remained unconfirmed, their existence must continue doubtful. Perhaps they were coprozoic forms.

Whether represented in man or not, species of *Nyctotherus* are known from a number of hosts, among them frogs, toads, fish, and insects as diverse as termites, cockroaches, and crickets. There is reason to think that Leeuwenhoek was the first to see *Nyctotherus* when he examined the fecal contents of the frog.

Ciliates belonging to this genus may be recognized by the deep laterally placed peristome ending in a cytostome and gullet. The peristome begins at the anterior end and is lined with a row of long cilia. Lying in the cytoplasm is a large reniform macronucleus with a minute micronucleus lying in its concavity. *Nyctotherus cordiformis*, best known of the species occurring in frogs and very common among them, is about the size of *Balantidium coli*, to which it bears a considerable superficial resemblance. Other species found in frogs are even larger.

Unlike many species of parasites, those belonging to both genera just discussed show little structural modification as the result of parasitism. Presumably this is because both *Balantidium* and *Nyctotherus* live in the intestinal lumen and are primarily dependent on solid food, just as were their free-living ancestors, and seldom take to invading or ingesting the tissues of their hosts. In parasites of both genera, the mouth is a very conspicuous part of the anatomy, and the resemblance to such free-living forms as *Bursaria* is quite striking—so much so, indeed, that Ehrenberg, who studied *Nyctotherus cordiformis* in 1838, called it *Bursaria cordiformis*. Probably these organisms live largely on bacteria and undigested dietary residues in the intestine, although in cultures, at least, *Balantidium coli* seems to be a vegetarian, living chiefly on starch. This appears to be a gross neglect of opportunities, in view of the leucocytes and even erythrocytes so readily available from its host. Surprising, too, is its ability to tolerate equally well (Schumaker, 1931) an aerobic and anaerobic environment, since conditions in the intestine are strictly anaerobic. Thus we may perhaps conclude that the physiological changes that have taken place as the result of parasitism are not very great either.

Other Ciliates of Mammals

Next to such ciliate genera as *Balantidium* and *Nyctotherus*, species of which are or have been claimed to be parasites of man, probably the best known are the numerous genera represented in the rumen fauna of cattle. This part of the digestive tract normally teems with such Protozoa. Related forms occur in other hoofed herbivores, particularly horses, camels, goats, and sheep, and species are even known from the gut of anthropoid apes. Apparently none of these parasites (and there are many species) plays any significant part in the digestive activities of the host, or in any other way affects its vital economy. In short, they seem to have become perfect commensals. Yet even so they may be symbionts of a sort, for it is very difficult to show that they are never of any benefit to the host. They may, for example, aid somewhat in the digestion of the host's food, or even contribute to satisfying its vitamin requirements, although it seems more likely that the rumen bacteria do the latter and that the Protozoa may even need to use a part of the product, especially of B vitamins, for their own needs (Oxford, 1955). Hungate (1955) has suggested that these ciliates may even supply the host with as much as one-fifth of its protein and energy sources, and there is some evidence that this may be so. Remarkably enough, the same suggestion was made more than a century ago, though it was then nothing more than a shrewd guess. Nevertheless it is certain that these Protozoa do nothing for the host which it cannot do for itself, for Becker (1932) long ago showed that defaunated animals, or those protected from primary infection, did quite as well as their inhabited stable mates.

These ciliates are interesting not only because of the unsolved problems still surrounding their host relationships, but also because of the remarkable specialization in structure which many of them exhibit—oddities of form for which it is often difficult to suggest a use or rational explanation. Some of these parasites are bizarre enough in appearance to rival prehistoric monsters, were they of comparable size. Such a one is *Troglodytella abrassarti*, a ciliate common in the colon of chimpanzees (Fig. 22.1*B*).

On the whole, these Protozoa are notable for their strong ciliation, possessing large membranelles which doubtless facilitate local travel in the viscid contents of the digestive tract. However, many species also are equipped with an apparatus of skeletal plates for which it is much harder to see any utility, and there is often a complex arrangement of posterior spinous processes which scarcely seems even ornamental. *Diplodinium* and *Ento-*

dinium are good examples of these highly evolved forms, and there are numerous other genera (Fig. 22.2). The structure of *Eudiplodinium* is shown in Figure 22.3. Dividing forms are shown in Figure 22.4.

In addition to oligotrichs such as these, there are also some genera of holotrichous ciliates occurring in the same environment, less spectacular in their morphology but often very numerous. It has been estimated that there may be as many as 200,000 per ml in the rumen of sheep and goats (Ferber, quoted

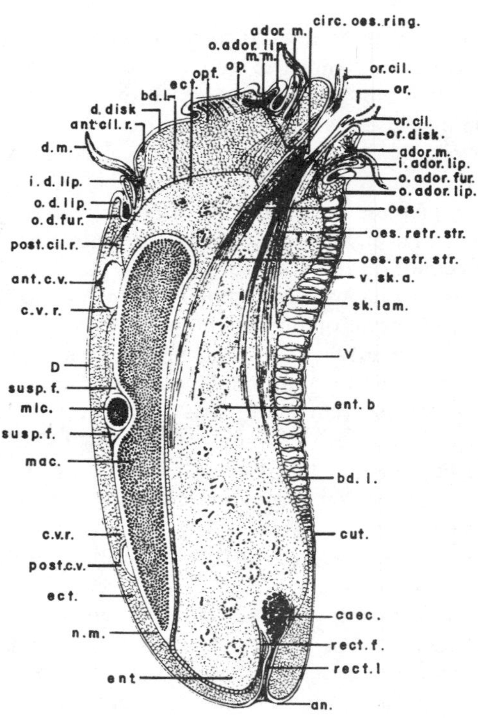

FIG. 22.2. *Diplodinium ecaudatum*, a cattle ciliate. (A schematic longitudinal section, as worked out by Sharp.) ador. m., adoral membranelles; an., anus; ant. cil. r., anterior ciliary roots; ant. c. v., anterior contractile vacuole; bd. l., boundary layer (ectoplasmic); cir. oes. r., circumesophageal ring; caec., caecum; cut., cuticle; c. v. r., region about contractile vacuole; D., dorsal side of body; d. disk, dorsal disk; d. fur., dorsal furrow; d. m. str., dorsal motor strand; d. m., dorsal membranelles; ect., ectoplasm; ent,. endoplasm; fd. vac., food vacuoles; i. ador, lip, inner adoral lip; i. d. lip, inner dorsal lip; L., left side of body; l. sk. a., left skeletal area; mac., macronucleus; mic., micronucleus; m. m., motor mass (motorium); o. ador. fur., outer adoral furrow; o. ador. lip, outer adoral lip; o. d. fur., outer dorsal furrow; o. d. lip, outer dorsal lip; oes., esophagus or cytopharynx; oes. f., esophageal fibers; oes. retr. str., esophageal retractor strands; op., operculum; op. f., opercular fibers; or., oral opening, mouth, or cytostome; or. cil., oral cilia; or. disk, oral disk; post. cil. r., posterior ciliary roots; post. c. v., posterior contractile vacuole; R., right side of body; rect., rectum; rect. f., rectal fibers; r. sk. a., right skeletal area; sk. lam., skeletal laminae; susp. f., suspensory fibers; V., ventral side of body; sk. lam., skeletal laminae; v. sk. a., ventral skeletal area; n. m., nuclear membrane.

by Hungate, 1955). Forms such as species of *Isotricha* and *Dasytricha*, superficially resembling free-living infusoria such as paramecium, are very common denizens of the stomach of cattle and sheep; others are encountered frequently enough in the gut of horses and even of guinea pigs.

At least in kids, transmission of these parasites appears to occur through

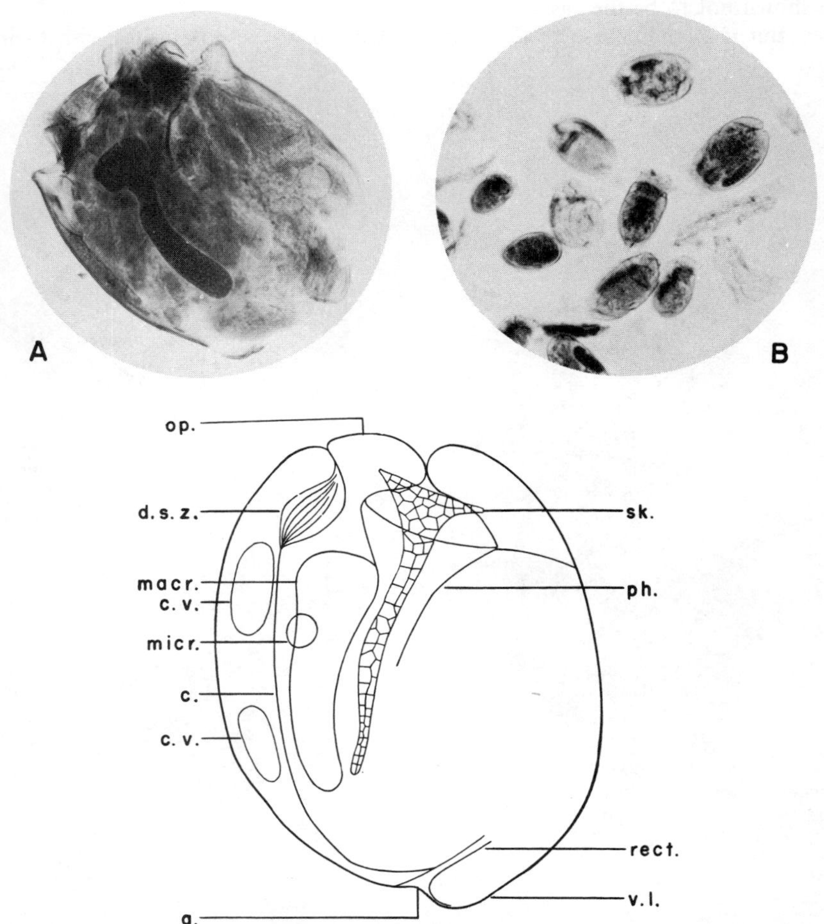

FIG. 22.3. Cattle ciliates. *A. Eudiplodinium maggii*, ×257. Note the adoral and dorsal zones of cilia at the anterior end, and the peculiar hooklike anterior end of the rod-shaped macronucleus. Just to the right of the latter the skeletal plate is visible as a slightly darker, narrow strip. *B.* A field chosen at random showing a densely packed group of ciliates from the rumen of cattle. Several species are probably represented. ×257. *Below.* Structure of the ciliate *Eudiplodinium neglectum*, from the stomach of the moose. a., anus; c., longitudinal cuticular line; c.v., contractile vacuole; d.s.z., dorsal synciliary zone; macr., macronucleus; micr., micronucleus; op., operculum; ph., pharynx or gullet; rect., rectum; sk., skeletal plate; v.l., ventral lobe. (After Kraschenninikow, *J. Protozool.*, 2: 124.)

the ingestion of contaminated food and water. Young animals doubtless initially acquire such internal guests with almost their first drink. Until the work of Becker and associates (1930; 1932) it had long been suspected that these ciliates might be as essential to the welfare of their hosts as are the gut flagellates of termites and woodroaches, but this was quite conclusively shown not to be the case.

But if in the course of the long association of these parasites with their

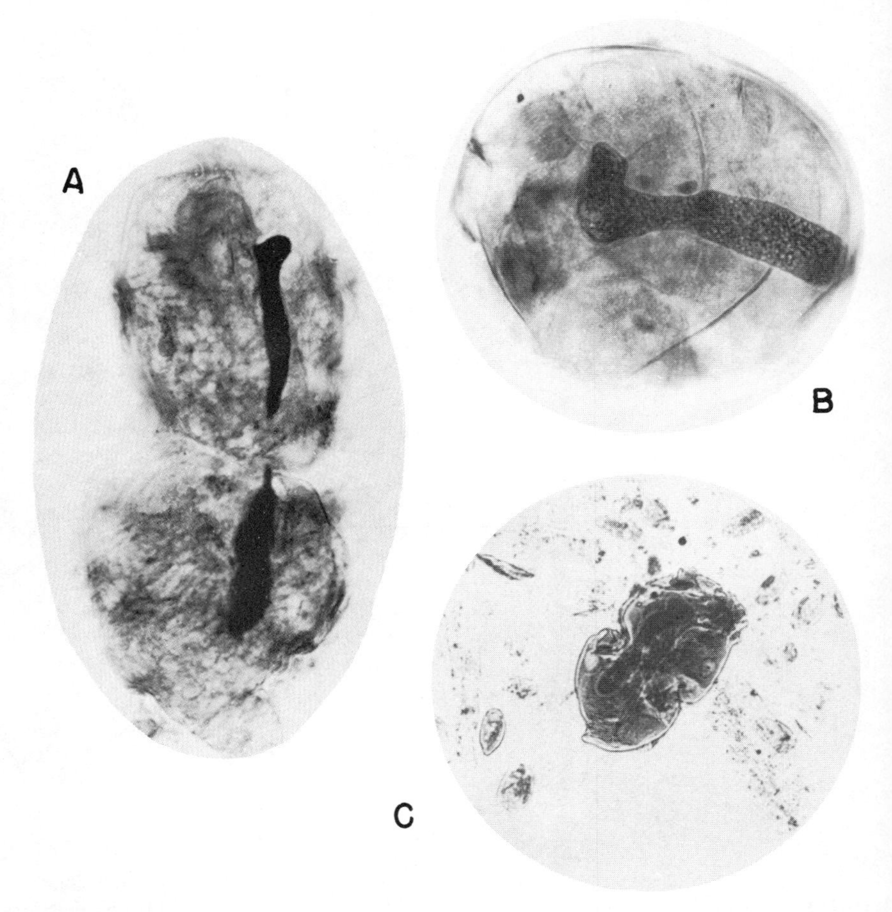

FIG. 22.4. *Troglodytella* and *Eudiplodinium* in division. *A. Troglodytella abrassarti* in telophase. The new macronuclei, just separated, can be clearly seen. Less distinct are the circlets of cilia. ×385. *B. Eudiplodinium maggii* in a very early stage of division, the micronucleus clearly visible in telophase. ×385. *C. Eudiplodinium maggii* in a late stage of division, the elongated macronucleus and new skeletal plates next to it visible. Note zone of constriction where the separation of the parent organism into two daughter individuals will take place. ×118.

hosts, the latter have been unable to find more than an incidental use for them (and even this is not certain), the case is otherwise for the parasites. They have become so well adapted to their peculiar environment that apparently no other will do. Their living standards seem to rise and fall with those of the host. Becker (1932) remarked that the ciliate population of the rumen of a goat fed only on hay numbered only 118,000 per ml, but when the ration was supplemented by a liberal allowance of grain the count rose to about 3 million per ml, and that they then amounted to almost 30 per cent by volume of the rumen contents. Despite the fecundity of the parasites in their natural surroundings, cultivation *in vitro* has proved a difficult problem, although maintenance in the laboratory for a time is possible. For this purpose Hungate has devised buffer solutions in which the ciliates remain alive long enough for some physiological studies. Species of *Diplodinium* and *Entodinium* may be cultured on media containing cellulose, dried grass, and sometimes starch (Hungate, 1955), but so far nothing remotely approaching axenic cultivation has been achieved. All the rumen ciliates are strict anaerobes.

The oligotrich ciliates of the mammalian gut appear to constitute a natural biological group, without obvious relationships to any other known free-living Protozoa. Within the group, however, there are indications of evolutionary relationships among the genera, and a genealogical tree showing such kinship, as worked out by the late Russian zoologist Dogiel (1946), is reproduced on p. 444. In Dogiel's opinion, both the mode of nutrition and the simpler structure of the genera *Entodinium* and *Anoplodinium* justifies their position at the base of the tree, and he suggests that the divergence which resulted in the evolutionary development of other genera in the group must have occurred at least as long ago as the Eocene—or more than 50 million years before our time.

Opalinidae

A second natural group of ciliates (though, as we shall see, their ciliate affinities are questionable) is the Opalinidae. Virtually all known species of the family are parasites in the gut of the Anura, or tailless Amphibia, and the incidence of infection is so high among them that an Opalina-less frog is extremely rare. Like the cattle ciliates, these parasites seem neither to harm nor to benefit their hosts.

The opalinids are remarkable forms. Externally, they show little structure other than numerous parallel rows of cilia. There is the usual differentiation of cytoplasm into ecto- and endoplasm. A mouth and therefore, of course, food vacuoles are lacking. Instead of a single nucleus, or the two types of nuclei typical of other ciliates, many very small nuclei (except in a few binucleated species) are scattered through the cytoplasm, as are numerous even smaller ovoidal bodies of chromatinlike material. Most opalinids are more or less leaflike in form, although species of the genera *Protoopalina* and

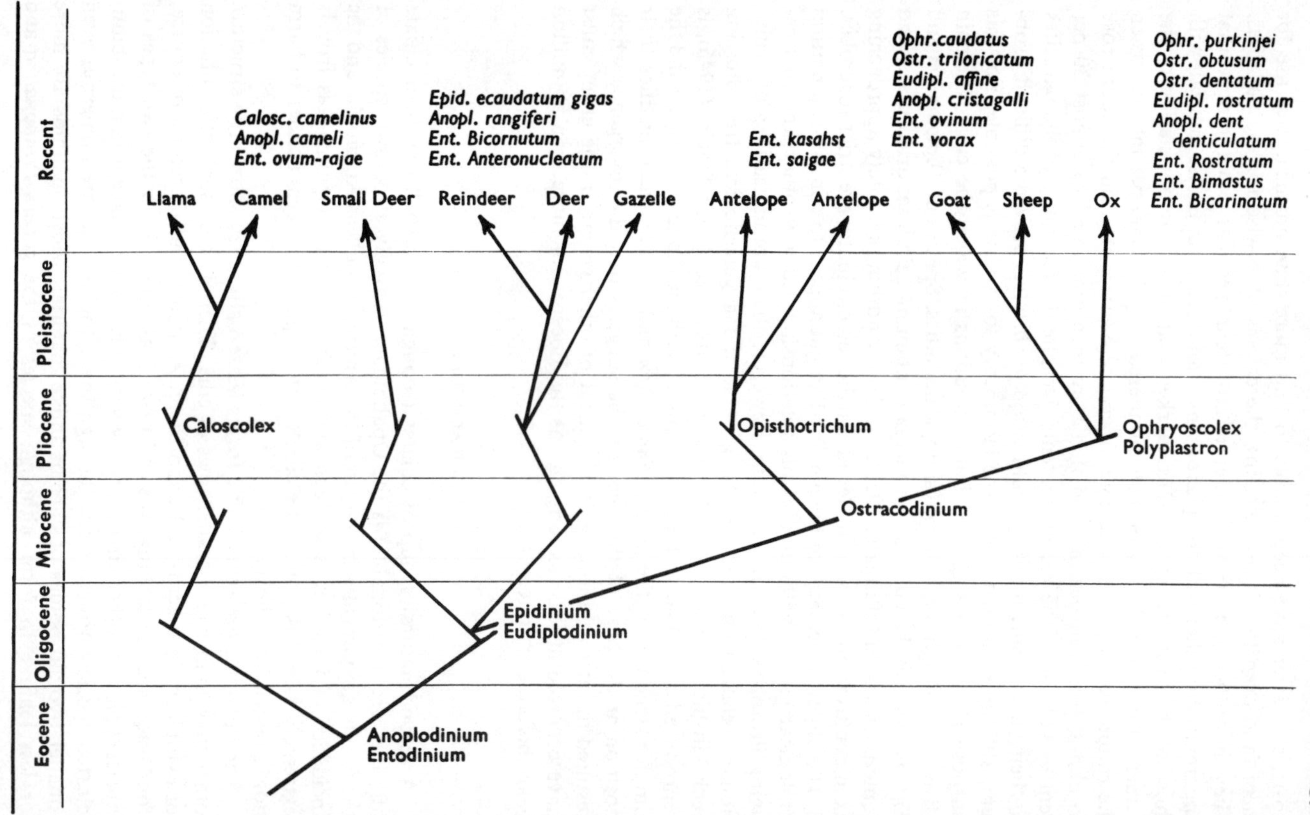
Recent
Pleistocene
Pliocene
Miocene
Oligocene
Eocene
Calosc. camelinus
Anopl. cameli
Ent. ovum-rajae
Epid. ecaudatum gigas
Anopl. rangiferi
Ent. Bicornutum
Ent. Anteronucleatum
Ent. kasahst
Ent. saigae
Ophr.caudatus
Ostr. triloricatum
Eudipl. affine
Anopl. cristagalli
Ent. ovinum
Ent. vorax
Ophr. purkinjei
Ostr. obtusum
Ostr. dentatum
Eudipl. rostratum
Anopl. dent
denticulatum
Ent. Rostratum
Ent. Bimastus
Ent. Bicarinatum
Llama
Camel
Small Deer
Reindeer
Deer
Gazelle
Antelope
Antelope
Goat
Sheep
Ox
Caloscolex
Opisthotrichum
Ophryoscolex
Polyplastron
Ostracodinium
Epidinium
Eudiplodinium
Anoplodinium
Entodinium

Cepedea may be cylindrical. Some reach a considerable size, even a millimeter in length, and are easy to see even without a microscope. *Opalina ranarum* is one of the better known common species (Fig. 22.1*A*).

The life histories of some of the 200 or more known species of the Opalinidae have been studied and described in detail. Division involves prior mitosis of the nuclei, followed at length by binary fission of the organism, the plane of division being parallel to the rows of cilia. Encystment occurs, and as in most protozoan parasites, the cyst is the infective stage. Reproduction does not occur in the cysts.

A remarkable feature of the life history is the manner in which the parasites respond to physiological events within the host. Just prior to egg-laying in the early spring, nature, with what seems like admirable foresight, arranges things so that division and encystment of the ciliates is greatly accelerated, with the result that the pond bottom is of course well seeded with cysts infective for the coming tadpole generation. Doubtless in the long course of evolution the parasites have become nicely adjusted to the hormonal mechanism of the host, just as in the case of the flagellates of termites and woodroaches (though meiosis among these flagellates is not always correlated with events in the host's life cycle).

From the cysts picked up by the foraging tadpoles, minute opalinids soon emerge in the course of passage through the gut, and the excysted individuals begin a series of divisions culminating in the production of gametes, the male being, as is usually the case, considerably smaller than the female. Sexual fusion soon takes place, and the zygote grows and initiates a period of active division, presumably ending only when the opalinid population has reached the limit of available food supply. The whole cycle is shown graphically in Figure 22.5.

From the foregoing account we see that the Opalinidae differ in several important ways from the true ciliates. Metcalf (1923), who studied the group intensively, was aware of these peculiarities and interpreted them as indicating the opalinids were more primitive than other ciliates. Thus he proposed the creation of a separate subclass, the "Protociliata," for this one family, and most parasitologists since have followed his taxonomic usage. However, Grassé (1952), Corliss (1955), and others have suggested that the true affinities of the opalinids are with the animal-like flagellates, and Grassé has proposed a new superorder, the "Opalinina," for their reception. Their argument is based on the absence of a micronucleus, or rather the absence of the two types of nuclei typical of true ciliates, the occurrence of division in a plane parallel to the rows of cilia (or "flagella," if their theory is accepted), and sexual reproduction unlike typical ciliate conjugation. Metcalf himself seriously considered the possibility of opalinid-flagellate kinship, for he remarked (Metcalf, 1923) "Among the flagellates the aberrant genus *Trichonympha** most nearly approaches the Opalinidae in structure . . .

**Trichonympha* is a genus of symbiotic flagellates inhabiting the gut of termites and woodroaches.

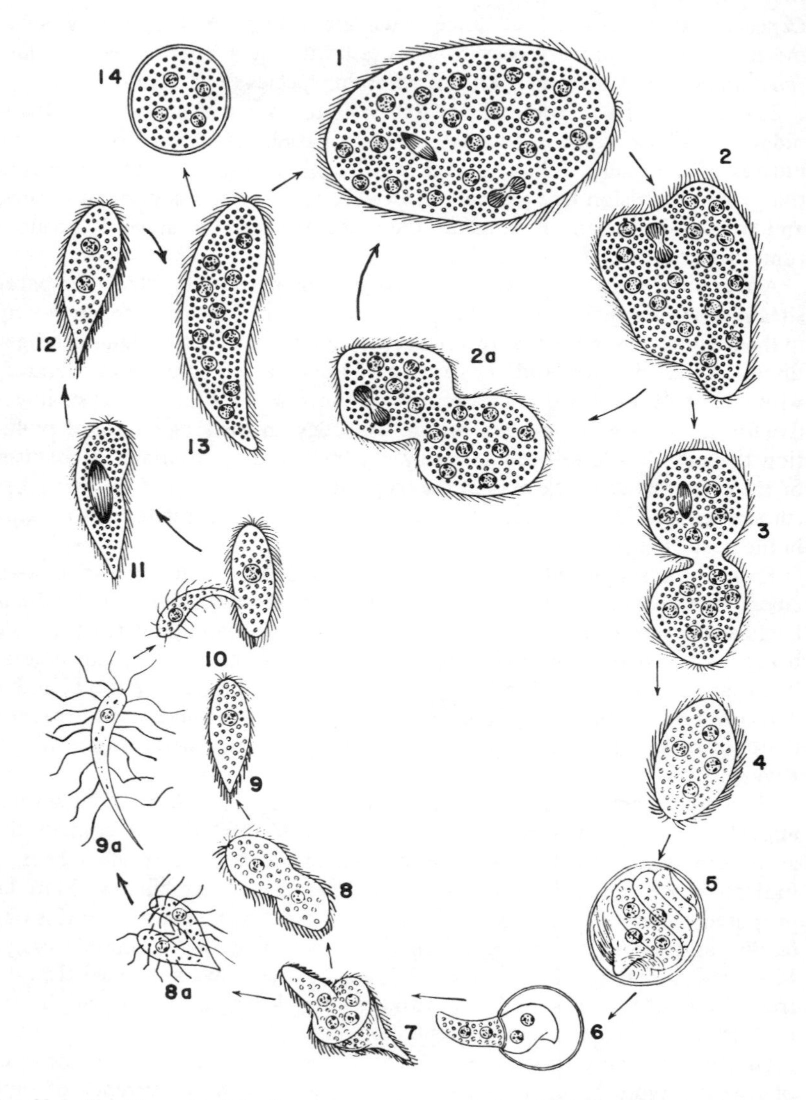

FIG. 22.5. The life cycle of *Opalina ranarum*. 1, trophozoite with 2 nuclei in mitosis; 2 and 2a, plasmotomy; 3, precystic plasmotomy; 4, precystic individual with few nuclei; 5, *Opalina* which has just encysted and undergone torsion; 6, excystment; 7, individual leaving the cyst; 8, macrogametocyte with 2 nuclei; 9, macrogamete; 8a, end of microgametic plasmotomy; 9a, microgamete; 10, fertilization; 11, first mitosis of the zygote; 12, "Protoopalina" stage 13, stage with few nuclei; 14, cyst formed in tadpole; (After Grassé, *Traité de Zoologie*.)

it seems not improbable that the Opalinidae and *Trichonympha* may have arisen from similar ancestors."

To this author, it seems better, at least for the time being, to continue to regard the opalinids as "protociliates," even though the connotation implies a degree of relationship to the true ciliates which these curious Protozoa do not have. It is doubtful whether our present knowledge of these organisms is sufficient to establish their real biological relationships with any certainty. As Corliss himself remarks, ". . . their organization seems to show a very high degree of differentiation and specialization indicating a long evolutionary history of their own far removed from the main line of development of any other protozoan group."

Ciliates of Echinoderms

Another group of ciliates infest echinoderms, particularly the alimentary tract of sea urchins. Unlike the protozoan parasites of ungulates and the Anura, those found in echinoids have little in common except host type and habitat, and thus do not constitute a natural biological group. Some are holotrichs, some hypotrichs, and some peritrichs, and some genera include both parasitic and free-living species. Although the number of ciliate species parasitizing echinoids is already known to be large, it is almost certain that many more exist, and much also remains to be learned about the nature of the host-parasite relationship.

These ciliates apparently infest sea urchins all over the world and the incidence of infection is often very high. For example, the green sea urchin *Strongylocentrotus dröbachiensis*, which occurs commonly along the North American and European Atlantic coasts, was found by Beers (1948) to show virtually 100 per cent infection with three of the seven ciliate species it harbors. Fortunately for the urchins, no harm seemed to result from these "inquilines," as parasites in such a relationship to the host are often called.

Oddly enough, some of these ciliates have a considerably more limited geographical distribution than do their host species. Powers (1937) suggests that the peculiarities of the ocean currents in which the latter live may account for this discrepancy.

In spite of the absence of any close kinship among echinoid ciliates, there is reason to think that this host-parasite association is an old one. It can hardly be otherwise when the infestation is so widespread. Although Power's theory has been recently questioned, his statement is worth quoting if for no other reason than that it shows how findings in one science often bear on another, sometimes in quite unexpected ways. He says, "Certain it is that the similarity between the infestation at Tortugas [Gulf of Mexico] and Acapulco [Mexican Gulf Coast] points to the same common origin for each of these now well-isolated groups—a common origin well established long before the present land masses of Central America had separated the Atlantic and

Pacific oceans. The time which has elapsed during this period is well demonstrated through the general topographical changes. . . ."

Ancient though the ciliate-sea urchin association may be, the evolutionary changes in these Protozoa have been chiefly physiological. That they have become well adapted to conditions in their host's gut is shown by the enormous numbers often found there. Beers (1948) counted as many as 10,000 per ml in the enteric fluid of some individuals.

Relatively little is known about the life cycles of these ciliates. The few that have been investigated seem similar to the cycles of related free-living species. Division in some of the former, however, appears to be cyclical with relatively long intervals in which no reproduction occurs; perhaps, since few ciliates are lost from the host in the fecal pellets, some form of protozoan "birth control" may be necessary. It would be decidedly interesting to know what factors inhibit or trigger division. An unusual (though not unique) feature of the division of one of the species, *Cyclidium stercoris*, is the discarding of a portion of the macronuclear substance as the daughter macronuclei separate in the telophase, with absorption of the spent chromatin into the cytoplasm.

Genera represented among the ciliates of sea urchins include, among others, *Entodiscus*, *Madsenia*, *Biggaria*, *Anophrys*, *Cryptochilidium*, *Trichodina*, *Colpoda*, *Cyclidium*, *Colpidium*, *Plagiopyla*, *Strobilidium*, *Metopus*, and *Euplotes*. The last seven include a number of free-living species, and those inhabiting the sea urchin gut indicate little morphological modification due to parasitism. Nevertheless, evolution has clearly resulted in physiological changes, for the parasitic species have a limited tolerance for sea water; apparently they can live and multiply only in the environment furnished by their hosts. Since the ciliates seem not to encyst, transmission probably depends on cannibalism—an old habit among sea urchins.

Ciliate Parasites of Fish

Among ciliates parasitizing the lower animals, another genus is especially significant because of the injury its species often do to fish. How many such species there may be is uncertain; the best known, and probably the most important, is *Ichthyophthirius multifiliis*, discovered in Europe in 1869. In this country it is responsible for annual losses estimated at a million dollars (Herman *et al.*, 1959). These ciliates resemble considerably the common free-living form *Holophrya*, a genus well represented in both fresh and salt water. *Holophrya*, however, lacks the cytostomal membranes possessed by *Ichthyophthirius*, a hymenostome ciliate. The life cycle of the latter is also very different, having some very peculiar features doubtless developed as adaptations to its parasitic mode of life.

The adult stage is oval or sometimes nearly spherical, with a mouth at the anterior end from which many longitudinal rows of cilia radiate posteriorly.

Within its cytoplasm are a number of contractile vacuoles and a kidney-shaped macronucleus with a nearly hidden micronucleus in its concavity (Fig. 22.6). Size varies, but the larger specimens may reach a length of almost a millimeter.

Many species of fresh-water fish are subject to attack and death may result from heavy infections. The ciliates actively seek out their victims and penetrate the epithelium, on the cells of which they feed; the resulting irritation causes the formation of a pustule about each parasite. Within the pustule the protozoan grows for several days and then drops off without previous division, gradually settling to the bottom of the pond or lake.

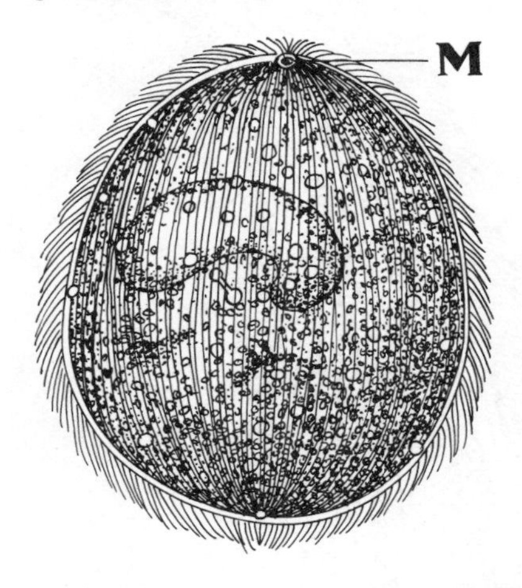

FIG. 22.6. *Ichthyophthirius multifiliis*, a highly pathogenic parasite of fish. The mouth (M) is said to have oral membranes organized much as in *Tetrahymena*. (After Bütschli.)

There it secretes about itself a double-walled cyst, and then begins a period of active division which ends only when anywhere from a hundred to more than a thousand minute "ciliospores" or "tomites" have been produced. Each is a replica of the adult, except for size and the absence of a mouth. Nevertheless, reorganization within the cyst is extensive, involving first a dedifferentiation of most of the ciliate structures, followed by equally complete redifferentiation.

Finally, the numerous offspring of this prolific parent (and it is this characteristic responsible for the species name "multifiliis") by their active boring movements break through the cyst wall, and are then on their own. The search for a fish is a big undertaking for so small an animal; it is only about $40\mu \times 10\mu$, sometimes less than 0.0001 the volume of the larger individuals of its parental generation. It has but two or three days to achieve success, and failure means death. Once the young parasite finds its piscine quarry, the life cycle of course begins all over again. *Ichthyophthirius* differs sharply from most parasites in undergoing the entire reproductive portion of its cycle during its free-living phase. More research is needed on this phase. Little is known about the effects of various environmental factors, and Herman *et al.* (1959) have recently suggested that low temperatures may be a factor in encystment; they were also able to demonstrate that cysts would survive for at least a week at 4° C., although when intracystic division occurred only a few daughter individuals resulted.

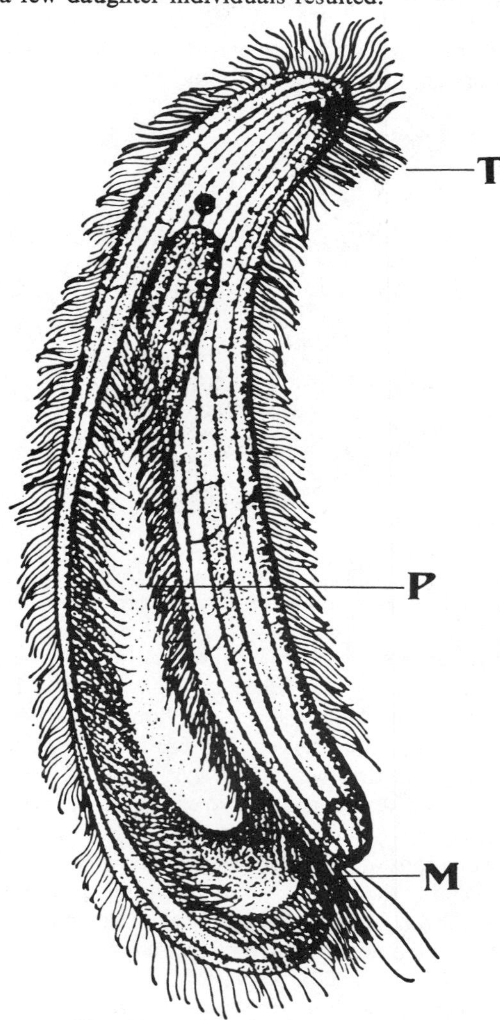

FIG. 22.7. *Ancistruma mytili*, a thigmotrichous ciliate parasitizing the clam *Mytilus edulis* (lateral view). T, tuft of straight tactile cilia at the anterior end; P, peristomal groove; M, mouth. (After Kidder, *Biol. Bull.*)

THIGMOTRICHIDA

Ciliates of the order Thigmotrichida, though of no known importance, are biologically interesting, partly because they illustrate in different degrees the modifications due to parasitism. The group receives its name from the presence of a ciliary field especially sensitive to contact with solid objects, to which the ciliates cling. The object is, in this case, appropriate tissues—usually the epithelium of the mantle cavity, gills, or palps—of their molluscan hosts. Such ciliary areas are not restricted to the Thigmotrichida, since they occur in some free-living species, as, for example, *Chilodonella*, but they are especially strongly developed in this group; often most of the other body cilia have become much reduced or lost altogether. The thigmotactic ciliary area is at what is thought to be the anterior end and with its gradual evolutionary development has gone a change in the position of the cytostome, which has moved posteriorly until in some forms it has become terminal and in others it has disappeared altogether.

Apparently the thigmotrichs are quite content as long as they can remain in contact with the tissues of their host. They remain quiet until dislodged, whereupon they then swim about slowly or actively. After all, they have little to be concerned about as long as they are assured of a square meal, which many of them get by sucking out the contents of their host's epithelial cells, one after another, with their suctorial tentacles. Such a form is *Ancistruma mytili* (Fig. 22.7), which lives on the palps and gills of certain clams. But there are species which even live on other Protozoa: *Hypocoma parasitica* (Fig. 22.8) preys on *Zoothamnium* (a colonial vorticellid), from which it sucks the protoplasm until the luckless victim dies, and there are other species with protozoan hosts in the genus *Hypocoma*.

Members of the family Hypocomidae, to which this genus belongs, present several characteristics of special interest. They are obligatory parasites, and all seem to be markedly host-specific. It is said that only one parasite of a species is usually found on a given host. The mouth has disappeared, but some vestiges of the adoral zone may remain. Within the tentacle a canal extending into the cytoplasm performs the functions of a cytostome and gullet. Two other peculiarities of thigmotrichs are worth comment. In some (e.g., *Ancistrocoma*) conjugants have the curious habit of joining by their posterior ends rather than in the conventional manner, and members of the Sphenophryidae are ciliated only as embryos, in which stage they resemble ciliates of another thigmotrich family, the Ancistrocomidae.

This order is a fairly large one, containing six (or seven) families, and numerous genera and species. Although the most notable contributions to what we know of the group are those of Chatton and Lwoff, French protozoologists also widely known for their work on other Protozoa, the best general account is probably that of Kirby (1941).

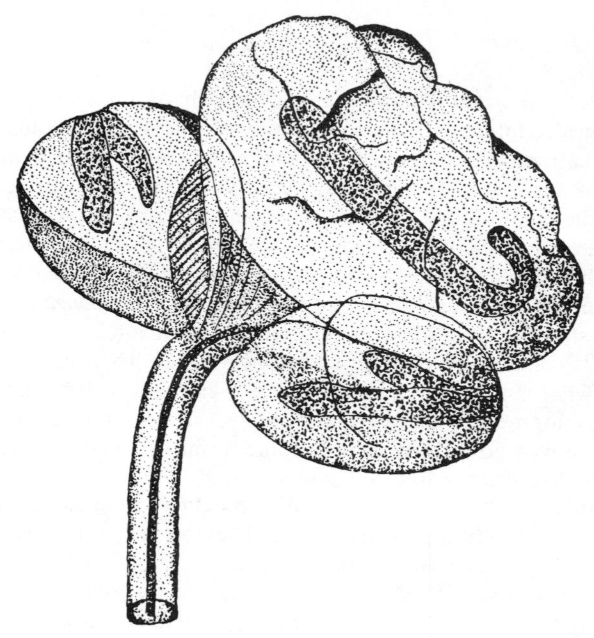

FIG. 22.8. Two thigmotrichs of the genus *Hypocoma* parasitizing *Zoothamnium*. The parasites are shown on each side of the host's stalk. (After Plate.)

ASTOMATIDA AND APOSTOMATIDA

Of the many other parasitic ciliates occurring in vertebrate and invertebrate hosts, two of them deserve special mention. These are the two orders, Apostomatida and Astomatida, comprising forms with a very small mouth and no mouth at all, respectively. The great majority of apostomes are parasites of crustacea, while the astomes exhibit more varied tastes in their choice of hosts. Many of the latter occur in annelids, particularly the oligochetes (earthworms and their near relatives). But numerous other invertebrates have been found infected, including groups as diverse as flatworms, bryozoa ("moss animals," a largely extinct and almost wholly marine phylum), molluscs, and crustacea. One genus of astomes even occurs in newts.

Little is known about the life cycles and mechanisms of transmission of these ciliates (though some form cysts), or of their host-parasite relationships, but most species are thought to be commensals. Their relationships to other ciliate groups are equally obscure. The astomes probably do not constitute a natural biological group, at least in the sense that the opalinids do, but they have undoubtedly been parasites for a long time. Many of them have developed suckers or other organelles of attachment, the better to maintain their positions in the host's alimentary tract. A few reproduce by budding off daughter individuals which may remain attached to the parent for some time, thus forming chains reminiscent of proglottid formation in tapeworms. This habit may lessen the danger of the younger generation being prematurely swept out of house and home.

The apostomes have been intensively investigated in recent years, especially by the French zoologists Chatton and Lwoff. Like the astomes, in nature they are not important parasites, but much basic knowledge has come to light from a study of their development (Fig. 22.9). This is particularly true with regard to the mechanisms underlying differentiation. In these ciliates (and, as further work has proved, in ciliates generally) the kinetosomes play a fundamental role in differentiation, and behind the manifold transformations in the life cycles of these remarkable forms lies a complex behavior

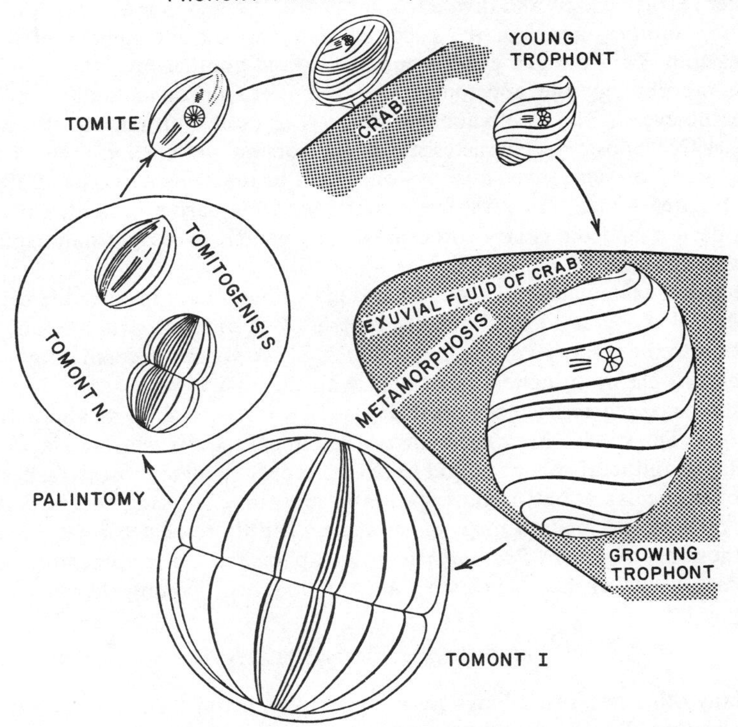

FIG. 22.9. The life cycle of *Gymnodinioides inkystans*, a ciliate parasitizing the hermit crab and belonging to the Apostomatida. The trophont is the actively growing stage; from it a cyst may be formed, and the protomont emerging later undergoes repeated transverse fission to produce a number of tomites. These encyst on a crab, and this last stage is known as the phoront. Excystation occurs only when the host moults, the excysted organism being the trophont. (After Lwoff, *Problems of Morphogenesis in Ciliates*.)

pattern of the kinetosomes. As Lwoff (1950) himself remarked, "All [apostomes] have very remarkable and unique properties" (p. 9). And later, "Differentiation in ciliates . . . represents essentially the movements of the kinetosomes. Two new 'daughter patterns' are formed anew at each generation from one or many kinetosomes."

The life cycle of *Synophrya hypertrophica*, a parasite of crabs, is ample evidence of Lwoff's first statement. This peculiar ciliate attunes its own reproductive behavior very delicately to that of its host. Individuals which have encysted on the integument of the crab excyst and enter its blood stream, where they feed and grow actively. Finally, while the host is mating, these parasites—by now enormously swollen in size, hence the name "hypertrophica"—initiate a period of rapid division. The resulting myriads of tiny daughter ciliates leave the host during the moult which soon follows, and begin feeding in the discarded exoskeleton. Apparently tiring of so quiet an existence, they encyst and divide within the cyst. But soon they abandon this life for another, again in the encysted state, on the integument of some other crab. Thus one life cycle is completed, and another one begun.

In morphology the apostomes are even more unconventional than in their life cycles. The rosettelike cytostome is in itself unusual, but the most remarkable apostome characteristic is the torsion or twisting seen in the "trophont" or vegetative stage. As one event in the cycle succeeds another, and tomites—small, infective, free-swimming forms—are produced, a detorsion occurs, and the ciliary rows, previously wound in a right-hand spiral, become straight.

Detailed examination of these ciliates at various stages in their life cycles has revealed a great deal about the genesis of organismal structure and its relation to the behavior of the kinetosomes. These are self-perpetuating units of the cytoplasm intimately associated with the cilia.*

Though largely restricted to crustacea, some apostomes show an alternation of hosts. The species *Foettingeria actiniarum* undergoes its phoront stages on crustacea and its trophont period in sea anemones. Unlike most parasites, it is not particular about its host species at any time, showing a remarkable lack of host specificity. Other apostomes with little respect for convention are those of the genera *Photorophrya*, which parasitize other apostomes, and *Pericaryon*, which has as hosts certain ctenophores ("comb-jellies").

Trichodina and its Relatives

Many other parasitic ciliates are known, and most of them are apparently unimportant. But *Trichodina pediculus*, first described almost two centuries ago, causes serious injury to fresh-water fish (Fig. 22.10). Mueller (1937)

*For the kinetosomes and their associated structures the term "infraciliature" has been proposed. A consideration of the possible broader significance of the behavior of kinetosomes in ciliate development may be found in Chapter 9.

found numerous Florida bass with their gills so heavily infested as to be on "the verge of suffocation." There are several genera in the family Urceolariidae, to which *Trichodina* belongs, all of them either commensal or potential pathogens. Some are ectoparasites, like those seen by Leeuwenhoek, but others prefer the security and privacy of life within their hosts. Such is the case with *Vauchomia nephritica*, described by Mueller from the urinary tract of the muskelonge; this ciliate has close relatives living in the bladder of a number of species of amphibia and fish. The ciliates of this family are all

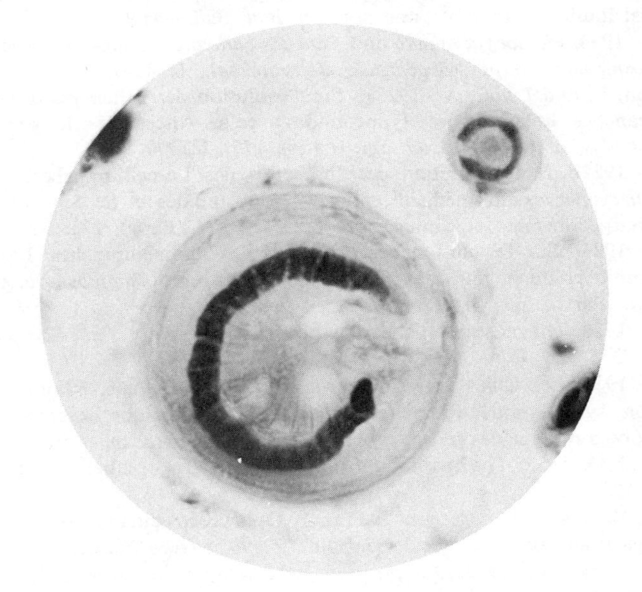

FIG. 22.10. Two species of *Trichodina*, ×570. The larger is *T. pediculus*, common on *Hydra* and also found on amphibian larvae. The large horseshoe-shaped macronucleus is clearly visible and the corona, consisting of radially arranged segments extending across the macronucleus toward the center, is also visible. The micronucleus, which is embedded in the macronucleus, cannot be seen.

equipped with a strong attaching organelle or sucker at the aboral end, by which they maintain their positions externally or escape expulsion when the bladder contracts.

Doubtless careful and painstaking study of other parasitic ciliates would reveal further important facts. Not only might we learn much of interest about the organisms themselves and their relations to their hosts, but some might be found to be important in the delicate balances so often seen in nature.

REFERENCES

Becker, E. R. 1932. The present status of problems relating to the ciliates of ruminants and Equidae. *Quart. Rev. Biol.*, 7: 282–97.

———, Schulz, J. A., and Emmerson, M. A. 1930. Experiments on the physiological relationships between the stomach Infusoria of ruminants and their hosts, with a bibliography. *Iowa State J. Sci.*, 4: 215–51.

Beers, C. D. 1948. The ciliates of *Strongylocentrotus dröbachiensis:* incidence, distribution in the host, and division. *Biol. Bull.*, 94: 99–112.

——— 1954. *Plagiopyla minuta* and *Euplotes balteatus*, ciliates of the sea urchin *Strongylocentrotus dröbachiensis. J. Protozool.*, 1: 86–92.

Chatton, E. and Lwoff, A. 1922a. Sur l'évolution des Infusoires des Lamellibranches, Relations des Hypocomides avec les Ancistridés. Le genre *Hypocomides*, n. gen. *C. Acad. Sci.* (Paris), 175: 787–90.

——— 1922b. Sur l'évolution des Infusoires des Lamellibranches. Le genre *Pelecyophrya*, intermédiaire entre les Hypocomidés et les Sphénophryidés. Bourgeonnement et conjugasion. *C. Acad. Sci.* (Paris), 175: 915–7.

——— 1923. Sur l'évolution des Infusoires des Lamellibranches. Les formes primitives du phylum des Thigmotriches; le genre *Thigmophrya. C. Acad. Sci.* (Paris), 177: 81–3.

——— 1926. Diagnoses de Ciliés thigmotriches nouveaux. *Bull. Soc. Zool. Fr.*, 51: 345–52.

——— 1935. Les Ciliés apostomes. Morphologie, cytologie, éthologie, évolution, systématique. Première partie. Aperçu historique et général. Etude monographique des genres et des espèces. *Arch. zool. exp. et gén.*, 77: 1–453.

Corliss, J. O. 1955. The opalinid infusorians: flagellates or ciliates? *J. Protozool.*, 2: 107–14.

Dogiel, V. 1946. Phylogeny of the family Ophryoscolecidae from a palaeontological and parasitological viewpoint. *Zoologicheskii Zhurnal*, 25: 395–402.

Fauré-Fremiet, E. 1955. La position systématique du genre *Balantidium. J. Protozool.*, 2: 54–8.

Grassé, P.-P. 1952. Super-ordre des Opalines (Opalinina, n. n.). In *Traité de Zoologie*. T. 1, Fasc. 1, Paris: Masson.

Herman, R., Nigrelli, R. F., and McLaughlin, J. J. A. 1959. Some *In Vitro* studies on *Ichthyophthirius filiis. J. Protozool.* 6(Suppl.): 25–6.

Hungate, R. E. 1955. Mutualistic intestinal Protozoa. *Biochemistry and Physiology of Protozoa*, vol. 2, S. H. Hutner and A. Lwoff (eds). New York: Academic Press.

Kirby, H. H. 1941. Relationships between certain Protozoa and other animals. In *Protozoa and Biological Research*, G. N. Calkins and F. M. Summers (eds). New York: Columbia University Press.

Levine, N. 1940. Changes in the dimensions of *Balantidium* from swine upon cultivation. *Am. J. Hyg.*, (Section C), 32: 1–7.

Lwoff, A. 1950. *Problems of Morphogenesis in Ciliates*. New York: Wiley.

McDonald, J. D. 1922. On *Balantidium coli* (Malmsten) and *B. suis* (sp. nov.), with an account of their neuromotor apparatus. *Univ. Calif. Pub. Zool.*, 20: 243–300.

Metcalf, M. M. 1923. The opalinid ciliate infusorians. Smithsonian Inst., U.S. Nat. Museum, *Bull. 120*.

——— 1940. Further studies on the opalinid infusorians and their hosts. *Proc. U.S. Nat. Museum*, 87: 465–634.

Mueller, J. F. 1937. Some species of *Trichodina* (Ciliata) from fresh-water fishes. *Trans. Am. Mic. Soc.*, 56: 177–84.

Netik, J., Lariviere, M., and Quenum, G. 1958. Un cas de dysenterie balantidienne mortelle. *Bull. Med. de l'A.O.F.*, 3: 136–40.

Oxford, A. E. 1955. The rumen ciliate Protozoa: their chemical composition, metabolism, requirements for maintenance and culture, and physiological significance for the host. *Exp. Parasitol.*, 4: 569–605.

Powers, P. B. A. 1933. Studies on the ciliates of sea urchins. 1. General taxonomy. *Biol. Bull.*, 65: 106–21.

——— 1936–7. Studies on the ciliates of sea urchins. *Ann. Rep. of Tortugas Laboratory*, Carnegie Inst. Wash., pp. 101–3.

Schumaker, E. 1931. The cultivation of *Balantidium coli*. *Am. J. Hyg.*, 13: 281–95.

Scott, M. J. 1927. Studies on the *Balantidium* from the guinea pig. I. Morphological studies. II. Studies on fission and conjugation. *J. Morphol. and Physiol.*, 44: 417–53.

Westphal, A. 1957. Experimentelle Infektionen des Meerschweinchens mit *Balantidium coli*. *Ztschr. f. Tropenmed. u. Parasit.*, 8: 288–94.

Wichterman, R. 1938. The present state of knowledge concerning the existence of species of *Nyctotherus* (Ciliata) living in man. *Am. J. Trop. Med.*, 18: 67–76.

Young, M. D. 1950. Attempts to transmit human *Balantidium coli*. *Am. J. Trop. Med.*, 30: 71–2.

ADDITIONAL REFERENCES FOR CHAPTER 20

Hoare, C. A. 1964. Morphological and taxonomic studies on mammalian trypanosomes. X. Revision of the systematics. *J. Protozool.*, 11: 200–7.

Lincicome, D. R., Rossman, R. N., and Jones, W. C. 1963. Growth of rats infected with *Trypanosoma lewisi*. *Exp. Parasit.*, *14*: 54–65.

——— and Shepperson, J. R. 1965. Experimental evidence for molecular exchanges between a dependent trypanosome cell and its host. *Exp. Parasit.*, *17*: 148–67.

23

Sporozoa Other Than the Coccidia and the Malaria Parasites

"Many important diseases have been supposed to originate from parasitic animals and vegetables . . . but they are considered to be animalcules so small that they cannot be discovered even with the highest powers of the microscope. . . . To assert under these circumstances, that there are spores and animalcules capable of giving rise to epidemics, but not discernible by any means at our command, is absurd. . . ." Joseph Leidy, *A Flora and Fauna Within Living Animals*, 1851.

The Sporozoa are a large and rather heterogeneous assemblage of Protozoa, sometimes considered a subphylum and sometimes a class. They are neither closely related among themselves nor to any free-living Protozoa. Although, like other parasites, their ancestors must have been free-living, all of them long ago evolved into parasites. Neither in their morphologies nor their life histories is there much hint of what their ancestors were like. Cleveland (1949) has suggested that the flagellates of the woodroach *Cryptocercus* and some of the Sporozoa have sexual cycles with much in common, and that they may therefore have descended from a "single group." Possibly both groups were well evolved by the "beginning of the Tertiary, if not much earlier"—70 million years ago. Since there are no fossil remains of the Sporozoa, and we have no hope of ever finding any, evidence based on life cycles, such as Cleveland's, is about all we shall ever have.

Because of the antiquity of the Sporozoa and because the different major groups in the class have evolved independently, it is difficult to form any opinion about biological or evolutionary relationships among them. Most protozoologists regard the class as consisting of the four subclasses listed in Chapter 17, but some would considerably restrict its size (e.g., Grassé, 1953); others would raise the taxonomic rank of some of its subgroups.

The common arrangement given below is that used by Kudo (1954) and, essentially, by Hall (1953); it has been given in skeleton form in Chapter 17.

THE TELOSPORIDIA

Without doubt the parasites embraced in this group, variously regarded as a class or (as we do) a subclass, are more important to man than any others among the Protozoa. For they include the coccidia and the malaria parasites, each of which is given a separate chapter (Chaps. 24 and 25) because they cause so much economic loss and disease. The gregarines, which constitute the first of the two orders, are of little practical importance but of much biological interest. They are considered below.

The Gregarines

Although the gregarines (Gregarinida) are a numerous group, they are, to paraphrase the noted nematologist Cobb, neither beautiful nor good to eat, nor valuable to man in any other way. They have, however, been a good deal studied.

Most gregarines are parasites of arthropods, particularly of insects, although they are not uncommon in echinoderms and a few species are known from molluscs and annelids. The primitive chordates known as Ascidia also harbor some of these organisms. For so peculiar and uneven a host distribution there is at present no plausible explanation. All these host types are segmented or of segmented stock (recently even a segmented mollusc has been discovered) and this may have evolutionary significance in the origin of parasitism among the gregarines. For example, gregarines may have parasitized the common ancestral stock from which segmented animals were derived, if such existed, as Patten believed. But this theory does not account for the numerous species among these groups which do not harbor gregarines, and for the fact that chordates above the ascidian level seem not to be parasitized by them. Possibly more thorough study of the gregarines will show that their hosts are more widely distributed among the major taxonomic groups of animals than is now apparent.

Relationships. The origin of the gregarines, like that of other Sporozoa, is obscure. Grassé remarks that a euglenid which had lost its flagellum and its pigment might easily be taken for a gregarine, and recalls that this mistake was actually made with *Astasia mobilis*, parasitic in the small crustacean *Cyclops*. However, peculiarities of the life cycles of certain species suggest a relationship to the coccidia, and it may be that coccidia and gregarines arose from a common ancestral stem.

Sporozoa of both groups are typically parasites of the epithelia of body cavities, most frequently of the digestive tract. Usually infection is accomplished by ingesting the oocysts (often spoken of as "sporocysts" in the case of the gregarines), from which the imprisoned sporozoites emerge in the gut and attack the epithelial cells (Fig. 23.1). The young gregarine grows and may reach a considerable size. At first it usually leads an intracellular existence, but eventually leaves the host cell, although it may remain attached for a time to its remnants. After a period of youthful wandering about its host's gut, it picks a partner and the pair become encysted in a primary protective envelope known as a "gametocyst." Then a series of nuclear divisions ensues, culminating at last in a cytoplasmic division and a numerous progeny of young gametes, plus usually a residual mass of cytoplasm. All offspring of a

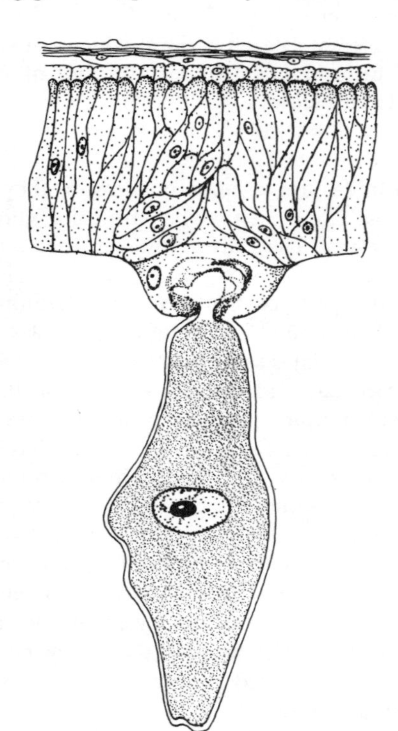

FIG. 23.1. A gregarine (*Lecudina fluctus*) from the oligochaete worm *Urechus unicinctus*, shown attached to the lining of the midintestine. Note the epimerite imbedded in an epithelial cell, apparently causing some hypertrophy. (After Iitsuka.)

given parent are believed to be of the same sex. They may or may not differ visibly from gametes of the other sex, although careful observation can generally distinguish such differences.

In a life cycle of this type there is no period of asexual reproduction, and thus each sporozoite is a potential gametocyte. Since there is no schizogony and the host is not overwhelmed by parasitic invasion of its tissues, as often happens with the coccidia, injury is limited to those host cells actually attacked by the sporozoites.

Within the gametocyst each of the gametes derived from one of the parent cells unites with a gamete from the other, and the zygotes then secrete secondary cysts (sporocysts or, more properly, oocysts) about themselves; three divisions promptly follow, with the production of eight sporozoites. In the meantime the gametocyst, with its cargo of future parasites, is passed from the host with its feces, and it begins a patient wait for the next host.

There is much variety of form in the gametocysts and oocysts of different species, but the sporozoites in each oocyst are almost invariably eight in number, rod or sicklelike in shape, and capable of the gliding movement typical of gregarines.

Life cycles of the Eugregarinina (Fig. 23.2), to which suborder most gregarines belong, are quite similar and follow closely the pattern just outlined. In a minority of species, however, collectively put into the suborder Schizogregarinina, multiplication occurs in both the sexual and asexual phases, just as with the coccidia, and this may be evidence of a kinship of the two groups. Or it may be that schizogony in the gregarines has been a secondary development. In the belief that the latter is indeed the explanation, Grassé (1953) has suggested that the Schizogregarinina should be placed in a new order, to be called the "Neogregarinida."

Physical Characteristics and Structure. In their physical characteristics the gregarines show great variety. Not only is there extreme range in size among the species, the largest reaching the gargantuan length of 16 mm. (more than half an inch), but different stages in the life cycle of even the same species may be of widely varying dimensions. In form there is also much diversity.

The body usually exhibits a rather sharp differentiation into ectoplasm and endoplasm, the first of these being typically further divided into three layers: the outer or "epicyte," the middle or "sarcocyte," and the innermost or "myocyte." The last contains contractile fibrils or "myonemes" which are believed to be the means by which gregarines achieve their remarkable variety of movement. In at least certain species, other structures, such as fibrils interpreted as having a skeletal function, apparently also exist.

The endoplasm, besides containing such organelles as mitochondria and a Golgi apparatus, is likely to be crowded with granules of reserve food material in the form of paraglycogen (a substance which has been regarded as a special form of glycogen, or animal starch), fat globules, and sometimes volutin (a protein rich in phosphorus, and probably consisting largely of nucleic acid).

Among the Eugregarines there are two major structural patterns, although in minor ways there is a good deal of variation. The acephaline ("headless") group is characterized by a relatively simple plan, without clearly differentiated body regions, although there is in the adult stage usually an organ of attachment at the anterior end. *Monocystis*, a gregarine genus very common in earthworms, is a good example.

The cephaline ("with a head") gregarines are, by comparison, very complex in their somatic architecture, and they are also the most frequently

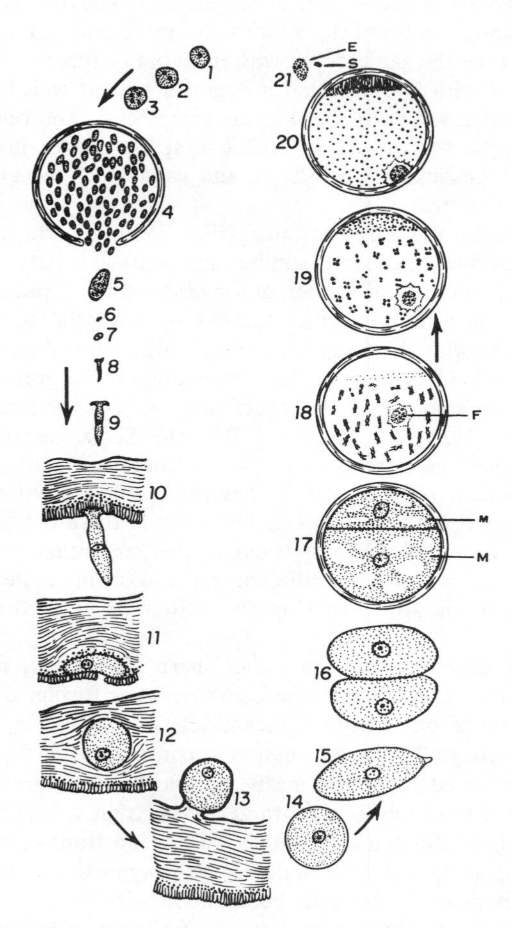

FIG. 23.2. Life cycle of the eugregarine *Filipodium ozakii*, from the sipunculoid worm *Siphonosoma cumaense*. *1–3*, zygote; *4*, formation of sporoblasts; *5*, sporozoite; *7–9*, formation of the trophozoite; *10*, trophozoite attached to intestinal wall; *11–12*, penetration of wall by trophozoite; *13*, trophozoite leaving the gut wall to enter coelomic cavity and become the sporont; *14*, sporont; *15*, change in shape of sporont to spindle form; *16*, pseudoconjugation, resulting from the union of two sporonts upon mutual contact; *17*, formation of cyst wall by pseudoconjugants; *18–20*, formation by a series of mitotic divisions, culminating in meiosis, of the micro- and macrogametes (E); *21*, fertilization. Key: E, macrogamete or egg; S, microgamete or sperm; F, residual fat body; m, portion of encysted pair destined to produce microgametes; M, portion destined to produce macrogametes. (After Hukui.)

encountered. The adult stage (often called a "cephalont") has three clearly visible body regions as a rule. The most anterior of these is the rostrum; then follow the "protomerite" and the "deutomerite." At the extremity of the first of these is an organelle known as the "epimerite," which may be modified in the most remarkable ways, the better to achieve its function of securely anchoring the parasite to its host cell. In the later stages of its existence, when the parasite becomes a nomad (otherwise known as a "sporadin") in its host's interior, the epimerite, its usefulness lost, is discarded. In many gregarines a "septum" separates the rostrum from the rest of the organism. The posterior portion of the parasite, or deutomerite, may also be further subdivided by partitions into two or more regions having the appearance of segments. Where no division exists the organism is said to be "monocystid"; otherwise it is called "polycystid."

The nucleus, often very large, is situated in the deutomerite, and is of the vesicular type. In later stages of parasite growth it develops a nucleolus which in certain species is said to pulsate like a contractile vacuole. Although gregarines are always uninucleate, they often appear multinucleate during the multiplicative phase of the life cycle.

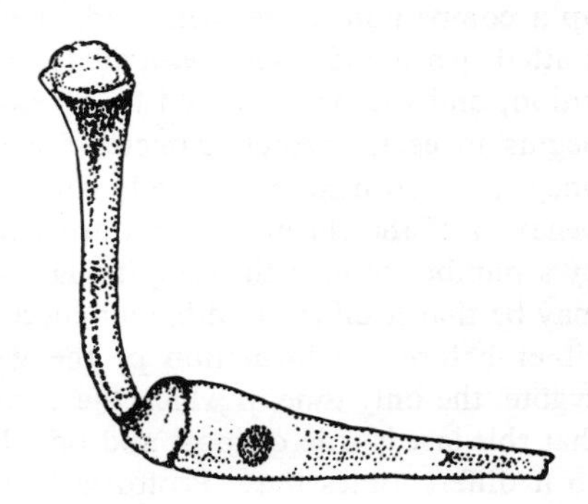

FIG. 23.3. *Carcinoecetes bermudensis*, a gregarine parasitizing the gut of certain Bermuda crustacea. Two individuals are shown in syzygy. (After Ball.)

Schizogregarines, like the acephalines, are relatively simple in structure, but they offer other complications. As its name suggests, life cycles of the group include schizogony as well as sporogony. Some species are extremely small, for example, the genus *Lipotropha*, containing species which are intracellular parasites of the fat bodies of certain dipterous larvae. Others are extremely large. Shape varies also, the adults ("gregarine phase") being sometimes vermicular, or conical, or even, in one genus (*Spirocystis*) coiled like a clockspring.

The gregarine phase is so named because it is in this stage of the life cycle that that peculiar form of gregariousness known as "syzygy" is observed, an association of a varying number of individuals end to end (posterior to anterior, or anterior to anterior), or even laterally (Fig. 23.3).

Life Cycles. Although some features of gregarine life cycles have already been briefly considered in connection with the problem of relationships of the group, they may best be understood by considering representative species in some detail, although known life cycles are still only a fraction of the whole.

Monocystis, many species of which are commonly seen in the seminal vesicles of earthworms, is a genus of acephalines examples of which are easily found for laboratory study. Since monocystids are usually relatively large (cysts of some species attain a diameter of nearly a millimeter), they are readily spotted, even by the inexperienced.

Infection occurs when the worm ingests the oocysts. Presumably these are scattered through the soil in various ways: they may be liberated by the death of parasitized worms, or dispersed by birds which have eaten such worms and later voided the resistant cysts. These germinate in the gut of the newly infected worm, and the released sporozoites wander about until they find the seminal vesicles. Here they become intracellular parasites of the sperm mother cells.

After a period of growth the matured gregarine, or sporadin, leaves the host cell (by this time reduced to a mere shell of protoplasm), and hunts up a companion of its own kind. It has acquired a camouflage of densely matted sperm and resembles a ciliate, but this seems no bar to mutual recognition, and the pair proceed to develop a gametocyst. Nuclear fission then begins in each partner, gametes are produced, sexual union follows, and oocysts are formed, each finally containing the usual eight sporozoites. This behavior of the chromosomes during these stages has been carefully studied by a number of investigators, though with somewhat variable results, and it may be that it differs in different species. Meiosis regularly occurs, however, either before the formation of the gamete or in the first division of the zygote, the only time in which the chromosome number is diploid. It is said that this last is true of most and possibly all coccidia and gregarines, and if so it differentiates these Protozoa from all other groups of animals.

The cephaline gregarine *Lankesteria culicis* is very common in *Aedes aegypti* mosquitoes. Since the mosquitoes are notorious vectors of yellow fever it is unfortunate that their gregarine parasites are not pathogenic. This species of mosquito has a nearly world-wide distribution. Sir Ronald Ross, who earned a knighthood for his discovery of the mosquito transmission of malaria, first observed the parasite in Indian aedine mosquitoes in 1895.

The infection is acquired when oocysts left in the water by adult mosquitoes are ingested by larvae. After liberation from the oocyst in the gut, the sporozoites penetrate cells of the stomach lining and develop to maturity there. A few days later, while the host pupates, they leave the invaded gut cells and migrate to the neighboring Malpighian tubules, there to associate in pairs preparatory to the formation of a gametocyst. Gametes then develop, much as in *Monocystis*, and sexual union follows, with subsequent production

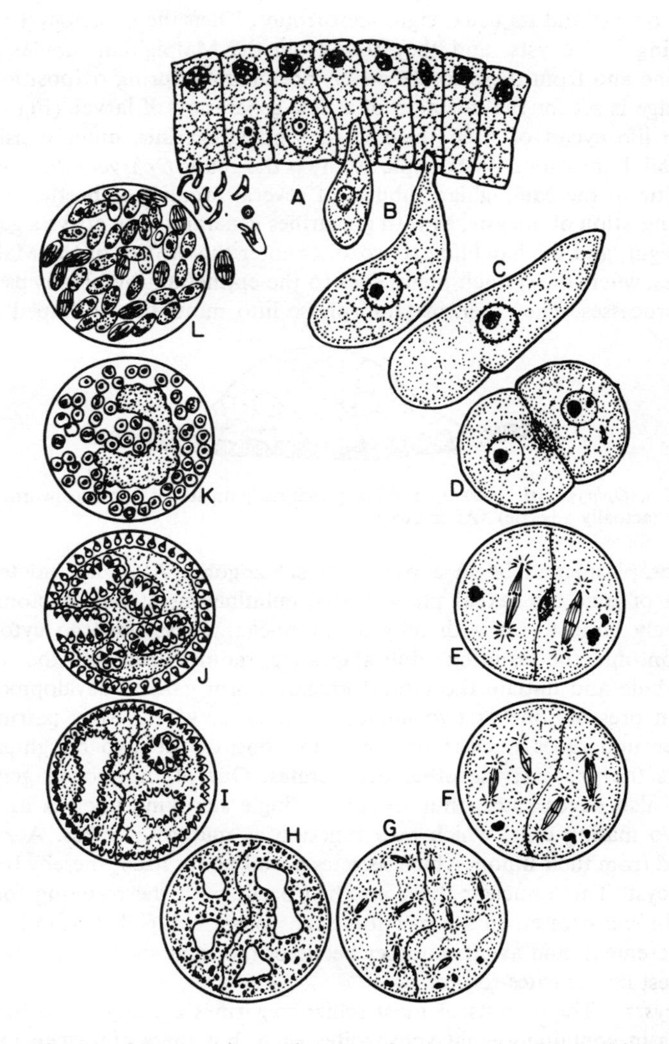

FIG. 23.4. Life cycle of the gregarine *Lankesteria culicis*, a parasite of the culicine mosquito *Aedes argenteus*. *A*. Escape of sporozoites (eight) from oocyst, and invasion of the gut cells of the mosquito larva. *B*. Growth of young gregarine until it outgrows its quarters in the host cell. *C*. Gregarine free in intestinal lumen, from which it migrates to the Malpighian tubules. *D*. Pairing in these tubules. *E–G*. Nuclear division within the gametocyst. *H*. Nuclear division completed, with gamete nuclei arranged about the periphery of the body and vacuoles within. *I*. Start of gamete formation by budding. *J*. Gametes formed and ranged about residual body of cytoplasm. *K*. Conjugation of gametes, which are of two kinds with respect to size of nuclei; *L*. Zygotes elongated and encysted within oocysts, where they form eight sporozoites. (After Wenyon, *Protozoology*.)

of an oocyst and its usual eight sporozoites. Then the gametocyst ruptures, releasing its oocysts, and they pass from the Malpighian tubules into the intestine and from there to the water, presumably during oviposition. Thus the stage is set for the infection of a new generation of larvae (Fig. 23.4).

The life cycles of even the known schizogregarines differ considerably in detail, but as good an example as any is the genus *Ophryocystis* (Fig. 23.5), parasitic in the Malpighian tubules of several families of beetles. Infection is by ingestion of oocysts, as with gregarines generally. The oocysts germinate in the gut, and the hatchling sporozoites enter the orifices of the Malpighian tubules, where they attach themselves to the epithelium by brushy pseudopodial processes. They then metamorphose into minute cone-shaped mounds

FIG. 23.5. *Ophryocystis mesnili*, a schizogregarine parasite of the mealworm *Tenebrio molitor* (actually a beetle). (After Léger.)

of protoplasm and begin a period of schizogony. After an indeterminate length of time the sexual phase starts, culminating in the development of relatively large forms with only a few nuclei. These undergo cytoplasmic division, and the resulting adult stages (sporadins) drop into the lumen of the tubule and initiate the typical gregariniform gamete development. The pattern presents one or two unique features, however. After pairing, each partner undergoes nuclear division, and one of the two daughter nuclei repeats the process; the other degenerates. One of the second generation nuclei also degenerates, thus leaving a single surviving nucleus in each of the two mating cells, which now function as mature gametes. A zygote is formed from their union, which secretes a wall about itself, thereby becoming an oocyst. Three nuclear divisions then occur with the resulting formation of eight sporozoites. In the meantime the oocyst is expelled from its host in the excrement, and awaits another beetle of the right species hungry enough to ingest its parasites-to-be.

Oocysts. The oocysts of most schizogregarines are similar to the eugregarines in containing eight sporozoites each, but there are exceptions. For example, *Spirocystis* (which is not always included with the schizogregarines) produces oocysts with only one sporozoite. The genera vary, however, in the number of oocysts characteristic of each gametocyst. *Ophryocystis* gametocysts contain a single oocyst each; *Lipocystis* is more generous, producing gametocysts with up to 200 oocysts. The oocysts of most species are boat-shaped, leading von Siebold, one of the pioneers in protozoology, to dub them "pseudonavicellae" when he first saw them in 1839. Some exhibit remarkable specialization in structure, as in *Syncystis.* Here the oocysts are provided with an armament of several spines at bow and stern, reminiscent of the ornamental beaks which used to be seen on sailing ships.

Pathogenicity. Unless infection is massive, gregarines seldom seem to do their hosts much harm. Perhaps this is added evidence that such association is a very ancient one. *Monocystis* may somewhat reduce the fertility of the earthworm, but infection is so common that little injury, even of this sort, seems to result. The cells of the intestinal epithelium which the gregarines of insects usually parasitize are destroyed, but epithelial cells are so rapidly regenerated that the host easily repairs such damage. When infections are intense enough local concentrations of phagocytes may attempt to ingest the parasites, destroying some, but the parasites seem quite resistant to phagocytic digestion. Occasionally they may become sufficiently numerous to cause intestinal blockage or the irritation from them may result in diarrhea. The desquamation of the intestinal lining which occurs during molting probably aids in getting rid of many of the parasites when infection of insect larvae takes place.

Classification. Classifying the gregarines presents a problem. No logical and biologically defensible taxonomic scheme has yet been constructed. The group itself has been variously accorded the rank of class, subclass, and order, depending on the prejudices of the zoologist. The scheme below is a modification of Grell's (1955).

Order GREGARINIDA Lankester.

- Suborder 1. EUGREGARININA Léger. Gregarines in which reproduction is limited to the sexual phase, without schizogony (merogony). Mostly parasitic in annelids and arthropods. A very large group.
  - Tribe 1. ACEPHALINA Kölliker. Body not differentiated into clearly defined regions. Mainly occurring in coelome. Early stages intracellular; later ones may be extracellular or not. About ten families.
  - Tribe 2. CEPHALINA Delage. Body differentiated into two main regions: the protomerite and the deutomerite (anterior and posterior respectively) with a septum between. Deutomerite may be further subdivided in some species. About twelve families.
- Suborder 2. SCHIZOGREGARININA Léger. Gregarines with reproduction in both asexual and sexual phases of the life cycle. Often intracellular in the former phase. A relatively small group. Two families, more or less (there is considerable difference of opinion as to the detailed taxonomy of this group).

  Parasites of insects, in which a few species (such as *Mattesia grandis* of the boll weevil) are pathogens.

THE CNIDOSPORIDIA

The Cnidosporidia are a somewhat heterogeneous assortment of organisms, and we do not know how many species there are. Because none is known to parasitize man they have been less studied than the Telosporidia. And yet a number of Cnidosporidia cause serious disease in animals used as sources of human food, such as fish and honey bees. The silkworm disease pébrine,

Pasteur's studies of which laid the foundation for his later researches in the biology of infectious disease, is caused by a cnidosporidian, *Nosema bombycis* (p. 479).

Variously regarded as a class or subclass of the Sporozoa, the Cnidosporidia received their name from the peculiar structure of their spores. The polar capsules and filaments that these spores contain closely resemble the nematocysts or stinging cells of the Coelenterata (and also of a few other animals). Cnidosporidia, loosely translated, means "nettle spore formers" (from the Greek *knide*, "nettle," and *sporos*, "seed").

The cnidosporidian spore is a truly remarkable structure. Although used for different purposes, the polar capsules and their spirally wound, extrusible

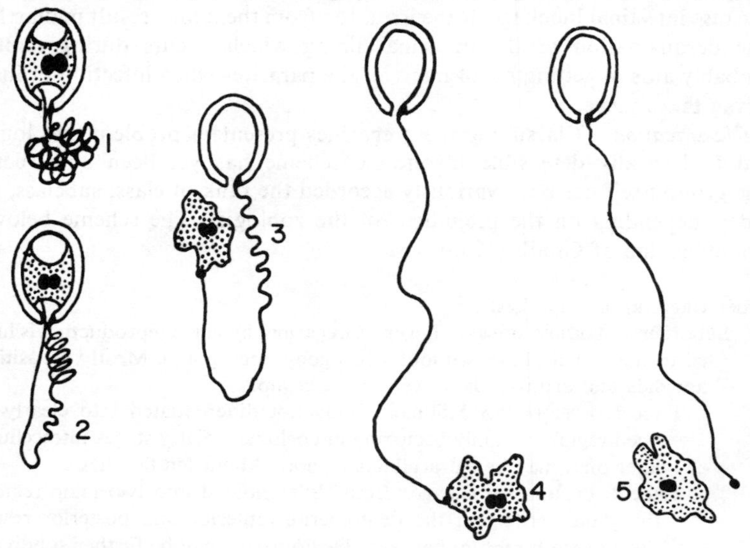

FIG. 23.6. Diagrammatic figure showing the emergence of filament and sporoplasm from a microsporidan spore. The filament appears to emerge first and its distal end brings out the sporoplasm, shown at the tip in *3*, *4*, and *5*. The spore case is left completely empty. (After Dissanaike and Canning, *Parasitol.*)

filaments seem to function much as do nematocysts (Fig. 23.6). Some biologists therefore believe that the Cnidosporidia and the Coelenterata may be related, the former being, as Poisson (1953) aptly put it, perhaps "very degraded Metazoa." As a matter of fact, it is even said that they resemble in some respects the larvae of certain medusae which have a parasitic phase. Additional evidence is the fact that cnidosporidian spores—unlike those of other Protozoa—are apparently formed from a number of cells, rather than from a single one. Thus, several such cells may give rise to the valves, and others may produce the polar capsular apparatus.

The use to which these spores are put is still undetermined. One suggestion is that they are discharged, probably as the result of the stimulation of gastric secretions, in the gut of the prospective host, where they serve as organelles of attachment in the gut wall until the sporoplasm can emerge. The still unopened spore is thus protected from absorption and possible passage through the intestinal wall (which could easily happen, since the spores are often only a few micra in diameter), or from being swept out with the intestinal contents. According to another theory, the filament is actually a tubule through which the sporoplasm, in the form of an amoebula, makes its way from spore case to host cell. But it seems unlikely that it can have this function in the many cnidosporidian species which are coelozoic in habitat.

The Cnidosporidia are customarily divided into three orders according to spore characteristics, as indicated in the classification in Chapter 18. Sometimes a fourth order is added, the Helicosporida, but it contains only a single genus and species, *Helicosporidium parasiticum.* The hosts of this parasite are Diptera and mites that live in rotting elm and horse chestnut trees. Its minute spore, with a diameter of only a few micra, is nevertheless able to contain a polar filament 60 or more micra in length when extruded, and a micron in width.

The Myxosporida

The Myxosporida, the first of the three orders of Cnidosporidia, are a moderately large group of Sporozoa which most laymen and perhaps some biologists have never heard of. It is an important group nonetheless, for it includes a good many species pathogenic to fish. Some species are often the cause of serious epizootics with a heavy mortality, resulting of course in less food for human consumption and higher fish prices.

Number of Species. No one can say at present how many species the order contains but there are certainly many. When Labbé wrote his famous monograph on the Sporozoa in 1899, he listed a total of "47 certain" and "20 uncertain" species of Myxosporida. According to Kudo (1933) the number had grown to 112 by 1910, and had almost doubled again in the next decade. By 1933 there were 402 known species. The number is undoubtedly much larger today, and it will continue to grow as an increasing realization of the importance of the group stimulates more intensive study.

Hosts and Habitat. Although the great majority of Myxosporida parasitize fish, host species being about equally divided between marine and fresh-water types, amphibia and reptiles are also occasionally attacked. Some myxosporidan species are tissue dwellers, living in organs such as the kidneys or in the muscles or integument of the body or gills; others are coelozoic, preferring residence in cavities such as the bladder. Most of the injury from myxosporidan infection is done by histozoic species. Externally they often provoke the development of large tumorlike masses, which anglers frequently suppose to be due to "worms." As Hegner once remarked,

". . . certainly no fish could possibly feel as badly as such a specimen looks." Internally, the cytoplasmic processes of the parasite may be so intimately intermingled with the tissues of the host that it may be impossible to tell where one ends and the other begins.

Fortunately, most of the many coelozoic species appear to do their hosts little harm. These forms seem content to reside in the hollow organs, especially the urinary and gall bladders, either freely floating about in their contents or attached to the lining epithelium by pseudopodia. Remarkably enough, no species of the group has yet been found inhabiting the alimentary tract.

Spores. The Myxosporida are chiefly characterized by the production of relatively large, bivalved spores with from one to four polar capsules (Fig. 23.7). The latter, unlike those of other orders in the group, are easily visible in the living condition without previous special treatment. The spores themselves are formed, usually in pairs, from a structure known as a pansporoblast, which in turn lies within a multinucleated protoplasmic mass or plasmodium.

Although the spores vary greatly in appearance—peculiarities of their shape and structure thus afford the best basis for the detailed taxonomy of the order—they are all essentially alike in architecture. Enclosed within the two valves, whose edges are applied to each other like the rims of a pair of watch glasses, lie the polar capsules (most often two) and the binucleated sporoplasm, as the infective germ is called. The two nuclei unite autogamously in the process of maturation, so that the mature spore has only one.

Apparently dispersal of the spores takes place when the external tumor-like masses ("myxosporidan cysts") burst, or when the body of the host disintegrates after death. Sometimes the parasite causes liquefaction of the tissues adjacent to the lesions, thus facilitating dissemination of the spores.

Once in the water, the spores await ingestion by fish or other susceptible vertebrates. Little is known of their longevity, but it apparently does not exceed several weeks. Since some species of Myxosporida attack both marine and fresh-water fish, their spores must be more than usually resistant to rather profound changes in salinity.

How the parasites, other than the few species inhabiting the alimentary tract, reach their final site of infection in the host is unknown. The young amoebula probably either manages to make its way to a connecting body cavity, such as the gall bladder (a preferred site for many species) or to the blood stream, by which it is passively carried to its destination. Considered collectively, the Myxosporida may attack virtually any part of their host's bodies.

Life Cycle. What happens after the young parasite establishes itself depends on the species of parasite and perhaps also on the species of host, since host reactions probably differ.

In the case of *Myxobolus* (one of the largest genera, comprising some 70 species typically histozoic in fresh-water fish), the amoebula invades a host cell and begins a period of active growth. At first only nuclear division occurs, with resulting formation of a plasmodium. This eventually fragments into many young uninucleate parasites, which in turn invade other cells. These repeat the process and eventually become extracellular. But instead of breaking up into uninucleate forms plasmotomy may ensue, with the

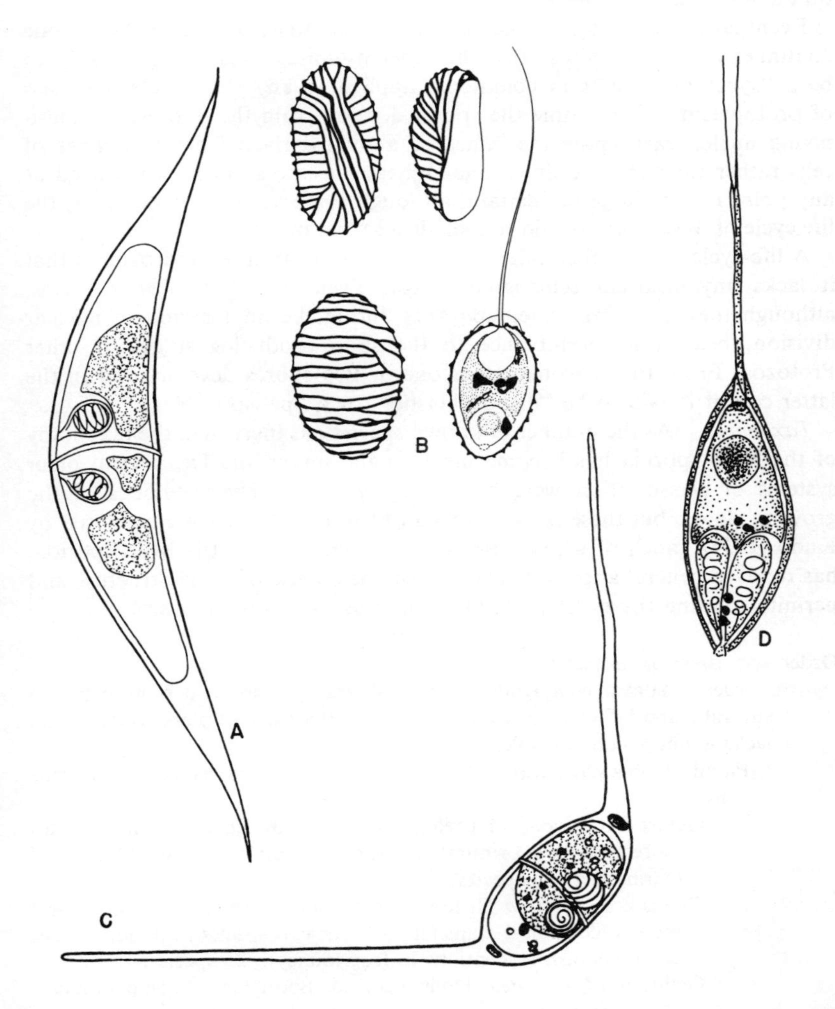

FIG. 23.7. Myxosporidan spores. *A. Ceratomyxa spherulosa* (after Kudo). Note the double (disporous) structure. *B. Myxidium serotinum* of frogs (Kudo and Sprague). Above: a lateral view and a valve; below: front view showing the complex ridge pattern characteristic of the species, and an exploded spore with a partially extruded polar filament. *C. Ceratomyxa spinosa* (after Kudo). *D. Henneguya macropodi*, a parasite of the paradise fish (after Shiba.)

production of a number of smaller but still multinucleated organisms. In some cases budding or gemmation may take place.

In the meantime the host may react by encapsulating the parasitized area, thus producing the typical "myxosporidan cyst." These masses may reach a considerable size, sometimes as big as a hen's egg. No wonder the effects on the host may be serious.

Eventually the nuclei of the parasite differentiate into two types: one continues to divide amitotically, the other by mitosis. This stage is said to be a "syncytium," a term commonly applied to any multinucleated mass of protoplasm. After a time the spores develop from the mitotically multiplying nuclei, each spore (as indicated above) derived from a number of cells rather than from a single one. Whether a sexual stage is involved at any point is not altogether certain, although some authorities think so; the life cycle of *Myxobolus* is said to include a sexual phase.

A life cycle such as this differs from those of most other Sporozoa in that it lacks any clear-cut schizogonic stage. Even the multiplicative forms, although they are often called sporonts during the final period of nuclear division, bear little resemblance to the spore-producing stages of other Protozoa. From the sporonts either one or two spores develop, and in the latter case it is said to be "disporoblastic," or a "pansporoblast."

Taxonomy. As the number of known species has increased, the taxonomy of the Myxosporida has become increasingly important. The several older systems of classification were based largely on the characteristics of the growing stages, but these proved difficult to use, and a scheme proposed by Kudo (1933), much of whose research has been done on the Myxosporida, has received general approval. This system makes use of spore structure and permits splitting the order into three suborders and nine families.

Order MYXOSPORIDA Bütschli.

Suborder 1. EURYSPOREA Kudo. Largest diameter of spore at right angles to sutural plane.* Two polar capsules, one on either side of plane. Sporoplasm lacks iodinophilous vacuole.

Family 1. CERATOMYXIDAE Doflein. Mostly coelozoic parasites of marine fish.

Genus 1. *Ceratomyxa* Thélohan emend. Arched spore with breadth more than twice sutural diameter. In gall or urinary bladder of marine fish. 67 species.

Genus 2. *Leptotheca* Thélohan 1895 emend. Breadth of spore less than twice sutural diameter. All but two species in organ cavities of marine fish (exceptions in frog kidney). 23 species.

Genus 3. *Myxoproteus* Doflein emend. Spore tends to be pyramidal;

*The "sutural line" is the line of union of the two halves, or valves, of the spore. The length of the spore is its greatest diameter, and its width the diameter at right angles to the sutural axis. The thickness is also at right angles to the former, but in the same plane. The "iodinophilous vacuole" is presumably a mass of glycogen, since it takes a deep brown stain with iodine.

may have distinct processes at base of pyramid. Urinary bladder of marine fish. 3 species.

Family 2. WARDIIDAE Kudo. Histozoic or coelozoic parasites of fresh-water fish.

Genus 4. *Wardia* Kudo. Spore in shape of isosceles triangle with 2 convex sides; oval profile. Shell valves with ridges and fringelike posterior processes. Large spherical polar capsules. 2 species.

Genus 5. *Mitraspora* Fujita emend. Circular or ovoidal spore (front view). Pyriform polar capsules. Shell valves finely striated longitudinally; with or without long posterior filaments. In kidneys of host. 4 species.

Suborder 2. SPHAEROPHRYA Kudo emend. Spherical spore, with 1, 2, or 4 polar capsules. Sporoplasm lacks iodinophilous vacuole.

Family 3. CHLOROMYXIDAE Thélohan. Four polar capsules.

Genus 6. *Chloromyxum* Mingazzini. Polar capsules at anterior end. Mostly coelozoic (though some are histozoic) in marine and fresh-water fish, and in amphibians. 41 species.

Family 4. SPHAEROSPORIDAE Davis. Two polar capsules.

Genus 7. *Sphaerospora* Thélohan. Sutural line straight. Polar capsules anterior. Coelozoic or histozoic in marine and fresh-water fish. 14 species.

Genus 8. *Sinuolinea* Davis. Sinuous sutural line. Two spherical polar capsules in middle. Coelozoic in marine fish. 8 species.

Family 5. UNICAPSULIDAE Kudo. One polar capsule.

Genus 9. *Unicapsula* Davis. Shell valves asymmetrical. Histozoic in marine fish. 1 species.

Suborder 3. PLATYSPOREA Kudo. Sutural plane coincides, or is at an acute angle to longest diameter. 1, 2, or 4 polar capsules. Sporoplasm may or may not show iodinophilous vacuole.

Family 6. MYXIDIIDAE Thélohan. Fusiform or semicircular spores; single polar capsule at each end. Sporoplasm lacks iodinophilous vacuole.

Genus 10. *Myxidium* Bütschli (=*Cysticercus* Lutz). Spore fusiform with pointed ends. Polar capsules typically pyriform, and long and slender polar filaments. Sutural line straight, coinciding with, or at an acute angle with axis of spore. Usually coelozoic, but sometimes histozoic in fresh-water fish, amphibia, and reptiles. 60 species.

Genus 11. *Spheromyxa* Thélohan. Spore fusiform; ends truncate. Pyriform polar capsules; short and thick polar filaments. Coelozoic in marine fish. 10 species.

Genus 12. *Zschokkella* Auerbach. Spore semicircular in front view; ellipsoidal profile. Pointed ends, with much curved sutural line. Large spherical polar capsules, with long, fine polar filaments. Usually coelozoic in marine and fresh-water fish. 8 species.

Family 7. COCCOMYXIDAE Léger and Hesse. Spore with single polar capsule; sporoplasm lacks iodinophilous vacuole.

Genus 13. *Coccomyxa* Léger and Hesse. Ellipsoidal spore, circular in cross-section. Coelozoic in marine fish. 2 species.

Family 8. MYXOSOMATIDAE Poche emend. Spore with either two or four polar capsules. Sporoplasm has no iodinophilous vacuole.

Genus 14. *Myxosoma* Thélohan (=*Lentospora Plehn*). Spore ovoidal and flattened; two pyriform polar capsules at anterior end. Histozoic in fresh-water and marine fish. 23 species.

Genus 15. *Agarella* Dunkerly. Ovoidal, flattened spore, with four polar capsules at anterior end. Long posterior processes. Histozoic in fresh-water fish. 1 species.

Family 9. MYXOBOLIDAE Thélohan. With one or two polar capsules at anterior end of spore, and iodinophilous vacuole in sporoplasm.

Genus 16. *Myxobolus* Bütschli. Spore flattened and ovoidal, without posterior process. Two pyriform polar capsules anteriorly. Iodinophilous vacuole. Histozoic in fresh-water fish. 70 species.

Genus 17. *Thelohanellus* Kudo. Pyriform, flattened spore, with single polar capsule anteriorly. Iodinophilous vacuole, histozoic in fresh-water fish. 11 species.

Genus 18. *Henneguya* Thélohan. Ovoidal, flattened spore; single or double caudal prolongations. Two pyriform polar capsules anteriorly. Iodinophilous vacuole. Histozoic in fresh-water fish. 52 species.

Genus 19. *Hoferellus* Berg. Pyramidal spore, with two posterior spinous processes arising from lateral faces. Parasitic in kidneys of fresh-water fish. 2 species.

The Actinomyxida

The Actinomyxida, the second of the orders of Cnidosporidia, is also one of the smallest orders of Sporozoa, and is interesting chiefly to the protozoologist making it his specialty. According to the Indian biologist Naidu, only 16 species were known when he added another one to the list in 1956. All of them are parasitic in annelids, but the host range in this phylum is rather wide. It includes the fresh-water Tubificidae (tube-dwelling oligochetes, familiar to anyone who has collected study material from the bottom of ponds), and the curious degenerate marine worms known as Sipunculoidea. The species described by Naidu, *Triactinomyxon naidanum*, adds not only another species to the order but another family, the Naididae, to the host list. Like the Tubificidae, the naids are common fresh-water dwellers.

Despite the very small size of the group, the Actinomyxida may not be wholly lacking in biological importance, since they may have some pathogenicity for their hosts. These are often of some significance in the food chains of larger organisms, among them fish used for human food. On this point, however, there is very little information. Naidu observed that infected naids were "less active than uninfected ones and conspicuously lacked the fission zones present in practically every uninfected worm." Against this statement may be set the apparent rarity of infection, at least in the case of some species. MacKinnon and Adam (1924), in the case of another species of the genus *Triactinomyxon*, examined 1250 worms from the mud of the Thames river before they found five infected specimens.

Little is known about the geographical distribution of the Actinomyxida. All studies on the group other than Naidu's have been done in Europe. It is likely, though, that these curious Sporozoa occur all over the world and that the total of species is much larger than the number now known.

The majority of Actinomyxida exhibit remarkable spore structure. The spores consist of three valves, giving them a kind of ternary radial symmetry. Though a few have the shape of simple tetrahedrons, most have three long, often branched, spinelike processes. The simplest arrangements of these may give the spore an anchorlike appearance (Fig. 23.8), but some species boast extremely intricate architecture which would inspire the keenest disciple of abstract art. Yet despite their complex design, these spores may be extremely small.

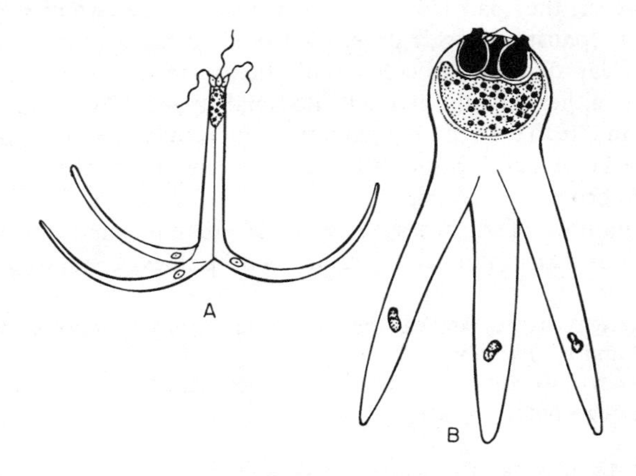

FIG. 23.8. Spores of two species of *Actinomyxida*. *A. Triactinomyxon ignotum*, occurring in the tube-building oligochaete annelid *Tubifex*. *B. Guyenotia sphaerulosa* in the same host. (After Léger and Naville respectively.)

The use of these peculiar branching processes can only be speculated. It has been suggested that they may enable the spore to float on the surface film, and floating spores have been observed. But Naidu noted that spores of the species he studied rested on the bottom, anchor prongs in the mud and the axial portion of the spore containing the sporoplasms directed upward. Perhaps the peculiar specialization of structure of the actinomyxidan spore is simply a characteristic, like freckles, conferring neither special advantage nor handicap on its possessor.

As in the Myxosporida, the spores are actually multicellular, each of the three valves being the product of a single cell, the three polar capsules likewise, and the sporoplasm (or sporoplasms) developing from the last, or seventh, cell. This complexity of origin led Štolc, the Czech biologist who sixty years ago pioneered in studying the Actinomyxida, to suggest that these peculiar parasites might not be Protozoa at all, but "Mesozoa." As we noted in the preceding section, a similar possibility exists for the Myxosporida.

Infection of a new host is presumably the result of ingesting spores, which then proceed to germinate in the intestine. But it also appears that autoinfection may be common, since the parasite population of the host may become very large. The sporoplasms, which are often multinucleate, escape through a pore at the triradiate end of the spore axis in which they are contained. Sometimes they break up into amoebulae before egress, sometimes after. In at least one species the number of amoebulae may exceed a hundred. In appearance and behavior these tiny amoebae act like any other such organism, having ectoplasm and endoplasm, moving about with pseudopodia, and ingesting food in typical amoeba-fashion.

Soon, however, they pair (sometimes while still in the parent spore) and a cyst stage, or "pansporocyst," develops from the binucleate form. Within this cyst nuclear division proceeds until there are a variable number of daughter cells either equally divided into smaller and larger ones or indistinguishable in size. These are the gametes, which then fuse to form zygotes, and from the latter develop the cells which collectively make up the spore. Details of the process, of course, show species differences.

Since the number of known species in the order is small, the taxonomy of the Actinomyxida is simple. There are two families, as follows:

Family 1. TETRACTINOMYXIDAE Poisson. One genus, *Tetractinomyxon*, parasitic in sipunculoid and naid worms.

Family 2. SYNACTINOMYXIDAE Poisson. Six genera (according to Poisson), the hosts being oligochete annelids.

Species of the first family produce tetrahedral spores without projecting processes, and the sporoplasm is at first uninucleate. Those of the second family are characterized by conspicuously developed spinelike processes, and contain a multinucleated sporoplasm from which may arise a variable number of amoebulae.

The Microsporida

The Microsporida, the third of the cnidosporidian orders, may well be more significant to man than the Myxosporida. Many microsporidan species are extremely common and pathogenic for their favored hosts, the insects. But their host range is even wider than the insects, for they infect a great variety of cold-blooded animals, and some species are pathogenic fish parasites.

Number of Species. Labbé in his famous 1899 monograph on the Sporozoa listed in the order Microsporida only "a single family with 3 genera, 9 certain species, and 44 uncertain." Kudo, in a very thorough monograph appearing in 1924, listed 178 "including the doubtful forms." Within the next twelve years, according to Jirovec (1936), 59 more had been added. It thus appears that the group is a relatively large one, today comprising well over 200 known species. Yet this figure probably does not even approximate the actual number; many possible hosts have never been carefully studied for microsporidal infection, and little is known about the host specificity of many of the species that have been described. Among certain Microsporida parasitizing mosquito larvae, host specificity seems to be quite strict, for only one of a number of larval species living in the same pool may be found with infection. On the other hand, the same species of microsporidan may attack numerous host species.

Definition. Outdated though his estimate of the size of the order may be, Labbé's definition of the group is still essentially correct: Sporozoa producing "spores with a single polar capsule invisible in the fresh state without the use of reagents. Spores very small. [Parasites] inhabiting the tissues and especially the muscles of certain fish and numerous invertebrates." We can now add that a few species are known which produce spores with two polar capsules, that the Microsporida are intracellular parasites, and that all of them have life cycles involving a single host, i.e., they lack vectors.

Spore Characteristics. Spore characteristics form the basis for generic and species identification and for general taxonomy. Although only a few micra in diameter as a rule (and in occasional species, hardly larger than many bacteria), the spore contains within its small space a sporoplasm and a very long coiled filament. In some species this filament reaches a length of several hundred micra (perhaps fifty times the diameter of the spore) when discharged. Its function is uncertain. So also is the trigger mechanism operating under natural conditions. Pressure on the spores and chemical stimulation causes extrusion of the filament. Kudo (1918) found hydrogen peroxide the most effective chemical stimulus; he suggested that the spore might contain a peroxidase which by liberating gaseous oxygen generated a sudden pressure resulting in the explosive discharge of the thread. Doubtless substances naturally present in the host's digestive tract exert a similar effect.

Infection may in some species occur through the egg, as Pasteur showed in his famous experiments with *Nosema bombycis* of the silkworm. But more often it is the result of ingesting spores. Like bacterial spores, these are highly resistant and may retain their viability for considerable periods of time. Little is known about the chemical nature of either the spore wall or the filaments. Kudo believed the former may be of chitin. Normally the spores pass from the body through the anus or are liberated from the host's body when it disintegrates after death. But in some cases at least they apparently germinate while the host is still alive and thus permit a continued build-up of the infection.

Pathogenicity. The effects of microsporidiosis vary greatly with the species. In some, disability and death are the rule; in others little harm is done.

Nosema apis kills honey bees the world over. *Thelohania opacita*, from our point of view a more useful parasite, often kills mosquito larvae. But probably, like most other parasites, the majority of microsporidan species do not greatly injure their hosts.

The mechanism of injury undoubtedly varies with the species also. We have little evidence of the production of toxins, but the effects of the parasite are often so marked on host and neighboring cells that it is probably safe to assume that toxic substances are produced. These hypothetical chemical irritants may well be closely related to certain carcinogens (cancer-producing compounds), for they cause hypertrophy and overgrowth of the affected tissues (Fig. 23.12) and irregularities of mitosis. Such parasite-induced alterations of normal tissue deserve more careful study than they have yet received.

Not all Microsporida exert such effects, however, and in many cases the injury seems to be simply the result of the progressive invasion and destruction of cells, sometimes on a massive scale. Certain species appear to have a strong predilection for given tissues, but others are less particular and attack almost every organ in the body.

Nosema bombycis. Such a nonselective species is *Nosema bombycis* of the silkworm. As Pasteur first showed, infection occurs via either the egg or from ingesting the spores ("corpuscles") left on mulberry leaves by other infected larvae when they defecate. Apparently such spores germinate in the gut of the new host within approximately 24 hours, and the emerged amoebulae promptly begin a period of rapid growth and multiplication by binary fission or budding. At first they remain in the gut, but they soon begin to wander and are now called "planonts" (from the Greek for "wanderers"). Some enter lining cells of the intestinal tract; others reach the blood stream and thus succeed in gaining access to virtually every tissue in the body, which they soon invade. Whatever the type of host cell, the young parasite (now known as a "meront," or dividing form) proceeds to undergo further binary fission, or it may reproduce by schizogony. In either case multinucleated forms eventually result and each produces a single spore, certain nuclei contributing to the formation of the sporoplasm, the spore envelope, and the polar capsule, much as in the Myxosporida. The life cycle is shown in diagrammatic form in Figure 23.9. Different investigators do not agree on all details of the process, which are difficult to work out because many stages are extremely minute.

Lightly infected silkworms may show little evidence of injury, but more heavily infected individuals eat little, develop slowly, move sluggishly, and may not live to pupate. If they do form cocoons, the silk is of poor strength and quality. Infected worms and moths often show numerous dark spots (hence the name pébrine from the French *pébré*, patois for "pepper"). Moths emerging from infected cocoons are often deformed. Meticulous cleanliness in the silkworm nursery and the use of healthy moths as sources of eggs are necessary to control the disease, and Pasteur taught the growers how to distinguish between healthy and infected moths. To certain skeptical growers who objected to the use of the microscope for this Pasteur remarked, "There is in my laboratory a little girl eight years old who has learned to use it without difficulty." As his biographer, René Dubos, goes on to say, the little girl was his daughter, Marie-Louise.

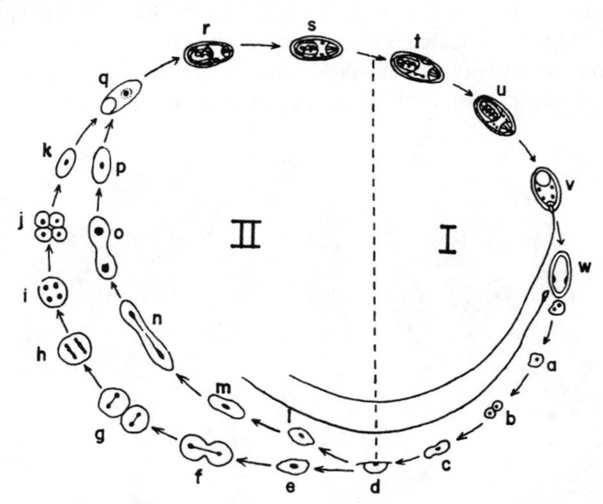

FIG. 23.9. The developmental cycle of *Nosema bombycis,* according to Stempell. *I.* Extracellular stages. *II.* Intracellular stages. a–c, planonts; d–p, meronts; q–s, sporulation stages; t–v, contamination stages; w, coming out of binucleated sporoplasm. (After Naville.)

Nosema apis. Since man-made fibers have reduced the importance of the silk industry, probably "Nosema disease" of bees is today the cause of greater losses than pébrine. The species causing it, *Nosema apis*, has already been mentioned. The characteristic spores of this parasite were first observed in 1857, but their protozoan, not fungal, nature was realized only some sixty years later when the German scientist Zander studied the disease. To him we also owe the species name *apis.*

Infection seems usually the result of ingesting spores with food and water or in licking the bodies of sick bees. The parasites, unlike those of pébrine, normally occur only in the midgut, the epithelial cells of which they invade and destroy (Fig. 23.10). At this stage the antibiotic fumagillin is effective against the organisms; it may be fed to the bees in syrup. Treatment is, however, no substitute for prevention based on good sanitation of the apiary.

Bees with Nosema disease may be recognized from the appearance of their stomachs, which are generally whitish and grossly distended. Severely affected individuals show symptoms suggestive of partial paralysis, and may be able to fly only short distances or not at all. Although some bees apparently recover, they seem to remain infected and are prone to relapse under adverse conditions.

Somewhat surprisingly, only the adult bees are susceptible to infection and workers and drones suffer more than the queens. Since the parasites are usually confined to the midgut, ovarian transmission does not occur, although eggs may be contaminated with the spores. The larvae, or brood, remain healthy except as they suffer from the inability of diseased workers to give them care. Lightly infected colonies may therefore survive, but heavily infected ones eventually perish. And since weakened hives are subject to raiding by stronger ones, such hives may spread the infection.

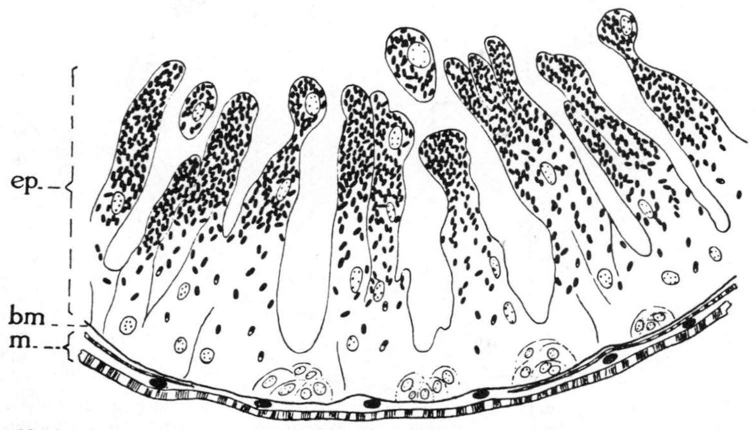

FIG. 23.10. *Nosema apis* in stomach epithelium of the honey bee; the black masses are the spores of the parasite. A few of the younger spores are shown in the basal portion of the gut lining; they are characterized by a lighter unstained area at one end. ep, epithelium; bm, basement membrane; m, muscular portion of the stomach wall. (After White.)

Although only the honey bee, *Apis mellifica*, is known to be a natural host of *Nosema apis*, other species of bees have been found experimentally susceptible. So also have insects belonging to several other orders, such as butterflies (Lepidoptera), blow flies, crane flies, and sheep keds (Diptera). Bumblebees are also subject to parasitism by another species of the genus, *Nosema bombi*. Like its close generic relation, *N. apis*, it is decidedly pathogenic and must be regarded as of some economic importance in view of the active role of the bumblebee as a pollenizer.

The life history of *Nosema apis* does not differ significantly from that of *N. bombycis*, and the two species closely resemble each other even in the characteristics of the spores (Fig. 23.11). Those of *Nosema apis* are passed from the host with excreta, and are relatively resistant to change in temperature, desiccation, and the chemicals commonly used as disinfectants. They are, however, quickly destroyed by direct exposure to sunlight and soon die in the presence of putrefaction. How long they live under natural conditions outside the host is unknown: certainly for some weeks, perhaps for several months, and by some estimates, as long as a year.

Other Pathogenic Species. A number of other pathogenic species of Microsporida is known, and the list is still growing. Species of *Thelohania* often kill their insect hosts, and one, *T. contejeani*, has caused destructive epizootics among river crayfish in France. Thomson (1957) has described a species of *Perezia* (*P. fumiferanae*) which infects some 40 per cent of the spruce budworm population of Ontario forests, and kills a substantial proportion of the larvae. Since the budworm is a very destructive parasite of

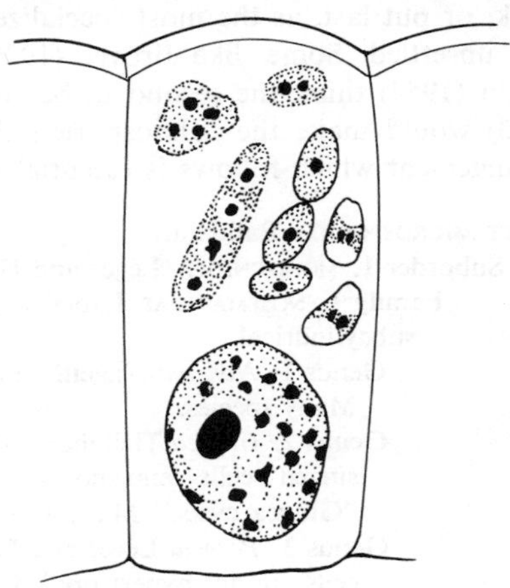

FIG. 23.11. An epithelial cell of the gut of the bee containing various stages of *Nosema apis*. To the right are two young spores; later stages are shown at the top, and at the bottom is the nucleus of the invaded cell. (After Jirovec.)

spruce, any aid in its natural control is beneficial. Not the least interesting fact about *Perezia* is that it is transmitted not only *per os* and through the egg, but occasionally at least by the male when mating.

Some species of *Glugea* afflict numerous species of fresh-water and marine fish, in whose flesh they cause cysts to develop (Fig. 23.12). These cysts, which may appear to be tumors, contain actively reproducing parasites in various stages of development, and eventually great numbers of spores. A few species of *Glugea* have amphibia, reptiles, and even crustacea for hosts.

There are also a few Microsporida which have carried parasitism almost to its nadir by taking other parasites as their hosts. One such is known from

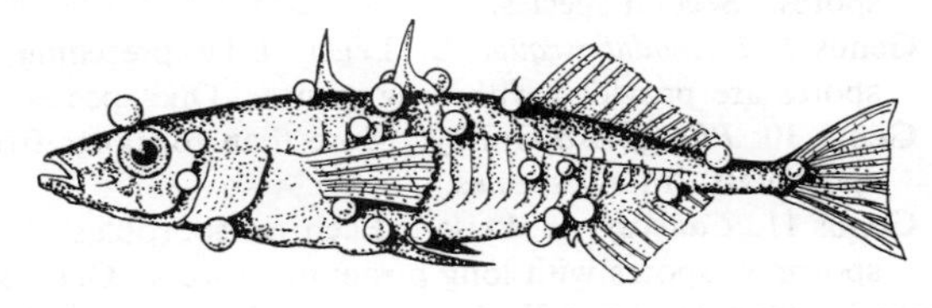

FIG. 23.12. A stickleback heavily parasitized with *Glugea anomala*, a pathogen causing hypertrophy of the invaded cells and the resulting tumorlike masses known as *Glugea* cysts. (After Poisson.)

a leech, several from the flukes and tapeworms, and two from gregarines. A species of microsporidan has as its host a *Balantidium* from a toad. All these parasites observe the rules of the game quite strictly, and ask no hospitality from their host's host either in food or shelter. But one microsporidan, oddly enough, has for a host a myxosporidan, and the Myxosporida are undoubtedly the closest relatives of the Microsporida.

Taxonomy. Opinion differs among protozoologists about the classification of the Microsporida. The group is usually given the rank of order, but whether it should be accorded first place, thus implying a lower evolutionary rank, or put last, as the most specialized of the Cnidosporidia, is a matter still unsettled. Some, like Jirovec (1936), urge the first view; others, e.g., Kudo (1954) think the second to be correct. Still others (such as Poisson, 1953) would make the Microsporida the second of the three orders. The arrangement which follows is essentially that of Kudo.

Order MICROSPORIDA Balbiani.

Suborder 1. MONOCNIDEA Léger and Hesse. Spores with a single polar capsule.

Family 1. NOSEMATIDAE Labbé. Spores ovoid, pyriform; occasionally subcylindrical.

Genus 1. *Nosema* Nageli. Each sporont produces a single spore. Many species.

Genus 2. *Glugea* Thélohan. Each sporont produces 2 spores. Parasitized cells enormously hypertrophied, with formation of "Glugea cysts." Many species.

Genus 3. *Perezia* Léger and Duboscq. Like *Glugea*, except that host cells are not hypertrophied. A number of species.

Genus 4. *Gurleya* Doflein. From each pansporoblast 4 sporoblasts develop, and finally 4 spores. Few species.

Genus 5. *Pyrotheca* Hesse. Spores with one end swollen, and the other somewhat pointed. Usually 4 spores from each pansporoblast. A number of species.

Genus 6. *Thelohania* Henneguy. Usually 8 sporoblasts and 8 spores (but up to 32) from each pansporoblast. Spores often of 2 sizes (macrospores and microspores). Many species.

Genus 7. *Stempellia* Léger and Hesse. Pansporoblast may produce 1, 2, 4, or 8 spores, according to circumstances. Two species known.

Genus 8. *Duboscquia* Perez. Each pansporoblast gives rise to 16 spores. Several species.

Genus 9. *Trichoduboscquia* L. Léger. Like preceding, except that spores are provided with long spines. One species.

Genus 10. *Plistophora* Gurley. More than 16 spores from each pansporoblast. Many species.

Genus 11. *Caudospora* Weiser. Each pansporoblast produces many sporonts; spores with long posterior process. One species known.

Family 2. COCCOSPORIDAE Kudo. Spores spherical, or subspherical.

Genus 12. *Coccospora* Kudo. Characters of the family. Several species.

Family 3. MRAZEKIIDAE Léger and Hesse. Spores cylindrical; about 5 times as long as broad.

Genus 13. *Mrazekia* Léger and Hesse. Spore straight with caudal process. Several species.

Genus 14. *Bacillidium* Janda. Spores elongate and narrow, without caudal appendages. Several species.

Family 4. COUGOURDELLIDAE Poisson. Spores rectilinear; sometimes curved or twisted.

Genus 15. *Cougourdella* Hesse. One end of spore swollen, and the other collarlike. Several species.

Genus 16. *Octosporea* Flu. Spores cylindrical and curved; ends nearly alike. Six species.

Genus 17. *Spiroglugea* Léger and Hesse. Tubular, helically wound, and flattened on axial side. Large polar capsule. One species.

Genus 18. *Toxoglugea* Léger and Hesse. Minute, curved or semicircular spores. Several species.

Suborder 2. DICNIDEA Léger and Hesse. Spores with a polar capsule at each end.

Family 5. TELOMYXIDAE Léger and Hesse. Characters of order.

Genus 19. *Telomyxa* Léger and Hesse. Each pansporoblast produces 8, 16, or *n* spores. Several species.

The Haplosporidia

By some protozoologists the Haplosporidia are regarded as an order in the subclass (or class) Acnidosporidia, and the Sarcosporidia may be treated as a second order. There is, however, some question whether the latter are Protozoa at all (Chap. 26). We prefer to regard the Haplosporidia as a subclass. It is a small group whose distinguishing characteristic is the production of spores without a polar capsule or filament. One genus, *Coelosporidium*, a parasite of the cockroach, produces spores lacking the filament but otherwise resembling microsporidan spores; it is of a doubtful taxonomic position.

The spores of Haplosporidia are apparently minimally equipped to survive the interval between leaving one host and finding another, for they consist of only a protective envelope with a sporoplasm inside (Fig. 23.13).

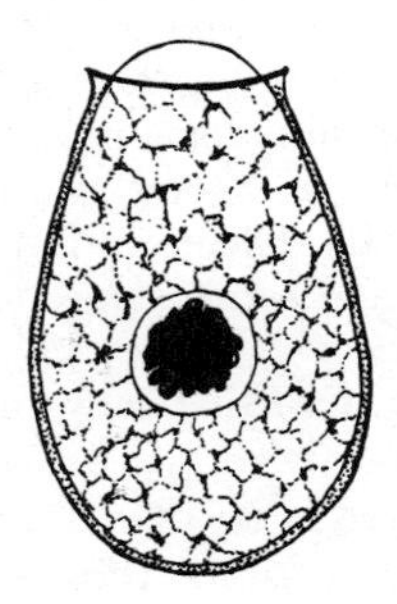

FIG. 23.13. The spore of a haplosporidian, *Haplosporidium cernosvitovi*, a parasite of oligochaete annelids. (After Jirovec.)

Definitive hosts are invertebrates and fish, and as far as known there are no intermediate hosts.

The Haplosporidia are still one of the least known groups of Sporozoa, even though the order was created in 1899. The distinguishing characteristics of these protozoan parasites are the simplicity of spore structure (the name "Haplosporidia" comes from the Greek *haploos* meaning "simple"), and the formation of multinucleated aggregations of protoplasm, or plasmodia, at one stage.

The host range is wide, including flatworms, molluscs, annelids, crustacea, insects, certain primitive chordates, and possibly fish. Haplosporidia have also been described from a variety of other animal hosts, and even from the Protozoa. Fuller study of many of these organisms, however, has shown some of them to be fungi, for example, *Rhinosporidium*, a supposed haplosporidian parasite of man.

Caullery and Mesnil, both noted French biologists, founded the group to receive a small assemblage of parasites belonging chiefly to three genera: *Haplosporidium*, *Urosporidium*, and *Anurosporidium*, occurring in annelid and trematode hosts. Attempts have since been made to attach numerous other genera to the group, but Caullery (1953) still believes that only five genera (the original three, plus two others) should properly be included, and these would embrace a total of only some 20 known species. He concedes, however, that further research will probably uncover additional species, and he appends a list of 30 other genera, comprising some 50 species, which have been assigned to the subclass (or order) by various biologists.

Just as the simplicity of spore structure separates the Haplosporidia from other corresponding groups, so peculiarities of shape and appearance of the spores afford the major basis for the identification of genera, as indicated below:

Haplosporidium. Spores of simple structure, although in some species the outer membrane may present bizarre modifications. Twelve or 13 species known.

Urosporidium. Spores with the outer envelope drawn out into a long and conspicuous platelike structure. One species.

Anurosporidium. Spores spherical, lacking a lid over the opening. The only known species is parasitic in the sporocyst of a fluke, which itself has a bivalve mollusc for a host.

Nephridiophaga. Spores said to be similar to those of *Haplosporidium*, but with certain differences in life history. Occurs in the Malpighian tubules of bees, and perhaps in certain roaches.

Physcosporidium. Spores said to be bivalved. A single species is known, parasitic in a free-living flatworm (Turbellaria).

One other genus is often included with the Haplosporidia—*Icthyosporidium*, parasitic in fish. Although Caullery and the late and eminent French biologist Mesnil were its creators, Caullery regards its affinities as doubtful, and it may properly belong with the Microsporida. In any event, some of its species apparently cause disease in fish.

One of the few species of Haplosporidia for which the life cycle is known in detail is *Haplosporidium heterocirri*. It parasitizes certain marine annelids, invading the intestinal epithelium. Here the minute spores begin a series of mitotic divisions, culminating in the production of relatively large, multinucleate forms, which then break up into cysts. From these, spores with remarkable flagelliform prolongations develop. But the occurrence of sexual stages is doubtful. In some species, as apparently in this one, binucleated stages are produced with subsequent nuclear recombination (autogamy). The existence of true gametes has been claimed for the life cycles of some species but these remain questionable.

REFERENCES

Bond, F. F. 1938. Resistance of myxosporidian spores to conditions outside the host. *J. Parasitol.*, 24: 470–1.

Calkins, G. N. 1933. Special morphology and taxonomy of the Sporozoa. In *Biology of the Protozoa*. Philadelphia: Lea & Febiger.

Caullery, M. 1953. Appendice aux Sporozoaires. Classe des Haplosporidies. In Grassé, P.–P., *Traité de Zoologie*, T. 1, Fasc. 2. Paris: Masson.

Cleveland, L. R. 1949. Hormone-induced sexual cycles of flagellates. I. Gametogenesis, fertilization, and meiosis in *Trichonympha*. *J. Morphol.*, 85: 197–296.

Grassé, P.-P. 1953. Sous-embranchement des Sporozoaires. In Grassé, P.-P. *Traité de Zoologie*, T. 1, Fasc. 2. Paris: Masson.

Grell, K. G. 1956. *Protozoologie*. Berlin: Springer-Verlag.

Hall, R. P. 1953. *Protozoology*. New York: Prentice-Hall.

Jirovec, O. 1936. Studien über Microsporidien. *Mémoires de la Soc. Zool. Tchecosl. Prague.*, 4: 1–74.

Kudo, R. R. Experiments on the extrusion of polar filaments of Cnidoporidian spores. *J. Parasitol.*, 4: 141–7.

——— 1920. Studies on the Myxosporidia. A synopsis of genera and species of Myxosporidia. *Ill. Biol. Monog.*, p. 265, Cont. 158.

——— 1924. A biologic and taxonomic study of the Microsporidia. *Ill. Biol. Monog.*, p. 268, Cont. 246.

——— 1933. A taxonomic consideration of the Myxosporidia. *Trans. Am. Mic. Soc.*, 52: 193–216.

——— 1954. *Protozoology*, 4th ed. Springfield: Charles C Thomas.

Labbé, A. 1899. Sporozoa. In *Das Tierreich*, O Bütschli, (ed.) Berlin: Verlag R. Friedlander und Sohn.

Léger, L. and Duboscq, O. 1902. Les grégarines et l'epithelium intestinal chez les trachéates. *Arch. de Parasit.*, 6: 377–473.

MacKinnon, D. L. and Adam, D. I. 1924. Notes on Sporozoa parasitic in *Tubifex*. I. The life history of *Triactinomyxon* Stolc. *Quart. J. Mic. Sci.*, 68: 187–210.

Naidu, K. V. 1956. A new species of actinomyxid parasitic in a fresh-water oligochete. *J. Protozool.*, 3: 209–10.

Naville, A. 1931. Les Sporozoaires. *Mém. Soc. Physique et d'Hist. Nat., Genève*, 41(1): 1–223.

Poisson, R. 1953. Sous-embranchement des Cnidosporidies. In Grassé, *Traité de Zoologie*, T. 1, Fasc. 2. Paris: Masson.

Steinhaus, E. A. 1947. *Insect Microbiology*. Ithaca, New York: Comstock.

Thomson, H. W. 1957. *Perezia fumiferanae* Thom., a protozoan parasite of the spruce budworm *Choristoneura fumiferanae* (Clem.). Ph.D. thesis, McGill University.

Weiser, J. 1955. A new classification of the Schizogregarina. *J. Protozool.*, 2: 6–12.

——— 1956. Protozoen Infektionen im Kampfe gegen Insekten. *Z. Pflanzenkr. u. Pflanzensch.*, 63: 625–38.

Wenyon, C. M. 1926. *Protozoology*, vol. 2. New York: Wood.

White, G. F. 1919. *Nosema* disease. *U.S. Dept. Agr. Bull. 780*, p. 59.

24

Malaria Plasmodia and their Relatives

> ". . . SECRETISSIMA CVLTRI PVDICITIAE HONESTAE. FAMAE QUAE VIXIT SINE FEBRIBUS. ANNIS XXXX D(iebus) XXVI Q(uintus) PVLLAENIVS. . . . ."
> Epitaph on a Roman tombstone commemorating a woman known "for her virtue and honesty, whom rumor states lived without fever (malaria) for 40 years." (*Notice sur l'Institut Pasteur d'Algerie*, December 31, 1934.)

Malaria has virtually disappeared from the United States and Canada, but it is still widespread in Africa, Asia, parts of Central and South America, and many other parts of the world. As recently as 1937, there were at least a million cases of malaria annually in the United States, and in the early part of the century it was one of the commonest diseases in the Deep South, In 1964 the estimated total number of cases in the world was 200 million each year and the resulting mortality at least 2 million. Furthermore, the disability and economic loss caused by malaria is beyond computation, and the World Health Organization has placed the abolition of malaria, now in theory quite possible, among its chief aims.

Malaria and History

Malaria has played an important role in shaping history. It was a major factor in the fall of the ancient Greek and Roman civilizations. In both the Civil War and the Spanish-American War malaria was responsible for more than a quarter of all hospital admissions for disease—and disease in these wars was more destructive than shot and shell. The historian Forry remarked in 1840, "By far the most fatal disease of Louisiana, whether in our city or the low land of the country, is the congested form of fevers, or, as it is called here the cold plague (malaria). . . ." He goes on to give a vivid account of the disease:

> Along the frontiers of Florida, as in our Southern states generally, may be seen

deplorable examples of the physical, and perhaps moral abjection, induced by marsh miasmata. In earliest infancy, the complexion becomes sallow, and the eye assumes a bilious tint. Advancing toward the years of maturity, the growth is arrested, and the limbs become attenuated, and the viscera engorged. Boys of fifteen years may be seen bowed down with premature old age—a mere vegetating being, with an obstructed, bloated, and dropsical system, subject to periodical fevers, passive hemorrhages, and those other forms of disease which follow in the train of malaria.

During World War II malaria seriously threatened our success in the Near and Far East. Although at present it seems to have retired from the stage of current history, any great social upheaval, such as another major war, would serve as an effective curtain call. Indeed, in Italy during the last war, malaria, apparently extinguished in many areas by years of public health work, reappeared in areas devastated by the advancing armies. Let no one talk of the "good old days!"

Etiology of Malaria

For a long time malaria in man was believed to result from living near marshes or breathing night air—hence the name "malaria," (*mal*, "bad" plus *aria*, "air"). Now we know that the mosquitoes associated with swamp conditions are the cause, rather than swampy emanations and darkness. But the direct cause of the malady is infection with one or more species of malaria parasites, and mosquitoes are only transmitters.

All species of malaria are usually placed in the genus *Plasmodium*, to which the genera *Haemoproteus*, *Leucocytozoon*, and *Hepatocystis* are closely related. There is, however, a tendency among malariologists to split the genus *Plasmodium* into several subgenera. Both *Haemoproteus* and *Leucocytozoon* are large genera and occur very commonly in birds; a few exceptional species are quite pathogenic and thus economically important. Some species of *Haemoproteus* also occur in reptiles. At present somewhat more than 60 species of *Plasmodium* are known, and here again birds are favored hosts; the infection has been found in several hundred species, the majority being perching birds (passerines). But the host range is a fairly wide one, and includes reptiles, chiefly lizards, and a moderate number of mammalian species, especially rodents and primates.

Fortunately, man is subject to infection with only four species* of malaria parasites: *Plasmodium falciparum*, *vivax*, *malariae*, and *ovale* (Fig. 24.1).

*It has recently been shown that man is also experimentally susceptible to infection with *Plasmodium cynomolgi bastianelli* of rhesus monkeys (Eyles, D. E., Coatney, G. R., and Getz, M. E. 1960. *Science*, 131: 1812) and to several other species of simian malaria. Transmission was accomplished by the bite of anopheline mosquitoes, and it is therefore possible that infection with these species may occur, at least occasionally, in nature. They include *Plasmodium knowlesi*, long thought transmissible to man only by blood inoculation.

Laveran, the French army surgeon who in 1880 first saw the causal relation of these tiny parasites in red cells to the disease, was probably dealing with *Plasmodium falciparum* when he gave it the name *Oscillaria malariae*. In the years that followed, it was realized that at least two other kinds of malaria were caused by the two species next on the list, but *Plasmodium ovale* was not discovered until 1922. After Danielewski's discovery of malaria in birds in 1885, the malaria parasites of birds and man were thought to be the same, but fortunately this proved untrue; if it were, human malaria would be much harder to control.

Plasmodium vivax is the most widespread species occurring in man, and it is found in most habitable regions of the world although rarely in the northern and southern hemispheres on the polar sides of the 42nd parallel. Not a very pathogenic parasite, it causes a disabling illness prone to relapse affecting millions of people.

Of the other three species, *Plasmodium falciparum* is by far the most important. Largely restricted to tropical and subtropical regions, it also afflicts large numbers of people and has a mortality which sometimes reaches 25 per cent. *Plasmodium malariae* causes a particularly chronic type of malaria, but it is common in only a few areas. *Plasmodium ovale* is the most benign and also the rarest species of the four. It is most common in tropical Africa, although cases have been recorded from various parts of the world. Possibly the genus *Plasmodium* evolved in Africa and radiated from there to the far corners of the earth. Since the evidence is growing that man also originated in that great continent, malaria may have been one of his chief afflictions through all his long history.

Life Cycle

All species of malaria plasmodia undergo a life cycle involving two hosts, a terrestrial vertebrate and, in every species yet determined, a mosquito. Only recently have certain gaps been finally filled in, so that the complete cycle, for certain species at least, is now known (Fig. 24.2).

A new host is infected when a mosquito carrying sporozoites injects them into the body of her victim, along with the salivary (anticoagulant) secretion. The minute parasites are quickly carried to remote parts of the body, for mosquitoes generally insert their probosces directly into the capillaries when they bite, and thus the malaria organisms are injected into the blood stream as effectively as if placed there by a hypodermic needle. The next stages in the cycle of monkey and human malarial parasites occur in the liver, as shown by recent studies of the English scientists, Shortt, Garnham, Hawking, and others. In *Plasmodium gallinaceum*, a malaria parasite of chickens (and the first species whose life cycle was worked out in complete detail), Huff and Coulston (1944) showed that the sporozoites entered cells of the lymphoid-macrophage (reticuloendothelial) system within 30 minutes after being

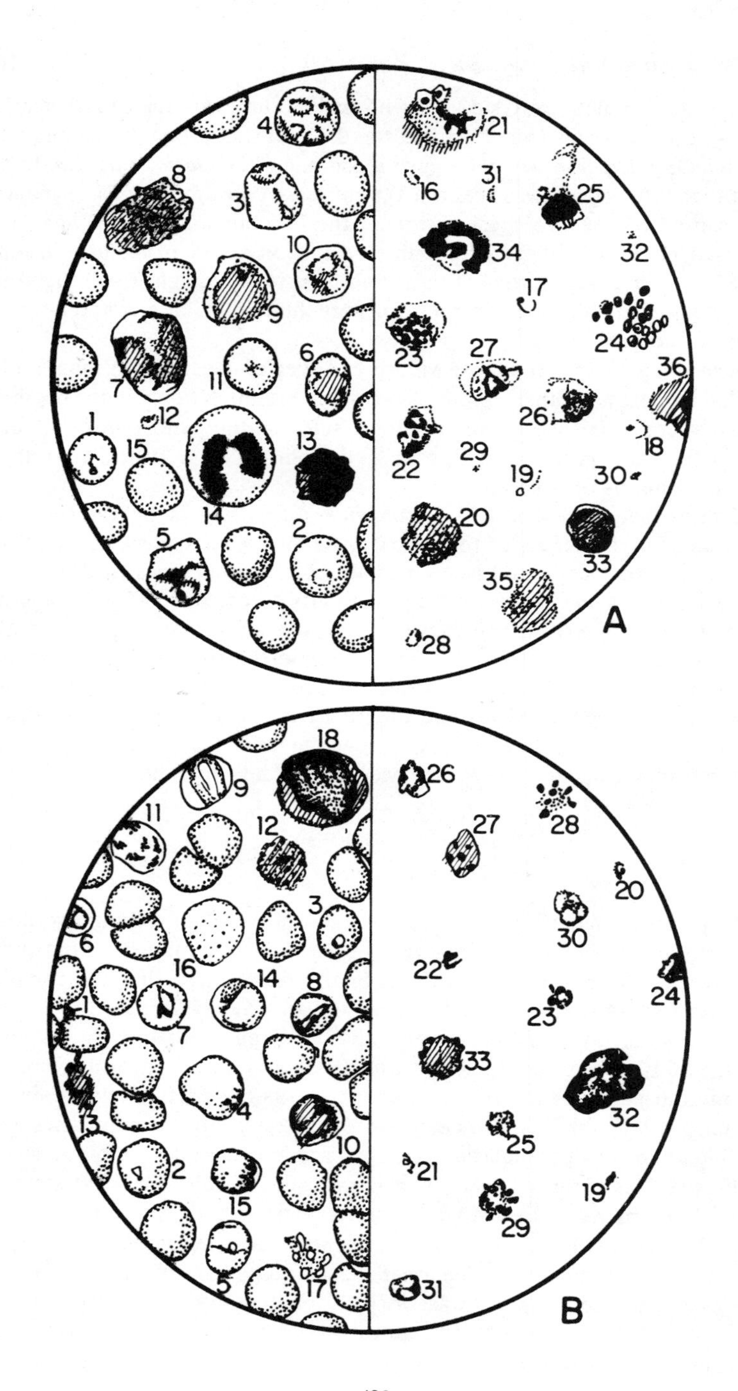
4
21
8
31
16
25
3
10
34
32
17
9
24
7
6
23
27
11
36
26
1
12
18
15
14
13
29
22
19
30
20
5
2
33
35
A
28
18
26
9
11
27
28
12
20
3
6
30
22
16
14
8
24
1
23
7
33
32
4
13
25
10
2
21
19
15
29
5
17
31
B

[Figure 24.1 A and B is reproduced in color on the inside front cover.]

FIG. 24.1. *A. Plasmodium vivax. Left.* Thin smear (Giemsa-stained). *1.* Very young trophozoite, with a few Schüffner's dots already in evidence. *2.* Ring stage, with host cell somewhat enlarged but not yet stippled. *3, 4.* Erythrocytes with 3 and 4 ring stages (respectively). In *3* two of these forms are marginal, simulating somewhat *P. falciparum. 5.* Trophozoite exhibiting characteristic amoeboid form. *6.* Trophozoite (probably a young gametocyte). *7.* Schizont with two nuclei. *8.* Segmenter with 16 massed merozoites. *9.* Macrogametocyte. *10.* Microgametocyte (somewhat smaller than usual). *11.* Erythrocyte with superimposed platelet. *12.* Platelet. *13.* Lymphocyte. *14.* Polymorphonuclear leucocyte. *15.* Normal red cell. (Compare with the increased size of those containing parasites.) *Right.* Thick smear (Giemsa-stained). *16.* Young trophozoite. *17, 18, 19.* Ring stages (note detached chromatin). *20.* Trophozoite (note characteristic amoeboid form). *21.* Trophozoite (note shadow about the parasite. When seen it is diagnostic of *P. vivax*, possibly representing the remains of the reticulum of the host cell, or of Schüffner's dots, or both). *22.* Presegmenter. *23.* Nearly mature segmenter. *24.* Group of merozoites, with mass of pigment. *25, 26.* Large trophozoites, or macrogametocytes (these may be difficult to tell apart in thick smears). *27.* Remains of a microgametocyte. *28–32.* Platelets. *33.* Lymphocyte. *34.* Polymorphonuclear leucocyte. *35, 36.* "Blue clouds," sometimes mistaken for parasites, probably the reticulum of young erythrocytes, often numerous in malaria.

B. Plasmodium malariae. Left. Thin smear (Giemsa-stained). *1.* A young merozoite. *2.* An erythrocyte containing a very young parasite. *3.* A young ring. *4.* A trophozoite simulating the marginal or "appliqué" forms so often seen in falciparum malaria. *5.* Trophozoite assuming the band form especially characteristic of *P. malariae. 6.* Moderately advanced trophozoite. Note small size of the host cell and the absence of pseudopodial processes. *7.* A slightly older trophozoite. *8.* Trophozoite illustrating the typical band form often assumed by the younger stages of this plasmodium. 9. An older parasite still exhibiting the band form. *10.* A schizont with the chromatin already divided more than once, but the pigment still scattered. *11.* A more advanced schizont. *12.* A segmenter with the eight merozoites (still not quite separated) typical of this species. *13.* A group of five free merozoites. *14.* A macrogametocyte. Note that it virtually fills the host cell, the cytoplasm takes a rather strong blue stain in Romanowsky-stained preparations, and there is a sharply stained marginal mass of chromatin. Pigment (in both sexual and asexual forms) is relatively dark and abundant. *15.* A microgametocyte. Pigment is abundant, rather coarse, and the cytoplasm takes a lighter stain than in the female. The chromatin stains lightly and is quite diffuse. *16.* An erythrocyte exhibiting basophilic stippling, often abundant in the blood of malarious persons. *17.* A mass of platelets, sometimes mistaken for a group of free merozoites. *18.* A large monocyte. *Right.* Thick smear (Giemsa-stained). *19, 20, 21.* Ring forms. In the thick smear, however, the ring shape is often lost, and the chromatin and cytoplasm are likely to appear slightly separated. *22.* A young parasite. *23, 24, 25.* Somewhat older trophozoites. In the thick smear little structure is likely to be visible, and the parasite has a somewhat condensed appearance which, together with the pigment, causes it to appear very dark. *26.* A schizont, with two masses of chromatin and conspicuous pigment. *27.* A segmenter. *28, 29.* A still more advanced segmenter. The eight merozoites so characteristic of this species of malaria are almost diagnostic. *30, 31.* Probably macrogametocytes, but in the thick smear it is not always easy to distinguish the sexual stages from the large trophozoites. Their small size and heavily stained appearance, however, should make them recognizable as quartan malaria parasites. *32.* The nucleus of a leucocyte. *33.* A "blue cloud."

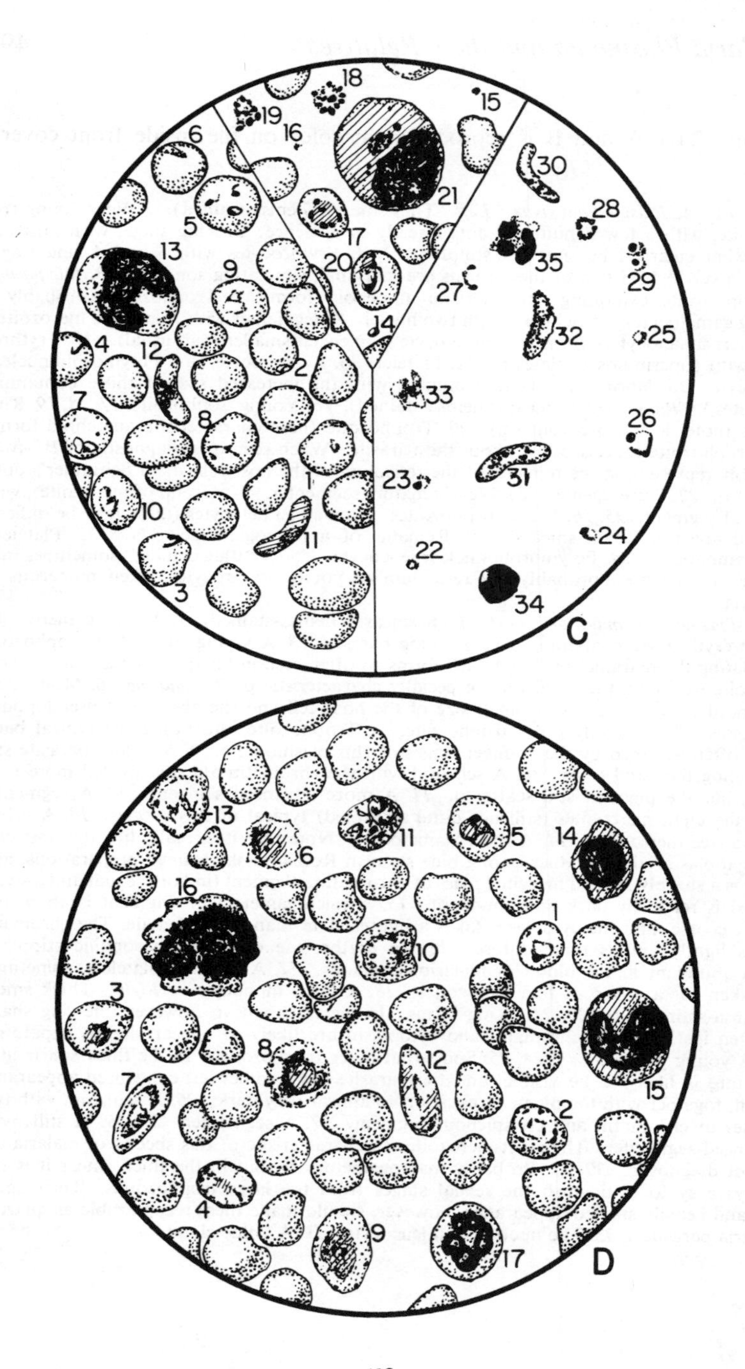
C
1
2
3
4
5
6
7
8
9
10
11
12
13
14
15
16
17
18
19
20
21
22
23
24
25
26
27
28
29
30
31
32
33
34
35
D
1
2
3
4
5
6
7
8
9
10
11
12
13
14
15
16
17

[Figure 24.1 C and D is reproduced in color on the inside back cover.]

C. Plasmodium falciparum. Left. Thin smear (Giemsa-stained). *1.* A young ring in an erythrocyte containing a trace of basophilic stippling. *2.* Slightly older ring in the marginal position frequently assumed by young parasites of this species. *3.* A slightly older ring. *4.* The so-called "*accolé*" form characteristic of this species. *5.* Multiple infection of the erythrocyte is frequent in falciparum malaria. *6.* A young trophozoite, marginal in position, but not of the "*accolé*" type. *7.* A more than usually amoeboid trophozoite. The erythrocyte also exhibits Maurer's dots. *8.* Another very amoeboid trophozoite, but one which has not caused the stippling of the host cell. *9.* Slightly older parasite than in *8* above. It is even more amoeboid, and the Maurer's dots are very conspicuous. *10.* This trophozoite exhibits the two chromatin masses which are often seen in the rings of *Plasmodium falciparum. 11.* A macrogametocyte. Characteristically sausage-shaped, with both chromatin and pigment condensed near the center, it takes a bluer stain than the male cell in Romanowsky preparations. *12.* A microgametocyte, showing the relatively dispersed chromatin and pigment, the greater breadth and more blunt ends, and the lighter stain, as compared to the female (above). *13.* A large mononuclear leucocyte containing malarial pigment. *Upper Center.* Thin smear of placental blood (Giemsa-stained). *14, 15, 16.* Young parasites lying free in blood. *17.* A schizont. *18, 19.* Two segmenters. No trace remains of the host cell. *20.* A half-grown crescent. *21.* A mononuclear which has ingested a segmenter. *Right.* Thick smear (Giemsa-stained). *22, 23, 24, 25.* Four very small rings, characteristically of delicate appearance and generally smaller than the corresponding stages of other species. *26.* Before becoming concentrated in the visceral blood. A moderately large falciparum ring. *27, 28.* Two rings, each with a double chromatin dot. *29.* Although there is nothing left of the host cell in the thick smear, one may see indications of the multiple infection so frequent in this species. These two rings were no doubt contained in the same red cell. *30, 31.* Two macrogametocytes. The crescents are not greatly changed in a thick smear. *32.* Apparently a microgametocyte. *33.* Some crescents may be considerably altered in the making of a thick drop preparation. *34.* The nucleus of a leucocyte. (Lymphocyte?) *35.* The nucleus of a granulocyte.

D. Plasmodium ovale. Thin smear (Leishman-stained). *1.* A young trophozoite. It has already produced some stippling of the host cell. *2.* A slightly older stage; the parasite has assumed a ring form. *3.* A parasite of about the same age as in *2*, but with a smaller central vacuole. The stippling is more prominent and the erythrocyte appears somewhat enlarged. *4.* Sometimes the smaller stages of *P. ovale* assume a strap or band form. *5.* Trophozoite that may have started to divide precociously. The lack of any pseudopodial processes is characteristic of this species. *6.* A stage similar to that in *5*, but the invaded erythrocyte exhibits the fimbriated edges often characteristic of *P. ovale*. *7.* Apparently a schizont, showing the oval distortion of the host cell which gives this species its name. *8.* A stage which probably corresponds to that just described. The host cell shows the intense stippling and the fimbriated edge. *9.* A schizont with four masses of chromatin. Note that the host cell has a somewhat oval shape. *10.* A presegmenter with six chromatin dots. This stage resembles *P. malariae* more than *P. vivax*. *11.* A segmenter with six merozoites. *12.* A macrogametocyte which has produced the typical oval distortion of the host erythrocyte. This distortion, however, appears in only a fraction of the invaded cells. *13.* A microgametocyte with a typical large, marginal chromatin mass. *14.* A lymphocyte. *15.* A mononuclear leucocyte. *16.* Also a mononuclear. *17.* A neutrophil polymorphonuclear leucocyte. (After Russell, West, Manwell.)

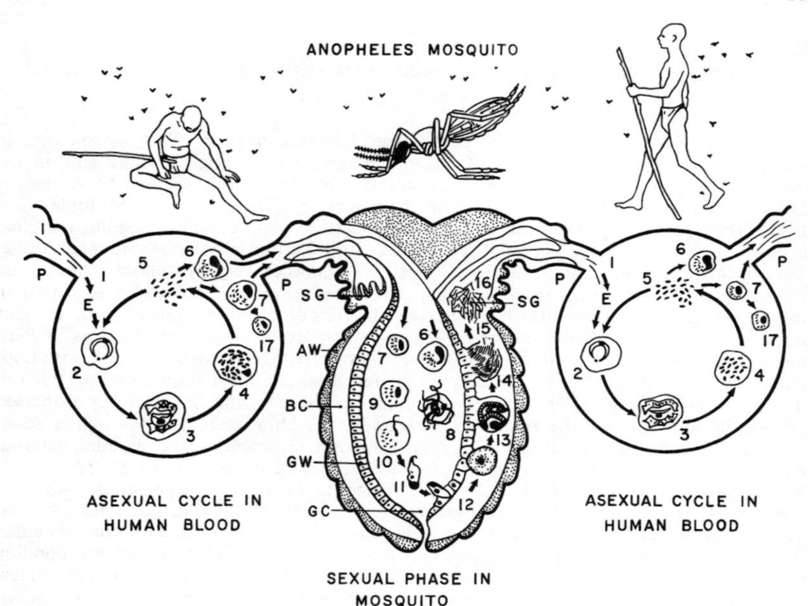

FIG. 24.2. The cycle of malaria transmission. An infected *Anopheles* mosquito inoculates man who becomes a gametocyte carrier; he infects *Anopheles* mosquito which in turn inoculates another man; the latter's blood becomes infectious to *Anopheles* mosquito. P, mosquito's proboscis; SG, salivary glands; AW, abdominal wall; BC, body cavity; GW, gut wall; GC, gut cavity. *1.* Sporozoites; E, exoerythrocytic stages. *2.* Trophozoite in ring stage. *3.* Schizont. *4.* Segmenter. *5.* Merozoites. *6.* Male or microgametocyte. *7.* Female or macrogametocyte. *8.* Exflagellating microgametocyte. *9.* Mature macrogametocyte. *10.* Flagellum or microgamete penetrating macrogamete. *11.* Ookinete. *12, 13, 14.* Developing oocyst. *15.* Bursting oocyst liberating sporozoites. *16.* Sporozoites in salivary gland and salivary duct. *17.* Gametocyte which, unless it leaves the blood stream, does not develop further. (After Russell, West, and Manwell.)

deposited in the body. After this short initial period, the blood remains free of parasites, in man as well as in chickens, for four or five days, while the tissue stages of the plasmodia complete what is known as the "pre-erythrocytic cycle." The first generation of this cycle is known as the "cryptozoite" stage (the Greek *krypte* means "hidden") and the several generations which follow, prior to invasion of the blood stream, are known as "metacryptozoites" (from the Greek *meta*, "after," and *zoion*, "animal"). Figures 24.3 and 24.4 show some of these stages as they appear in birds.

Finally some of the tiny offspring of these forms spill over into the blood stream and infect red cells. In the avian plasmodia and some species occurring in mammalian hosts the tissue cycle probably continues, perhaps for as long as the infection lasts, which is often for years. These tissue stages are then known as "phanerozoites" (from the Greek *phaneros*, "visible," since they are fairly easy to demonstrate). The persistence of the tissue stages may

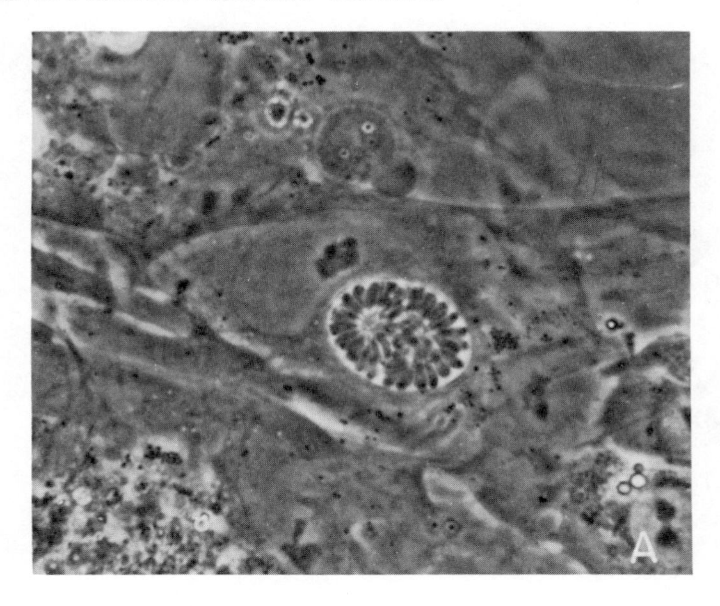

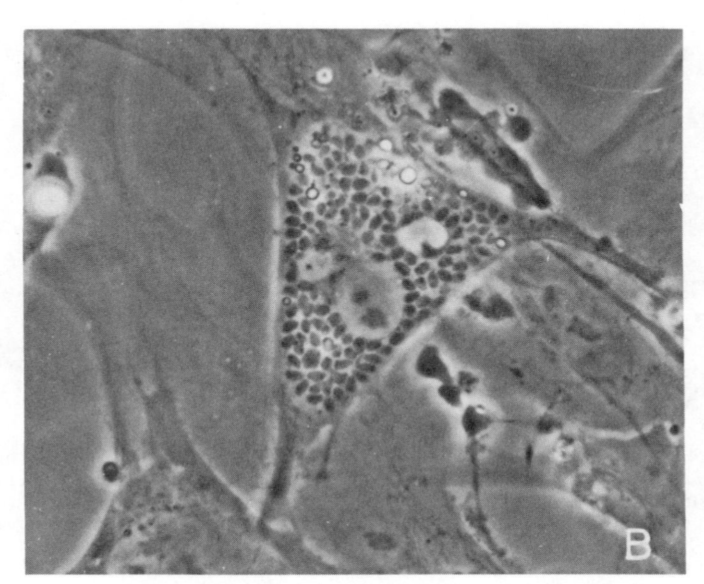

FIG. 24.3. *A*. A characteristic daisy-shaped exoerythrocytic segmenter of *Plasmodium fallax*, as seen in a cell (perhaps a fibroblast) in tissue culture. *B*. A cell, also in tissue culture, containing a large number of trophozoites of *Plasmodium fallax*. ×1000. (Huff. Naval Medical Research Institute.)

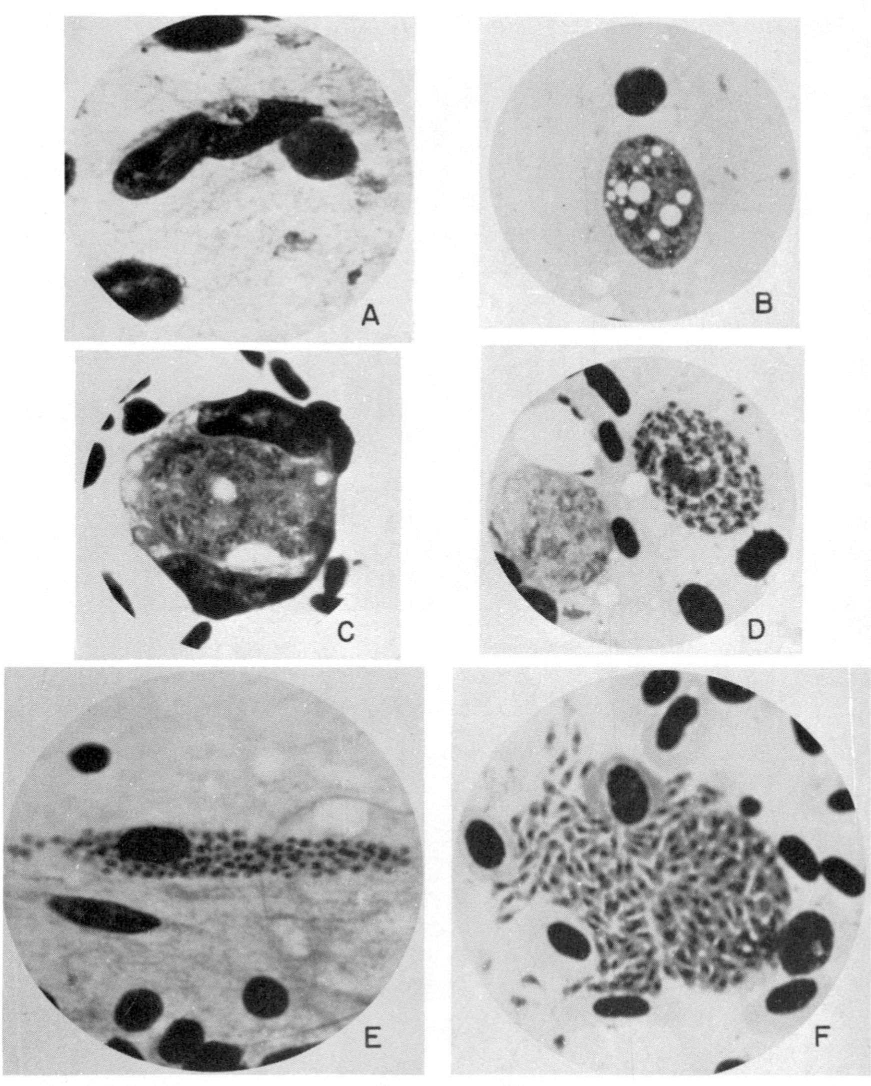

FIG. 24.4. Exoerythrocytic stages of avian malaria parasites, ×1200. *A.* A young stage of *Plasmodium hexamerium*, a parasite chiefly of passerine birds. The host cell is presumably an endothelial cell of a brain capillary of the orange crowned warbler. *B.* A medium-sized schizont of *Plasmodium relictum* var. *matutinum*, a morning-segmenting species common in robins. The vacuolization is typical of this species of malaria. *C.* A somewhat larger schizont of *Plasmodium circumflexum*, a common species of avian malaria, seen in the lung of a canary. *D.* Segmenting exoerythrocytic stages of *Plasmodium circumflexum* look like this form seen in the lung of a canary. *E.* Exoerythrocytic stages of *Plasmodium circumflexum* (host, canary) in the endothelial cells of brain capillaries. *F.* A freshly liberated brood of merozoites from an exoerythrocytic segmenter of *Plasmodium circumflexum*, seen in the lung of a canary. Note the elongate shape of most of the young parasites.

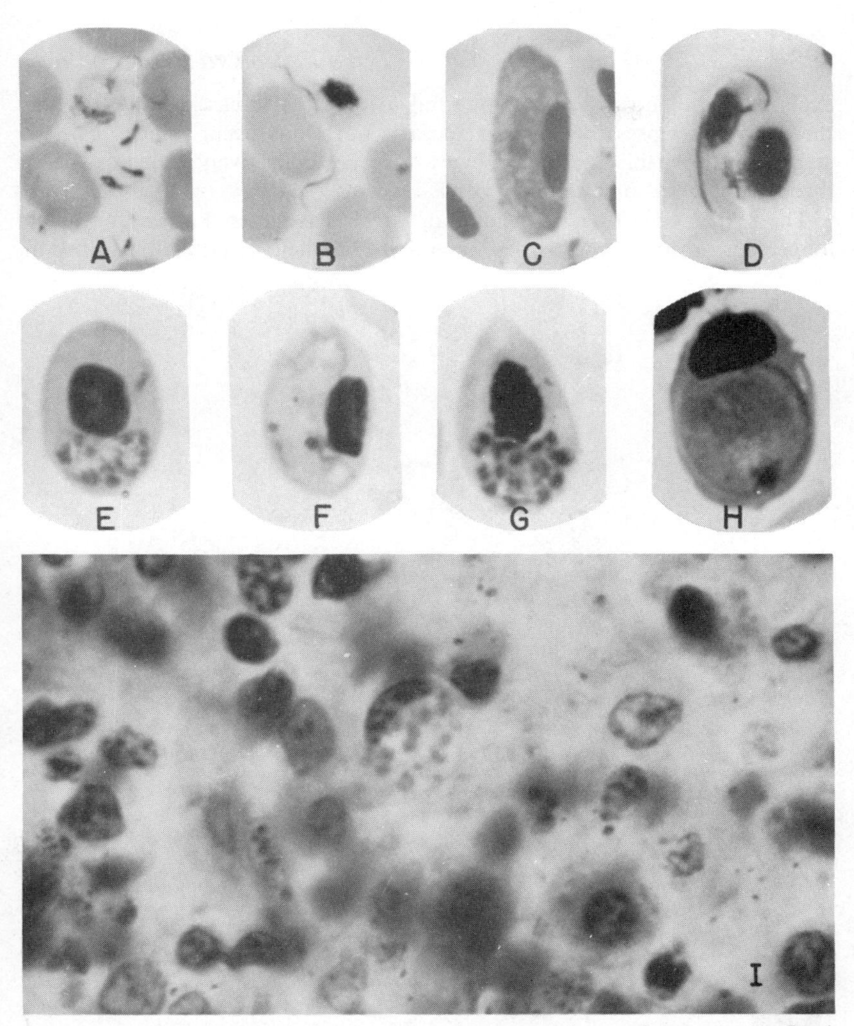

FIG. 24.5. Malaria and malaria-related parasites. *A*. Microgametogenesis ("exflagellation") in bat malaria (although here the parasite concerned is probably not a true plasmodium, but a species of the closely related genus *Hepatocystis*). *B*. Two microgametes, or "flagella," of the parasite above. Although it may not be a true malaria parasite, the microgametes of this and all the species of plasmodia appear to look alike. *C*. *Haemoproteus sacharovi* of the mourning dove. Probably a macrogametocyte, as evidenced by the relatively heavily stained cytoplasm. *D*. Haemogregarine of the turtle. Although the gametocytes of haemogregarines parasitize red cells, as do those of malaria and *Haemoproteus*, they exhibit no pigment. *E*. A segmenter of *Plasmodium floridense*, a malaria parasite occurring in lizards. *F*. A gametocyte (male ?) of the same species. *G*. A segmenter of *Plasmodium mexicanum*, a species of plasmodium occurring in lizards. *H*. A gametocyte (macro ?) of *P*. *mexicanum*. *I*. A spleen preparation from a lizard infected with *P*. *mexicanum*, showing exoerythrocytic forms. Note the group of merozoites in the upper center, and the less advanced schizont near the upper left margin. This species exhibits both types of exoerythrocytic schizogony seen in *Plasmodium elongatum* and *P*. *gallinaceum*. ×1800. (Army Medical Museum, Washington, D.C.)

explain why such drugs as quinine fail to cure the disease completely and the delayed relapses for which malaria is notorious occur. Since all these stages occur outside the erythrocytes, they are collectively called "exoerythrocytic stages," or often simply "E-E forms."

The stages occurring in the blood, which have been known for much longer than those in the tissues, essentially parallel the latter, but differ in

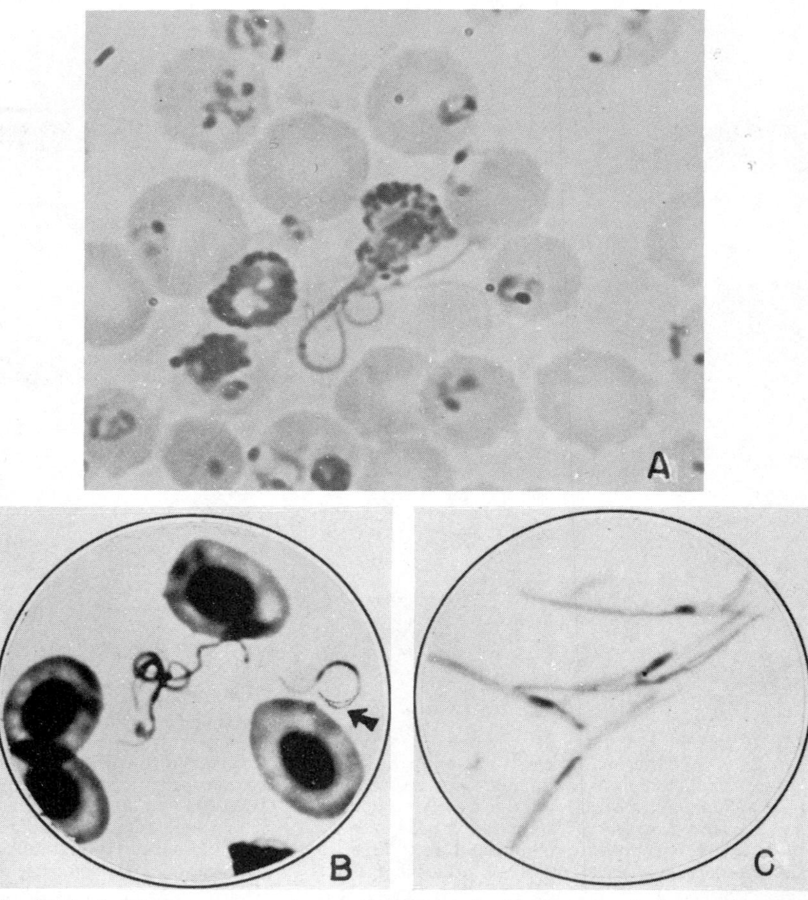

FIG. 24.6. *A.* Microgametogenesis or exflagellation in the simian malaria parasite *Plasmodium knowlesi.* The sinuously curved filaments are the microgametes, still attached to the parent microgametocyte but struggling to get away. (Pasteur Institute, Coonoor, South India; Oxford University Press.) *B.* Microgametes of the avian malaria parasite *Plasmodium lophurae.* The structure of the microgamete appears to be the same in all species. The arrow indicates the delicate adherent filament near the end of the microgamete on the right. (Stella Zimmer, Oxford University Press.) *C.* Sporozoites of the avian malaria parasite *Plasmodium cathemerium.* The sporozoites of all species of malaria are so similar as to be indistinguishable. (Army Medical Museum.)

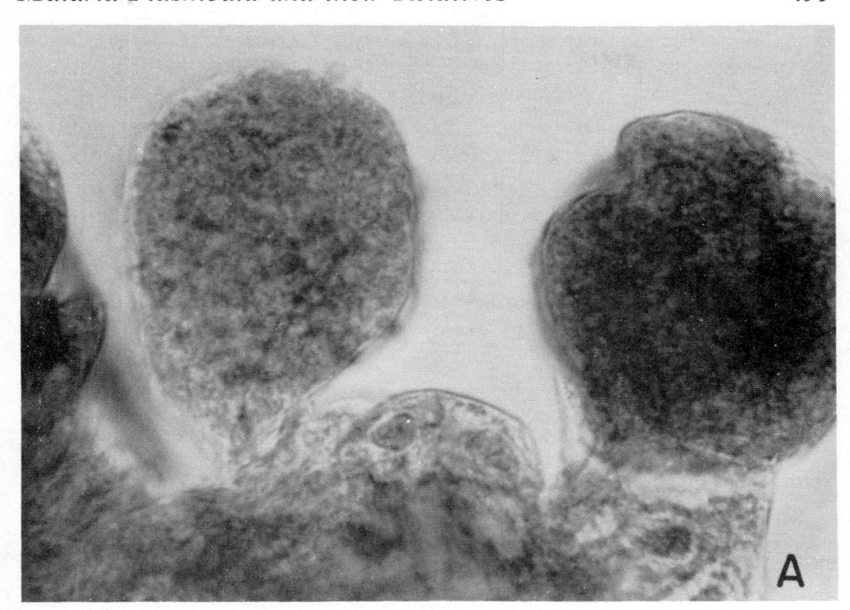

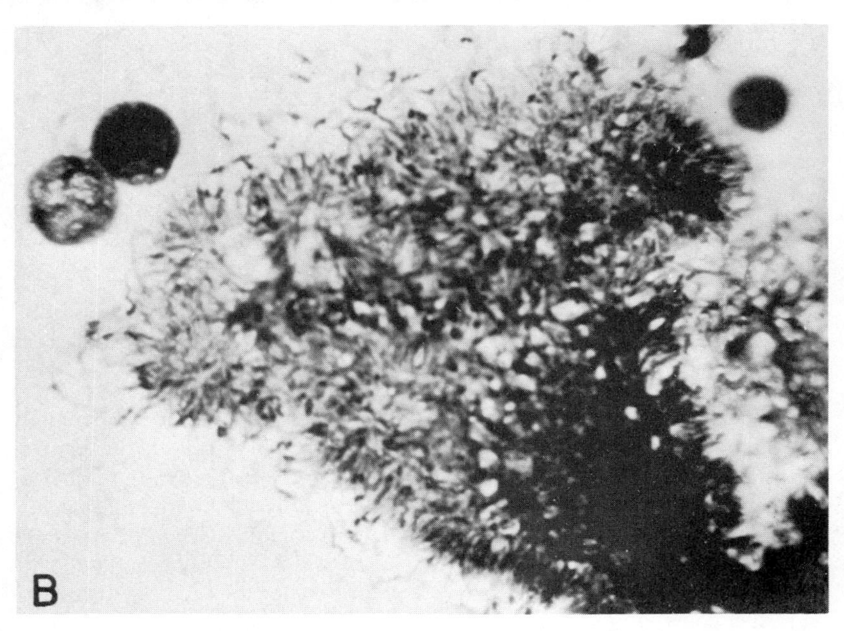

FIG. 24.7. Mosquito phase of malaria life cycle. *A*. Several oöcysts of *Plasmodium vivax* as they appear on the outer wall of the gut of the vector *Anopheles*. *B*. A swarm of sporozoites just after liberation from the parent oöcyst. ×1200.

that some of the parasites are incapable of reproduction, destined to continue life in the mosquito, if anywhere. These are the sexual stages, or gametocytes. The few lucky enough to be picked up by some hungry culicine or anopheline (the usual transmitters of bird and mammalian malaria, respectively) speedily mature into gametes, and fertilization follows. Maturation of the male cells is known as "exflagellation" (*A* in Figs. 24.5 and 24.6). Then the zygote (oökinete) makes its way to the outer layer of the mosquito's gut, and develops into a tumorlike mass, from which in a few days a myriad of sporozoites (Fig. 24.7) escape, some finding their way into the salivary glands. We can only marvel at a cycle so complex and speculate about the steps and time required for its evolution. The physiological cycle is even more complex (Fig. 24.8).

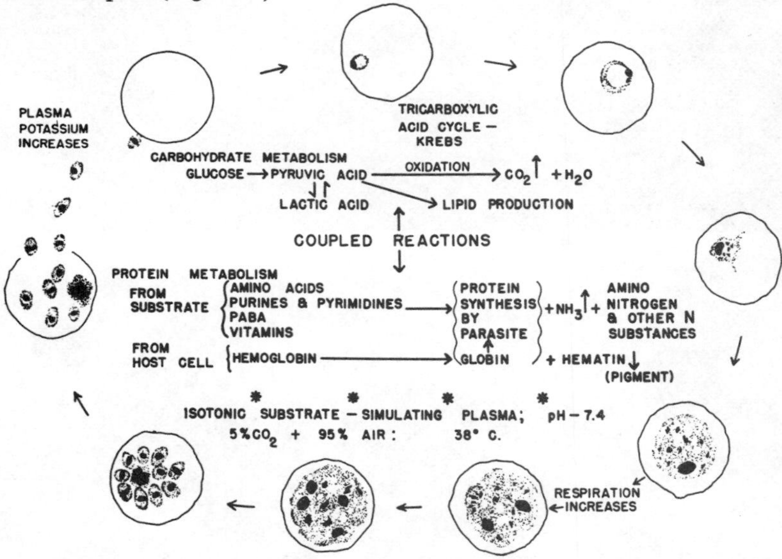

FIG. 24.8. Growth and metabolism of malarial parasites: probable sequence. (Geiman and McKee, 1948.)

In many species of malaria, the cycle in the red cells requires a definitely fixed time, generally some multiple of 24 hours, varying with the kind of parasite. Thus, in *vivax* (benign tertian), *falciparum* (aestivo-autumnal or subtertian) malaria and in *Plasmodium ovale* it is 48 hours. *Plasmodium malariae*, however, requires 72. The Italian scientist Golgi observed 80 years ago that the symptomatology of the disease was nicely correlated with events in the life cycle of the parasite, the liberation of each new brood of parasites coinciding with successive chills and fever.

The periodic nature of the cycle is even more sharply defined in some of the avian malarias than in the human types (although in the former there are no chills and fever), the peak of reproduction occurring not only regularly but at almost the same hour each day. Thus, *Plasmodium cathemerium* brings forth each new crop of merozoites at about 6:00 P.M. Here there are obviously two factors concerned: one is genetic, since other species of malaria do not reproduce similarly even though in a similar environment; the other is the cyclic nature of the host's physiological processes, for when the latter are changed, as they can be by placing the bird in a room artificially illuminated and darkened at regular intervals, the parasite alters in corresponding fashion the length of its reproductive cycle.

TABLE 24.1

DIFFERENTIAL CHARACTERISTICS OF HUMAN PLASMODIA

Character compared	*Plasmodium vivax* (Grassi and Feletti, 1890)	*Plasmodium malariae* (Grassi and Feletti, 1890)	*Plasmodium ovale* (Stephens, 1922)	*Plasmodium falciparum* (Welch, 1897)
Rings	——— not easily distinguished ———			delicate, small; often 2 chromatin dots; appliqué forms frequent; often very numerous
Trophozoites	highly amoeboid	not amoeboid; strap-shaped forms frequent	not amoeboid	rarely seen in peripheral blood
Schizonts	amoeboid; large	not amoeboid	not amoeboid	rarely seen in peripheral blood
Segmenters	large; usually 16 merozoites (12–24)	small; usually 8 merozoites (6–12)	smallish, usually 8 merozoites (6–12)	rarely seen in peripheral blood; usually 6–12 merozoites (often more)
Gametocytes	round or irregular	like *vivax*, but smaller	*like vivax*, but smaller	sausage-or crescent-shaped
Pigment	yellow-brown	brownish-black; abundant	like *vivax*	dark and quite abundant
Effect on host cell	marked	slight	marked	marked (especially in case of gametocyte)
Enlarged?	yes (markedly)	no (often seems shrunken)	yes (somewhat)	no
Distorted?	rather frequently	no	often oval	no (but often reduced to envelope about gametocyte)
Decolorized?	yes	no	somewhat	no
Stippled?	Schüffner's dots	no	Schüffner's dots (especially abundant)	Maurer's (Stephens and Christopher's) dots or clefts
Preferred erythrocyte type	reticulocyte	senescent	unknown	indifferent
Cycle	48 hours	72 hours	48 hours	48 hours

TABLE 24.2

COMPARATIVE TABLE OF THE AVIAN PLASMODIA

A. SPECIES PRODUCING ELONGATE GAMETOCYTES

Species Plasmodium	*Blood Cell Types Infected*	*Effect on Erythrocyte (Asexual Forms)*	*Merozoites per Segmenter: Mean-Range*	*Pigment (In larger asexual forms)*	*Cycle Length (peak)*	*Natural Occurrence*
			Larger Species			
elongatum (Huff, 1930)	"all blood and blood forming cells"	nucleus displaced	9 (4–20) (segmentation chiefly in bone marrow)		24 hours (8–10 AM)	in many passerine birds (common)
circumflexum (Kikuth, 1931)	erythrocytes only	unaltered (surrounds nucleus)	19 (13–30)		48 hours (4–6 PM)	many passerine birds (common)
**fallax* (Schwetz, 1930)	erythrocytes only	unaltered	"as many as 16"			known only from 2 African species (owl and guinea fowl)
lophurae (Coggeshall, 1938)	erythrocytes only	unaltered (resembles *circumflexum*)	(8–18)		24 hours	in fire-backed pheasant (Borneo)
			Smaller Species			
polare (Manwell, 1935)	erythrocytes only	unaltered (parasites usually polar)	10.5 (8–14)			Chiefly cliff swallows (uncommon)
nucleophilum (Manwell, 1935)	erythrocytes only	unaltered (parasites tend to cling to nucleus)	6 (4–10)	dark (several granules)	"probably 24 hours"	catbird and other species (uncommon)
hexamerium (Huff, 1935)	erythrocytes only	unaltered	6 (4–8)	brown (several granules)	"probably 24 hours" (8–9 AM)	various passerine species (fairly common)
vaughani (Novy and MacNeal, 1904)	erythrocytes only	unaltered	4 (4–8)	dark (characteristic refractile granules)		very common in robins; rare otherwise
rouxi (Sergent, Sergent, and Catanei, 1928)	erythrocytes only	unaltered	4 (always)	dark (usually 2 granules)		known only in sparrows of Near East

(TABLE 24.2 contd.)

B. SPECIES PRODUCING ROUND OR IRREGULAR GAMETOCYTES

*Species Plasmodium**	*Blood Cell Types Infected*	*Effect on Erythrocyte*	*Merozoites per Segmenter: Mean-Range*	*Pigment*	*Cycle Length (peak)*	*Natural Occurrence*
			Larger Species			
gallinaceum (Brumpt, 1935)	erythrocytes only	nucleus displaced	(8–30)	Few and large granules in gam.	36 hours (alternately 12 M. and 12 N.)	domestic fowl in Orient (much used experimentally)
relictum (Grassi & Feletti, 1891)	erythrocytes only	nucleus displaced (often expelled by gam.)	11 (8 to 15 in American strain; 16 to 32 in German strain)	Fine, point-like granules in gam.	12, 27, 30 & 36 hours in different strains	many species of birds chiefly passerines
relictum var. *matutinum* (Huff, 1937)	erythrocytes only	nucleus displaced (often expelled by gam.)	14 (10–22)	Spherical granules, often of larger size (in gam.)	24 hours (about 9 AM)	as above
cathemerium (Hartman, 1927)	erythrocytes only	nucleus displaced (often expelled by gam.)	15 (6–24)	Rod-like especially in gam.	24 hours (about 6 PM)	as above
			Smaller Species			
†*durae* (Herman, 1941)	erythrocytes only	nucleus often displaced by gam.	8 (6–14)	Large, round, black granules	24 hours (9–10 AM)	African turkeys
juxtanucleare (Versiani and Gomes, 1941)	erythrocytes only	nucleus displaced	3.5 (2–5)	one to several course dark granules	24 hours (3–7 AM) synchronicity low	domestic fowl in Brazil and Mexico

*Several other species of avian *Plasmodium* have been recently described, but are not included here because little is yet known about them other than their morphology.

†The gametocytes of *Plasmodium fallax* are said to be remarkably like those of *Haemoproteus;* schizonts possess conspicuous vacuoles. It is infective for chicks and pigeons. *Plasmodium durae* produces gametocytes which are generally more or less elongate, but they are also often amoeboid in appearance. For this reason and because the morphology of the asexual forms and the character of the infection it causes seem to ally it more closely with species in this group, it is placed in Part B of the table.

Morphology and Diagnosis

Species of malaria are usually recognized by peculiarities in the morphology of the stages occurring in the erythrocytes, and by differences in their effects on the host cell. For example, *Plasmodium vivax* is highly amoeboid (hence the name from the Latin *vivo*, meaning "live") and is larger than the other plasmodia of man. *Plasmodium falciparum* produces gametocytes which are sausage- or crescent-shaped (giving rise to the species name, from the Latin *falx*, "sickle," and *par*, "equal"). Other characteristics used by the taxonomist include the number of merozoites per parent cell, peculiarities of the pigment, or haemozoin, and the number and arrangement of the chromatin masses in the youngest stages, which because of their form are usually called "rings."

Likewise, alterations produced by the parasite in the size, shape, and haemoglobin of the host cell, the appearance of dots ("stippling") in erythrocytes stained with the usual Romanowsky stains, and displacement of the nucleus (except in the mammalian malarias) all aid in species identification. These characteristics are summed up, for human and most avian species of malaria, in Tables 24.1 and 24.2.

Although Laveran was able to recognize the malaria parasites when he saw them in unstained blood films, in which they are readily visible, they are much more easily studied in stained preparations. The living plasmodia appear as colorless globules, containing perhaps a few grains of dark malarial pigment (the so-called haemozoin, derived from the chief food of the organisms, haemoglobin), against the faint orange background of the red cell. In Romanowsky-stained preparations (so termed because the Russian physiologist Romanowsky originated the process), the parasites appear as sharply defined bodies with bluish cytoplasm and lilac-colored chromatin.

For many years little more could be learned about the structure of the organisms because of the limitations of the staining method. Haematoxylin, probably the best of the cytological stains, gives poor results with the malaria plasmodia and related blood parasites. It is a "basic stain" and the nuclear material of the parasites apparently varies from that of most other cells during much of its life cycle. This is shown by negative results obtained with the Feulgen stain (a delicate test for thymonucleic acid or DNA) when used on the malarial organisms, except when they are in schizogony. In general, the parasites stain well only with derivatives of methylene blue and eosin (an "acid" dye), which together form the base of the Romanowsky stains.

Phase and electron microscopy, however, have recently revealed much about the structure of these organisms that was previously unsuspected. Though only a few species have been studied, it now appears that the genus is characterized by two chromosomes. These are of equal length in human and simian plasmodia, and unequal in bird and reptilian species (Wolcott,

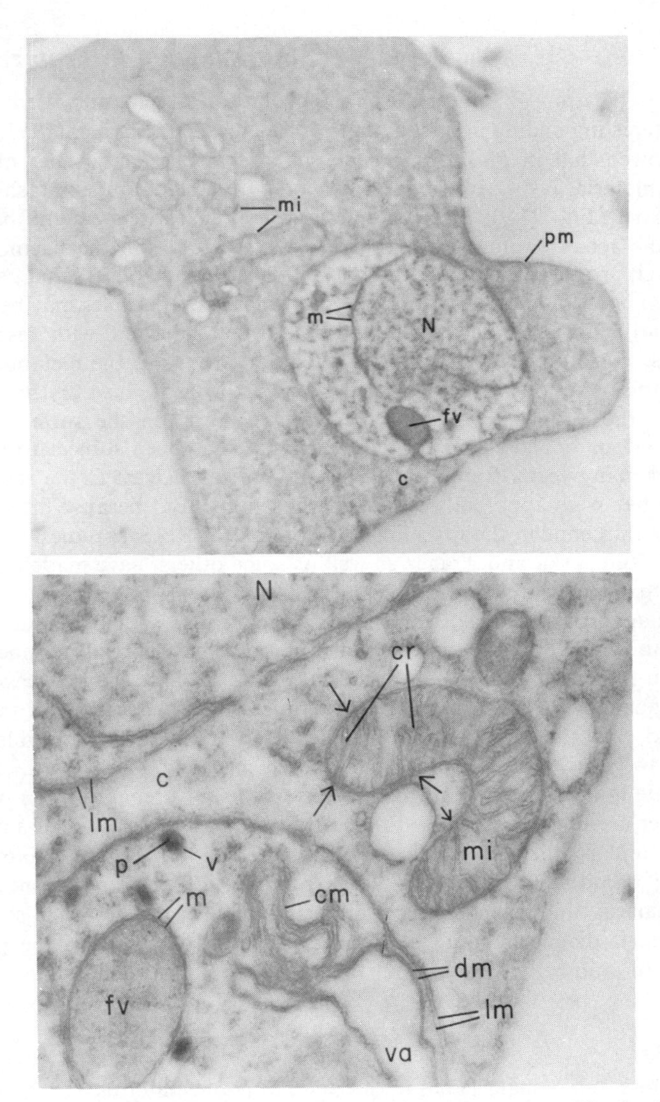

FIG. 24.9. *Plasmodium berghei* under the electron microscope. *Above*. The food vacuole (fv) is just being formed. Both the membrane surrounding the nucleus (N) and that enclosing the parasite are double. ×18,000. *Below*. Two parasites appear, although little more than the nucleus (N) of one is visible. In the second (lower left), a food vacuole (fv), a pigment granule within a vacuole (p, v), and an organelle of unknown function (cm) surrounded by concentric double membranes, are seen. Notice the U-shaped mitachondrion of the host cell, with its double-layered cristae (cr) criss-crossing it. ×34,200. Key: c, cytoplasm of host cell; cm, structure composed of double concentric membranes; cr, cristae mitochondriales; dm, double membrane of vacuole; fv, food vacuole; lm, double limiting membrane; m, double membrane surrounding food vacuole; mi, mitochondria; N, nucleus; p, pigment hematin; v, vesicle; va, vacuole. (Rudzinska and Trager, *J. Biophys. Biochem. Cytol.*)

1955; 1957), strongly suggesting close evolutionary relations. Perhaps the most interesting finding of all was that of Rudzinska and Trager (1957; 1959), who showed that in *Plasmodium lophurae* and *P. berghei* (species of avian and rat malaria, respectively), food is actually ingested, amoeba-fashion, by the parasite (Fig. 24.9). Electron microscope preparations caught the parasite in the act of swallowing bits of its host's cytoplasm, the haemoglobin of which it proceeded to digest in food vacuoles. Now it appears that parasites may even possess a minute cytostome into which food particles may be taken. Apparently, however, the parasites are not self-sufficient with respect to enzymes. Certain chemical transformations, necessary to the metabolism of the organisms, must be accomplished by enzymes of the host erythrocyte.

Where the parasites really are, crawling perhaps on the surface of the erythrocyte or imbedded in the cell substance, has been subject to controversy for many years. Ratcliffe (1927) showed that at least in the species he studied they were apparently within the host cells. But because his sections were not thin enough the question was not settled. Recent studies, however, those of Rudzinska and Trager as well as some others, have made it almost certain that malaria parasites are indeed intracellular.

For diagnostic purposes two kinds of preparations are in general use. In one ("thin film"), the blood cells constitute an evenly spread, monocellular layer; in the other ("thick film"), several drops of blood may cover an area about equal to that of a dime, and the processing is such that all the cells are laked, only the parasites and the leucocyte nuclei remaining visible. Thin films show parasite morphology much more accurately, but using thick films saves time in diagnostic laboratories (see methods of preparation on p. 587). The latter are of course useless for the study of blood parasites of any kind in birds and reptiles, in which red cells are nucleated. Since making and studying blood films from very large numbers of persons are necessary in antimalaria campaigns, attempts are being made to develop slide-scanning machines. If diagnosis could be done by an electronic reader, the present struggle to eradicate malaria from the world would be facilitated.

Animal Malaria as a Research Tool

Most research in malaria has in the past been done on birds, partly because of the frequent occurrence of bird malaria in nature, and partly because of the ease with which convenient host species, like canaries, ducklings, and chickens, can be maintained in the laboratory (Fig. 24.10). But in 1948 a species of *Plasmodium*, *P. berghei*, was discovered in African tree rats and proved readily transmissible to other species of rats, as well as to mice and hamsters. As a result *berghei* malaria has now become a popular research tool, especially for experimental chemotherapy and the study of immunity. Many newly elaborated and highly successful antimalarial drugs were developed only after thorough testing on malarious chickens, ducklings, and

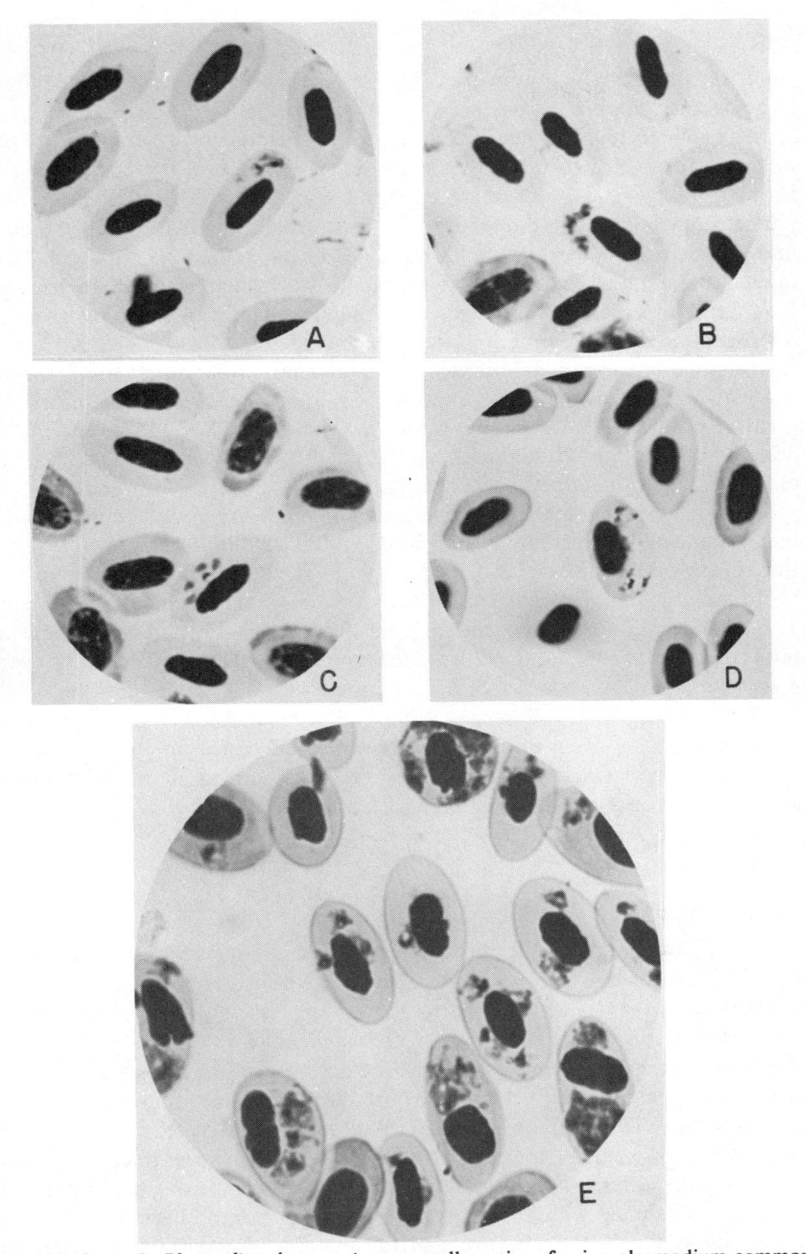

FIG. 24.10. *A–D. Plasmodium hexamerium*, a small species of avian plasmodium common in birds. *A*, trophozoite. *B*, schizont. *C*, segmenter with four merozoites (although six is the more common number). *D*, gametocyte (probably male). The dark granules are pigment. *E*, *Plasmodium gallinaceum*. Virtually every erythrocyte is parasitized. ×1440.

canaries. But sparrows were used by Sir Ronald Ross, the English Nobel Prize winner, in his epochal researches on the mosquito transmission of malaria, which in 1898 finally culminated in the successful transfer of the infection from bird to bird. Avian malaria is still a tool for basic research in malariology, and it is convenient for the beginner in protozoology and parasitology to study in the laboratory. It is usually easier to find plasmodia and compare their various stages on slides of avian malaria than on those made from human cases. This is especially true when we wish to become familiar at first hand with the exoerythrocytic stages and those involved in microgametogenesis (formation of the male gamete, or "exflagellation"). The former are often abundant and easily found in sections or contact preparations (p. 589) made from the brain, spleen, liver, and bone marrow of malarious birds. Exflagellation is readily demonstrated in fresh blood preparations mixed with enough citrated saline solution to prevent clotting, and it may also be studied in stained preparations.

When infected birds are available (and infections are easily secured from a malarious English sparrow by injecting a little of its blood into a canary), the course of the disease and its pathology (greatly enlarged dark spleen, profound anemia, hyperplastic bone marrow) make a fascinating study. The dramatic effects of the antimalarial drugs, such as quinine, quinacrine (atebrine), camoquin, and chloroquine are easily shown. In almost every respect, the avian malarias closely parallel those of man, although they exhibit one important difference: When human malaria is induced by blood inoculation, as is often done in the treatment of neurosyphilis, no exoerythrocytic stages develop, though they regularly do so in many kinds of avian malaria, As a result, this type of human malaria is easily treated, and little prone to relapse.

EPIDEMIOLOGY

The epidemiology of the disease is much the same in bird and human populations. Certain species of birds, such as robins, seem especially prone to malaria; others, like English sparrows, are less so. In man, whites are more susceptible to most strains of *Plasmodium vivax* than Negroes. The greater resistance of Negroes seems to be accounted for, in part at least, by the frequent occurrence in them of a peculiarity of the red blood cells known as "sickling." This trait is inherited, and when homozygous may cause a serious type of anemia. But persons heterozygous for it seem to enjoy normal health and to suffer less from malaria. This may in turn be due to the presence in such individuals of a different type of haemoglobin, perhaps less adequate nutritionally for the requirements of the malaria parasites. But geography makes a difference, as we might expect. Human malaria has always been more prevalent in the warmer parts of the world, largely because mosquitoes in such places have a longer breeding season, although the cause of aestivo-autumnal malaria, *Plasmodium falciparum*, develops more readily in the infected mosquito at higher temperatures. Avian malaria also seems to be more prevalent in these areas.

English sparrows, which do not migrate, appear to have a considerably lower incidence of malarial infection in the northern states than in the Deep South. Yet one species of malaria has been found to be not uncommon in a number of species of penguins, also nonmigrating (since they are flightless) and resident in or close to the Antarctic. The author has also found Canada Jays ill with malaria in the High Rockies, where summers are short and cool and winters severe. Since the Canada Jay is also nonmigratory, it must have acquired its disease in that cool climate. Human malaria has never been able to establish itself in climates as cold as these. In malarious regions, very young children are most commonly and severely affected. Among birds malaria is also likely to be a disease of nestlings and juveniles, causing a significant mortality among them (although actual evidence is naturally very difficult to get); thus it may play a part in maintaining biological balances, the result of which is the survival of the fittest. Under more primitive conditions, disease played an exactly similar role in the evolution of the human species, as well as in limiting the population to the number a given area could support. Thus in Ceylon, one of the most densely populated parts of the world, malaria has been almost completely controlled in the last decade, and with the near disappearance of this disease has gone a burgeoning of the number of mouths the island must feed.

Under natural conditions, the continued existence of malaria, or of any other infectious or parasitic disease, depends on a delicate equilibrium between those factors favoring the agent of infection and those against it. Thus a warm and moist climate favors the multiplication of mosquitoes and tends to lengthen their life span. A high birth rate insures a population of malaria-susceptible individuals, and the likelihood that they will be bitten by infected mosquitoes is all the greater because heat and humidity make much clothing uncomfortable and encourage open windows and outdoor life. On the other hand, individuals tend to develop a relative immunity to malaria after prolonged exposure, with the result that gametocytes in their blood are few and mosquitoes biting them are less likely to become infected. A dry and cool season may also cause a sharp drop in active cases of malaria. Indeed, were it not for the chronic nature of this disease and its tendency to relapse months or even years after the original infection, it might often disappear spontaneously in a region without control measures. It is likely that the reservoir of infection is always man (or the vertebrate host, for species of malaria other than the human); mosquitoes may retain an infection as long as they live, but their life spans are short—a few weeks, often much less. Exceptionally they may survive the winter (otherwise there would, of course, be no mosquitoes the next season with most species), but the proportion of such Methuselahs among the mosquito population is small indeed.

Malaria-Related Genera

Haemoproteus and Leucocytozoon

The two malaria-related genera, *Haemoproteus* and *Leucocytozoon*, have already been briefly mentioned. The former is closely akin to the true malaria parasites in that the gametocytes, which occur in erythrocytes, contain the typical malaria pigment, haemozoin. Although this substance varies somewhat in color and the shape of granules seen in different species of *Plasmodium* and *Haemoproteus*, it is probably nearly the same in chemical composition wherever it occurs. It seems likely not to be hematin, as was long thought, but more complex in composition.

The reproductive stages of the species of *Haemoproteus* so far carefully studied occur only in the internal organs, particularly in the lungs (as in the common *Haemoproteus columbae* of the pigeon, and in the species most often seen in the song sparrow). Here they exist as relatively huge masses of protoplasm, eventually breaking up into hundreds or even thousands of minute uninucleated merozoites (Fig. 24.11). These multiplicative stages are no doubt comparable to the exoerythrocytic forms seen in malaria, which in birds also show a strong predilection for the lungs. Nothing is known about the vectors of most species of *Haemoproteus;* the majority have been thought to be transmitted by hippoboscid flies, as is *H. columbae* of the pigeon, but a *Haemoproteus* of ducks has been recently shown to be transmissible by biting midges of the genus *Culicoides* (Fallis and Wood, 1957).

As the name *Leucocytozoon* suggests, the gametocytes of this species are characteristically seen in leucocytes (usually lymphocytes) which their presence grossly distorts, so that only the host cell nucleus and a thin film of cytoplasm enveloping the parasite remain at maturity. The parasite itself is generally spherical in shape (Fig. 24.11*C*), although some species are elongate, and, as in *Haemoproteus*, few peculiarities of morphology can be used in the description of species. As a result, no one has any idea of how many species there are in either of these genera. Each of them is, however, extremely common in birds, and few avian species have been found uninfected when such parasitism was diligently looked for. Infection with either genus confers no protection against the other, as far as known, nor does cross protection exist to the avian malarias.

Leucocytozoon, like *Haemoproteus*, reproduces in the internal organs, often the liver and kidneys. Its multiplicative stages, while they differ from those of the malaria organisms and *Haemoproteus* in appearance (Fig. 24.11), are likewise irregularly shaped masses of protoplasm of relatively large size, and give rise to multitudes of merozoites. These in turn may develop into successive generations of reproductive stages, or they invade lymphocytes and become sexual forms. For the latter there is no future except as they are taken up by some species of vector fly, apparently usually a species of *Simulium* ("black fly"). In them, as in the vectors of *Haemoproteus*, the cycle is almost identical with that of the malarial organisms in mosquitoes.

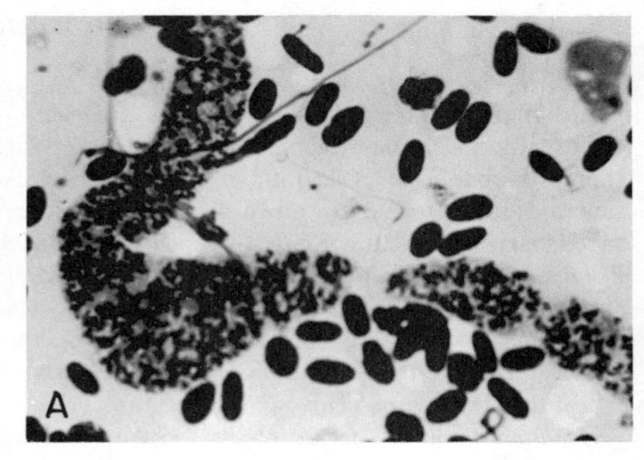

A. Reproducing stage of the *Haemoproteus* of the song sparrow. These stages occur in the capillaries of the lung. From such forms come both the gametocytes seen in the erythrocytes of the peripheral blood, and progeny destined to continue asexual reproduction. ×1125.

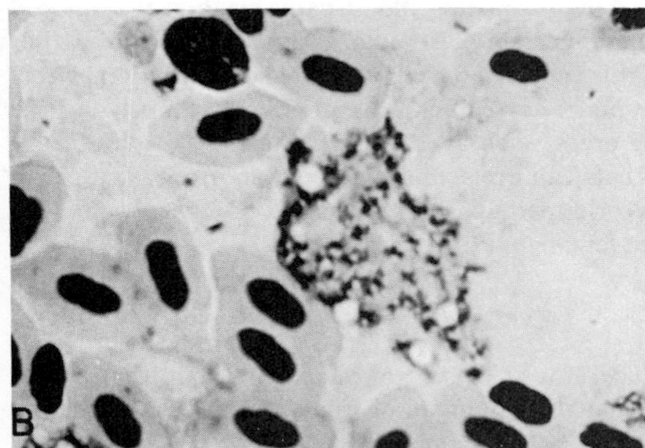

B. Leucocytozoon, like *Haemoproteus*, reproduces exclusively in the tissues of the vertebrate host. In this figure may be seen a schizont of the *Leucocytozoon* of the grackle in its preferred host organ, the kidney. Some species of *Leucocytozoon* pass their reproductive stages in the liver. ×1350.

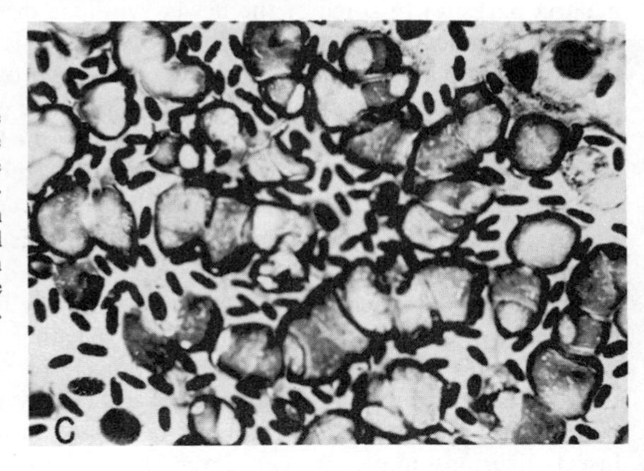

C. Apparently most of the offspring of the tissue stages of *Leucocytozoon* are destined to become gametocytes. Here may be seen 50 or more young sexual stages crowded together in a contact preparation made from the grackle kidney. ×590.

FIG. 24.11. Malaria-related parasites: *Haemoproteus* and *Leucocytozoon.*

Hepatocystis

Although the three genera, *Plasmodium*, *Haemoproteus*, and *Leucocytozoon*, are usually said to be the only ones in the family Plasmodiidae, other forms should probably be placed there. One of them is *Hepatocystis kochi*, a simian parasite formerly known as *Plasmodium kochi;* perhaps all the so-called malarial parasites of bats, (except *P. roussetti*) and one from antelopes, *P. limnotragi*, also belong in the new genus *Hepatocystis*. Species of *Hepatocystis*, like those of *Haemoproteus*, have their sexual stages in red cells where they produce pigment, and their reproductive stages in the liver. In this last respect they resemble the true malarial parasites of mammals, except that there is no schizogony in erythrocytes.

Hepatocystis is of special interest because it is believed to have evolved very early in the history of the pigmented blood parasites, and may indeed have been ancestral to the mammalian plasmodia. Its apparent lack of pathogenicity is consistent with this view. On the whole, its host distribution closely parallels that of the true malarias, though it is unknown in man and the anthropoid apes. In some areas, such as Taiwan (Formosa), it is the commonest blood parasite of bats, squirrels, and macacque monkeys, but seems to be absent in the New World, perhaps because species of *Culicoides* (so far its only known vector) are few there (*Manwell, 1965).

Evolution

And so, even though malaria is at present of diminishing importance and may conceivably vanish from the catalogue of human scourges, reports of growing resistance of mosquitoes to commonly used insecticides, and of the malaria parasites to some of the newer synthetic drugs make attainment of complete eradication seem somewhat doubtful. It is increasingly clear that much remains to be learned about the great group of parasites to which the organisms belong. We can at present only speculate on the evolutionary history of the group. Although it seems probable that the Plasmodiidae must have arisen from the coccidia, a large and important assemblage of parasites occurring in both vertebrate and invertebrate (chiefly insect) hosts, this must remain uncertain since fossil malaria parasites are and must remain unknown. The plasmodia and coccidia have much in common, differing mainly in that nearly all coccidia are directly transmitted from one host to another, without the agency of a vector such as a mosquito. Whether these ancestral coccidia were parasites of insects, adopting a life cycle of the two-host type when their hosts discovered the advantages of a diet of vertebrate blood, or were perhaps first parasitic in vertebrates, from which they entered blood-sucking insects like mosquitoes, we can never know. We can be reasonably certain that the epochal discovery of the convenience of having one host find the next one was made long ago, for mosquitoes are known as fossils from Eocene time, almost 70 million years ago. The many unsolved or only partly solved problems in malariology include many relating to the physiology of the parasites, their cultivation, their epidemiology, and their occurrence in nature.

Despite the fact that we now know enough about the parasites and the disease in man to eradicate it, if our resources permitted us to apply what we know, many frontiers in malariology remain awaiting the researcher.

REFERENCES

Bray, R. S. 1957. Studies on the Exoerythrocytic Cycle in the genus *Plasmodium*, London Sch. Hyg. and Trop. Med., *Memoir* 12. London: H. K. Lewis.

Fallis, A. M., Davies, D. M., and Vickers, M. A. 1951. Life history of *Leucocytozoon simondi* Mathis and Leger in natural and experimental infections and blood changes produced in the avian host. *Can. J. Zool.*, 29: 305–28.

——— and Wood, D. M. 1957. Biting midges (Diptera: Ceratopogonidae) as intermediate hosts for *Haemoproteus* of ducks. *Can J. Zool.*, 35: 425–35.

Faust, E. C. 1951. The history of malaria in the United States. *Am. Sci.*, 39: 121–30.

Forry, S. 1840. Statistical report on the sickness and mortality in the army of the United States from January 1819 to January 1839. Prepared under the direction of Thomas Lawson, Surgeon General, U.S. Army. Washington, D.C.: Jacob Gildeon, Jr.

Garnham, P. C. C. 1948. The development cycle of *Hepatocystis* (*Plasmodium*) *kochi* in the monkey host. *Trans. Roy. Soc. Med. and Hyg.*, 41: 601–16.

——— 1951. Patterns of exoerythrocytic schizogony. *Brit. Med. Bull.*, 8: 10–15.

———, Bray, R. S., Cooper, W., Lainson, R., Awad, F. I., and Williamson, J. 1955. The preerythrocytic stage of *Plasmodium ovale. Trans. Roy. Soc. Trop. Med. and Hyg.*, 49: 158–67.

Geiman, Q. M. and McKee, R. W. 1948. Malarial parasites and their mode of life. *Sci. Month.*, 57: 217–25.

Hewitt, R. 1940. *Bird Malaria.* Baltimore: John Hopkins Press.

Huff, C. G. 1938. Studies on the evolution of some disease-producing organisms. *Quart. Rev. Biol.*, 13: 196–202.

——— 1942. Schizogony and gametocyte development in *Leucocytozoon simondi*, and comparisons with *Plasmodium* and *Haemoproteus. J. Inf. Dis.*, 71: 18–32.

——— and Coulston, F. 1944. The development of *Plasmodium gallinaceum* from sporozoite to exoerythrocytic trophozoite. *J. Inf. Dis.*, 75: 231–9.

James, S. P. and Tate, P. 1937. New knowledge of the life-cycle of the malaria parasites. *Nature*, 139: 545.

Jeffery, G. M., Wolcott, G. B., Young, M. D., and Williams, D. Jr. 1952. Exoerythrocytic stages of *Plasmodium falciparum. Am. J. Trop. Med. and Hyg.*, 1: 917–26.

Jones, W. H. S. 1909. *Malaria and Greek History.* Manchester: Manchester University Press.

Manwell, R. D. 1938. Identification of the avian malarias. *Am. J. Trop. Med.*, 18: 565–75.

——— 1949. Malaria, birds, and war. *Am. Sci.*, 37: 60–8.

——— 1955. Some evolutionary possibilities in the history of the malaria parasites. *Ind. J. Malariol.*, 9: 247–53.

McKee, R. W. 1951. Biochemistry of *Plasmodium* and the influence of antimalarials. In *Biochemistry and Physiology of the Protozoa*, Hutner and Lwoff (eds.). New York: Academic Press.

Moulder, J. W. 1948. The metabolism of the malarial parasites. *Ann. Rev. Microbiol.*, pp. 101–20.

Moulton, F. R. (ed). 1941. Human malaria. A.A.A.S., *Pub. No. 15.*

O'Roke, E. C. 1930. The morphology, transmission, and life-history of *Haemoproteus lophortyx* O'Roke, a blood parasite of California Valley Quail. *Univ. Calif. Pub. Zool.*, 36: 1–50.

Pampana, E. J. and Russell, P. F. 1955. Malaria—a world problem. World Health Organization, Geneva.

Ratcliffe, H. L. 1927. The relation of *Plasmodium praecox* and *Plasmodium vivax* to the red cells of their respective hosts as determined by sections of blood cells. *Am. J. Trop. Med.*, 7: 383–8.

Rudzinska, M. A. and Trager, W. 1957. Intracellular phagotrophy by malaria parasites: an electron microscope study of *Plasmodium lophurae. J. Protozool.*, 4: 190–9.

——— 1959. 1959. Phagotrophy and two new structures in the malaria parasite *Plasmodium berghei. J. Biophys. and Biochem. Cytol.*, 6: 103–12.

Russell, P. F. 1952. The eradication of malaria. *Sci. Am.*, 186: 22–5.

———, West, L. S., Manwell, R. D., and MacDonald, G. 1963. *Practical Malariology*. 2nd ed. New York: Oxford University Press.

Shortt, H. E. and Garnham, P. C. C. 1948. The pre-erythrocytic development of *Plasmodium cynomolgi* and *Plasmodium vivax. Trans. Roy. Soc. Trop. Med. and Hyg.*, 41: 785–94.

———, Fairley, N., Hamilton, C. G., Shute, P. G., and Garnham, P. C. C. 1951. The pre-erythrocytic stage of *Plasmodium falciparum. Trans. Roy. Soc. Trop. Med. and Hyg.*, 44: 405–19.

Wolcott, G. B. 1955. Chromosomes of the four species of human malaria. *J. Heredity*, 46: 53–7.

——— 1957. Chromosome studies in the Genus *Plasmodium. J. Protozool.*, 4: 48–51.

ADDITIONAL REFERENCES

Garnham, P. C. C. 1966. *Malaria Parasites and Other Haemosporida.* Oxford: Blackwell Scientific Publications.

Manwell, R. D. 1965. The lesser *Haemosporidina. J. Protozool.*, 12: 1–9.

25

Coccidia

> "Further, I examined the bile from three old rabbits. The first had a very few small globules, but very many oval corpuscles* of a figure like those that, as I have said, I saw in the bile of a cow." Leeuwenhoek, Letter 7, October 19, 1674.

The coccidia are a very important group of parasitic Protozoa. Although only two coccidian species parasitize man, many species parasitize domestic and wild animals. They are responsible for losses totaling millions of dollars each year, and the coccidioses of wild animals help maintain the biological balances so necessary in nature as well as causing illness or death in numerous species prized by the sportsman.

There is little doubt that the coccidia were the forbears of the malaria parasites, the cause of a most important human disease. In general, the life cycles of the vertebrate coccidia (most of which belong to the superfamily Eimerioidea) resemble considerably those of their malarial cousins. The most important differences (with exceptions to be noted) are the absence of vectors for the coccidia, and in their place, direct infection by ingestion of oocysts with food and water.

The Parasites and their Life Cycles

The coccidia are Sporozoa most of which have a single host. The best known are parasites of vertebrates, and it is likely that most species have vertebrate hosts. They may have arisen from flagellate ancestors. The life cycle of the dinoflagellate *Coccidinium* suggests this (Fig. 18.2). Most often they inhabit the alimentary tract or organs associated with it, such as the liver (as in Leeuwenhoek's rabbits). The asexual portion of the life cycle

*The "oval corpuscles" were, in all probability, oocysts of the coccidian *Eimeria stiedae*, a common parasite of rabbits.

occurs in tissue cells, such as those of the intestinal mucosa, and culminates in the production of gametocytes from which arise gametes and then the infective stage, the oocyst.

Coccidia with invertebrate hosts, of which there are a considerable number (comprising forms as diverse as centipedes, water scorpions, worms, and cuttle fishes), also show much variety in their life cycles.

Infection in the vertebrate usually begins with the ingestion of the oocyst. The oocyst contains the actual infective forms, the sporozoites, which leave it somewhere in the alimentary tract and actively penetrate cells of the intestinal mucosa.

The asexual stages, which in the first instance develop from the sporozoites, generally undergo several generations or more of reproduction, producing large numbers of merozoites. These, like the sporozoites, are motile and invade new host cells. Finally gametocytes are formed, and it is generally thought that their appearance marks the end of the multiplicative cycle, at

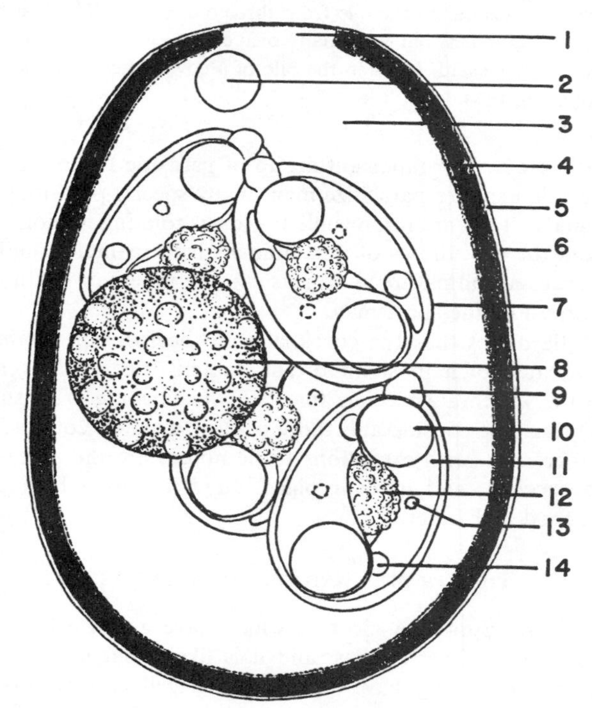

FIG. 25.1. Oocyst of the genus *Eimeria* (diagrammatic). *1*, micropyle; *2*, polar inclusion; *3*, oocyst jelly; *4*, endomembrane of cyst wall; *5*, middle or "granular layer" of cyst wall; *6*, exomembrane of cyst wall; *7*, sporocyst or envelope of a spore; *8*, oocystic or extraresidual body; *9*, stieda body of spore; *10*, refractile globule of sporozoite; *11*, sporozoite; *12*, sporocystic or intraresidual body; *13*, nucleus of sporozoite; *14*, small refractile globule at more attenuated end of sporozoite. (After Becker, *Coccidia and Coccidiosis*.)

least in many species. There must be many exceptions to this rule, however, since oocysts are often passed by hosts for long periods of time.

With that foresight so frequently exhibited by nature, the macrogametes grow to a considerable size and accumulate large reserves of food materials. The microgametes, much the more numerous of the two cell types in the Eimerioidea, and also able to move actively because of their two flagella and minute size, then begin a search for the immobile macrogametes. Fertilization ensues, and the zygote becomes an oocyst, which soon develops one or more sporocysts, each with several sporozoites. Thus the cycle is completed. The number of sporocysts each oocyst contains, and the number of sporozoites typical of each sporocyst, are diagnostic for the different genera and species, especially in the case of the family Eimeriidae (Fig. 25.1).

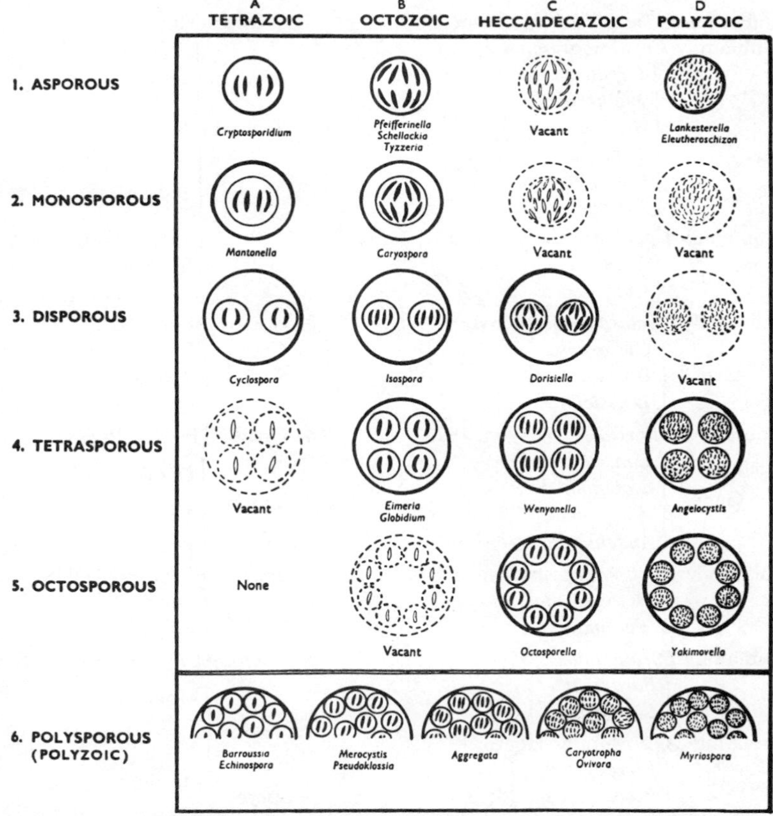

FIG. 25.2. Schema for the classification of the Eimeriidae into genera, based on the number of sporocysts (spores) within the oocyst, and the number of sporozoites within the sporocyst. Individual diagrams in each column illustrate the characteristic oocyst of each genus. Those in dotted lines show oocysts with arrangements of sporocysts and sporozoites theoretically possible, but not yet discovered. (After Hoare, *Rev. Brasileira de Malariol.*)

Taxonomy

Just how inclusive the family Eimeriidae should be (it is considered first because of its great importance) is still debated among protozoologists. The English protozoologist Hoare (1933; 1957), whose opinion in matters coccidial is still respected, suggested that it should embrace 7 subfamilies and 25 genera. His classification follows, and may be nicely shown in diagrammatic form (Fig. 25.2). As the schema shows, there are several vacancies in the arrangement which we hope, as in the periodic table of the elements, will eventually be filled in as our knowledge of the coccidia grows.

Classification of Coccidina, Eimeriidae

Family *Eimeriidae* (Minchin) Poche, 1913.	Diagnostic Characters
1. Subfamily *Cryptosporidiinae* Hoare, 1933. Genus: *Cryptosporidium*; *Pfeifferinella*, *Schellackia*, *Tyzzeria*; *Lankesterella*, *Eleutheroschizon*	1. Oocyst: ASPOROUS Tetrazoic Octozoic Polyzoic
2. Subfamily *Caryosporinae* Wenyon, 1926. Genus: *Mantonella*, *Caryospora*	2. Oocyst: MONOSPOROUS Tetrazoic Octozoic
3. Subfamily *Cyclosporinae* Wenyon, 1926. Genus: *Cyclospora*, *Isospora*, *Dorisiella*	3. Oocyst DISPOROUS Spores: Dizoic, Tetrazoic, Octozoic
4. Subfamily *Eimeriinae* Wenyon, 1926. Genus: *Eimeria*, *Globidium*; *Wenyonella*; *Angeiocystis*	4. Oocyst TETRASPOROUS Spores: Dizoic; Tetrazoic; Polyzoic
5. Subfamily *Yakimovellinae* Gousseff, 1937. Genus: *Octosporella*, *Yakimovella*	5. Oocyst OCTOSPOROUS Spores: Dizoic, Polyzoic
6. Subfamily *Baroussiinae* Genus: *Baroussia*, *Echinospora*	6, 7. Oocyst POLYSPOROUS Spores: Monozoic
7. Subfamily *Aggregatinae* Hoare, 1933. Genus: *Merocystis*, *Pseudoklossia*; *Aggregata*; *Caryotropha*, *Ovivora*; *Myriospora*	Spores: Dizoic Trizoic Dodecazoic Ployzoic

Of these rather numerous genera, *Eimeria*, *Isospora*, *Cryptosporidium*, *Caryospora*, and *Cyclospora* contain the great majority of the species parasitizing birds and mammals.

Coccidia of Man

For a long time it was thought that there was only one species of coccidian parasitizing man, but now it is fairly certain that there are actually at least two. Other species have been described but these were probably based on finding in human feces oocysts of species of coccidia occurring in the lower animals. Coccidian oocysts are so resistant that they often survive passage through the gut of man and other animals in which they are unable to set up an infection.

Of these two species, *Isospora belli* is much more common, although, curiously *Isospora hominis* was apparently discovered first. The great German pathologist Virchow noted in 1860 that a fellow scientist, Kjellberg of Stockholm, had seen parasites in the intestinal villi in the course of an autopsy, and it now seems quite certain that these were the asexual stages of *Isospora hominis*.

Infection is diagnosed by finding the occysts in the stools. Those of *Isospora belli* (Fig. 25.3) are immature when passed, and range in length between 25μ and 33μ, and in width from 13μ to 16μ. The oocysts of *Isospora hominis*

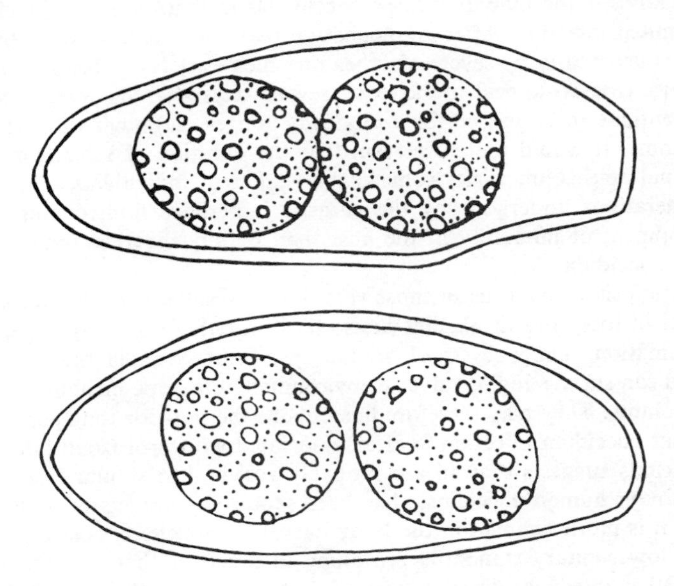

Fig. 25.3. Mature oocysts of (probably) *Isospora belli*. These two species differ chiefly in the size of their oocysts. (After Beltrán.)

are smaller, with corresponding dimensions of about 16μ and 10μ, and are said to be fully developed when they leave the body. Like other members of the genus, both species produce oocysts containing two sporocysts and four sporozoites ("disporous" and "octozoic," as indicated in Hoare's schema). In shape the oocysts resemble minute eggs, which is no doubt as it should be.

Stages of these parasites other than the oocysts have been little studied, since the relatively benign nature of the disease precludes much opportunity for post mortem examination. Virchow, who performed many thousand autopsies during a long life, is said by Wenyon (1926) to have seen asexual stages in the villi of the small intestine, and remarked that they looked exactly like some forms, which were undoubtedly coccidia, that he had previously observed in a dog. From a scientific point of view, Virchow and Kjellberg were lucky in their choice of patients for post mortems.

Were it not for several quite unusual characteristics shared by *Isospora hominis* with only a few other members of the genus, it might not be possible to say with any assurance that Virchow had actually seen this coccidian. These characteristics (as listed by Becker, 1954) include asexual development in the cores of the villi during the chronic phase, though the entire epithelium of the small intestine may be invaded in the acute stage of the infection; completion of sporulation while the parasites are still intracellular; and the production of oocysts with very thin walls, sometimes slightly constricted between the two sporocysts within.

Luckily for the human species, coccidiosis in man is usually a mild and self-limited infection. After an incubation period of about a week, the acute stage is ushered in by severe diarrhea and moderate fever, lasting for about ten days. Oocysts appear in the stools several days after the onset of diarrhea and continue to be passed for a month or longer, although without further symptoms. It would therefore seem that the duration of schizogony in the intestinal epithelium varies somewhat in different individuals. The number of generations undergone by the parasite is perhaps limited more by the development of immunity in the host than by any inherent genetic factors in the coccidian.

Coccidiosis in man, as in most vertebrates, results from the ingestion of oocysts in food and drink, but house flies probably also play a part in its dissemination. The oocysts of all the species of coccidia that have been studied can survive highly adverse environmental factors, including ordinary disinfectants. They may therefore live outside the host for long periods. The fact that coccidian infection is often associated with protozoan infection of other kinds suggests that all are probably acquired in similar fashion.

Although human coccidiosis has been reported from many parts of the world, it is more frequent in the Near East than elsewhere, possibly because of the low sanitary standards prevailing in much of that poverty-stricken area. Other intestinal diseases, as well as tuberculosis, thrive under similar conditions and they are common there too.

Coccidiosis of Animals

Canine and Feline Coccidiosis

Canine and feline coccidiosis may be of more human concern the world over than the disease in man himself, for the latter (at least in a clinical form), although cosmopolitan, seems uncommon in most areas and is not a very serious disease anyway. However, the frequency of infection in cats and dogs is not widely known.

Reports of the incidence in the United States of coccidial infection in dogs vary all the way from 8 per cent to 79 per cent in different areas. The disease is probably no less common in cats. Some (e.g., Morgan and Hawkins, 1948) believe that the species of coccidia occurring in cats and dogs may also cause disease in the "fox, mink, and probably other closely related carnivores," but this remains uncertain; on the whole, host specificity is very marked in coccidial infections (particularly in those due to *Eimeria*), much more so than with some other protozoan parasites.

Coccidiosis behaves in dogs and cats essentially as it does in man, except that it is often a much more serious disease. Apparently the mechanism of transmission is also the same. After a short incubation period there is bloody diarrhea, rapid loss of weight, and general weakness. Death follows in severe cases, and no treatment avails. Recovery is said to result in immunity to reinfection.

The number of species of coccidia to which dogs and cats are subject is somewhat uncertain. There are at least three of *Isospora* (*I. felis*, *bigemina*, *rivolta*) and two or three of *Eimeria* (*E. canis*, *felina*, *cati*). These differ among themselves in the shape and size of their oocysts, as well as in the number of sporozoites and sporocysts (*Eimeria* oocysts, as already indicated, contain sporocysts each with two sporozoites.) *Isospora bigemina* takes its name from the slight constriction which may appear between the two sporocysts within the oocyst, thus giving them the appearance of "twins." It was this species Virchow probably saw in dogs, and which caused him to remark on their similarity to forms he had seen in a human case. Oocysts of some of the species listed above are shown in Figure 25.4. Although life cycles of these species doubtless show minor differences, that of *Isospora felis* is probably typical. Figure 25.5 shows the various stages as seen by Wenyon (1923) in the intestinal epithelium.

Coccidiosis of Sheep and Goats

It is of interest that the first mention of coccidiosis in sheep seems to have been a note in 1879 by Leuckart, the great German pioneer in zoology and parasitology. The treatise in which it appeared was *Die Parasiten des Menschen*, of which the historian Nordenskiold said that on it was based "all subsequent research in this field." Both sheep and goats often succumb

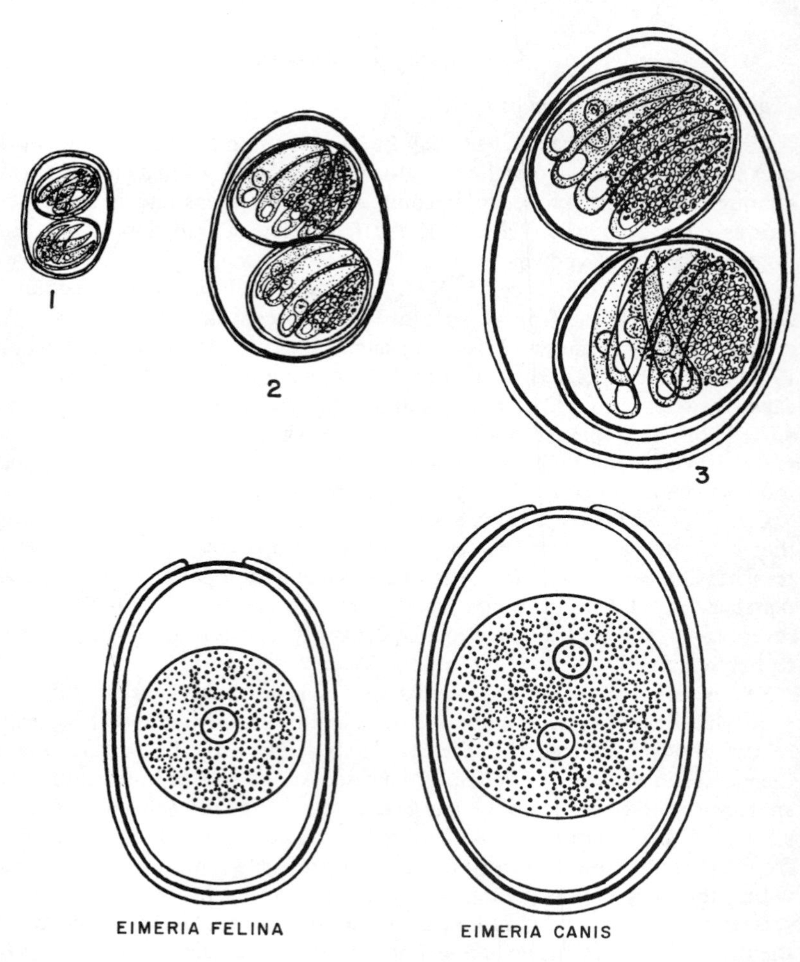

FIG. 25.4. *Above*. Mature oocysts of coccidia commonly parasitizing dogs and cats (after Becker). *1*, *Isospora bigemina*; *2*, *I. rivolta*; *3*, *I. felis*. All three have also been found in foxes, leopards, and jungle cats. *Below*. Oocysts *Eimeria felina* and *E. canis*, also dog and cat coccidia, although both rare. (After Morgan and Hawkins, *Veterinary Protozoology*.)

to infection with one or more of possibly nine different species of coccidia. Apparently coccidiosis of these hosts occurs throughout the world. Kids and lambs are especially susceptible.

There is little detailed knowledge of the life cycles of any of these species, all of which belong to the genus *Eimeria*, but there is no reason to think that they differ significantly from the general pattern. The parasites undergo schizogony and gametogenesis in the mucosa and submucosa of the small

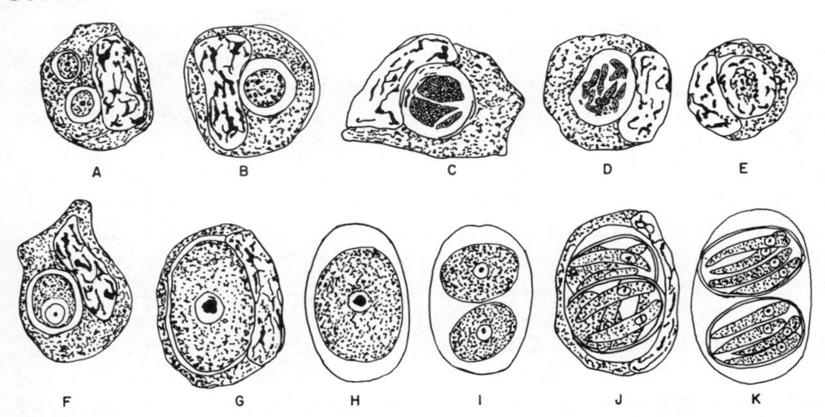

FIG. 25.5. Life cycle of *Isospora bigemina*. *A–D*. Schizogony. *E*. Possible microgametocyte with microgametes. *F–G*. Growth of macrogametocyte. *H–K*. Development of oocyst in tissues of villi. (After Wenyon, *Protozoology*.)

intestine, and the resulting injury to the intestinal lining results in severe diarrhea and weakness, with death in some cases. Mortality figures are variously given: some say it seldom exceeds 5 per cent and others that a "large proportion of the flock" may die (Becker, 1934). In any case, the over-all losses are large.

As with species of coccidia already discussed, infection results from picking up the oocysts where they have been deposited in the droppings of infected animals. Such oocysts are not immediately infective; the oocysts of all species of *Eimeria* occurring in warm-blooded animals are passed in an immature condition and must have about two or three days to ripen. How long they are viable outside the host is uncertain, but apparently a contaminated range does not remain so for more than a single season. Persisting maternal infections are believed to constitute the reservoir from which the young animals acquire their parasites.

Perhaps the chief problem of coccidiosis in sheep and goats is a method of effective therapy. But there are also questions such as which species are responsible for recurring outbreaks in various localities, and even which are the most pathogenic. Of the nine species, more or less, *Eimeria faurei*, *E. arloingi*, and *E. parva* are said to be the most common in the United States. The first of the three species was also the first to be described, by the French authors Moussu and Marotel in 1901.

Coccidiosis in Cattle

Perhaps of even greater importance than sheep and goat coccidiosis is coccidial infection in cattle. When it is recalled that cattle raising is a worldwide industry, and that bovine coccidiosis is at least as important in other parts of the world as it is in this country where it causes an annual loss of at least 10 million dollars (Foster, 1949), the magnitude of the problem stands out in bold relief. What it means, of course, is that the people of the world, many parts of which are already overcrowded, have less to eat and less footwear. It likewise means that all of us pay more for beef and leather.

Cattle are subject to attack by at least twelve species (Morgan and Hawkins, 1948) instead of a paltry nine or less. The parasitic nature of the disease was first noted by the German Zurn, for which one of the more pathogenic species, *Eimeria zurnii*, was named by the Italian biologist Rivolta (who called it *Cytospermium Zurnii*) in the same year. As in the case of the coccidia of sheep and goats, all the coccidia of cattle belong to the genus *Eimeria*, and probably have a world-wide distribution.

Much of what has been said of coccidiosis in other animals applies equally to the bovine variety. Calves are more susceptible than older animals, and suffer more severely from its effects; some die of it. Those that recover are believed to have at least a transient immunity to reinfection with the same species of coccidian. But it appears that immunity sufficient to protect the host from injury by the parasites may not be enough to cause their complete elimination, and thus the carrier state may supervene. It is therefore dangerous to introduce newly acquired calves into an old herd. Overcrowding adds to the risk.

Coccidiosis of Swine

Much less important than bovine coccidiosis, but still a matter of economic significance, is coccidiosis of swine. Like the bovine variety, it occurs throughout the world. There are apparently five species (four of *Eimeria* and one of *Isospora*) that infect swine. Of these, *Eimeria debliecki* and *E. scabra* are said to be the most pathogenic. The course and symptoms of the acute phase of the malady resemble those of coccidiosis in other animals, and young animals here also are the most susceptible. The infection may kill, but when it does not, weight and general health may be impaired for an extended period of time.

Coccidiosis in Rabbits

Rather surprisingly, rabbit coccidiosis is of greater importance than coccidiosis in swine. Just how expensive it is to the rabbit fancier no one can say, but Foster (1949) remarked that annual losses of as much as a million dollars had been claimed, though "the true figure is probably very much lower." In any case, it is a very common disease among rabbits and "one of the main causes of mortality" (Morgan and Hawkins, 1948). Figures for losses apply, of course, only to domestic rabbits; coccidiosis in the wild rabbit population is also common, and is considered beneficial by everyone except the backyard sportsmen who like to while away an afternoon taking potshots at them. For rabbits take an appreciable toll of farm produce, as all gardeners know. After myxomatosis, an introduced virus disease, virtually wiped out the rabbits in the English countryside, the production of a number of crops is said to have increased some 10 per cent.

Rabbits are host to five species of *Eimeria.* One of these species (*E. stiedae*) was apparently first seen by Leeuwenhoek in 1674, and thus the honor of having been the discoverer of the coccidia, as well as of so many other things, is his. The "oval corpuscles" (oocysts), which reach a maximum of about 40μ in length by 25μ in breadth, and are characteristically reddish pink in color, must have taxed the power of his crude single-lens microscope.

Unlike most species of vertebrate coccidia, *Eimeria stiedae* resides in the liver, where it gives rise to tumorlike white nodules that may reach the size of small marbles. The sporozoites are liberated from the ingested oocysts in the intestine, and reach the liver apparently by penetrating the capillaries of the intestinal wall and riding the portal blood stream to their ultimate destination.

Perhaps because the rabbit lends itself so readily to use as a laboratory animal, more has been learned about rabbit coccidiosis than about the disease in most other species. The course of the infection with *Eimeria stiedae* has been thoroughly studied (Fig. 25.6). After a period of asexual reproduction in which the schizonts give rise to broods of about eight merozoites each, another type of schizont is said to develop, which forms four merozoites. These in turn enter other cells, and eventually become gametocytes. The female gametocyte then matures into a macrogamete, which proceeds to secrete an oocyst with a tiny opening (micropyle) at one end for the entry of the microgamete. The microgamete is a biflagellated cell developed from the microgametocyte, and one of a numerous brood. The oocyst with the zygote within is then (if it is lucky) carried down the bile ducts and into the gall bladder, which is often crowded with these parasites. From there it passes through the common duct into the gut. Final development of sporozoites is said to require about three days or a little less.

The affected rabbit is meanwhile likely to be decidedly ill, and the disease is frequently fatal if a large number of oocysts have been ingested. The acute stage is marked by severe diarrhea and great enlargement of the liver. If death does not occur the disease subsides in a few weeks, although it is said that among wild rabbits a chronic stage often follows, and no doubt it is the oocysts passed by these animals that are primarily responsible for keeping the disease going.

Although *Eimeria stiedae* causes much loss to the owners of rabbitries, *E. perforans* may be of equal or even greater importance (Fig. 25.7). The incidence of this type of coccidiosis among California rabbits is said to be about 30 per cent (Kessel and Jankiewicz, 1931). This species, and the other three (about which less is known) attack the intestinal epithelium, and parasites are sometimes so numerous as to crowd almost every cell in a villus, with resulting extensive destruction of the mucosa. The mechanism of injury is

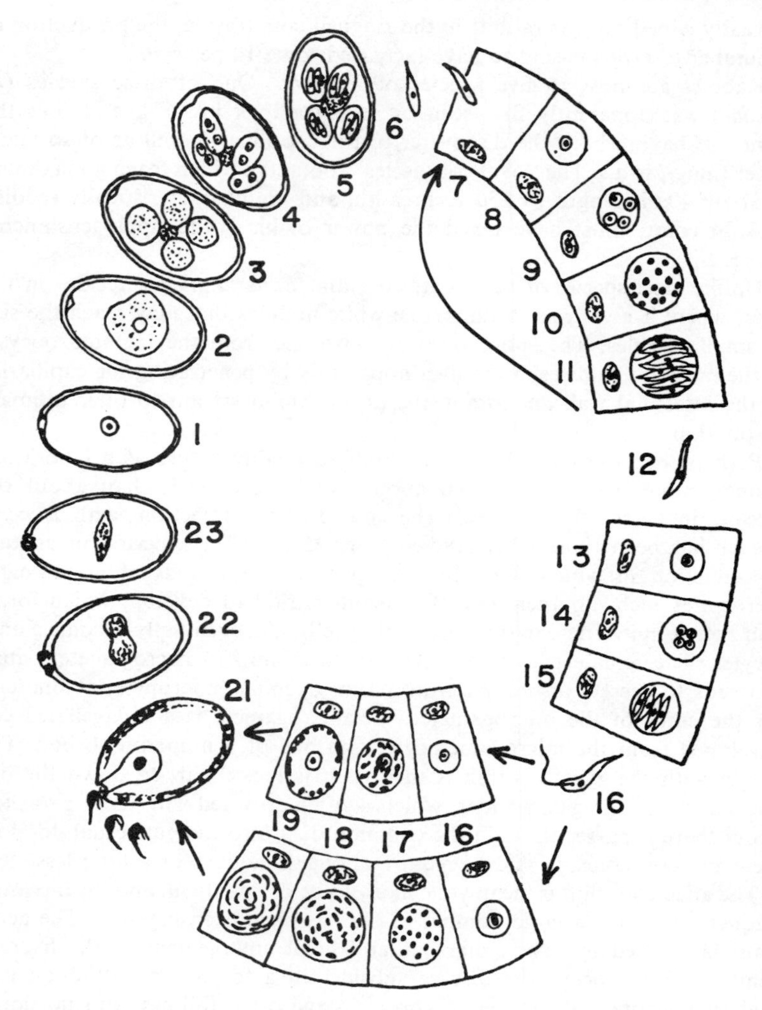

FIG. 25.6. The developmental cycle of *Eimeria stiedae*, a parasite of the biliary passages of the rabbit. *1–12*, schizogony; *13–16*, gamogony; *16–19*, formation of gametes; *21–23*, fertilization; *1–5*, sporogony. (After Reich; reproduced from Naville.)

believed to be largely mechanical, with possible secondary toxemia, due to absorption of poisonous substances of bacterial origin.

Coccidiosis in Fowl

Much more serious than the losses due to rabbit coccidiosis are those chalked up to coccidiosis of domestic fowl. Eight species are involved, of

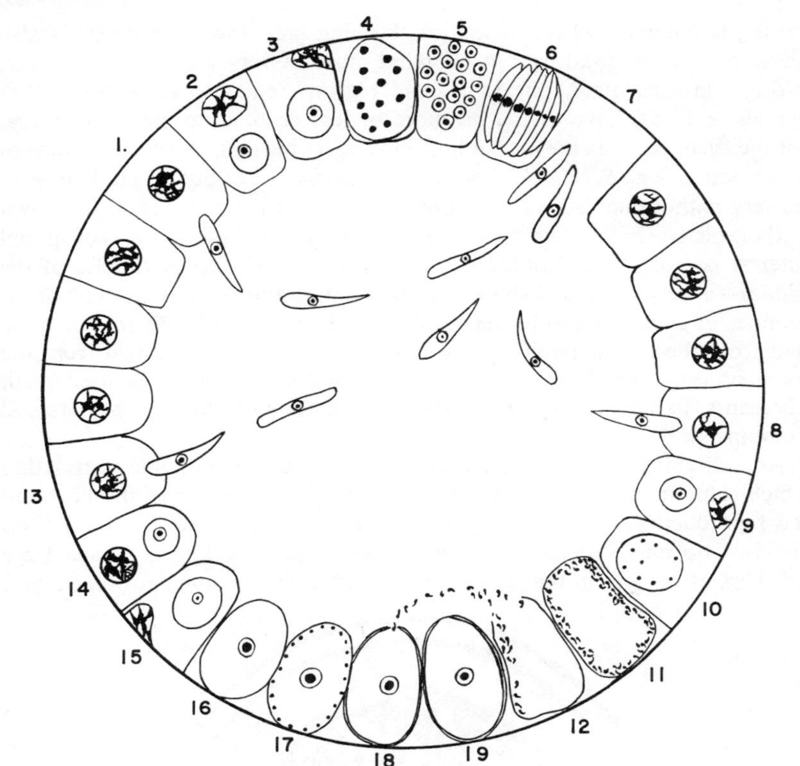

FIG. 25.7. The life cycle of *Eimeria perforans*, a parasite of rabbits (after Thomson and Robertson, *Protozoology*). *1–7*. Schizogony; the invasion of an epithelial cell by a sporozoite is shown in *1*; the formation of merozoites is shown in *5* and *6*. *8–12*. Gametogony; formation of the microgametocytes, from which develop the microgametes (*11* and *12*). *13–19*. Gametogony, with formation of the macrogametocyte, followed by fertilization (*18*) and development of the oocyst (*19*).

which two are much more pathogenic than the others. *Eimeria tenella* (Fig. 25.8) and *E. necatrix* are often the cause of severe epizootics affecting a large proportion of the birds in a flock, resulting in heavy mortality. Even when recovery occurs, weight gain and egg production may be considerably curtailed.

The size of the nation's bill for coccidiosis in chickens is uncertain, but it is thought to at least equal the 10-million-dollar loss credited to the bovine variety of the disease (Foster, 1949). The mortality directly due to poultry coccidiosis is said to amount to about 10 per cent of that from all causes, and to this figure must be added lessened egg production, lessened weight gain per pound of food consumed, and the cost of medicine and preventive measures, as well as a variety of indirect costs. Since poultry raising is common wherever there is farming, and fowl coccidiosis is also cosmopolitan, its total cost to mankind must be very great. Geese, ducks, turkeys, grouse, quail, pheasants, and pigeons (and probably most other species of birds) have their own more or less peculiar species of coccidia, for the avian coccidia are probably in most cases highly host-specific. Some of these species (e.g., *Schellackia*, or *Tyzzeria, perniciosa* of ducks) are known to be very pathogenic; others seem not to be, and still others are of unknown pathogenicity. One of these species, *Eimeria truncata* of the goose, is of special interest because it attacks the epithelium of the uriniferous tubules of the kidney—a very unusual habit for a coccidian. Some idea of the very large number of coccidial species reported from domestic and wild fowl may be had from the review published by Levine (1953) of those known from the three orders Anseriformes (ducks and geese), Galliformes (grouse, quail, pheasants, partridges, chickens, turkeys), and Charadriiformes (plovers and sandpipers).

An intriguing aspect of avian coccidiosis is the remarkable correlation which sometimes exists between the physiological activities of the host and the reproductive cycle of the parasite. This was first noted by Boughton (1933), who observed that sparrows infected with coccidiosis passed large numbers of oocysts in the afternoon, and often few or none at other times.

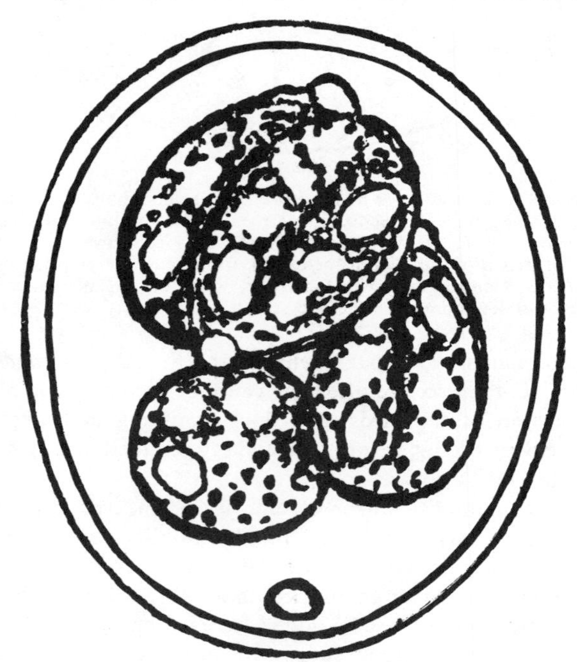

FIG. 25.8. Mature cyst of *Eimeria tenella*. (After Tyzzer, *Am. J. Hyg.*)

When the birds were exposed to "days" of artificially lesser or greater length than the normal 24 hours, the parasites, after a short initial lag, adjusted their cycles accordingly. Later it was found that certain species of coccidia infecting the pigeon behaved similarly, but in neither case was it possible to discover just what physiological changes of the host were responsible. In all likelihood the host-parasite dependence in these cases is fundamentally similar to that seen in malaria infections, in which the reproduction of the plasmodia often exhibits like periodicity. Here also the cycle of the parasite has been shown to be subject to artificial change in the length of the "day" to which the host is exposed.

An incidental result of Boughton's study was the finding that English sparrows displayed an almost 100 per cent incidence of coccidiosis, provided only that oocysts were sought for in feces passed at the proper time of day. Most such infections were chronic, instead of being rather sharply self-limited as they apparently are in many other host species. These findings suggest that coccidiosis may prove to be equally common in many another species of bird, if only it is looked for with equal diligence and regard for the behavior of the parasites.

Since many species of coccidia are highly pathogenic, and the character of the English sparrow is hardly such as to make it many friends, one might wish that its coccidian parasites were of a sort to aid in its natural control. Unfortunately this seems not to be the case, even among nestlings. Whether species of coccidia infecting other species of wild birds are a cause of significant mortality is in most cases unknown, but experience with coccidiosis in domestic fowl suggests that at least in some cases young birds may suffer considerably. This is known to be true of young quail chicks raised on state-owned game farms in California (Herman, 1949). Probably there are not only pathogenic species of coccidia, but relatively virulent strains even among some normally nonpathogenic species.

Coccidia of Other Vertebrates

Coccidial infection is also common in many species of wild mammals, and in other vertebrates. Species of the so-called "bigemina" group of the genus *Isospora* are known from some marsupials and certain reptiles, and several occur in frogs. Rather surprisingly, none has been found in vertebrates lower in the evolutionary scale. So peculiar a host distribution is difficult to account for, but it may indicate that coccidia of this group broke away from the main line of *Isospora* descent a long time ago. The hosts, as Becker notes, share few characteristics other than a liking for a flesh diet.

In some species of coccidia infecting the lower vertebrates the parasites are introduced when the prospective host is bitten by some blood-sucking invertebrate, such as a leech or a mite. This occurs with all members of the family Haemogregarinidae. In general very little is known of the actual incidence of coccidiosis among the lower vertebrates, and still less about its possible role as a cause of disability or death.

Coccidia of Invertebrates

Among invertebrates the picture is equally clouded. Numerous species of coccidia have been described from a great variety of hosts, but the part assumed by these parasites in the maintenance of biological balances is entirely unknown. Remarkably enough, in view of the much greater importance of coccidial infection among the higher vertebrates, a species of coccidian (*Eimeria schubergi*) infecting a centipede was the first for which a life cycle was completely worked out. This was done by the great German biologist Fritz Schaudinn in 1900.*

Schaudinn's account of the life cycle of *Eimeria schubergi* (Fig. 25.9), which he called *Coccidium schubergi*, has proved to be essentially correct, and it differs only in minor degree from the cycles of other species of the genus. Schaudinn also proposed a terminology for the different stages of the life cycle which has been almost universally adopted. The growing form is known as a "trophozoite" and it becomes a "schizont" with the onset of changes leading to fission. When the steps preceding division are complete, the parasite is called a "segmenter," and its daughters are "merozoites." Asexual multiplication is "schizogony," and sexual reproduction "sporogony." The fertilized egg cell, or zygote, may also be spoken of as the "synkaryon," but since this term is now widely used in a somewhat different sense in ciliate genetics it is better not to use it to describe a stage in the coccidian life cycle.

The new host is infected by ingesting oocysts, and these germinate in the gut of the centipede, usually a common garden variety known as *Lithobius forficatus*. The sporozoites emerge from the ruptured sporocyst and then pass through a tiny opening in the oocyst (the "micropyle"), and promptly begin an active search for susceptible epithelial cells. These they penetrate by a peculiar sinuous gliding movement though they lack any locomotor organelles. Once inside the doomed cell of the gut lining, they round up and begin a period of active growth culminating after about 24 hours in multiple fission and the birth of a brood of several dozen young parasites.

Like the sporozoites, the young parasites are motile. They invade neighboring epithelial cells and the process of growth and multiple fission is repeated. Finally, after several generations of such asexual reproduction, gametocytes develop, essentially as described above for the rabbit coccidian *Eimeria stiedae*, and these undergo maturation with the formation of gametes. The macrogamete is relatively large and well provided with food material

*Schaudinn is usually best remembered for his discovery of *Treponema pallidum*, the spirochete responsible for syphilis. He also worked on *Entamoeba histolytica*, a frequent cause of dysentery in man (see p. 416), with the result that he acquired the disease himself and is said to have paid for his scientific curiosity with his life.

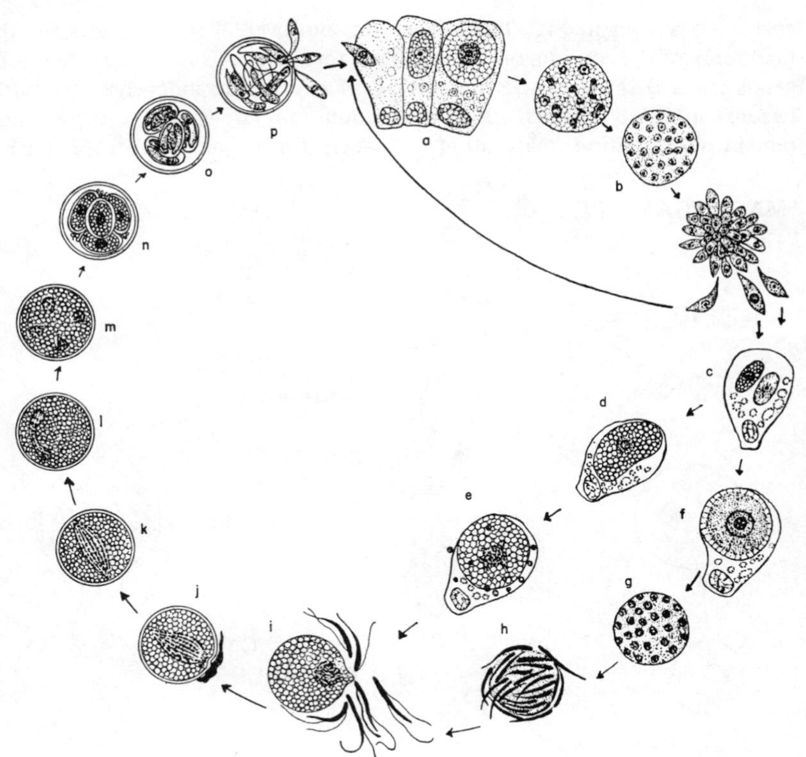

FIG. 25.9. Life cycle of *Eimeria schubergi*, a parasite of the centipede (Schaudinn). *a*, invasion of gut epithelial cell by sporozoite; *b*, schizogony; macro- and microgametocyte; *d–e*, development of macrogamete; *f–h*, development of male gametes; *i*, ripe gametes; *j–k*, fertilization; *l–n*, formation of spores; *o*, mature oocyst with four spores; *p*, spore germination in gut.

in the form of numerous granules. The microgamete is filamentous with two trailing flagella, although one arises anteriorly.

Fertilization ensues, and the flagella of the male cell are discarded, while the nuclei of the two cells fuse and form a spindle; in the mean time an oocyst is secreted. Then comes a division in which the doubled number of chromosomes is believed to be reduced to the normal figure, and this division is succeeded by a second one. Each daughter nucleus then appropriates a little cytoplasm, and secretes a wall, the sporocyst, about itself. Development is completed by a final division which leaves two sporozoites and a mass of residual cytoplasm in each sporocyst.

Although the pattern of the coccidial life cycle varies among the different families, genera, and species, such individual peculiarities may be regarded as minor and will not be discussed here. One such cycle, that of *Aggregata eberthi*, is shown in Fig. 25.10. They do, however, form the basis for the taxonomy of the coccidia (given in outline form on p. 536). In some (e.g., the genus *Adelea*) the gametocytes, instead of developing independently, early become attached to each other in a condition known as "syzygy," and remain thus during maturation. In such forms the number of micro-

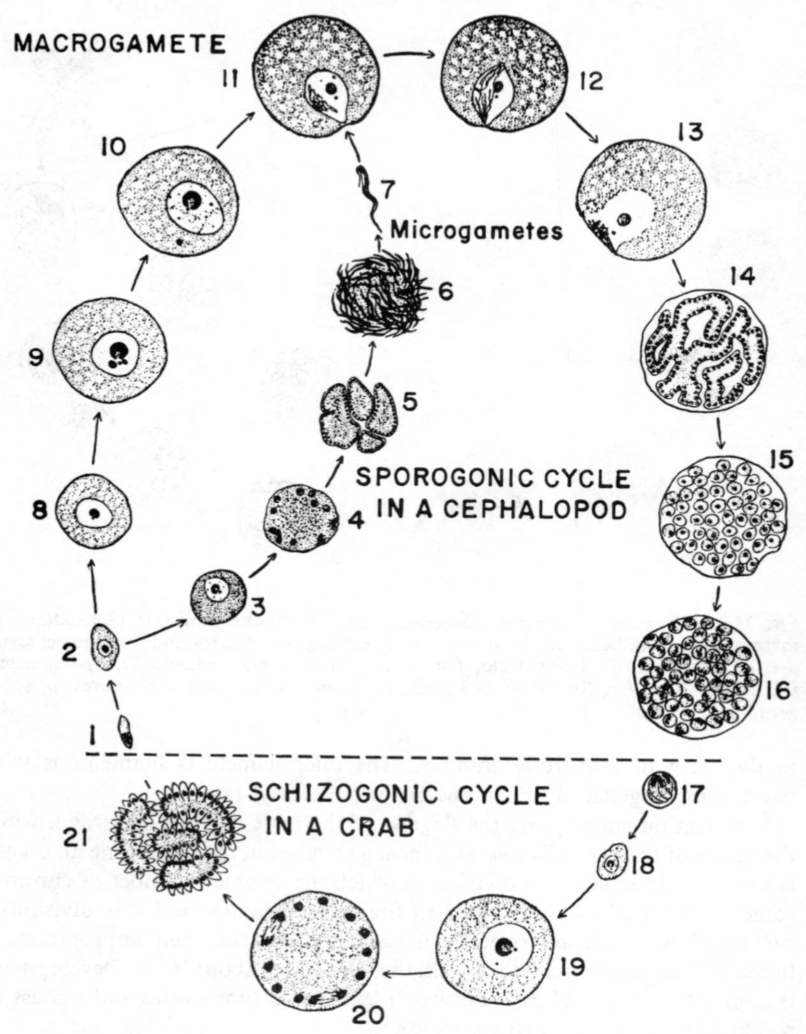

FIG. 25.10. Life cycle of *Aggregata eberthi*, a coccidian which alternates between the crab and the cuttlefish as hosts. *1–2*, young undifferentiated schizont; *3–7*, microgametogenesis; *8–11*, macrogametogenesis; *12*, fertilization; *13–17*, sporogenesis; *18–24*, schizogony. (After Naville.)

gametes is small, often not more than from two to four. Actual fusion does not take place until the process of ripening is completed. There are differences, too, in the number of asexual generations preceding the appearance of gametocytes. Indeed in *Eimeria bovis* there is but a single generation of asexual reproduction, though since each segmenting parasite produces about 100,000 offspring, ample provision seems to be made for the continuance of the species.

Coccidian and Malaria Life Cycles Compared

Several features of this rather complicated train of events are also found in the developmental pattern of the malaria parasites (see Chap. 24), and constitute reasons for regarding as closely related the true coccidia and these most important agents of human disease. In both, the sporozoites are the actual infective forms, and in both they have a somewhat similar shape. The merozoites in both also tend to be falciform, although this is not always true in all species of malaria plasmodia. Asexual multiplication is initiated by a number of generations in tissue cells (exoerythrocytic stages in the malaria parasites) before gametocytes are produced, and the structure of the gametes, particularly in the case of the microgamete, is much alike. A few species parasitize even red blood cells, such as *Haemogregarina stepanowi* (Fig. 25.11).

To be sure, there are differences. The Protozoa of malaria reproduce in blood cells as well as in tissue cells; except in the case of the microgametes and the ookinete, or zygote, they lack motility. Development of the malarial oocyst does not involve the formation of a resistant envelope, or of sporocysts. While the coccidian sporozoites are imprisoned within the walls of the oocyst until swallowed by the prospective host those of malaria are free and are liberated within the body cavity of the mosquito, some of them managing to reach the salivary glands, from which they pass to the new host when the insect sucks blood. But these differences are not fundamental, and have probably evolved because the sheltered situation of the malaria parasites makes the protection of a tough oocyst wall unnecessary. And even among the true coccidia there are a few species in which sporocysts are lacking within the oocyst.

Mode of Injury to Host

In this connection, we may well ask what so great an influx of invading parasites may do to the host. *Eimeria bovis*, like many species of coccidia, is highly pathogenic. How do coccidia injure their hosts? To this question no final answer can be given now. Obviously, there must be rather gross destruction of cells, since all invaded cells ultimately die. Yet mucosal epithelium is quickly regenerated, and this kind of damage is therefore ephemeral. Tyzzer (1929) made the interesting observation that cells attacked by certain

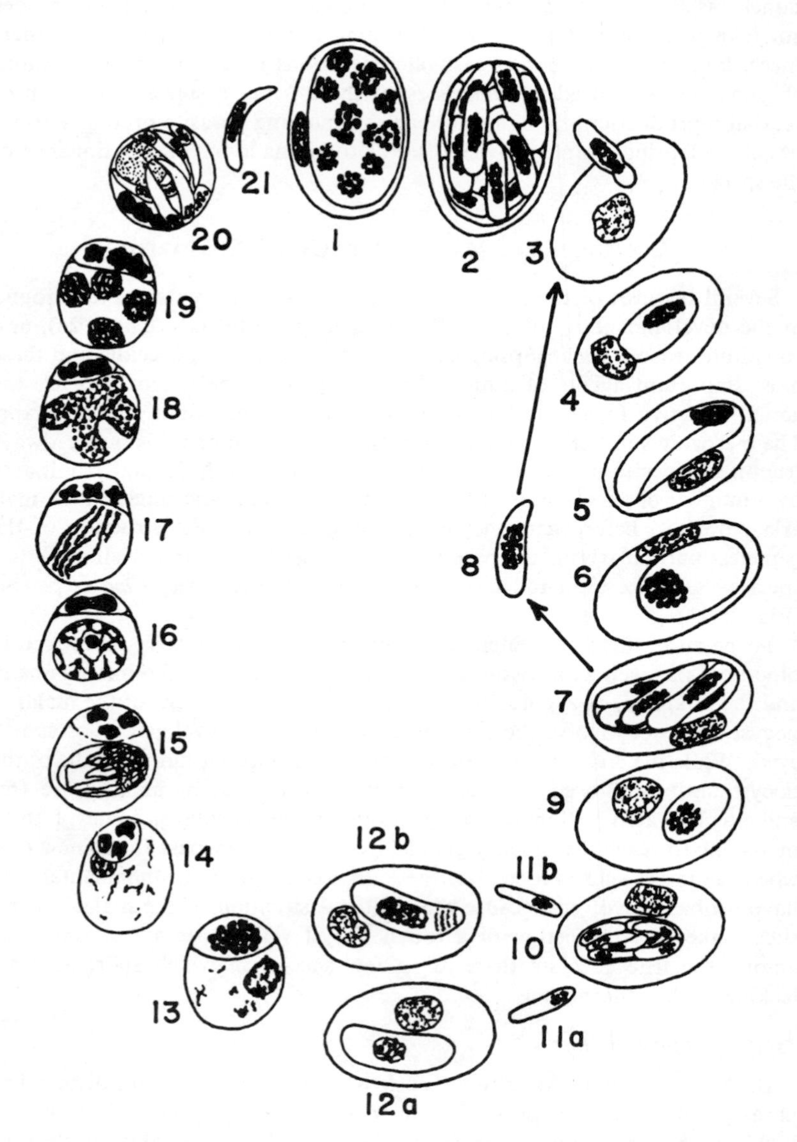

FIG. 25.11. The developmental cycle of *Haemogregarina stepanowi*, a blood parasite of the turtle transmitted by a leech. *1–7*, schizogony; *8–11*, bipolar gamegony; *12–15*, gametogenesis and syzygy; *16*, zygote; *17–20*, formation of sporozoites. (After Reichenow; reproduced from Naville.)

species of chicken coccidia alter their normal behavior, growing to a great size and becoming invasive of adjacent tissue. As a result local pressure necrosis occurs, with severe haemorrhage. Blood loss and intestinal irritation may be great enough to cause immediate death, but even when recovery ensues there may be extensive formation of scar tissue. Thus, both during the acute stage of the malady and later, the functioning of the alimentary tract may be so much impaired that food absorption fails to take place normally, and anemia and weight loss result.

We would also like to know why some species of coccidia are so much more pathogenic than others. So far there is no evidence that any coccidia elaborate toxins. Doubtless there are fundamental differences in the relationships of pathogenic and less pathogenic species to their hosts, but about such relationships there is as yet little real knowledge.

Many other unsolved problems remain in the field of "coccidiology." The proper criterion for the definition of species, for example, and indeed how many species there are. Reasons largely of convenience have dictated the choice of characteristics used in the description and identification of species—oocyst and sporocyst morphology, assumed strict host-parasite specificity, and life cycle peculiarities. Degree of pathogenicity and the site of tissue invasion have also been regarded as useful in the blue-printing of species. But the work of Doran (1953) suggests that host-parasite specificity may not always be as rigid as is usually supposed, and we have little information about the amount of intraspecific variation there may be.

With regard to the number of species which exist, Levine and Becker (1933) published a review of the genus *Eimeria* listing 220 species from hosts as varied as annelids, centipedes, insects, and a great assortment of vertebrates. *Isospora*, a much smaller genus, nevertheless embraced at least 48 species, according to Becker (1934), who reviewed it about the same time. Since then the number of recognized species of *Eimeria* and *Isospora* has doubled (Hardcastle, 1943; Becker, 1956; Pellerdi, 1956) and it will doubtless continue to grow. Most of the other genera have been less studied, but we may be sure that they will also continue to increase in number of included species as they receive more attention from protozoologists and parasitologists.

Thus, it is clear that there is still much opportunity for exploration in the field of coccidia. Problems in the physiology, pathology, taxonomy, genetics, and evolution of this important and interesting group of Protozoa will challenge the biologist for a long time. And we still have much to learn about the control and treatment of coccidiosis, which remains a major threat to the poultryman and cattleman, and hence to the food supply of us all.

The Classification of the Coccidia

The coccidia, like other groups of Sporozoa, are not easy to classify, for a good scheme of classification is based on biological relationships, and among the coccidia these are often obscure. The elucidation of probable evolutionary relationships depends on our knowledge of life cycles and morphology, and information about many species is lacking. More fundamentally, our difficulty is caused by the profound alterations in structure and developmental patterns which long-continued parasitism has entailed. Historically, the coccidia have been arranged in various ways by taxonomists, and a cursory examination of contemporary works in protozoology shows that this is still true.

The scheme which follows is essentially Hall's (1953) except for taxonomic rank of the larger groups. The coccidia constitute one of two suborders of the order Coccidiomorphida, subclass Telosporidia. Sporozoa of this subclass have life cycles with sexual and asexual phases, reproduction occurring in both. In the former phase sporozoites are the end product ("Telosporidia" derives from the Greek *telos* and *spora* meaning "end" and "seed," or "spores at the end"). The spores are distinctive in lacking the polar capsules seen in Sporozoa of the subclass Cnidosporidia, and are also quite unlike the spores of the fourth subclass, the Haplosporidia.

The coccidia are typically parasites of the epithelium, generally of the gut. Hosts may be either invertebrates or vertebrates, or sometimes both. In forms having but a single host, the infective stage is usually the oocyst with its contained sporozoites, and the host does the rest when it picks up these forms with its food and water.

Subclass. TELOSPORIDIA.

Order. COCCIDIOMORPHIDA.

Suborder. COCCIDINA.

Superfamily. ADELEOIDEA. Gametocytes exhibit syzygy; few microgametes produced.

Series 1. ADELEIA. Zygote inactive; typical oocyst formed.

Family 1. ADELEIDAE. Oocyst contains sporocysts; life cycle typical of order.

Family 2. DOBELLIIDAE. No sporocysts; numerous microgametes from microgametocyte. One genus.

Family 3. KLOSSIELLIDAE. Oocysts each contain numerous sporocysts, within each of which are many sporozoites. One genus.

Family 4. LEGERELLIDAE. Oocysts with many sporozoites but no sporocysts. One genus.

Series 2. HAEMOGREGARIA. Zygote motile ("ookinete"); life cycle involves two hosts.

Family 1. HAEMOGREGARINIDAE. Oocysts contain no sporocysts. Sexual cycle in leeches, and asexual cycle in turtles or other of lower vertebrates. One genus: *Haemogregarina*.

Family 2. HEPATOZOOIDAE. Oocysts rather large, with many sporocysts, each containing numerous sporozoites. Gametocytes in avian or mammalian leucocytes; vectors may be ticks, mites, or biting insects. One genus: *Hepatozoon*.

Family 3. KARYOLYSIDAE. Sporokinetes (motile forms) penetrate egg of invertebrate host (mite), where they become sporocysts. The vertebrate host (lizard) harbors the gametocytes in its red cells, the nuclei of which are attacked (hence the name "Karyolysidae," from two Greek words meaning "nucleus dissolver"). One genus: *Karyolysus.*

Superfamily 2. EIMERIOIDEA. No syzygy of gametocytes. Many microgametes.

Family 1. AGGREGATIDAE. Few to many sporocysts (up to several hundred) in single oocyst. Life cycle may be completed in one host, or require two.

Family 2. CARYTROPHIDAE. Oocyst with many sporocysts, each containing numerous sporozoites. One genus.

Family 3. CRYPTOSPORIDIIDAE. Exceptional in that development is extracellular in gut mucus, rather than in underlying epithelium. Oocysts without sporocysts, but with four sporozoites. One genus.

Family 4. EIMERIIDAE. Largest family of order. Oocysts with one to many sporocysts, each with one to a number of sporozoites. Family characteristics less well defined than in other families. Gametocyte development independent; only one host.

Family 5. LANKESTERELLIDAE. Sporozoites invade erythrocytes of vertebrate host, with mechanical transmission to new vertebrate host by vector such as mite or leech.

Family 6. SELENOCOCCIDIIDAE Growth stages occur, for both sexual and asexual forms, extracellularly in intestine of lobsters, although development must be completed in tissue cells. One genus.

REFERENCES

Becker, E. R. 1934. *Coccidia and Coccidiosis of Domesticated, Game and Laboratory Animals, and of Man.* Ames, Iowa: Collegiate Press.

——— 1954. The host affinities of *Isospora bigemina* type Coccidia. *Proc. Iowa Acad. Sci.*, 61: 463–7.

——— 1956. Catalog of Eimeriidae in genera occurring in vertebrates and not requiring intermediate hosts. *Iowa State Coll. J. Sci.*, 31: 85–139.

Boughton, D. C. 1933. Diurnal gametic periodicity in avian *Isospora. Am. J. Hyg.*, 18: 161–84.

Brackett, S. and Bliznick, A. 1950. The occurrence and economic importance of coccidiosis in chickens. Lederle Lab.

Carvalho, J. C. M. 1943. The coccidia of wild rabbits of Iowa. I. Taxonomy and host specificity. *Iowa State Coll. J. Sci.*, 18: 103–35.

Connal, A. 1922. Observations on the pathogenicity of *Isospora hominis* Rivolta, emend. Dobell, based on a second case of human coccidiosis in Nigeria; with remarks on the significance of Charcot-Leyden crystals in the feces. *Trans. Roy. Soc. Trop. Med. and Hyg.*, 16: 223–45.

Craig, C. F. and Faust, E. C. 1951. The Sporozoa. The Coccidia of man. In *Clinical Parasitology*, 5th ed. Philadelphia: Lea & Febiger.

Doran, D. J. 1953. Coccidiosis in the kangaroo rats of California. *Univ. Calif. Pub. Zool.*, 59: 31–58.

Foster, A. O. 1949. The economic losses due to coccidiosis. *Ann. N.Y. Acad. Sci.*, 52: 434–42.

Hall, R. P. 1953. *Protozoology*. New York: Prentice-Hall.

Hardcastle, A. B. 1943. A check list and host-index of the species of the protozoan genus *Eimeria*. *Proc. Helm. Soc. Wash.*, 10: 35–69.

Herman, C. M. 1949. Coccidiosis in native California Valley quail, and problems of control, *Ann. N.Y. Acad. Sci.*, 52: 621–3.

Hoare, C. A. 1933. Studies on some new ophidian and avian coccidia from Uganda, with a revision of the classification of the Eimeriidae. *Parasitology*, 25: 359–88.

——— 1957. Classification of Coccidia Eimeriidae in a "Periodic System": Homologous Genera. *Rev. Brasil. de Malariol.*, 8: 197–202.

Kessel, J. F. and Jankiewicz, H. A. 1931. Species differentiation of the domestic rabbit. *Am. J. Hyg.*, 14: 304–24.

Kudo, R. R. 1954. *Protozoology*. 4th ed. Springfield: Charles C Thomas.

Levine, N. D. 1953. A review of the coccidia from the avian orders Galliformes, Anseriformes, and Charadriiformes, with descriptions of three new species. *Am. Midl. Nat.*, 49: 696–719.

——— and Becker, E. R. 1933. A catalog and host-index of the species of the coccidian genus *Eimeria*. *Iowa State Coll. J. Sci.*, 8: 83–106.

Liebow, A. A., Milliken, N. T., and Hannum, C. A. 1948. *Isospora* infections in man. *Am. J. Trop. Med.*, 28: 261–74.

Morgan, B. B. and Hawkins, P. A. 1948. *Veterinary Protozoology*. Minn.: Burgess.

Pellerdi, E. 1956. Catalog of the genus *Eimeria* (Protozoa: Eimeriidae). *Acta. Vet., Acad. Sci. Hungarieae* (Fasc. 1), 6: 75–102.

Schaudinn, F. 1900. Untersuchungen über den Generationswechsel bei Coccidien. *Zool. Jahrb. Abt. Morph.*, 13: 197–292.

Swales, W. E. et al. 1948. Parasitology (committee report). *J.A.V.M.A.*, 113: 235–9.

Tyzzer, E. E. 1929. Coccidiosis in gallinaceous birds. *Am. J. Hyg.*, 10: 1–115.

Virchow, R. 1860. Helminthologische Notizen. 4. Zur Kenntnis der Wurmknoten. *Arch. f. Pathol., Anat.*, 18: 523.

Wenyon, C. M. 1923. Coccidiosis of cats and dogs and the status of the *Isospora* of man. *Ann. Trop. Med. and Parasit.*, 17: 231–38.

——— 1926. *Protozoology*, vol. 2. New York: Wood.

26

Parasites of Doubtful Status

"The list of described species of parasitic animals and plants to which man is liable is already a long one, but nevertheless, in different parts of the world, others will yet be discovered." Joseph Leidy, *A Flora and Fauna Within Living Animals*, 1851.

Biologists are always concerned with the relationships of living things, and when these cannot be determined proper classification may be difficult. Any system of classification should fulfill two requirements: it must be convenient to be of practical value, and it ought to reflect kinship. The microbiologist who aspires to be a taxonomist faces special difficulties. Comparative morphology, so important in illuminating the possible common origins of many multicellular organisms, is less easily applied to microscopic animals. Palaeontology, which tells so much about evolution of the larger forms of life, gives us little information about the Protozoa. Embryology, when applied to living things that apparently consist of a single cell, is even harder to work out and interpret.

Yet closely related to embryology is the series of developmental stages which collectively make up the life cycles of organisms. A knowledge of the life histories of the Protozoa greatly aids the taxonomist. But nature has exercised a most exasperating degree of ingenuity in devising complicated life cycles that are difficult to work out. Thus many of them are still incompletely known, and the classification of many species of organisms is a matter of controversy and doubt.

Toxoplasma gondii

Toxoplasma gondii, in many ways a most fascinating organism, concretely illustrates the dilemmas that taxonomists often face. First described by the French biologists Nicolle and Manceaux in 1909 from the African gondi (a rodent), its true position in the world of living things is still undetermined. Although many scientists have since studied this organism, almost everything about it is still unclear.

Toxoplasma gondii is a crescent-shaped organism about 6 to 12 micra long and half as wide. One end is usually more pointed than the other. Few details of structure are visible under the ordinary light microscope except for a centrally placed nucleus (Fig. 26.1). Though the organisms are capable of independent movement they appear to possess no locomotor organelles, and how they accomplish the short trips they make is still something of a mystery. Under the electron microscope the morphology appears more complex, and it is possible to discern a system of some 15 longitudinal fibrils ("toxonemes") arising in the median portion of the cytoplasm and extending to the anterior, pointed end. From the latter projects a tiny knob, which is

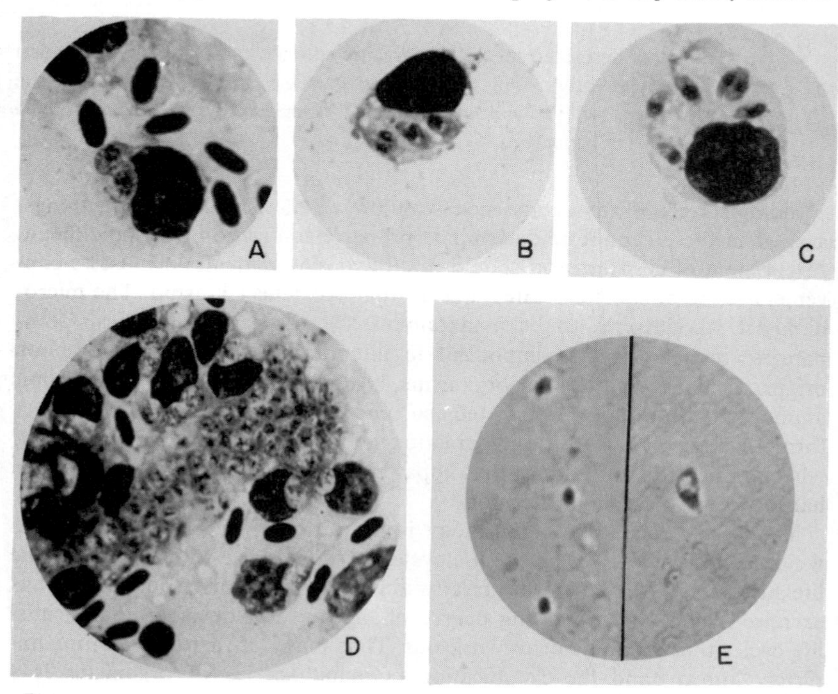

FIG. 26.1. *Toxoplasma* and *Atoxoplasma*. *A*. *Atoxoplasma* seen in a contact preparation from the spleen of an English sparrow. The host cell nucleus (lymphocyte ?) is indented by the organisms. ×1050. *B*. *Toxoplasma gondii* in a lymphocyte (?) in the spleen of a mouse. Note the organism on the lower edge, apparently undergoing longitudinal binary fission. ×1050. *C*. A cell from mouse spleen parasitized by *Toxoplasma*. Note the paired organisms, apparently from a just completed fission. ×1050. *D*. *Atoxoplasma* in the liver of an English sparrow. Probably *Atoxoplasma* multiplies by longitudinal binary fission; note some of the intracellular forms. ×1050. *E*. *Toxoplasma* seen in a Sabin-Feldman dye test. When positive (right), the parasites remain unstained in a mixture of immune serum, a heat labile factor present in normal human serum, and methylene blue. Otherwise (a negative test, left) they stain readily. ×980.

said to be the truncated extremity of a cone (called a "conoid") of unknown function (Gustafson, *et al.*, 1954).

The host range of *Toxoplasma* in nature is extremely wide and includes many species of mammals, among them man, and some birds, particularly the pigeon. Experimentally, the host spectrum is even wider, and the great majority of warm-blooded animals have been found susceptible. Whether cold-blooded forms are ever infected is questionable, although toxoplasmosis has been reported from them also. Since the source of most human infections is unknown, the frequency of toxoplasmosis in animals is of considerable potential importance. Cook (1959) remarks, "The studies of infections in animals (of Australia) led us to expect that there would probably be many infections in the human population. The results reported here have confirmed that expectation. . . ."

The effects of the parasite on the host vary greatly, but on the whole the organism is not very pathogenic. Natural infections in mammals and birds are usually demonstrable only by subinoculation of tissue suspensions (preferably brain) into laboratory animals, of which the white mouse is probably the most susceptible (p. 578). Positive results are indicated by the development of an acute infection, usually within a few days and with a fatal termination.

Just how the organism injures its host is uncertain. Despite the name,* which suggests the ability to produce a toxin, no toxin has yet been demonstrated. *Toxoplasma* is normally intracellular, and seems to prefer cells of the reticulo-endothelial system. During the acute stage of the disease, and perhaps at infrequent intervals thereafter, it is present in the circulating blood, probably in the leucocytes. Apparently it has a special predilection for the nervous system, and here it may survive for a long time in what have been called "pseudocysts," since it was thought that they might represent aggregations of parasites surrounded by the remains of the invaded cell. Recently, however, van der Waaij (1959) has adduced evidence that the parasites themselves form the cyst, and suggests that some of them from time to time leave the cyst and invade neighboring cells, there to multiply and eventually give rise to new cysts. Such cyst formation is especially common in the brain, which seems to have less ability to resist the infection than do other tissues. It is said that the cell nucleus is especially susceptible to injury by the parasites.

Toxoplasmosis in the clinical form is rare in man, but asymptomatic infections are common. Surveys in various parts of the world for the presence of antibodies have shown that the proportion of persons reacting positively increases with age, until in middle life it reaches or exceeds 40 per cent in some areas. Thus it seems that we are constantly exposed to infection by the parasites, and that many of us either develop subclinical toxoplasmosis (as we do tuberculosis and fungal diseases such as histoplasmosis), or we

*The name "Toxoplasma" actually derives from a Greek word meaning "bow" and refers to the crescentic shape of the organism.

are stimulated to produce antibodies because of frequent contact with the causative organisms in our environment.

Probably clinical toxoplasmosis results only when a nonimmune individual is exposed to an unusually large dose of parasites, or a genetically induced predisposition may be necessary. In general, the most severe cases of toxoplasmosis are seen in infants during the first few weeks of life. In them the disease is congenital, apparently acquired *in utero* or at birth from the mother—although the mothers of these children almost always appear to be in good health. No effective treatment is known, nor any practical way of predicting when congenital toxoplasmosis will occur. Fortunately, such cases are uncommon, for even when not fatal permanent damage to the brain and eyes often results. The actual frequency of ocular toxoplasmosis is uncertain, for the role of *Toxoplasma* in the production of eye disease has only recently been recognized. Cook (1959), in a study of *Toxoplasma* in Queensland, states that there is probably "an appreciable amount of ocular and glandular toxoplasmosis in the community, particularly in the young" and there is reason to think that Australia does not differ particularly in this respect from the rest of the world.

Aside from congenital transmission, due probably to parasites in the circulating blood—perhaps liberated for unknown reasons from some pseudocyst—the means by which *Toxoplasma* is spread in nature is unknown. Nurseling mice may acquire the disease from their mother's milk if she has an acute infection. Experimental infections in various animals may be produced occasionally by feeding flesh containing the organisms. It has been suggested that human infections may originate from eating infected pork (pigs are often infected), but toxoplasmosis occurs in persons (e.g., orthodox Jews) who deny ever having consumed pork. Even close contact between sick and healthy animals in the same cage usually fails to transmit the parasite, at least under laboratory conditions. The difficulty of accounting for the frequency of toxoplasmosis in nature is increased by the apparent fragility of the organisms. They are quickly destroyed by physical agents such as heat and drying, and even by mild disinfectants. Spore stages seem wholly absent, and the cysts or "pseudocysts," though able to survive in tissues considerably longer than the vegetative stages, are still much less resistant than protozoan cysts usually are.

Once in the body, *Toxoplasma* reproduces by a process generally agreed to be longitudinal binary fission, but claimed by Goldman *et al.* (1958) to be internal budding stemming in two progeny. Since they regard this method of multiplication as different from anything previously described for the Protozoa they have suggested that it be called "endodyogeny." Although schizogony has also been reported, there is no convincing evidence that it ever occurs. Nor are stages known other than the typical crescents.

It also appears that there are no intermediate hosts. This makes the naturally acquired cases of toxoplasmosis often seen in herbivorous animals (such as sheep, in which the disease may be common) especially hard to explain. Experimental attempts to infect various blood-sucking arthropods have shown that the parasites may survive a few hours, a few days, or even as long as a month in certain ticks. Yet there is little actual evidence that *Toxoplasma* ever reproduces in them, and no real reason to think that they ever transmit the infection in nature.

Successful cultivation of parasitic microorganisms has often led to important discoveries about their physiology and life cycles, but so far *Toxoplasma* has been grown only in tissue culture. Nothing like true cultivation on artificial media has yet been achieved. Since it is probably an obligate intracellular parasite (even though often seen free in tissue fluids such as peritoneal exudate), it may be some time yet before a suitable culture medium is devised.

Thus what we really know about *Toxoplasma* is largely negative, and the organism clearly fits nowhere in the conventional scheme of protozoan classification. Since it does not have locomotor organelles we cannot place it among the Sarcodina, Mastigophora, or Ciliophora. And with an apparently simple life history it hardly belongs with the Sporozoa—particularly if binary fission is its only method of reproduction. Even if the mode of multiplication should turn out to be the very simple type of internal budding dubbed endodyogeny, *Toxoplasma* will remain a very atypical sporozoan, if indeed it is one at all. Its remarkable lack of host-specificity is typical of the fungi, but the fungi, like most Sporozoa, usually have complex life cycles. Possibly its life history is more complex than present knowledge indicates, but intensive search has so far failed to reveal it, and its discovery is still a parasitologist's challenge.

Sarcocystis

Perhaps related to *Toxoplasma* is the genus *Sarcocystis*. The Sabin-Feldman dye test to detect antibodies to the former may also react positively to the presence of the latter. Like *Toxoplasma*, *Sarcocystis* is a parasite whose biological nature remains in doubt, although there is some evidence, still unconfirmed, that it is in reality a fungus. It is, however, usually accorded a place in most protozoology texts, sometimes to the extent of being placed in an order, Sarcosporidia, all by itself.

Sarcocystis is a genus of muscle parasites having a good deal in common with *Toxoplasma*. Both genera are small, and may contain a single species each (*Toxoplasma gondii* and *Sarcocystis miescheriana*, Kuhn, 1865); this, however, is less probable in *Sarcocystis*. Both show a conspicuous lack of host specificity. Indeed, *Sarcocystis* infects not only many species of warm-blooded animals (including, though rarely, man), but some reptiles. Mice make good laboratory hosts, and are quickly killed by it.

Sarcocystis invades both skeletal and cardiac muscle, forming tubes ("Miescher's tubes") large enough to be faintly visible to the unaided eye as whitish streaks (Fig. 26.2). These filamentous tubes are filled with myriads of minute banana-shaped spores called "Rainey's corpuscles." The spores somewhat resemble, both in appearance and behavior, the crescents of *Toxoplasma*.

Despite the occasional presence of almost incredible numbers of parasites (Scott mentions "an old lean ewe in which it was estimated there were 88,000 sarcocysts per cubic centimeter, or over 1,400,000 per cubic inch of heart muscle"), sarcosporidiosis ordinarily results in no great harm to the

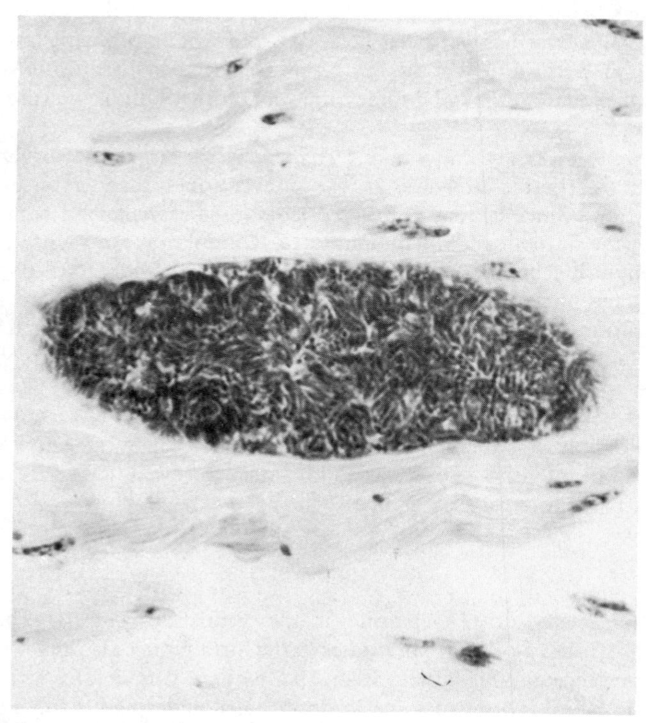

FIG. 26.2. *Sarcocystis* from the monkey, ×570. The large black body in the center is a cross section of a cyst ("Miescher's tube") filled with many minute crescentic spores. Sometimes known as "Rainey's corpuscles," the spores are contained in compartments into which the interior of the cyst is divided; these may also be faintly seen in the microphotograph.

host. Nevertheless, it does have some economic importance, particularly among sheep on the western plains of the United States. "Any flock that is not doing well is almost sure to be heavily infected with sarcocysts" (Scott, 1920). Apparently, the infections are acquired when spore-contaminated food and water are ingested.

The life history of *Sarcocystis* is not yet fully known. According to Scott

(1943), who made a long-term study of the genus, the spores of the *Sarcocystis* of the sheep become tiny amoeboid forms ("sporoblasts") after invading a muscle fibril. There they multiply by binary fission, and their presence stimulates the neighboring tissues to produce a wall about them, leaving the parasites within a structure known as a "sarcocyst." Within this envelope reproduction continues, and secondary walls produced by the parasites further divide the cyst into chambers. When the cyst walls break down, as apparently occasionally happens, crops of spores are liberated. Many of them eventually reach the blood stream and then the intestinal lumen, from which they pass with the host's feces.

Spindler and Zimmerman (1945) give a different account of the life cycle of the *Sarcocystis* of the pig. If confirmed, it means that this parasite is a fungus allied to the common mold *Aspergillus*. They observed the development of mycelia and hyphae—typical fungoid structures—from spores placed in sterile dextrose culture solution, and they then produced typical sarcosporidiosis in young pigs fed or injected with the cultured parasites.

Thus *Sarcocystis* presents the biologist with problems of much the same kind as does *Toxoplasma*. Is there one species, or perhaps many? What is is its life cycle? If it is simple, and if reproduction is by binary fission, it hardly belongs with the Sporozoa. Yet it has no locomotor organelles, even though motile, and certainly does not fit in any other major group of Protozoa.

Globidium and Besnoitia

Perhaps allied closely to the two genera we have been discussing are two others, *Globidium* and *Besnoitia*. Parasites found in the intestinal villi of a horse were described under the name of *Globidium* by Flesch in 1884, and since then species from a number of hosts, among them the kangaroo, wallaby, wombat, and even watersnake, have been added. The genus *Besnoitia* was created in 1913 by Henry to receive certain parasites found in dermal and internal lesions of cattle in the French Pyrenees. These organisms had at first been thought to be a species of *Sarcocystis*, and at various times since have been assigned to the genera *Gastrocystis* and *Globidium*, as well as to *Besnoitia*, where they are presently placed (Jellison, 1956).

Globidia occur in the skin as well as in the internal organs of their hosts. The cysts formed can cause enormous hypertrophy of the host cells, sometimes resulting in a thousandfold increase in their volume. By some (e.g., Jirovec, 1953; Grassé, 1953; Frenkel, 1956), *Globidium* is regarded as a genus of coccidia. But others, (Jellison *et al.*, 1956) group it with *Toxoplasma*, *Sarcocystis*, and *Besnoitia*. Wenyon (1926) comments on the similarity of the spores of *Sarcocystis* and *Globidium*. Here again what is needed is more knowledge of the life cycles of the various species assigned to the genus.

Recently a new species of *Besnoitia* has been described from an Idaho field mouse. Fortunately, it is easily transmissible to white mice, and therefore conveniently studied in the laboratory. In this host its morphology and behavior are strikingly like *Toxoplasma*, as Frenkel (1953) noted; it is possible that *Besnoitia* is responsible for some of the naturally occurring infections in animals ascribed to *Toxoplasma*. Like the latter parasite, *Besnoitia jellisoni* (as the species from the Idaho field mouse is called) produces severe and frequently fatal infections in laboratory mice; whether it also does so in their wild Idaho counterparts is not known. Hamsters, rats, and some other rodents are also susceptible to experimental infection; in hamsters ocular lesions are often a complication, as in toxoplasmosis.

The original and still the only other species of *Besnoitia*, *B. besnoiti*, apparently has a wide distribution. It has been found not only in France but in other parts of Europe, in North and South America, and in Australia. Jellison (1956) refers to it as "an important parasite of domestic cattle," although Morgan and Hawkins (1948) fail to mention it at all in their *Veterinary Protozoology*. In cattle and horses it gives rise to a generalized infection.

Dissemination presumably occurs through dispersal of the feces of infected animals. The intestinal lesions are doubtless a ready source for such contamination.

Despite the great similarity of the reproductive forms of *Toxoplasma* and *Besnoitia jellisoni* (and hence, doubtless, *B. besnoiti*) they are stated by Frenkel (1956) to be "serologically and immunologically distinct." The pseudocysts of the former also differ from the cysts of parasites of the latter genus. Since both *Toxoplasma* and *Besnoitia* multiply by binary fission (or, according to Goldman *et al.*, by internal budding, with two offspring per parent cell), and *Globidium* is said to reproduce by schizogony, real and important differences separate the two genera.

Encephalitozoon and Pneumocystis

Two other genera about which there has been and still is much confusion are *Encephalitozoon* and *Pneumocystis*. The first was created for certain supposed microorganisms found in the brains of rabbits suffering from encephalitis. They occurred in groups, each surrounded by an envelope thought to be the remains of the host cell, and bearing a strong resemblance to the pseudocysts characteristic of *Toxoplasma*. To the second genus, *Pneumocystis*, have been assigned a number of species of Sporozoa observed in a variety of mammalian hosts, among them, rodents, rabbits, sheep, and man. It is so named because the organisms were first seen in cystlike aggregations in the lungs of Parisian rats.

There is at present considerable doubt whether the bodies given the generic name *Encephalitozoon* are really parasites at all, and an even greater skepticism as to their protozoan or sporozoan nature. They have been provisionally classed with the Cnidosporidia, but without a convincing demonstration of the characteristic spores fitted with polar filaments. Grassé (1953) suggests that the supposed organisms may actually represent cytoplasmic abnormalities caused by the presence of a virus. There has also been occasional confusion as to the possible identity of *Encephalitozoon* and *Hepatozoon*, but the latter genus is generally accepted as a member of the coccidian group Adeleoidea.

Pneumocystis is an interesting genus of parasites presenting much the same problems as *Toxoplasma*. Like the latter, it was first described from a rodent (in this case the guinea pig), in which it occurred in the form of pulmonary cysts. Chagas, the eminent Brazilian biologist who made the discovery in 1909, thought he was seeing stages in the life cycle of *Trypanosoma cruzi*, with which his guinea pigs were also infected. He considered his opinion confirmed two years later when he made a similar observation on the tissues of a patient dead of Chagas' disease. But the parasites have since been seen in a number of other species of mammals, though not always in the lungs, and they were finally recognized by the Delanoës of France as a distinct species, to which was given the name *Pneumocystis carinii*.

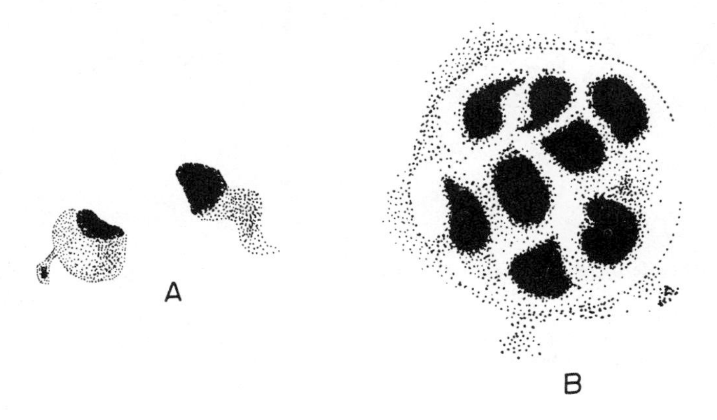

FIG. 26.3. *A*. Two isolated organisms, *Pneumocystis carinii*, from a lung smear. *B*. A cyst with eight spores. Note the clear space around the latter indicating the cyst wall. All but the smallest forms are surrounded with mucoid capsules. (Redrawn from a microphotograph after Vaněk, Jirovec, and Lukeš, *Ann. Paed.*, 180: 1.)

Pneumocystis is important, for it was recently shown to be a frequent cause of pneumonia, especially in premature infants. Only rarely has it been diagnosed in adults. In fatal cases, the alveoli are found filled with foamy material at autopsy, with most of the lung completely airless. Aggregations of parasites occur in the vacuoles of the foamy masses, usually as small cysts, each with eight spores (Fig. 26.3). According to Jirovec and others, reproduction of the individual parasites is always by binary fission. The organism secretes a mucoid envelope about itself and divides. Eventually an additional cyst wall is secreted, and after three divisions the characteristic cyst with its eight spores results. Figure 26.4 shows a hypothetical life cycle as suggested by Vaněk *et al.* (1953).

Although Grassé (1953) regards *Pneumocystis carinii* as a species which should probably be assigned to the coccidian genus *Klossiella*, there is as yet too little evidence for such disposition. Unlike the coccidia, all stages are apparently extracellular, and schizogony does not occur; moreover, sexual forms are unknown. Jirovec places it provisionally with the Haplosporidia, but as he himself remarks, this is itself an ill-defined group.

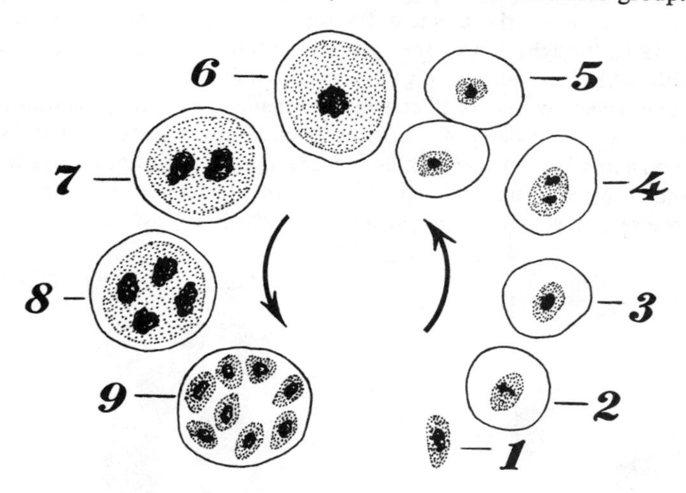

FIG. 26.4. The life cycle of *Pneumocystis carinii* (after Vaněk, Jirovec, and Lukeš). *1–5.* Stages multiplying by binary fission; during this period the parasites are surrounded by mucoid envelopes which contribute to the foamy material filling the alveoli and bronchioles of the victim of *Pneumocystis* pneumonia. *6–9.* Sporogony with the production of a cyst containing eight spores. (After Vaněk, Jirovec, and Lukeš, *Ann. Paed.*, 180:1.)

ATOXOPLASMA

Still another genus that somewhat resembles *Toxoplasma* and was for a long time thought to be a species or variety of it, is now known as *Atoxoplasma** (Garnham, 1950). It seems to occur exclusively in birds, and was formerly called "avian *Toxoplasma*" (Fig. 26.1). It shares with *Toxoplasma* a crescentic shape and an intracellular habitat, being most often seen in leucocytes, but it differs in not being transmissible either to other birds or to mammals.

Natural infections of this parasite are usually very light, only an occasional

*Lainson (1959) believes that *Atoxoplasma* is the same parasite described by Labbé in 1899 under the name *Lankesterella*. It would thus be a coccidian. He has also described what he believes to be the life cycle as seen in the English sparrow (*J. Protozool.*, 6: 360).

parasite being seen in blood or lung smears, but parasitemia may rise to rather high levels in birds kept in captivity for a time, for example, English sparrows.

Nothing whatever is known about the life cycle of *Atoxoplasma*, nor of its mode of transmission. Reproduction appears to be by binary fission, but it is not certain that this is the only method. Both its geographic and host range seem to be wide. Coulston (1942) suggested that these parasites might indeed represent coccidia which had invaded the blood stream from their primary focus in the digestive tract, but he had little real evidence other than the coexistence of the two types of infection; Garnham noted that atoxoplasmosis might exist without coccidiosis. However, Box (still unpublished) has apparently shown that Coulston's earlier hypothesis is correct, and that the parasites called *Atoxoplasma* are actually stages of *Isospora*.

SERGENTELLA

The genus *Sergentella*, containing certain little-known human parasites, illustrates how problems of life cycle, mode of reproduction, and even biological kinship may remain unsolved for a long time. This genus was created some fifty years ago by the French parasitologist Emile Brumpt to receive certain peculiar parasites discovered in the blood of an Algerian patient by the brothers Sergent, medical friends of Brumpt's at the Pasteur Institute. The organisms resembled minute round worms more than anything else, but they showed little structure and even lacked a clearly defined nucleus. Each organism took the form of a rather closely coiled spiral (Fig. 26.5), and was no longer than 40 micra and no wider than 1.5 or 2 micra.

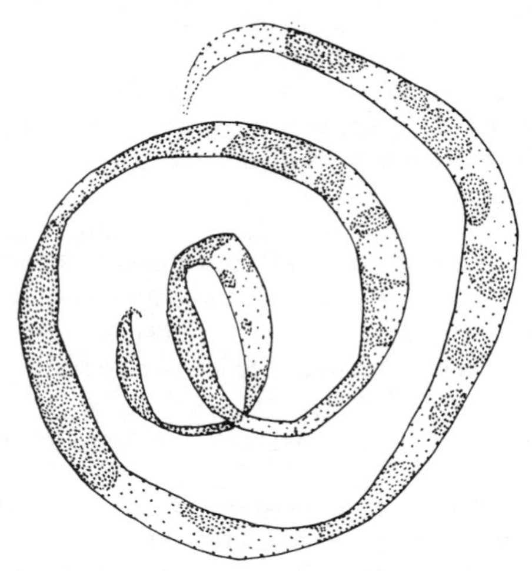

FIG. 26.5. *Sergentella spiroides*, a human parasite found in the pus from abcesses and in sputum. It varies in length (48–116μ) and in breadth (1.2–2μ). (After Jirovec.)

Little more than these bare facts about its morphology have been learned in the half-century since the original observation, but the organism has been seen in a number of human cases, all in central Europe, and also in a chimpanzee. Three species names have been proposed: *Sergentella Yakimoffi* Perekropoff, 1925, and *S. spiroides* Jirovec, 1953, for the parasites seen in man, and *S. anthropitheci* Deschiens, 1927, for those of the chimpanzee. Whether these species are in fact distinct is uncertain.

The microorganisms are evidently not pathogens. The patients in whom they were seen had a variety of maladies, such as malaria, tuberculosis, other febrile conditions, and anemias. Usually only one or a few parasites were seen in the blood and pus. Conceivably the organisms may have been contaminants of some sort, or secondary invaders. "In any case," says Jirovec (1953), who has made a more intensive study of them than anyone else, "it seems to us that this organism deserves the attention of microbiologists as well as clinicians."

The biological position of *Sergentella* is more obscure than that of the parasites discussed previously, for less is known about the organism. Certainly its morphology gives few clues, although it is possible we are dealing with an aberrant spirochete.

Spirochetes

Thus we come to the question of the protozoan affinities, if any, of the spirochetes and the Protozoa. Both bacteriologists and protozoologists have claimed the spirochetes, which were first seen by Leeuwenhoek. A part of the confusion undoubtedly stems from the work of the justly famous, though often mistaken, German microbiologist Schaudinn, who was convinced that the spirochetes were Protozoa. Wenyon (1926) devoted a section to the spirochetes in his monumental *Protozoology*, despite his conviction (p.1235) that "It is perfectly clear that the spirochetes are not Protozoa. . . ."

Yet it is also clear that the spirochetes, though they differ greatly from most Protozoa, do share with them certain attributes. They are motile, generally difficult to cultivate (as contrasted with most bacteria), and form no spores. The parasitic types among them stimulate the development of an immunity of the "premunition" type, at least in their vertebrate hosts, just as do most parasitic Protozoa. These types are also in many cases transmitted by arthropod vectors, unlike the great majority of bacterial agents of infection.

Nevertheless, the spirochetes, though exhibiting much variety, seem to have more in common with the bacteria; they lack, as the latter do, visible organelles (although the electron microscope has revealed flagella in some species), antero-posterior polarity and dorso-ventral differentiation, proved sexual stages, and a definite nucleus. Reproduction is by simple transverse fission. In general the life history seems to be simple, though there may sometimes be produced minute granular bodies from which (as from spores) the normal vegetative forms may be reconstituted.

Although it is not difficult to list characteristics the spirochetes lack, it is less easy to give a description fitting the entire group. Most are flexible spirals, the organism having a cylindrical cross section, but in some the helix consists of many tightly wound turns, while in others they are few and loose. The length of the coils ranges from a few to several hundred micra. There may or may not be a longitudinal membrane ("crista") or an axial filament, or septa dividing the organism transversely. This last characteristic has led to some speculation that the spirochetes may have been derived from the septate algae.

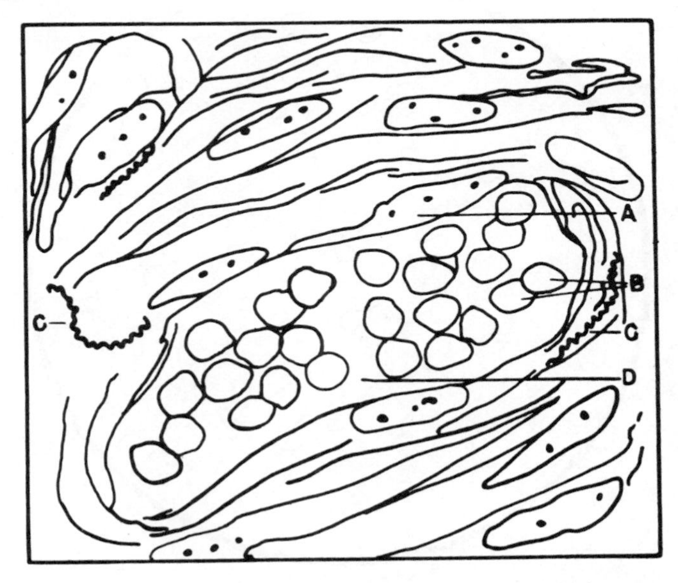

FIG. 26.6. *Trepanema pallidum*, syphilis spirochete localized in wall of mouse blood vessels. *A*, nucleus of endothelial cell; *B*, red cells in the vessel lumen; *C*, spirochetes; *D*, vessel lumen. (After Vaisman, *Annales de Dermatologie et de Syphiligraphie*, 10: 651.)

Spirochetes are extremely common in nature, and the free-living species are often found associated with Protozoa in stagnant waters containing organic material. Parasitic species probably constitute a minority, yet spirochetal infection is common in both vertebrate and invertebrate hosts (man included). Syphilis, caused by the spirochete *Treponema pallidum* (Fig. 26.6), is venereally or congenitally transmitted and occurs throughout the world. Yaws, caused by a closely related spirochete *Treponema pertenue* is a scourge in many tropical countries. It is acquired through direct contact with an infected individual, or from the bite of a small fly of the genus

Hippelates, which acts as a mechanical carrier. The relapsing fevers, the etiological agents of which may be any one of a number of different species of *Borrelia*, are transmitted by ticks (usually of the genus *Ornithodorus*), or by body lice. All these species of *Borrelia* (Fig. 26.7) are believed to have originated as tick parasites, becoming secondarily adapted to life in man and other vertebrates.

Some 30 known species (sometimes called varieties or "serotypes," since serological tests are usually required to distinguish between them) normally occur in the lower animals but are capable of causing disease in man. Some

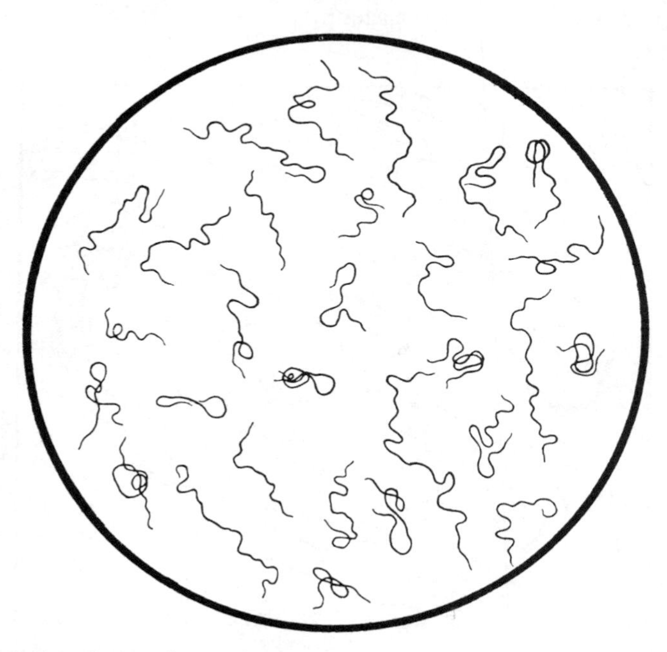

FIG. 26.7. *Spirochaeta cobayae* of the guinea pig as seen in a blood film. The blood spirochetes form a large group, and the various species often strongly resemble one another in morphology though they differ in host specificity and immune reactions of the host. *Spirochaeta cobayae* is said to be indistinguishable in appearance from *S. recurrentis* (also known as *Borrelia recurrentis*), the cause of relapsing fever in man. (After Knowles and Basu, *Ind. J. Med. Res.*)

of these are listed in Table 26.1 below. Others have their natural habitats in fresh or salt water, in mud and in sewage. Doubtless minimal physiological change enabled such naturally occurring saprophytes to transfer their residence to the body fluids and tissues of vertebrates, where food and shelter are more easily obtained. Many parasitic leptospiras can survive, for some weeks at least after leaving their hosts, in urine-contaminated water and mud. Apparently human infections are acquired by contact with such sources of contagion, the organisms entering the body through any minute abrasion, or possibly even through the unbroken skin or mucous membranes.

TABLE 26.1

SPIROCHETES OF ANIMALS TRANSMISSIBLE TO MAN*

Spirochetes (Leptospira)	*Animal Hosts*	*Human Disease*
L. autumnalis	rats	Hasami fever
L. bovis	cattle	Cattle leptospirosis
L. canicola	dogs	Canicola fever
L. grippotyphosa	mice; other rodents	Swamp fever; mud or slime fever
L. hebdomadis	mice; other rodents	Seven-day fever
L. icterohaemorrhagiae	rats	Weil's disease
L. pomona	cattle; horses; swine	Swineherd's disease

*To this list should perhaps be added *Spirillum minus* (which is, however, a spiral bacterium, distinguishable from true spirochetes by its rigid body form). This organism is transmitted to man when infected rats bite.

A species of spirochete known as *Treponema vincenti* seems incapable of causing disease except in association with certain peculiar bacteria known as "fusiform bacilli." The two kinds of organisms frequently set up an invidious partnership, with resulting severe mouth infections.

But spirochetes are significant not only because many of them are responsible for human illness. Some of the animal diseases listed in Table 26.2 are

TABLE 26.2

COMMON SPIROCHETAL DISEASES OF ANIMALS*

Spirochetes	*Animal Hosts*	*Disease*
Leptospira canicola	Dog	Stuttgart disease
Leptospira icterohaemorrhagiae	Dog; rat; cattle	Yellows (in dog)
Leptospira pomona	Cattle; pigs	Bovine leptospirosis Porcine leptospirosis
Leptospira bovis	Cattle	Bovine leptospirosis
Leptospira bataviae	Cat; rat; mouse	
Spirochaeta cobayae	Guinea pig	
Borrelia anserina (many strains)	Chickens, turkeys, ducks, geese, grouse, and others	Fowl spirochetosis

*Some of these spirochetes also have invertebrate hosts, particularly ticks. Fowl spirochetosis, for example, is commonly transmitted by the tick *Argas persicus*, but some epidemics have apparently been due to mosquitoes.

of economic importance and these are only a few of many. Among them are the avian spirochetoses, highly fatal to ducks, chickens, and geese. Wild animals also often fall a prey to spirochetal infections.

Quite apart from their considerable role in the etiology of human and animal disease, the whole group of spiral organisms loosely termed spirochetes is a challenge to microbiologists. Clearly these little living corkscrews occupy a unique position in nature, differing from bacteria and Protozoa alike, and yet having something in common with each. Yet the spirochetes exhibit so much variety among themselves that perhaps they constitute a rather artificial grouping.

BABESIIDAE, THEILERIIDAE, AND ANAPLASMIDAE

Babesias, theilerias, and anaplasms are all of importance, because of the serious diseases many species cause in domestic and wild mammals all over the world. Among cold-blooded vertebrates a perhaps related genus, *Dactylosoma*, is not uncommon in frogs and toads, and occurs also in fish.

The first two of the three families are quite certainly Protozoa, but there is question about the nature of the anaplasms. However, all infect the red cells, and often cause severe anemia—frequently resulting in hemoglobinuria. All are also transmitted by ticks, in which some may pass from one generation to the next in the egg.

The babesias (or piroplasms, as they are often called) are relatively large organisms, though still minute. Some species rather resemble the malaria organisms, but do not produce pigment from hemoglobin as do the latter. Both reproductive stages and gametocytes occur in the erythrocytes.

The theilerias, intermediate in size between babesias and anaplasms, reproduce in the endothelium of blood vessels of the internal organs, and occur in red cells only as gametocytes; these also produce no pigment.

Anaplasms are very minute. In Romanowsky-stained preparations viewed under the light microscope, little structure is visible; they seem entirely devoid of cytoplasm—hence the genus name. Usually they appear as tiny, roughly circular bodies less than a micron in diameter, located near the periphery of the infected red cell.

Because of their apparent simplicity, the anaplasms have been thought to be bacteria, Protozoa, or (because of the possible presence of viruses within the cell) cellular inclusions. They have also been considered abnormalities of erythrocytes resulting from associated disease. Nevertheless, they have in the past generally been assigned to the Protozoa.

Little is known of their life history. Reproduction often appears to be by binary fission, but under an electron microscope De Roberts and Epstein (1951) saw them "breaking up" into numerous small particles. Perhaps we might call this "submicroscopic schizogony."

Babesias appear as piriform (from the Latin *pirum*, meaning "pear") bodies within erythrocytes. Reproduction is apparently by fission, but recent study with the electron microscope indicates that budding may also occur (* Rudzinska and Trager, 1962). As in the case of malaria parasites, minute fragments of the host cell are ingested by phagotrophy.

Babesias are of historic importance because Smith and Kilborne in 1893 showed that *B. bigemina* is transmitted by ticks (see p. 600), the first demonstration of such a role by arthropods. Within the tick some of the parasites are destroyed by phagocytosis, but others (gametocytes) give rise to gametes, and fertilization follows. The zygote (ookinete) penetrates the intestinal wall, and then the ovule. Subsequent development occurs in the embryonic stages of the tick after hatching, and eventually the parasites enter the salivary glands of the adult; these are infective for cattle, the definitive host.

Theilerias (if *T. dispar* is typical) undergo schizogony in cells of the reticulo-endothelial and lymphatic systems, and then enter the endothelium of blood vessels. Forms destined to be gametocytes invade the red cells, and if ingested by ticks, continue development much as described for *Babesia*.

Much still remains unknown about all of these groups. The life-cycles of very few have been adequately studied. There is no doubt that many species remain to be discovered, and host-specificity (potentially important) has been little investigated.

REFERENCES

Cook, I. 1959. *Toxoplasma* in Queensland. IV. Serological surveys of the human population. *Aust. J. Exp. Biol. and Med. Sci.*, 37: 581–92.

Coulston, F. 1942. The coccidial nature of "avian *Toxoplasma*." *J. Parasitol.*, 28: 16.

Davis, G. E. 1948. The spirochetes. *Ann. Rev. Microbiol.*, 2: 305–34.

De Roberts, E. and Epstein, B. 1951. Electron microscope study of anaplasmosis in bovine red cells. *Proc. Soc. Exp. Biol. and Med.*, 77: 254–8.

Frenkel, J. K. 1953. Infections with organisms resembling *Toxoplasma*, together with the description of a new organism *Besnoitia jellisoni*. *Atti VI Cong. Intern. Microbiol.* (Rome) 5: 426–34.

——— 1956. Pathogenesis of toxoplasmosis and of infections with organisms resembling *Toxoplasma*. *Ann. N.Y. Acad. Sci.*, 64 (Art. 2): 215–51.

Garnham, P. C. C. 1950. Blood parasites of East African vertebrates, with a brief description of exoerythrocytic schizogony in *Plasmodium pitmani*. *Parasitology*, 40: 328–37.

Goldman, M., Carver, R. K., and Sulzer, A. J. 1957. Similar internal morphology of *Toxoplasma gondii* and *Besnoitia jellisoni* stained with silver protein. *J. Parasitol.*, 43: 490–1.

——— 1958. Reproduction of *Toxoplasma gondii* by internal budding. *J. Parasitol.*, 44: 161–71.

Gustafson, P. V., Agar, H. D., and Cramer, D. I. 1954. An electron microscope study of *Toxoplasma*. *Am. J. Trop. Med. and Hyg.*, 3: 1008–22.

Jellison, W. L. 1956. On the nomenclature of *Besnoitia besnoiti*, a protozoan parasite. *Ann. N.Y. Acad. Sci.*, 6 (Art. 2): 268–70.

———, Fullerton, W. J., and Parker, H. 1956. Transmission of the protozoan *Besnoitia jellisoni* by ingestion. *Ann. N.Y. Acad. Sci.*, 64 (Art. 2): 271-4.

Jirovec, O. 1953. La pneumocystose, une nouvelle maladie des nourrisons causée par le parasite *Pneumocystis carinii* Delanoë 1912. *Atti VI Cong. Intern. Microbiol.* (Rome) 5: 327–36.

——— 1953. *Sergentella spiroides*, parasite nouveau de l'homme. *Atti VI Cong. Intern. Microbiol.* (Rome) 5: 337–40.

Leidy, J. 1853. A flora and fauna within living animals. *Smithsonian Contrib. Knowl.*, 5: 1–67.

Manwell, R. D. and Drobeck, H. P. 1953a. Toxoplasmosis. *Sci. Am.*, 188: 86–92.

——— 1953b. The behavior of *Toxoplasma*, with notes on its taxonomic status. *J. Parasitol.*, 39: 577–84.

Morgan, B. B. and Hawkins, P. A. 1948. *Veterinary Protozoology*. Minn.: Burgess.

Poisson, R. 1953. Sporozoaires incertaines. Superfamille des Babesioidea nov. In Grassé, P.-P., *Traité de Zoologie*, Paris: Masson.

——— 1953. Protistes parasites, intra- ou extracellulaires, d'affinités incertaines. In Grassé, P.-P., *Traité de Zoologie*. Paris: Masson.

Scott, J. W. 1920. *Sarcocystis tenella*, the muscle parasite of sheep. *Univ. Wyoming Agr. Exp. Station. Bull. 124*, pp. 69–94.

——— 1930. The Sarcosporidia: a critical review. *J. Parasitol.*, 16: 111–30.

——— 1943. Life history of the Sarcosporidia, with particular reference to *Sarcocystis tenella. Univ. Wyoming Agr. Exp. Station Bull. 259*, pp. 1–63.

Spindler, L. A. and Zimmerman, H. E., Jr. 1945. The biological status of *Sarcocystis. J. Parasitol.*, 31(Suppl.): 13.

van der Meer, G. and Brug, S. L. 1942. Infection avec *Pneumocystis* chez l'homme et chez les animaux. *Ann. Soc. Belge Med. Trop.*, 22: 1–7.

van der Waaij, D. 1959. Formation, growth and multiplication of *Toxoplasma gondii* cysts in mouse brains. *Trop. Geog. Med.*, 11: 345–60.

Vanek, J., Jirovec, O., and Lukes, J. 1953. Interstitial plasma cell pneumonia in infants. *Ann. Pediatrici*, 180: 1–21.

Wenyon, C. M. 1926. *Protozoology*. New York: Wood.

ADDITIONAL REFERENCE

Rudzinska, M. A., and Trager, W. 1962. Intracellular phagotrophy in *Babesia rodhaini* as revealed by the electron microscope. *J. Protozool.*, 9: 279–88.

27

Methods and Techniques

"In the summer I divers times applied myself to the study of the waters lying about our town. . . ." Leeuenhoek, Letter, January 2, 1700.

Collecting and Examining Material

Luckily for the biologist studying the Protozoa, most free-living species occur throughout the world in comparable environments, and none is protected by hunting and fishing laws. The richest sources of free-living Protozoa are bogs and shallow pools containing water plants; tidal pools with vegetation support marine forms. Many species, including those tolerant of gradual changes in salinity, inhabit the brackish marshes found along sea coasts.

Some protozoan species, however, are so adapted to special environments that they are seldom found in any other. For example, Foraminiferida are almost exclusively marine, and although some species occur in brackish estuaries, they are best collected in beach sands, which sometimes contain many thousand shells per ounce. Radiolarida are also marine, but less likely to be found in sands along the shore since they tend to be pelagic. The ooze on the ocean floor often contains their shells in great quantity.

Some species of ciliates are characteristic of the intertidal zones of beaches, and other ciliates and flagellates are particularly abundant in sewage. Soils likewise contain their more or less typical protozoan fauna, although some species in this fauna are among those commonly found in sewage.

How many species of Protozoa may one expect to find in a given situation? Even though thousands of species exist, many of them cosmopolitan, a cursory study of a given sample usually discloses that some species are repeatedly and frequently encountered whereas others are harder to find. Not many localities have been carefully surveyed for their protozoan fauna, but a fresh-water pond near the Mt. Desert Island Biological Laboratory on the Maine coast is fairly typical. Here Unger (1941) counted 84 species, of which 35 were flagellates, 15 Sarcodina, and 42 ciliates. Had he not limited his study to the month of July he would doubtless have found more, and the total would have been greater had he extended his census of species to other ponds and bogs in the neighborhood. The protozoologist in search of variety will find it if he is willing to put in the necessary time and effort.

In collecting Protozoa always include some vegetation and even a little soil with each sample, and of course label the container as to source. The protozoan fauna changes gradually, both in the dominant types and the relative abundance of different species, with length of time in the laboratory. It is also affected by physical factors such as temperature and light, both in the laboratory and in natural bodies of water. Containers should remain uncovered or covered only with wire gauze or mosquito netting.

To demonstrate soil Protozoa by direct microscopic examination may not be easy, despite their nearly universal presence in the upper few inches, although patience will reveal the shelled forms. To reveal other types, culturing is usually required, and for this purpose nutrient agar and dilute hay infusion may be used (Sandon, 1927; Skinner, 1939). Some species develop rapidly in these media, some quite slowly (especially the testate forms), and there are doubtless others for which conditions are quite unsuitable.

The observation of living Protozoa is best done without resort to quieting agents, since organisms under such conditions are in an abnormal environment. Most of what we know about the behavior of these little animals has been learned by studying them in conditions as nearly as possible like those of their natural habitat. Take a drop or two of water, preferably from the surface, edge, or bottom of the sample, or close to vegetation, with a pipette or medicine dropper, place it on a slide and cover, and examine under the low power objective. The larger Protozoa and many smaller ones are then easily seen. More exact observation requires the high dry objective, but it is seldom desirable to use the oil immersion objective with a wet preparation; the viscosity of the oil tends to move the cover slip and hence to make focusing and keeping the organism in the field difficult. If it seems important to slow rapidly moving individuals (always a strong temptation) introduce a little methyl cellulose, which is relatively nontoxic, under the edge of the cover slip.

Success in studying the Protozoa demands skill with the microscope. The first essential is absolutely clean lenses. Since good resolution is more important than high magnification, always use the lowest power consistent with seeing the desired structures. Adequate illumination is necessary, and to get it requires some experimentation. Although much can be seen without a condenser, it is desirable to have an instrument equipped with one; a mechanical stage also facilitates slide manipulation. Definition is usually improved by cutting down the light somewhat, but this should be done with the iris rather than by dropping the condenser.

For some purposes, phase contrast microscopy is an advantage. Structures such as nuclei, often difficult to distinguish with the living animal, may thus become visible. But only experiment will enable such observation to be done to best advantage. Dark-field observation will also often reveal detail not readily discernible otherwise. But to be effective both these techniques require special equipment and some experience.

CULTIVATING PROTOZOA

Free-living Species

Attempts to grow free-living Protozoa in the laboratory probably began when the French amateur biologist Maupas, studying conjugation in ciliates, established mass cultures in the 1880's. But even in media thought to have been sterilized, during earlier investigations of spontaneous generation, "infusoria" appeared and this may also be called artificial cultivation, though unplanned.

Most modern researches on growth, genetics, and the physiology of Protozoa involve the use of cultures started from a single individual, a technique generally credited to Calkins. Several methods exist for such isolation, but all begin with finding individuals of the desired species in material collected in the field, and maintaining them in the form of "mass cultures" for a time thereafter in the laboratory.

The method most often used is selecting from material in a small depression slide or watch glass one or more individuals of the kind being sought, and then picking them up singly with a fine pipette under a wide-field binocular. After a little practice individual Protozoa can be fairly easily isolated this way, and you can make sure that they get into the container of culture medium by watching for them as the pipette is slowly discharged.

Suitable pipettes are made by drawing out a glass capillary in a Bunsen burner, in the manner of a melting point tube. The glass tube is allowed to soften to just the right point and is then slowly drawn out into a capillary of approximately the calibre desired. It may then be fitted with a rubber bulb or mouthpiece so that the operation can be controlled with the tongue. Johnson (1928) suggests another method of handling Protozoa, useful under special circumstances. A minute platinum spoon is made from an ordinary bacteriological needle. The first step is to flatten the end of the needle to form a circular disk about 0.2 mm. thick and 0.5 to 1 mm. in diameter, and then indent it with some blunt instrument to make a concavity just deep enough so that organisms picked up with it are covered with a film of water. Finally bend the end of the needle so that the spoon bowl is approximately a right angle with the stem. The device is used as follows:

> "A specimen desired for isolation is selected and if lying on the bottom of the dish, the instrument is lightly passed beneath it, then with a jerk it is brought to the surface by the currents produced; the spoon is now placed directly under the individual and gently raised through the surface film. In this way one can transfer active as well as sluggish animals, e.g., didinia, paramecia, amebae, etc. With a little practice specimens as small as 150 micra can be isolated under the low and high powers of the microscope."

This instrument is strong and easily sterilized, and organisms handled with it are transferred with a minimum of fluid or culture medium.

The utensils used in actual cultivation are varied, depending on the preferences of the investigator and the kind of problem. Syracuse and other types of watch glasses have been employed; the former especially are conveniently handled and take little space when stacked. Counts of Protozoa in them can be quickly made under a wide-field binocular.

Mirror plate depression slides are smaller. Among the most convenient sizes are those of plate glass, 3″ x 1″ x ¼″, with a concave depression about ⅛″ deep. Because of their limited capacity they must be kept in a container with a tight cover, which itself encloses an open water dish, so that humidity will remain high enough to control significant evaporation. Cultures maintained in slides of this kind are especially convenient for counting organisms under the binocular, and hence for keeping an accurate check of the reproductive rate.

Studies of protozoan physiology, however, usually require the use of "axenic" cultures—cultures in which the species being grown is the only organism present. For this purpose bacterial and protozoan contamination from the outside must be rigorously excluded, and the medium must be as nearly chemically defined as possible. Hence test tubes, capped either with cotton plugs or covers (both plastic and steel have been used), are employed. The latter have the advantage of preventing evaporation. *Tetrahymena*, which bids fair soon to rival *Drosophila* in its use in research, and for which a medium of precisely known composition is available, is grown in this way.

Protozoa to be grown axenically must first be freed from bacteria, fungi, and other Protozoa—the organisms of any sort usually associated with them in nature. This isolation is often rather difficult, and various methods have been used, of which that of Parpart (1928) is still one of the best.

This method requires a box large enough to hold a wide-field binocular and to allow manipulating of culture dishes and organisms. The oculars of the binocular project through an opening otherwise sealed, to exclude the contaminating dust. An opening on one side of the box is covered by a cloth curtain (or provided with loose sleeves) through which the operator can reach the interior with his hands. The Protozoa are freed from bacteria by repeated washing, being transferred from one dish of sterile medium to another by a pipette observed through the microscope. Parpart found that about ten transfers are necessary to insure sterility of the organisms, for many washings allow the Protozoa to eject and leave behind any bacterial spores they might have ingested.

A modification of Parpart's method was suggested by Kidder, Lilly, and Claff (1940). They found manipulating the watch glasses used for washing Protozoa more convenient when the former were enclosed in cellophane bags. Thus wrapped, the watch glasses could be sterilized. Observation of the Protozoa during transfer and washing was easily possible, after the organisms had been introduced into the dish through an opening in the side of the bag just large enough to accommodate the necessary pipette.

Another method of growing Protozoa involves the use in cultures of penicillin or other antibiotics, which for most species are of low toxicity, although some previous experimentation may be necessary to establish this. Streptomycin, for example, may bleach the green flagellates.

The media used for the culture of free-living Protozoa were for a long time empirical, but they are now carefully and scientifically compounded, although the older formulae (Needham *et al.*, 1937) are often still of value.

* Mugard and Rouyer's (1966) method of cultivating the ciliate *Ophryoglena mucifera* axenically is an interesting example of the modern approach. They were able to grow it in well aerated flasks, using as food the brains of *freshly* killed mice. Bacteria-free organisms to start the culture were obtained by the use of forms just excysted (which contain no bacteria), freed of surface bacteria by repeated washing, centrifugation, and migration.

Nutritionally speaking, Protozoa fall into several roughly defined groups. The autotrophs normally employ photosynthesis for the manufacture of carbohydrate and (indirectly) of other organic compounds; the saprozoic (saprophytic) species absorb needed nutrients in solution; and the holozoic forms ingest solid food, usually other organisms. Many Protozoa are capable of more than one of these modes of nutrition, theoretically making the problem of cultivation easier. An example of such a nutritionally versatile species is *Tetrahymena pyriformis*, in nature mainly a bacterial feeder, but easily grown on an artificial medium of known chemical composition (Table 27.1).

TABLE 27.1

MINIMAL CHEMICALLY DEFINED MEDIUM FOR *TETRAHYMENA**

Amino Acids			*Growth Factors*	
	L	*DL*		
Arginine	150 mg/l.	—	Thiamine·HCl	1.00 mg/1.
Histidine	110	—	Riboflavin	0.10
Isoleucine	50	100 mg/1.	Ca pantothenate	0.10
Leucine	70	—	Niacin	0.10
Lysine	35	—	Pyridoxine·HCl	2.00
Methionine	35	35	Folic Acid	0.01
Phenylalanine	50	100	Thioctic acid	0.001
Serine	90	180		(1000 units/1.)
Threonine	90	180		
Tryptophane	20	40	*Inorganic Salts*	
Valine	30	60	K_2HPO_4	100.0 mg/l.
			$MgSO_4 \cdot 7H_2O$	10.0
Carbon Source			$Zn(NO_3)_2 \cdot 6H_2O$	5.0
Glucose	1000 mg/l.		$FeSO_4 \cdot 7H_2O$	0.5
Na acetate	1000		$CuCl_2 \cdot 2H_2O$	0.5
Nucleic Acid				
Guanylic acid	25 mg/l.			
Adenylic acid	25			
Cytidylic acid	25			
Uracil	25			

*This medium will support the growth of nearly all strains of *Tetrahymena pyriformis*. The medium worked out by Kidder and associates is somewhat more complex, and includes additional compounds not essential but useful as growth stimulants. Formulation furnished by A. M. Elliott.

On the whole, the greatest success in cultivating Protozoa has been achieved with the green flagellates. Many species have been cultured on chemically defined media, the composition of several of which is given below.

Although these media may support growth and multiplication, determining the precise nutritional requirements of a protozoan species is often difficult. Impurities in reagents, trace elements from the culture tubes, and growth essentials in minute but adequate amounts from the cotton plugs closing the culture tubes complicate the interpretation of experimental results. Nature may also introduce other complexities such as differing nutritional requirements in strains within a species. These varieties in protozoan tastes may explain the need for modifications in media said to be adequate by different researchers. It is beyond the scope of this work to give detailed formulae of this sort. Fortunately the Protozoa are often adaptable and as a practical matter may be grown successfully in media of relatively simple composition, at least from a strictly empirical point of view.

Protozoa with special dietary requirements present a problem of another sort. To grow *Didinium*, for instance, we must first have vigorously multiplying cultures of paramecia. Most carnivorous Protozoa are less fastidious than *Didinium*, but they demand strictly fresh food, which means that it must be alive when eaten.

The media whose formulae are given below have been used with success by various investigators for growing protozoan species. Among the genera represented are *Euglena*, *Colacium* and *Chlorogonium* (media in Table 27.2), *Chilomonas*, *Polytoma*, and *Astasia* (media in Table 27.3), and *Paramecium*, *Colpidium* and *Glaucoma* (media in Table 27.4).

TABLE 27.2

HALL'S MEDIUM FOR CHLOROPHYLL-BEARING FLAGELLATES
(Hall, 1937)

	Media*					
Constituents	A	B	C	D	E	F
NH_4NO_3 ..	0.5 gm		1.0 gm			
KNO_3 ..		0.5 gm			0.5 gm	
Tryptone ..		2.5 gm		2.0 gm	5.0 gm	
Glycine ..						1.996 gm
K_2HPO_4 ..						0.209 gm
KH_2PO_4 ..	0.5 gm	0.5 gm	0.2 gm	0.25 gm	0.5 gm	
$MgSO_4$ ..	0.1 gm	0.1 gm	0.2 gm	0.25 gm	0.25 gm	0.048 gm
NaCl ..	0.1 gm	0.1 gm			0.1 gm	
KCl ..			0.2 gm	0.25 gm		
$FeCl_3$ ..	trace		trace	trace	trace	
Sodium acetate		2.5 gm		2.0 gm		1.48 gm
Dextrose ..		2.0 gm				
Distilled water	1.0 liter	1.0 liter	1.0 lite	1.0 liter	1.0 liter	1.0 liter

*In general, these media work best when adjusted to pH 7.00.

TABLE 27.3

MEDIUM FOR COLORLESS FLAGELLATES

(Hall, 1937)

Constituents	Media			
	G	H	I	J
Glycine	2.0 gm			
Peptone		2.0 gm	10.0 gm	10.0 gm
Gelatin		150.0 gm		
NH_4NO_3			0.5 gm	
Dextrose	2.0 gm			
Sodium acetate	2.0 gm	2.0 gm	2.0 gm	
K_2HPO_4	0.2 gm			2.0 gm
KH_2PO_4		0.25 gm	1.5 gm	
K_2CO_3	2.0 gm			
KCl		0.25 gm		
$MgSO_4$	0.1 gm	0.25 gm	0.25 gm	
Distilled water	1.0 liter	1.0 liter	1.0 liter	
Tap water				1.0 liter

TABLE 27.4

MEDIUM FOR CILIATES

(Hall, 1937)

Constituents	Media					
	K	L	M	N	O	P
Peptone*	10.0 gm	10.0 gm	10–30 gm		5.0 gm	
KH_2PO_4			2.0 gm			
K_2HPO_4						0.02 gm
Na_2HPO_4	0.01 gm					
NaCl	0.5 gm	4.0 gm			0.003 gm	0.02 gm
KCl	0.01 gm					
$MgSO_4$	0.01 gm			0.025 gm	0.0045 gm	0.02 gm
$CaCl_2$	0.01 gm			0.001 gm		
KNO_3				0.5 gm	0.0013 gm	
$(NH_4)_2PO_4$				0.05 gm		
$FeCl_3$				0.001 gm	0.0002 gm	
$CaSO_4$					0.015 gm	
$Ca(NO_3)_2$						0.2 gm
$FeSO_4$						trace
Distilled water	1.0 liter	1.0 liter	1.0 liter	1.0 liter	1.0 liter	1.0 liter

*Difco tryptone is said to be the most generally satisfactory of the available peptones, although Difco proteose-peptone may sometimes be substituted with good results.

Recently it has been possible to cultivate one species of *Paramecium* (*P. multimicronucleatum*) on a medium whose composition is chemically known, except for a single ingredient. This one substance is the nondialysable yeast fraction ("N.D.F."). *Paramecium* is a ciliate genus of great biological interest and the formula for this medium (Johnson, 1956) is therefore reproduced in Table 27.5 below. The medium includes some three dozen ingredients and is a further illustration of the difficulty of problems involving nutritional requirements of the more highly evolved Protozoa.

TABLE 27.5

MEDIUM FOR *PARAMECIUM MULTIMICRONUCLEATUM*
(Johnson and Miller, 1956)

	Final Conc., ug./ml		Final Conc., ug./ml
$MgSO_4 \cdot 7H_2O$	40.0	Stigmasterol	2
$Fe(NH_4)_2(SO_4)_2 \cdot 6H_2O$	10.0	Ca pantothenate	2
$MnCl_2 \cdot 4H_2O$	0.1	Nicotinamide	4
$ZnCl_2$	0.02	Pyridoxal·HCl	4
$CaCl_2 \cdot 2H_2O$	20.0	Riboflavin	4
$CuCl_2 \cdot 2H_2O$	2.0	Folic acid	2
$FeCl_3 \cdot 6H_2O$	0.5	Thiamine	12
K_2HPO_4	570.0	Amino acids[a], ovalbumin, or casein	
KH_2PO_4	570.0		
Ethyldiamine tetraacetic acid	16.0	N.D.F. (nondialysable yeast fraction)[b]	
Sodium acetate	570.0		
Sodium pyruvate			

[a]Amino acids were incorporated in the medium as follows:

	Final Conc., ug./ml		Final Conc., ug./ml
L–tyrosine	50.0	L–arginine	100.0
L–phenylalanine	75.0	DL–valine	75.0
L–tryptophane	50.0	DL–serine	200.0
DL–methionine	150.0	L–proline	50.0
DL–threonine	150.0	L–alanine	25.0
DL–isoleucine	150.0	L–aspartic acid	50.0
L–leucine	150.0	L–glutamic acid	75.0
L–lysine	125.0	glycine	25.0
L–histidine	50.0		

[b]N.D.F. yeast fraction prepared as follows: A 1:4 (w/v) suspension of Fleischmann's "active" yeast (dry) was made by continuously stirring the yeast in hot (50°–70°C.) distilled water for 2–3 hours to effect a slow extraction. The suspension was then autoclaved to coagulate the proteins. The mixture was separated on a Sorvall high-speed centrifuge at maximum speed (about 14,000 r.p.m.) and a clear amber supernatant was collected. The supernatant was again autoclaved in a section of cellulose dialyzer tubing with one end left open to be tied off immediately after sterilization. The nondialyzable fraction (N.D.F.) contains the required substances which together with a defined basal medium supports sterile growth of *Paramecium*.

TABLE 27.6

MEDIUM FOR *LABYRINTHULA*[a]
(Vishniac and Watson, 1953)

	Per 100 Ml. Distilled Water		Per 100 Ml. Distilled Water
NaCl	2.5 gm	Thiamine	0.2 mg
$MgSO_4 \cdot 7H_2O$	0.5 gm	Nicotinic acid	0.1 mg
K_2HPO_4	0.01 gm	Ca pantothenate	0.1 mg
Ethylene diamine tetraacetic acid[b]	0.05 gm	Pyridoxin·HCl	0.04 mg
		Pyridoxamin·2HCl	0.02 mg
$CaCO_3$	0.025 gm	p–aminobenzoic acid	0.01 mg
		Biotin	0.5 ug.
Mn (as manganous sulfate)	2.0 mg	Cobalamin (vit. B_{12})	0.05 ug.
Fe (as ferrous sulfate)	0.2 mg	Folic acid	2.5 ug.
Gelatin hydrolysate	0.1 gm	Agar	0.2 gm

[a]The medium is boiled for 20 minutes, allowed to cool to room temperature, and the pH adjusted to 8.0–8.2 with KOH. The agar is added to furnish the physical support needed by the organism.

[b]'Versene'-free acid (Bersworth Chemical Co., Framingham, Mass.).

On the whole, less attention has been given to the problem of devising chemically defined media for the Sarcodina than for the flagellates and ciliates. Vishniac and Watson, however, have developed such a medium for the rhizopod *Labyrinthula.* One species in this genus is notorious as the parasitic agent responsible for the destruction of eel grass in widespread epidemics occurring along the Atlantic coast. This is a shallow-water grass important in the ecology of a variety of forms of coastal wildlife, and as insulating, packing, and upholstering material. The composition of the medium is given in Table 27.6. Its nutritional requirements, although exacting, are rather less so than those of *Tetrahymena* and *Paramecium*, since it demands less variety in vitamins and amino acids.

Cultures of the more common free-living Protozoa available for laboratory study are more easily achieved. Satisfactory media may be prepared from pond water and a nutrient source, such as timothy hay, water weeds, (e.g., *Elodea* and *Ceratophyllum*, often used in balanced aquaria), or grain, especially wheat and rice. Pond water is better than tap water which usually contains chlorine, or distilled water which lacks the salts and other substances needed for growth, often in minute amounts, by many Protozoa.

The exact proportions of nutrient and water are of no great consequence, except that in general too little of the former is better than too much. Three or four dozen one-inch shreds of timothy hay or a few heads boiled in a liter of water make a suitable infusion. A tea may be prepared in much the same way from shredded water weeds. If wheat or rice grains are used, a dozen or so per liter usually proves ample. They should also be boiled, but not until they burst.

Libbie H. Hyman of the American Museum of Natural History has suggested boiled lettuce leaves as a base of media in which a variety of small animals, Protozoa included, can be grown. The boiled leaves are added after the container has been filled with several inches of tap water, at the rate of about one leaf to each square foot of bottom surface, additional leaves replacing the old ones as they disintegrate and sink to the bottom.

Battery jars make convenient containers for such mass cultures of Protozoa. The medium is allowed to age for a few days or perhaps a week, so that the bacterial flora on which the Protozoa depend for food will have time to develop, and the jars are then seeded with the organisms under cultivation. Cultures should be kept at room temperature and protected from exposure to too much direct sunlight. However, the preferences of Protozoa for degrees of light and temperature vary considerably with the species.

For larger amoebae, such as *Amoeba proteus* and *A. dubia*, Dawson (1928) suggested the use of finger bowls. Material collected from the field is placed in a number of bowls and diluted somewhat each day until the original volume of the medium is about doubled. The contents of the various bowls are examined periodically for amoebae; if present, they usually become numerous enough for easy detection within two weeks, unless prevented by their ruthless enemies, the smaller crustacea.

When amoebae are present in numbers it is desirable to transfer them to Syracuse watch glasses, in which they are more conveniently observed. To each watch glass is added a few ml of distilled or sterilized pond water, plus about 0.02 gm of sterilized timothy hay. Since the development of amoebae in such cultures may at first be highly variable, it is best to start a number of cultures initially. After two weeks or so the more thriving ones may be transferred back to finger bowls with enough additional water to bring the volume of medium in each to about 30 ml, plus 0.06 gm (dry weight) sterilized timothy hay and a wheat and oat grain. Sterile water is added every few days thereafter until the total in each bowl reaches 65–70 ml, and now and then a little sterilized hay (but not more than 0.06 gm at any one time) is added. Cultures maintained in this way often last for a month or more, sometimes for many months. Apparently the condition and continued life of the cultures depend primarily on a relatively low concentration of nutrient material and the presence of ample numbers of *Chilomonas paramecium*, the ubiquitous flagellate on which the amoebae chiefly subsist.

Another species of free-living amoeba, the much less common but perhaps more interesting *Pelomyxa palustris*, may be readily cultivated in Carrel flasks containing filtered lake water and shreds of the alga *Spirogyra* (Kudo, 1957). Such cultures are kept in moist chambers exposed to diffuse daylight. They do well at room temperature (18°–24°C.).

For the cultivation of the two often encountered species of minute and colorless chrysomonads belonging to the genera *Monas* and *Oikomonas*, Hutner *et al.* (1953) found the following medium very satisfactory for routine maintenance:

powdered skim milk	0.2%
Trypticase (Balto. Biol. Lab.)	0.05%
sucrose	0.1%

These are dissolved in the proper amount of distilled water, and the pH adjusted to 6.4–5.9. Other tryptic digests of casein may be used instead of Trypticase. Cultures are kept at room temperature, and often contain numerous flagellates even after months.

Another colorless flagellate, the voracious euglenoid *Peranema trichophorum*, is fond of milk—an acquired taste! Storm and Hutner (1953) found that it would thrive on the following medium:

liquid whole milk	1.0 ml
soil extract	10.0 ml
distilled water, to make	100.0 ml

The soil extract is prepared by autoclaving garden soil 15 or 20 minutes with an equal weight of water, cooling and filtering. Cultures of *Peranema* maintained on this medium are said to be good for several months.

Two other protozoan species with dairy tastes are the swanlike *Lacrymaria olor* and highly evolved hypotrichous ciliate *Euplotes patella*. The former does well on a medium consisting of a 0.003 per cent solution of malted milk, while the latter prefers it stronger, 0.25–0.50 per cent (Hyman, 1931, quoting Ibara, 1926, and Yocom, 1918).

Hyman also states that *Paramecium* does extremely well in a medium made by boiling very briefly 60 to 70 grains of wheat in a liter of water, and allowing it to stand for a few days before the ciliates are added.

Stentors may be grown in a less concentrated medium (20 wheat grains per liter) of the same sort. Aside from bacteria, it is desirable to have rotifers and small Protozoa present since *Stentor* has somewhat omnivorous tastes.

Vorticellas may also do well in this medium, though usually only for limited periods, but in Hyman's experience water in which "a boiled spray of water plant, or a wisp of boiled hay" per liter served as nutrient material proved better.

The colonial vorticellids, such as *Epistylis* and *Carchesium*, often do well in balanced aquaria. A few grains of boiled wheat may be added to the water occasionally, but the plants essential to maintaining a balanced condition should not be allowed to decay, nor should bacterial growth ever become noticeably abundant.

Hypotrichous ciliates, such as *Oxytricha*, *Pleurotricha*, and their close relatives, will usually do well in any environment supporting heavy bacterial growth. Hyman (quoting Woodruff, 1905) suggests a medium made by boiling for one minute three grams of hay or fresh grass in 200 ml of water.

Cultivating Parasitic Species

Cultivating parasitic Protozoa has until recently received more attention than growing the free-living forms because of the greater practical importance of many parasites. The aim has usually been to discover a medium on which the organisms could be grown rather than determining their exact nutritional requirements.

The first successful medium for the *in vitro* cultivation of an important group of protozoan parasites was devised by Novy and MacNeal in 1903 and modified by Nicolle in 1908 for use with the trypanosomes and leishmanias. Known as the NNN medium in their honor, the formula is as follows:

agar	14 gm
sodium chloride (sea salt)	6 gm
distilled water	900 ml

The agar and salt are mixed with distilled water, the whole brought to a boil, tubed (about 6 ml each), and autoclaved. The tubes are allowed to cool to about 48° C., often most conveniently done in a water bath. About 2 ml of sterile, defibrinated rabbit blood is then added to each tube and thoroughly mixed. Great care must be taken to avoid bacterial contamination, which would prevent growth of the flagellates. The tubes are then slanted; further cooling is best done in a refrigerator, thus increasing the amount of condensate in which the flagellates chiefly grow. It is a curious fact for which there is no satisfactory explanation that, even though the medium has a blood base, the stages observed in culture are those which would normally develop only in the insect host. Although the NNN medium is not equally suitable for all species of trypanosomes and leishmanias, and others better adapted for certain species have been devised, it is still probably the best for general use. For a thorough discussion of mammalian trypanosome cultivation see *Tobie (1964).

The intestinal flagellates may be cultivated on the same media used for growing the amoebae of the alimentary tract. Boeck and Drbohlav's (also known as the Locke-egg-serum, or L.E.S. medium) is one of the oldest of such media and still one of the best, although *Endolimax nana* does not grow well on it and *Giardia* not at all.

This medium is prepared from the following ingredients:

4 eggs
Locke's solution (see below)
inactivated human blood serum

The eggs are washed in water and then in 70 per cent alcohol, and broken into a sterile flask containing glass beads and 50 ml of Locke's solution. The flask is then shaken until its contents are well mixed, after which the suspension is transferred to test tubes, each containing about 5 ml. These are slanted, exposed to a temperature of about 70° C. (158°F.) over a water bath until solidified, and sterilized for 20 minutes at 15 lbs. in an autoclave.

Then enough of a mixture of Locke's solution and inactivated human serum (in equal sterile proportions) is added to each tube to cover the slant to a depth of about 1 cm. (To insure the sterility of the mixture, it should be first passed through a Berkefeldt, asbestos, or fritted glass filter, and then tested by incubation).

It is also possible to substitute for the blood serum a 1 per cent solution of crystallized egg albumen in Locke's solution, sterilized by filtration, as above. Locke's solution consists of:

NaCl	9.0 gm
$CaCl_2$	0.2 gm
KCl	0.4 gm
$NaHCO_3$	0.2 gm
dextrose	2.5 gm
distilled water	1000.0 ml

Sterilization is best done in the autoclave.

Instead of Locke's solution, sterile Ringer's solution may be used. Its formula follows:

NaCl	8.0 gm
KCl	0.2 gm
$CaCl_2$	0.2 gm
$MgCl_2$	0.1 gm
NaH_2PO_4	0.1 gm
$NaHCO_3$	0.4 gm
distilled water	1000.0 ml

A convenient substitution for the mixture of Locke's solution and human blood serum is one liter of Ringer's solution to which has been added 0.25 gm of Loeffler's Dehydrated Blood Serum (Difco Laboratories, Detroit). This mixture is boiled an hour to facilitate solution, filtered, and autoclaved 20 minutes at 15 lbs.

A little sterile Chinese rice flour* added to each tube is also advantageous. Sterilize by placing a few grams in a cotton-plugged test tube and exposing to 90°C. (dry heat) for 12 hours, preferably in three installments of 4 hours each. Overheating is indicated by a change from the natural white color. The flour is introduced into the tubes from a 1-ml wide-mouthed pipette by gentle tapping. The tubes should be subsequently incubated as a check on their continued sterility.

Initial inoculation of cultures is effected by adding a small amount of feces, about the size of a pea or perhaps half a ml, if the stool is liquid, to the culture tube. Sterile precautions are of course not necessary for this.

Once started, cultures of *Entamoeba histolytica* should be transferred every 48 hours, every 72 hours for *E. coli*. In examining cultures for amoebae,

*If the medium is to be used for the cultivation of *Trichomonas hominis* and *Chilomastix mesnili*, omit the rice flour.

remember that organisms are most likely to be found at the bottom of the liquid portion of the medium, and that their numbers are often not very great.

Although the intestinal flagellates other than *Giardia* can be grown on media designed for amoebae of the alimentary tract, media of simpler composition suffice. Trichomonads are especially easily cultivated. Hogue's medium is a good one and is made from the following:

physiological salt solution (0.7 per cent)	600 ml
whites of 6 hen's eggs	

After shaking the egg whites thoroughly in a flask containing glass beads, the salt solution is added and the whole heated over a water bath for half an hour, being further shaken at frequent intervals. The mixture is then filtered through a cheese cloth, and the filtrate is passed through cotton under suction. It is then tubed, about 5 ml in each, and sterilized in the autoclave, as directed for the modified Ringer's.

Balantidium coli may be cultivated on Schumaker's medium, consisting of Ringer's solution and horse serum in the proportions of 9:1. Rice starch is essential to the growth of the ciliate and the amount used is 7 mg to 10 ml. Or the cecal content medium may be employed, in which instead of the serum a cecal content preparation, made up as described below, is used in the same proportion. Material from a pig's cecum (easily obtained at slaughter houses) is filtered, first through a sieve and then through a funnel lined with cotton. The filtrate is conveniently kept in a flask fitted with a siphon, and may be stored at room temperature, though a refrigerator is better. The pH rises gradually, reaching an end point of about 8.00. This medium is said to present less of a bacterial growth problem than when serum is used, the growth of the ciliates being limited chiefly by a decrease in pH. Below pH 6.00 multiplication is slowed, and pH 5.00 is lethal.

To study a given species of protozoan without isolating it yourself, the Culture Index of the Society of Protozoologists is useful (*J. Protozool.*, 5: 1–38, 1958). It lists the names of laboratories and investigators using different species of Protozoa and gives information regarding their maintenance. Usually cultures of these species, or in the case of some parasitic forms animals infected with them, may be had on request. For certain pathogens, federal permits may be required. These are obtainable from the Animal Inspection and Quarantine Branch, Agricultural Research Service, United States Department of Agriculture, Washington 25, D.C. In some cases, a permit must also be secured from the Division of Foreign Quarantine of the Department of Health, Education, and Welfare.

Preserving Parasitic Protozoa by Low Temperature Freezing

Low temperature deep freezing is especially useful for preserving some species of parasitic Protozoa that otherwise require frequent passage in laboratory animals or regular transfer in cultures for their maintenance. Not all parasitic species of Protozoa survive such treatment, but trypanosomes, malaria plasmodia, and trichomonads have been kept in the frozen state for long periods. Many of the organisms are killed, either in the freezing or subsequent thawing, but enough remain viable for the successful infection of laboratory animals or the seeding of cultures. The lethal effects of freezing are believed to be due to the formation of minute ice crystals in the protoplasm, but they also vary markedly with the particular species or strain of parasite.

Methods vary with the investigator, but they usually involve the immersion of thin-walled Pyrex or plastic tubes containing the organisms in blood or culture medium, in alcohol chilled to about −76°C. by the addition of dry ice. Electric deep-freeze cabinets are also used, as is immersion in liquid nitrogen. In many cases, e.g., trypanosomes and malaria parasites, mortality seems to vary inversely with the speed of the freezing and thawing process; in others, as with the trichomonads (Levine and Marquardt, 1955), slow freezing is essential, and the organisms must be further protected by the addition of about 10 per cent glycerol. Speed of freezing and thawing—thawing is accomplished by immersing the tube containing the organisms in a beaker of warm water (37°–40°C.)—is promoted by rotation, either manual or mechanical (Manwell and Edgett, 1943). Mechanical rotation is achieved by fitting the tubes to a motor-driven drink mixer, on which the agitator has been replaced by a rubber stopper in the manner of a chuck.

Success in the preservation of parasitic Protozoa depends not only on factors such as speed of freezing and thawing, and (often) the addition of glycerol, but also on maintaining a relatively constant low temperature—as close to −76°C. as possible.

The Use of Animals

Experimental animals are necessary for anything more than a superficial study of protozoan parasites, for they furnish the only natural environment of these organisms. An increasing number of parasitic Protozoa can be cultivated *in vitro*, but it is almost impossible exactly to duplicate conditions within the host animal or within the body cells which may form the immediate environment of the parasite.

Furthermore, laboratory infections are intrinsically important: they make possible firsthand observation of the course of the infection. We can thus determine the lengths of the incubation and prepatent period, the acute stage, and the latent period that usually follows in protozoan diseases. The pathological changes in the blood and organs are easily observed, as is also the manner in which the host responds to the presence of the parasite. Sometimes we can learn how the latter responds to changes associated with developing immunity in the host.

Not all species of protozoan parasites can be successfully maintained in experimental animals, and frequently it is necessary to study related species which can be so maintained. Thus it is with the malaria parasites: since human malaria can not be transmitted to any of the common laboratory animals, species of *Plasmodium* restricted in nature to other animals, such as birds, rats, and monkeys, must be used instead.

For most experimental purposes, rats, mice, hamsters, guinea pigs, and sometimes cats, dogs, and rabbits, are well suited as laboratory hosts. Birds, such as canaries, pigeons, chickens, and ducklings, are also used. Animals used for experimentation must be well treated, adequately housed without crowding, and kept under conditions preventing as far as possible chance transmission of the organisms being studied or others which might cause intercurrent infections.

Anaesthesia is not usually necessary when only injections or blood examinations are planned, but it should be used whenever pain or discomfort is likely. As a rule mammals are easily anesthetized with ether; nembutal may also be injected in appropriate amounts. Birds usually tolerate ether very poorly and for them nembutal is better. Only experience enables one to administer anesthetics with uniform success, but the skill is easily acquired.

To interpret results correctly, controls are of course absolutely necessary. So also is accurate and complete record-keeping. A good system is essential, for it saves time and permits proper evaluation of the outcome of experiments. Mimeographed report sheets are often convenient; in some recent work IBM cards have been used. To study the pathological changes resulting from parasitic infection autopsies are often necessary. Since these are of no value if post mortem changes are marked, experimental animals should be frequently observed, any that have died should be either immediately examined, or if this cannot be done, kept in the refrigerator until they can be.

Malaria in the Living Animal

Studying malaria in the living animal is more rewarding than from stained slides alone. *Plasmodium berghei* in the rat or mouse, or one of the numerous species of avian malaria, is well adapted to laboratory experimentation. Animals infected with malaria can occasionally be secured from someone using them for research, but often it may be necessary to resort to naturally occurring infections in wild birds.

Trapping wild birds is simple but generally requires a government permit. English sparrows, however, are common almost everywhere, and are often infected with malaria; they are also easily trapped and no permit is needed. For capturing them a double funnel trap (see Farmer's Bulletin 493, Dept. of Agriculture) is easily made of wire cloth, or a drop trap may be employed.

Blood smears are first made and stained (using Giemsa, Wright's, or J.S.B.), and examined under oil. If malaria parasites are seen in a few minutes of observation, 0.1 ml of blood may be withdrawn from the tarso-metatarsal vein on the inside of one of the legs, mixed with perhaps 0.05 ml of citrated saline (consisting of 0.7 gm NaCl and 2 gm sodium citrate in 100 ml of water) to prevent clotting, and injected into a canary. The most convenient site for injection is the breast muscle. An active malarial infection usually develops in a week or 10 days, although if the parasite dosage is small or the species or strain not very virulent, two weeks may be required.

Such an infection makes excellent material for the study of the behavior of the parasite, or for making slides to illustrate the life cycle; it may also be used for the demonstration of periodicity, which is so characteristic of some species of malaria parasites.

The infection itself varies considerably with the species of the *Plasmodium. Plasmodium cathemerium*, *P. relictum*, and *P. circumflexum* are common in passerine birds, including the English sparrow. Ordinarily they produce infections in which the acute stage lasts a week or ten days followed by a chronic phase (punctuated sometimes by relapses) which may endure for the life of the bird.

If you wish to follow the rise and fall in the level of parasitemia, daily slides may be made of the peripheral blood, stained, and the number of parasites per unit number of red cells (e.g., per 10,000) determined. The counts may then be plotted.

The immunity developed in malaria is of the "premunition" type, depending on the continued presence of parasites in the body. Should parasites completely disappear, little residual immunity persists. Thus the failure of attempted reinfection is evidence of a chronic infection. This is easily demonstrated by the injection (preferably intravenous) of a heavy challenge dose of parasites from an acutely infected bird. Controls develop a typical attack of malaria, while the experimental (chronic) birds show an extremely transitory parasitemia.

To maintain the strain of avian malaria, it is usually only necessary to keep a few birds with chronic infections. Active (acute) infections may then be induced in healthy birds at any time by the injection of approximately 0.25 ml of blood from a chronic case. The incubation period is likely to be shortened and positive results more certain if the parasites are introduced intravenously. For this, the tarso-metatarsal vein is usually the most convenient site. A No. 26 or 27 needle fitted to a 1 ml hypodermic syringe is required, and it often works better if the point is first bent somewhat in a microflame. This makes it possible to hold the syringe nearly upright while the needle enters the vein.

It is also possible to infect mosquitoes and study the stages occurring in them. Oocysts are readily seen on the outer surface of the gut, and sporozoites are easily demonstrated in smear preparations of the salivary glands. *Culex pipiens*, a widely distributed species, is susceptible to infection with *Plasmodium cathemerium* and *P. relictum*. It breeds in rain barrels and can be maintained in large aquaria for oviposition and the aquarium itself is covered with wire gauze. A sleeve of mosquito netting affixed to the wire gauze enables insertion of the hand for manipulating a catching tube, collecting egg masses, etc. Birds to be exposed to mosquitoes may be confined in a tiny wire screen cage and left for any time desired.

Although the account above applies particularly to avian malaria, much of it can be adapted to handling *Plasmodium berghei*. Either rats or mice are suitable as experimental animals, but the disease is not readily maintained in mice since it usually kills them. It is also not possible to secure the parasites by isolation from naturally occurring cases, since the disease is native to Africa. Avian malaria is therefore much more easily come by.

Trypanosoma lewisi

Trypanosoma lewisi is especially suitable for laboratory study. Common in the ordinary rat found about houses, city dumps, and the like, it is easily established in the white rat by intraperitoneal inoculation with a little blood from such a source. More conveniently, infected rats can often be secured from a biological supply house or from a laboratory where the trypanosomes are being used for experimental purposes.

If trypanosomes are at all abundant in the blood, a drop or two, plus enough citrated physiological saline solution to prevent clotting, is ample to produce infection when injected intraperitoneally. The incubation period lasts only three or four days, and there is then a rapid rise in parasitemia for several days, followed by an equally precipitous fall. Very seldom does the rat appear to suffer any ill effects. In nature young rats acquire the infection perhaps as an almost normal event, just as children get mumps. Rats six weeks old or less are more susceptible than older ones when laboratory infections are required.

For microscopic examination, a drop of blood is taken from the tail after snipping off a bit with a pair of sharp scissors. It is smeared on a clean slide, stained with some Romanowsky stain (e.g., Giemsa, J.S.B., Wright's) and placed under the oil immersion objective. Trypanosomes are readily spotted if present. They may also be sought in fresh preparations, made with a drop or two of blood plus enough physiological citrated saline to prevent clotting and to facilitate vision through the film.

Parasite counts can be made from stained preparations, using the average number per field or per unit number of red cells, if it is desired to study the behavior and course of the infection.

Since the immunity which *Trypanosoma lewisi* induces is strong and permanent, it is not possible to maintain the parasite in animals with chronic infections, as in the case of malaria. But the immunity is itself worth study, since its mechanism is in part an unusual one, depending first on the elaboration of a reproduction-inhibiting substance (ablastin), followed by the appearance of lytic substances which kill off the parasites.

All this can be easily demonstrated by making a series of length measurements from stained blood films prepared daily during the course of the infection. Such measurements are most conveniently done from camera lucida drawings by the use of dividers, the magnification being first known, so that the equivalent in micra can be calculated. The parasites are extremely variable in size as long as reproduction is actively going on, but when the crisis of the infection has been reached with its inhibition of further parasite multiplication, they become nearly uniform in length.

To demonstrate the existence of a very strong immunity at this stage of the infection, all that is necessary is the injection of a heavy challenge dose of trypanosomes from a rat in the acute phase. A follow-up series of blood films will be found completely negative, although controls develop a typical infection.

If desired, transmission of the parasites can also be accomplished through the agency of the rat flea, *Ceratophyllus fasciatus* (easily obtainable from wild rats), which in nature serves as the vector. In such fleas the trypanosomes undergo a rather complicated developmental cycle, with intracellular, crithidial, and metacyclic stages, the initial stages being in the cells of the stomach lining and the final ones in the lumen of the rectum (the so-called "posterior station"). The new host acquires its infection by licking the feces of the flea from the site of the wound, or by swallowing the flea itself.

Trypanosoma cruzi

Of the various species of trypanosomes of direct or indirect importance to man, *Trypanosoma cruzi* is one of the most interesting because of its relative lack of host specificity and the peculiarities of its life history. Fortunately it is easily maintained in laboratory animals and in cultures. Mice are readily infected, and their blood remains positive for a considerable length of time, but nurseling rats are more susceptible and for them the parasites are often quite pathogenic. After the age of about four weeks, however, they develop a strong age resistance and can then no longer be infected.

Since *Trypanosoma cruzi* occurs naturally only in South and Central America, and in rodents in southwestern United States, the parasites are obtained originally from a laboratory where they are used for study. Thereafter the infection may be transferred by intraperitoneal injection of parasitized blood into white mice or nurseling white rats (hooded ones are somewhat less susceptible). A drop or two of blood is easily obtained by snipping the tail or by cardiac puncture, although the latter requires some skill. An infection in animals may sometimes be induced by inoculation with the cultured trypanosomes, but this frequently fails to take. Whether the parasites are obtained from infected animals or from cultures, it is essential to remember that they are potentially pathogenic to man.

The length of the incubation period varies with the parasite dosage, but it is usually a week or less, especially in nurseling rats. If the latter are used, it is desirable to inoculate them within a day or two of birth.

Since *Trypanosoma cruzi* does not multiply in the blood, dividing forms will not be found in blood films, although two types of parasite are often to be seen there: the typical "C"-shaped trypanosomes, with a conspicuous terminal kinetoplast, and somewhat smaller, very slender organisms.

Crithidial and leishmania forms may be observed in heart smears or sections. The latter are better for the demonstration of intracellular stages.

Cultures are excellent for study of the stages normally occurring in the vertebrate host. *Trypanosoma cruzi* grows well on the usual media used for blood and tissue flagellates (see p. 569), and remains viable and active for two or three weeks without requiring transfer.

The Leishmanias

Since the leishmanias do not occur naturally in most areas, either cultures or infected animals must be procured from a laboratory using them for research. *Leishmania tropica* and *L. braziliensis* are not readily maintained in animals, but hamsters are very susceptible to infection with *Leishmania donovani*, although it is not always easy to induce the disease in them by inoculation from cultures. As in numerous other pathogens, prolonged cultivation often seems to reduce virulence. Nevertheless, in working with these parasites it is well to remember that they are potentially pathogenic to man.

Hamsters are not prone to develop age resistance to leishmanial infection, nor do they often show much natural immunity, although experimental inoculations do not always result successfully. The route of infection may be intraperitoneal (usually the most convenient), or intracardial. Other routes are of course possible. The inoculum may consist of a suspension or homogenate of spleen and/or liver in physiological saline solution, for the parasites are usually the most numerous in those organs. The actual volume of the inoculum is not important, especially if the injection is to be intraperitoneal, but it should contain enough tissue to insure the introduction of numerous organisms.

The infection in the hamster runs a varying course, depending on the susceptibility of the animal and the number of infective organisms and their virulence. With a heavy parasite dosage of a virulent strain, hamsters are likely to become gravely ill within a month or even less. Autopsy will reveal a greatly enlarged and dark spleen, and a somewhat hypertrophied liver. Tissue (contact) preparations of these organs show very numerous organisms both intracellular and lying free (although doubtless almost all of them are normally within cells).

Less virulent strains of leishmania may produce a slowly developing infection requiring as much as six months to reach a fatal termination, but natural recovery probably does not occur.

Experimental Amoebiasis

Entamoeba histolytica occurs naturally in several species of the lower mammals, especially in monkeys, apes, dogs, and rats. Infection with similar species of amoebae is also common in many other host species. Thus only individuals known to be free from pre-existing infection with amoebae should be used as experimental animals.

Although monkeys, dogs, and rats have been employed for the study of amoebiasis in the laboratory, kittens are still probably the best experimental hosts. Cheap and easily obtained, they apparently are not subject to natural infection. Kittens are much more susceptible than adult cats, although the resistance acquired with advancing age is not complete. When cats have passed an age of about twelve weeks, or a weight of about 600 grams, experimental infections are less easily obtained, and when they do develop likely to be rather benign.

The amoebae (which may be from cultures) are injected rectally into a weaned kitten, using a long soft rubber catheter attached to a syringe. The animals should first be anesthetized. The organisms may also be given orally, but probably cysts are the only infective stage in this method. It is important to remember, in handling material which may contain them, that they are also infective to man.

Acute amoebiasis develops within a week or ten days after introduction of the parasites, and is characterized by diarrhea or dysentery and the passage of bloody mucus. In some animals the infection results fatally. Examination of the colons of these animals will reveal the typical ulcerated patches in the mucosa and submucosa. Smear preparations of the fecal mucus of acutely ill kittens, or tissue sections of the lesions, reveal numerous amoebae. They are also easily recovered by inoculating cultures with a bit of fecal material.

Kittens vary considerably in their response to the presence of the amoebae. Some develop light infections, and recovery is probably complete with the passage of time, when the acute stage does not result in death.

Experimental Toxoplasmosis

Although Toxoplasma may not be a protozoan it is usually regarded as one. It is the subject of much research because of the many unsolved problems relating to its transmission in nature, its life cycle, and therapy. Since it has not yet been cultivated (except in tissue culture), it must be maintained in animals, and subinoculation of tissue from a suspected case affords the only absolute proof of the presence of parasites.

Apparently most species of warm-blooded animals are susceptible, but white mice are the most suitable experimental hosts. They are easily infected by intraperitoneal inoculation of material containing the parasites. Peritoneal exudate from a mouse in the acute stage of the disease, or an organ suspension (preferably of brain in physiological saline) from an animal in the chronic phase is a good source. For maintaining the parasites over an extended period, pigeons are the most suitable host, since infections are likely to become chronic and seldom kill the bird during the acute stage. Probably a very few parasites are sufficient to infect a mouse, perhaps even a single organism.

Since toxoplasmosis in the mouse is usually fatal in a few days, frequent transmission is necessary to maintain the parasites, although they may retain their viability in infected tissue for a week or longer in the refrigerator. Chronic infections of long duration are possible in pigeons.

In testing wild animals for the incidence of naturally acquired toxoplasmosis, brain suspensions are the best inoculum for mice, since the parasites seem to persist longest in the brain. Because some degree of adaptation may be required from the parasite when introduced into a new host species, the first series of test mice may give apparently negative results. Therefore it is desirable to subinoculate from them into a second series (for which purpose a suspension of several organs, such as brain, lungs, spleen, liver may be used), and from the second series even into a third, if negative results persist. Mice developing acute toxoplasmosis usually indicate it by a generally unthrifty appearance and a much swelled abdomen. If the peritoneal exudate is aspirated with a syringe and examined under the microscope it will be found teeming with the minute crescentic parasites. They may be readily seen under the high power, either in a fresh preparation or after staining with one of the Romanowsky stains.

Although toxoplasmosis seems to be transmissible only with difficulty in nature (despite the great number of naturally acquired infections in man and animals), it is important to remember that it is, after all, a pathogenic parasite. An accidental prick with an infected hypodermic needle used in inoculating mice might result in a laboratory-acquired infection in the operator.

Staining

Free-living Forms and Parasites other than Blood Protozoa

Heidenhain's Haematoxylin. Despite the increasing employment of newer techniques, such as the electron microscope to elucidate minute structure, the more conventional and older staining methods are still useful. There are many such, but for most Protozoa other than blood parasites staining with Heidenhain's haematoxylin is one of the best. Preparations made in this way reveal cytological detail with great clarity, and if properly mounted will last for many years without fading.

The method involves the use of an albumen fixative to insure adhesion of the organisms to the cover glass or slide while it is being passed through the various solutions. Two reagents are required for the actual staining:

> Mayer's Albumen Fixative: Beat the white of an egg and transfer to a cylinder to allow any suspended material to rise to the surface. Skim this off, and add an equal amount of glycerine, and about 1 gm of sodium salicylate or a crystal of thymol to prevent spoilage.

Solution 1: Dissolve 3.0 gm of clear violet ferric alum crystals in 100 ml of distilled water.

Solution 2: Dissolve 0.5 gm haematoxylin in 100 ml of distilled water, and allow to ripen for several weeks. Or prepare the desired amount of aqueous solution by adding a suitable quantity of a 10 per cent alcoholic stock solution to distilled water. The stock solution (consisting of 1 gm haematoxylin in 10 ml absolute ethyl alcohol) ripens slowly on the shelf, and aqueous solutions made from it are therefore ready for immediate use.

Some killing or fixing agent is also required, and Schaudinn's among many others is a favorite. It has the following formula:

saturated solution of corrosive sublimate in distilled water	100 ml
absolute ethyl alcohol	50 ml
glacial acetic acid	5 drops

The acetic acid may be omitted, depending on the nature of the structures to be differentiated. Chromosomes and the details of mitosis, for example, are shown very well without it. The fixative may be used hot (60°C.) or cold.

The organisms to be stained are picked up with a capillary pipette and discharged on the albumen-smeared surface of a slide or cover slip. (Each has certain advantages from the point of view of handling). It is a good idea to watch the discharge of the culture medium from the pipette under a wide-field binocular to make sure that the organisms are not left behind on the surface of the glass; if a given individual is to be stained this precaution is very necessary. If a cover slip is used, its manipulation is best done with cover-glass forceps. When a number of cover-glass preparations are being stained at one time, cover-glass racks, such as those of Chen (1942) and Wichterman (1946), will be found very convenient.

The organisms are now fixed by flooding with Schaudinn's fluid for 5 minutes or more (fixation is probably almost instantaneous, but some very good technicians prefer fixation periods of 15 minutes or more). A schedule for subsequent processing of the preparation follows:

1.	Immerse in 70% ethyl alcohol plus iodine	2 hours
2.	" " 50% " " " "	5 minutes
3.	" " tap water	5 minutes
4.	" " mordant (Sol. 1)	4–12 hours
5.	" " tap water	5 minutes
6.	" " stain (Sol. 2)	4–24 hours
7.	" " tap water	5 minutes
8.	Destain under microscope while in weak iron alum solution (Sol. 1 diluted with equal amount of distilled water), until differentiation is sharp.	
9.	Immerse in tap water	2 hours

10.	Immerse in 50% alcohol	5 minutes
11.	" " 70% "	5 minutes
12.	" " 95% "	5 minutes
13.	" " absolute alcohol	5 minutes
14.	" " xylol	5 minutes

15. Mount in neutral Canada balsam ("neutral" balsam, though purchased as such, is not always neutral. Any degree of acidity is likely to cause fading in time; as a precaution, a piece of marble may be kept in the balsam.)

The schedule above is an empirical one, and most technicians have their favorite and time-tested variants. In particular it is possible to reduce the times (even to as little as 10 minutes) in which the preparation is left in the mordant and the stain. There is reason to think, however, that staining is then less critical and less permanent. Differentiation is seldom improved by counterstaining.

For special purposes it may be desirable to use fixatives other than Schaudinn's or to use Schaudinn's in a modified form. Calkins, for example, employed a saturated solution of corrosive sublimate in absolute alcohol, with or without acetic acid. Another valuable modification is a solution consisting of equal parts of 2 per cent chromic acid and the original formula (2 parts saturated aqueous $HgCl_2$ plus 1 part of absolute alcohol), to which is added 5 per cent of glacial acetic acid.

Other good fixatives for Protozoa are the much used Bouin's and Zenker's. Osmium oxide is also often employed, either as the vapor, or in Champy's solution:

1% chromic acid	7 ml
3% $K_2Cr_2O_7$	7 ml
2% OsO_4	4 ml

Borrel's Method. Another good staining method is that of Borrel. It gives beautifully differentiated preparations that, in the author's experience, have resisted fading for many years. For critical cytological work, however, Borrel-stained slides are not equal to those made with haematoxylin. As used by Borrel, fixation was first accomplished by exposure of the organisms for 24 hours to a solution of the following:

OsO_4	2 gm
$PtCl_2$	2 gm
H_2CrO_4	3 gm
CH_3COOH (glacial)	20 ml
distilled water	350 ml

Staining was done by immersion in a 1 per cent aqueous solution of magenta for 1 hour, followed by 5 minutes in a mixture of 2 parts of saturated aqueous solution of indigo carmine and 1 part of a saturated solution of picric acid.

Preparations Involving Many Individuals; Sectioning. If preparations showing many individuals are desired, or if sections of Protozoa *en masse* are to be made, large numbers may be concentrated from mixtures by centrifuging. The concentrate is processed in the same manner as single preparations, with, of course, repeated centrifugation followed by pipetting off the supernatant at each step. After final dehydration and transfer to xylol, the tube containing the suspended organisms is placed in a beaker of warm water (65°–70°C.), then again centrifuged, the supernatant xylol pipetted off, and melted hard paraffin added. After allowing the mixture of organisms and paraffin to stand in a paraffin oven for 15 minutes, centrifuging is repeated, the melted wax removed and more added, and the whole process is carried out once more. A final 15 minutes in the oven is followed by centrifuging, and the tube is then placed in cold water to hasten hardening of the paraffin. A small wire loop has previously been inserted into the wax to facilitate its removal, which is easily accomplished after briefly immersing the tube in hot water. Sections may be cut at four micra.

Another method of handling Protozoa to be sectioned is that of Beers (1937). A small bag made from mammalian mesentery (e.g., cat or rabbit) is formed over a paraffin mold, fixed and hardened by exposure to Bouin's fluid for 24 hours, and then freed of the wax after dehydration by exposure to xylol. It is then reimmersed in absolute alcohol, followed by a day in 80 per cent alcohol. A small basket of fine mesh wire gauze for holding the bag is then made, and the basket with the bag inside is placed in a watch glass of 95 per cent alcohol, the opening of the bag at the top.

To get the Protozoa into the bag, they are first placed in a watch glass containing 80 per cent alcohol. Then "with a pipette having a straight, slender tip about 2 cm long, transfer the Protozoa to the bag under the dissecting binocular. They need merely to be released from the pipette directly above and near to the opening in the bag; they will then drop or stream into it because of the greater specific gravity of the 80 per cent alcohol." The bag is then closed with a No. 80 black cotton thread previously placed in a position to serve as a draw string, and "the entire bag plus contents is now to be dehydrated, cleared, infiltrated, imbedded, and sectioned, though it is advisable to stain it in alcoholic eosin to facilitate orientation." Since the thread cannot be sectioned, cutting is started "tangentially to the deepest part of the bag" and "at right angles to (its) longitudinal axis" (Beers, 1937). This method has the advantage of greatly reducing the loss of organisms during processing, which is always a problem, even when the manipulation of slides and cover slips carrying whole mounts is involved.

Feulgen's Method. Feulgen's method of staining is less a method for the demonstration of minute structure than it is a delicate microchemical test for the presence of desoxyribonucleic (thymonucleic) acid, or DNA. Ribonucleic acid (RNA) is not revealed by the Feulgen reaction, but may be demonstrated by other methods. DNA is a constituent of chromosomes and believed to be active in protein synthesis. The reaction of Feulgen really depends on the hydrolysis of chromatin, followed by the use of Schiff's reaction to demonstrate aldehydes resulting from such hydrolysis. A positive test is indicated by a delicate pink or violet coloration of chromatin structures, the cytoplasm remaining colorless (unless counter stains are applied). The process as given below is a slight modification of the original, and is said by Rafalko (1946) to be better adapted for use with small organisms such as Protozoa. Two reagents are used: leuco-fuchsin and sulfurous acid. Their preparation and a staining schedule follow (quoted from Rafalko):

By gentle bubbling of sulfur dioxide gas from a small aperture in glass tubing into 100 ml of 0.5% basic fuchsin solution, decolorization takes place in about an hour and the reagent is then ready for use. Distilled water is similarly saturated for the sulfurous acid rinsing bath and may be stored for weeks in a tightly corked flask. The sulfur dioxide gas can be produced easily by a simple flask-and-funnel generator using sodium bisulfite and dilute sulfuric acid.

The reagents thus made were then used according to the slightly modified standard procedure as follows:

1. Fixation (see below), mostly 2 to 20 minutes. (Time is for smear preparations).
2. Washing, minimum time according to the fixative used (not over 20 minutes).
3. Distilled water, 2 minutes.
4. Normal HCl at room temperature, 2 minutes.
5. Normal HCl at 60°C., 8 to 10 minutes.
6. Normal HCl at room temperature, rinse.
7. Distilled water, rinse.
8. Sulfurous acid, 2 minutes.
9. Leuco-basic fuchsin, 1½ to 2 hours.
10. Sulfurous acid bath, for sufficient time to remove the free unreacted leuco-basic fuchsin (usually two or three 1-minute changes).
11. Tap water, 10–15 minutes.

Counterstaining may be done with fast green in aqueous or alcoholic solutions if desired. Dehydration through the alcohols, immersion in xylol, and mounting are done as in other staining techniques.

Not all fixatives are good for use with Feulgen's method. Lee (1937) lists those which may be safely used with the optimum time. Corrosive sublimate mixtures are usually employed, the most popular being probably sublimate acetic (6 per cent aqueous corrosive sublimate solution 98 parts, glacial acetic acid 2 parts.)

Silver Staining. To demonstrate the "silver-line system" characteristic of many ciliates, the organisms are fixed by drying in a thin film on a slide or cover slip, or by treatment with a fixative such as Schaudinn's, exposed to light after preliminary treatment with silver nitrate, and mounted. Klein (1927), the originator of the method, fixed his ciliates by drying, impregnated them with silver nitrate for 6–8 minutes, exposed them to strong light for some hours, and mounted them after microscopic observation showed sufficient differentiation.

Others have modified the method somewhat. Lund (1933) found the following schedule best for *Paramecium:*

1. Concentrate the organisms with the centrifuge.
2. Fix for 3 minutes in corrosive sublimate-formalin mixture (95 parts concentrated aqueous solution of $HgCl_2$ to 5 parts formalin).
3. Wash twice in filtered water.
4. Impregnate 3 minutes in 2 per cent solution of $AgNO_3$.
5. "Without washing, reduce in distilled water by direct sunlight for 8 minutes." (Lund found a stream of air, introduced into the centrifuge tube through a fine glass capillary, useful to keep the organisms in agitation, and thus more uniformly exposed to light.)
6. Wash 5 times in distilled water.
7. Dehydrate slowly, by the slow addition of 9 ml of 95 per cent alcohol to the 1 ml aqueous suspension of organisms.
8. Complete dehydration by passage through absolute alcohol, clear in xylol, and mount in balsam.

Vital staining. Vital staining is a technique by which living protoplasm takes up dye so that structures not ordinarily visible can be distinguished. Theoretically it is superior to ordinary stained preparations that reveal only what may, in a sense, be regarded as artifacts, since fixation and staining inevitably cause profound chemical and physical changes. Unfortunately, however, even the least toxic dyes when incorporated into the substance of the cell seem to be more or less inconsistent with its continued normal functioning. Either the organism survives only a limited time in the stained condition, or it metabolizes the dye and soon reverts to its normal dyeless state. Fortunately, modern phase microscopy has made it possible to see more structure in the living cell than could ever be seen with the conventional light microscope, and has greatly reduced our dependence on staining.

Despite its limitations, vital staining is nevertheless often still useful in studying Protozoa. The pH of vacuoles may be tested by employing dyes such as neutral red, which appears cherry red or yellow, according to whether it is in an acid or alkaline medium. India ink, which is a fine suspension of carbon particles and is also relatively nontoxic, may be used to demonstrate the mechanics of ingestion and gastric vacuole formation.

Basic dyes seem to be much less toxic than acid ones and are therefore more valuable in vital staining. Some of them, such as Bismarck brown, are not only capable of staining the cytoplasm but may be retained in the protozoan for long periods without evident harm. But any staining of the nucleus usually soon results in death. Cytoplasmic staining may injure the cell by sensitizing it to light, even when other effects seem minor.

The number of dyes which have been tried for vital staining properties is large. Ball (1927), who experimented with a number of them on *Para-*

mecium caudatum, found them useful and least toxic in the concentrations tabulated below. Probably results on other ciliates, if not on Protozoa generally, would be comparable.

TABLE 27.7

DYES STAINING CYTOPLASM OF NORMAL LIVING *PARAMECIUM*

Dye	Minimal Concentration for Cytoplasmic Staining	Mortality in 1 Hour, per cent	Hours Required for Cytoplasmic Destaining
Bismarck brown	1/150,000	0	7
Methylene blue	1/100,000	5	7
Methylene green	1/37,500	5	4
Neutral red	1/150,000	3	9
Toluidin blue	1/105,000	5	9
Basic fuchsin	1/25,000	30	9
Safranin	1/9,000	30	1½
Aniline yellow	1/5,500	0	1
Methyl violet	1/500,000	20	2
Janus green	1/180,000	40	7

Vital staining may be done either by the addition of dyes in suitable amounts to the culture medium, or by placing a little of the substance in alcoholic solution on a slide or cover slip, allowing it to evaporate, and then putting a drop or two of medium containing the organisms on the film of dye.

Staining Blood Protozoa

Staining Thin Films. Since many species of blood Protozoa are often encountered and may be of considerable importance, the study of stained blood films is often required. Usually one of the rather numerous Romanowsky stains is employed. They are all based on methylene blue and eosin and give essentially similar results. The fixing agent (methyl alcohol) may be used separately or incorporated in the staining solution. Methylene blue is not used as such, but is first subjected to acid or alkaline decomposition, with the resulting formation of derivative dyes known as azures. One of these (Azure B) is the most active coloring agent. Eosin may be in solution with the methylene blue complex, or it may serve as a counter stain into which the preparation is introduced after immersion in the latter.

The most commonly used Romanowsky stains are Wright's and Giemsa's, although Leishman's is popular abroad, and the "J.S.B." stain (the letters are the initials of the two Indian scientists, Jaswant Singh and L. M. Bhattacharji who devised the process) is fast and cheap, which has led to its increasing adoption.

For most purposes thin blood films (often referred to as "smears") are employed, although thick films are very useful to diagnose malaria and other blood infections.

The first step in making a thin film is to spread a drop of blood thinly and evenly over a glass slide (see Fig. 27.1) by using the edge of a second slide held at an angle of about 45°. Slides must be clean and free of any trace of grease. To avoid fingerprinting them, they should be held only by their

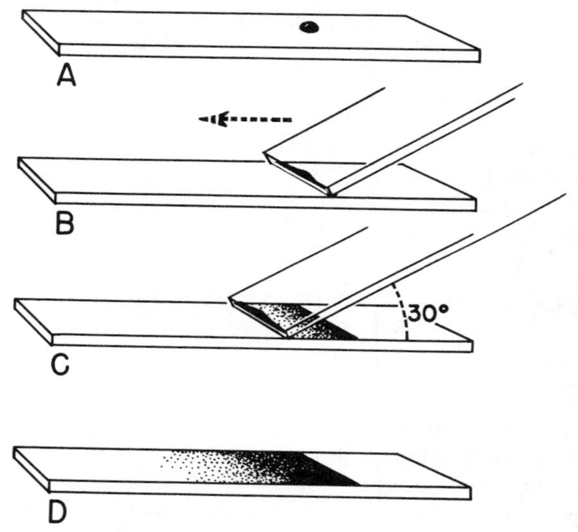

FIG. 27.1. Steps in making a thin blood film. *1.* Place a small drop of blood near one end of the slide. *2.* A second slide is brought back toward it until it touches, and the blood spreads along its edge. *3.* The slide is then moved gently forward at an angle of about 30° in such a way as to leave an even blood film behind it. *4.* The completed film. (After Russell, West, and Manwell.)

edges. After the film is made, it is allowed to dry and fixed in absolute methyl alcohol (except when using a stain such as Wright's, which already contains the fixing agent). The staining procedure differs with the stain used, and schedules are given below for Wright's, Giemsa's, and J.S.B., when used with thin smears.

1. Wright's Stain
   a. Flood with undiluted Wright's for 1½ minutes.
   b. Add an equal amount of water (buffered to pH 6.8 to 7.0) and allow to stand for three minutes.
   c. Drain, wash in tap water and stand on end to dry. (Slide may be pressed between two sheets of clean blotting paper to save time.)
2. Giemsa's Stain
   a. Fix dried film in absolute methyl alcohol (dipping is sufficient).

b. Place in diluted Giemsa's (dilution may be of various strengths: one part of stock solution to 15 of water, buffered to about pH 7.2 is satisfactory, but good results may be had with a dilution up to 1 in 50, if staining time is increased about 50 per cent) and leave about 20 minutes for the minimal dilution.
c. Wash in tap water just enough to remove excess stain, and stand up to dry. (Or blot, as indicated above).

3. J.S.B. Stain
   a. Fix in absolute methyl alcohol, as in "2a" above.
   b. Immerse in Solution A for 30 seconds.
   c. Wash for about 10 seconds, by immersion in distilled water buffered to a pH of from 6.2 to 6.6. (It is better to use two washing jars, dipping slide first in one and then the other.)
   d. Dip in Solution B (eosin).
   e. Wash for about 10 seconds, as in "c" above.
   f. Immerse in Solution A for 30 seconds.
   g. Wash again for 10 seconds, as in "c," and stand up to dry.

Staining Thick Films. To make a thick film (which, of course, is only good for examining mammalian blood, since the red cells of all other vertebrates are nucleated), several drops of blood are spread on a slide over an area approximately equal to a dime and allowed to dry thoroughly. If the preparation is not wholly dry it will not adhere to the slide during processing; if it is too thick it will be difficult to distinguish parasites against the background of leucocyte nuclei and red cell debris. It should be possible to distinguish newsprint through a preparation of this type when of the right density. *Do not fix.* If Wright's stain is used, the dried blood must first be laked in distilled water; with stains in aqueous solution (e.g., Giemsa's, J.S.B.), laking and staining take place at the same time. The actual staining process is the same for both thick and thin films and the schedules above may be used for both. It is often an advantage to have thick and thin films on the same slide. Labeling can be done either with a diamond pencil or a lead pencil, the latter being used as a stylus to write on the thin film.

Success in blood staining is easily learned. But good differentiation of blood cells and parasites depends as much on the quality of reagents as on the care in processing. A small variation in the pH of staining and washing solutions may prevent the appearance of structures such as the Schüffners dots so important in the diagnosis of *Plasmodium vivax* and *P. ovale*. Careless exposure of the slides to dust before staining may result in artifacts puzzling to the novice, and preparations not promptly fixed may be very quickly injured by stray flies, ants, and roaches. Slow drying, occurring under conditions of heat and high humidity, is also a frequent cause of poor staining. It may be hastened by exposure to the draft from an electric fan or hair dryer. Once fixed, however, blood films may be kept for some time without serious deterioration even though unstained, and may be shipped without damage to the film if first individually wrapped in tissue.

It is of course unnecessary to cover stained blood preparations unless they are to be observed with dry objectives. If covered, a neutral medium such as diaphane should be used, since fading may otherwise be very rapid. Diaphane-mounted Romanowsky-stained blood films will keep for years without much deterioration.

Preparation of Blood Stains. Wright's Stain is made up in a stock solution consisting of the following ingredients:

0.3 gm powdered stain
3.0 ml glycerine (C.P.)
97.0 ml methyl alcohol (acetone-free)

The powdered dye is ground up with the glycerine, and the mixture is added to the methyl alcohol and mixed well. The solution should then be kept for several weeks and filtered before use. It is best kept in a brown bottle, and if it is to be stored for any length of time it should be placed in a cool spot (Wright's often fails to give good results in the tropics).

Giemsa's Stain is prepared from these ingredients:

1.0 gm powdered stain
66.0 ml glycerine (C.P.)
66.0 ml methyl alcohol (acetone-free)

Grind the stain and glycerine together, mix well, and expose the suspension to heat over a water bath held at a temperature of 55° to 64°C. until solution is complete. After it is cooled, add the methyl alcohol and allow it to stand for several weeks. Make the dilutions indicated in the schedule above only when the stain is to be used. Since the staining qualities deteriorate rather rapidly after dilution such dilutions should not be used after a lapse of more than a few hours. Stain from different sources may vary (*Saal, 1964).

Two solutions ("A" and "B") are used in the J.S.B. process. Solution A is made up either as directed below, or from the dried stain, by dissolving 100 mg in 100 ml of M/20 Na_2HPO_4. Solution B is prepared by dissolving water soluble eosin in water in the proportion of 200 mg per 100 ml. Both solutions keep well at room temperature, and thus are ready for use whenever required. They should stain well for weeks unless used too intensively.

The dye used in Solution A is prepared from these ingredients:

Medicinal methylene blue	0.5 gm
Potassium dichromate	0.5 gm
Sulfuric acid	3.0 ml
Water	500.0 ml

The methylene blue is first dissolved in the water; then the acid is added, and after thorough mixing, the dichromate. This results in the formation of a heavy, amorphous, purple precipitate. The mixture is then heated over a boiling water bath for about three hours, by which time polychroming is usually complete, as indicated by a blue color. Should the color remain greenish, some further heating may be required. After cooling at room temperature, the precipitate that forms is filtered out and dissolved in 500 ml of M/20 Na_2HPO_4. The solution is ready for use after a maturation period of about 48 hours.

If desired, the precipitate may be allowed to dry at room temperature (a vacuum dessicator is convenient for this purpose) and kept for later use. The dried stain keeps well almost indefinitely.

Tissue Contact Preparations. For certain purposes, tissue contact preparations have real advantages. They are easy to make and the entire operation takes little longer than the autopsy of the animal to be examined.

To make the preparation the cut surface of the organ (e.g., lung, spleen, liver) is gently touched to a clean slide, and then fixed and stained as with a blood film. Romanowsky stains are used here also. If the slide is to be kept for future use, it may be covered after mounting in diaphane.

To check the brain for the presence of exoerythrocytic stages of the malaria parasites, the process is somewhat different. A small bit of the pulp is removed and placed on a slide. Then it is smeared much as with a thin blood film, except that the slide with which the tissue spreading is done is moved the other way, allowing the film to grow as the slide edge recedes. A little practice may be needed to enable one to spread the tissue to just about the right thickness; if too thick, it will be hard to make out much detail, and if too thin, there will be little detail to make out. Fixing and staining are done as directed above, but the slide may have to remain longer in the dye to insure good differentiation.

Contact preparations, although of little value for the study of cell structure, are very useful to diagnose protozoan tissue parasites such as the leishmanias, and to demonstrate the exoerythrocytic stages of malaria.

Examination of Fecal Samples for Intestinal Protozoa

Microscopic Examination of Fresh Stool

Infection with intestinal Protozoa occurs throughout the world and is not confined to man, and therefore fecal samples must often be examined. Since the behavior and appearance of the organisms change rapidly after they leave the body, examination should be made as soon as possible after the stool is passed. Trophozoites may be unrecognizable after thirty minutes, though cysts may retain their morphology for some time.

A positive diagnosis is usually based on (1) microscopic demonstration of trophozoites or cysts in fresh fecal samples, or (2) in stained preparations, (3) finding the organisms after the use of concentration methods, or (4) positive cultures.

To examine a stool specimen microscopically, pick up a bit of the sample, preferably with a little mucus, using a wooden applicator. Mix thoroughly with several drops of physiological saline solution (or Ringer's) on a slide and cover. The suspension should not be too thin nor so thick that ordinary newsprint cannot be fairly clearly distinguished through it. Examine under the microscope, first with the low power and then under the high dry objective. Look for amoebae, which are often of a glassy appearance, for their cysts, and for evidences of moving organisms. But beware of mistaking motion due to currents in the fluid, or Brownian movement of particles for true protozoan locomotion. And remember that bacteria, many of them actively motile, are always extremely numerous in colon contents. The diagnosis of intestinal Protozoa is not easy, but with practice it may be successfully done even with the low powers of the microscope. (Diagnostic characteristics for the various intestinal Protozoa of man are given in Chap. 21.)

To improve visibility of glycogen bodies and the nuclei of cysts, a little d'Antoni's solution* may be run under the cover glass. Glycogen masses are then stained a dark brown, and it is usually possible to distinguish with considerable clarity not only the nuclei but even their karyosomes, when present.

Of course, since the amount of feces which can be examined in an ordinary smear preparation is extremely small, negative results are not really significant. Repeated examinations of stools passed on successive days are often desirable.

Concentration Methods

To increase the probability of finding parasitic Protozoa concentration techniques have been devised, although these are useful only for demonstrating cysts. One of the best involves the use of centrifugation and flotation of the cysts in a zinc sulphate solution of high specific gravity. It is carried out as follows:

1. Mix a small sample of feces with about 10 parts of water.
2. Strain about 10 ml of this suspension through wet cheesecloth.
3. Centrifuge at about 2500 R.P.M. for about a minute, pour off the supernatant liquid, replace with water after breaking up the sediment by shaking or tapping, and centrifuge. Repeat the whole process until the supernatant fluid is essentially clear.
4. After the last supernatant liquid is poured off, add a little 33 per cent zinc sulphate solution, break up the sediment, and then fill the tube nearly to the top with more solution. Centrifuge again for about a minute.
5. A few loopfuls of the surface film may now be removed, placed on a slide with a drop or two of iodine solution, and examined microscopically.

Staining other than with iodine is in general not very useful in diagnosis of parasitism by intestinal Protozoa. This is partly because of the added time

*Other iodine solutions may be used, such as one consisting of a 1 per cent solution of potassium iodide saturated with iodine by the addition of iodine crystals. But d'Antoni's solution has the advantage of standardization and excellent keeping qualities.

and labor it requires, but chiefly because equally accurate results can usually be had without it.

However, stained preparations, if such are desired, can be made in much the same way as for other Protozoa, except that previous coating of the slide or cover slip with albumen fixative is not required. A bit of feces is simply smeared on the surface of the glass with a wooden applicator, care being taken not to make it too thick, the preparation is fixed in hot Schaudinn's solution, and the schedule for staining in Heidenhain's haematoxylin is then followed. (If the slide is to be used for diagnostic purposes only, staining and mordanting times may be shortened to ten minutes, more or less.)

Various modifications of this method have been suggested. One involves the use of warm (37°C.) saturated aqueous solution of picric acid as a mordant, the smears being exposed to its action for five hours "or more," with subsequent staining in a 0.5 per cent aqueous solution of haematoxylin for "from 30 minutes to several hours, depending on the ripeness of the stain."* The advantage of this method is said to be that it eliminates the need for microscopic differentiation.

Polyvinyl Alcohol as a Fixative

Polyvinyl alcohol ("PVA" or "Elvanol") has recently come into wide use as an ingredient of a combined fixative and mounting medium for fecal preparations, especially when for any reason they must be kept for some time before laboratory examination. The solution has the following formula:

Polyvinyl alcohol	5 gm
Glacial acetic acid	5 ml
Glycerol	1.5 ml
Schaudinn's solution, to make	100 ml

Stir under gentle heat (75°C.) until clear.

To use, mix about three parts of PVA solution with one of feces. Slides may be made immediately, or whenever convenient. Smearing is done in the usual way and the preparation dried thoroughly (for some hours at 37°C.). Before staining in the customary manner, all traces of corrosive sublimate must be removed by 10 or 15 minutes immersion in a 70 per cent alcohol-iodine mixture.

According to Goldman and Brooke (1953), the use of PVA fixation for trophozoites and zinc sulphate flotation method for cysts in combination reveal more infections than is possible in any other way.

Cultivation

Diagnosis of the presence of intestinal Protozoa by cultivation is a method useful chiefly as an adjunct to direct microscopic demonstration. Several

*Quoted from mimeographed material distributed by the Communicable Disease Center of the U.S. Public Health Service.

culture tubes should be inoculated with amounts of feces about the size of a pea, and incubated for at least 48 hours. Results are evaluated by taking a little fluid and sediment from the bottom of the tube with a wide-mouthed 1 ml pipette, placing several drops on a slide, covering and examining under the microscope. It must always be remembered that organisms will never be as numerous as in bacterial cultures, and may indeed be relatively few.

Preparation of d'Antoni's Solution

D'Antoni's solution is prepared as follows (the directions are taken largely from Craig and Faust, 1951):

One hundred grams of potassium iodide are placed in a chemically clean 1000 ml volumetric flask, and enough distilled water is added to bring the volume up to one liter.

Then weigh a 25 ml clean and dry volumetric flask to the fourth place, fill to the mark with the above solution, and again find the weight to the fourth place. Subtract the weight of the empty flask from that of the full one to get the actual weight of the 25 ml of KI solution (in theory 26.925 gm, but likely to be less, due to the deliquescence of the salt). Divide the difference in weights by the theoretical weight and subtract the quotient (expressed in percentage) from 10 to get the actual percentage of the solution. Now use the formula below, solving for "x" (grams of KI):

100: (actual percentage of tested solution) = x: 10 per cent

Subtracting 100 from "x" gives the number of grams of potassium iodide to be added to secure a true standardized 10 per cent solution.

The staining solution is made by preparing 100 ml of a 1 per cent solution from the stock solution above and adding 1.5 gm powdered iodine crystals. Filter and age for four days before use. If tightly stoppered the solution will keep well for a long time.

Fossil Foraminiferida in Oil Well Cores

Since recent Foraminiferida are not easily secured at points distant from the ocean, and in any event do not differ very greatly from fossil types, the latter are an acceptable substitute for laboratory study. Sometimes they may be numerous in locally occurring limestones, but more often fossiliferous material is most easily obtained from oil companies, which are frequently quite willing to supply it on request. Their laboratories receive great quantities of it from the field, since determination of the foraminiferidan fossils it contains allows the driller to identify strata penetrated by the drill.

Core samples are taken at different depths. Some are rich in fossils, while others contain few or none. It is the association of certain species, rather than the occurrence of given key species, which is usually important in recognizing one or another rock stratum.

To examine a core for fossils in the laboratory, the simplest method is to mix a very small fragment of material with water in a watch glass, making a light suspension. A drop or two of this is then transferred to a slide, covered, and placed under the microscope. The sample should be free of large grains and thin enough so that newsprint can be discerned rather clearly through it. Fossil Foraminiferida are usually easily spotted under the low power objective. Many of them will be observed to contain black carbonaceous material.

Although the method above is satisfactory for the qualitative demonstration of foraminiferidan fossils, the procedure used in oil company laboratories is a little different. It involves allowing the mixture of core material and water to stand for 24 hours, so that a smoother preparation can be had. Then it is placed on a 200 mesh sieve to facilitate removal of clay and silt (which is carried off by a water spray), and the residue remaining on the sieve is then subjected to microscopic examination.

REFERENCES

Ball, G. H. 1927. Studies on *Paramecium*. III. The effects of vital dyes on *Paramecium caudatum*. *Biol. Bull.*, 52: 68–78.

Beers, C. D. 1937. A method for sectioning of Protozoa en masse. *Science*, 86: 381–2.

Chen, Tze-Tuan. 1942. A staining rack for handling cover-glass preparations. *Stain Tech.*, 17: 129–30.

Craig, C. F. and Faust, E. C. 1951. *Clinical Parasitology*, 5th ed. Philadelphia: Lea & Febiger.

Dawson, J. 1928. The culture of large free-living amoebae. *Am. Nat.*, 62: 453–66.

Goldman, M. and Brooke, M. M. 1953. Protozoans in stools unpreserved and preserved in PVA-fixative. *Pub. Health Rep.*, 68: 703–6.

Guyer, M. F. 1927. *Animal Micrology*. University of Chicago Press.

Hall, R. P. 1937. Growth of free-living Protozoa in pure cultures. In *Culture Methods for Invertebrate Animals*, Needham *et al.* (eds). Ithaca, New York: Comstock.

Hutner, S. H., Provasoli, L., and Filfus, J. 1953. Nutrition of some phagotrophic fresh-water chrysomonads. *Ann. N.Y. Acad. Sci.*, 56: 852–62.

Hyman, L. H. 1925. Methods of securing and cultivating Protozoa. I. General statement and methods. *Trans. Am. Mic. Soc.*, 44: 216–21.

——— 1931. Methods of securing and cultivating Protozoa. II. *Paramecium* and other ciliates. *Trans. Am. Mic. Soc.*, 50: 50–7.

——— 1941. Small animal cultures maintained on lettuce leaves. *Educational Focus* (Bausch & Lomb Opt. Co.), 12: 14–9.

Johnson, P. L. 1928. A platinum spoon for isolating and transferring Protozoa. *Science*, 67: 299.

Johnson, W. H. and Miller, C. A. 1956. A further analysis of the nutrition of *Paramecium*. *J. Protozool.*, 3: 221–6.

Kidder, G. W. and Dewey, V. C. 1949. Studies on the biochemistry of *Tetra-*

hymena. XI. Components of factor II of known chemical nature. *Arch. Biochem.*, 20: 433–43.

Klein, B. M. 1926. Über eine neue Eigentümlichkeit der Pellicula von *Chilodon uncinatus. Ehrbg. Zool. Anz.*, 67: 160–2.

Kudo, R. R. 1957. *Pelomyxa palustris* Greeff. I. Cultivation and general observations. *J. Protozool.*, 4: 154–64.

Lee, B. 1937. *The Microtomists' Vade-Mecum*, 10th ed. J. B. Gatenby and T. S. Painter (eds). Philadelphia: Blakiston.

Levine, N. and Marquardt, W. C. 1955. The effect of glycerol and related compounds on survival of *Tritrichomonas foetus* at freezing temperatures. *J. Protozool.*, 2: 100–7.

Lund, E. E. 1933. A correlation of the silver-line and neuromotor systems of *Paramecium. Univ. Calif. Pub. Zool.*, 39: 35–76.

Manwell, R. D. and Edgett, R. 1943. The relative importance of certain factors in the low-temperature preservation of malaria parasites. *Am. J. Trop. Med.*, 23: 551–7.

——— and Feigelson, P. 1948. A modified method of preparing the J.S.B. stain. *J. Lab. and Clin. Med.*, 33: 777–82.

McClung, C. E. 1950. *Handbook of Microscopical Technique*, 3d ed. (R. M. Jones, ed.) New York: Hoeber.

Miller, C. A. and Johnson, W. H. 1957. A purine and pyrimidine requirement for *Paramecium multimicronucleatum. J. Protozool.*, 4: 200-4.

Needham, J. G. *et al.* 1937. *Culture Methods for Invertebrate Animals.* Ithaca, New York: Comstock.

Nicolle, C. 1908. Culture du parasite du Bouton d'Orient. *C. Acad. Sci.*, 146: 842–3.

Novy, F. G. and MacNeal, W. J. 1903. On the cultivation of *T. lewisi*. Contributions to Medical Research dedicated to V. V. Vaughn.

Pantin, C. F. A. 1946. *Notes on Microscopical Technique for Zoologists.* New York: Columbia University Press.

Parpart, A. K. 1928. The bacteriological sterilization of *Paramecium. Biol. Bull.*, 55: 113–20.

Rafalko, J. S. 1946. A modified Feulgen technique for small and diffuse chromatin elements. *Stain Tech.*, 21: 91–3.

Rice, N. E. 1947. The culture of *Volvox aureus* Ehrenberg. *Biol. Bull.*, 92: 200–09.

Rudzinska, M. A. 1955. A simple method for paraffin and plastic embedding of the Protozoa. *J. Protozool.*, 2: 188–90.

Sandon, H. 1927. *The Composition and Distribution of the Protozoan Fauna of the Soil.* Edinburgh: Oliver & Boyd.

Singh, J. and Bhattacharji, L. M. 1944. Rapid staining of malarial parasites by a water-soluble stain. *Ind. Med. Gaz.*, 79: 102.

Skinner, C. E. 1939. Soil Protozoa for classroom demonstrations. *Turtox News*, 17: 129.

Sonneborn, T. M. 1950. Methods in the general biology and genetics of *Paramecium aurelia. J. Exp. Zool.*, 113: 87–148.

Storm, J. and Hutner, S. H. 1953. Nutrition of *Peranema. Ann. N.Y. Acad. Sci.*, 56: 901–9.

Unger, W. B. 1941. A preliminary survey of the Protozoa of Beaver Lake near Salsbury Cove, Maine. *Bull. Mt. Desert Island Biological Laboratory.*

Vishniac, H. S. and Watson, S. W. 1953. The steroid requirements of *Labyrinthula vitellina* var. *pacifica. J. Gen. Microbiol.*, 8: 248–55.

Wells, M. M. and Gamble, D. L. 1928. Protozoan cultures. *Gen. Biol. Sup. House.*

Wenrich, D. and Diller, W. F. 1950. Methods of Protozoology. In *McClung's Handbook of Microscopical Technique*, 3rd ed. Ruth McClung Jones (ed). New York: Hoeber.

Wichterman, R. 1946. A new glass device for staining cover-glass preparations. *Science*, 103: 23–4.

ADDITIONAL REFERENCES

Mugard, H. and Rouyer, M. 1966. Cultures axeniques d'*Ophryoglena mucifera*, infusoire, hymenostome histiophage. *Protistologica*, 2(4): 53–5.

Saal, J. R. 1964. Giemsa stain for the diagnosis of bovine babesiosis. I. Staining properties of commercial samples and their component dyes. *J. Protozool.*, 11: 573–82.

Tobie, E. J. 1964. Cultivation of the mammalian trypanosomes. *J. Protozool.*, 11: 418–23.

28

Protozoology in America

"There appear to be but trifling steps from the oscillating particle of organic matter to a Bacterium; from this to a Vibrio; thence to a Monad, and so gradually up to the highest orders of life! The most ancient rocks containing remains of living beings, indicate the contemporaneous existence of the more complex as well as the simplest of organic forms; but, nevertheless, life may have been ushered upon earth, through oceans of the lowest types, long previously to the deposit of the oldest palaeozoic rocks known to us!" Joseph Leidy, *A Flora and Fauna Within Living Animals*, 1851.

Protozoology is a young science, especially on this side of the Atlantic. This is partly because of the differing goals of American and European universities in the last century, when protozoology was developing as an independent science. American universities tended to be parochial in view and were largely interested in teaching*; most of the leading private institutions were church-founded and originally dedicated to turning out a better educated ministry. The universities of Europe, and particularly of Germany, were devoted to research and creative scholarship. Therefore talented young men interested in graduate work in science usually matriculated in European universities, most often in Germany. Others were inspired to creative work and independent investigation by contacts with European scholars; thus it

*American universities may have been slow to encourage work in the laboratory sciences because for a long time good equipment could only be had from abroad. First-class microscopes became available in America only when Charles A. Spencer began their manufacture about 1850. According to Woodruff (1943), Yale University was probably the first American institution to procure a microscope; it was imported in 1734 at about 15 dollars, from "Matthew Loft, Maker of the Golden Spectacles the Backside of the Royal Exchange, London." It was 1806, however, before a second instrument was acquired. Protozoological investigation remained relatively inexpensive for some time thereafter. Leidy tells us that he made all his observations with a microscope that cost only 50 dollars and that the Geological Survey, which underwrote his studies in the west, had to invest only 222 dollars!

was with Leidy, of whom we have already learned. While still in his twenties, he spent a summer abroad touring Europe and meeting some of its most eminent scholars. The first American university to be organized on the European model was Johns Hopkins, opened in 1876. The wealthy Baltimore merchant Johns Hopkins, its founder, directed that it should equal the best in Europe. In later years its faculty included a number of scholars who left their mark on American protozoology; the most notable were undoubtedly Herbert Spencer Jennings and Robert Hegner.

Rather surprisingly, the earliest contributors to protozoology in the United States, and in the Americas, were more often than not physicians. Both Leidy and Stokes were nineteenth-century medical men for whom the Protozoa were more fascinating than their patients. Indeed Leidy was so captivated by his study of fossilized vertebrates of the remote past, of parasites of contemporary man and animals, and of the microscopic world that he very early gave up the practice of the healing art and never returned to it. Stokes, too, spent more of his time with his microscope than advising patients. Samuel G. Morton, a pioneer in the study of the Foraminiferida in the first half of the nineteenth century, was also a physician; so was William Thomas Councilman, who wrote important works on malaria and amoebic dysentery toward the end of the century, though his chief interest lay in pathology. Like Leidy, he received much of his initial scientific inspiration during a sojourn in Europe, and especially in Austria and Germany. As he remarked in the final lecture of his teaching career at Johns Hopkins, "I came back from Europe very full of all the things which I had learned . . . both at the University and at the Hospital there was that wonderful happiness in work."

Gary N. Calkins (1869–1943) (Fig. 28.1) wrote the first authoritative text in protozoology to appear in this country, which was also one of the first

FIG. 28.1. Gary N. Calkins (1869-1943).

to appear in English. For many years his research centered on conjugation in ciliates, but he also found time to investigate other aspects of the biology of the Protozoa and to write a text in general biology. Furthermore, Calkins trained a large group of graduate students, some of whom in turn achieved eminence as protozoologists. It was the writer's good fortune to receive his introduction to protozoology through Calkins and one of his students, Mary Stuart MacDougall, whose own contributions to the field were considerable.

Enrique Beltran, a distinguished Mexican biologist on the faculty of the University of Mexico and himself the author of a number of studies dealing with the Protozoa (particularly parasites), remarked in a biographical sketch that it was difficult to express adequately protozoology's debt to Calkins. His contributions included not only much writing and research but also the establishment of the first course in protozoology at an American university. Beginning at Columbia, it was also given for a number of years during the summer at the great Marine Biological Laboratory at Woods Hole, Massachusetts. It may be worth recalling that this laboratory is itself the lineal descendant of one started in 1873 on Penikese Island, in Buzzards Bay, by the eminent Swiss-born biologist Louis Agassiz. Inspiration for the founding of this laboratory undoubtedly had its source in Agassiz's early training in Swiss and German universities. When we remember that only a quarter of a century before the inauguration of Calkins' course in protozoology Agassiz was teaching that Protozoa were not distinctive enough to constitute a phylum, Calkin's pioneering becomes especially remarkable. To this day there are few American universities with chairs of protozoology, and to Calkins belongs the honor of having occupied the first one.

Of Calkins' students, L. L. Woodruff (1879–1947), long professor of protozoology, was certainly one of the most notable. He became as interested as Calkins in the significance of conjugation in the ciliates, but for a long time took quite a different view of it. To him, it was not essentially a rejuvenating process (though he eventually came to believe that this was one of its effects), but rather nature's method of promoting survival through the greater production of variations. He also showed that even asexual reproduction results in populations of significantly varying individuals.

In contrast to Calkins and Woodruff, the chief interests of Jennings were genetic; he thought of himself as a geneticist though his research was mainly with the Protozoa, which he saw as tools for research into the broader problems of biology. Like Calkins, he had a number of graduate students and among them was T. M. Sonneborn of Indiana University, widely known for his studies of mating types. Jennings was for many years director of the Zoological Laboratory at Hopkins.

Robert Hegner (1880–1942) was one of Jennings' contemporaries at Johns Hopkins, though in another division of the university. He became primarily interested, however, in the parasitic Protozoa rather than the free-living ones.

His earliest adventure into the realm of protozoology was a study of inheritance in *Arcella* and *Difflugia*, but thereafter he wrote a long series of papers on parasitic species as well as a text on the Protozoa of man. He had a keen sense of humor and one of his later books on parasitic Protozoa was illustrated with numerous cartoons (Fig. 28.2). His *College Zoology* was a zoological Bible for generations of students, and a number of other books from

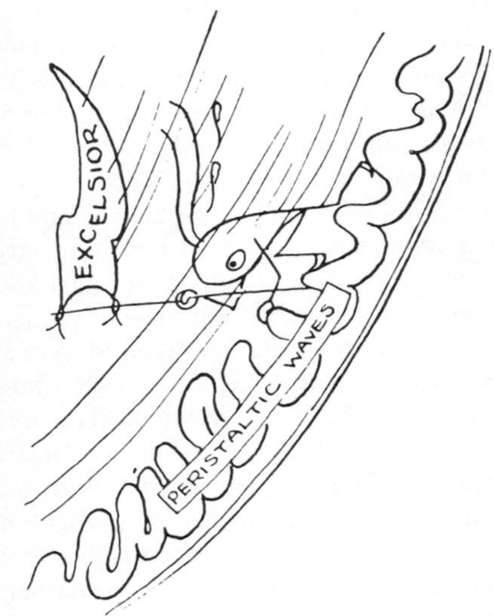

FIG. 28.2. *Left*. Robert Hegner (1880-1942). *Right*. *Giardia lamblia*, a flagellate parasitic in the human duodenum, "riding the peristaltic waves" of the intestine. (Cartoon from Hegner, *Big Fleas Have Little Fleas*.)

his pen aimed to make biology interesting to the general public. Graduate students came to him for training, as they did to Calkins and Jennings, and among them were some now well known for their own work with the parasitic Protozoa.

S. O. Mast (1878–1947) and M. M. Metcalf (1868–1940) were also members of what might be called the Hopkins group. Mast's studies dealt with a variety of problems relating to the free-living Protozoa, but he is probably best known for his work on amoeboid movement. Metcalf spent most of his scientific life delving into the opalinids, and was perhaps the first to suggest that their real affinities are with the flagellates rather than the ciliates.

C. A. Kofoid (1865–1947) served as chairman of the Department of Zoology at the University of California (Berkeley) for more than a quarter of a century. The minute organisms of the microscopic world were his chief interest, though his initial research concerned the early stages of slug eggs, and for a time he was actively engaged in studying the biology of shipworms. Both the free-living and parasitic Protozoa received his attention, but the former particularly fascinated him. Perhaps his greatest single work was the monumental treatise on *The Free-living Unarmored Dinoflagellata*, written with the collaboration of his colleague Olive Swezey. Almost as complete were his studies of the great assemblage of marine ciliates known as the tintinnids, some of which were published with his colleague Arthur Campbell. Graduate students flocked to him during his long career, as they do to most great scholars, and among them was Harold Kirby (1900–52), who himself later achieved recognition as an able protozoologist, especially for his researches on trichomonads and the protozoan parasites of termites.

Most of these pioneers are remembered chiefly for their work with the free-living Protozoa, though several did much in the field of protozoan parasitology. Hegner and Kofoid especially were influential in directing the attention of American physicians and biologists toward the parasitic species, the importance of which had been largely overlooked in this country. We enjoy a climate which, over most of the nation, tends to discourage the spread of the more pathogenic parasites and we have no colonies in the tropics. Both Hegner and Kofoid were among the organizers of the American Society of Parasitologists in 1926 and both were elected to its presidency. The society has since grown until its present membership exceeds a thousand; it now publishes the *Journal of Parasitology*, founded by Henry Baldwin Ward of the University of Illinois, in which appear numerous papers dealing with protozoan parasites as well as with parasites in general.

But though Leidy may be called the father of protozoology in America, and Hegner and Kofoid the godfathers, others were also endowed with a pioneering spirit in the field. Among them (and here the author has drawn heavily on an interesting sketch by Wenrich) were six who should not be forgotten. Daniel E. Salmon (1850–1914) and Theobald Smith (1859–1934) are remembered for their work with *Babesia* (or *Piroplasma*) *bigemina*, the causative agent of Texas cattle fever. Smith, in collaboration with Kilborne in 1893, demonstrated the role of the tick in the dissemination of this disease, then a scourge which threatened the cattle industry. The mode of transmission established, the way was paved for control of the malady.

George Dock (1860–1951) was another American physician for whom Protozoa held a special interest. With the great clinician Sir William Osler, he was one of the first to confirm in this country the role of the plasmodia in causing human malaria, and he was also one of the first to find clear-cut cases of infection with *Entamoeba histolytica* and *Trichomonas vaginalis*.

Charles W. Stiles (1867–1941), like so many of the earlier American scientists, received his graduate training in Germany; both his master's and doctor's degrees were earned at the University of Leipzig. Had it not been for the counsel of the famous German zoologist Leuckart, Stiles would probably have completed the course in medicine, on which he was already well started, and protozoology would have lost the powerful impetus he later gave it in educating the medical and biological public to the importance of animal parasites as causes of human disease. It is said of Leuckart, when Stiles asked his advice, that he advised against medicine saying "Why, you will kill all your patients by experimenting upon them. The physician must cure his patients, not feed them worm eggs as you do laboratory animals. The place for a venturesome spirit like yours is in the laboratory. There are lots of men well fitted to give potions to dowagers, but not many with enough imagination to make progress through experimentation. Science, like navigation, requires a venturesome spirit."*

Colonel Charles F. Craig (1872–1950), who made a career of army medicine, had ample opportunity in his assignments in many parts of the world to observe the role of parasites as causes of human and animal disease, and he made numerous important studies of malaria and amoebiasis. Instrumental in organizing a department of tropical medicine at the Medical School of Tulane University in New Orleans, one of the first such departments in any American university, he was also the author of several textbooks in this and related subjects.

William G. MacCallum (1874–1944), who later became one of the best-known pathologists in the world (though he always insisted, when introduced as "the greatest of living pathologists," that he was "only second best"), is remembered by protozoologists for his discovery, while still a medical student at Johns Hopkins, of the sexual cycle in the malaria-related parasite *Haemoproteus* of crows. He immediately appreciated the significance of the exflagellating gametocytes, and applied it to the life cycle of the malaria parasites, in which such forms had been often observed but not understood.

One more of MacCallum's contemporaries should also be mentioned. Ernest Edward Tyzzer (1875–   ) is also a pathologist, but he early developed an interest in the parasitic Protozoa, about which he published a number of papers, including several important ones on coccidia and on blackhead on turkeys.

Of course there have been others whose pioneering studies have made protozoological history in America. Some of the earliest investigations dealt not with living forms but with foraminiferidan and radiolaridan fossils. Among these early students of the Protozoa were men in professions as varied as chemistry, geology, and the ministry. Undoubtedly the most illustrious of them all was Joseph A. Cushman (1881–1949). After taking his doctorate in zoology at Harvard in 1909, he became fascinated by micropalaeontology, and eventually proved to oil drillers so convincingly the utility of a knowledge of foraminiferidan fossils that he made himself wealthy as well as famous. (Undoubtedly he is the only man who ever made a fortune from protozoology.)

A decisive step in protozoology's "coming of age" as a recognized branch of biology was taken in 1947, when a small group organized the Society of

*Quoted from "Charles Wardell Stiles, Intrepid Scientist," by F. G. Brooks, *Bios*, 18: 139, 1947.

Protozoologists. Even at its inception the Society was international in outlook, and thus membership is open to all protozoologists and graduate students in the field, whatever their country of origin. The society publishes the *Journal of Protozoology*, in which any acceptable study dealing with the Protozoa, whether free-living or parasitic, may be published. There are of course other journals (see p. 603) open to the protozoologist, but no other is limited to studies in this field.

The maturity of American protozoology is also evident in the number of colleges and universities that offer courses in protozoology. Common in undergraduate curricula, they are not at all infrequent at the graduate level. Many institutions also support research in protozoological problems. These include not only institutions of higher learning, but also privately and publicly supported agencies, such as the Rockefeller Institute, the Haskins Laboratories in New York City, the National Institutes of Health, National Science Foundation, and the U.S. Bureau of Animal Industry. The annual cost of such research runs into millions of dollars.

Now, scarcely half a century after Calkins offered the first course in protozoology in an American university, instruction in the subject and research on problems involving the Protozoa are being actively prosecuted throughout the land. Protozoology has indeed come of age.

REFERENCES

Three American Microscope Builders. 1945. American Optical Company.

Ball, G. H. and Hall, R. P. 1953. In Memoriam: Harold Kirby. *J. Parasitol.*, 39: 110–12.

Beltran, E. 1943. La deuda de la protozoologia con Gary N. Calkins. *Rev. Soc. Mex. Hist. Nat.*, 4: 97–114.

——— 1947. Lorande Loss Woodruff (1879–1947), miembro honorario de la Sociedad Mexicana de Historia Natural, y sus investigaciones protozoologicas. *Rev. Soc. Mex. Hist. Nat.*, 8: 15–28.

Brooks, F. G. 1947. Charles Wardell Stiles, intrepid scientist. *Bios*, 18: 139–69.

Cushing, H. 1933. William Thomas Councilman. *Science*, 77: 613–8.

Faust, E. C. 1951. Charles Franklin Craig. An appreciation. *Am. J. Trop. Med.*, 31: 267–9.

Kirby, H. 1947. Charles Atwood Kofoid. *Science*, 106: 462–3.

Nicholas, J. S. 1954. Lorande Loss Woodruff. *J. Protozool.*, 1: 4–5.

Osborne, H. F. 1913. Sketch of Joseph Leidy. In *Biographical Memoirs*, published by National Academy of Sciences, 7: 339–96.

Sonneborn, T. M. 1948. Herbert Spencer Jennings (1867–1947). *Genetics*, 33: 1–4.

Wenrich, D. H. 1955. Some early Philadelphia zoologists. *Proc. Penn. Acad. Sci.*, 29: 22–35.

——— 1956. Some American pioneers in protozoology. *J. Protozool.*, 3: 1–7.

Woodruff, L. L. 1943. The advent of the microscope at Yale College. *Am. Sci.*, 31: 241–5.

SOURCES OF INFORMATION

The following list of periodicals and books includes those in which most of the work in protozoology is published. There are others published in Central and Eastern Europe in which valuable papers appear, but which are seldom available in this country. Journals chiefly useful as sources of information about parasitic Protozoa are marked with an asterisk.

PERIODICALS

Acta Protozoologica, Warsaw.
Anais da Academia Brasileira de Ciencias, Rio de Janeiro.
**American Journal of Hygiene*, Baltimore.
**American Journal of Tropical Medicine and Hygiene*, Baltimore.
American Midland Naturalist, Notre Dame.
American Naturalist, Tempe, Ariz.
American Scientist, Easton, Penn.
Annales de l'Institut Pasteur, Paris.
Annals of the New York Academy of Sciences, New York.
**Annales de Parasitologie Humaine et Comparée*, Paris.
Annales de Protistologie, Paris.
Annales des Sciences Naturelles, Paris.
**Annals of Tropical Medicine and Parasitology*, Liverpool.
Annual Review of Microbiology, Stanford.
Archives d'Anatomie Microscopique et de Morphologie Expérimentales, Paris.
**Archives de l'Institut Pasteur d'Algérie*, Algiers.
Archives de l'Institut Pasteur, Paris.
**Archives de parasitologie*, Paris.
Archiv für Protistenkunde, Jena.
Archives de Zoologie expérimentale et génerale, Paris.
Biological Abstracts, Philadelphia.
Biological Bulletin, Lancaster, Penn.
Biological Reviews, London.
Biologische Zentralblatt, Leipzig.
Bulletin Biologique de la France et de la Belgique. Paris.
Bulletin de la Société Française de Microscopie, Paris.
**Bulletin de la Société de pathologie exotique*, Paris.
Bulletin de la Société zoologique de France, Paris.
Canadian Journal of Zoology, Ottawa.
Comptes rendus des séances de l'Académie des Sciences, Paris.
Comptes rendus des séances de la Société de Biologie, Paris.
Cytologia, Tokyo.
Ecology, Durham, N.C.
Experimental Cell Research, New York.
**Experimental Parasitology*, New York.
Genetics, Austin, Texas.
Hydrobiologia, The Hague.
Illinois Biological Monographs, Urbana, Ill.

**Indian Journal of Malariology*, Delhi.
**Journal of the American Medical Association*, Chicago.
Journal of Cellular and Comparative Physiology, Philadelphia.
**Journal of Experimental Medicine*, Baltimore.
Journal of Experimental Zoology, Philadelphia.
Journal of General Microbiology, London.
Journal of Heredity, Baltimore.
Journal of Infectious Diseases, Chicago.
Journal of Morphology, Philadelphia.
**Journal of Parasitology*, Lancaster, Penn.
Journal of Protozoology, Utica.
Journal of the Royal Microscopical Society, London.
Nature, London.
Naturwissenschaften, Berlin.
Ohio Journal of Science, Columbus.
**Parasitology*, Cambridge, England.
Physiological Zoology, Chicago.
Proceedings of the American Philosophical Society, Philadelphia.
Proceedings of the National Academy of Sciences, Washington, D.C.
Proceedings of the Society for Experimental Biology and Medicine, Utica.
Protistologica, Paris.
Protoplasma, Vienna.
Protoplasmatologia, Vienna.
Quarterly Journal of Microscopical Science, Oxford.
Quarterly Review of Biology, Baltimore.
Revue Suisse de Zoologie, Geneva.
**Rivista di Malariologia*, Rome.
**Rivista di Parassitologia*, Rome.
Science, Washington, D.C.
Scientific American, New York.
Scientific Monthly (now incorporated into *Science*), Washington, D.C.
Systematic Zoology, Baltimore.
Transactions of the American Microscopical Society, Columbus.
Transactions of the New York Academy of Sciences, New York.
**Transactions of the Royal Society of Tropical Medicine and Hygiene*, London.
**Tropical Diseases Bulletin*, London.
University of California Publications in Zoology, Berkeley.
Verhandlungen der Deutschen Zoologischen Gesellschaft, Leipzig.
Zeitschrift für Naturforschung. Ausgabe B. Chemie, Biochemie, Biophysik, Biologie und verwandte Gebiete, Tübingen.
**Zeitschrift für Parasitenkunde*, Berlin.
Zeitschrift für wissenschaftliche Mikroscopie und für Mikroskopische Technik, Stuttgart.
**Zentralblatt für Bakteriologie, Parasitenkunde und Infektionskrankheiten*, Jena.

Books

Calkins, G. N. 1901. *The Protozoa*. New York: Macmillan.
——— 1933. *Biology of the Protozoa*. Philadelphia: Lea & Febiger.
——— and Summers, F. M. 1941. *Protozoa in Biological Research*. New York: Columbia University Press.

Chandler, A. C. and Reid, C. 1961. *Introduction to Parasitology*, 10th ed. New York: Wiley.

Corliss, J. O. 1961. *The Ciliated Protozoa*. New York: Pergamon Press.

Craig, C. F. 1926. *A Manual of the Parasitic Protozoa of Man*. Philadelphia: Lippincott.

——— 1942. *Laboratory Diagnosis of Protozoan Diseases*. Philadelphia: Lea & Febiger.

——— and Faust, E. C. 1957. *Clinical Parasitology*, 6th ed. (rev. by E. C. Faust and P. F. Russell). Philadelphia: Lea & Febiger.

Doflein, F. (rev. and cont. by Eduard Reichenow). 1953. *Lehrbuch der Protozoenkunde*. Teil 1. Allgemeine Naturgeschichte der Protozoen. Teil 2. 1. Hälfte: Mastigophora und Rhizopoden. Teil 2. 2. Hälfte. Sporozoa und Ciliophora. pp. 436; 411–776; 777–1213.

Grassé, P.-P. (ed). 1952. *Traité de Zoologie*. Tome 1. Fasc. 1. Phylogénie. Protozoaires: Généralité. Flagellés. Paris: Masson.

——— 1953. *Traité de Zoologie*. Tome 1. Fasc. 2. Rhizopodes, Actinopodes, Sporozoaires, Cnidosporidies. Paris: Masson.

Grell, K. G. 1956. *Protozoologie*. Berlin: Springer Verlag.

Hall, R. P. 1953. *Protozoology*. New York: Prentice-Hall.

Hegner, R. and Andrews, J. (eds). 1930. *Problems and Methods of Research in Protozoology*. New York: Macmillan.

——— and Taliaferro, W. H. 1925. *Human Protozoology*. New York: Macmillan.

Hutner, S. H. and Lwoff, A. (eds). 1955. *Biochemistry and Physiology of Protozoa*, vol. 2. New York: Academic Press.

Jahn, T. L. and Jahn, F. F. 1949. *How to Know the Protozoa*. Dubuque: Brown.

Jirovec, O. (et al.). 1953. *Protozoologie*. Prague: Nakladat, Ceskosl. Akad. Ved.

Kent, W. S. 1880–1. *Manual of the Infusoria*. London: David Bogue.

Kudo, R. R. 1944. *Manual of Human Protozoa*. Springfield: Charles C Thomas.

——— 1966. *Protozoology*. 5th ed. Springfield: Charles C Thomas.

Lankester, E. R. 1903–9. *A Treatise on Zoology*. London: Black.

Lwoff, A. (ed). 1951. *Biochemistry and Physiology of the Protozoa*. Vol. 1. New York: Academic Press.

Minchin, E. A. 1912. *An Introduction to the Study of the Protozoa*. London: Edward Arnold

Morgan, B. B. and Hawkins, P. A. 1948. *Veterinary Protozoology*. Minn.: Burgess.

von Prowazek, S. 1920. *Handbuch der Pathogenen Protozoen*. Leipzig: Johann Ambrosius Barth.

v. Stein, F. R. 1859–83. *Der Organismus Infusionsthiere*. Leipzig: Verlag von Wilhelm Engelmann.

Tartar, V. 1961. *Biology of Stentor*. New York: Pergamon Press.

Thomson, J. G. and Robertson, A. 1929. *Protozoology*. New York: Wood.

Wenyon, C. M. 1926. *Protozoology*. New York: Wood.

Additional Books

Chen, T. T. 1963. *Research in Protozoology*. New York: Pergamon Press.

Hutner, S. H. (ed.). 1964. *Biochemistry and Physiology of the Protozoa*. Vol. 3. New York: Academic Press.

Levine, N. D. 1961. *Protozoan Parasites of Domestic Animals and Man*. Minneapolis: Burgess.

Lundin, F. C. and West, L. S. 1963. *The Free-living Protozoa of the Upper Peninsula of Michigan.* Monographic Series No. 1, Marquette: Northern Michigan College Press.

MacKinnon, D. L. and Hawes, R. S. J. *An Introduction to the Study of Protozoa.* London: Oxford University Press.

Richardson, U. F. and Kendall, S. B. 1957. *Veterinary Protozoology.* Rev. ed. Edinburgh and London: Oliver and Boyd.

GLOSSARY

ABLASTIN: a reproduction-inhibiting antibody produced by the rat in response to infection with *Trypanosoma lewisi.*

ACEPHALINE: "without a head" (designating a group of gregarines).

ACTINOPODIUM: see AXOPODIUM.

ADORAL ZONE: the specialized zone of membranelles leading to the cytostome in ciliates.

AEROBIC: able to live only in the presence of free oxygen.

ALLELE: one of a pair (or, sometimes, of a number) of genes affecting the same trait; in a given individual there would never be more than two.

ALLELOCATALYTIC effect: the accelerating effect on reproduction in some Protozoa resulting from population increase.

AMICRONUCLEATE: without a micronucleus.

AMITOSIS: nuclear division without the formation of a spindle; only certainly known to occur in the macronucleus of ciliates.

AMOEBULA: a small amoeboid stage occurring in the life cycles of organisms such as the slime molds; a small amoeba.

AMPHIMIXIS: the union of two nuclei, each derived from a different organism (as in the wandering and stationary pronucleus commonly observed in most conjugating ciliates).

AMPHINUCLEUS: the fertilization nucleus (e.g., the product of the wandering and stationary pronucleus of conjugating ciliates); now more commonly known as the synkaryon.

ANAEROBIC: able to live only in the absence of free oxygen.

ANARCHIC FIELD: unorganized group of kinetosomes destined to migrate and determine the development of certain structures in the course of ciliate differentiation.

ANISOGAMETES: gametes in which the males are morphologically unlike the females.

ANTERIOR STATION: the anterior portion of the gut of insect vectors of trypanosomiasis.

ANTIBODY: a substance produced by the body in response to the presence of some foreign substance (especially proteins or polysaccharides).

ANTIGEN: a substance, such as a protein or polysaccharide, capable of stimulating the body to produce antibodies.

ARENACEOUS: shells incorporating foreign bodies, such as sand grains.

AUTOGAMY: reorganization involving the union of two nuclei, both derived from a single parent nucleus; when applied to ciliates (e.g., *Paramecium*), reorganization very similar to conjugation but involving a single individual.

AUTOTROPHIC: able to manufacture food from simple inorganic substances (e.g., as in the chlorophyll-bearing flagellates).

AXENIC: cultures containing only a single species of organism.

AXIAL FILAMENT: the central filament of an axopodium or a flagellum (but should be distinguished from the eleven longitudinal filaments revealed by electron microscopy).

AXONEME: the fibril extending from the base of a flagellum to its blepharoplast.

AXOPODIUM: a pseudopodium containing an axial filament (as in the Heliozoida).

AXOSTYLE: an elongate structure running longitudinally in flagellates such as the trichomonads, and usually projecting posteriorly.

BASAL DISC: anchoring organelle at aboral pole of sessile ciliates.

BASAL GRANULE: a minute body underlying a cilium (the kinetosome); believed to be homologous with the blepharoplast of the flagellum.

BINARY FISSION: division into two offspring.

BLEPHAROPLAST: basal granule of the flagellum (thought to be homologous with the kinetosome); gives rise to axoneme.

BUCCAL CAVITY: a depression associated with the mouth of ciliates and with organelles of ciliary origin, such as undulating membranes and membranelles.

CARYONIDE (or KARYONIDE): a ciliate stock derived from any member of a clone in which cellular reorganization has occurred (thus a single clone may give rise over a period of time to a number of caryonides).

CENTROBLEPHAROPLAST: an organelle having both flagellar and mitotic functions characteristic of flagellates such as *Trichomonas* and of some Mycetozoa and Actinopoda (where it is associated with axopodia).

CEPHALIN: an epimerite-bearing gregarine trophozoite.

CEPHALINE: having a head.

CHEMOAUTOTROPHIC: able to elaborate energy sources from inorganic substrates.

CHITIN: a horny substance, like that of arthropod exoskeletons; actually said to be a monosaccharide, d-glucosamine. Cysts of some Protozoa have been claimed to consist of chitin, but this is now questioned.

CHOANOCYTES: collared cells with flagella typical of the gastral (inner layer) of sponges and very similar to the collared flagellates known as the Choanoflagellata Kent; (also placed in the family Codosigidae).

CHONDRIOSOME: a term often used for the mitochondria of the Protozoa.

CHROMATOID (or CHROMATOIDAL BAR): body commonly occurring in the young cysts of intestinal amoebae and taking basophilic stains, such as haemotoxylin; probably consists of reserve foodstuffs.

CHROMATOPHORE: a discrete body of chlorophyll (or other pigment); a chloroplast.

CHROMIDIA: cytoplasmic granules taking basophilic stains (e.g., haemotoxylin), and for a long time thought to consist of extranuclear chromatin.

CILIOSPORE: a minute ciliated stage occurring in the life cycles of ciliates such as the fish parasite *Icthyophthirius*: small replicas of the adult but lacking a mouth.

CILIUM: a whiplike locomotor organelle, fundamentally similar to a flagellum, but typically much more numerous and not restricted to the anterior region as flagella usually are.

CIRCUMFLUENCE: the engulfing of food particles by means of the cytoplasm flowing around them, as in certain amoebae.

CIRCUMVALLATION: the pseudopodial capture of food by surrounding it, but without actual contact of pseudopodia and food particles.

CIRRUS: an organelle resulting from the fusion of a tuft of cilia, used chiefly for locomotion; characteristic of spirotrich ciliates.

CLONE: a stock of organisms all derived from the same ancestor, hence usually having an identical genetic constitution.

COCCOLITH: a type of skeletal platelet characteristic of the Coccolithithina, a suborder of Chrysomonadida; such platelets are often found as fossils in chalk.

COELOZOIC: living in one of the cavities of the host.

COENOBIUM: a colony of flagellates (as in *Volvox*).

COLONIAL: used for Protozoa which live as an integrated group (as distinguished from a community or simple aggregation); in a true colony the number of constituent organisms is often fixed and there is likely to be some division of labor.

COMMENSAL: living within or on some other organism, but without doing it harm.

COMPACT: a term applied to protozoan nuclei containing a relatively large amount of chromatin (as opposed to nucleoplasm); e.g., the macronucleus of ciliates.

CONJUGATION: a type of sexual union typical of the ciliates; usually temporary.

CONTRACTILE VACUOLE: a pulsating vacuole, especially characteristic of free-living forms; regarded as hydrostatic in function.

COPROZOIC: applied to organisms found living in feces, or in water polluted with feces.

CORTEX: the outer layer of a protozoan (also known as ectoplasm or ectosarc).

COSTA: a rodlike supporting organelle of trichomonads; extends along undulating membrane.

CRESTA: a marginal, bandlike structure of varying size, easily demonstrable with haemotoxylin, in the descovinid flagellates of termites.

CRISTA: an ultramicroscopic membrane within a mitochondrion.

CRYPTOZOITE: the stage developing from the malarial sporozoite.

CYCLOSIS: movement of cytoplasm within a cell.

CYTOGAMY: a kind of cellular reorganization occurring in ciliates such as *Paramecium*, involving fusion of gametic nuclei produced within each member of a conjugating pair (rather than a mutual exchange of pronuclei).

CYTOPHARYNX: that part of the food canal lying just back of the cytostome; nonciliated and often very short.

CYTOPROCT: cell anus (also known as a cytopyge).

CYTOPYGE: cell anus.

CYTOSTOME: cell mouth.

DAUERMODIFIKATION: a term coined by the German protozoologist Jollos to designate enduring modifications or adaptations developed by Protozoa exposed to abnormal environments or circumstances.

DEFINITIVE HOST: the host in which a parasite undergoes its sexual cycle.

DEUTOMERITE: the more posterior of the two body regions in a cephaline gregarine, the anterior being the protomerite.

DIASTOLE: the period of relaxation, or filling, of the contractile vacuole.

DINOCARYON: a type of nucleus in which the chromatin appears in rows of granules, apparently corresponding to the chromosomes, in the "interphasic" or resting stage (as in the dinoflagellates and some coccidia).

ECTOCYST: outer layer of cyst wall.

ECTOPARASITE: a parasite living on the outside of its host.

ECTOPLASM: outer or covering layer of cytoplasm.

ECTOSARC: outer layer of cytoplasm (ectoplasm).

ENDOCYST: inner layer of cyst wall.

ENDOMIXIS: a term coined by Woodruff and Erdmann to designate a periodic asexual reorganization in *Paramecium*, without synkaryon formation.

ENDOPLASM: internal cytoplasm (i.e., between nucleus and ectoplasm or cortical zone).

ENDOSARC: inner cytoplasm (endoplasm).

ENDOSOME: intranuclear body (often called a karyosome).

EPICYTE: thick outer layer of gregarine body.

EPIMERITE: anchoring organelle of cephaline gregarines.

EXFLAGELLATION: microgametogenesis in such blood Protozoa as the malaria parasites and their close relatives.

EXOERYTHROCYTIC: living in cells other than erythrocytes (used especially for the tissue stages of the malaria parasites).

FACULTATIVE ANAEROBE: an organism capable of living both in the presence and absence of free oxygen (although preferring the latter).

FEULGEN STAIN: a stain serving as a test for DNA (also called thymonucleic acid).

FILOPODIUM: pseudopod producing filamentous (but not anastomosing) branches.

FLAGELLUM: a whiplike organelle.

FLIMMER-FLAGELLUM: flagellum with numerous lateral branches (often called a ciliary flagellum).

FORAMEN (pl. FORAMINA): minute openings through which pseudopodia may be extruded in the shells of Foraminiferida or, more correctly, openings allowing cytoplasmic communication between the chambers of such shells. Also used for any minute opening, as that in the spores of Myxosporida.

GAMETOCYST: cystlike envelope formed by gregarines immediately after pairing.

GAMETOCYTE: parental stage from which gametes or mature sex cells develop.

GAMOGONY: reproduction leading to the formation of gametes.

GASTRIC VACUOLE: food vacuole.

GEMMATION: formation of buds (gemmules).

GEMMULE: a bud from which a young organism develops (by gemmation).

GENOTYPE: genetic constitution of an individual.

GIRDLE: transverse groove characteristic of dinoflagellates, and containing one of the flagella.

GLYCOGEN: a carbohydrate of very high molecular weight used by animals as a reserve foodstuff; built up of numerous molecules of glucose.

GOLGI APPARATUS: a canalicular system within cells made visible by osmic acid fixation and subsequent staining (existence doubtful in Protozoa).

HAEMATOCHROME: a bright red pigment occurring in some plantlike flagellates; allied to chlorophyll.

HAEMOZOIN: malarial pigment; by-product of haemoglobin metabolism.

HEMIXIS: nuclear reorganization involving the macronucleus only; so far known only in a few ciliate species (e.g. *Paramecium*).

HETEROAUTOTROPHIC: designates a type of nutrition in which the energy source is organic (e.g., from acetate).

HETEROTROPHIC: applied to a mode of nutrition used by colorless organisms, where an organic source of carbon is required.

HISTOZOIC: living in tissue.

HOLOPHYTIC: see AUTOTROPHIC.

HOLOZOIC: used for animal-like nutrition, in which the organism ingests living prey.

HYPERPARASITE: a parasite of a parasite.

IMPORT: the taking of food upon contact, but with little movement on the part of the amoeba (coined by Rhumbler).

INFRACILIATURE: the complex system of granules and fibrils underlying cilia; located beneath the pellicle.

INQUILINES: commensals.

INTERMEDIATE HOST: the host in which the parasite undergoes its asexual cycle.

INVAGINATION: a folding in of ectoplasm, as when an amoeba uses this method of engulfing food (coined by Rhumbler).

INVERSION: one of the stages in the maturation of a *Volvox* colony, in which the inner flagellated layer becomes the outer one.

ISOGAMETE: used for gametes morphologically alike in both sexes.

KAPPA: particles resembling genes in their behavior (though equally resembling virus particles) occurring in the cytoplasm of certain species of *Paramecium*, and determining the production of a substance (*paramecin*) which, when liberated into the medium, kills paramecia lacking *kappa*.

KARYOSOME: a discrete body within the nucleus.

KATHAROBIC: an environment rich in oxygen and essentially free of pollution.

KERNSPALT: a transverse nuclear fissure appearing in the macronucleus of some ciliates (particularly hypotrichs) during division; it moves from one end of the nucleus to the other as division progresses.

KINETODESMA: a fibril lying to the right of and under each kinetosomal row.

KINETOPLAST: the parabasal body of trypanosomes and related forms.

KINETOSOME: a minute body (basal granule) underlying the cilium, capable of independent reproduction, and giving rise to cilia, trichocysts, and trichites.

KINETY: a row of kinetosomes with their associated fibril, the kinetodesma.

LOBOPODIUM: a pseudopodium of fingerlike (blunt) character.

LORICA: an envelope or covering fitting the protozoon somewhat loosely.

MACROGAMETE: female sex cell.

MACROGAMETOGENESIS: maturation of the macrogametocyte culminating in the formation of the macrogamete.

MACRONUCLEUS: the larger or vegetative of the two types of nucleus of ciliates; now believed to be polyploid.

MASTIGONEMES: laterally branching fibrils ("flimmer") of the flagella of some species.

MASTIGONT: a term applied to the complex of kinetic elements underlying the flagellum.

MATING TYPE: a population of ciliates which do not conjugate among themselves, but with those of a complementary group; individuals of like mating type appear to have the same macronuclear genetic constitution.

MEGALOSPHERIC: the uninucleate form of two assumed by Foraminiferida; initially larger than the second or microspheric form.

MEIOSIS: nuclear division with a reduction of chromosome number to the haploid condition.

MEMBRANELLE: an organelle, triangular, rectangular, or ribbonlike in cross section, formed by fusion of adjacent filia in the peristomal area; its ciliary base is two to four granules across.

MERIDIAN: a row of cilia more or less encircling the organism.

MERONT: the schizont of a microsporidan (but used occasionally as a synonym for schizont without regard to the sporozoan group).

MEROZOITE: a daughter organism resulting from multiple fission or schizogony.

MESOCYST: middle cyst wall (not always present in cysts).

MESOMITOSIS: a type of moderately advanced mitosis seen in some Protozoa (infrequently used today).

MESOSAPROBIC: an environment in which oxidation and decomposition are only moderately active.

METACRYPTOZOITE: second exoerythrocyte (post-sporozoite) generation of the malaria parasite.

METACYCLIC: individuals which have recently completed some phase of the life cycle (e g., trypanosomes which have reached the infective stage in the vector).

METACYSTIC: a freshly excysted individual.

METAMITOSIS: the more advanced types of mitosis seen in Protozoa.

MICROGAMETE: male sex cell.

MICROGAMETOGENESIS: maturation of the microgametocyte resulting in microgamete formation.

MICRONUCLEUS: the smaller (generative) of the two types of ciliate nuclei.

MICROSPHERIC: one of the two forms assumed by Foraminiferida; initially smaller than the other (macrospheric) form, and often multinucleate.

MITOCHONDRIA (sing. MITOCHONDRION): cytoplasmic structures concerned with metabolic, especially respiratory, activities of the cell, and almost universally present in the Protozoa.

MIXOTROPHIC NUTRITION: nutrition of both animal and vegetal types occurring at the same or different times in the same individual (synonym, HETEROTROPHIC).

MONILIFORM: beadlike (e.g., the macronuclei of some ciliates).

MONOCYSTID: the nonseptate gregarines.

MONOENERGID: with one nucleus.

MOTORIUM: an organelle of numerous species of ciliates, which, with its connecting fibrils, is believed to have a motor coordinating function; located anteriorly usually, often near cytostome.

MULTIPLE FISSION: splitting of the parent organism into numerous offspring (the "merozoites" or "schizozoites").

MUTUALISM: a relationship between two species of organisms for the good of both (SYMBIOSIS).

MYOCYTE: innermost layer of the gregarine body envelope.

MYONEME: a contractile filament occurring in the cytoplasm of numerous species of Protozoa (e.g., in the stalk of *Vorticella*).

NEMATOCYSTS: stinging cells typical of the coelenterates, and occurring also in some dinoflagellates.

NEUROID: a fibril associated with the myonemes of certain ciliates (e.g., *Stentor*).

NEUROMOTOR APPARATUS: a fibrillar system believed to have a motor coordinating function in many species of ciliates; its center is the motorium.

OCELLUS: eyespot of a peculiar type found in the dinoflagellate family Pouchetiidae.

OLIGOSAPROBIC: a mineral-rich environment, with little or no decomposition.

OOCYST: a cyst containing the zygote; often a stage in the sporozoan life cycle.

OOKINETE: the motile zygote of the parasite of malaria and of its relatives.

OOSPHERES: ripe macrogametes of *Volvox*.

OPERCULUM: the minute lid or cover closing the opening of cysts of some Protozoa.

OPISTHE: the posterior of two daughter cells resulting from binary fision in the ciliates.

ORAL APPARATUS: the mouth parts of a ciliate considered collectively.

ORGANELLES: the protozoan analogue of a metazoan organ.

ORTHOMITOSIS: type of mitosis in which there are spindle fibers and a typical metaphase.

OSMOTROPHY: a type of nutrition in which food is absorbed in solution.

PAEDOGAMY: union of two nuclei one or two generations removed from their single gametocyte ancestor (as in microsporidan spores).

PALMELLA: a stage in the life cycle of many flagellates involving the aggregation of numerous individuals in a matrix of mucus, simulating a colony.

PANSPOROBLAST: the multinucleated sporont from which two spores usually develop (as in the myxosporidian).

PANSPOROCYST: the parent organism within which the spores of Actinomyxidia are formed.

PARABASAL BODY: a conspicuously staining body lying at or near the base of the flagellum in numerous parasitic flagellates (e.g., those of termites and wood-roaches).

PARAGLYCOGEN: a type of reserve foodstuff of starchlike character especially characteristic of gregarines.

PARAMECIN: an antibiotic secreted by *kappa*-bearing paramecia; at least two varieties of this substance occur.

PARAMYLUM (PARAMYLON): a starch-related carbohydrate forming the reserve foodstuff of many phytoflagellates.

PARTHENOGENESIS: development of an egg without fertilization, or reorganization in a protozoan with changes simulating those following fertilization (as in autogamy).

PATHOGEN: a parasite or infectious agent which harms its host.

PEDUNCLE: a stalklike organelle of attachment (as in sessile peritrichs).

PELAGIC: drifting or floating freely about in the ocean.

PELLICLE: the relatively stiff outer envelope of some Protozoa (but not to be confused with a secreted covering or lorica).

PELTA: a somewhat crescentic membranelike structure occurring in trichomonads, lying to the right of the blepharoplast complex, and best demonstrated with silver staining.

PENICULUS (also spelled PENNICULUS): a compound ciliary organelle present in the buccal cavity of certain hymenostomes (two or more occur).

PERISTOME: the area (usually specialized) immediately about the cytostome; more properly restricted to the buccal cavity.

PHAGOTROPHY: the ingestion of food in solid form.

PHANEROZOITE: the tissue (exoerythrocytic) stage of the malaria parasite, derived from the metacryptozoites.

PHARYNGEAL BASKET: a reinforcing structure of trichites about the cytostome and gullet (as in *Chilodonella*).

PHENOTYPE: the apparent genetic characteristics of an individual (as distinct from the actual inherited gene pattern, or *genotype*).

PHIALOPORE: the small opening in a young *Volvox* colony through which the organism may turn itself inside out (i.e., accomplish inversion).

PHORONT: encysted stage of apostomatous ciliates in which they are attached to the crustacean host and undergo reorganization.

PHOTOTAXIS: response of an organism to light.

PHOTOTROPHIC: able to synthesize foodstuffs by light (synonym "HOLOPHYTIC").

PHOTOTROPISM: response to light (almost synonomous with *phototaxis*, but implies attraction).

PLANKTON: the aggregate of the drifting or feebly swimming organisms in any body of water (but refers especially to the algae and Protozoa).

PLANONT: an amoebula from a freshly germinated microsporidan spore.

PLASMAGEL: the stiffer layer lying between plasmalemma and plasmasol of an amoeboid organism.

PLASMAGENE: a hypothetical cytoplasmic bearer of hereditary traits.

PLASMALEMMA: the elastic outer layer of an amoeboid organism.

PLASMASOL: the inner and more fluid portion of an amoeboid organism.

PLASMODIUM: an aggregation of a variable number of organisms (as in the slime molds) with cytoplasmic fusion, thus making a multinucleate mass (synonym, *syncytium*); not to be confused with the genus *Plasmodium*, to which the malaria parasites belong.

PLASMOTOMY: a form of binary fission occurring in some multinucleated Protozoa, in which each of the offspring receives half the nuclei, more or less; although there is cytoplasmic fission the nuclei do not divide.

PLASTIDS: cytoplasmic living structures in cells (e.g., the pyrenoids of flagellates).

PLEUROMITOSIS: a type of mitosis in which the chromosomes are inserted on the nuclear membrane or at the base of a centrosomal fibrillar cone, and in which a typical metaphase is lacking.

POLAR CAPSULE: a peculiar structure occurring in Sporozoa of certain groups, such as the Cnidosporidia, from which a long filament may be explosively discharged.

POLAR FILAMENT: the extrusible filament within a polar capsule.

POLYCYSTID: the septate gregarines.

POLYENERGID: multinucleate.

POLYMORPHIC: taking on many forms, or highly variable in form, as with many trypanosomes.

POLYSAPROBIC: an environment poor in oxygen and rich in decomposition products.

POSTERIOR STATION: the posterior portion of the gut of insect vectors of trypanosomiasis.

PREMUNITION: a type of immune response dependent on the continued presence of the infectious agent; occurs especially in protozoan and spirochetal infections.

PROLOCULUM: the initially formed chamber in the foraminiferidan shell.

PROMITOSIS: a simple type of mitosis (e.g., seen in many amoebae).

PRONUCLEUS: the final product of the maturation or pregamic divisions of the micronuclei in ciliates.

PROTER: the anterior of the daughter organisms resulting from binary fission in ciliates.

PROTOMERITE: the anterior region of the body of cephaline gregarines.

PSAMMOPHILOUS: organisms preferring a beach sand environment.

PSEUDOCHITIN: a substance resembling chitin, such as the base material of foraminiferidan shells.

PSEUDOPODIUM: a temporary cytoplasmic process used for locomotion or food-getting (especially characteristic of the Sarcodina).

PULSELLUM: a flagellum which propels by pushing (whether locomotion of this sort ever exists in Protozoa is disputed).

PUSULE: a reservoir or vacuole of unknown function, generally filled with a rose-colored liquid, occurring in dinoflagellates.

PYLEA: the apertures in the central capsule of a radiolaridan shell, permitting cytoplasmic connection between ectoplasm and endoplasm.

PYRENOID: a body associated with the chromatophore or chloroplast, apparently functioning in starch formation.

RADIATING (FEEDER) CANALS: radially arranged canals emptying into contractile vacuoles of ciliates such as *Paramecium.*

REGENERATION: the ability of an organism to repair injury by the reproduction of lost parts.

RESERVOIR: cavity lying at the base of the anterior canal, or so-called gullet, of the euglenid flagellates.

RHIZOPLAST: a connecting filament, such as that between the blepharoplast and the nucleus of flagellates.

RHIZOPODIUM: pseudopodium which both branches and shows union or anastomosis of branches (synonyms MYXOPODIUM, RETICULOPODIUM).

RING: youngest erythrocytic stage of the malaria parasite.

RODS: see TRICHITES.

ROSETTE: a specialized organelle occurring near the cytostome of apostome ciliates.

ROSTRUM: attaching organ of monocystid gregarines (corresponds to *epimerite* of cephaline gregarines).

SAPROBIC: environment in which decomposition is active.

SAPROPHYTIC: a type of nutrition in which foodstuffs are absorbed in solution (synonym SAPROZOIC).

SARCOCYTE: the layer immediately underlying the epicyte or outer envelope of gregarines.

SCHIZOGONY: reproduction by multiple fission.

SCHIZONT: an organism in which multiple fission has started, as usually evidenced by division of the nucleus.

SCOPULA: an organelle, itself derived from cilia, from which originates the stalk of certain ciliates (e.g., Suctorida).

SEGMENTER: a ripe schizont of the erythrocytic cycle of the malaria parasite.

SELFING: conjugation between individuals of the same clone, thus indicating a change in mating type of at least one of the pair, due to cellular reorganization.

SENSITIVES: paramecia which are susceptible to the lethal action of *paramecin* from *kappa*-bearing individuals of the same species.

SENSORY BRISTLES: minute cilia occurring dorsally on hypotrichs, and non-motile cilia of other ciliates, supposedly sensory.

SEPTUM: in protozoology, the dividing wall between protomerite and deutomerite of cephaline gregarines.

SESSILE: fixed or attached to the substrate.

SILVER-LINE SYSTEM: a complex system of surface markings or network of superficial fibrils revealed by silver impregnation (Klein's method, or some modification thereof).

SOMATELLA: a kind of plasmodium formed by certain flagellates (e.g., of plants and insects), in which both nuclei and kinetoplasts are present in equal numbers.

SPORADINS: adult stage in the gregarine life cycle.
SPORE: in protozoology, the resistant stage of a sporozoan developing from the zygote, and containing the infective forms (sporozoites).
SPOROCYST: the sporozoite-containing stage of many Sporozoa.
SPOROGONY: spore formation (reproduction in the sexual cycle, as in the malaria parasites, culminating in the production of sporozoites).
SPORONT: trophozoite destined to produce gametocytes or gametes (also sometimes used for stages developing from the zygote).
SPOROPLASM: an amoeboid organism within a spore (e.g., of the Cnidosporidia).
SPOROZOITE: the final stage in the sexual cycle of Sporozoa, usually the infective one.
SPORULATION: reproduction by multiple fission such as occurs in the sporozoan life cycle.
STIGMA: the light-perceiving organelle of many phytoflagellates; usually orange-red.
SULCUS: the longitudinal groove in which one of the flagella of the dinoflagellates lies.
SUTURAL LINE: the line of union of the valves, or major component parts, of a myxosporidan spore (sometimes used in a similar sense for other organisms).
SUTURE: any line of apparent union of two parts.
SWARMERS: minute motile products of multiple fission; known in the life cycles of a variety of Protozoa, among them, the Radiolarida, and some ciliates.
SYMBIOSIS: an association between two species to their mutual benefit (synonym MUTUALISM); the word is also used in a somewhat broader sense for any such association irrespective of mutual advantage.
SYNCYTIUM: a multinucleated mass of cytoplasm (synonym PLASMODIUM).
SYNGEN: a "genetic species" of ciliate, or population of compatible mating types.
SYNKARYON: fertilization nucleus (*amphinucleus* in older literature).
SYSTOLE: contraction, especially of the contractile vacuole.
SYZYGY: various types of association of gamonts; seen especially in the gregarines, in which it is often end to end.
TECTIN: see PSEUDOCHITIN.
TELOTROCH: a freely motile stage of the vorticellids in which the bell develops a posterior ring of cilia and swims away, leaving its stalk behind; one of two daughter cells from ordinary peritrich division may also be a telotroch.
TEST: the "house" or dwelling of a protozoan; synonym SHELL.
THECA: a closely fitting skeletal envelope, often of cellulose.
THERMOCLINE: a point in a lake in which the temperature change with increasing depth is especially sharp.
THIGMOTROPISM: response to a solid object.
TOMITE: the small ciliated stage resulting from division of the apostome TOMONT; free-swimming but nonfeeding. Also used for individuals resulting from rapid fission within a cystic membrane of some ciliates.
TRACTELLUM: a flagellum which propels by pulling.
TOMONT: Stage in the life cycles of some ciliates in which a cystic membrane is secreted before beginning fission destined to produce the tomites.
TRICHITE: an internal skeletal structure, usually rodlike in form, reinforcing the mouth and gullet of many ciliates and also in some flagellates (e.g., *Peranema*).
TRICHOCYST: an organelle underlying the surface of many ciliates and some flagellates capable of sudden discharge; of uncertain function.

TROPHONT: the growth stage of ciliates such as apostomes and others (e.g., hymenostomes).

TROPHOZOITE: the growing or vegetative stage in the protozoan (and especially the sporozoan) life cycle.

TROPISM: a reaction of constant character to some type of stimulus.

UNDULATING MEMBRANE: a delicate membrane consisting of a fused row (or rows) of cilia in the cytostomal neighborhood of ciliates; also a fold of the surface membrane enveloping a flagellum from its point of origin to its free extremity (e.g., as in trypanosomes and trichomonads).

VACUOME: originally used for vacuolar system system of plant cells, but also applied to certain cytoplasmic inclusions of some Protozoa (not necessarily homologous with the vacuome of plant cells).

VALVES: the main parts or divisions of the spores of certain Sporozoa (e.g. Myxosporida).

VECTOR: the invertebrate (transmitting) host of a parasite (e.g., tsetse fly for African sleeping sickness).

VERMICULE: like a worm (as in the ookinete of Haemosporidina).

VESICULAR: having small cavities or vesicles within (in the literal sense); usually used for a type of nucleus having a relatively large amount of nucleoplasm (as opposed to chromatin) and an endosome.

VESTIBULUM (VESTIBULE): depression or invaginated area, typically anterior or on ventral surface, leading to cytostome and cytopharynx and fitted with rather simple ciliature.

VOLUTIN: a substance taking nuclear stains, probably containing ribonucleic acid, and constituting one of the types of cytoplasmic inclusions of some Protozoa.

ZOONOSIS: a disease of animals transmissible under some conditions to man (e.g., malaria in monkeys caused by *Plasmodium knowlesi* may occasionally be transmitted to man by infected mosquitoes).

ZYGOTE: the fertilized egg or, among the Protozoa, the stage regarded as its homologue.

Index of Names

Subject Index

Numbers in italics refer to pages on which figures appear.

A CATALOGUE OF SELECTED DOVER BOOKS
IN ALL FIELDS OF INTEREST

A CATALOGUE OF SELECTED DOVER BOOKS IN ALL FIELDS OF INTEREST

JOHANN SEBASTIAN BACH, Philipp Spitta. One of the great classics of musicology, this definitive analysis of Bach's music (and life) has never been surpassed. Lucid, nontechnical analyses of hundreds of pieces (30 pages devoted to St. Matthew Passion, 26 to B Minor Mass). Also includes major analysis of 18th-century music. 450 musical examples. 40-page musical supplement. Total of xx + 1799pp.
(EUK) 22278-0, 22279-9 Two volumes, Clothbound $15.00

MOZART AND HIS PIANO CONCERTOS, Cuthbert Girdlestone. The only full-length study of an important area of Mozart's creativity. Provides detailed analyses of all 23 concertos, traces inspirational sources. 417 musical examples. Second edition. 509pp. (USO) 21271-8 Paperbound $3.50

THE PERFECT WAGNERITE: A COMMENTARY ON THE NIBLUNG'S RING, George Bernard Shaw. Brilliant and still relevant criticism in remarkable essays on Wagner's Ring cycle, Shaw's ideas on political and social ideology behind the plots, role of Leitmotifs, vocal requisites, etc. Prefaces. xxi + 136pp.
21707-8 Paperbound $1.50

DON GIOVANNI, W. A. Mozart. Complete libretto, modern English translation; biographies of composer and librettist; accounts of early performances and critical reaction. Lavishly illustrated. All the material you need to understand and appreciate this great work. Dover Opera Guide and Libretto Series; translated and introduced by Ellen Bleiler. 92 illustrations. 209pp.
21134-7 Paperbound $1.50

HIGH FIDELITY SYSTEMS: A LAYMAN'S GUIDE, Roy F. Allison. All the basic information you need for setting up your own audio system: high fidelity and stereo record players, tape records, F.M. Connections, adjusting tone arm, cartridge, checking needle alignment, positioning speakers, phasing speakers, adjusting hums, trouble-shooting, maintenance, and similar topics. Enlarged 1965 edition. More than 50 charts, diagrams, photos. iv + 91pp. 21514-8 Paperbound $1.25

REPRODUCTION OF SOUND, Edgar Villchur. Thorough coverage for laymen of high fidelity systems, reproducing systems in general, needles, amplifiers, preamps, loudspeakers, feedback, explaining physical background. "A rare talent for making technicalities vividly comprehensible," R. Darrell, *High Fidelity.* 69 figures. iv + 92pp. 21515-6 Paperbound $1.00

HEAR ME TALKIN' TO YA: THE STORY OF JAZZ AS TOLD BY THE MEN WHO MADE IT, Nat Shapiro and Nat Hentoff. Louis Armstrong, Fats Waller, Jo Jones, Clarence Williams, Billy Holiday, Duke Ellington, Jelly Roll Morton and dozens of other jazz greats tell how it was in Chicago's South Side, New Orleans, depression Harlem and the modern West Coast as jazz was born and grew. xvi + 429pp.
21726-4 Paperbound $2.50

FABLES OF AESOP, translated by Sir Roger L'Estrange. A reproduction of the very rare 1931 Paris edition; a selection of the most interesting fables, together with 50 imaginative drawings by Alexander Calder. v + 128pp. 6½x9¼.
21780-9 Paperbound $1.25

TWO LITTLE SAVAGES; BEING THE ADVENTURES OF TWO BOYS WHO LIVED AS INDIANS AND WHAT THEY LEARNED, Ernest Thompson Seton. Great classic of nature and boyhood provides a vast range of woodlore in most palatable form, a genuinely entertaining story. Two farm boys build a teepee in woods and live in it for a month, working out Indian solutions to living problems, star lore, birds and animals, plants, etc. 293 illustrations. vii + 286pp.
20985-7 Paperbound $2.50

PETER PIPER'S PRACTICAL PRINCIPLES OF PLAIN & PERFECT PRONUNCIATION. Alliterative jingles and tongue-twisters of surprising charm, that made their first appearance in America about 1830. Republished in full with the spirited woodcut illustrations from this earliest American edition. 32pp. 4½ x 6⅜.
22560-7 Paperbound $1.00

SCIENCE EXPERIMENTS AND AMUSEMENTS FOR CHILDREN, Charles Vivian. 73 easy experiments, requiring only materials found at home or easily available, such as candles, coins, steel wool, etc.; illustrate basic phenomena like vacuum, simple chemical reaction, etc. All safe. Modern, well-planned. Formerly *Science Games for Children*. 102 photos, numerous drawings. 96pp. 6⅛ x 9¼.
21856-2 Paperbound $1.25

AN INTRODUCTION TO CHESS MOVES AND TACTICS SIMPLY EXPLAINED, Leonard Barden. Informal intermediate introduction, quite strong in explaining reasons for moves. Covers basic material, tactics, important openings, traps, positional play in middle game, end game. Attempts to isolate patterns and recurrent configurations. Formerly *Chess*. 58 figures. 102pp. (USO) 21210-6 Paperbound $1.25

LASKER'S MANUAL OF CHESS, Dr. Emanuel Lasker. Lasker was not only one of the five great World Champions, he was also one of the ablest expositors, theorists, and analysts. In many ways, his Manual, permeated with his philosophy of battle, filled with keen insights, is one of the greatest works ever written on chess. Filled with analyzed games by the great players. A single-volume library that will profit almost any chess player, beginner or master. 308 diagrams. xli x 349pp.
20640-8 Paperbound $2.75

THE MASTER BOOK OF MATHEMATICAL RECREATIONS, Fred Schuh. In opinion of many the finest work ever prepared on mathematical puzzles, stunts, recreations; exhaustively thorough explanations of mathematics involved, analysis of effects, citation of puzzles and games. Mathematics involved is elementary. Translated by F. Göbel. 194 figures. xxiv + 430pp. 22134-2 Paperbound $3.00

MATHEMATICS, MAGIC AND MYSTERY, Martin Gardner. Puzzle editor for Scientific American explains mathematics behind various mystifying tricks: card tricks, stage "mind reading," coin and match tricks, counting out games, geometric dissections, etc. Probability sets, theory of numbers clearly explained. Also provides more than 400 tricks, guaranteed to work, that you can do. 135 illustrations. xii + 176pp.
20338-2 Paperbound $1.50

THE PRINCIPLES OF PSYCHOLOGY, William James. The famous long course, complete and unabridged. Stream of thought, time perception, memory, experimental methods—these are only some of the concerns of a work that was years ahead of its time and still valid, interesting, useful. 94 figures. Total of xviii + 1391pp.
20381-6, 20382-4 Two volumes, Paperbound $8.00

THE STRANGE STORY OF THE QUANTUM, Banesh Hoffmann. Non-mathematical but thorough explanation of work of Planck, Einstein, Bohr, Pauli, de Broglie, Schrödinger, Heisenberg, Dirac, Feynman, etc. No technical background needed. "Of books attempting such an account, this is the best," Henry Margenau, Yale. 40-page "Postscript 1959." xii + 285pp. 20518-5 Paperbound $2.00

THE RISE OF THE NEW PHYSICS, A. d'Abro. Most thorough explanation in print of central core of mathematical physics, both classical and modern; from Newton to Dirac and Heisenberg. Both history and exposition; philosophy of science, causality, explanations of higher mathematics, analytical mechanics, electromagnetism, thermodynamics, phase rule, special and general relativity, matrices. No higher mathematics needed to follow exposition, though treatment is elementary to intermediate in level. Recommended to serious student who wishes verbal understanding. 97 illustrations. xvii + 982pp. 20003-5, 20004-3 Two volumes, Paperbound $6.00

GREAT IDEAS OF OPERATIONS RESEARCH, Jagjit Singh. Easily followed non-technical explanation of mathematical tools, aims, results: statistics, linear programming, game theory, queueing theory, Monte Carlo simulation, etc. Uses only elementary mathematics. Many case studies, several analyzed in detail. Clarity, breadth make this excellent for specialist in another field who wishes background. 41 figures. x + 228pp. 21886-4 Paperbound $2.50

GREAT IDEAS OF MODERN MATHEMATICS: THEIR NATURE AND USE, Jagjit Singh. Internationally famous expositor, winner of Unesco's Kalinga Award for science popularization explains verbally such topics as differential equations, matrices, groups, sets, transformations, mathematical logic and other important modern mathematics, as well as use in physics, astrophysics, and similar fields. Superb exposition for layman, scientist in other areas. viii + 312pp.
20587-8 Paperbound $2.50

GREAT IDEAS IN INFORMATION THEORY, LANGUAGE AND CYBERNETICS, Jagjit Singh. The analog and digital computers, how they work, how they are like and unlike the human brain, the men who developed them, their future applications, computer terminology. An essential book for today, even for readers with little math. Some mathematical demonstrations included for more advanced readers. 118 figures. Tables. ix + 338pp. 21694-2 Paperbound $2.50

CHANCE, LUCK AND STATISTICS, Horace C. Levinson. Non-mathematical presentation of fundamentals of probability theory and science of statistics and their applications. Games of chance, betting odds, misuse of statistics, normal and skew distributions, birth rates, stock speculation, insurance. Enlarged edition. Formerly "The Science of Chance." xiii + 357pp. 21007-3 Paperbound $2.50